TOWN AND COUNTRY PLANNING IN THE UK

Thirteenth Edition

This extensively revised edition of *Town and Country Planning in the UK* retains and enhances its reputation as the bible of British planning. The book now covers the whole of the UK and gives a critical discussion of current issues and problems. It provides an explanation of the nature of planning, the institutions and organisations involved, the plans and other tools used by planners, the system of controlling development and land use change, and planning policies pursued. Detailed consideration is given to:

- The nature of planning and its historical evolution
- Central and local government, the EU and other agencies
- The framework of plans and other planning instruments
- Development control
- Land policy and planning gain
- Environmental and countryside planning
- Sustainable development, waste and pollution
- Heritage and transport planning
- Urban policies and regeneration
- Planning, the profession and the public

This thirteenth edition has been completely revised to take into account the many changes to the planning system and policies introduced by the Labour government. The devolution of Scotland, Wales and Northern Ireland, the new instruments of regional and strategic planning, new area-based urban policy initiatives, innovation in planning for sustainable development and the rapidly expanding role of the European Union in spatial planning and environmental policy are all given comprehensive treatment in the new edition. Each chapter ends with notes on further reading and there are lists of official publications and an extensive bibliography at the end of the book.

Barry Cullingworth has held academic posts at the Universities of Manchester, Durham, Glasgow, Birmingham and Toronto and is Emeritus Professor of Urban Affairs and Public Policy at the University of Delaware.
Vincent Nadin is Director of the Centre for Environment and Planning and Reader in the Faculty of the Built Environment, University of the West of England, Bristol.

TOWN AND COUNTRY PLANNING IN THE UK

Thirteenth Edition

Barry Cullingworth
and
Vincent Nadin

London and New York

First published 1964
Thirteenth edition published 2002
by Routledge
11 New Fetter Lane, London EC4P 4EE

Simultaneously published in the USA and Canada
by Routledge
29 West 35th Street, New York NY 10001

Routledge is an imprint of the Taylor & Francis Group

Typeset in Garamond by Keystroke, Jacaranda Lodge, Wolverhampton
Printed and bound in Great Britain by
St Edmundsbury Press, Bury St Edmunds, Suffolk

British Library Cataloguing in Publication Data
A catalogue record for this book is available from the British Library

Library of Congress Cataloging in Publication Data
Cullingworth, J. B.
Town and country planning in the UK / J. Barry Cullingworth and Vincent Nadin.– 13th ed.
p.   cm.
Includes bibliographical references and index.
1. City planning–Great Britain. 2. Regional planning–Great Britain. I. Nadin, Vincent.
HT 169.G7 C8 2001
361.6'0941–dc21   2001019932

ISBN 0–415–21774–1 (hbk)
ISBN 0–415–21775–X (pbk)

CONTENTS

FIGURES

TABLES

BOXES

PREFACE

Since 1963, when the first edition of this book was drafted, it has become increasingly uncertain what should be included under the title of *Town and Country Planning*. At one time it could be largely defined by reference to a limited number of Acts of Parliament. Such a convenient benchmark no longer exists: planning policies are now far broader, and there is general acceptance of the important interrelationships that exist with other spheres of policy (though catering for these has proved exceptionally difficult). It is therefore not easy (or even useful) to define the boundaries of town and country planning. No longer can it be claimed (as it was in the first edition) that this book provides 'an outline of town and country planning and the problems with which it is faced'. Such an enterprise would now take several volumes. Successive editions have been expanded to incorporate the increasing range of issues with which planners are concerned but, beyond basic statutory and administrative matters, selection has become increasingly arbitrary. The problems have increased with the uncertainties about new features in the legislative landscape, particularly the Human Rights Act. This could have major implications for the planning system.

The current edition follows the pragmatic course of updating its predecessors – adding in some parts; deleting in others. The result does not satisfy the authors: too many compromises have had to be made, and too much has had to be omitted. But, like practising planners, the authors have had to operate within constraints which are externally determined.

As in the last edition, each chapter ends with a guide to further reading. These guides are intended to assist students who wish to follow up the discussion in the text. However, they are only an introductory pointer to some of the useful available material: they are in no way comprehensive. Though there may well be a need for an annotated bibliography of planning literature, this is not the place to provide it. The literature is now so vast that the selection of titles for recommendation is inevitably a personal (and, to some extent, an arbitrary) matter. However, it is not, we hope, idiosyncratic, though no doubt other teachers may prefer alternatives. The bibliography has been expanded, and the list of official publications has been completely reorganised by broad subject matter (rather than by source). It is hoped that this will make it more manageable and useful.

In the last edition endnotes were omitted, mainly because they seemed to be getting out of hand. However, we were encouraged to reinstate the idea since the notes provide a way of referring to additional useful sources, wider points of interest, and other relevant matter which could not be included in the body of the book without overburdening the text. Acknowledgement is made with sincere thanks to Sally Jones who helped with the figures; to the many people who have supplied information; and to the Editors at Routledge for their patience.

Barry Cullingworth

Vincent Nadin

ACRONYMS AND ABBREVIATIONS

Acronyms and abbreviations are a major growth area in public policy. The following list includes all that are used in the text and others that readers will come across in the planning literature. No claim is made for comprehensiveness.

AAI	area of archaeological importance
ACC	Association of County Councils
ACCORD	assistance for coordinated rural development
ACO	Association of Conservation Officers
ACOST	Advisory Council on Science and Technology
ACRE	Action with Communities in Rural England (formerly Association of Community Councils in Rural England)
ADAS	agricultural development and advice service
ADC	Association of District Councils
AESOP	Association of European Schools of Planning
AGR	advanced gas-cooled reactor
AIS	agricultural improvement scheme
ALA	Association of London Authorities (now the Association of London Government)
ALG	Association of London Government
ALNI	Association of Local Authorities in Northern Ireland
ALURE (1)	alternative land use and rural economy
ALURE (2)	Action with Communities in Rural England
AMA	Association of Metropolitan Authorities
ANPA	Association of National Park Authorities
AONB	area of outstanding natural beauty
AOSP	areas of special protection for birds
APRS	Association for the Protection of Rural Scotland
ARC	Action Resource Centre
ASAC	area of special advertisement control
ASNW	area of semi-natural woodland
ASSI	area of special scientific interest (Northern Ireland)
ATB	Agricultural Training Board
BACMI	British Aggregate Construction Materials Industries
BATNEEC	best available techniques not entailing excessive cost (Environmental Protection Act 1990, s.7(10))
BIC	Business in the Community
BNFL	British Nuclear Fuels Ltd
BPEO	best practicable environmental option
BPF	British Property Federation
bpk	billion passenger-kilometres
BPM	best practicable means

BR	British Rail
BRE	Building Research Establishment
BRF	British Road Federation
BSI	British Standards Institution
BTA	British Tourist Authority
BUFT	British Upland Footpath Trust
BVPI	best value performance indicator
BWB	British Waterways Board
CABE	Commission for Architecture and the Built Environment
Cadw	Not an acronym, but the Welsh name for the Welsh Historic Monuments Agency. The word means 'to keep', 'to preserve'.)
CAF	Coalfields Area Fund
CAP	Common Agricultural Policy
CAT	City Action Team
CBI	Confederation of British Industry
CC	Countryside Commission
CCS	Countryside Commission for Scotland (now Scottish Natural Heritage)
CCT	compulsory competitive tendering
CCW	Countryside Council for Wales
CDA	comprehensive development area
CDP	community development project
CEGB	Central Electricity Generating Board
CEMAT	Conférence Européen des Ministres d'Aménagement du Territoire (European Conference of Ministers responsible for Regional Planning)
CEMR	Council of European Municipalities and Regions
CES	Centre for Environmental Studies
CFC	chlorofluorocarbon
CHAC	Central Housing Advisory Committee
CHP	combined heat and power
CIA	commercial improvement area
CIEH	Chartered Institute of Environmental Health
CIPFA	Chartered Institute of Public Finance and Accountancy
CIT	Commission for Integrated Transport
CITES	Convention on International Trade in Endangered Species

CLRAE	Conference of Local and Regional Authorities of Europe (Council of Europe)
CLES	Centre for Local Economic Strategies
CLEUD	certificate of lawfulness of existing use or development
CLOPUD	certificate of lawfulness of a proposed use or development
CMS	countryside management system
CNCC	Council for Nature Conservation and Countryside (Northern Ireland)
COBA	cost–benefit analysis
COE	Council of Europe
COI	Central Office of Information
COMARE	Committee on Medical Aspects of Radiation in the Environment
COPA	Control of Pollution Act (1974)
COR	Committee of the Regions (EU)
COREPER	Council of Permanent Representatives
CoSIRA	Council for Small Industries in Rural Areas
CORINE	Community Information System on the State of the Environment (EU)
COSLA	Convention of Scottish Local Authorities
CPO	compulsory purchase order
CPOS	County Planning Officers' Society
CPRE	Council for the Protection of Rural England
CPRS	Central Policy Review Staff
CPRW	Campaign (formerly Council) for the Protection of Rural Wales
CRE	Commission for Racial Equality
CRRAG	Countryside Recreation Research Advisory Group
CSD (1)	Commission on Sustainable Development (UN)
CSD (2)	Committee on Spatial Development (EU)
CSERGE	Centre for Social and Economic Research on the Global Environment
CSF	Community Support Framework
CSO	Central Statistical Office
CWI	Controlled Waste Inspectorate

DAFS	Department of Agriculture and Fisheries for Scotland		EBRD	European Bank for Reconstruction and Development
DATAR	Délégation à l'aménagement du territoire et à l'action régionale (French national planning agency)		EC	European Community
			ECMT	European Conference of Ministers of Transport
DBFO	design, build, finance, and operate (roads by the private sector)		ECOSOC	Economic and Social Council (United Nations)
DBRW	Development Board for Rural Wales		ECS	Economic and Social Committee (EU)
DC	development corporation		ECSC	European Coal and Steel Community
DCAN	Development control advice note (Northern Ireland)		ECTP	European Council of Town Planners
			Ecu	European currency unit (no longer in use)
DCC	Docklands Consultative Committee		EEA (1)	European Economic Area (the EU plus Iceland, Liechtenstein, Norway, and Switzerland)
DCMS	Department for Culture, Media and Sport			
DEA	Department of Economic Affairs		EEA (2)	European Environmental Agency
DEFRA	Department for Environment, Food and Rural Affairs		EEC	European Economic Community
			EFTA	European Free Trade Association
DETR	Department of Transport, Environment and the Regions		EIA	environmental impact assessment
			EIB	European Investment Bank
DG	Directorate General (of the European Commission)		EIF	European Investment Fund
			EIP	examination in public
DHSS	Department of Health and Social Security		EIS	environmental impact statement
			EMAS	eco-management and audit scheme
DLG	derelict land grant		EMP	environmental management area
DLGA	derelict land grant advice note		EMU	European Monetary Union
DLT	development land tax		EN	English Nature
DNH	Department of National Heritage		EP	English Partnerships
DoE	Department of the Environment		EPA (1)	educational priority area
DoENI	Department of the Environment Northern Ireland		EPA (2)	Environmental Protection Act (1990)
			EPC	economic planning council
DoT	Department of Transport (formerly DTp)		ERDF	European Regional Development Fund
DPOS	District Planning Officers' Society		ERP	electronic road pricing
DRIVE	dedicated road infrastructure for vehicle safety in Europe		ES	environmental statement (UK)
			ESA	environmentally sensitive area
DTI	Department of Trade and Industry		ESDP	European Spatial Development Perspective
DTLR	Department for Transport, Local Government and the Regions			
			ESF	European Social Fund
DWI	Drinking Water Inspectorate		ESPRID	European Spatial Planning Research Information Database
EA	environmental assessment		ESRC	Economic and Social Research Council
EAGGF	European Agricultural Guidance and Guarantee Fund			
			ETB	English Tourist Board
EAZ	education action zone		ETC	English Tourism Council

EU	European Union
EUCC	European Union for Coastal Conservation
EURATOM	European Atomic Energy Community
EZ (1)	employment zone
EZ (2)	enterprise zone
FA	Forestry Authority
FC	Forestry Commission
FCGS	Farm and Conservation Grant Scheme
FEOGA	Fonds Européan d'Orientation et de Garantie Agricole (European Agricultural Guidance and Guarantee Fund)
FIFG	Financial Instrument for Fisheries Guidance
FIG	Financial Institutions Group
FMI	financial management initiative
FOE	Friends of the Earth
FTA	Freight Transport Association
FWAG	Farming and Wildlife Advisory Group
FWGS	farm woodland grant scheme
FWPS	farm woodland premium scheme
GATT	General Agreement on Tariffs and Trade
GCR	Geological Conservation Review
GDO	General Development Order
GDP	Gross domestic product
GDPO	General Development Procedure Order
GEAR	Glasgow Eastern Area Renewal (scheme)
GIA	general improvement area
GLC	Greater London Council
GLDP	Greater London Development Plan
GOR	Government Offices for the Regions:
GO-EM	Government Office for the East Midlands
GO-ER	Government Office for Eastern Region
GO-L	Government Office for London
GO-M	Government Office for Merseyside
GO-NE	Government Office for the North East

GO-NW	Government Office for the North West
GO-SE	Government Office for the South East
GO-SW	Government Office for the South West
GO-WM	Government Office for the West Midlands
GO-YH	Government Office for Yorkshire and Humberside
GPDO	General Permitted Development Order
HAA	housing action area
HAG	housing association grant
HAT	housing action trust
HAZ	health action zone
HBF	House Builders' Federation
HBMC	Historic Buildings and Monuments Commission
HC	House of Commons
HCiS	Housing Corporation in Scotland
HERS	Heritage Economic Regeneration Schemes (English Heritage)
HIDB	Highlands and Islands Development Board (now Highlands, and Islands Enterprise)
HIE	Highlands and Islands Enterprise
HIP	housing investment programme
HL	House of Lords
HLCA	hill livestock compensatory allowances
HLF	Heritage Lottery Fund
HLW	high-level waste
HMIP	Her Majesty's Inspectorate of Pollution
HMIPI	Her Majesty's Industrial Pollution Inspectorate (Scotland)
HMNII	Her Majesty's Nuclear Installation Inspectorate
HMO	hedgerow management order
HMSO	Her Majesty's Stationery Office
HRF	Housing Research Foundation
HSA	Hazardous Substances Authority
HSE	Health and Safety Executive
HWI	Hazardous Waste Inspectorate

IACGEC — Inter-Agency Committee on Global Environmental Change

IAEA — International Atomic Energy Agency

IAP — inner area programmes

IAPI — Industrial Air Pollution Inspectorate

IAURIF — Institut d'aménagement du territoire et d'urbanisme de la région d'Île de France

ICE — Institution of Civil Engineers

ICOMOS — International Council on Monuments and Sites

ICT — information and communications technology

ICZM — Integrated Coastal Zone Management

IDC — industrial development certificate

IEEP — Institute for European Environmental Policy

IIA — industrial improvement area

ILW — intermediate-level waste

INLOGOV — Institute of Local Government Studies (University of Birmingham)

IPC — integrated pollution control

IPCC — Intergovernmental Panel on Climate Change

IPPC — integrated pollution, prevention, and control

IRD — Integrated rural development (Peak District)

ISOCARP — International Society of City and Regional Planners

IWA — Inland Waterways Association

IWAAC — Inland Waterways Amenity Advisory Committee

JNCC — Joint Nature Conservation Committee

JPL — *Journal of Planning and Environment Law* (formerly *Journal of Planning and Property Law*)

LA21 — Local Agenda 21 (UNCED)

LAAPC — Local Authority Air Pollution Control

LAW — Land Authority for Wales

LAWDC — local authority waste disposal company

LBA — London Boroughs Association (now the Association of London Government)

LCO — landscape conservation order

LDDC — London Docklands Development Corporation

LEAP — local environmental agency plan

LEC — local enterprise company (Scotland)

LEGUP — local enterprise grants for urban projects (Scotland)

LETS — local employment and trading systems

LFA — less favoured area (agriculture)

LGA — Local Government Association

LGC — Local Government Commission for England

LGMB — Local Government Management Board

LLW — low-level waste

LNR — local nature reserve

LOTS — living over the shop

LPA — local planning authority

LPAC — London Planning Advisory Committee

LSP — local strategic partnership

LSPU — London Strategic Policy Unit

LTP — local transport plan

LTS — local transport strategy (Scotland)

LULU — locally unwanted land use

LWRA — London Waste Registration Authority

MAFF — Ministry of Agriculture, Fisheries and Food

MARS — Monuments at Risk Survey

MCC — metropolitan country council

MEA — Manual of Environmental Assessment (for trunk roads)

MEHRA — marine environmental high-risk area

MEP — Member of the European Parliament

MHLG — Ministry of Housing and Local Government

MINIS — management information system for ministers

MLGP — Ministry of Local Government and Planning

MMS — multi-modal study

MNR — marine nature reserve

MOD	Ministry of Defence	NNR	national nature reserve
MPA	mineral planning authority	NPA	national park authority
MPG	minerals planning guidance note	NPF	National Planning Forum
MPOS	Metropolitan Planning Officers Society	NPG	national planning guideline (Scotland)
MSC	Manpower Services Commission	NPPG	national planning policy guideline (Scotland)
MTCP	Ministry of Town and Country Planning	NRA	National Rivers Authority
		NRTF	national road traffic forecasts (GB)
NACRT	National Agricultural Centre Rural Trust	nSA (1)	National scenic area (Scotland)
		NSA (2)	nitrate-sensitive area
NAO	National Audit Office	NUTS	the nomenclature of statistical territorial units; designates levels of regional subdivision in the EU
NAQS	National Air Quality Strategy		
NARIS	national roads information system		
NATA	new approach to road appraisal	NVZ	nitrate-vulnerable zone
NAW	National Assembly for Wales	NTDC	new town development corporation
NBN	National Biodiversity Network		
NCB	National Coal Board	OECD	Organisation for Economic Cooperation and Development
NCBOE	National Coal Board Opencast Executive		
		OEEC	Organisation for European Economic Cooperation
NCC	Nature Conservancy Council		
NCCI	National Committee for Commonwealth Immigrants	OFLOT	Office of the National Lottery
		OJ	Official Journal of the European Communities
NCCS	Nature Conservancy Council for Scotland (now Scottish Natural Heritage)		
		ONS	Office for National Statistics
		OOPEC	Office for Official Publications of the European Communities
NCR	New commitment to regeneration		
NCVO	National Council of Voluntary Organisations	OPCS	Office of Population Censuses and Surveys (now part of ONS)
NDC	New Deal for Communities		
NDPB	non-departmental public body	PAG (1)	Planning Advisory Group (1965)
NEC	noise exposure category	PAG (2)	Property Advisory Group
NEDC	National Economic Development Council	PAN	planning advice note (Scotland)
		PAT	policy action team
NEDO	National Economic Development Office	PDO (1)	Permitted Development Order
		PDO (2)	potentially damaging operation (in a site of special scientific interest)
NERC	National Environment Research Council		
		PDR	permitted development right
NFFO	non-fossil fuel obligation	PEP	Political and Economic Planning (now the Policy Studies Institute)
NGO	non-governmental organisation		
NHA	Natural heritage area (Scotland)	PFI	Private Finance Initiative
NHMF	National Heritage Memorial Fund	PI	Planning Inspectorate
NII	Nuclear Installations Inspectorate	PIC	Planning Inquiry Commission
NIO	Northern Ireland Office	PLI	public local inquiry
NIREX	Nuclear Industries Radioactive Waste Executive	PPA	priority partnership area (Scotland)
		PPG	planning policy guidance note

PPP	polluter pays principle
PPS	planning policy statement (Northern Ireland)
PRIDE	programmes for rural initiatives and developments (Scotland)
PSA	Property Services Agency
PSI	Policy Studies Institute
PTA	passenger transport authority
PTE	passenger transport executive
PTRC	Planning and Transport Research and Computation
PVC	polyvinyl chloride
PWR	pressurised water reactor
quango	quasi-autonomous non-governmental organisation
RA	renewal area
RAC	Royal Automobile Club
RAWP	regional aggregates working parties
RBMP	river basin management plan
RCC	rural community council
RCEP	Royal Commission on Environmental Pollution
RCHME	Royal Commission on the Historical Monuments of England
RCI	Radiochemical Inspectorate
RCU	road construction unit
RDA (1)	regional development agency
RDA (2)	rural development area
RDC	Rural Development Commission
RDG	regional development grant
RDP	rural development programme
REG	regional enterprise grant
RIBA	Royal Institute of British Architects
RICS	Royal Institution of Chartered Surveyors
RIGS	regionally important geological/geomorphological sites
RPB	regional planning body
RPG	regional planning guidance (note)
RSA (1)	regional selective assistance
RSA (2)	Regional Studies Association
RSDF	regional sustainable development framework

RSPB	Royal Society for the Protection of Birds
RTB (1)	right to buy (public-sector housing)
RTB (2)	regional tourist board
RTPI	Royal Town Planning Institute
RTS	regional transport strategy
RWMAC	Radioactive Waste Management Advisory Committee
SAC	special area of conservation (habitats)
SACTRA	Standing Advisory Committee on Trunk Road Assessment
SAGA	Sand and Gravel Association
SCEALA	Standing Conference of East Anglian Local Authorities
SCLSERP	Standing Conference on London and South East Regional Planning (see also SERPLAN)
SDA	Scottish Development Agency (now Scottish Enterprise)
SDC	Sustainable Development Commission
SDO	Special Development Order
SDS	Spatial Development Strategy (London)
SEA	strategic environmental assessment
SEDD	Scottish Executive Development Department
SEED	Scottish Executive Education Department
SEEDA	South East England Development Agency
SEEDS	South East Economic Development Strategy
SEELLD	Scottish Executive Enterprise and Lifelong Learning Department
SEERA	South East England Regional Assembly
SEHD	Scottish Executive Health Department
SEM	Single European market
SEPA	Scottish Environment Protection Agency
SERAD	Scottish Executive Rural Affairs Department

SERC	Science and Engineering Research Council	TEN	Trans-European Network(s)
SERPLAN	London and South East Regional Planning Conference	TEST	Transport and Environment Studies
		TfL	Transport for London
SHAC	Scottish Housing Advisory Committee	THORP	thermal oxide reprocessing plant
		TPO	tree preservation order
SI	statutory instrument	TPPs	transport policies and programmes
SIC	social inclusion partnerships (Scotland)	TRL	transport Research Laboratory (formerly Transport and Road Research Laboratory)
SINC	site of importance for nature conservation	TSG	transport supplementary grant
SLF	Scottish Landowners Federation	TSO	The Stationery Office
SME	small and medium-sized enterprises (Europe)	TUC	Trades Union Congress
SNAP	Shelter Neighbourhood Action Project	UCO	Use Classes Order
		UDA	urban development area
SNH	Scottish Natural Heritage	UDC	urban development corporation
SMR	sites and monuments records (counties)	UDG	urban development grant
		UDP	unitary development plan
SO	Scottish Office	UEI	urban exchange initiative
SOAEFD	Scottish Office Agriculture, Environment and Fisheries Department	UKAEA	United Kingdom Atomic Energy Authority
SODD	Scottish Office Development Department	UNCED	United Nations Conference on Environment and Development ('Earth Summit', Rio, 1992)
SOEnD	Scottish Office Environment Department (now SOAEFD)	UNCSD	United Nations Commission on Sustainable Development
SOID	Scottish Office Industry Department	UNCTAD	United Nations Conference on Trade and Development
SOIRU	Scottish Office Inquiry Reporters Unit	UNECE	United Nations Economic Commission for Europe
SPA	Special protection area (for birds) (EU)	UNEP	United Nations Environment Programme
SPG	supplementary planning guidance		
SPZ	Simplified planning zone	UNESCO	United Nations Educational, Scientific and Cultural Organisation
SRB	Single Regeneration Budget		
SSHA	Scottish Special Housing Association	UP (1)	urban programme
SSSI	site of special scientific interest	UP (2)	urban partnership (Scotland)
STB	Scottish Tourist Board	URA	Urban Regeneration Agency
TAN	technical advice note (Wales)	VFM	value for money
TCPA	Town and Country Planning Association	VISEGRAD	four former communist countires: Poland, the Czech Republic, Slovakia, and Hungary
TCPSS	Town and Country Planning Summer School		
TEC	training and enterprise council	WCA	waste collection authority

WCED	World Commission on Environment and Development	WMO	World Meteorological Organisation
WDA (1)	Welsh Development Agency	WOAD	Welsh Office Agriculture Department
WDA (2)	waste disposal authority	WQO	water quality objectives
WDP	waste disposal plan	WTO	World Trade Organisation
WMEB	West Midlands Enterprise Board	WWF	World Wide Fund for Nature (formerly the World Wildlife Fund)
WO	Welsh Office		
WRA	waste regulation authority		
WRAP	waste reduction always pays	YTS	youth training scheme
WTB	Welsh Tourist Board		

Encyclopedia refers to Malcolm Grant's *Encyclopedia of Planning Law and Practice*, London: Sweet & Maxwell, loose-leaf, regularly updated by supplements.

INTERNET RESOURCES

The last two editions of this book have made extensive use of resources available on the Internet, the global computer network that links millions of computers world-wide. There are many hundreds of web sites that are relevant to town and country planning and hypertext links among them make it relatively easy to find many sites of interest. Reference is made to relevant web sites throughout this book. This section lists some of the main sites that provide useful starting points for surfing (All the URLs start with *http://*). Users should be aware that web pages and their addresses are constantly changing. If in doubt use one of the many search engines.

UK GENERAL RESOURCES AND DIRECTORIES

Online planning resources by Hugh McLintock
    Extensive list of planning related web sites
www.nottingham.ac.uk/sbe/planbiblios/contacts/web.html

Planweb
    Internet resource for UK planners by Peter Thorpe
www.planweb.co.uk

Planning Online
    The weekly newspaper with links to documents mentioned
www.planning.haynet.com

Index of national and local government, and government agencies in the UK
www.open.gov.uk

Guide to Official Information in the UK
    The Stationery Office web site
www.clicktso.com

Online Planning
    Site dedicated to the impact of the web on planning
www.onlineplanning.org

National Statistics for the UK, including the Office for National Statistics
www.statistics.gov.uk

The Planning Exchange
www.planex.co.uk

Urban Design Resources
www.rudi.herts.ac.uk

Regeneration network
www.regen.net

INTERNATIONAL GENERAL RESOURCES AND DIRECTORIES

Cyburbia (USA)   www.cyburbia.org
   Internet resources for the built environment

Environmental organisations web directory   www.webdirectory.com

Planners Forum (USA)   www.asu.edu/caed/onlineplanner
   Directory of planning related web content

Planum (Europe)   www.planum.net

EUROPEAN GOVERNMENT AND SOURCES

Europa   europa.eu.int/index-en.htm
   The main site of the European Union

The European Commission Office in the UK   www.cec.org.uk/index.htm

Council of the European Union   ue.eu.int

EU Environment Directorate   europa.eu.int/comm/dg11/index_en.htm

EU Regional Policy Directorate   europa.eu.int/comm/dg16

Inforegio   www.inforegio.cec.eu.int/dg16_en.htm
   Regional policy and EU Structural Funds

Eurostat – EU Statistics   europa.eu.int/comm/eurostat

European Parliament   www.europarl.eu.int

Committee of the Regions (EU)   www.cor.eu.int

Council of Europe   www.coe.int

European Environment Agency   www.eea.dk

EU Special Study on European Spatial Planning on   www.nordregio.se
the site of Nordregio

European Spatial Planning Research and Information   www.data.ncl.ac.uk/esprid
Database (ESPRID)

Catalogue of Free Publications from the EU   www.cec.org.uk/sourcedi/catalog1.htm

Office for the Official Publications of the European   eur-op.eu.int/general/en/index.htm
Union

A Spatial Vision for North-west Europe   www.uwe.ac.uk/fbe/vision

Local Government International Bureau   www.lgib.gov.uk

European Environment Bureau   www.eeb.org

| Rural Europe | www.rural-europe.aeidl.be/rural-en/index.html |
| European Centre for Nature Conservation | www.ecnc.nl |

ENGLAND GOVERNMENT, AGENCIES AND BODIES

UK Online	www.ukonline.gov.uk
General web site for access to government related bodies and documents	
Department of Transport, Local Government and the Regions	www.dtlr.gov.uk
Department of Environment, Food and Rural Affairs	www.defra.gov.uk
Department of Culture, Media and Sport	www.culture.gov.uk
Environment Agency	www.environment-agency.gov.uk
Royal Commission on Environmental Pollution	www.rcep.org.uk
The New Deal for Communities	www.newdeal.gov.uk
English Partnerships	www.englishpartnerships.co.uk
English Nature	www.english-nature.org.uk
Heritage Lottery Fund	www.hlf.org.uk
English Tourism Council	www.englishtourism.org.uk
English Heritage	www.english-heritage.org.uk
National Land Use Database	www.nlud.org.uk
Countryside Agency	www.countryside.gov.uk
Home Office Human Rights Unit	www.homeoffice.gov.uk/hract
The Planning Inspectorate	www.planning-inspectorate.gov.uk

NORTHERN IRELAND GOVERNMENT AND AGENCIES

Northern Ireland Assembly	www.ni-assembly.gov.uk
Department of the Environment Northern Ireland	www.doeni.gov.uk
The Planning Service	www.doeni.gov.uk/planning/index.htm
Northern Ireland Executive	www.nics.gov.uk/index.htm
Environment and Heritage Service Northern Ireland	www.ehsni.gov.uk

SCOTLAND GOVERNMENT AND AGENCIES

The Scottish Executive   www.scotland.gov.uk

Scottish Natural Heritage   www.snh.org.uk

Historic Scotland   www.historic-scotland.gov.uk/sw-frame.htm

Scottish Environment Protection Agency   www.sepa.org.uk

Scottish Homes   www.scot-homes.gov.uk

WALES GOVERNMENT AND AGENCIES

National Assembly for Wales   www.wales.gov.uk/index_e.html

Welsh Development Agency   www.networks.co.uk/wda

Countryside Council for Wales   www.commonwealth-planners.org

REGIONAL AND LOCAL GOVERNMENT

Directory of regional bodies and local authorities   www.open.gov.uk/index/orgindex.htm

Local Government Association   www.lga.gov.uk

Improvement and Development Agency (IDeA)   www.idea.gov.uk

Greater London Authority   www.london.gov.uk

INTERNATIONAL ORGANISATIONS

United Nations   www.un.org

UN Environment Programme   unep.frw.uva.nl

UN Commission for Sustainable Development   www.un.org/esa/sustdev/csd.htm

UN Economic Commission for Europe   www.unece.org

UN Sustainable Cities Programme   www.unchs.org

HABITAT +5 Summit   www.istanbul5.org

RIO +10 2001 Summit   www.johannesburgsummit.org

OECD Territorial Development Service   www.oecd.org/tds

PLANNING ASSOCIATIONS, PROFESSIONAL BODIES AND NETWORKS

Royal Town Planning Institute	www.rtpi.org.uk
Town and Country Planning Association	www.tcpa.org.uk
Planning Officers' Society	www.barnsley.gov.uk/planning/index.html
British Urban Regeneration Association	www.bura.org.uk
Royal Institute of British Architects	www.architecture.com
European Council of Town Planners	www.ceu-ectp.org
Commonwealth Association of Planners	www.commonwealth-planners.org
American Planning Association	www.planning.org/abtapa/abtapa.html
International Federation of Housing and Planning	www.ifhp.org
International Urban Development Association (INTA)	www.inta-aivn.org
ISOCARP International Society of City and Regional Planners	www.soc.titech.ac.ip/titsoc/higuchi-lab-isocarp/index
METREX The European network of metropolitan planning authorities	www.metrex.dis.strath.ac.uk/en/intro.html
European Sustainable Cities and Towns Campaign	www.sustainable-cities.org
Eurocities Network of 100+ European cities	www.eurocities.org
World Town Planning Day	planning.org/abtaicp/world.htm

NON-GOVERNMENTAL ORGANISATIONS AND SOURCES

Association of National Park Sites	www.anpa.gov.uk
Brownfield land	www.brownfieldsites.com
Centre for Local Economic Strategies	www.cles.org.uk
Civic Trust	www.civictrust.org.uk
Commission for Architecture and the Built Environment (CABE)	www.cabe.org.uk
Council for the Protection of Rural England	www.cpre.org.uk

European Local Transport Information Service (ELTIS)   www.eltis.org

Forum for the Future   www.forumforthefuture.org.uk/new2/index.htm

Friends of the Earth, UK   www.foe.co.uk

Greenpeace   www.greenpeace.org

House Builders' Federation   www.hbf.co.uk

Local Agenda 21 in the UK   www.la21-uk.org.uk

Local Sustainability
    European good practice information service   cities21.com/europractice

National Retail Planning Forum   www.nrpf.org

National Trust   www.nationaltrust.org.uk

Royal Society for the Protection of Birds   www.rspb.org.uk/education

The Association for the Protection of Rural Scotland   www.aprs.org.uk

The Campaign for the Protection of Rural Wales   www.cprw.org.uk

Transport 2000   www.transport2000.org.uk

Sustainable Development Commission   www.sd-commission.gov.uk/index/index.htm

EDUCATION AND RESEARCH

Association of Collegiate Schools of Planning (ACSP)   www.uwm.edu/org/acsp

The Association of European Schools of Planning (AESOP)   www.ncl.ac.uk/aesop

Asian Planning Schools Association (APSA)   www.hku.hk/cupem/apsa

London Research Centre   www.london-research.gov.uk/et/planhpweb.htm

PlaNet
    Network of European planning students   www.planningnetwork.org

The Joseph Rowntree Foundation – summaries of research papers   www.jrf.org.uk

Town and Country Planning Summer School   www.tcpss.co.uk

Transport Research Laboratory   www.trl.co.uk/ctopics.htm

UK PLANNING SCHOOLS (many provide links to other planning resources)

Bristol, West of England   www.uwe.ac.uk/fbe

Cardiff	www.cf.ac.uk/uwcc
Cheltenham and Gloucestershire	www.chelt.ac.uk/el/soe
Dublin	www.ucd.ie/~engineer
Dundee	www.trp.dundee.ac.uk
Edinburgh, Heriot Watt	www.eca.ac.uk/plan-hous/index.html
Leeds Metropolitan	www.lmu.ac.uk/hen/benv/index.htm
Liverpool John Moores	cwis.livjm.ac.uk/blt
Liverpool	www.liv.ac.uk/civdes/civdes.html
London, University College (The Bartlett)	www.bartlett.ucl.ac.uk/planning
London, South Bank	pisces.sbu.ac.uk/BE/UES
Manchester	www.art.man.ac.uk/planning
Newcastle-upon-Tyne	www.npl.ncl.ac.uk
Nottingham (see online planning resources)	www.nottingham.ac.uk/sbe
Oxford Brookes	www.brookes.ac.uk/schools/planning
Reading	www.rdg.ac.uk/AcaDepts/le/SPS
Sheffield Hallam	www.shu.ac.uk/schools/sed
Sheffield	www.shef.ac.uk/uni/academic/R-Z/trp
Strathclyde	www.strath.ac.uk/department/architecture/index.html

PUBLICATIONS SEARCHES AND BUYING

British Library Public Catalogue	blpc.bl.uk
Zetoc British Libray Contents of Journals	zetoc.mimas.ac.uk
The Stationery Office Bookstore	www.clicktso.com

1

THE NATURE OF PLANNING

If planning were judged by results, that is, by whether life followed the dictates of the plan, then planning has failed everywhere it has been tried. No one, it turns out, has the knowledge to predict sequences of actions and reactions across the realm of public policy, and no one has the power to compel obedience.

(Wildavsky 1987)

The challenge for planning in the 1990s is to 'adapt' not only to new substantive agendas about the environment and how to manage it, but to address new ways of thinking about the relation of state and market and state and citizen, in the field of land use and environmental change.

(Healey 1992b)

INTRODUCTION

It is the purpose of this chapter to give a general introduction to the character of land use planning. Since this is so much a product of culture, it differs among countries. The understanding of one system is helped by comparing it with others and, for this reason, some international comparisons are introduced. The chapter presents a broad discussion of some basic features of the UK planning system, which is essentially a means for reconciling conflicting interests in land use.

Many of the arguments about planning revolve around the relationships between theory and practice. Planning theories (along with related theories on management, government, and other facets of human interaction) have often been based on abstract models stemming from notions of rationality, defined in normative terms. There are difficulties with the concept of rationality. Some of these stem from the fact that planning operates within an economic system which has a 'market rationality' which can differ from, and conflict with, the rationality that is espoused in some planning theories. But the crucial issue is that

the concept of rationality cannot be divorced from objectives, ambitions, and interests – as well as place and time. These variables are the very stuff of planning: disputes and conflict arise not because of irrationalities (though these may be present), but because different interests are rationally seeking different objectives. A brief discussion of these matters leads into issues such as those of incrementalism and implementation, both of which present their own rationales for behaviour and attitudes.

A notable feature of the UK system is the unusual extent to which it embraces discretion. This allows for flexibility in interpreting the public interest. It is in sharp contrast to other systems which, more typically, explicitly aim at reducing such uncertainty. The European and US systems, for example, eschew flexibility, and lay emphasis on the protection of property rights. Flexibility is highly regarded in the UK because it enables the planning system to meet diverse requirements and the constantly changing nature of the problems with which it attempts to deal. Shifts in planning policy have been dramatic, and seem to be accelerating – with a greater concern for market forces, though only within circumscribed limits, and with surges in public and political support for and against development and conservation.

CONFLICT AND DISPUTES

Land use planning is a process concerned with the determination of land uses, the general objectives of which are set out in legislation or in some document of legal or accepted standing. The nature of this process will depend in part on the objectives which it is to serve. The broad objective of the UK system is to 'regulate the development and use of land in the public interest' (PPG 1: 2). Like all such policy statements, this is a very wide formulation of purpose. It is not, however, empty of content: it enshrines the essential character of the UK planning system. Its significance is highlighted when it is compared with possible alternatives. These might be 'to encourage the development and use of land' or 'to facilitate the development of land by private persons and corporations'. Other alternatives include 'to plan the use of land to ensure that private property rights are protected and that the public interest is served' and (an example from Indiana) 'to guide the development of a consensus land use and circulation plan'. These scene-setting statements convey the overall philosophy or principles which are to guide the planning system. They are important for that reason, and they are of direct concern in disputes about the validity or appropriateness of policies which are elaborated within their framework. They are called upon in support of arguments about specific policies.

Politics, conflict, and dispute are at the centre of land use planning. Conflict arises because of the competing demands for the use of land, because of the externality effects that arise when the use of land changes, and because of the uneven distribution of costs and benefits which result from development. If there were no conflicts, there would be no need for planning. Indeed, planning might usefully be defined as the process by which government resolves disputes about land uses.

Alternatives arise at every level of planning – from the highest (supranational) level to the lowest (site) level. The planning system is the machinery by which these levels of choice are managed – from plan-making to development control. Though planning systems vary among countries, they can all be analysed in these terms. The processes involved encompass the determination of objectives, policy-making, consultation and participation, formal dispute resolution, development control, implementation, and the evaluation of outcomes.

The explicit function of the processes is to ensure that the wide variety of interests at stake are considered and that outcomes are in the general 'public interest'. In reality there are very many interests that might be served. Four main groups of participants are politicians serving various levels of government, the development industry, landowners, and 'the public'. The latter is a highly diverse group which is achieving an increased role (not always meaningfully) by way of pressure group and public involvement. Governments usually argue that a reasonable balance is being achieved between the different interests. Critics argue that intervention through land use planning serves to maintain the dominance of particular interests. Evaluations of planning suggest that those with a property interest are more influential and get more out of the planning system, but organised interest groups and even some individuals have had success in individual cases, so the outcomes are by no means certain.

One of the reasons for the increased importance attached to planning processes, and public involvement in them (apart from questions of democracy), lies in the belief that they are effective in reducing the scope for later conflict. The clearer and firmer the policy, and the wider its support, the less room there is for arguing about its application and implementation. Thus for the managers of the system efficiency is increased. But there are limits to this: there is no way that conflict can be planned away.

A central problem for the planning system is to devise a means for predicting likely future changes that may impact on the system. In fact, this is extremely difficult, and past attempts have demonstrated that there is a severe limit to prediction. This is one of the reasons why discretion has to be built into the processes: without this, it is difficult to take account of changing circumstances. A second, more immediate reason for discretion is the impossibility of devising a process which can be applied automatically to the enormous variety of circumstances that come to light when action is being taken. Plans and other policy documents provide a reference point for what has been agreed through the planning process, and against which proposals will be measured. Professional research and analysis, together with opportunities for consultation, public participation, and formal objection and adoption by political representatives, give such documents legitimacy. But they cannot be blueprints. The implementation of a plan always differs from what is anticipated.

There are several reasons for this. For the individuals concerned, the actuality of implementation may appear different from the perceived promise of a plan. For the landowners involved, the market implications may prove to be unwelcome (whether or not market conditions have changed). There will generally be those whose objections at the plan-making stage were rejected in favour of alternatives, and who will naturally take advantage of any opportunity to repeat their objections at the stage of implementation: the passage of time and the changes that it has wrought provide that opportunity. The change in conditions may be so great that the plan is outdated or even counter-productive. Where this is the case, there is a clear need for a revised plan, but problems mount in the meantime and (since the process for elaborating a new plan has still to be completed) the areas of dispute multiply. In addition to these general issues, there is always scope for dispute on the detailed application of policy to individual cases: no plan can be so detailed as to be self-implementing. Finally, there are the cases where there is no policy; or where the policy is simply not relevant to the action that needs to be taken. For these and similar reasons, there has to be machinery for settling disputes concerning implementation.

Adjudication of disputes may be the responsibility of an administrative system (which is theoretically subservient to the political system), the courts, or an *ad hoc* machine. The courts have a major role in countries where planning involves issues which are subject to constitutional safeguards (of property rights or due process), or where plans have the force of law. Where there are no such complications, as is typical in the UK, matters of dispute are more likely to be dealt with by an administrative appeal system. However, there is no hard and fast rule about this, and recent years have witnessed an increasing role for the UK courts. Nevertheless, there remains a huge difference in the role of the courts in the UK compared with the US and most European countries.

PLANNING, THE MARKET AND THE DEVELOPMENT PROCESS

In the early years of the less sophisticated immediate postwar period, plans were drawn up in a vacuum which blissfully ignored the manner in which the property market and development processes worked. Land was allocated to uses which seemed sensible in planning terms, but with little regard to the market. Indeed, market considerations were often explicitly expressed in terms which cast them as subservient to needs. Given the positive role which was envisaged for public development, this had some semblance of logic, but it rapidly disintegrated in the face of the realities of public finance and the incapacity of public authorities to take on the primary development role.

There is today a better (though very incomplete) understanding of how land and property markets work, and a greater appreciation of the need to take account of market trends (even if they have to be subjected to public control or influence). There is also a greater willingness on the part of both the public and the private sectors to pool their efforts and resources: the word 'partnership' is an important addition to the planning lexicon. Of course, this has not ushered in a new era of sweet harmonious cooperation: there are inherent conflicts of interest between (and within) the two sectors. The planning system provides an important mechanism for mediating among these conflicts.

There have been serious difficulties here (by no means resolved) which stem, in part, from a mutual ignorance between the planning and the development sectors. However, attitudes have changed somewhat as a result of a miscellany of forces, ranging from an increased concern on the part of local authorities to promote economic development, to the changed fortunes of the constituent parts of the development industry. Since there is no prospect of a future period of calm stability, attitudes will continue to change. This hardly promises a good base for traditional-type planning (as caricatured by the term 'end state planning'); rather, it promises an even greater role for flexibility and discretion.

A mention of some crucial features of the development process highlights the nature of the conflicts with which planning is concerned:

- Developers are concerned with investment and profits, particularly in a time-frame considerably shorter than is typical in the planning world, where the preoccupation is with long-term land use.
- Developers need to act quickly in response to market opportunities and the cost of capital. Planners operate in a different time scale.
- Development is much easier on greenfield sites than on inner-city locations: to developers, projects are risky enough without being burdened with 'extraneous' problems. Problems which are 'extraneous' to developers can be central to the concerns and objectives of planners.
- Markets are very diverse: and one location is not as good as another. They are, moreover, dynamic: timing may be the crucial factor in the feasibility of a development. Markets are frequently simply not understood by planners: their concern is more generally with the unfolding of a long-term plan. Any pressures they experience are more likely to be political than economic in origin.
- Developers are concerned with the particular; for planners, the particular is only one among many which add up to the general policy matters with which they are concerned.

Given such major differences between the two sectors, it is not surprising that their relationships can be difficult. The problem for planning is that full consideration for developers' concerns can quickly lead to *ad hoc* responses which undermine planning policy.

The very vagueness of policy statements and the high degree of discretion in the system increase the likelihood of this. The dilemma is inherent: there is no simple solution.

It is not surprising that the comprehensive planning philosophy which was dominant in the early postwar period is now discredited. It relates to a world which does not exist. Thus, planning has moved from a preoccupation with grand plans to a concern for finding ways of reconciling the conflicting interests which are affected by development. This leads away from development control to negotiation and mediation. Paradoxically, this is happening at the same time that the central government is attempting to secure a greater degree of certainty through a plan-led system. Perhaps the circle will be squared if it is found that mediation leads to greater certainty? US experience shows how developers can work more easily under a negotiatory system than within a regulatory framework: they know how to operate it to their benefit.

RATIONALITY AND COMPREHENSIVE PLANNING

Central to planning is the concept of rationality. Since rationality requires all relevant matters to be taken into account, the use of the concept readily leads to a comprehensive conception of planning. This stems from two simple ideas: the first (valid) is that in the real world, everything is related to everything else; the second (invalid) is that the planning of one sector cannot properly proceed without coordinated planning of others. Rationality also requires the determination of objectives (and therefore, though not always explicitly, of values), the definition of the problems to be solved, the formulation of alternative solutions to these problems, the evaluation of these alternatives, and the choice of the optimum policy.

This application of the idea of rational scientific method to policy-making has been subject to scrutiny from many different perspectives, giving rise to a wide range of ideas about the nature of planning. First, there are those who have criticised the simple notion of rationality noted above but have continued to maintain that the task for planning theorists is to elaborate the notion of planning as a set of procedures by which decisions are made (Davidoff and Reiner 1962; Faludi 1987). Second, some people reject the objectivity implied by the simple rational approach and instead focus on the role that planning plays in the distribution of resources among different interests in society. Part of this criticism has included the development of a body of thought variously described as social, community, or equity planning, where planning is promoted as a tool that can redress inequalities and work to benefit minority and disadvantaged groups (Gans 1991). Third, there are more fundamental criticisms from those who have used a neo-Marxist critique to draw attention to more fundamental divisions of power in the political and economic structure of capitalist society (Paris 1982). This view asserts that irrespective of its explicit intentions, planning will inevitably 'serve those interests it seeks to regulate' (Ambrose 1986).

The persuasiveness of the concept of comprehensive rational planning is seen at every stage of the plan-making process, from the initial production of goals and objectives, to definitions of problems, and proposed solutions. But all this is done in the context of the politics of the place and the time, and against the background of public opinion and the acceptability or otherwise of governmental action. Some important issues may be regarded not as problems capable of solution but as powerful economic trends which cannot be reversed. Others may be of a nature for which possible solutions are conceivable but untried, too costly, too administratively difficult, too uncertain, or even dangerous to the long-term future of the area. And, as will be apparent from later chapters, these acutely difficult problems (of urbanisation, congestion, inner-city decay, for example) have continually proved beyond the powers of governments to solve, at least in the short run; and the long run is unpredictable. Big differences of opinion exist among experts, politicians, and electors on these matters. As a result, there are severe constraints operating on the planning process, and there is little resembling a logical, calm set of procedures informed by intellectual debate. Certainly, the process is far from scientific or rational.

Practitioners are quick to point out that planning involves deciding between opposing interests and objectives: personal gain versus sectional advantage or public benefit, short-term profit versus long-term gain, efficiency versus cheapness, to name but a few. It entails mediation among different groups, and compromise among the conflicting desires of individual interests. Above all, it necessitates the balancing of a range of individual and community concerns, costs, and rights. It is essentially a political as distinct from a technical or legal process, though it embraces important elements of all these aspects.

Currently, one of the most difficult planning issues is concerned with reconciling the implications of the growth in traffic with traditional ideas about town centres and urban growth. The policy response has been as confused as the issues are complex and politically daunting. So far, the main focus has been on controlling the number of additional out-of-town shopping centres, and directing development back to town centres and brownfield sites. Conflicts that arise here include the apparently irrepressible demand for car ownership and use, the traditional view that road space is free and should not incur any type of congestion charge, the desire of town centre businesses to maintain their custom and to avoid the risk of losing it because of tighter parking restrictions, the financial difficulties facing public transport, and so on. Any one of these issues on its own would be difficult enough; all of them together constitutes a planning witches' brew. And, as so often happens, ends and means become intertwined in a hopelessly confusing way: protecting city centres, safeguarding inner-city jobs, conserving the countryside, reducing pollution, facilitating ease of access for the car-less as well as the car traveller, providing greater choice for the shopper – which is the objective and which the solution? These issues are examined in later chapters; they are listed here to underline the essentially political nature of planning. Grand phrases about rational planning to 'coordinate land uses' crumble against the stark reality of the complex real world.

The concept of comprehensive planning in theory may be contrasted with the narrowly focused planning which takes place in practice. Each administrative agency takes its decisions within its particular sphere of interest, understanding, resources, and competence. How can it be otherwise? The task of any agency is to undertake the task for which it is established, not to take on the complicating and possibly conflicting responsibilities of others (which in any case would be resistant to a take-over). Thus, a conservation agency will take decisions of a very different character from an economic development agency: they have separate and potentially conflicting goals. The idea that there is some level of planning which can rise above the narrow sectionalism of individual agencies is not only inconceivable in terms of implementation: it also assumes that an overriding objective can be identified and articulated. This is typically expressed in terms of the public interest; yet there are very many 'publics'. They have conflicting interests which are represented by, or reflected in, different agencies of government. This simple point is worth emphasising at a time when planning is promoted as a means of sectoral policy integration and achieving 'joined-up government'.

Change takes place not only in physical terms but also socially, economically, institutionally, and indeed in many other ways. The spatial restructuring is the most dramatic visually but, in terms of the quality of everyday life, other dimensions are of greater importance: income and income security, employment, health services, and education, and also matters relating to race, handicap, age, and gender. Each of these has its own brand of planning, and it is sometimes suggested that there should be an overarching planning system which coordinates all of these. This is an extreme form of comprehensive planning which, even if inconceivable, frequently arises in discussions of the limitations of purely physical planning.

Even coordination among the various agencies of planning is difficult and, not surprisingly, rare. Planners have made more claims for comprehensiveness than other professionals; indeed, the search for this has been their distinguishing feature. The fact that they have neither the responsibilities nor the resources for such ambitious aims has not, in the past, prevented them from being articulated. A classic example is the Greater London Development Plan, which was subject to a searching inquiry in 1970. The plan dealt not only with the land use issues which fell within the remit of the Greater London Council, but also with a wide range of policy areas including employment, education, transport, health, and income distribution. There was no doubt about the importance of these and similar issues, but they did not fall within the responsibility of the Council; indeed, it had no way of exerting any influence in these fields (Centre for Environmental Studies 1970).

The experience of the Greater London Development Plan cast a shadow over the hopes of the more ambitious planners of the time, and it affected the attitude of central government to the definition of matters which were relevant to an official development plan. It has taken a long time for central government to feel able to countenance giving official blessing to the need even to take account of 'social needs and problems'. That it has now done so (in PPG 12, quoted later) suggests that the fear of over-ambitious plan-making has receded: local planning authorities are now too experienced in the implementation of plans to seek the impossible. Indeed, the contemporary problem may be the opposite one of believing that too little is possible. The apparent impossibility of tackling problems raised by increased traffic, rising housing demands, and suchlike could lead to a virtual demise of all planning that is not simply regulatory: planning could degenerate into nothing more than a sophisticated form of building control. This is probably an exaggeration – but it is not impossible.

INCREMENTALISM

The obvious failure of comprehensive planning to attain desired goals has led to a number of alternative models of decision-making processes. Many of these revolve around the problem of making planning effective in a world where values, attitudes, and aspirations differ, where market and political forces predominate, and where uncertainty prevails. Lindblom (1959) dismissed rational comprehensive planning as an impractical ideal. In his view, it is necessary to accept the realities of the processes by which planning decisions are taken: for this he outlined a 'science of muddling through'. Essentially, this incrementalist approach replaces grand plans by a modest step-by-step approach which aims at realisable improvements on an existing situation. This is a method of 'successive limited comparisons' of circumscribed problems and actions to deal with them. Lindblom argues that this is what happens in the real world: rather than attempt major change to achieve lofty ends, planners are compelled by reality to limit themselves to acceptable modifications of the status quo. On this argument, it is impossible to take all relevant factors into account or to separate means from ends. Rather than attempting to reform the world, the planner should be concerned with incremental practicable improvements. There has been much debate on Lindblom's ideas (a good selection is given in Faludi 1973); here it is necessary only to make two points.

First, incrementalism is theoretically different from opportunism: it is a rational and realistic approach to dealing with problems. It rejects a comprehensive analysis of all the available options, and concentrates on what appears practicable and sensible given the constraints of time and resources. The classic illustration of the infeasibility of its opposite is the zero-base budget, which, instead of being based on a previous year's figures, rejects history and questions the justification for every individual item. (The term comes from the baseline of the new budget – zero.) As Wildavsky (1978) has demonstrated, this is a completely unmanageable approach: it overwhelms, frustrates, and finally exhausts those who try it. (Of course, selected items may with benefit be isolated for such treatment; but that is a very different matter.)

Second, incrementalism is more a practical necessity than a desirable model to be followed. All policies need thorough review at times – particularly those embedded in an established development plan. Without the occasional upheaval (and that is what zero-base budgeting or policy-making implies), policies can continue well after they have served their purpose: they may even have become counter-productive. Indeed, incrementalism can lead to disasters, wars often being dramatic illustrations of the point. (The escalation of the Vietnam War is a horrendous example.) The ease with which incrementalism continues does in fact make a break in the continuity difficult; and often both the political and the administrative systems are averse to change. Nevertheless, changes in direction are sometimes essential; and events (particularly unexpected ones) may create the basis for a change, despite any fears about uncertain outcomes. The difficulties are well illustrated by the current heart-searching about transport policy. Here, a reversal of trends is necessary, and is increasingly being recognised; but how change should be brought about, and what its character should be, are highly problematic.

IMPLEMENTATION

The rational model of planning embraces the simplistic view that there is a logical progression through successive stages of 'planning', culminating in implementation. The beguiling logic does not translate into reality. On the contrary, it is highly misleading and dangerous to separate policy and implementation matters. In fact, sometimes policy emanates from ideas about implementation rather than the other way round. Thus a policy of slum clearance or redevelopment focuses on the clearly indicated types of action. The implementation becomes the policy, and the underlying purpose is left in doubt. If the objective is to improve the living conditions of those living in slum areas, there might be better ways of doing this, such as rehabilitation, or area improvement through local citizen action. With such an approach, demolition might be merely an incidental element in the local programme. With clearance as a policy, however, there is a danger that quite different objectives might be served, such as commercial development, or provision for roads and car parking.

Unfortunately, the difficulties involved here are even greater than this suggests, since clearly focused efforts are not enough. For instance, a policy of improving a low-income area by environmental improvements may be explicitly intended to benefit the existing inhabitants, but the added attraction of the area may become reflected in higher rents and prices which could lead to gentrification, thus benefiting a very different group. Similarly, a policy of providing grants to industrialists to move to an area of unemployment may result in the substitution of capital for labour, or the influx of workers with skills not possessed by the local people. A policy of preserving historic buildings by prohibiting demolition or alteration may lead to accelerated deterioration as owners seek ways of circumventing the regulations (and, in the period before the prohibition comes into effect, a rash of demolitions – as with the Firestone factory on the Great West Road, London; which was demolished on a holiday weekend). A policy of reducing urban congestion by controlling growth through the designation of green belts may result in 'leapfrogging' of development, increased commuting, and thus increased urban congestion. Examples could easily be multiplied (Derthick 1972; Hall 1980; Kingdon 1984).

To confuse matters further, arguments about such effects are often complicated by differing views on what the objectives of the policy really are. The green belt case is a particularly good illustration of the point, since defenders (and there are many) can slip from one objective to another with ease. If the green belt does not reduce urban congestion, it provides 'opportunities for access to the open countryside for the urban population' and 'opportunities for outdoor sport and outdoor recreation near urban areas'; and if it does not do this, it does 'retain attractive landscapes and enhance landscapes, near to where people live'. Other objectives (all listed in PPG 2) are 'to improve damaged and derelict land around towns'; 'to secure nature conservation interest'; and 'to retain land in agricultural, forestry and related uses'. There is nothing unique in such a long list of miscellaneous policy objectives (though this one is unusual in the manner in which it is conveniently assembled and articulated). It would be a very sad policy indeed that was unable to meet any objectives in such a list! (To quote Wildavsky's pithy observation, 'objectives are kept vague and multiple to expand the range within which observed behaviour fits'; Wildavsky 1987: 35).

It should be added that sometimes policies have unintended good side effects. Unfortunately, it is often difficult to relate cause and effect, but one example is the imposition by the US federal government of a 55 mile per hour speed limit. This was introduced to reduce petrol consumption, but a welcome effect was a reduction in road accidents: this, for a time, became the basis of a powerful argument for retaining the speed limit after the fuel crisis had passed.

The points do not need labouring: the certainty which is required for the type of rational planning envisaged in some traditional theories is impossible. The underlying assumptions, relevance, and political support can change dramatically; and the outcomes of policy are difficult to predict, are frequently different from expectations, are hard to identify and to separate from all the other forces at work, and are rarely clear. Thus, not only is planning a hazardous exercise, with serious likelihood of failure; it is also an exercise whose outcomes are remarkably difficult to evaluate, even when they are felt to be a resounding success.

It is perhaps unsurprising that most planners have neither the time nor the remit to examine what went wrong with the last plan: they have moved on to the next one! It is, however, a matter of some surprise that there have been so few analyses of the (UK) planning scene to fill the vacuum. The wealth of US studies indicates how valuable this can be. Perhaps it is another cultural characteristic that there is little interest in learning why things go wrong?

THE BRITISH PLANNING SYSTEM IN COMPARATIVE PERSPECTIVE

Since it is easier to understand one planning system by comparing it with another, it is worth exploring a little further the differences between the UK, the USA and other European systems. Three features are of particular interest: first, the extent to which a planning system operates within a framework of constitutionally protected rights; second, the degree to which a system embodies discretion; third, the importance of history and culture.

In many countries the constitution limits governmental action in relation to land and property. In the USA, the Bill of Rights provides that 'no person shall . . . be deprived of life, liberty or property without due process of law; nor shall private property be taken without just compensation'. These words mean much more than is apparent to the casual (non-American) reader. Since land use regulations affect property rights, they are subject to constitutional challenge. They can be disputed not only on the basis of their effect on a particular property owner (i.e. *as applied*), but also in principle: a regulation can be challenged on the argument that, in itself, it violates the constitution (this is described in the legal jargon as being *facially* unconstitutional). Moreover, the constitution protects against arbitrary government actions, and this further limits what can be done in the name of land use planning. No such restraints exist in the UK system. Indeed, the UK does not have a codified constitution of the type common to most other countries (Yardley 1995).

Constitutions can influence the system in more subtle ways. In some European countries, including Italy, the Netherlands, and Spain, the constitution provides that all citizens have the right to a decent home. This may limit planning action, but may also influence policy priorities and provide legitimacy for intervention. In Finland and Portugal, landowners are granted the constitutional right to build on their land. This presents obvious difficulty in pursuing policies of restraining urban growth. Constitutions also often allocate powers to different tiers of government, which effectively ensures a minimum degree of autonomy for regional and local governments. Again, there is no such constitutional safeguard in the UK. As a result, the Thatcher government was able to abolish a whole tier of metropolitan local government in England and, in consequence, that part of the planning system that went with it. Such haughty action would be inconceivable in most countries. In the USA, for example, there is little to compare with the central power which is exercised by the national government in Britain. Plan-making and implementation are essentially local issues, even though the federal government has become active in highways, water, and environmental matters, and in recent years a number of states have become involved in land use planning. So local is the responsibility that even the decision on whether to operate land use controls is a local one; and many US local governments have only minimal systems.

Lack of constitutional constraint allows for a wide degree of discretion in the UK planning system. In determining applications for planning permission, a local authority is guided by the development plan, but is not bound by it: other 'material considerations' are taken into account. In most of the rest of the world, plans become legally binding documents. Indeed, they are part of the law, and the act of giving a permit is no more than a certification that a proposal is in accordance with the plan. In practice, there are mechanisms that allow for variations from the provisions of a plan but, since these are by definition contrary to the law, they may entail lengthy procedures, and perhaps an amendment to the plan.

This discretion is further enlarged by the fact that the preparation of a local plan is carried out by the same local authority that implements it. This is so much a part of the tradition of British planning that no one comments on it. The US situation is different, with great emphasis being placed on the separation of powers. (Typically the plan is prepared by the legislative body – the local authority – but administered by a separate board.) The British system has the advantage of relating policy and administration (and easily accommodating policy changes) but, to US eyes, 'this institutional framework blurs the distinction between policy making and policy applying, and so enlarges the role of the administrator who has to decide a specific case' (Mandelker 1962: 4). The Human Rights Convention also focuses attention on the separation of powers since it provides for the right to appeal to an independent body against actions of government. Though there is a limited right of appeal to the courts in the UK (which are independent), most appeals are heard by the government or its representatives. There may need to be changes to the planning system to meet the requirements of the Convention.

Above all, in comparing planning systems, there are fundamental differences in the philosophy that underpins them. Thus, put simply (and therefore rather exaggeratedly), US planning is largely a matter of anticipating trends, while in the UK there is a conscious effort to bend them in publicly desirable directions. In France, *aménagement du territoire* (the term often incorrectly used as a translation for town and country planning) deals with the planning of the activities of different government sectors to meet common social and economic goals, while in the UK, town and country planning is about the management of land use, albeit taking into account social and economic concerns.

Planning systems are rooted in the particular historical, legal, and physical conditions of individual countries and regions. In the UK, some of the many important factors which have shaped the system are the strong land preservation ethic, epitomised in the work of the Council for the Protection of Rural England (and its Scottish and Welsh counterparts) and, of longer standing, the husbandry of the land-owning class. Added (but not unrelated) to this are the popular attitudes to the preservation of the countryside and the containment of urban sprawl, which in turn are related to the early industrialisation of the UK; the small size of the country; the long history of parliamentary government; and the power of the civil service in central government and the professions in local government.

In comparison, land in the USA has historically been a replaceable commodity that could and should be parcelled out for individual control and development; and if one person saw fit to destroy the environment of his or her valley in pursuit of profit, well, why not? There was always another valley over the next hill. Thus the seller's concept of property rights in land came to include the right of owners to earn a profit from their land, and indeed to change the very essence of the land, if necessary to obtain that profit.

In the Mediterranean countries of Greece, Italy, Portugal, and Spain there has been a short history of democratic government, and planning regulation has enjoyed little general public support. Controlling land use has been much less a political priority than housing the population. In large parts of these countries rapid urbanisation has proceeded with little regard to regulations or plans. The historic cores of cities, meanwhile, have not until recently felt the scale of pressure for redevelopment which has been the norm in northern Europe.

However, in all countries land for development is becoming more valuable, and the problem of coping with land use conflicts is of increasing importance. In Europe this has led to the growth of a conservationist ethic, with the restraint of urban growth being a top priority. In the USA this has happened to a limited extent, particularly with environmentally valuable resources, but a major effect was in the opposite direction: to increase the attractiveness of land as a source of profit. Speculation has never been frowned upon in the USA. In many countries, land is regarded as different from commodities: it is something to be preserved and husbanded. In the USA, the dominant ethic regards land as a commodity, no different from any other. Though there is much rhetoric to the contrary, actions speak louder than words.

The contrast in the operation of planning in different countries is abundantly clear to anyone who travels.

ACCOMMODATING CHANGE

Having drawn the comparison, we immediately need to qualify it: times and attitudes change, sometimes slowly, sometimes dramatically. The largest postwar change in the UK has been the move from 'positive planning' to a more market-conscious (and sometimes market-led) approach. The elements of this (which range from the abolition of development charges to the embrace of property-led urban regeneration) are discussed later, but it needs to be stressed that the extent of the change in planning attitudes towards market forces has been dramatic. The limits of the possible have been redefined in the light of experience and a recognition of the character of the forces at work in the modern world.

Governments are responsive to shifts in electoral opinion, particularly when changes can be made painlessly. The UK planning system provides a route by which change can be implemented not only without pain, but also with little effort. Indeed, the ease with which it can accommodate change is quite remarkable. There has, for instance, been a see-saw in the extent to which economic development, social needs, and environmental concerns have had a high profile. In the

1980s, economic efficiency rose to prime place in the government's order of priorities. (This was the time when the planning system was attacked for its restrictive character: 'locking away jobs in filing cabinets'.) Environmental concerns later became salient – a result of a fascinating combination of conservative forces ranging from green belt voters keen to protect the belt, to a younger generation of protestors who had less to lose but saw more to protect. Social considerations have for long been regarded by central government as being outside the legitimate responsibility of the planning system (a curiously British myopia). Major arguments have raged between the centre and the localities on what is, and what is not, appropriate for inclusion in a development plan. After many years of pressing local authorities to exclude 'social factors', the central government made a curious about-turn in a planning policy guidance note of 1992: 'authorities will wish to consider the relationship of planning policies and proposals to social needs and problems' (PPG 12: para. 5.48). The most recent version of this guidance, published in 1999, continues in the same way.

This flexibility (another aspect of the discretionary nature of the system) is a built-in feature. The statutory framework is essentially procedural; it is almost devoid of substantive content. Local authorities are given the duty to prepare development plans (a rare case where no discretion is allowed; unlike their US counterparts, a local authority is not free to decide not to have a plan). What goes into the plan, however, is very imprecise: 'general policies in respect of the conservation of the natural beauty and amenity of the land; the improvement of the physical environment; and the management of traffic' (Town and Country Planning Act 1990, section 13.3A). More detailed requirements are, of course, spelled out in a range of directions and advice from the central government. But that is the point: the content is added separately, and can be changed in line with what 'the Secretary of State may prescribe'.

Yet changes are often not easy to evaluate, even if only because the implementation of planning policy rests with local authorities and, despite much bandying of words, central government powers over them are limited. There are, of course, various control mechanisms and default powers, but these are cumbersome to use, and they carry political risks. Moreover, central government's understanding of how local government works, and its awareness of what happens in practice, is even more circumscribed. These depths of ignorance have had surprisingly little academic light shed on them: few studies have been undertaken of the actual working of the planning machine. (Note the surprise which was expressed when the report on development control in North Cornwall (Lees 1993) revealed that the local councillors gave favourable consideration to the personal circumstances of local applicants for planning permission.)

Given such considerations, it can be difficult to chart (or even to be aware of) important changes. Legislative amendments and new policy statements are more apparent, but they may not be as important as trends which emerge over time. (For example, it *may* be that one of the most significant operational changes has been the way in which local authorities and housebuilders have evolved a system for negotiating housing land allocations; perhaps in time this model might be followed in other development sectors?) Moreover, major political statements and new laws typically follow rather than precede changes in attitudes and perceptions. The picture is also confused by grand claims for new approaches which seldom last far beyond an initial flurry. Much is obscured by political debate and the use of fuzzy jargon. Changes are more easily seen in retrospect than contemporaneously.

PLANNING QUESTIONS

Planning is an imperative: only the form it takes is optional. At a minimum, some system is required to provide infrastructure – preferably in the right place at the right time. Something more than this is generally accepted to be necessary (and general acceptance is the bedrock of any form of effective planning). But there is no way of determining the extent to which a planning system ought to go in determining 'how much of what should go where and when'. The decision is a political one, even if it is taken by default (i.e. with no effective opposition to its growth). Hence, as stressed earlier, cultural influences are crucial. However, this does not mean that a planning system is hallowed or immune from review and radical change. It may be that the UK system has reached precisely the stage at which this is required, though this is not the place to elaborate such an argument. It is, nevertheless, appropriate to point to some issues which need addressing if the planning system is to adapt to conditions which are very different from those that existed half a century ago when it was introduced.

First, the UK planning system is highly effective in stopping development: it is much less effective in facilitating it. Comparative research on property markets in Europe (Williams and Wood 1994) underlines the lack of 'positive planning'. There are serious weaknesses in anticipating needs and allocating sufficient land for these to be met; in the assembling and acquisition of land (especially in inner cities); and in integrating the planning of infrastructure with new development. Powers exist for such important planning actions, but they are underused since there is insufficient relationship between the (public) planning process and the (largely private) development process. In the 1947 Act it was envisaged that 'positive' planning would be undertaken directly by the public sector. This proved infeasible; and alternative mechanisms are underdeveloped.

Second, the most difficult issue facing any policy is defining the right questions. A mechanism is needed to facilitate this. It could be argued that current UK debates are focused on the wrong questions. Too many are concerned with the 'how' of planning policy rather than the 'why'. Why is the countryside to be protected? Why are city centres to be rehabilitated? Why are additional facilities for travel to be provided? There are many such questions, and though they do not have simple (or readily acceptable) answers, debate upon them would provide a firmer base for policy than exists at present. The debate would, however, raise a further level of policy questions. Thus, it might be asked where retail outlets should be located to maximise convenience, service, and profitability (or whatever other criteria are to be employed), rather than posing the questions in terms of safeguarding existing patterns of development (particularly existing town centres). Instead of asking where the best locations are for housing an additional x million households, argument rages over protecting the countryside from housing development and concentrating new housing in urban areas.

Third, planning deals with a highly complex series of interrelated processes which are imperfectly understood. Though better understanding should be high on the research agenda, these processes will inevitably remain beyond the comprehension needed for fully competent land use planning. It follows that planning must proceed on the basis of either a high degree of ignorance, or belief in the efficacy of some over-riding political or economic philosophy. In practical terms, this implies debating how far the planning process should ally itself to market forces (or socio-economic trends, if that term is preferred).

These issues arise throughout this book. It should be evident from this introductory discussion that they are not easily resolved. Indeed, much 'planning' effort is spent on wrestling with them. There seems no doubt that this will continue.

FURTHER READING

Good starting points on the nature of planning are Taylor (1998) *Urban Planning Theory since 1945* and the older but still relevant chapter 2 of Healey *et al.* (1982) *Planning Theory: Prospects for the 1980s*, and Ravetz (1986) *The Government of Space* (which contains a chapter on theoretical perspectives). A useful collection of early articles is contained in Faludi (1973) *A Reader in Planning Theory*. (This includes the paper by Lindblom referred to in the chapter, together with important papers by writers such as Davidoff, Etzioni, Friedmann, and Meyerson). Later collections contain more recent writings: Healey *et al.* (1982) *Planning Theory: Prospects for the 1980s*, Campbell and Fainstein (1996) *Readings in Planning Theory*, and Fainstein and Campbell (1996) *Readings in Urban Theory*. A helpful analysis of arguments for and against planning is given by Klosterman (1985). For an insightful, succinct discussion of the constant flood of changes which besets planning, see Batty (1990) 'How can we best respond to changing fashions in urban and regional planning?'. Sillince (1986) *A Theory of Planning* gives useful summaries of rational comprehensive and incremental models of procedural planning theory. A fuller account is provided by Alexander (1992) *Approaches to Planning*.

A particularly useful introduction to the analysis of policy issues is Kingdon (1984) *Agendas, Alternatives and Public Policies*. A clear and succinct account of policy processes is given in Ham and Hill (1993) *The Policy Process in the Modern Capitalist State*. There is a good range of readings in an accompanying volume edited by Hill (1993) *The Policy Process*. There are a number of useful papers in Tewdwr-Jones (1996) *British Planning Policy in Transition*. Hall (1980) *Great Planning Disasters* is required reading for all planners, as well as for non-planners who want to know why planning is so difficult. A more complex, but fascinating, study focused on the operation of US federal policy in one urban area is Pressman and Wildavsky (1984) *Implementation*. Very interesting as well as revealing is Derthick (1972) *New Towns In-Town: Why a Federal Program Failed*. Such case studies are much more common in the USA than in the UK (a reflection of the cultural differences in the openness of government). Among the small number of British studies, see Muchnick (1970) *Urban Renewal in Liverpool*; Levin (1976) *Government and the Planning Process* (which focuses on the new and expanded towns); Healey (1983) *Local Plans in British Land Use Planning*; and Blowers (1984) *Something in the Air: Corporate Power and the Environment*. Simmie (1993) *Planning at the Crossroads* summarises research findings on the impact of planning in the UK. A radical critique of the role of planning in society is given by Ambrose (1986) *Whatever Happened to Planning?* See also his *Urban Process and Power* (1994).

For a comparative study of 'certainty and discretion' in planning, see Booth (1996) *Controlling Development*. Discretion is discussed at length (comparing the UK and the USA) by Cullingworth (1993) *The Political Culture of Planning*. A broader discussion of the two countries is given by Vogel (1986) *National Styles of Regulation*.

The challenge that Thatcherism and 'the market' brought to ideas of planning has been addressed in many studies – notably Thornley (1993) *Urban Planning under Thatcherism: The Challenge of the Market*; Allmendinger and Thomas (1998) *Urban Planning and the British New Right*; and Brindley *et al.* (1996) *Remaking Planning*.

Communication, negotiation, and argumentation through planning have dominated discussions about planning theory during the 1990s. Early contributions are by Forester (1982) 'Planning in the face of power' and other papers brought together in the book *Planning in the Face of Power* (1989). Later contributions include those of Innes (1995) 'Planning theory's emerging paradigm: communicative action and interactive practice'; and Healey (1997) *Collaborative Planning*; (1998) 'Collaborative planning in a stakeholder society'; and (1992c) 'Planning through debate: the communicative turn in planning theory'. Consequently, aspects of planning practice have been investigated using these ideas; for example, Healey (1993) 'The communicative work of development plans'; Davoudi *et al.* (1997) 'Rhetoric and reality in British structure planning in Lancashire: 1993–95'; and Tait and Campbell (2000) 'The politics of communication between planning officers and politicians: the exercise of power through discourse'. For a critique, see Tewdwr-Jones and Allmendinger (1998) 'Deconstructing communicative rationality: a critique of Habermasian collaborative planning'.

For some amusing homespun philosophy on planning, see Zucker (1999) *What Your Planning Professors Forgot to Tell You*.

2

THE EVOLUTION OF TOWN AND COUNTRY PLANNING

> The first assumption that we have made is that national planning is intended to be a reality and a permanent feature of the internal affairs of this country.
>
> (Uthwatt Report 1942)

THE PUBLIC HEALTH ORIGINS

Town and country planning as a task of government has developed from public health and housing policies. The nineteenth-century increase in population and, even more significant, the growth of towns led to public health problems which demanded a new role for government. Together with the growth of medical knowledge, the realisation that overcrowded, insanitary urban areas resulted in an economic cost (which had to be borne at least in part by the local ratepayers), and the fear of social unrest, this new urban growth eventually resulted in an appreciation of the necessity for interfering with market forces and private property rights in the interest of social well-being.

The nineteenth-century public health legislation was directed at the creation of adequate sanitary conditions. Among the measures taken to achieve these were powers for local authorities to make and enforce building by-laws for controlling street widths and the height, structure, and layout of buildings. Limited and defective though these powers proved to be, they represented a marked advance in social control and paved the way for more imaginative measures. The physical impact of by-law control on British towns is depressingly still very much in evidence; and it did not escape the attention of contemporary social reformers. In the words of Unwin,

much good work has been done. In the ample supply of pure water, in the drainage and removal of waste matter, in the paving, lighting and cleansing of streets, and in many other such ways, probably our towns are as well served as, or even better than, those elsewhere. Moreover, by means of our much abused bye-laws, the worst excesses of overcrowding have been restrained; a certain minimum standard of air-space, light and ventilation has been secured; while in the more modern parts of towns, a fairly high degree of sanitation, of immunity from fire, and general stability of construction have been maintained, the importance of which can hardly be exaggerated. We have, indeed, in all these matters laid a good foundation and have secured many of the necessary elements for a healthy condition of life; and yet the remarkable fact remains that there are growing up around our big towns vast districts, under these very bye-laws, which for dreariness and sheer ugliness it is difficult to match anywhere, and compared with which many of the old unhealthy slums are, from the point of view of picturesqueness and beauty, infinitely more attractive.

(Unwin 1909: 3)

It was on this point that public health and architecture met. The enlightened experiments at Saltaire (1853), Bournville (1878), Port Sunlight (1887), and elsewhere had provided object lessons. Ebenezer Howard and the garden city movement were now exerting considerable influence on contemporary thought. The National Housing Reform Council (later the National Housing and Town Planning Council) was campaigning for the introduction of town planning. Even more significant was a similar demand from local government and professional associations such as the Association of Municipal Corporations, the Royal Institute of British Architects, the Surveyors' Institute and the Association of Municipal and County Engineers. As Ashworth has pointed out,

> the support of many of these bodies was particularly important because it showed that the demand for town planning was arising not simply out of theoretical pre-occupations but out of the everyday practical experience of local administration. The demand was coming in part from those who would be responsible for the execution of town planning if it were introduced.
>
> (Ashworth 1954: 180)

THE FIRST PLANNING ACT

The movement for the extension of sanitary policy into town planning was uniting diverse interests. These were nicely summarised by John Burns, president of the Local Government Board, when he introduced the first legislation bearing the term 'town planning' – the Housing, Town Planning, Etc. Act 1909:

> The object of the bill is to provide a domestic condition for the people in which their physical health, their morals, their character and their whole social condition can be improved by what we hope to secure in this bill. The bill aims in broad outline at, and hopes to secure, the home healthy, the house beautiful, the town pleasant, the city dignified and the suburb salubrious.

The new powers provided by the Act were for the preparation of 'schemes' by local authorities for controlling the development of new housing areas. Though novel, these powers were logically a simple extension of existing ones. It is significant that this first legislative acceptance of town planning came in an Act dealing with health and housing. The gradual development and the accumulated experience of public health and housing measures facilitated a general acceptance of the principles of town planning:

> Housing reform had gradually been conceived in terms of larger and larger units. Torrens' Act (Artizans and Labourers Dwellings Act, 1868) had made a beginning with individual houses; Cross's Act (Artizans and Labourers Dwellings Improvement Act, 1875) had

introduced an element of town planning by concerning itself with the reconstruction of insanitary areas; the framing of bylaws in accordance with the Public Health Act of 1875 had accustomed local authorities to the imposition of at least a minimum of regulation on new building, and such a measure as the London Building Act of 1894 brought into the scope of public control the formation and widening of streets, the lines of buildings frontage, the extent of open space around buildings, and the height of buildings. Town planning was therefore not altogether a leap in the dark, but could be represented as a logical extension, in accordance with changing aims and conditions, of earlier legislation concerned with housing and public health.

> (Ashworth 1954: 181)

The 'changing conditions' were predominantly the rapid growth of suburban development: a factor which increased in importance in the following decades:

> In fifteen years 500,000 acres of land have been abstracted from the agricultural domain for houses, factories, workshops and railways . . . If we go on in the next fifteen years abstracting another half a million from the agricultural domain, and we go on rearing in green fields slums, in many respects, considering their situation, more squalid than those which are found in Liverpool, London and Glasgow, posterity will blame us for not taking this matter in hand in a scientific spirit. Every two and a half years there is a County of London converted into urban life from rural conditions and agricultural land. It represents an enormous amount of building land which we have no right to allow to go unregulated.
>
> (*Parliamentary Debates*, 12 May 1908)

The emphasis was entirely on raising the standards of *new* development. The Act permitted local authorities (after obtaining the permission of the Local Government Board) to prepare town planning schemes with the general object of 'securing proper sanitary conditions, amenity and convenience', but only for land which was being developed or appeared likely to be developed.

Strangely, it was not at all clear what town planning involved. It certainly did not include 'the remodelling of the existing town, the replanning of badly planned areas, the driving of new roads through old parts of a town – all these are beyond the scope of the new planning powers' (Aldridge 1915: 459). The Act itself provided no definition; indeed, it merely listed nine-teen 'matters to be dealt with by general provisions prescribed by the Local Government Board'. The restricted and vague nature of this first legislation was associated in part with the lack of experience of the problems involved.

Nevertheless, the cumbersome administrative procedure devised by the Local Government Board (in order to give all interested parties 'full opportunity of considering all the proposals at all stages') might well have been intended to deter all but the most ardent of local authorities. The land taxes threatened by the 1910 Finance Act, and then the world war, added to the difficulties. It can be no surprise that very few schemes were actually completed under the 1909 Act.

INTERWAR LEGISLATION

The first revision of town planning legislation which took place after the war (the Housing and Town Planning Act of 1919) did little in practice to broaden the basis of town planning. The preparation of schemes was made obligatory on all borough and urban districts having a population of 20,000 or more, but the time limit (January 1926) was first extended (by the Housing Act 1923) and finally abolished (by the Town and Country Planning Act 1932). Some of the procedural difficulties were removed, but no change in concept appeared. Despite lip-service to the idea of town planning, the major advances made at this time were in the field of housing rather than planning.

It was the 1919 Act which began what Marion Bowley (1945: 15) has called 'the series of experiments in State intervention to increase the supply of working-class houses'. The 1919 Act accepted the principle of state subsidies for housing and thus began the nationwide growth of council house estates. Equally significant was the entirely new standard of working-class housing provided: the three-bedroom house with kitchen, bath, and garden, built at the density recommended by the Tudor Walters Committee (1918) of not more than twelve houses to the acre. At these new standards, development could generally take place only on virgin land on the periphery of towns, and municipal estates grew alongside the private

suburbs: 'the basic social products of the twentieth century', as Asa Briggs (1952 vol. 2: 228) has termed them.

This suburbanisation was greatly accelerated by rapid developments in transportation – developments with which the young planning machine could not keep pace. The ideas of Howard (1898) and the garden city movement, of Geddes (1915), and of those who, like Warren and Davidge (1930), saw town planning not just as a technique for controlling the layout and design of residential areas, but as part of a policy of national economic and social planning, were receiving increasing attention, but in practice town planning typically meant little more than an extension of the old public health and housing controls.

Various attempts were made to deal with the increasing difficulties. Of particular significance were the Town and Country Planning Act of 1932, which extended planning powers to almost any type of land, whether built-up or undeveloped, and the Restriction of Ribbon Development Act 1935, which, as its name suggests, was designed to control the spread of development along major roads. But these and similar measures were inadequate. For instance, under the 1932 Act, planning schemes took about three years to prepare and pass through all their stages. Final approval had to be given by Parliament, and schemes then had the force of law, as a result of which variations or amendments were not possible except by a repetition of the whole procedure. Interim development control operated during the time between the passing of a resolution to prepare a scheme and its date of operation (as approved by Parliament). This enabled, but did not require, developers to apply for planning permission. If they did not obtain planning permission, and the development was not in conformity with the scheme when approved, the planning authority could require the owner (without compensation) to remove or alter the development.

All too often, however, developers preferred to take a chance that no scheme would ever come into force, or that if it did, no local authority would face pulling down existing buildings. The damage was therefore done before the planning authorities had a chance to intervene (Wood 1949: 45). Once a planning scheme was approved, on the other hand, the local authority ceased to have any planning control over individual developments. The scheme was in fact a zoning plan: land was zoned for particular uses such as residential or industrial, though provision could be made for such controls as limiting the number of buildings and the space around them. In fact, so long as developers did not try to introduce a non-conforming use, they were fairly safe. Furthermore, most schemes did little more than accept and ratify existing trends of development, since any attempt at a more radical solution would have involved the planning authority in compensation which it could not afford to pay. In most cases the zones were so widely drawn as to place hardly more restriction on the developer than if there had been no scheme at all. Indeed, in the half of the country covered by draft planning schemes in 1937 there was sufficient land zoned for housing to accommodate 291 million people (Barlow Report 1940: para. 241).

A major weakness was, of course, the administrative structure itself. District and county borough councils were generally small and weak. They were unlikely to turn down proposals for development on locational grounds if compensation was involved or if they would thereby be deprived of rate income. The compensation paid either for planning restrictions or for compulsory acquisition had to be determined in relation to the most profitable use of the land, even if it was unlikely that the land would be so developed, and without regard to the fact that the prohibition of development on one site usually resulted in the development value (which had been purchased at high cost) shifting to another site. Consequently, in the words of the Uthwatt Committee,

> an examination of the town planning maps of some of our most important built-up areas reveals that in many cases they are little more than photographs of existing users and existing lay-outs, which, to avoid the necessity of paying compensation, become perpetuated by incorporation in a statutory scheme irrespective of their suitability or desirability.

These problems increased as the housing boom of the 1930s developed: 2,700,000 houses were built in England and Wales between 1930 and 1940. At the outbreak of war, one-third of all the houses in England

and Wales had been built since 1918. The implications for urbanisation were obvious, particularly in the London area. Between 1919 and 1939 the population of Greater London rose by about two million, of which three-quarters of a million was due to natural increase and over one and a quarter million to migration (Abercrombie 1945). This growth of the metropolis was a force which existing powers were incapable of halting, despite the large body of opinion favouring some degree of control.

THE DEPRESSED AREAS

The crux of the matter was that the problem of London was closely allied to that of the declining areas of the North and of south Wales, and both were part of the much wider problem of industrial location. In the South-East the insured employed population rose by 44 per cent between 1923 and 1934, but in the North-East it fell by 5.5 per cent and in Wales by 26 per cent. In 1934, 8.6 per cent of insured workers in Greater London were unemployed, but in Workington the proportion was 36.3 per cent, in Gateshead 44.2 per cent, and in Jarrow 67.8 per cent. In the early stages of political action these two problems were divorced. For London, various advisory committees were set up and a series of reports issued.[1]

For the depressed areas, attention was first concentrated on encouraging migration, on training, and on schemes for establishing the unemployed in small-holdings. Increasing unemployment accompanied by rising public concern necessitated further action.[2] Special Commissioners were appointed for England and Wales, and for Scotland, with very wide powers for 'the initiation, organisation, prosecution and assistance of measures to facilitate the economic development and social improvement' of the special areas. The areas were defined in the Act and included the north-east coast, west Cumberland, industrial south Wales, and the industrial area around Glasgow. The Commissioners' main task – the attraction of new industry – proved to be extraordinarily difficult, and in his second report Sir Malcolm Stewart, the Commissioner for England and Wales, concluded that 'there is little prospect of the special areas being assisted by the spontaneous action of industrialists now located outside these areas'. On the other hand, the attempt actively to attract new industry by the development of trading estates achieved considerable success, which at least warranted the comment of the Scottish Commissioner that there had been 'sufficient progress to dispel the fallacy that the areas are incapable of expanding their light industries'.

Nevertheless, there were still 300,000 unemployed in the special areas at the end of 1938, and although 123 factories had been opened between 1937 and 1938 in the special areas, 372 had been opened in the London area. Sir Malcolm Stewart concluded, in his third annual report, that 'the further expansion of industry should be controlled to secure a more evenly distributed production'. Such thinking might have been in harmony with the current increasing recognition of the need for national planning, but it called for political action of a character which would have been sensational. Furthermore, as Neville Chamberlain (then Chancellor of the Exchequer) pointed out, even if new factories were excluded from London, it did not necessarily follow that they would forthwith spring up in south Wales or west Cumberland. The immediate answer of the government was to appoint the Royal Commission on the Distribution of the Industrial Population.

THE BARLOW REPORT

The Barlow Report (1940) is of significance not merely because it is an important historical landmark, but also because some of its major recommendations were for long accepted as a basis for planning policy.

The terms of reference of the Commission were:

to inquire into the causes which have influenced the present geographical distribution of the industrial population of Great Britain and the probable direction of any change in the distribution in the future; to consider what social, economic or strategic disadvantages arise from the concentration of industries or of the industrial population in large towns or in particular areas of the country; and to report what remedial measures if any should be taken in the national interest.

These very wide terms of reference represented, as the Commission pointed out, 'an important step forward in contemporary thinking' and, after reviewing the evidence, it concluded that

the disadvantages in many, if not most of the great industrial concentrations, alike on the strategical, the social and the economic side, do constitute serious handicaps and even in some respects dangers to the nation's life and development, and we are of opinion that definite action should be taken by the government towards remedying them.

The advantages of more urban concentration at that time were clear: proximity to markets, reduction of transport costs, and availability of a supply of suitable labour. But these, in the Commission's view, were accompanied by serious disadvantages such as heavy charges, on account mainly of high site values, loss of time through street traffic congestion, and the risk of adverse effects on efficiency due to long and fatiguing journeys to work. The Commission maintained that the development of garden cities, satellite towns, and trading estates could make a useful contribution towards the solution of the problems of urban congestion.

The London area, of course, presented the largest problem, not simply because of its huge size, but also because 'the trend of migration to London and the Home Counties is on so large a scale and of so serious a character that it can hardly fail to increase in the future the disadvantages already shown to exist'. The problems of London were thus in part related to the problems of the depressed areas:

It is not in the national interest, economically, socially or strategically, that a quarter, or even a larger, proportion of the population of Great Britain should be concentrated within twenty to thirty miles or so of Central London. On the other hand, a policy:

(i) of balanced distribution of industry and the industrial population so far as possible throughout the different areas or regions in Great Britain;
(ii) of appropriate diversification of industries in those areas or regions

would tend to make the best national use of the resources of the country, and at the same time would go far to secure for each region or area, through diversification of industry, and variety of employment, some safeguard against severe and persistent depression, such as attacks an area dependent mainly on one industry when that industry is struck by bad times.

Such policies could not be carried out by the existing administrative machinery: it was no part of statutory planning to check or to encourage a local or regional growth of population. Planning was essentially on a local basis; it did not, and was not intended to, influence the geographical distribution of the population as between one locality or another. The Commission unanimously agreed that the problems were national in character and required 'a central authority' to deal with them. They argued that the activities of this authority ought to be distinct from and extend beyond those of any existing government department. It should be responsible for formulating a plan for dispersal from congested urban areas – determining in which areas dispersal was desirable and whether and where dispersal could be effected by developing garden cities or garden suburbs, satellite towns, trading estates, or the expansion of existing small towns or regional centres. It should be given the right to inspect town planning schemes and 'to consider, where necessary, in cooperation with the government departments concerned, the modification or correlation of existing or future plans in the national interest'. It should study the location of industry throughout the country with a view to anticipating cases where depression might probably occur in the future and encouraging industrial or public development before a depression actually occurred.

Whatever form this central agency might take (a matter on which the Commission could not agree), it was essential that the government should adopt a much more positive role: control should be exercised over new factory-building, at least in London and the Home Counties, that dispersal from the larger conurbations should be facilitated, and that measures should be taken to anticipate regional economic depression.

THE IMPACT OF WAR

The Barlow Report was published in January 1940, some four months after the start of the Second World War. The problem which precipitated the decision to set up the Barlow Commission, that of the depressed areas, rapidly disappeared. The unemployed of the depressed areas now became a powerful national asset. A considerable share of the new factories built to provide munitions or to replace bombed factories were located in these areas. By the end of 1940, 'an extraordinary scramble for factory space had developed'; and out of all this 'grew a wartime, an extempore, location of industry policy covering the country as a whole' (Meynell 1959). This emergency wartime policy, paralleled in other fields, such as hospitals, not only provided some thirteen million square feet of munitions factory space in the depressed areas which could be adapted for civilian industry after the end of the war, but also provided experience in dispersing industry and in controlling industrial location which showed the practicability (under wartime conditions at least) of such policies. The Board of Trade became a central clearing-house of information on industrial sites. During the debates on the 1945 Distribution of Industry Bill, its spokesman stressed:

> We have collected a great deal of information regarding the relative advantage of different sites in different parts of the country, and of the facilities available there with regard to local supply, housing accommodation, transport facilities, electricity, gas, water, drainage and so on. . . . We are now able to offer to industrialists a service of information regarding location which has never been available before.

Hence, though the Barlow Report 'lay inanimate in the iron lung of war',[3] it seemed that the conditions for the acceptance of its views on the control of industrial location were becoming very propitious: there is nothing better than successful experience for demonstrating the practicability of a policy.

The war thus provided a great stimulus to the extension of regional planning into the sphere of industrial location. This was not the only stimulus it provided: the destruction wrought by bombing transformed 'the rebuilding of Britain' from a socially desirable but somewhat visionary and vague ideal into a matter of practical and clear necessity. Nor was this all: the very fact that rebuilding would be taking a large scale provided an unprecedented opportunity for comprehensive planning of the bombed areas and a stimulus to overall town planning. In the Exeter Plan, Thomas Sharp (1947: 10) urged that

> to rebuild the city on the old lines . . . would be a dreadful mistake. It would be an exact repetition of what happened in the rebuilding of London after the Fire – and the results, in regret at lost opportunity, will be the same. While, therefore, the arrangements for rebuilding to the new plan should proceed with all possible speed, some patience and discipline will be necessary if the new-built city is to be a city that is really renewed.

Lutyens and Abercrombie (1945: 1) argued that in Hull, there is now both the opportunity and the necessity for an overhaul of the urban structure before undertaking this second refounding of the great Port on the Humber. Due consideration, however urgent the desire to get back to working conditions, must be given to every aspect of town existence.

This was the social climate of the war and early postwar years. There was an enthusiasm and a determination to undertake 'social reconstruction' (i.e. public-sector intervention) on a scale hitherto considered utopian. The catalyst was, of course, the war itself. At one and the same time war occasions a mass support for the way of life which is being fought for and a critical appraisal of the inadequacies of that way of life. Modern total warfare demands the unification of national effort and a breaking down of social barriers and differences. As Titmuss (1958: 85) noted, it 'presupposes and imposes a great increase in social discipline; moreover, this discipline is tolerable if, and only if, social inequalities are not intolerable'. On no occasion was this more true than in the Second World War. A new and better Britain was to be built. The feeling was one of intense optimism and confidence. Not only would the war be won, but it would be followed by a similar campaign against the forces of want. That there was much that was inadequate, even intolerable, in prewar Britain had been generally accepted. What was new was the belief that the problems could be tackled in the same way as a military operation. What supreme confidence was evidenced by the setting up in 1941 of committees to consider postwar reconstruction problems: the Uthwatt Committee on Compensation and Better-

ment, the Scott Committee on Land Utilisation in Rural Areas, and the Beveridge Committee on Social Insurance and Allied Services. Perhaps it was Beveridge (1942: 170) who most clearly summed up the spirit of the time, and the philosophy which was to underlie postwar social policy:

> The Plan for Social Security is put forward as part of a general programme of social policy. It is one part only of an attack upon five great evils: upon the physical Want with which it is directly concerned, upon Disease which often causes Want and brings many other troubles in its train, upon Ignorance which no democracy can afford among its citizens, upon Squalor which arises mainly through haphazard distribution of industry and population, and upon Idleness which destroys wealth and corrupts men, whether they are well fed or not, when they are idle.

It was within this framework of a newly acquired confidence to tackle long-standing social and economic problems that postwar town and country planning policy was conceived. No longer was this to be restricted to town planning 'schemes' or regulatory measures. There was now to be the same breadth in official thinking as had permeated the Barlow Report. The attack on squalor was conceived as part of a comprehensive series of plans for social amelioration. To quote the 1944 White Paper *The Control of Land Use*, 'provision for the right use of land, in accordance with a considered policy, is an essential requirement of the government's programme of postwar reconstruction'.

THE NEW PLANNING SYSTEM

The prewar system of planning was defective in several ways. It was optional on local authorities; planning powers were essentially regulatory and restrictive; such planning as was achieved was purely local in character; the central government had no effective powers of initiative, or of coordinating local plans; and the 'compensation bogey', with which local authorities had to cope without any Exchequer assistance, bedevilled the efforts of all who attempted to make the cumbersome planning machinery work.

By 1942, 73 per cent of the land in England and 36 per cent of the land in Wales had become subject to interim development control, but only 5 per cent of England and 1 per cent of Wales were actually subject to operative schemes (Uthwatt Report 1942: 9); and there were several important towns and cities as well as some large country districts for which not even the preliminary stages of a planning scheme had been carried out. Administration was highly fragmented and was essentially a matter for the lower-tier authorities: in 1944 there were over 1,400 planning authorities. Some attempts to solve the problems to which this gave rise were made by the (voluntary) grouping of planning authorities in joint committees for formulating schemes over wide areas, but, though an improvement, this was not sufficiently effective.

The new conception of town and country planning underlined the inadequacies. It was generally (and uncritically) accepted that the growth of the large cities should be restricted. Regional plans for London, Lancashire, the Clyde Valley, and south Wales all stressed the necessity of large-scale overspill to new and expanded towns. Government pronouncements echoed the enthusiasm which permeated these plans. Large cities were no longer to be allowed to continue their unchecked sprawl over the countryside. The explosive forces generated by the desire for better living and working conditions would no longer run riot. Suburban dormitories were a thing of the past. Overspill would be steered into new and expanded towns which could provide the conditions people wanted, without the disadvantages inherent in satellite suburban development. When the problems of reconstructing blitzed areas, redeveloping blighted areas, securing a 'proper distribution' of industry, developing national parks, and so on are added to the list, there was a clear need for a new and more positive role for the central government, a transfer of powers from the smaller to the larger authorities, a considerable extension of these powers and, most difficult of all, a solution to the compensation–betterment problem.

The necessary machinery was provided in the main by the Town and Country Planning Acts, the Distribution of Industry Acts, the National Parks and Access to the Countryside Act, the New Towns Act and, later, the Town Development Acts.

The 1947 Town and Country Planning Act brought almost all development under control by making it subject to planning permission. Planning was to be no longer merely a regulative function. Development plans were to be prepared for every area in the country. These were to outline the way in which each area was to be developed or, where desirable, preserved. In accordance with the wider concepts of planning, powers were transferred from district councils to county councils. The smallest planning units thereby became the counties and the county boroughs. Co-ordination of local plans was to be effected by the new Ministry of Town and Country Planning. Development rights in land and the associated development values were nationalised. All the owners were thus placed in the position of owning only the existing (1947) use rights and values in their land. Compensation for development rights was to be paid 'once and for all' out of a national fund, and developers were to pay a development charge amounting to 100 per cent of the increase in the value of land resulting from the development. The 'compensation bogey' was thus at last to be completely abolished: henceforth development would take place according to 'good planning principles'.

Responsibility for securing a 'proper distribution of industry' was given to the Board of Trade. New industrial projects (above a minimum size) would require the board's certification that the development would be consistent with the proper distribution of industry. More positively, the Board was given powers to attract industries to development areas by loans and grants, and by the erection of factories.

New towns were to be developed by *ad hoc* development corporations financed by the Treasury. Somewhat later, new powers were provided for the planned expansion of towns by local authorities. The designation of national parks and 'areas of outstanding natural beauty' was entrusted to a new National Parks Commission, and local authorities were given wider powers for securing public access to the countryside. A Nature Conservancy was set up to provide scientific advice on the conservation and control of natural flora and fauna, and to establish and manage nature reserves. New powers were granted for preserving amenity, trees, historic buildings, and ancient monuments.

Later controls were introduced over river and air pollution, litter, and noise. Indeed, there has been a steady flow of legislation, partly because of increased experience, partly because of changing political perspectives, but perhaps above all because of the changing social and economic climate within which town and country planning operates.

The ways in which the various parts of this web of policies operated, and the ways in which both the policies and the machinery have developed since 1947 are summarised in the following chapters. Here a brief overview sets the scene.

THE EARLY YEARS OF THE NEW PLANNING SYSTEM

The early years of the new system were years of austerity. This was a truly regulatory era, with controls operating over an even wider range of matters than during the war. It had not been expected that there would be any surge in pressures for private development, but even if there were, it was envisaged that these would be subject to the new controls. Additionally, private building was regulated by a licensing system, which was another brake on the private market. Building resources were channelled to local authorities, and (after an initial uncontrolled spurt of private housebuilding) council housebuilding became the major part of the housing programme.

The sluggish economy made it relatively easy to operate regulatory controls (since there was little to regulate), but it certainly was not favourable to 'positive planning'. It had been assumed that most of this positive planning would take the form of public investment, particularly by local authorities and new town development corporations. Housing, town centre renewal, and other forms of 'comprehensive development' were seen as essentially public enterprises. This might have been practicable had resources been plentiful, but they were not, and both new building and redevelopment proceeded slowly. Thus, neither the public nor the private sectors made much progress in 'rebuilding Britain' (to use one of the slogans which had been popular at the end of the war).

The founders of the postwar planning system foresaw modest economic growth, little population increase (except an anticipated short postwar 'baby boom'), little migration either internally or from abroad, a balance in economic activity among the regions, and a generally manageable administrative task in maintaining controls. Problems of social security and the initiation of a wide range of social services were at the forefront of attention: welfare for all rather than prosperity for a few was the aim. There was little expectation that incomes would rise, that car ownership would spread, and that economic growth would make it politically possible to declare (as Harold Macmillan later did) that 'you have never had it so good'. The plans for the new towns were almost lavish in providing one garage for every four houses.

The making of plans went ahead at a steady pace, frequently in isolation from wider planning considerations, though the regional offices of the planning ministry made a valiant attempt at coordination; but even here, progress was much slower than expected, and it soon became clear that comprehensive planning would have to be postponed for the sake of immediate development requirements.

For a time, the early economic and social assumptions seemed to be borne out, but during the 1950s dramatic changes took place, some of which were the result of the release of pent-up demand that followed the return of the Conservative government in 1951 – a government which was wedded to a 'bonfire of controls'. One of the first acts of this government in the planning sphere was a symbolic one: a change in the name of the planning ministry – from 'local government and planning' to 'housing and local government'. This reflected the political primacy of housing and the lack of support for 'planning' (now viewed, with justification, as restrictive). The regional offices of the planning ministry were abolished, thus saving a small amount of public funds, but also dismantling the machinery for coordination. Though this machinery was modest in scope (and in resources), it was important because there was no other regional organisation to carry out this function.

The first change to the 1947 system came in 1953 when, instead of amending the development charge in the light of experience (as the Labour government had been about to do), the Conservatives abolished it. At about the same time, all building licensing was scrapped. Private housebuilding boomed; and curiously, so did council housebuilding, since the high building targets set by the Conservative government could be met only by an all-out effort by both private and public sectors. The birthrate (which – as expected – dropped steadily from 1948 to the mid-1950s) suddenly started a large and continuing rise.

The new towns programme went ahead at a slow pace, accompanied by a constant battle for resources which, so the Treasury argued, were just as urgently needed in the old towns. (The provision of 'amenities' was a particular focus of the arguments.) By contrast, public housing estates and private suburban developments mushroomed. Indeed, there was soon a concern that prewar patterns of urban growth were to be repeated. The conflict between town and country moved to centre stage. This was a more difficult matter for the Conservative government than the abolition of building controls, development charges, and other restrictive measures. New policies were forged, foremost of which was the control over the urban fringes of the conurbations and other large cities where an acrimonious war was waged between conservative counties seeking to safeguard undeveloped land and the urban areas in great need of more land for their expanding housebuilding programmes. On the side of the counties was the high priority attached to maintaining good-quality land in agricultural production. On the side of the urban areas was a huge backlog of housing need. The war reached epic proportions in the Liverpool and Manchester areas, where Cheshire fought bitterly 'to prevent Cheshire becoming another Lancashire'.

Similar arguments were used in the west Midlands, where a campaign for new towns (led by the Midland New Towns Society) was complicated by the government view that Birmingham was a rich area from which to move industry to the depressed areas. London, of course, had its ring of new towns, but these were inevitably slow in providing houses for needy Londoners, particularly since tenants were selected partly on the basis of their suitability for the jobs which had been attracted to the towns. The London County Council therefore, like its provincial counterparts, built houses for 'overspill' in what were then called 'out county estates'. Similarly, Glasgow and Edinburgh built their 'peripheral estates'.

The pressures for development grew as households increased even more rapidly than population – a little-understood phenomenon at the time (Cullingworth 1960a) – and as car ownership spread (the number of cars doubled in the 1950s and doubled again in the following decade). Increased mobility and suburban growth reinforced each other, and new road building began to make its own contribution to the centrifugal forces.

Working in the opposite direction was the implacable opposition of the counties. They received a powerful new weapon when Duncan Sandys initiated the green belt circular of 1955. Green belts no longer even had to be green: their function was to halt urban development. Hope that all interests could be appeased was raised by the Town Development Acts (1952 in England, 1957 in Scotland). These provided a neat mechanism for housing urban 'overspill' and, at the same time, rejuvenating declining small towns and minimising the loss of agricultural land. But though a number of schemes were (slowly) successful, the local government machinery was generally not equal to such a major regional task.

It was this local machinery which was at the root of many of the difficulties. Few politicians wanted to embark on the unpopular task of reforming local government, and even those who appreciated the need for change could not agree on why it was wanted – whether to resolve the urban–rural conflict, to facilitate a more efficient delivery of services, or to provide a system of more effective political units. These and similar issues were grist to the academic mill, but a treacherous area for politicians. Perhaps the biggest surprise here was the decision to go ahead with the reorganisation of London government. The legislation was passed in 1963: this followed (in sequence but not in content) a wide-ranging inquiry. The surprise was not that the recommendations were altered by the political process, but that anything was done at all. One important factor in the politics of the situation was the desire to abolish the socialist London County Council (though ironically the hoped-for guarantee of a permanent Conservative GLC was dashed by the success of the peripheral districts in maintaining their independence).

One effect of the London reorganisation was that further changes elsewhere were taken very seriously. The writing was now on the wall for local government in the rest of the country, and campaigns and counter-campaigns proliferated. Three inquiries (for England, Scotland, and Wales) were established by the Labour government which assumed office in 1964. These reported in 1969, but implementation fell to its successor Conservative government. For Scotland, the recommendations were generally accepted (with a two-tier system of regions and districts over most of the country). The city-region recommendations for England, however, were unacceptable, and a slimmer two-tier system was adopted. Wales was treated in the same way. The result south of the Border was that the boundaries for the urban–rural strife, though amended in detail, were basically unchanged in character. It would be only a matter of time before a further reorganisation was seen to be necessary. What followed is discussed in Chapter 3, but it is clear that the story is a continuing one.

ADDITIONAL NEW TOWNS

In the meantime, truly alarming population projections had appeared which transformed the planning horizon. The population at the end of the century had been projected in 1960 at 64 million; by 1965 the projection had increased to 75 million. At the same time, migration and household formation had added to the pressures for development and the need for an alternative to expanding suburbs and 'peripheral estates'. It seemed abundantly clear that a second generation of new towns was required.

Between 1961 and 1971, fourteen additional new towns were designated. Some, like Skelmersdale and Redditch, were 'traditional' in the sense that their purpose was to house people from the conurbations. Others, such as Livingston and Irvine, had the additional

function of being growth points in a comprehensive regional programme for central Scotland. One of the most striking characteristics of the last new towns to be designated was their huge size. In comparison with the Reith Committee's optimum of 30,000 to 50,000, Central Lancashire's 500,000 seemed massive. But size was not the only striking feature. Another was the fact that four of them were based on substantial existing towns: Northampton, Peterborough, Warrington, and Central Lancashire (Preston-Leyland). Of course, town-building had been going on for a long time in Britain, and all the best sites may have already been taken by what had become old towns. The time was bound to come when the only places left for new towns were the sites of existing towns.

There were, however, other important factors. First, the older towns were in need of rejuvenation and a share in the limited capital investment programme. Second, there was the established argument that nothing succeeds like success; or, to be more precise, a major development with a population base of 80,000 to 130,000 or more had a flying start over one with a mere 5,000 to 10,000. A wide range of facilities was already available, and (it was hoped) could be readily expanded at the margin.

No sooner had all this been settled than the population projections were drastically revised downwards. It was too late to reverse the new new towns programme, though it was decided not to go ahead with Ipswich (and Stonehouse was killed by the opposition of Strathclyde because of its irrelevance to the problems of the rapidly declining economy of Clydeside). However, the reduced population growth prevented some problems becoming worse, though little respite was apparent at the time. Household formation continued apace, as did car ownership and migration. The resulting pressures on the South-East were severe – and have remained so, with little resolution of the difficulties of 'land allocation'.

THE REDISCOVERY OF POVERTY

While much political energy was spent on dealing with urban growth, even more intractable problems of urban decay forced their attention on government. Every generation, it seems, has to rediscover poverty for itself, and the postwar British realisation came in the late 1960s (Sinfield 1973). As usual, there were several strands: the reaction against inhuman slum clearance and high-density redevelopment; the impact of these and of urban motorways on communities ('get us out of this hell', cried the families living alongside the elevated M4 (Goodman 1972)); and fear of racial unrest (inflamed by the speeches of Enoch Powell). These issues went far beyond even the most ambitious definition of 'planning', and they posed perplexing problems of the coordination of policies and programmes. Not surprisingly, the response was anything but coordinated, and programmes proliferated in confusion.

Housing policy was the clearest field of policy development. Slum clearance had been abruptly halted at the beginning of the Second World War, when demolitions were running at the rate of 90,000 a year. It was resumed in the mid-1950s, and steadily rose towards its prewar peak. Both the scale of this clearance and its insensitivity to community concerns, as well as the inadequate character of some redevelopment schemes, led to an increasing demand for a reappraisal of the policy. Added force was given to this by the growing realisation that demolition alone could not possibly cope with the huge amount of inadequate housing – and the continuing deterioration of basically sound housing. Rent control had played a part in this tide of decay, and halting steps were taken to ameliorate its worst effects, though not with much success.

More effective was the introduction of policies to improve, rehabilitate, and renovate older housing: changing terminology reflected constant refinements of policy. Increasingly, it was realised that *ad hoc* improvements to individual houses were of limited impact: area rehabilitation paid far higher dividends, particularly in encouraging individual improvement efforts. A succession of area programmes have made a significant impact on some older urban neighbourhoods, but a considerable problem remains; and it is debatable whether the overall position is improving or deteriorating.

Housing policies have typically been aimed at the physical fabric of housing and the residential environment. Their impact on people generally, and the poor in particular, has been less than housing reformers had hoped (the lessons of earlier times being ignored). This realisation followed a spate of social inquiries, of which *The Poor and the Poorest* by Brian Abel-Smith and Peter Townsend (published in 1965) was a landmark in raising public concern. A bewildering rush of programmes was promoted by the Home Office (including the urban programme in 1968, community development projects in 1969, and comprehensive community programmes in 1974), the DoE (urban guidelines in 1972, area management trials in 1974, and 'the policy for the inner cities' in 1977), the Department of Education (educational priority areas in 1968), and the DHSS (cycle of deprivation studies in 1973). This list is by no means complete, but it demonstrates the almost frantic search for effective policies in fields which had hitherto largely been left to local effort.

Despite all this, the problems of the 'inner cities' (a misnomer, since some of the deprived areas were on the periphery of cities) grew apace. The most important factors were the rapid rate of deindustrialisation and the massive movement of people and jobs to outer areas and beyond. Unlike in the interwar years, the problems were not restricted to the 'depressed areas': the South-East, previously the source for moving employment to the north, was badly affected. In absolute (rather than percentage) terms, London suffered severely (losing three-quarters of a million manufacturing jobs between 1961 and 1984; Hall 1992: 150).

There was initially little difference between the two main political parties here: both were searching for solutions which continued to evade them. Lessons from the USA indicated that more money alone was not sufficient, and academic writers pointed to the need for societal changes, but there were few politically helpful ideas around. Following a period in which the problems were seen in terms of social pathology, attention was increasingly directed to 'structural' issues, particularly of the local economy. In the 1980s the Conservative government put its faith in releasing enterprise, though it was never clear how this would benefit the poor. New initiatives included urban development corporations, modelled on the new town development corporations but with a different, private enterprise ethic. The London Docklands UDC seemed almost determined to ignore, if not override, the community in which it was located, but this attitude eventually changed, and both the LDDC and later UDCs became more attuned to local needs and feelings. Indeed, later policies are characterised by an attempt to be much more sensitive to human needs, with an emphasis on 'bottom-up' planning.

LAND VALUES

The issue of land values was addressed by both the two later Labour governments. In the 1964–70 administration, the Land Commission was established to buy development land at a price excluding a part of the development value and to levy a betterment charge on private sales. Its life and promise were cut short by the incoming Conservative government. Exactly the same happened with the community land scheme and the development land tax introduced by the 1974–79 Labour government. Thus, there were three postwar attempts to wrestle with the problem, and none was given an adequate chance to work. For a time, attention focused on land availability studies. These became a time-demanding ritual for planners, later transformed when increased household projections in the 1990s widened and intensified the debate.

The abandonment of attempts to solve 'the betterment problem' (which may no longer even be perceived as a problem) is more than a matter of land taxation or even equity. The so-called 'financial provisions' of the 1947 Act underpinned the whole system, and made positive planning a real possibility. Though it seems unlikely that the issue will return to the political agenda in the foreseeable future, it should not be forgotten that this vital piece of the planning machinery is missing. Planning is therefore essentially a servant of the market (in the sense that it comes into operation only when market operations are set in motion). This change, made some forty years ago, is far more

fundamental than the high-profile changes made under the Thatcher regime. Whether 'planning gain' can be made the basis of a new approach remains to be seen.

ENTREPRENEURIAL PLANNING

The theme of the Conservative era which began in 1979 was a commitment to 'releasing enterprise'. This was translated into a miscellany of policies which had little in the way of a coherent underlying philosophy, but which could be characterised in terms of removing particular barriers that were identified as holding back initiative. The identified problems ranged from inner-city landholding by public bodies (dealt with by requiring publicising of the vacant land, which would thereby automatically trigger a market use); to the 'wasteful' and 'unnecessary' tier of metropolitan government in London and the provincial conurbations (simply abolished).

Many areas of public activity were privatised, large parts of government were hived off to executive agencies, and compulsory competitive tendering was imposed on local government. The emphasis on 'market orientation' and the concerted attack (regrettably, the word is not an exaggeration) on local government had some strange results. More power was vested in central government and its agencies. Public participation was reduced. But, though the planning system was affected in tangible ways (Thornley 1993), in no sense was it dismantled, or even changed in any really significant way. True, it was bypassed (by UDCs); its procedures were modified (by government circulars, and changes in the General Development and Use Classes Orders); development plans were, for a time, downgraded, and threatened with severe curtailment; and simplified planning zones were introduced: a system in which 'simplification' meant less planning control, but might involve even more human resources in negotiation.

The list can be extended, but the rhetoric which preceded and accompanied the changes was harsher than the changes themselves. Moreover, the language of confrontation which the politicians employed disguised the fact that previous governments had done similar things, even if more *sotto voce*. The development corporation initiative, for example, was essentially the brainchild of a much earlier period, and indeed, as applied to redevelopment (as distinct from new town development), had for long been proposed by socialists as a means of assisting local authorities. Some of the early days of the UDC flagship; the London Docklands Development Corporation, were characterised by an excess of zeal, a lack of understanding of the way in which the administration of government is different from the administration of business, and an authoritarian style which was widely – and justifiably – criticised. Time, however, mellowed misplaced enthusiasm, and brought about a better understanding of the inherent slowness of democratic government. There was also a keener awareness of the need to pay attention to the 'social' issues of the locality as well as its physical regeneration.

More generally, an old lesson was relearned: it is extremely difficult for one level of government to impose its will on another unless it has some broad and powerful support from outside, as well as willing co-operation inside. (There is, however, the draconian alternative of simply abolishing a wayward layer of local government, as was done with the Greater London Council and the metropolitan county councils.)

An about-turn on structure plans illustrates the pragmatic nature (what some call the flexibility and others the inconsistency) of the Conservative government's thinking. The initial decision to abolish them was one option for dealing with a problem which dates back to 1947: how to ensure that plans provide (without overwhelming detail) sufficient guidance for the land use planning of an area, while being adaptable to unforeseen changing circumstances. The option actually adopted was a 'streamlining' – not unlike earlier attempts. The 1965 Planning Advisory Group report had highlighted the problem: 'It has proved extremely difficult to keep these plans not only up to date but forward looking and responsive to the demands of change.'

Twenty years later, the 1985 White Paper *Lifting the Burden* was in a similar key: 'There is cause for concern that this process of plan review and up-dating is becoming too slow and cumbersome.' More effective structure plans require a framework of regional policy.

In the 1990s, this began to be accepted and, following the election of the Blair government, regional policy moved to centre stage.

THE ENVIRONMENT

All governments operate with some degree of pragmatism: electoral politics force this upon them. So it was with the 1979–97 Conservative government. After many years of relegating environmental issues to a low level of concern, there was a sudden conversion to environmentalism in 1988. This was heralded in a remarkable speech by Mrs Thatcher in which she declared that Conservatives were the guardians and trustees of the earth. At base, this reflected a heightening of public concern for the environment which is partly local and partly global.

The action which followed looks impressive (though critics have been less impressed by the results). A 1990 White Paper, *This Common Inheritance*, spelled out the government's environmental strategy over a comprehensive range of policy areas (untypically, this covered the whole of the UK). Environmental protection legislation was passed, 'integrated pollution control' is being implemented, 'green ministers' have been appointed to oversee the environmental implications of their departmental functions, and new environmental regulation agencies have been established. The latter follow a spate of organisational changes which remind one of the old saying: 'when in doubt, re-organise'. But there are difficult issues here which, though they include organisational matters, go much deeper. Questions about the protection of the environment underline a perhaps (to the layperson) surprising ignorance of the workings of ecosystems at the local, national, and global levels. Additionally, new questions of ethics have come to the fore. Difficult problems of deciding among alternative courses of action are rendered ever more complex. Cost–benefit analysis is of little help: indeed, all forms of economic reasoning are being challenged. International pressures have played a role here, as, of course, has the coming of age of the EU. This has added a new dimension to the politics of the environment (and much else as well).

Concern for historic preservation (now embraced in the term 'heritage') is of much longer standing. Though many historic buildings were destroyed during the Second World War, the more effective stimulus to preservation came from the clearance, redevelopment, renewal, and road-building policies which got under way in the 1950s and accelerated rapidly. As with housing, the emphasis has been mainly on individual historic structures, but a conservation area policy was ushered in by the Civic Amenities Act 1967, sponsored by a private member (Duncan Sandys), though with wide support. This proved a popular measure, and there are now over 9,000 of them. Indeed, there has been mounting concern that too many areas are designated, and too few resources applied to their upkeep and management.

The 1983 National Heritage Act bore a modern name that signified a new and wider appreciation of the historical legacy. A new executive agency, English Heritage (formally called the Historic Buildings and Monuments Commission for England), was established and took over many of the functions previously housed within the DoE. In Northern Ireland, Scotland, and Wales, rather different administrative solutions were devised, as befits the distinctive character of these parts of the UK.

Unfortunately, the new environmental and historical awareness was too late in raising sufficient concern about transport to bring about any significant change from a preoccupation with catering for the car.

ROAD-BUILDING POLICIES

Transport policy has traditionally been largely equated with road-building policy, and protests that alternatives need to be considered have been unavailing until recently. On a number of issues, however, the protests could not be ignored. One has already been mentioned: the brutal impact of urban motorways on the communities through which they passed. The outcry against this led to a reassessment of both the location of urban roads and their necessity. Compensation for 'distress' caused by new roads was increased as part of a policy labelled (in a 1972 White Paper) 'putting people first'. Closely related was a growing concern about the inadequacy of the road inquiry process, which resulted in a significant improvement of the provisions for public participation. These and other changes curbed but did not allay the concerns: indeed, they are still being voiced. The turning point came in 1989 when new forecasts of huge increases in car ownership and use were published. It was widely considered to be impossible to accommodate the forecast amount of traffic satisfactorily. The results of a change in attitude were working their way through the political system before the era of 'integrated transport planning'. Traffic calming became part of the contemporary vocabulary (and is now statutorily enshrined); road pricing moved onto the agenda for serious discussion (but little action); and road-building was slashed. This extraordinary reversal of the long standing policy of building roads to meet the demand for them started under the Conservative government. It reflected that government's interpretation of public attitudes to road-building, which nicely attuned with the political objective of reducing tax-related expenditure. In this area at least, a bankruptcy of political ideas (for which persuasive alternatives were sadly in short supply) led firmly into the doldrums.

THE COUNTRYSIDE

The countryside has always been dear to hearts of conservatives, though support for the protection and enjoyment of the countryside has traditionally cut across party and class lines. Increasing concern for the rural landscape, growing use of the countryside for recreation (and investment), and huge changes in the fortunes of the agricultural industry have transformed the arena of debate of rural land use. At the end of the Second World War, and for many years afterwards, the greatest importance was attached to the promotion of agriculture. There were, however, established movements for countryside conservation and recreation, some of which came together with the National Parks and Access to the Countryside Act of 1949 (but a separate Nature Conservancy Council was also established, thus dividing the conservancy function).

The pressures for conservation and for recreation have varied over time, and the balance between them is inevitably an ongoing problem, particularly in areas of easy access (which now include most of the country). Limited budgets held back incipient pressures in the early postwar years, but increasing real incomes and mobility led to mounting pressures which were acknowledged in the 1966 White Paper *Leisure in the Countryside* and the 1968 Countryside Act. This replaced the National Parks Commission with a Countryside Commission which was given wider powers and improved finance. At the same time, the powers of local authorities were expanded to include, for instance, the provision of country parks. Unlike national parks, these were not necessarily places of beauty, but were intended primarily for enjoyment. They were also seen as having the added advantage of taking some of the pressure off the national parks and similar areas where added protection was needed.

The 1972 reorganisation of local government was accompanied by a requirement that local authorities which were responsible for national parks should establish a separate committee and appoint a park planning officer. The modesty of this provision was clearly a compromise between concerns for local government and for the planning of national parks. It was a step forward, but an enduring case for *ad hoc* park authorities remained. Local authorities had too many local interests to satisfy to give adequate resources for national parks – whose very name indicated their much wider role. The growth of pressures on the parks continued, and the administrative knot was finally cut when the Environment Act 1995 provided for the establishment of *ad hoc* national park authorities for all the parks.

More widely, a long-standing debate continued on the divided organisational arrangements for nature conservation and amenity, and for scientific conservation and wildlife. In England that separation continues (on the basis of arguments which are not easy to follow), but in Scotland and Wales the responsibilities are now vested in single bodies: Scottish Natural Heritage and the Countryside Council for Wales. Of particular note was the first outcome of Scottish thinking on integrated countryside planning, which built upon the simple (but rarely used) notion that all countryside activities 'are based on use, in one way or another, of the natural heritage'. This thought has passed into the realm of 'ideas in good currency', and it is echoed in three highly coloured White Papers on the countryside issued in 1995 and 1996.

THE BLAIR GOVERNMENT 1997–99

The flood of proposals, discussion documents, consultation documents, and legislation from the Blair government would justify a separate book, rather than a note towards the end of this chapter. However, important matters are discussed, or at least mentioned, at appropriate points later. Here a note is made of some of the outstanding features, in so far as they relate to town and country planning (generously defined).

It is on constitutional matters that the most dramatic changes have been made. Not only have devolution proposals been made, but they have been passed into law, and both the Scottish Parliament and the National Assembly for Wales are operational. It is too early to comment on what the impacts on planning may be, though some preliminary indications are discussed in relevant chapters. Scotland in particular is engaged on some thoroughgoing reviews, while Wales (where the advent of devolution was uncertain) has already produced a number of planning statements. Sadly, the Northern Ireland situation has proved too problematic for resolution and, at the time of writing, it is unclear whether the devolution plans will go ahead.[4]

One of the unknowns in these constitutional changes is their impact on England. There was already a consensus that regional planning needed more direction than it was getting through the regional planning guidance system. The Blair government rapidly made moves on two fronts. First, regional planning guidance was given a new lease of life, with a 'bottom-up' involvement of local government and other 'regional stakeholders'. The aim now is to produce a more comprehensive and integrated spatial strategy, with coverage of a wider range of issues, including regeneration and transportation. Regional planning guidance has also 'gone public': examinations in public were held in 1999 on the first of the new generation of regional planning guidance (East Anglia and the South-East). Secondly, Regional Development Agencies were established and are already producing regional strategies. Though these bodies are appointed by and responsible to the Secretary of State, regional assemblies are developing which will provide 'stakeholder' input, and, it is to be hoped, a much-needed link between economic planning and land use planning. Things are happening rapidly on this front, and the outcome is by no means clear, but it seems that the regional planning dimensions are now becoming central to both land use and economic policies (even though the relationships between the two are problematic).

A possible resolution may lie in a political move towards regionalism. The possibility of regional devolution is explicitly embraced in the Labour Party manifesto, and there are signs of a reawakening and strengthening of regional consciousness. There is also an expressed concern that, whereas the Scottish members of the House of Commons can vote on English domestic affairs, English members cannot do likewise on devolved Scottish affairs. There is no simple answer to this question (which Gladstone unsuccessfully wrestled with in the controversy on Irish Home Rule a century ago), but it remains to be seen which way political currents go.[5] The position is further complicated by the appointment of an elected mayor for London, and the possibility that some provincial cities may seek the same, particularly since there is legislation in the pipeline which would provide precisely for this.[6] The course of events will be interesting to follow.

The most problematic political issue in regional planning is the allocation of land for new housing. This was a very troublesome issue for the Conservative government of 1979–97, and it is proving no less so for its successor. A major commitment has been made to increase the proportion of 'brownfield' sites, with an aspirational target of 60 per cent. Such targets have little rationale or credibility at the national level, but they concentrate effort and they also have political value. A revised PPG 3 on housing includes a 'sequential' method for identifying housing sites. It is the government's hope that the new regional planning guidance system will create the framework for agreement on housing figures without too much intervention by the Secretary of State. It is unclear whether this hope may be fulfilled. How far it will be possible to increase (and accelerate) the development of brownfield sites is equally unclear despite the stimulation of Lord Rogers' Urban Task Force Report of 1999.[7]

Devolution is not the only constitutional issue with which the Blair government has dealt. The European Convention on Human Rights has been incorporated into British law (which the previous government refused to do). However, though it passed the Human Rights Act (see 1997 White Paper *Rights Brought Home*), it has not yet accepted the need for a Human Rights Commission to advise and monitor the legislation.[8] The Act guarantees a number of basic rights and freedoms, including freedom from discrimination and the right to the peaceful enjoyment of property. In effect, the Act marks an increase in the power of the courts over Parliament. Judges will be looking beyond the letter of the law to its substance. There will be a greater role for judicial review, with a concern for the merits of a decision rather than the fairness of the way in which it was reached.[9]

The power of Britain's domestic government is also curtailed by membership of the European Union. The direct impacts of membership on town and country planning are limited so far, though the indirect influence of Community competences in regional policy, environment, transport, and other fields is important (Nadin 1999). Environmental policy, in particular, owes much to cooperation with other EU countries and Community legislation. The regional debate too is now strongly influenced by Community policies. The new regional planning guidance is adopting the notion of spatial planning, embracing a wider concern for the integration of sectoral policies around a territorial strategy. This is not unknown in the UK, of course, but it is now strongly advocated in the *European Spatial Development Perspective*, presented by the Committee on Spatial Development to the meeting of ministers of regional/spatial planning at Potsdam in 1999. Financial incentives through Community initiatives are tempting more planners to experiment with cross-national planning and exchange of experience, and government departments are looking to other countries for ideas for the 'Modernising Planning' agenda. Increasing interdependence among the EU states may mean that transnational planning strategies (now commonplace elsewhere in Europe) will become accepted for the UK.

Yet another change has a major constitutional character: a strikingly new attitude to local government. In a 'commitment to the fundamental principles of local government', the government ratified in 1998 the European Charter of Local Self-Government (which the previous government rejected). It has legislated for a Greater London Authority with an elected mayor; it has abolished compulsory competitive tendering; and it has published a stream of proposals for modernising the structures to which local government has so long adhered (including new provisions for the consideration of planning applications). It has also promised a new duty (not just a power) for local authorities to promote the economic, social, and environmental well-being of their areas. (On the other hand, exactly the opposite philosophy underlies the centralisation of education policy and even of teaching methods. Large areas of educational practice are now the subject of detailed direction.)

This paradox is not evident in Scotland, where the Executive has gone even further in its 'renewal' of local government. It has, for example, embraced local government as a partner 'in a shared duty to serve the public' (McIntosh Report 1999, para. 11). It has also embarked on designing new concepts of 'community planning'. Devolution is freeing up thinking on a number of fronts, including the right of public access to the Scottish countryside, and a tentative start has been made on reducing the concentration of land ownership by way of enabling 'community bodies' to buy rural land.

Finally, in this selective list of initiatives, mention must be made of the commitment to an integrated transport policy, which has proved more elusive than expected. Rural policies have also presented serious difficulties as they have been beset by political controversy over hunting and the right to roam. A programme for 'Modernising Planning' is making more progress. Like many of the issues touched upon in this rapid survey, it is discussed in relevant chapters.

WHITHER PLANNING?

It is now well over half a century since the postwar planning system was put into place. Major changes have taken place during this time in society, the economy, and the political scene – some of which have been touched upon in this rapid overview. In these shifting sands, 'town and country planning' has grown into (or been submerged by) a series of different policy areas which defy description, let alone coordination. Yet 'planning' is nothing if not a coordinative function, and the frenetic activity in reorganising machinery which has absorbed so much energy in the past fifty years must, at some point, give way to substantive progress. The difficulty lies in determining the direction in which this lies.

One thing is clear: some of the most important underlying problems are well beyond any conceivable scope of 'planning': for example, much urban change has been due to global forces which are currently beyond *any* political control. Multinationals and international finance were not in the standard vocabulary in the early postwar years. Planners find it easier to think in terms of 'need'. In recent years they have been forced to recast some of their thinking in 'market' terms. But could they ever come to terms with the workings of the property investment market? As many studies have shown, 'the channelling of money to promote new urban development is determined not by need or demand, but by the relative profitability of alternative investments' (Bateman 1985: 32) – which may be in different sectors, such as industrial equities, or in quite different geographical locations. Much private-sector development is now 'driven more by investment demand and suppliers' decisions than by final user demand – and even less by any sort of final user needs' (Edwards 1990: 175).

This widening gap between land use development and 'needs' throws considerable doubt on the adequacy of a planning system which is based on the assumption that land uses can be predicted and appropriate amounts of land 'allocated' for specific types of use. Overriding all other pressing considerations, of course, is the state of the economy. (It is little comfort that so many other countries share the same problem.) One result has been a strengthening of the 'partnership' philosophy which has gradually grown over the past two decades. The term now means more than coordination of the efforts of different agencies: it implies that planning has to embrace the agents of the market, and adapt a regulatory system of planning to the need for negotiation. At the least, risks are shared.

The implications of all this are not clear. Though an obvious response may be to try harder to identify emerging trends, this is more difficult to do than ever before. Economic and social trends seem as unpredictable as the weather or the course of scientific inquiry. Comprehensive planning based on firm predictions of the future course of events is now clearly impossible. Incrementalism is the order of the day, and Burnham's famous aphorism ('make no little plans') has now been turned on its head: 'make no big plans'.

But planners have always strained for unattainable goals, whether they be frankly utopian or simply overenthusiastic. Contemporary plans are more practicable in this regard than many earlier ones. The plans prepared at the end of the Second World War were often quite unrealistic in the assumptions that were made about the availability (and control) of resources – though that did not prevent them being very influential in moulding planning ideas. As noted in the previous chapter, the Greater London Development Plan was replete with policies over which the GLC had no control (Centre for Environmental Studies 1970). It remains to be seen whether the lesson has been learned – or whether some currently unpredictable change will transform the future. Be that as it may, there seems little doubt that in the perpetual planning conflict between flexibility and certainty, the former is the clear winner.

FURTHER READING

Though the Barlow and Uthwatt reports are seldom read these days, they are well worth at least a perusal. Like other reports of the time (particularly Beveridge), they give an insight into the spirit of the times which produced the planning system. Hennessy (1992) narrates this wonderfully in *Never Again: Britain 1945–1951*. A little-known but insightful essay is Titmuss (1958) 'War and social policy'. An excellent account over a longer period (1890–1994) is given by Ward (1994) *Planning and Urban Change*. Two of Hall's books are also essential reading: *Cities of Tomorrow* (1988) and *Urban and Regional Planning* (1992). Ashworth (1954) *The Genesis of Modern British Town Planning* is a thorough account up to the passing of the 1947 Act. A clear exposition of the (original) 1947 Act is given by Wood (1949), a civil servant who was heavily involved in drafting the legislation. Cherry (1996) *Town Planning in Britain since 1900: The Rise and Fall of the Planning Ideal* carries the story up to date, while his *Evolution of British Town Planning* (1974) incorporates a history of the planning profession and its Institute. A review entitled *British Planning: 50 Years of Urban and Regional Policy* is edited by Cullingworth (1999).

A number of earlier writers are quoted in the text or in the endnotes, as are several of the wartime and postwar plans. LeGates has edited a useful selection of writings entitled *Early Urban Planning 1870–1940*.

Analyses and commentaries on the operation of the planning system rapidly become out of date. Among the books and articles published in the past fifteen years are Ambrose (1986) *Whatever Happened to Planning?*; Reade (1987) *British Town and Country Planning*; Healey *et al.* (1988) *Land Use Planning and the Mediation of Change*; Simmie (1994) *Planning London*; Adams (1994) *Urban Planning and the Development Process*; Thornley (1993) *Urban Planning under Thatcherism: The Challenge of the Market*; Ambrose (1994) *Urban Process and Power*; Allmendinger (1997) *Thatcherism and Planning*; Taylor (1998) *Urban Planning Theory since 1945*; Davies (1998) 'Continuity and change: the evolution of the British planning system 1947–97'; Allmendinger and Chapman (1999)

Planning beyond 2000; and Vigar *et al.* (2000) *Planning, Governance and Spatial Strategy in Britain*.

There are a large number of White Papers and consultation documents published by the Blair Government. Among these are *Modern Local Government: In Touch with the People* (Cm 4014, 1998); *Modernising Government* (Cm 4310, 1999); and *Local Leadership: Local Choice* (Cm 4298, 1999). Recent revisions of planning policy guidance are listed at the end of the book.

NOTES

1 Royal Commission on the Local Government of Greater London (1921–3), the London and Home Counties Traffic Advisory Committee (1924), the Greater London Regional Planning Committee (1927), the Standing Conference on London Regional Planning (1937), as well as *ad hoc* committees and inquiries, for example, on Greater London Drainage (1935) and a Highway Development Plan (Bressey Plan, 1938).

2 Government 'investigators' were appointed and, following their reports, the Depressed Areas Bill was introduced in November 1934, to pass (after the Lords had amended the title) as the Special Areas (Development and Improvement) Act.

3 The phrase was coined by Alix Meynell, a senior official in the Board of Trade; see Meynell (1959).

4 Following the Good Friday Agreement of April 1998, the UK and Irish governments passed legislation on referendums on the Agreement. These gave a clear endorsement by the electorates of both Northern Ireland and the Irish Republic. The agreement provides for devolution to an elected Assembly of legislative and executive powers for all matters which are currently the responsibility of the six Northern Ireland departments (thus including environmental and planning policies). Additionally, a North–South Ministerial Council will deal with matters of mutual interest.

5 In the North-East a Constitutional Convention (chaired by the Bishop of Durham) was established in April 1999. Inspired by Scottish experience, this has 'set itself the task of agreeing a scheme for an elected assembly and ensuring the widest possible agreement on the scheme'. In Yorkshire and Humberside a campaign for 'Yorkshire Democracy' was launched in March 1999. 'These regional campaigns have recently formed the Campaign for the English Regions, designed to coordinate and extend the efforts in the individual regions' (Tomaney and Mitchell, 1999). See also the material available through the English Regional Associations web site <www.somerset.gov.uk/era/>. Other references are given in the following chapter.

6 Bogdanor (1999: 275) has argued that 'such mayors would constitute a powerful force against devolution to the regions since regional assemblies would be likely to detract from their authority'.

7 *Towards an Urban Renaissance*. This is discussed in Chapter 6.

8 Such a body could also scrutinise proposed legislation, train lawyers, provide legal representation for test cases, and initiate its own cases. See Spencer and Bynoe (1998).

9 The Human Rights Act is currently thought unlikely to have much impact on the planning world. Nevertheless, there are areas where it *could* have significant impacts, particularly where discretion or personal liberty is involved. See Corner (1999) and Upton (1999). One author has speculated that the hearing and determination of local plan objections may well be in breach of the Convention's provisions relating to civil rights and the entitlement 'to a fair and a public hearing by an independent and impartial tribunal established by law'. See Kitson (1999), and also the succinct account by Johnston (1999b).

3

THE AGENCIES OF PLANNING

(A) EUROPEAN GOVERNMENT

What the Community is doing, and the ways in which it is developing, will affect planning in Britain, as it will affect planning in every other Member State of the Community; through its direct and indirect impact on planning policies and legislation . . . and in the constraints which it may impose, but, more significantly, in the opportunities which it will open up for the development of planning practice.

(Davies and Gosling 1994)

THE GROWING INFLUENCE OF EUROPE

The impact of the European Union (EU) on town and country planning has been felt predominantly in the field of environmental controls rather than in mainstream planning practice. The most striking example is environmental impact assessment. In coming years, the influence of Europe will increase, and especially so in cross-border and transnational spatial planning. Later chapters identify a range of agricultural, environmental, economic, and regional policies of the EU which are having an effect on parts of the British planning system. Chapter 4 includes a note on supranational and cross-border planning instruments and policies that are being introduced at the European level. Here a brief and more general account is given of important EU institutions and actions.

BRITAIN IN THE EU

The UK was not an enthusiastic supporter of the postwar moves towards a federal Europe. Though it favoured intergovernmental cooperation through such bodies as the Organisation for European Economic Cooperation (set up 1948) and the Council of Europe (1949), it was opposed to the establishment of organisations which would facilitate functional cooperation alongside nation-states. It therefore did not join the European Coal and Steel Community (1952), nor was it a signatory to the 1955 Treaty of Rome, which established the European Economic Community and the European Atomic Energy Community. However, along with the other members of the Organisation for European Economic Cooperation, it formed the European Free Trade Association (EFTA) in 1960. Britain envisaged that EFTA would form the base for the development of stronger links with Europe. When it became clear that this was not viable, Britain applied for membership of the European Community. This was opposed by France and, since membership requires the unanimous approval of existing members, negotiations broke down. The opposition continued until a political change took place in France in 1969. Renewed negotiations led finally to membership at the beginning of 1973.

The Treaty of Accession provided for transitional arrangements for the implementation of the Treaty of Rome, which Britain agreed to accept in its entirety. The objectives include the elimination of customs duties between member states and of restrictions on the free movement of goods; the free movement of people, services, and capital between member states; the adoption of common agricultural and transport policies; and the approximation of the laws of member states to the extent required for the proper functioning of the common market. These objectives are often referred to as the 'four freedoms': the free movement of goods, people, services, and capital.

There are fifteen members of the EU: Austria, Belgium, Denmark, Finland, France, Germany, Greece, the Republic of Ireland, Italy, Luxembourg, the Netherlands, Portugal, Spain, Sweden and the United Kingdom. Thirteen more countries have applied for membership.[1]

The EU has a population of 363 million and a land area of over 3.2 million square kilometres. Compared to the USA, this is about 50 per cent more people living on just over a third of the space. The EU is also easily the world's largest trading bloc, having a share of exports more than three times its nearest rivals of the USA and Japan.

The organisational and political structure of the EU is complex and, as with all such bodies, its actual workings are somewhat different from the formal organisation chart. The main institutions of the EU and their elements which are of particular interest to planning are shown in Figure 3.1. In brief, there is an elected Parliament which operates as an advisory body, and a Council of Ministers which is the legislature and makes policy largely on the basis of proposals made by the European Commission, which is the executive. There is also a Court of Justice which adjudicates matters of legal interpretation and alleged violations of Community law. The distribution between Parliament and the executive is very different from that in most national governments.

THE EUROPEAN COUNCIL

A summit of heads of state or government of the member states, together with the President of the European Commission, provides general political direction for the European Union, considers fundamental questions related to the 'constitution' and construction of the EU, and makes decisions on the most contentious issues (Dinan 1998). It is not the legislature – that is the function of the Council of the European Union. Decisions which require legislation have to go through the normal EU legislation process, but agreements and declarations reached in the European Council are binding on the EU institutions, and have been critical in shaping the evolution of the EU. The Council is usually held twice a year, once under each six-month Presidency.

THE COUNCIL OF THE EUROPEAN UNION (COUNCIL OF MINISTERS)

The main policy-making body of the EU is the Council of Ministers. This is the legislature of the Community, a task it shares for some matters with the European Parliament. Unlike most other legislatures it is indirectly elected – being composed of representatives elected in the member states – and it deliberates in private. These characteristics have given rise to the criticism of 'democratic deficit' in comparison to national legislatures and the European Parliament, which is directly elected and debates in public. But the characteristics reflect the fundamental nature of the EU as a pooling of national sovereignties and legislative powers. These require complex negotiation among the member states, rather than a federal structure with a unitary legislature.

The Council has many committees and working groups with various functions and membership, such as the Council of Ministers for the Environment or for Agriculture. There is a Council meeting of some sort every working week, often lasting for three days, and 100,000 documents are produced by the Council each year (Dinan 1998: 106). Representatives are usually senior ministers of national government, although

THE EUROPEAN COUNCIL
Meeting of the Heads of State

Gives broad guidance and impetus to action

COREPER
Committee of Permanent Representatives - civil servants from the member states who manage the work of the Council.

THE COUNCIL OF THE EU
Meetings of ministers (one from each member state). The Council meets in a different form depending on the issue e.g.
The Environment Council
The Transport Council

The Presidency of the Council rotates every six months.

2000	Portugal	France
2001	Sweden	Belgium
2002	Spain	Denmark
2003	Greece	

Legislature (on some matters shared with the European Parliament)

INFORMAL MEETING OF MINISTERS OF SPATIAL PLANNING
Generally meets once during each Presidency.
It is not a formal council and has limited powers.

COMMITTEE ON SPATIAL DEVELOPMENT
An intergovernmental meeting of civil servants from the member states.

EUROPEAN COMMISSION
20 Commissioners
36 Directorates, (departments) including :

TRANSPORT

ENVIRONMENT

REGIONAL POLICY
(includes spatial planning and ERDF)

Applies the Treaties by initiating legislation and implements policy as executive body.

EUROPEAN PARLIAMENT
626 elected members (87 from UK)
20 standing committees, including :

**Agriculture and Rural Development
Regional Policy; Transport and Tourism
Environment, Public health and Consumer Policy**

Political driving force, supervising and questioning the Council and Commission. Joint power to adopt legislation with Council. Supervises appointment of Commission.

ECONOMIC AND SOCIAL COMMITTEE
222 nominated members from empoyers, workers and other interests
(24 from UK).
6 sections, including :

**Agriculture, Rural Development and the Environment
Transport, Energy, Infrastructure and the Information Society**

Is consulted and delivers opinions on proposed legislation.

COMMITTEE OF THE REGIONS
222 members representing regional and local government (24 from UK)
8 Commissions, including :

**Commission 1 : Regional Policy
Commission 2 : Agriculture and Rural Development
Commission 3 : Trans-European Networks and Transport
Commission 4 : Spatial Planning, Urban Issues, Energy and the Environment**

Is consulted and delivers opinions where regional interests are involved.

THE COURT OF JUSTICE & COURT OF FIRST INSTANCE
15 judges & 9 advocates general   15 judges

Interpret the Treaties and apply judgements and penalties in cases of non-compliance.

Figure 3.1 Institutions of the European Union and Spatial Planning

regional ministers may also represent the country concerned, a point which may become more significant for the UK as devolution starts to bite. The criterion is that the representative must be authorised to commit the member state to the decisions made.

There is no formal Council of Ministers responsible for planning but, since 1991, there have been biannual informal meetings of ministers responsible for spatial planning. A Committee on Spatial Development (CSD) consists of officials representing the planning ministries of member countries operating under the auspices of the informal meeting of ministers.[2] The UK has been represented on the CSD by the Department of the Environment, Transport and the Regions (DETR), the Department of Trade and Industry (DTI), and the Scottish Office. It is the CSD that has taken the most important action on European planning in preparing the European Spatial Development Perspective, which is discussed in Chapter 4.

The Council (in some cases in cooperation with the European Parliament) can make three main types of legislation. *Regulations* have direct effect and are binding throughout the EU. They require no additional implementing legislation in the member states and are used mostly for detailed matters of a financial nature or for the technical aspects of (for example) administering the Common Agricultural Policy (CAP). By contrast, *directives* provide framework legislation which, though equally binding, is implemented by national legislation. This leaves a degree of choice over the method of implementation to the member states. Environmental matters are typically dealt with in this way. The Council can also issue *decisions* which are binding on the member state, organisation, firm, or individual to whom they are addressed. Finally, there are *recommendations* and *opinions*, which have no binding force.

The work of the Council of Ministers is supported by officials in the Council of Permanent Representatives (COREPER). These are civil servants of the member states. Indeed, it has been argued that COREPER is where the real decisions are made. It is the officials who conduct often very lengthy negotiations to reach agreement about measures among the member states before proposals are put before the ministers.

THE EUROPEAN COMMISSION

The main work of the EU is undertaken by the executive of the Community, the European Commission. The Commission is a major driving force within the EU because it has the primary right to initiate legislation. It prepares proposals for decision by the Council, and oversees their implementation. (Only rarely can the Council of Ministers make a policy decision without a proposal from the Commission.) The Community's decision-making process is dominated by the search for consensus among the member states, and this gives the Commission a crucially important role in mediation and conciliation. Of the same nature is the ethos of achieving compromise and of progressing in an incremental way. In promoting action at the EU level, the key reference for the Commission is the European Treaties.

Among the Commission's powers is that of dealing with infringements of Community law. If it finds that an infringement has occurred, it serves a formal notice on the state concerned requiring discontinuance or comments with a specified period (usually two months). If the matter is not resolved in this way, the Commission issues a reasoned opinion, requiring the state to comply by a given deadline. As a last resort, the Commission can refer a matter to the Court of Justice, whose judgement is legally binding. Most matters are dealt with informally, but Britain has been subject to reasoned opinions on environmental matters (Haigh 1990: 153 and 160).

During the 1990s the influence of the Commission waned under fierce criticism of its perceived greed for power and the acquisition of national competencies. Its attempts at harmonisation of standards in the pursuit of the Single Market, though often well founded, have sometimes been inept, giving an impression of remoteness and arrogance, exacerbated by its poor control of Community funds. However, much of the popular criticism is misconceived. Thus, for example, to label the 17,000 officers of the Commission (including 3,000 translators and more than 3,000 scientists) as a massive bureaucracy is a gross exaggeration. (The DETR had a total of 15,000 staff.) Nevertheless, the media have harried the Commission on its interference in national affairs, and the cronyism of the Commissioners. Protecting national competencies in the face of expanding EU powers was a prime objective of the Thatcher administration, but other member states too have grown wary of the expanding competence of the Community. The European Parliament has taken the Commission to task on poor management. The Council took action during the 1990s to reduce unnecessary interference from the Commission, citing the principles of subsidiarity and proportionality. The effect has been dramatic, with a considerable fall in the amount of Community legislation, and, less obviously, a weakening of its influence.

The culmination of mounting criticism came at the end of 1998 when the European Parliament, to which the Commission is accountable, threatened to sack all Commissioners. Although the proposal was defeated, the debate fuelled popular antagonism against the Commission, and in March 1999 the Commissioners resigned en bloc. A new Commission was approved by the European Parliament in September with major reforms to its organisational structure and procedures. These were designed, in Mr Prodi's words, 'to win back ordinary people's confidence in Europe and the European vision . . . [and] to transform the Commission into a modern, efficient administration'.[4] A 'blueprint for reform' is expected imminently. In the meantime, the Commission's thirty Directorates-General ('DGs') and numerous other services have been reorganised into thirty-six departments that will be known by their names rather than the old and frustrating system of numbers. The main departments are each headed by a Director-General, but considerable influence over the work of the Directorate is exercised by the personal 'cabinet' of the Commissioner, and in particular by the chair (who is known as the *chef du cabinet*).

The departments with an interest in town and country planning or the broader concept of spatial planning are Regional Policy (known as DG Regio, and whose main responsibility is for the Structural Funds), Environment, and Transport.

THE EUROPEAN PARLIAMENT

The European Parliament is a directly elected body consisting of 626 members, who are elected every five years. Britain has eighty-seven representatives, known as MEPs: Members of the European Parliament. The Single European Act and Treaty on European Union extended the powers of the Parliament, and the Amsterdam Treaty (which came into force in May 1999) has again increased its role in joint decision-making with the Council and its supervisory powers over the Commission. The Parliament is consulted on all major Community decisions, and it has powers in relation to the budget which it shares with the Council, and in approving the appointment of the Commission. The assent of Parliament is needed also for accession of new members and international agreements. However, it is important to note that the Parliament was established essentially as an advisory and supervisory body, while the Council of Ministers is the legislature. One reflection of the lack of legislative power is that the Parliament sits in plenary session for only three days each month and bizarrely continues to divide its sittings between two locations: Brussels and Strasbourg.

Parliament is organised along party political (not national) lines. The political groups have their own secretariats and are the 'prime determiners of tactics and voting patterns' (Nugent 1999: 130). Much of their work is carried out by standing committees and through questions to the Commission and Council. The Regional Policy, Transport and Tourism Committee considers matters related to spatial development, including European regional planning policy and the common transport policy.

In 1995 the European Parliament established the office of the European Ombudsman, charged with improving the quality of Community relations with the public. The Ombudsman can investigate complaints within all the Community institutions except for the courts acting in their judicial role. Complaints can be made by anyone living in the European Union, and 1,372 were received in 1998. Almost 70 per cent of complaints are outside the mandate of the Ombudsman, and many of these are about the application of

Community law within the member states. Three-quarters of admissible complaints were made against the European Commission, and the highest proportion, one-third, were related to access to information.

THE COMMITTEE OF THE REGIONS

The Committee of the Regions (COR) is the youngest European institution, set up following the Treaty of European Union, and holding its first session in March 1994. It is intended to give a voice to the regions and local authorities in European Union debates and decision-making. It has 222 members representing the regions, including twenty-four from UK local authorities. (The UK representation is made up of fourteen from England, five from Scotland, three from Wales, and two from Northern Ireland.) The COR has taken a particular interest in regional planning and in advocating wider use of the principle of subsidiarity, so as to strengthen the role of regional and local authorities.

The Treaty identifies particular areas where the COR has to be consulted by the Commission, including trans-European networks, economic and social cohesion, and structural fund regulations. It can also offer opinions in other areas that it thinks appropriate, typically when an issue has a specifically regional dimension. It has issued many opinions on planning, urban, and environmental issues. A committee (confusingly known in the COR as a commission) has been established to deal exclusively with spatial planning, urban issues, energy, and the environment.

THE EUROPEAN COURTS

There are two main European courts: the European Court of Justice and the Court of First Instance. The European Court of Justice has thirteen judges. It decides on the legality of decisions of the Council and the Commission, interprets Community law, ensures its consistent application, and determines violations of Treaties. Cases can be brought before it by member states, organisations of the Community, and private firms and individuals. Since 1989 the Court of First Instance has dealt with most actions involving private applicants. It is organised on a similar basis to the European Court of Justice. The Courts have played an important part in extending the competencies of the European Union by confirming that actions by the Community are legal under the Treaties (Nadin and Shaw 1999), and by promoting harmonisation by ruling that certain actions are illegal (Nugent 1999: 263). These courts are quite separate from the European Court of Human Rights.

THE COUNCIL OF EUROPE

The Council of Europe is not to be confused with the EU. It was set up in 1949 with ten member countries to promote awareness of a common European identity, to protect human rights, and to standardise legal practices across Europe in order to achieve these aims. Since 1989, its main role has been to monitor human rights in the post-communist democracies, and to assist them in carrying out political, constitutional, and legal reform. It now has forty-one member countries (including sixteen countries that were formerly part of the communist bloc).

It has a three-tier structure with a Council and Ministers, a Parliamentary Assembly, and a Congress of Local and Regional Authorities. With an annual budget of less than £100 million, it is much less powerful than the EU (which has an annual budget of over £50 billion), but nevertheless it has played an important part in maintaining and establishing democracy on the Continent. It is best known for its Convention on European Human Rights. Anyone who feels that their rights under the Convention have been breached may take a case to the European Court on Human Rights for a decision which will be binding on those states that have signed up to the Convention.[5] The Convention is now incorporated into UK law,[5] and its impacts on planning are discussed in Chapter 12.

The Council has been active for many years in the field of regional planning and environment, and perhaps the most notable achievement is the Bern

Convention on Conservation of Wildlife and Habitats. It has published conference and other reports on the implications of sustainability for regional planning, the representation of women in urban and regional planning, and many other topics. A conference of ministers of spatial and regional planning (CEMAT) has been meeting since 1970, and its most important contribution has been the European regional/spatial planning charter, known as the Torremolinos Charter. This was adopted in 1983 and committed the Council to producing a 'regional planning concept' for the whole of the European territory. It has taken some time, but CEMAT has also now published *Guiding Principles for Sustainable Spatial Development of the European Continent* (2000).

The Council was also responsible for the European Campaign for Urban Renaissance (1980–82). This led to a programme of *ad hoc* conferences, various reports and 'resolutions' on such matters as health in towns, the regeneration of industrial towns, and community development. In 1992 the Conference adopted *The European Urban Charter*. This 'draws together into a single composite text, a series of principles on good urban management at local level'. The 'principles' relate to a wide range of issues, including transport and mobility, environment and nature in towns, the physical form of cities, and urban security and crime prevention.

(B) CENTRAL GOVERNMENT

We live in an age when most of the old dogmas that haunted governments in the past have been swept away. We know now that better government is about much more than whether public spending should go up or down, or whether organisations should be nationalised or privatised. Now that we are not hidebound by the old ways of government we can find new and better ones.

(White Paper, *Modernising Government*, 1999)

MODERNISING GOVERNMENT

The quotation illustrates the style as well as the zeal of the Blair government in its attempt to change the nature of the governmental system. It is not, of course, the first government to enter office with such flourishes; nor is it unique in proclaiming innovations which are recognisably in line with secular social and political changes. But 'the third way' (Giddens 1998) is in marked contrast to at least the rhetoric of the long-lived Conservative administration which was defeated in the election of 1997. Moreover, the first years of office have witnessed a continuing torrent of measures to bring both policy and the machinery of implementation in line with the philosophy of the new government. Changes continue to be introduced, and the following account represents a snap-shot picture in late 2000. First, however, reference needs to be made to a remarkable innovation in the strategic planning of public expenditure: the Comprehensive Spending Review announced by the Chief Secretary to the Treasury in June 1997. Instead of adjusting departmental budgets at the margin, priorities are being attained by the use of zero-based budgeting:

'Every department will scrutinise its spending plans in detail from a zero base, and ask, how does each item contribute to the Government's objective as set out in out manifesto? Why are we spending this money? Do we need to spend it? What is it achieving? How effective is it? How efficiently are we spending it? . . . Its conclusions will inform a new set of public spending plans for the rest of this Parliament – a set that reflects our priorities.'[6]

The outcome was a significant shift in spending priorities over a three-year period 'towards education, health and capital expenditure in transport and housing, and away from defence, agriculture, the diplomatic service, and the legal system'. The new arrangements represented 'the most ambitious re-engineering of the public expenditure system for several decades, shifting the emphasis away from annual negotiations and their emphasis on inputs, and towards objectives and outputs' (James 1999: 195).

This suggests that Whitehall departmentalism has become less rigid than previously. Certainly, Blair did not initially share Harold Wilson's experience in the early 1960s of the tardiness of the civil service in adapting to a Labour government after thirteen years of Conservative government. Indeed, he 'found a civil service almost startlingly keen to prove that they had not been politicised by eighteen years of Conservative rule'.[7] However, Blair's drive for change may now be facing some problems with the traditional departmentalism. He is reported as expressing frustration at civil servants 'defiantly defending their own departments: they are felt to oppose any structural changes to their fiefdom, particularly if it means ceding any territory'.[8] To combat this, he has asked Lord Simon (formerly of BP) 'to introduce a revolution in civil service culture, including stripping out layers of management and imposing payment by results'. In the words of Michael White,

> Performance-related pay, targeted objectives for departments and individuals, more inter-departmental cooperation, fast-track promotion for bright young things, above all a shift from being preoccupied with policy and process to a new focus on outcome and delivery are what it's all about.'[9]

Time will tell: these developments are (at the time of writing) much too recent to evaluate.

ORGANISATIONAL RESPONSIBILITIES

A large number of governmental departments and agencies are involved in town and country planning. Those having the main responsibility for the planning acts are the Scottish Executive Development Department (SEDD), the Transport, Planning and Environment Group of the Welsh Assembly, the Planning Service Executive Agency of the Department of the Environment for Northern Ireland (DoENI), and, for England, the Department for Transport, Local Government and the Regions (DTLR). There are, of course, many planning functions that fall to departments responsible for agriculture, the countryside, the human heritage, national heritage, nature conservation, and trade and industry. Additionally, an

increasing number of functions have been transferred from government departments to agencies and public bodies. Figure 3.2 shows the main institutional arrangements, and gives a flavour of their complexity.

Planning responsibilities have evolved over time and, though there have been numerous reorganisations, the machinery inevitably has a patchwork appearance. (As an example of the problems involved: in which department should questions of the rural economy be placed – the one concerned with agriculture, or natural resources, or economic development, or employment? Or should it form a separate department of its own?) The machinery is also unstable: changing perceptions, conditions, problems, and objectives demand new policy responses, which in turn can lead to organisational changes. For example, increased concern for environmental planning has resulted in the transfer of widespread environmental functions into new comprehensive environment agencies. Sometimes, different patterns emerge in different parts of the UK. Thus nature conservation and access to the countryside are the responsibility of one agency in Scotland (Scottish Natural Heritage) and in Wales (Countryside Council for Wales), but are divided between two in England (English Nature and the Countryside Agency).

THE DEPARTMENT FOR TRANSPORT, LOCAL GOVERNMENT AND THE REGIONS

The central planning department for England is the Department for Transport, Local Government and the Regions (DTLR). The ungainly title mirrors its vast range of responsibilities. It was formed in June 2001, when the environment responsibilities of the former Department of Environment, Transport and the Regions (DETR) were transferred to the newly formed Department for the Environment, Food and Rural Affairs (DEFRA). The Department's responsibilities include not only the three areas indicated by its title, but also housing, local government, regeneration, and of course planning.

The Secretary of State has final responsibility for all the functions of the Department. He or she is,

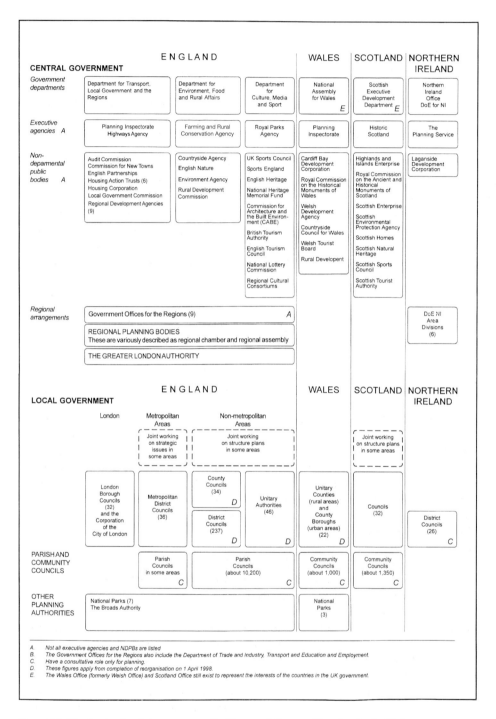

Figure 3.2 The Institutional Arrangements for Town and Country Planning in the UK

however, concerned primarily with strategic issues of policy and priority, including public expenditure, which determine the operations of the Department as a whole. The former DETR was a large department: in 1999, it had over 15,000 staff, of whom 3,900 were at the headquarters, 900 were in the Government Offices for the Regions, and around 10,500 were in eight executive agencies such as the Highways Agency and the Planning Inspectorate. The Department is organised in Groups and Directorates. For example, the Planning, Roads and Local Transport Group includes the Planning Directorate, the Integrated and Local Transport Directorate, and the Roads and Traffic Directorate.

The machinery of government is constantly changing. The original Department of the Environment (formed in 1970) took over transport responsibilities, but these were moved back to a separate Department of Transport in 1976 – a move which was reversed by the Labour government in 1997. Following the 1992 general election, a separate Department for National Heritage was created (superseded in 1997 by the Department for Culture, Media and Sport). Other changes have included the gathering together of the pollution regulation functions within Her Majesty's Inspectorate of Pollution (HMIP), and later the establishment of Environmental Agencies for England and Wales, and for Scotland; these have taken over the functions of the HMIP, the National Rivers Authority, and the waste regulation functions of local government. It has been a restless time in Whitehall.[10]

The Department's policy aims and objectives are set out in its annual report, and are under constant review. The overall aim of the former DETR was 'to improve the quality of life by promoting sustainable development at home and abroad, fostering economic prosperity and supporting local democracy'. The list of objectives refers to a wide range of issues including housing and social cohesion, integrated transport, rural areas, responsive elected local government, regeneration, and efficiency in the construction industry (Box 3.1). The range of responsibilities in the field of planning is shown in Box 3.2.

THE DEPARTMENT FOR CULTURE, MEDIA AND SPORT

The Department for Culture, Media and Sport was established in 1997 (following the Department of National Heritage). It has a wide range of responsibilities, including the arts, sport and recreation, libraries, museums, broadcasting, film, press freedom and regulation, heritage, and tourism. Its overall aim is 'to improve the quality of life for all through cultural and sporting activities, and to strengthen the creative industries'. There are now greatly enhanced resources for these worthy objectives by way of the National Lottery. The areas of 'good causes' for which Lottery funds provide support are sport, the arts, heritage, charities, millennium projects, health, education, and the environment. In 1998–99 a total of £3,192 million was awarded, of which £1,283 million went to the Heritage Lottery Fund, £148 million to sport, and £281 million to the National Lottery Charities Board (*Whitaker's Almanac 2000*: 611). (Funding heritage projects by the Lottery is discussed further in Chapter 8.)

The Department is responsible for over forty executive and advisory non-departmental public bodies, including the British Library, the British Tourist Authority, the Millennium Commission, the National Heritage Memorial fund, and English Heritage (which is discussed in Chapter 8). It has also established close relationships with the Local Government Association and is promoting local authority cultural strategies.[11]

THE DEPARTMENT FOR ENVIRONMENT, FOOD AND RURAL AFFAIRS

Other departments of government have special status in respect of town and country planning, notably the Department for Environment, Food and Rural Affairs (DEFRA). This new ministry, created in 2001, includes all functions of the former Ministry of Agriculture, Fisheries and Food (MAFF) and the environment, wildlife and countryside functions of the

---

BOX 3.1 GOVERNMENT OBJECTIVES FOR PLANNING, THE ENVIRONMENT, TRANSPORT AND LOCAL GOVERNMENT

Aim

The DETR aims to improve everyone's quality of life, now and for the future, through:

- thriving, prosperous regions and communities;
- better transport;
- better housing;
- a better environment;
- safer, healthier surroundings; and
- prudent use of natural resources.

Objectives

- To protect and improve the environment and to integrate the environment with other policies across government and internationally.
- To offer everyone the opportunity of a decent home and so promote social cohesion, well-being, and self-dependence.
- To promote modern and integrated transport for everyone and to reduce the impact of transport on the environment.

- To provide customer-focused regulatory and other transport services and to collect taxes fairly and efficiently.
- To enhance opportunity in rural areas, improve enjoyment of the countryside, and conserve and manage wildlife resources.
- To promote a sustainable pattern of physical development and land and property use in cities, towns, and the countryside.
- To promote a system of elected government in England which responds to the needs of communities.
- To enhance sustainable economic development and social cohesion throughout England through effective regional action and integrated local regeneration programmes.
- To secure an efficient market in the construction industry, with innovative and successful UK firms that meet the needs of clients and society and are competitive at home and abroad. To improve health and safety by reducing risks from work activity, travel, and the environment.

Source: DETR *Annual Report 2000* (p. 210)

---

former DETR. DEFRA has had to be consulted on important proposals, and MAFF classification of agricultural land quality remains a potentially important consideration in development control (PPG 7: Annex B). The influence of the former MAFF has waned somewhat in parallel with the decline of agriculture in the British economy, but it still has to be consulted on any planning proposal which involves a significant loss of high-quality agricultural land. Objections by MAFF have fallen considerably over recent years.[12] At the same time, it assumed an increasing responsibility for countryside protection functions such as the protection of environmentally sensitive areas (discussed in Chapter 9). In the words of MAFF's annual report, one of its objectives is

> to sustain and enhance the rural and marine environments and public enjoyment of the amenities they provide. The principles laid down in the UK's Sustainable Development Strategy, of pursuing a balance between economic development and environmental protection, are integrated into the development of MAFF's policies.

The Farming and Rural Conservation Agency is an executive agency of DEFRA (established in 1997), and, following a review in 1999, concentrates on 'the development and maintenance of strategic relationships with planning authorities, Government offices, and the

BOX 3.2 THE DTLR'S PLANNING RESPONSIBILITIES

- Issuing national planning guidance on the key areas of planning policy, such as housing, economic development, transport, retail and town centres, the countryside, green belts, sport and recreation, minerals and waste, and on the operation of the system at regional and local levels.
- Working with the government offices for the regions to ensure that regional planning guidance, local authorities' development plans and decisions on planning applications are in line with national policies.
- Seeking to improve the legislative framework

for planning, including the system of development plans and development control.
- Acting as the focal point within government for the property industry and liaising with the Local Government Association, the main representative bodies of business and environmental interests, professional groups, and other government departments on the development of policy.
- Representing the UK at international negotiations on planning; and commissioning research and statistical information to support policy development.

DETR *Annual Report* 2000

regional development agencies with a view to achieving fuller integration of agricultural considerations into wider government policies affecting the rural economy and environment'.[13]

EXECUTIVE AGENCIES

The proliferation of new government agencies is confusing. Essentially it has taken two main forms: executive agencies (of which the highly successful Driver and Vehicle Licensing Executive was the forerunner) and non-departmental public bodies (exemplified by the Housing Corporation, the Local Government Commission for England, and the former new town development corporations).

Executive agencies remain part of their Department, and their staffs are civil servants; but they have a wide degree of managerial freedom (set out in their individual 'framework' documents). They enjoy delegated responsibilities for financial, pay, and personnel matters. They work within a framework of objectives, targets, and resources agreed by ministers. They are accountable to ministers, but their chief executives are personally responsible for the day-to-day business

of the agency. Ministers remain accountable to Parliament. If this sounds somewhat confusing, that is because it is. However, in principle, the stated intention is to increase accountability. A distinction is drawn between responsibility, which can be delegated, and accountability, which remains a matter for ministers – a contention which is the subject of considerable controversy. Examples of executive agencies are the Planning Inspectorate (discussed further below) and Historic Scotland.

By contrast, non-departmental public bodies are a type of quango (quasi-autonomous non-governmental organisation). These bodies (of which there are over 1,300) range enormously in function, size, and importance. They all play a role in the process of national government, but are not government departments or parts of a department. There are three types of NDPB: executive bodies such as the Countryside Agency, the Environment Agency, and English Heritage; advisory bodies such as the Advisory Committee on Business and the Environment, the Radioactive Waste Management Committee, and the Royal Commission on the Environment; and tribunals such as the lands tribunals, rent assessment committees, and the Agricultural Land Tribunal.

THE PLANNING INSPECTORATE

The way in which the aims and objectives of executive agencies are cast is illustrated by the case of the Planning Inspectorate. This is a joint executive agency of the DTLR and the Welsh Office. In Scotland the equivalent is the Scottish Office Inquiry Reporters Unit (SOIRU). The major areas of work are the holding of local plan inquiries (thirty-six opened in 1999–2000, down from a high of ninety-one in 1995–96, but for which it seeks to recover full costs from the planning authorities concerned); determining planning appeals (12,619 in 1999–2000); and determining enforcement appeals (2,746 in 1999–2000).[14] During the wave of plan-making in the mid-1990s, local plan inquiries placed heavy demands on the Inspectorate. Not only did numbers increase, but recent inquiries have generally been longer and more complex, so reporting times inevitably lengthened.[15]

Other responsibilities of the Planning Inspectorate include highway inquiries, and footpath orders under the Highways, Town and Country Planning, and Wildlife and Countryside Acts.[16] Increasing resources are devoted to environmental matters such as inquiries under the Environmental Protection Act. The Inspectorate has an annual budget of £37 million (1999–2000), and a total staff of around 600, 400 of whom are inspectors. In line with current ideas about governmental administration, it has performance targets which include deciding 80 per cent of written representation appeals within eighteen weeks (achieved); and providing an inspector for local plan inquiries at the time the authority requested in 80 per cent of cases (achieved). The Inspectorate now also publishes its own journal.

The future of the Planning Inspectorate has been the subject of much debate over recent years. During 2000 the Inspectorate was subject to a major review by the DETR (a review is required every five years), and in the same year the Environment, Transport and Regional Affairs Select Committee conducted an inquiry into the Planning Inspectorate and public inquiries. On the first review the outcome was positive and recommended retention of the Inspectorate as an executive agency. Attention has now shifted to stage two of the DETR review, on how performance can be enhanced. This will take into account the findings of the Select Committee report, which had numerous recommendations. The report is generally positive: it notes that the 'the Inspectorate is held in high regard by most who come into contact with it' and that there had been a dramatic improvement in efficiency over recent years. The main problems seem to be in ensuring consistency and in the relatively small number of cases where complaints arise, where the Committee found an 'apparently high-handed attitude to people querying . . . decisions' (para. 29). The report makes other recommendations on the potential of more 'instant decisions'; the increasing need for inspectors to be able to provide specialist knowledge; and the difficulty of keeping up with incremental changes to government policy (or even knowing what government policy is), which is not so much an issue for the Inspectorate as for the government.

In the context of new legislation implementing the Human Rights Convention, there have been suggestions that the Inspectorate should be replaced by a system of environmental courts. The reason is that the Human Rights legislation requires that everyone be 'entitled to a fair and public hearing . . . by an independent and impartial tribunal established by law'. Advocates of the environmental court argue that the Inspectorate does not meet this requirement because it is an executive agency of government and thus not independent.[17] This matter is taken up in the discussion of inquiries and hearings in Chapter 12; suffice it to say here that the government (and the courts) have so far determined otherwise. The Select Committee came to the common-sense conclusion that if there is a need for more independence through a court system, the Inspectorate should be established as the first part of that, in effect a 'court of first instance'.

CENTRAL GOVERNMENT PLANNING FUNCTIONS

Relationships between central and local government vary significantly among various policy areas, 'reflecting, in part, the difference in weight and concern which the centre gives to items on its political agenda, and, in part, differences in the sets of actors involved in particular issue areas' (Goldsmith and Newton 1986: 103).

The 1943 Town and Country Planning Act (which preceded the legislation on the scope of the planning system) placed on the relevant minister the duty of 'securing consistency and continuity in the framing of a national policy with respect to the use and development of land'. Though this is no longer an explicit statutory duty, the spirit lives on, and the Secretary of State has extensive formal powers. These, in effect, give the Department the final say in all policy matters (subject, of course, to parliamentary control – though this is in practice very limited). For many matters, the Secretary of State is required or empowered to make regulations or orders. Though these are subject to varying levels of parliamentary control, many come into effect automatically. This delegated legislation covers a wide field, including the Use Classes Orders and the General Development Orders. These enable the Secretary of State to change the categories of development which require planning permission.

The formal powers over local authorities are wide-ranging. If a local authority fails to produce a 'satisfactory' plan, default powers can be used. The Secretary of State can require a local authority to make 'modifications' to a plan, or 'call in' a plan for 'determination'. Decisions of a local planning authority on applications for planning permission can, on appeal, be modified or revoked. Development proposals which the Secretary of State regards as being sufficiently important can be 'called in' for decision by the minister.

These powers are now frequently employed in the plan-making process, usually informally through the DETR regional offices.[18] In less interventionist times they were reserved for cases where there was a deadlock between local and central government. This can amount almost to a game of bluff as, for instance, when a local authority wants to make a political protest, or to demonstrate to its electors that it is being forced by central government to follow a policy which is unpopular. Thus, opposition in Surrey to the M25 was so strong that the county omitted it from its structure plan. The Secretary of State made a direction requiring it to be included. Another battle arose over the Islington unitary development plan, where the Secretary of State took strong objection to the stringent controls which the borough proposed (*inter alia*) for its thirty-four conservation areas. The Secretary of State issued a direction requiring most of these to be changed. The borough took the matter to court, which held that it had no power to intervene on the planning aspects of the case; and since the Secretary of State had not acted perversely or in conflict with his own policies, his action was quite legal. Judicial review cannot be used as an oblique appeal. It was therefore the responsibility of the borough and the Secretary of State to resolve their differences to the satisfaction of the Secretary of State (*JPL* 1995: 121–5).

A more recent case was a direction to Berkshire County Council to modify its proposed structure plan to increase the provision for new housing in the county (from 37,500 to 40,000) by the year 2006. This is a common issue of friction between central government and a number of county councils, particularly in the South-East. Perhaps the classic case of open political conflict was the North Southwark Local Plan, which was formally called in by the Secretary of State. A departmental press release explained the situation:

> We do not lightly call-in local plans, let alone reject them; indeed this is the first to be rejected and only the second to be called in. However, we will not hesitate to take action where such plans fly in the face of national policies for economic regeneration, especially of the inner city. The Southwark Local Plan is opposed to private investment and is hostile to the London Docklands Development Corporation. The Council has rejected the helpful comments of the inspector following the public local inquiry. The plan is quite unacceptable.
>
> (Read and Wood 1994: 11)

The interest of cases such as these lies in their exceptional status. It is very rare for a local authority to engage in a pitched battle with central government. Equally, it is seldom that central government will feel compelled to use its reserve powers. It is perhaps noteworthy that these two cases arose in the politically charged areas of inner London between radical Labour authorities and a Conservative government that had become openly hostile to local government.

The North Cornwall case was handled in a way more consistent with tradition. The case is discussed on p. 378: here it is sufficient to note that the local authority was giving planning permissions for development in the open countryside contrary to national policies and the approved county structure plan. Pressure was brought to bear upon the district council by way of a special inquiry carried out by an independent professional planner (Lees 1993). Normally, informal pressures are sufficient: the threat of strong action by the Secretary of State is typically as good as – if not better than – the action itself. With the enhanced position of development plans in the so-called plan-led system, attention now focuses on the provisions of draft plans. Regional officials pore over the wording of local policies to ensure that they accurately reflect those established at the national level. To the outsider, this plan scrutiny can develop into a game of words, sometimes taking on the character of academic hair-splitting. For instance, at one time the word 'normally' was acceptable, now it is taboo; 'a presumption against' is also out of order – unless this accords with specific national policy (DoE (1992) *Development Plans: A Good Practice Guide*: 87).

In spite of all this, it is not the function of the Secretary of State to decide detailed planning matters. In a recent ministerial statement, it was explained that

> It is the policy of the Secretary of State 'to be very selective about calling in planning applications. He will, in general, only take this step if planning issues of more than local importance are involved. Such cases may include, for example, those which, in his opinion,
>
> • may conflict with national policies on important matters;
> • could have significant effects beyond their immediate locality;
> • give rise to substantial regional or national controversy;
> • raise significant architectural or urban design issues; or
> • may involve the interests of national security or of foreign Governments.
>
> (HC Debates 16 June 1999, col 138)

This echoes with statements made by previous Secretaries of State: local planning decisions are normally the business of local planning authorities. The Secretary of State's function is to coordinate the work of individual local authorities and to ensure that their development plans and development control procedures are in harmony with broad planning policies. That this often involves rather closer relationships than might prima facie be supposed follows from the nature of the governmental processes. The line dividing policy from day-to-day administration is a fine one. Policy has to be translated into decisions on specific issues, and a series of decisions can amount to a change in policy. This is particularly important in the British planning system, where a large measure of administrative discretion is given to central and local government bodies. This is a distinctive feature of the planning system. There is little provision for external judicial review of local planning decisions (Scrase 1999; Keene 1999); instead, there is the system of appeals to the Secretary of State. The Department in effect operates both in a quasi-judicial capacity and as a developer of policy.

The Department's quasi-judicial role stems in part from the vagueness of planning policies. Even if policies are precisely worded, their application can raise problems. Since local authorities have such a wide area of discretion, and since the courts have only very limited powers of action, the Department has to act as arbiter over what is fair and reasonable. This is not, however, simply a judicial process. A decision is not taken on the basis of legal rules as in a court of law: it involves the exercise of a wide discretion in the balance of public and private interest within the framework of planning policies.

Appeals to the Secretary of State against (for example) the refusal of planning permission are normally decided by the Planning Inspectorate. Such decisions are the formal responsibility of the Secretary of State; and there is no right of appeal except on a question of law. Inspectors also consider objections made to local development plans, but in this case the recommendations are made to the local authority rather than the Secretary of State. The local authority is responsible for deciding how to act on the recommendations, but there are additional public safeguards allowing further objections and call-in by the Secretary of State.

Planning authorities, inspectors, and others are guided in their decisions and recommendations by government policy. Central government guidance on planning matters is issued by way of circulars and, since 1988, in policy guidance notes. There are four series of these: planning policy (PPG), regional planning policy (RPG), minerals policy (MPG), and derelict land grant advice (DLGA). In Scotland, Wales, and Northern Ireland there are similar 'national' policy statements, which are explained in Chapter 4.

Since the introduction of planning guidance documents, circulars have been concerned mainly with the explanation and elaboration of statutory procedures. PPGs deal with government policy in substantive areas, ranging from green belts to outdoor advertising. Circulars and guidance notes are generally subject to some consultation with local authorities and other organisations prior to final publication, but the Secretary of State has the final word.

Circulars and guidance notes are recognised as important sources of government policy and interpretation of the law, although they are not the authoritative interpretation of law (this is the role of the courts), nor are they generally legally binding. Indeed, advice can be conflicting, perhaps as a result of piecemeal revision at different times. Moreover, as is demonstrated repeatedly at public inquiries, differing interests can 'cherry pick' from the twenty-five PPGs to show how well their arguments meet the official guidance. Arguments for and against development in villages can be equally supported. While 'the overall strategy . . . should be to allocate the maximum number of houses to existing larger urban areas' (PPG 13), the building of houses in villages can help to sustain the local services which are necessary for their economic survival (PPG 7). Nevertheless, circulars and guidance notes command a great deal of respect and form an important framework for development planning and development control.

Policy, of course, has to be translated into action. This presents inevitable problems: policy is general, action is specific. In applying policy to particular cases, interpretation is required; and often there has to be a balancing of conflicting considerations – of which many examples are given throughout this book. Policies can never be formulated in terms which allow clear application in all cases, since more than one 'policy' is frequently at issue. Even the most hallowed of policies has to be flouted on occasion – as witness developments in the green belts, in protected sites of natural or historic importance, and in national parks. Such developments may be unusual (if only because they attract great opposition – of an increasingly strident nature); but they represent only the most obvious and the most public of the conflicts over land use.

Given the realities of land use controls, policies are usually couched in very general terms such as 'preserving amenity', 'sustaining the rural economy', 'enhancing the vitality of town centres' or 'restraining urban sprawl', and suchlike. This is a very different world from that of a zoning ordinance, which is the principal instrument of development regulation in many countries. Such an ordinance may provide (for example) that a building shall be set back at least five metres from the road, have a rear yard of six metres or more, and side yards of at least two and a half metres. Zoning was intended to be clear and precise, and subject to virtually no 'interpretation'. Indeed, it was hoped that it would be virtually self-executing. Though these hopes failed to materialise, it is fundamentally different in approach from the British planning system. Above all, the British system embraces discretion and general planning principles rather than certainty for the landowner and developer.

It is important to recognise that discretion means much more than 'making exceptions' in particular cases. The system requires that all cases be considered on their merits within the framework of relevant policies. Local authorities cannot simply follow the letter of the policy: they must consider the character of a particular proposal and decide how policies should apply to it. But they cannot depart from a policy unless there are good and justifiable planning reasons for so doing. The same applies to the Secretary of State, who is equally bound both by the formulated policies and by the merits of particular cases. The courts will look into this carefully in cases which come before them and, though they will not question the merits of a policy, they will ensure that the Secretary of State abides by it. Thus, in a curious way, discretion is limited. All material considerations must be taken into account and justified. Arbitrary action is unacceptable, as it is in the USA, which has written constitutional safeguards (Booth 1996; Purdue 1999).

(C) DEVOLVED AND REGIONAL GOVERNMENT

The Union will be strengthened by recognising the claims of Scotland, Wales and the regions with strong identities of their own. The Government's devolution proposals, by meeting those aspirations, will not only safeguard but also enhance the Union.

(White Paper, *Scotland's Parliament*, 1997)

DEVOLUTION TO SCOTLAND AND WALES

The campaign for devolution to Scotland and Wales failed at the end of the 1970s, but succeeded twenty years later. The aftermath of the earlier failure proved to be an important factor in the later success. The 1979 collapse of devolution led to the defeat of the Labour government and eighteen years of Conservative governments bent not on devolving power, but on centralising it. During this period the strength of the movement for devolution increased, particularly in Scotland, where the Thatcher government displayed a marked insensitivity to Scottish feelings. As Vernon Bogdanor has put it,

> Margaret Thatcher saw Scotland as an outpost of the dependency culture which she was determined to extirpate, while the 'very structure' of the Scottish Office 'added a layer of bureaucracy, standing in the way of the reforms which were paying such dividends in England'.
>
> (1993: 624 and 619)

The Thatcher Government's policies of competitive individualism were resented in both Scotland and Wales where they were seen as undermining traditional values of community solidarity; and policies such as privatisation and opting out from local authority control had little resonance there. But resented above all was the community charge, the poll tax. . . . Only devolution, so

it seemed, could protect Scotland and Wales against future outbursts of Thatcherism.

(1999: 195–6)

Following the publication of White Papers (*Scotland's Parliament* and *A Voice for Wales*), the Scotland Act and the Government of Wales Act were passed in 1998. The very titles of the White Papers point to a major difference between them. Scotland has a Parliament with legislative powers over all matters not reserved to the UK Parliament. Wales has only executive functions, but it does have full powers in relation to subordinate legislation. The latter include environmental, housing, local government, and planning functions. Thus Wales can change the provisions relating to the Use Classes Order, the General Development Order, the General Development Procedure Order, as well as the regulations concerning planning applications.

The devolution to Scotland and Wales is of importance to England for a variety of reasons. One of these is its effect on the possible pressure for devolution to English regions. This might be fostered if Scotland and Wales were perceived to be benefiting economically from devolution at the expense of the poorer regions of England. Encouragement might lie in the new regional machinery being established in England. As noted earlier, the Northern Region is already showing considerable interest in some degree of devolution to its area.[19]

THE SCOTTISH EXECUTIVE

Scotland has had a special position in the machinery of government since the 1707 Act of Union. It has maintained its independent legal and judicial systems, its Bar, its established Church (Presbyterian), and its heraldic authority (Lord Lyon King-at-Arms). The Scottish Office has a long history and, even before devolution, had a large degree of independence from Whitehall. Many years of responsibility for Scottish services, the relative geographical remoteness of Edinburgh (perhaps essentially psychological), the nature of the distribution of people and economic activity, the vast areas of open land, the close relationship between central and local administrators and politicians – such are the factors which gave Scottish administration a distinctive character. (This is not to mention the additional party political factors, such as the very small number of Conservative MPs: eleven out of seventy-two seats in 1996, and none in 1979!) Currently the departments include development (SEDD), education (SEED), enterprise and lifelong learning (SEELLD), health (SEHD), and rural affairs (SERAD). Following devolution, the ministers for these departments are members of the Scottish Parliament.

The role of the Scottish Executive and the Scottish Parliament in planning policy is as yet unclear, but it is unlikely to be a passive one. One preliminary indication is given in a 1999 Consultation Paper, *Land Use Planning under a Scottish Parliament*, issued by the Scottish Office:

> The form of any national planning policy guidance which emerges from the Scottish Executive could have significant implications for statutory development plans. . . . A national plan would almost certainly be perceived as unduly centralist and excessively rigid. However, guidance produced by the Scottish Parliament and Executive, bringing together the various National Planning Policy Guidelines and incorporating spatial issues more explicitly, might be attractive. This could inform future development in Scotland and provide some degree of consistency in the pursuit of sustainable development. It could be a vehicle for high level coordination of the objectives of the major agencies as they relate to development and land use. It could also prove attractive for those areas where progress with structure plans has been slow.

This gives some idea of early thinking on the way in which the new machinery might work. Later in the document there is a more certain statement: 'there is a clear expectation that all national strategic policy guidance will be subject to scrutiny by the Scottish Parliament'. Also revealing is the list of 'questions for consultees'. These include questions on whether there is a continuing need for structure plans, or whether an alternative should be subnational planning guidance. A further question asks whether there is a case for planning powers to be reduced by widening permitted development rights, or extended to cover

additional aspects, e.g. agriculture, forestry, and land management.

THE NATIONAL ASSEMBLY FOR WALES

In Wales, increasing responsibilities over a wide field were gradually transferred from Whitehall to the (former) Welsh Office. This transfer took many years to achieve. Welsh affairs were dealt with by the Home Secretary until 1960, with many services being administered directly by the departments which served England. There has been a minister responsible for Wales since 1951, but it was not until 1964 that the (Labour) government established the Welsh Office and a Secretary of State for Wales (Bogdanor 1999: 157–62).

Following devolution, the National Assembly for Wales has taken over responsibility for a wide range of functions.[20] Relevant to the fields covered in this book are culture, economic development, environment, historic buildings, housing, local government, tourism, town and country planning, and transport. All these functions are now transferred to the Assembly. Particularly important are the powers of secondary legislation. This is in contrast to the Scottish Parliament, which has the wider powers of primary legislation.[21] However, in the field of town and country planning, the effective difference is smaller than might at first sight appear. This is because of the particular character of the British planning legislation. This provides only a very general framework for the substantive measures which are enacted in secondary legislation such as the Use Classes Order, the General Development Order, the General Development Procedure Order, and a host of statutory rules and regulations (Bosworth and Shellens 1999). The latter deal with such matters as advertisements, development plans, environmental impact assessment, inquiries procedures, and planning obligations. Additionally, of course, plans are the responsibility of the local authorities, now subject both to Welsh planning guidance[22] and to approval by the Assembly. The Assembly is both an executive and a deliberative body.

The executive functions for planning are organised in a Transport, Planning and Environment Group. This comprises five management units, dealing with transport, highways, planning, property matters (estate division), and the environment. It also has oversight of the Welsh built heritage agency, Cadw. The Assembly is also responsible for the Countryside Council for Wales, the Welsh Development Agency (incorporating the former Land Authority for Wales and the Development Board for Rural Wales), and the Welsh Tourist Board. Some of these are discussed in later chapters.

NORTHERN IRELAND

Government in Northern Ireland has a unique character and structure. National government performs, either directly or through agencies, virtually all governmental functions; local government has few responsibilities. Though there are twenty-six elected district councils, their powers are limited to matters such as building regulations, consumer protection, litter prevention, refuse collection and disposal, and street cleansing. The councils nominate representatives on the various statutory bodies responsible for regional services such as education, health and personal social services, and the fire service. They also have a consultative status in relation to a number of services, including planning. All the major services, including countryside policies, heritage, pollution control, urban regeneration, transport, roads, and town and country planning, are administered directly by the Northern Ireland Office. The DoENI is the responsible department for these. Housing is administered by the Northern Ireland Housing Executive, which was formed in 1971 to take control of the local authority housing stock. Other departments include Agriculture and Economic Development.

Given the tragic history of Northern Ireland, the Office's priority aims are significantly different from those of other parts of the UK: 'to create the conditions for a peaceful, stable and prosperous society in which the people of Northern Ireland may have the opportunity of exercising greater control over their own affairs'. Planning has an important role in this, and is undertaken through an executive agency: the Planning Service Agency of the Northern Ireland Office. The general status of executive agencies is discussed below. The Agency's aim is 'to plan and manage development in ways which will contribute to a quality environment and seek to meet the economic and social aspirations of present and future generations' (Trimbos 1997).

The consultative role of the district councils is regarded as having great importance by the Agency, and it consults with them on a wider range of issues than is required by law. It is the government's intention that there should be a 'substantial democratic control of the planning process . . . as soon as politically possible by cross-party agreements in the context of a comprehensive political settlement' (*NI Planning Service Agency Annual Report 1998–99*: 81).

It is also intended to reorganise the departments in the new administration. A new Department for the Environment will be responsible for planning control, while a Department for Regional Development will be responsible for strategic planning. The Belfast Agreement gave a commitment to make rapid progress with a long-term regional strategy for consideration by the Assembly (once it has been established). This strategy will be a statutory document to which all Northern Ireland and UK Departments will be required 'to have regard'.[23] At the time of writing, any statutory action is on hold as a result of the political impasse.

TOWARDS REGIONAL GOVERNMENT IN ENGLAND?

The Labour Party manifesto proposal (Box 3.3) for elected regional governments in the regions where there was a popular demand for them was not for immediate action (particularly in areas where local government reorganisation into unitary authorities would be involved). In the meantime, regional chambers were to be developed to debate and formulate views about future policies. These chambers would vary according to regional wishes, but were expected to be based on existing regional local authority

BOX 3.3 LABOUR PARTY ELECTION MANIFESTO ON REGIONAL GOVERNMENT

Demand for directly-elected regional government so varies across England that it would be wrong to impose a uniform system. In time we will introduce legislation to allow the people, region by region, to decide in a referendum whether they want directly elected regional government. Only where clear popular consent is established will arrangements be made for elected regional assemblies. This would require a predominantly unitary system of local government, as presently exists in Scotland and Wales, and confirmation by independent auditors that no additional public expenditure would be involved. Our plans will not mean adding a new tier of government to the existing English system.

Source: Labour Party Election Manifesto 1997

organisations such as standing conferences and the like. As with these bodies, the regional chambers would be local authority led, but they would include representatives from other regional stakeholders – defined as 'any group or individual within the chamber's region with an established personal of professional interest in the issues of concern to the chamber'.[24] This proved an even more appealing idea than the government could have imagined, and all eight regions quickly established chambers (Table 3.1).

Concurrently with these initiatives, there has been considerable debate on the need for a more effective way of dealing with the regional dimensions of planning. The main approach until recently has been through the establishment and strengthening of the Government Offices for the Regions. These have built up a relationship with local government, and created a real governmental locus away from Whitehall. Set up in 1994, they have a range of functions within the remit of the DETR, the DTI, and the Department for Education and Employment. Their overall official role is to promote a coherent approach to competitiveness, sustainable economic development, and regeneration.

Table 3.1 Regional Development Agencies and Regional Chambers

Region	Regional development agencies	Regional chambers
Eastern	East of England	East of England Regional Chamber
East Midlands	East Midlands	East Midlands Regional Local Government Association
North-East	One North-East	North of England Assembly of Local Authorities
North-West	North-West	North West Regional Assembly
South-East	South-East	SERPLAN (from April 2001: South-East England Regional Assembly)
South-West	South-West	South-West Planning Conference
West Midlands	Advantage West Midlands	West Midlands Local Government Association
Yorkshire and Humberside	Yorkshire Forward	Regional Assembly for Yorkshire and Humberside

Attention was initially focused on the publication and revision of regional planning guidance, but this had tended to be rather bland statements of general central government policies. The Blair government quickly adopted a different approach with its campaign to 'modernise' both local government and the planning system. An early policy statement, *Modernising Planning*, pointed to the shortcomings of the regional planning guidance (RPG). Existing guidance had been criticised on the basis that it was too 'top-down', lacked regional focus and spent too much time reiterating national policies. It was also too narrowly land use oriented, took too long to produce, and did not 'command the confidence or commitment of regional stakeholders'. In short, RPG needed a major overhaul. Fundamentally, this was seen to include a strong 'bottom-up' approach:

> We propose a more inclusive process, involving the local authority conferences working with the Government Offices, business and other regional stakeholders, in producing drafts of the regional guidance itself. This would replace the current arrangement under which the regional planning conference merely provides 'advice' on the basis of which planning guidance is subsequently produced by the Government Office. In effect, this would replace the existing two stage process with a single stage of regional strategy preparation. If this works as it should, it will normally only be necessary for the Secretary of State to endorse the draft, subject to any specific reservations in relation to issues of national policy
>
> (*Modernising Planning* 1998: para. 18)

This was envisaged as being a second-best solution: 'statutory planning at the regional level will have to await a democratically-accountable statutory body to undertake it'. There are strong indications that this might be emerging. In London, of course, such a body is being established. Elsewhere, there is a generally positive response to the government's proposals.[25] A major stimulus has come from another policy area: regional development. The epigraph at the head of this section explicitly indicates the government's commitment to elected regional government in England, though its viability and nature is uncertain at this stage, particularly since it would involve yet another upheaval in local government. In the absence of elected regional authorities, a new regional organisation has

been established. Following the 1997 White Paper *Building Partnerships for Prosperity*, Regional Development Agencies (RDAs) were set up in each of the eight regions outside London. (The London Development Agency will be appointed by and responsible to the Mayor.)

REGIONAL DEVELOPMENT AGENCIES

RDAs are, as their name indicates, agencies to promote economic development in their regions. The Regional Development Agency Act 1998 requires each agency to formulate and keep under review a strategy for implementing its statutory responsibilities to further economic development and regeneration, to promote business efficiency, to promote employment, to enhance the development and application of skills, and to contribute to the achievement of sustainable development. Given the traditional emphasis on urban areas (as well as the political prominence of rural concerns), the RDAs are specifically required to give equal attention to rural areas.[26] Indeed, funding is to be separately allocated or earmarked for rural projects.[27]

The strategy is subject to any 'guidance' and directions issued by the Secretary of State. The first guidance was published in 1999 and dealt with regional strategies (Box 3.4). Guidance has also been issued or announced on rural policy, sustainable development, regeneration policy, education and skills issues, competitiveness, inward investment, performance indicators, state aid rules, and equal opportunities.

A particular difficulty arises where a planning area straddles more than one RDA, as is the case with Thames Gateway.[28] Even more problematic is the problem with the Eastern Region, where three counties (Cambridgeshire, Suffolk, and Norfolk) are covered by the Standing Conference of East Anglian Local Authorities (SCEALA), while three other counties (Bedfordshire, Essex, and Hertfordshire) are covered by the London and South East Regional Planning Conference (SERPLAN). In July 1999, proposals were put forward by SERPLAN and the

---

BOX 3.4 GUIDANCE AND DIRECTIONS TO REGIONAL DEVELOPMENT AGENCIES ON REGIONAL STRATEGIES

The Regional Development Agencies Act 1998 enables the Secretary of State to give guidance and directions to an RDA, in particular with respect to:

- the matters to be covered by the strategy;
- the issues to be taken into account in formulating the strategy;
- the strategy to be adopted in relation to any matter; and
- the updating of the strategy.

The agency's strategy should provide:

- a regional framework for economic development, skills, and regeneration which will ensure better strategy focus for and co-ordination of activity in the region whether by the agency or by other regional, subregional, or local organisations;
- a framework for the delivery of national and European programmes which may also influence the development of government policy: and
- the basis for detailed action plans for the agency's own work, setting out the wider aims and objectives for its annual corporate plan.

Source: DETR (1999) *RDAs Regional Strategies*

South East England Regional Assembly (SEERA) for a new joint committee to set out a broad strategy for the three regions of London, the South East and East Anglia.[29] This is regional planning on a truly grand scale!

An obvious problem arises on the relationship between the RDA economic strategy and RPG. Much has been written on this, and the House of Commons Select Committee report on Regional Development Agencies includes a range of views.[30] Some have argued that RDAs should be required to work within the framework of a 'comprehensive overarching strategy' to be prepared and approved by the appropriate regional bodies. The DETR rejected this on the grounds that the two strategies cover different issues, and that areas of mutual interest can be dealt with by constructive collaboration. What is not usually made explicit is the concern of central government that such an 'overarching' plan could be used to frustrate desirable economic development or housing provision. It is a nicely arguable question whether the requirements of regional land use planning should take

priority over economic or housing 'requirements'. Those who see the protection of the countryside as an overriding policy objective will have no doubts on the answer to this, but central government takes a different view. The case of biotechnology clusters is a case in point, as is illustrated by the decisions on development proposals in the rural area south of Cambridge.[31] Interestingly, though one of the proposals was not accepted in full (40,000 sq. m at the Wellcome Trust Human Genome Project at Hinxton Hall), the decision letter explicitly stated that 25,000 sq. m would be acceptable, despite the location in an area of restraint. (It was the size of the development, not the location, that was unacceptable.) Two other proposals, both in the Cambridge green belt, were accepted. Thus important economic developments can receive priority over countryside protection.

Nowhere has this been made more explicit than in the Public Examination on the RPG for the South-East. The Panel Report on this is scathing about the argument that economic and housing developments should be 'dampened down' in this region in an attempt to benefit other regions. The essence of this argument was that 'regional imbalances' should be tackled by preventing the 'economic magnetism of the 'overheated' South-East from 'draining away economic vitality and population from other UK regions'. The Panel castigated this view 'with its manifest overtones of postwar Barlow based industrial development policy'. Government policy was very different, with an emphasis on the economy being encouraged 'to go ahead at full speed on all engines'. The Panel Report continued:

> In our view it is high time that the ghost of Barlow (his report that is) be finally exorcised from regional strategy. Whereas in the 1940s and for some time thereafter, it may have been quite reasonable to consider the UK as the principal unit for economic planning, this is manifestly not the situation at the present time. Economic activity and investment discouraged from settling in the South East of England will not now find alternative landing places in the other UK regions; they are just as likely to go to other parts of the EU. The effect therefore of reducing 'development pressures' by 'dampening down' the economy of the South East would have little or no beneficial effect on the economies of the other regions of the UK. . . . The whole of the UK (and indeed the EU) has a vested interest in the economic success of the South East region as a core area for economic activity and a major source of capital and tax revenues. . . . It is an engine of growth for the whole country. . . . RPG needs to make it clear that there can be no question of doing anything but building on the success of the South East's economy with a view to recovering its premier status in the EU and world league.[32]

Needless to say, this argument is not shared by all! Nevertheless, it now seems clear that the Blair government has made the decision to back the RDAs in maximising their individual growth potentials, irrespective of the impact on migration from the less favoured regions. In this, it is following in the steps of the previous Conservative government.[33]

FUNDING OF THE RDAS

The RDAs are financed with the funds of the government programmes they have taken over, such as urban regeneration, industrial land improvement, and rural development. In total they have a budget of around £800 million and will be expected to lever another £1 billion or so of private money for programmes such as urban regeneration and the redevelopment of derelict land. Their staffs come mainly from the agencies they have subsumed, and number around 100 (*The Economist*: 'An England of regions', 27 March 1999). Some functions that might have been transferred to the RDAs have been retained by their departments – business support services by the Department of Trade and Industry, and skills training by the Department for Education and Employment. The RDAs report to the DTI. There thus appears to be scope for much bureaucracy. This could be put to the test if an RDA wishes to follow a policy which is in conflict with 'central government orthodoxy'. (For instance, 'the chief executive of the North-East agency is keen to spend more on improving the skills of those already in work. That would run counter to the government's belief in concentrating training on the unemployed'.[34] Such conflicts are familiar to federal systems, and will no doubt come to the fore in the new system. Regional government means not only giving power to the regions, but also taking some power away from central government.

Only time will tell whether elected regional government will emerge. Even in the areas (such as the North-East and Yorkshire) where it looks promising, there could be a contest between the forces for regional government and those for city mayoralties. There are interesting times ahead!

(D) LOCAL GOVERNMENT

> UK local authorities in the past two decades had to manage and adapt themselves to a scale of change and upheaval that has approached the revolutionary.
> (Wilson and Game 1998)

The Government was elected with a bold mandate to modernise Britain and build a fairer, more decent society. To do that and to deliver its key pledges it needs the support of local government. . . . At its best local government is brilliant and cannot be bettered. But to play its full part in modernising Britain local government itself needs to modernise.

(Blair 1998)

REORGANISING LOCAL GOVERNMENT

Reorganising local government has become a tradition. Its functions and its structure have been subject to frequent change. The pace of this change has become almost frenetic since the early 1960s when it became apparent from the government's decision to reorganise London government that there were serious prospects for the reorganisation of local government elsewhere in the country. Since then analysis, debate, legislation, review, and further legislation has been ceaseless.

In summary, a uniform two-tier system of local government was established by legislation passed in 1963 for London, 1972 for Scotland, and 1973 for the rest of England and Wales. The two-tier system in London and the metropolitan counties was subject to drastic change by the Thatcher government in 1986 under the banner of 'streamlining the cities'. The upper tier (the Greater London Council and the metropolitan county councils of Greater Manchester, Merseyside, Tyne and Wear, West Midlands, and West Yorkshire) were abolished, thus leaving a unitary system of local government in these areas.

Further reorganisation into unitary authorities took place in Scotland and Wales in 1996, and a number of unitary authorities were introduced in parts of non-metropolitan England between 1995 and 1997 (although much of the two-tier system remains). In Northern Ireland a unitary system of local government was set up in 1973. Thus, while Northern Ireland, Wales, and Scotland have a unitary system, England has a varied structure of local government, in which 115 areas have a unitary system and the remainder are two-tier, forming thirty-four counties and 238 districts.

THE ENGLISH LOCAL GOVERNMENT REVIEW

The anomalous structure of local government in England stems from the distinctive nature of the local government review that preceded it. A Local Government Commission was established to work within policy and procedural guidance published by the DoE. This 'guidance' proved to have greater power than the government expected: it had the effect of limiting the changes which could be made to the Commission's proposals. Moreover, because of the consultative way in which the Commission operated, these proposals were significantly influenced by the views of articulate local interests. The government's initial proposals for change were set out prior to the establishment of the Commission in a consultation paper, *The Structure of Local Government in England*. This made the argument, widely accepted across the political spectrum, for a unitary structure of local government in the shires (the pattern which had been put in place in the metropolitan counties). It was argued that a single tier would reduce bureaucracy and costs, and improve coordination. It would clarify responsibility for services and, since taxpayers would be able to relate their local tax bills more clearly to local services, would provide for greater accountability. In the early stages of the Commission's review, the Environment Secretary, John Gummer, had stated unequivocally that the aim was to produce a unitary structure in England, with the two-tier system remaining in only exceptional circumstances. The way this was to be achieved was left open, with several possibilities: existing districts might become unitary authorities, two or more authorities could be merged into larger ones, and wholly new authorities might be created. The main criteria for judging the need for change was responsiveness to local needs and 'sense of identity', as well as the ubiquitous 'cost-effectiveness'.

During the two years of the Commission's review, district and county authorities sought to justify their existence through an expensive and sometimes bitter propaganda war. In fourteen cases this led to challenge to the Commission's recommendations in the courts. There was also a legal challenge by the Association of

County Councils, which successfully prevented the Secretary of State from modifying the guidance he had previously given to the Commission in an attempt to strengthen the case for unitary authorities. Despite the government's wish to see a unitary structure, the eventual undoubted winners were the counties. The Commission found little evidence that change would improve service provision. In the main, changes were limited to renewing unitary status for former county boroughs, and abolishing new and contrived counties created in the 1974 reorganisation.

After much debate, the Commission recommended only fifty new unitary authorities. These were mostly former county boroughs (unitary authorities before the 1974 reforms), although a significant number of 'special cases' were included on the basis of 'substantial local support for change'. The Commission explained the modest extent of their recommendations as due to the 'weight of evidence from national organisations pointing to the problems and risks associated with a breaking up of county wide services' – a view that was strongly supported by local opinion. However, these arguments failed to satisfy the many districts which were not proposed for unitary status and which had campaigned for this. More significantly, it did not satisfy the government, which was concerned to further increase the number of unitary authorities. Following these disagreements, the chairman of the Commission resigned, and the new chairman was given the remit to review again the case of twenty-one districts where the government believed there was a strong case for unitary status. Further guidance was issued for this mini-review, stressing the potential benefits of unitary status particularly for areas needing economic regeneration (as in the Thames Gateway). It was argued that the 'single focus' of unitary local government would be more effective in promoting multi-agency programmes in these areas. This final review initially recommended unitary status for ten of the twenty-one districts, but this was reduced to eight after consultation. The new councils came into being in April 1997.

The process of reorganisation in the shires has been the subject of considerable criticism and, after three years of work, many commentators are asking whether it was worthwhile. Certainly, reorganisation seems to have been handled much more expeditiously in Wales and Scotland.

Parish councils (or community or town councils) can play a role in the democratic process by providing an effective voice for local interests and concerns. Unlike their counterparts in Scotland and Wales they have statutory functions, though these are very restricted. Of particular importance (and widely used) is their right to be consulted on planning applications in their areas. They can also play a part in the consultation process for the preparation of development plans.

LOCAL GOVERNMENT IN SCOTLAND

Even before devolution, the cultural history and physical conditions of Scotland have dictated that, to a degree, the administration of planning is distinctive. Changes to the law in Scotland require specific legislation, and the Scottish Office has for long had administrative discretion within which it could take account of the special circumstances that exist in parts of the country. Nevertheless, the broad thrust and impact of government policy have been much the same (Carmichael 1992).

In setting out to reorganise Scottish local government, the government was firmly committed to a single-tier structure, and the 1992 Scottish consultation paper provided options only on the number that were to be established.

There were, of course, some political factors involved in this decision: the problem of conflicting interests within the Conservative Party was much less in Scotland since only a handful of the sixty-five Scottish local authorities were in Conservative control. The consultation document in Scotland was also more forthright about the role of local government reform in direct service provision. While the government confirmed its commitment to 'a strong and effective local authority sector', it also argued that local authorities no longer needed to 'maintain a comprehensive range of expertise within their own organisation', since 'much could be done by outside contractors'.

In reviewing the possible number and size of the proposed unitary authorities, a consultation paper provided four illustrations showing structures ranging from fifteen to fifty one authorities. The choice between mainly small or mainly large authorities has important implications for the planning function, especially structure plans. Only the fifteen-authority option would have allowed for unitary authorities to prepare their own structure plans. Even then, special arrangements would have been needed for Glasgow to ensure effective strategic planning. The outcome of reorganisation in Scotland was thirty-two unitary councils, each of which has full planning powers for its area. The fragmentation of the strategic planning function across a larger number of authorities threatens a recognised strength of the Scottish system, and the need for special arrangements for strategic planning was acknowledged by the Scottish Office during the review. The country has been divided into seventeen structure planning areas, six of which require joint working between authorities. The plan framework is discussed further in Chapter 4.

Scottish legislation provides for the establishment of community councils where there is a demand for them, under schemes prepared by local authorities. As in England and Wales, their purpose is to represent the local community and 'to take such action in the interests of the community as appears to its members to be desirable and practicable'. A study of community councils concluded that 'in contemporary moves towards democratic renewal in local government, community councils are seen as having no special status or role by most local authorities', though some do accord them a distinctive role in consultation, and there is a wide variety throughout Scotland in their operations and effectiveness (Goodlad *et al.* 1999). Nevertheless, community councils may have a new role in the proposals of a working group of the Scottish Office and the Convention of Scottish Local Authorities (SO, *Report of the Community Planning Working Group*).

SCOTTISH LOCAL GOVERNMENT AND 'COMMUNITY PLANNING'

Though there has been less than overwhelming support for community councils in Scotland, the debate on their future has been transformed into an enthusiastic promotion of the idea of 'community planning'. This is defined as any process through which a local authority comes together with other organisations to plan, provide for, or promote the well-being of the communities they serve. The objectives are to improve levels of service and to increase the collective capacity of public-sector agencies to tackle problems which require action from more than one agency. Though much cooperation between agencies already exists, there was a need for a more systematic approach which would provide an overarching strategy. The working group recommend that the Scottish Parliament should signify the importance of 'community planning' by providing a statutory basis for it.

Similarly, the 1999 McIntosh Report on local government and the Scottish Parliament recommended both the retention of community councils (properly resourced) and the promotion of their role 'within the wider context of the area approach adopted by many councils, as a means of obtaining the fullest possible consultation at the local level'.

SCOTTISH LOCAL GOVERNMENT AND THE SCOTTISH PARLIAMENT

The establishment of the Scottish Parliament raises a host of questions concerning local government, some of which have long been of importance (such as public apathy and mistrust (Carole Millar Research 1999), some of which arise because of devolution (particularly relationships between Parliament and local government), while others have arisen on the tide of reform which devolution has created (such as the electoral system). Whatever the reason, there is a major endeavour to improve the system of governance in Scotland.

The Commission on Local Government and the Scottish Parliament (the McIntosh Report) has

Figure 3.3 Planning Authorities in the UK

Note: Divisional planning offices are shown for Northern Ireland

Non-metropolitan unitary councils

1	Darlington	17	Derby	32	Wokingham
2	Stockton-on-Tees	18	Nottingham	33	Slough
3	Hartlepool	19	Leicester	34	Windsor and Maidenhead
4	Middlesbrough	20	Rutland	35	Bracknell Forest
5	Redcar and Cleveland	21	Peterborough	36	Thurrock
6	York	22	Herefordshire	37	Southend
7	East Riding of Yorkshire	23	Milton Keynes	38	Gillingham and
8	Kingston upon Hull	24	Luton		Rochester upon Medway
9	North Lincolnshire	25	South Gloucestershire	39	Plymouth
10	North-East Lincolnshire	26	Bristol	40	Torbay
11	Blackpool	27	North-West Somerset	41	Poole
12	Blackburn	28	Bath and North-East	42	Bournemouth
13	Halton		Somerset	43	Southampton
14	Warrington	29	Thamesdown	44	Portsmouth
15	Stoke-on-Trent	30	Newbury	45	Isle of Wight
16	The Wrekin	31	Reading	46	Brighton and Hove

Metropolitan unitary district councils

Greater Manchester
Bolton
Bury
Manchester
Oldham
Rochdale
Salford
Stockport
Tameside
Trafford
Wigan

Merseyside
Knowsley
Liverpool

St Helens
Sefton
Wirral

South Yorkshire
Barnsley
Doncaster
Rotherham
Sheffield

Tyne & Wear
Gateshead
Newcastle upon Tyne
North Tyneside
South Tyneside
Sunderland

West Midlands
Birmingham
Coventry
Dudley
Sandwell
Solihull
Walsall
Wolverhampton

West Yorkshire
Bradford
Calderdale
Kirklees
Leeds
Wakefield

London borough councils

Barking and Dagenham
Barnet
Bexley
Brent
Bromley
Camden
City of Westminster
Croydon
Ealing
Enfield
Greenwich

Hackney
Hammersmith and Fulham
Haringey
Harrow
Havering
Hillingdon
Hounslow
Islington
Kensington and Chelsea
Kingston upon Thames
Lambeth

Lewisham
Merton
Newham
Redbridge
Richmond upon Thames
Southwark
Sutton
Tower Hamlets
Waltham Forest
Wandsworth

Wales

A	Flintshire	E	Vale of Glamorgan	I	Caerphilly
B	Wrexham	F	Rhondda, Cynon, Taff	J	Newport
C	Neath and Port Talbot	G	Cardiff	K	Torfaen
D	Bridgend	H	Merthyr Tydfil/Blaenau Gwent	L	Monmouthshire

Figure 3.3 continued

BOX 3.5 STIRLING LISTENING TO COMMUNITIES

Stirling Council has developed a range of ways of ensuring that it is attentive to the communities it serves, both in terms of policy development and service delivery. It has adopted a radical committee structure which focuses on people's needs rather than the Council's internal organisation (e.g. a children's committee, a care committee, and environmental committee, and a community and economic development committee). It has created a 'civic assembly' which meets four times a year to give citizens and community groups the opportunity to scrutinise services and make an input to future planning for the area. Community councils are elected by postal ballot. There has been a turnout of over 60 per cent in some areas. The Council goes out to discuss budget priorities with the communities, and over the last two years over 1,000 people have made their views on budget priorities known to the Council.

Area forums are progressively being established for different localities within the Council area, to enable local communities to have an input into the decisions which affect them and to strengthen the accountability between local Council managers (who have been given substantial budget authority) and the people they serve. A network of local offices is maintained to give people convenient access to Council services.

Source: Scottish Office (1999) *Report of the Community Planning Working Group*

---

recommended a number of wide-ranging proposals for reforming local government in the context of devolution. Its starting point is a declaration that

> relations between local government and the Parliament ought to be conducted on the basis of mutual respect and parity of esteem. . . . Councils, like Parliament, are democratically elected and consequently have their own legitimacy as part of the whole system of governance.
>
> (McIntosh Report 1999)

To play the role envisaged by the Commission, local government should take the initiative to respond to the challenges it now faces. It should review its procedures and 'renew itself'. This involves citizen participation, not merely by way of consultation but also in decision-making; 'open, transparent and intelligible' methods of conducting business; a focus on the consumer; and quality and cost-effectiveness in the delivery of services. Local government needs to develop new ways of working in partnership: it is uniquely placed to take an overview of local needs and to provide leadership in 'community planning'.

It is on the basis of this type of thinking (and a long list of specific recommendations) that the Commission stresses that 'the world is about to change for Scottish local government'. Among its recommendations are the ratification of a covenant between the Parliament and the thirty-two councils setting out their working relationship. This concept of a direct working relationship between local and central government is 'without parallel or precedent at Westminster', though it is in harmony with the European Charter of Local Self-Government, as well as the Hunt Report.[35]

Other proposals include a statutory power of general competence, thus freeing local authorities from the limitations imposed by the constitutional position that they can carry out only those specific powers granted by legislation; further study of the ways in which local authorities may become financially more independent; and a review of local government elections, with the introduction of proportional representation in 2002.

This selection of recommendations gives some flavour of the extent to which Scottish local government is under fundamental review.[36] In addition, there is a consultative process under way for determining a new ethical framework.[37] This includes a review of aspects of the planning process, such as the training of members for the work of a planning committee, and the introduction of 'best practice' (see the discussion in Chapter 12).

As with somewhat more modest English ideas for modernising local government, these are big aspirations which can be met only by major changes in the culture of local government (Brooks 1999: 43).

LOCAL GOVERNMENT IN WALES

The reorganisation of local government in Wales proceeded more quickly than in England. The review was carried out by the Welsh Office (rather than by an independent commission) and the country was considered as a whole (rather than by separate areas). After a two-year period of consultation, a White Paper, *Local Government in Wales: A Charter for the Future*, was published, in 1993, setting out detailed proposals. There was widespread agreement that the new structure should be unitary in character, and the debate was focused on the number and boundaries of the new local authorities.

The underlying thinking included a restoration of authorities which had been swept up in an earlier reorganisation (Cardiff, Swansea, Newport) and some of the traditional counties, such as Pembrokeshire and Anglesey. However, to fit into a unitary structure, the boundaries had to be stretched somewhat, and a number of counties had to be amalgamated. After consideration of proposals for thirteen, twenty, and twenty-four unitary councils, the final outcome was twenty-two authorities. (Reflecting their history, these have varied formal names such as county borough, city, and county council, but they all have the same functions.) They range in population from 66,000 in Cardiganshire to 318,000 in Cardiff.

In the White Paper, the unitary system was commended for its administrative simplicity, its roots in history, its familiarity, and the relative ease with which residents could identify 'with their own communities and localities'. The intention was to create 'good local government which is close to the communities it serves'. The White Paper continued:

> Its aims are to establish authorities which, so far as possible, are based on that strong sense of community identity that is such an important feature of Welsh life; which are clearly accessible to local people; which can, by

taking full advantage of the 'enabling' role of local government, operate in an efficient and responsive way; and which will work with each other, and with other agencies, to promote the well-being of those they serve.

These desirable objectives do not all work in the same direction, of course, and some compromise was inevitable. Some of the areas are very large. Powys, for example, has over 500,000 hectares: this is a very large area for local government. There is potential for the community councils to take on an increased role, but the Welsh Office has stressed that there is no intention of forming a second tier of local government.

WELSH LOCAL GOVERNMENT AND THE WELSH ASSEMBLY

Preparations for devolution in Wales were far less advanced than they were in Scotland, where there was a much firmer expectation that devolution would take place. The first steps included the mounting of a consultation exercise on the establishment of a Partnership Council with local government. This Council was mandated by the Government of Wales Act. It consists of twenty-five members: ten from the Assembly, ten from the county and county boroughs, two from the community councils, and one each from the police authorities, the fire authorities, and the national park authorities (NAW, *The Partnership Council: Preparing the Ground*, 1999).

The Assembly has also produced a paper on the development of planning policy (*The Approach to Future Land Use Planning Policy*, 1999) Unlike the English and Scottish planning policy guidance publications, the Welsh guidance is published in two guidance notes only: 'This enables the inter-relationships between policies to be clarified, and means that each revision has to be a full one across all policy topics, rather than piecemeal.' The guidance closely follows the English publications and has relied on DETR research and policy development. The question of its Welsh distinctiveness has been debated, and the paper concludes:

> Planning policy for Wales should no longer track DETR priorities slavishly, nor should it diverge from GB policies

unless this is for good reason. Both the process of developing planning policy, and its content, should be appropriate to Welsh circumstances, and be produced in a shorter time scale than hitherto.

It is stressed that in developing planning policies, there should be a partnership with local government, business, and the voluntary sector. To facilitate this, a forum is to be established. This will include representatives from a wide range of organisations, and will have the remit 'to inform planning research and policy development'. To ensure that Welsh needs are fully met, a Welsh Planning Research Programme is to be developed. A Research Scoping Study has been mounted to identify key research areas. These will include speeding up the preparation of development plans; planning for rural areas; planning and integrated transport; improving local authority development control performance; waste planning; making planning more responsive to business; and locational policy for renewable energy (e.g. wind farms).

LOCAL GOVERNMENT IN NORTHERN IRELAND

Local government in Northern Ireland was last re-organised in 1973, when thirty-eight authorities, made up of counties, county and municipal boroughs, and urban districts, were replaced by a single tier of twenty-six district councils. Although this reduced the enormous variation in the size of districts (previously ranging from 2,000 to over 400,000), there is still a wide variation, from Moyle with a population of some 15,000 to Belfast City with a population approaching 300,000. Planning powers were centralised under the then Northern Ireland Ministry of Development. Since the demise of the power-sharing Northern Ireland Assembly in 1974, planning, like all public services, has been subject to 'direct rule' under the supervision of the Secretary of State for Northern Ireland. The preparation of plans and the control of development are functions of the Department of the Environment for Northern Ireland (DoENI), which it exercises through the Planning Service (an executive agency).

Local government is consulted only on the preparation of plans and development control matters.

The lack of accountability through local government (described as the 'democratic deficit') obviously needs to be seen in the light of the very special circumstances, though it has been judged to have operated with a 'considerable measure of success' (Hendry 1992: 84). Nevertheless, local councillors have been able to attack planning and to 'represent themselves as the champions of the local electorate against the imposed rule of central government' (Hendry 1989: 121). Even when the central bureaucracy has made determined efforts to open decision-making and involve local people, it has been accused of having ulterior motives (Blackman 1991b).

The promise of a 'lasting peace' in Northern Ireland during the ill-fated ceasefire brought with it ideas for reform which are still on the agenda. Several possible scenarios have been suggested, including the continuation of a central planning authority accountable to an elected Assembly, devolution to joint regional boards, and complete delegation of powers to the Districts (RICS 1994). The relatively weak position of local government over many years and the dearth of skills and experience will not be put right quickly. It is likely that any reform will be introduced incrementally. The volatile political conditions in the Province make prediction impossible.

THE NEW GREATER LONDON AUTHORITY

The abolition of the Greater London Council in 1986 left a gap in the machinery of government which was cumbersome, inefficient, and indefensible. London became the only Western capital city without an elected city government. Some functions carried out by the GLC were transferred in part to the London boroughs, but many were taken over by a range of joint bodies, committees, *ad hoc* agencies, and suchlike (including the London Planning Advisory Committee, which prepared strategic planning guidance for the capital). The result was 'a degree of complexity that can be seen not so much as a "streamlining" as a return to the administrative "tangle" of the 19th century' (Wilson and Game 1998: 54). The Labour Party's election manifesto promised a referendum 'to confirm popular demand' for a strategic authority and mayor. The referendum was held in 1998, and though only a third of Londoners voted, there was an overall 72 per cent majority. The Greater London Authority Bill provides for an elected Mayor and an elected Greater London Assembly (GLA). The Mayor is to be no figurehead, but a highly influential leader. In the words of the 1998 White Paper *A Mayor and Assembly for London*, 'the Mayor will have a major role in improving the economic, social and environmental well-being of Londoners, and will be expected to do this by integrating key activities'. The main responsibilities include:

- The production of an integrated transport strategy for London (extending to transport issues for which the Mayor is not directly responsible) to be implemented by a new executive agency, Transport for London (TfL), which will have responsibility for a wide range of services including London's bus and light rail services, the Croydon Tramlink, the Docklands Light Railway, the Victoria Coach Station, taxis and minicabs, and river services. It will also acquire responsibility for a strategic London road network. Government funding will be paid in a single block grant, and capital investment schemes within the budget available will not require central government approval.
- Preparation of strategic planning guidance for London in the form of a new Spatial Development Strategy (SDS). The content of this will be for the Mayor to decide, but will include transport, economic development and regeneration, housing, retail development and town centres, leisure facilities, heritage, waste management, and guidance for particular parts of London such as the central area and the existing Thames Policy Area (there are also other strategies, including transport). The unitary development plans of the boroughs will be required to be 'in general conformity' with the SDS. Development control will remain with the boroughs, but the Mayor will be a statutory consultee for planning applications of strategic importance, and will have defined powers of intervention, which are already being used for significant applications.
- The setting of an economic development and regeneration strategy for London. A London Development Agency will be appointed by, and responsible to, the Mayor.
- Improvement of London's environment, the development of an air quality management strategic plan, and the production of a report every four years on the state of the environment in London.
- Appointment of half the members of a new independent Metropolitan Police Authority, and scrutiny of the policies of the Authority.
- Overall responsibility for a new London Fire and Civil Defence Authority, and appointment of the majority of its members.
- Preparation of a strategy for the development of the culture, media and leisure sectors, appointments and nominations to the key cultural organisations.

Clearly this is a highly significant change to the government of London, providing an eloquent indication of the government's commitment to a more effective and democratic system of government. The position of Mayor will not be an easy one, since it will involve extensive – and intensive – negotiation with the London boroughs and innumerable governmental bodies, as well as many professional and voluntary organisations. However, the arrangements have been carefully thought through, and there is a general commitment to make them work.

MANAGING PLANNING AT THE LOCAL LEVEL

For town and country planning, the apparent and seemingly paradoxical outcome of change in the 1980s and 1990s has been a larger and stronger body of planners with widened statutory functions. The indirect effect of market deregulation, the increasing complexity of development issues, and the growing emphasis on environmental protection was bound to lead to a greater demand for planning skills (Healey 1989). The concept of an 'enabling' local government also increases the need for strategic thinking and focuses attention on the corporate planning function (Carter *et al.* 1991). The direct impact on the way in which the planning service is delivered is less significant. So far, planning has been subjected to only minimal change in comparison to other services, and the concept of the local authority as 'enabler' requires more attention to strategy and in-house planning, rather than less. The conclusion has to be that, in contrast to most local services, planning as a statutory and regulatory function has been somewhat protected from the pressure for change. Nevertheless, there has been a clear increase in contracting out planning tasks, particularly those of a specialist nature, such as retail studies and environmental assessment.[38] Moreover, the spread of auditing and 'value for money' (VFM) has been given a new gloss by the Labour government with the concept of 'Best Value'.[39] These are not easy concepts to define for planning because of the difficulty of assessing quality in plans and planning decisions. The Audit Commission provides guidance for local authorities and district auditors on performance indicators for all services, including planning, but these have been criticised for the reliance on quantitative measures, the classic example being the proportion of applications decided within eight weeks.

Best Value requires that performance reviews are prepared looking ahead over a five-year period, starting with areas of work where there are problems. The reviews must challenge why and how the service is being provided; invite comparison with others' performance across a range of relevant indicators; involve consultation with local taxpayers, service users, and the wider business community in the setting of new performance targets; and embrace fair competition as a means of securing efficient and effective services. The reviews will produce new performance targets to be published in an annual local performance plan together with comparisons with other authorities (note that the District Audit Service is to stop producing the annual local authority reports based on CIPFA statistics); identification of forward targets for all services

annually and in the longer term (at least five years); and commentary on how the targets will be achieved, including proposed changes to procedures. Local performance plans will be audited.

The Audit Commission's 1992 report *Building in Quality* addressed criticisms of the accent on efficiency rather than effectiveness of the planning system in performance review. It made a real attempt to introduce a wider assessment, recognising that there were many ancillary tasks in providing advice and negotiating with applicants, and making 'complex professional and political judgments'. After consultation, the Audit Commission settled on six key indicators. As with the earlier version, these concentrate on matters of efficiency rather than on the effectiveness of the system, though the added breadth of performance review will be a significant improvement on previous practice. The consultation document on this lists seven 'best value performance indicators' (or BVPIs as they are inevitably called). They are reproduced in Box 3.6.

Much of this 'business' approach to local government stemmed from the Conservative government's concern to reduce the autonomy of local government. Rate capping and compulsory competitive tendering (CCT) were striking examples of this. The former has been greatly reduced by the Blair government as part of its programme for modernising and democratising local government.[40] Compulsory competitive tendering has been replaced by the principles of 'Best Value'. The 1998 White Paper *Modern Local Government: In Touch with the People* succinctly describes this as 'a duty to deliver services to clear standards – covering both cost and quality – by the most effective, economic and efficient means available'. This is essentially a positive recasting of the 'enabling' concept. (This is the norm in western Europe, with local government implementing its functions through a diverse range of agencies, often in partnership, but essentially seeing its role as prioritising community needs and acting as the focus for local political activity.) Best Value is also seen as an aid to local government 'to address the cross-cutting issues facing their citizens and communities, such as community safety or sustainable development, which are beyond the reach of a single service or service provider'. The 'very best-performing councils' will be

BOX 3.6 BEST VALUE PERFORMANCE INDICATORS FOR PLANNING

Percentage of new homes built on previously developed land.

Planning cost per head of population.

The number of advertised departures from the statutory plan approved by the authority as a percentage of total permissions granted.

Percentage of applications determined within 8 weeks.

Average time taken to determine all applications.

Score against a checklist of planning best practice. (The answer should be 'yes' to all questions.)

Do you have a development plan which was adopted in the last 5 years? If 'no',

For those adopted plans not adopted in the last 5 years, are there proposals on deposit for their alteration or replacement, and have you publicly adopted a timetable for adopting those alterations or the replacement plan?

Does your development plan contain a comprehensive set of indicators and targets and do you monitor your performance against these?

Has all supplementary planning guidance (including planning briefs) produced and adopted by you during the last year followed the guidance given in paragraphs of Planning Policy?

Do you provide for pre-application discussions with potential applicants on request?

Do you have a publicised charter which sets targets for handling the different stages of the development control process (including enforcement and compliance) and arrangements for monitoring your performance against these targets?

Is the percentage of appeals where the council's decision has been overturned lower than 40%?

Does your authority delegate 70% or more of its applications to officers?

In the last financial year, have you run your planning service in such a way that:
   a) you have not had any planning costs awarded against you;
   b) you have not had any adverse ombudsman's reports issued against you finding maladministration with or without injustice; and
   c) there have been no court findings against you under sections 287 and 288 of the Town and Country Planning Act 1990 or on judicial review?

Does your council operate a one stop shop service which includes the following characteristics:

   a) a single point giving initial advice to members of the public and other enquirers on all development-related consents operated by the authority;
   b) application discussions covering all the development-related consent regimes operated by the authority which are appropriate to each potential applicant; and
   c) a nominated case officer acting as the contact point within the authority for each planning applicant and coordinating the authority's response to the application including progress and advising on the application's relationship to other development-related consents.

Have you implemented a policy for ensuring that different groups have equal access to the planning process including, as necessary, the provision of advice in ethnic minority languages and in Braille/on tape based on consultation with relevant members of the community about the accessibility of the planning service, and do you have arrangements for keeping this policy under review?

eligible for 'Beacon' status, normally for particular services. Applicant councils will be chosen by an independent advisory panel, and will be rewarded by being given wider discretion in the operation of the beacon service. Streamlining planning decisions for business is one of seven priority service areas which will be judged.

A new statutory duty has been placed on local authorities to promote the economic, social, and environmental well-being of their areas. This is in line with the European Charter of Local Self-Government, which provides that local authorities can do anything to further their electors' interest unless prevented by statute. (The previous Conservative government refused to sign this charter, preferring to keep local authorities under central control (Jenkins 1996: 254–8).

These are some of the elements of the local government modernisation programme which, at the time of writing, is emerging from the Blair government. Others include proposals for a major recasting of the political structures of local authorities, with cabinet-style executives in place of the traditional committee system, and a separation of planning policy and development control functions (White Paper *Local Leadership, Local Choice*, 1999). Under this model, planning decisions would be taken by a quasi-judicial planning committee. Whether the separation of policy and its execution is viable is debatable. The RTPI has argued convincingly that it is impracticable.[41]

THE ETHICAL LOCAL AUTHORITY

Local government has for long had a reputation for probity, particularly in planning, where foreign observers are quick to point out the obvious opportunities (nay: temptations) for corruption. That the temptations have not always been resisted is now well known. It was in the 1970s that the Poulson scandal blew up: several local authority politicians and officials were found guilty of securing contracts for the architectural business of John Poulson. A number of well-known figures went to jail. It was an extreme case, and shocked the local government world. It led

to the setting up of a Royal Commission on Standards of Conduct in Public Life and to the introduction of a National Code of Local Government Conduct. There have been other cases of local government impropriety, particularly in the planning arena, of which the most recent were in Warwick, Bassetlaw (Nottinghamshire) and Newark.[42] Currently, there is considerable argument about the alleged misdoings of some Doncaster councillors and officials.[43]

Though only a small number of such cases have arisen, they are clearly unacceptable. In fact, the numbers involved were certainly fewer than the number of cases of Westminster 'sleaze' (a conveniently vague and all-embracing term) in the later years of the Conservative government.[44] This led to the appointment of a Committee on Standards in Public Life, under the chairmanship of Lord Nolan, which embarked upon a series of inquiries into various areas of public life. Its third report, devoted to local government, was published in 1997.[45]

The Nolan inquiry was concerned not to put local government on trial but to provide guidance on what standards of conduct should apply and how they could be maintained. The National Code was criticised for being inadequate, complicated, and, in parts, inconsistent and even impenetrable. Moreover, in the words of Standards of Conduct in Local Government (para. 56), it represented 'something that is done to local authorities, rather than done with them'. Building upon the report, the government proposed a 'new ethical framework' to govern the conduct of elected members and also local government employees (who were not covered by the National Code). A Code of Conduct, based on a national model, is required of all local authorities, together with a Standards Committee to oversee ethical issues and to provide guidance on the code and its implementation. An independent Standards Board will have the responsibility of investigating alleged breaches of the local authority's code.[46]

Planning was seen to require additional measures. Because of its complexity and the problems of dealing fairly and properly with planning law and its implementation, it was proposed that members of planning committees should be trained in the planning

THE AGENCIES OF PLANNING 69

system.[47] There should also be a greater degree of openness in the planning process; this would, among other things, assist in dealing with the problems facing local authorities in granting permission for their own proposed developments, and 'the potential for planning permission being bought and sold'.

In coming to these conclusions, the Nolan Report noted that in 1947 'the need for postwar reconstruction was clear. Development enjoyed broad public support.' Things have now changed.

> Development is now a term which has a pejorative ring, and the planning system is seen by many people as a way of preventing major changes to cherished townscapes and landscapes. If the system does not achieve this (and it is a role which it was not originally designed to perform), then the result can be public disillusionment.
>
> (Standards of Conduct in Local Government, para. 277)

In Scotland a 1998 consultation paper on the Nolan Report, 'A New Ethical Framework for Local Government in Scotland', broadly accepted its recommendations, but took issue with a number of them. It proposed a single code for all local governments (instead of a model code), and it favoured a national Standards Commission instead of local authority standards committees. It also argued that reasons should not be required for the granting of planning permission since such decisions are not subject to any appeal process, and it would not only add to the difficulties facing a planning committee but put permissions at increased risk of legal challenge on purely technical grounds. These and related matters are discussed in Chapters 5 (development permission for local authority developments), 6 (planning gain) and 12.

FURTHER READING

European Government

The institutions and policies of the EU are summarised in a series of free booklets, Europe on the Move, which are updated periodically (available from the UK Office of the European Commission at Jean Monnet House, 8 Storey's Gate, London SW1 – and a long list of free publications is available at <www.cec.org.uk/sourcedi/catalog1.htm>. For a summary of the history of the EU, see Borchardt (1995) *European Integration: The Origins and Growth of the European Community*.

There are a great many general accounts of the making of the European Union, including the very comprehensive *Encyclopedia of the European Union* edited by Dinan (1998); and Borchardt (1995) *European Integration: The Origins and Growth of the European Community*. For a more critical account, see Chisholm (1995) *Britain on the Edge of Europe*; and for a more theoretical view of European integration, Nelson and Stubb (1998) *The European Union*; and Emerson (1998) *Redrawing the Map of Europe*. There are also a great number of texts on EC Law and institutions, many of which are no more than reprints of official texts. Tillotson (2000) *European Union Law*, and Craig and de Búrca (1999) *EU Law* are useful in their own right, and these are also updated regularly and provide brief summaries of the history of the EU. The history of each policy area in which the Community has acted, including regional policy, environment and transport, is given in Moussis (1999) *Access to European Union Law, Economics, Policies*. Williams (1996) *European Union Spatial Policy and Planning* gives an account of the European institutions from a planning perspective.

A chronological review of how Europe has influenced planning is given in Nadin (1999) 'British planning in its European context'; and there are two DETR research reports that address the consequences of the development of European policies for planning in the UK: Nadin and Shaw (1999) *Subsidiarity and Proportionality in Spatial Planning Activities in the European Union*, and Wilkinson *et al.* (1998) *The Impact of the European Union on UK Planning Practice*. See also Bishop *et al.* (2000) 'From spatial to local: the impact of the European Union on local authority planning in the UK', and Shaw *et al.* (2000) *Regional Planning and Development in Europe*. Further references on European spatial planning are given at the end of Chapter 4.

On European comparative planning systems, including the organisation of government, see the *EU Compendium of Spatial Planning Systems and Policies*, which comprises individual volumes describing the systems and policies of spatial planning in each member state, and a comparative review; Davies *et al.* (1989b) *Planning Control in Western Europe*; and Seaton and Nadin (2000) *A Comparison of Environmental Planning Systems Legislation in Selected Countries*.

Central Government

Whitehall by Hennessy (1989) is an insightful analysis of the culture of the British civil service and the changes that have been imposed upon it. See also Drewry and Butcher (1991) *The Civil Service Today*. Osborne and Gaebler (1992) *Reinventing Government* give a coloured account of changes which they perceive in the US systems of government: resemblances with the British scene are striking, even if inconclusive. An invaluable overview, though now dated, is given by Dynes and Walker (1995) *The Times Guide to the New British State: The Government Machine in the 1990s*. A more up-to-date examination of the British system of government is the deservedly popular account by Birch (tenth edition 1998). Jones and Kavanagh (1998) provide a good introduction to British politics today in the sixth edition of their book. On Northern Ireland, see Connolly and Loughlin (1990) *Public Policy in Northern Ireland: Adoption or Adaptation?*.

An up-to-date summary description of government departments and their functions is given in the annual official handbook *Britain*, prepared by the Central Office of Information. Greater detail is given in the annual *Whitaker's Almanack*. A principal source of information on the work of government departments is the *Departmental Annual Reports*. These are the *Government's Expenditure Plans* for the forthcoming three years and are sometimes referenced in this way. They are now available on the Internet.

An excellent insight into the operation of central government in exercising its controls over local government is given in Read and Wood (1994) 'Policy, law and practice'.

Devolved and Regional Government

On devolution, the essential book is Bogdanor (1999) *Devolution in the United Kingdom* (which has been the main source for the discussion in the text, and which has an extensive bibliography). See also the excellent set of essays edited by Hazell (1999) *Constitutional Futures: A History of the Next Ten Years*; Connal and Scott (1999) 'The New Scottish Parliament: what will its impact be?' McCarthy and Newlands (1999) *Governing Scotland: Problems and Prospects – The Economic Impact of the Scottish Parliament*; and Bosworth and Shellens (1999) 'How the Welsh Assembly will affect planning'. The two White Papers were *Scotland's Parliament* (Cm 3658) and *A Voice for Wales* (Cm 3718). Important for the current debate are the various official publications, particularly DETR (1997) *Building Partnerships for Prosperity*; and DETR (1998) *The Future of Regional Planning Guidance*. Baker (1998) provides a useful statement of the position of regional land use planning in 'Planning for the English regions: a review of the Secretary of State's regional planning guidance'. Mawson (1996) reviews the re-emergence of the regional agenda in the English regions, while the history of the two strands of regional planning (inter-regional economic and intra-regional land use) and the move towards 'a more integrated and comprehensive approach' are well analysed by Roberts and Lloyd (1999) 'Institutional aspects of regional planning, management, and development: models and lessons from the English experience'. Roberts *et al.* (1999) provide an extensive discussion of Metropolitan planning in Britain, with case studies of nine British metropolitan regions.

The progress of the RDAs and their strategies is being monitored by various academic centres. See, for example, Roberts and Lloyd (1998) *Developing Regional Potential* (Centre for Planning Research, University of Dundee); Nathan *et al.* (1999) *Strategies for Success?* (Centre for Local Government Economic Strategies, Manchester); and Deas and Ward (2000) 'The song has ended but the melody lingers'. See also the statements and publications of the English Regional Associations (www.somerset.gov.uk/era), including the 1999 *Regional Governance: Statement of General Principles*. Additionally, there is Charter 88, whose publications include Tomaney and Mitchell (1999) *Empowering the English Regions*.

Hall's essay 'The regional dimension' (1999a) gives an overview of postwar regional economic policy, with a short comment on regional land use planning. Bradbury and Mawson (1997) *British Regionalism and Devolution: The Challenges of State Reform and European Integration* provide an informative analysis of the emergence of regionalism from a number of perspectives. Wannop (1995) *The Regional Imperative: Regional Planning and Governance in Britain, Europe and the United States* is an excellent account, by a knowledgeable practitioner, of the endeavours to plan on a regional scale, primarily in the UK, but also in Europe and the USA. Two different aspects of the development of regionalism are discussed in Mawson (1996) 'The re-emergence of the regional agenda in the English regions: new patterns of urban and regional governance', and Garmise (1997) 'The impact of European regional policy on the development of the regional tier in the UK'.

Local Government

The principal textbooks which give a general introduction to local government structure and organisation are Elcock (1994) *Local Government*, Chandler (1996) *Local Government Today*, and the second edition of Wilson and Game (1998) *Local Government in the United Kingdom*. For an overview of the politics of local government, including the roles and relationships between councillors, officers, and political parties, see Stoker (1991) *The Politics of Local Government*. On changing management approaches, see Stoker's volume of essays *The New Management of British Local Governance* (1999). See also Allen (1990) *Cultivating the Grass Roots: Why Local Government Matters*; and for an examination of the relationship between planning and changing government, see Vigar *et al.* (2000) *Planning, Governance and Spatial Strategy in Britain*.

European comparisons are given by Hirsch (1994) *A Positive Role for Local Government: Lessons for Britain from Other Countries* and Batley and Stoker (1991) *Local Government in Europe*. There are several reports of foreign experience and practice in local government prepared for the Commission on Local Government and the Scottish Parliament: Hughes *et al.* (1998) *The Constitutional Status of Local Government in Other Countries*; Hambleton (1998) *Local Government Political Management Arrangements: An International Perspective*; University of Edinburgh (1999) *Summary of Devolved Parliaments in the European Union*; and Centre for Scottish Public Policy (1999) *Parliamentary Practices in Devolved Parliaments*.

On Scotland and Wales, see Carmichael (1992) 'Is Scotland different?'; Boyne *et al.* (1995) *Local Government Reform: A Review of the Process in Scotland and Wales*; and Midwinter (1995) Local Government in Scotland. On devolution to Scotland and Wales, Bogdanor (1999) is essential reading. On Northern Ireland, see Bannon *et al.* (1989) *Planning: The Irish Experience 1920–1988*; Hendry (1992) 'Plans and planning policy for Belfast'; and a short paper by Lipman (1999) 'Difficult decisions in a rural balancing act'.

Parish councils are the subject of a survey by the Public Sector Management Research Centre (1992) *Parish Councils in England*. A Welsh Office consultation paper was issued in 1992: *The Role of Community and Town Councils in Wales*. A particularly interesting document is the Scottish Office (1999) *Report of the Community Planning Working Group*. On central–local government relations, see Cochrane (1993) *Whatever Happened to Local Government?*. A case for improving relations is made in Carter and John (1992) *A New Accord: Promoting Constructive Relations between Central and Local Government. Local Government in the 1990s* is a collection of essays edited by Stewart and Stoker (1995). See also Jones and Stewart (1983) *The Case for Local Government*. A lively critical analysis of the Conservative government's attitudes and policies in relation to local government is provided by Simon Jenkins (1995) in *Accountable to None: The Tory Nationalisation of Britain*. A useful postwar review is Young and Rao (1997) *Local Government since 1945*.

The rate of change in local government under the Blair government requires a perusal of the relevant journals, such as *Local Government Chronicle* and *Municipal Review*. See also Hambleton (2000) 'Modernising political management in local government'.

There have been a large number of official publications discussing and proposing changes in the operation of local government. Of particular importance are *Modern Local Government: In Touch with the People* (Cm 4014, 1998); *Local Leadership. Local Choice* (Cm 4298, 1999); and *A Mayor and Assembly for London* (Cm 3897, 1998). For references on 'the ethical local authority', see Chapter 12.

NOTES

1 The countries that have applied for membership of the EU are Bulgaria, Cyprus, the Czech Republic, Estonia, Hungary, Latvia, Lithuania, Malta, Poland, Romania, Slovakia, Slovenia, and Turkey. Agreement was reached for Norway to join in 1972 and 1995, but on both occasions membership was rejected by a referendum. Switzerland has also applied for accession; its application was accepted by the EU but rejected in a referendum in 1992. Iceland is the only significant western European country which has not sought accession to the EU.

2 The CSD is to be incorporated into the committee that provides advice on the Structural Funds, which is a formal EU committee.

3 The European Treaties lay the foundation for economic and political integration. The first was the European Coal and Steel Community (ECSC) of 1951, which was followed in 1957 by the European Atomic Energy Community (EAEC or Euratom) and the Treaty establishing the European Economic Community (later to become the Treaty establishing the European Community, TEC) both signed in Rome. The latter is generally referred to as the 'Treaty of Rome', and set objectives for the creation of the Economic Community and established the basis institutions which would achieve it. The Treaties have subsequently been amended by the Single European Act (SEA) of 1987, which made firm the commitment to creating the Single Market; the Treaty on European Union (TEU) of 1992, known as the Maastricht Treaty, which considerably widened the areas of cooperation of the Union into foreign affairs, defence, and justice; and the Amsterdam Treaty of 1997 (although not coming into force until 1999), which prepared for enlargement and made sustainable development an objective of the Union. During the process of preparing the Amsterdam Treaties it was decided to consolidate the Treaties, but in the event this was limited to a tidying up of the numbers of articles in the TEU and the TEC. The addition of numerous protocols (which have legal force) and declarations has increased the complexity, and navigating the treaties is very much a job for experts.

4 Romano Prodi's speech to the European Parliament, 14th September 1999, Strasbourg.

5 It came into full effect in October 2000.

6 Alistair Darling, HC Debates, 11 June 1997, col. 1144, quoted in James (1999: 194–5). See the White Paper *Modern Public Services in Britain: Investing in Reform* (Cm 4011, 1998).

7 James (1999: 42). Of course, many things had changed since Wilson's time, including those brought about by Mrs Thatcher. On a later page (p. 45), James comments that 'The Thatcher Government's biggest impact on Whitehall may have been to alter attitudes by proving how much can be achieved by sheer political will.'

8 Jill Sherman, 'Whitehall faces major revamp', *The Times* 12 July 1999.

9 Michael White, 'Whitehall warfare', *Guardian* 12 August 1999.

10 In 1999 there was press speculation that the DETR would be 'downsized', either by separating transport again (allegedly because of public dissatisfaction with inadequate transport policies) or by allocating responsibilities for the countryside to a new department of rural affairs (thus demonstrating the government's commitment in an area where there has been considerable controversy). In the event the DETR lost its responsibilities for rural and environment affairs to DEFRA, and for the government offices for the regions to the Cabinet Office, after the June 2001 election.

11 In 1999, the Department requested local authorities to volunteer to pilot local cultural strategies. Draft guidance on these strategies has been circulated for consultation. (The Greater London

Authority has a specific duty to produce a cultural strategy for London.)

12 Curiously, in 1999 the Ministry made the first ever use of its powers under section 43(6) of the Planning Act to veto the allocation of high-grade farmland in East Yorkshire for development. This was overruled by the Secretary of State for the Environment. See *Planning* 26 February 1999: 1.

13 The extract is from a statement issued by MAFF in July 1999. The statement continues: 'There will be less emphasis than hitherto on detailed input to local plans and on ad hoc planning applications. As far as possible, FRCA will complete discussions with local planning authorities and other bodies in respect of existing consultations, but it will no longer be possible to undertake detailed field work to establish the correct Agricultural Land Classification grading of proposed development sites. Instead, greater use will be made of desk studies and predictive mapping techniques when evaluating proposals from LPAs for allocations to development. . . . These changes should be fully in place by 31 March 2000.'

14 All figures refer to England only and are taken from the Planning Inspectorate (2000) *Statistical Report 1999–2000*, Bristol: PIEA.

15 Whereas in 1993–94 the duration from the opening of an average inquiry to the submission of the inspectors report was 40 weeks, by 1995–96 this had increased to 49 weeks.

16 The Inspectorate's statistical report suggests that it can be called upon to adjudicate in relation to about 200 categories of work.

17 See the DETR research report on the Environmental Court Project (2000) and the memorandum of evidence by Professor Malcolm Grant to the Select Committee Inquiry.

18 'The Department monitors all draft plans . . . this is a formidable task. In England it is undertaken by the Regional Offices of the Department. . . . For many years the Department has maintained a Planning Handbook which gives internal guidance to decision-making officers on procedural questions. This guides officers in monitoring development plans. The Handbook is now computerised and can be instantly updated. This is the primary tool for the coordination of Departmental action. It is not a public document. Officers are advised to check the wording of policies and proposals against current planning guidance. Realism is looked for, in particular whether sites said to be available can actually be developed. A most difficult task is to identify possible strategic implications of local policies. . . . The Department also encourages informal approaches to regional offices by development plan teams in the course of plan preparation' (Read and Wood 1994: 10).

19 See pp. 30–1 and 52–4.

20 The Welsh Office has been renamed Wales Office, but the term is applicable only in Whitehall. In Wales, the reference is to the Assembly for Wales (Welsh Affairs Committee, *The Role of the Secretary of State for Wales*, 26 October 1999, HC 854, para. 36).

21 The decision to grant legislative powers to Scotland, but only executive power to Wales, 'had as much to do with political compromise and accident as with any rational argument' (Osmond 1997: 149.)

22 Planning policy guidance in various forms has been issued by the Welsh Office for some years. See particularly *Unitary Development Plans* (1966) and *Planning Policy* (first revision 1999).

23 See Fourth Standing Committee on Delegated Legislation, 3 March 1999, *Draft Strategic Planning (Northern Ireland) Order 1999*.

24 See *Regional Chambers* (DETR 1999).

25 An encouraging insight is given by Kitchen (1999a).

26 In the inelegant language of the Act (section 4.2), 'A regional development agency's purposes apply as much in relation to the rural parts of its area as in relation to the non-rural parts of its area.'

27 The rural regeneration functions of the Rural Development Commission have been transferred to the RDAs. (Other functions have been merged in the new Countryside Agency.) An interesting

history of the RDC is A. Rogers (1999) *The Most Revolutionary Measure: A History of the Rural Development Commission 1909–1999*.

28   See DoE (1993) *East Thames Corridor: A Study of Development Capacity and Potential*; DoE (1995) *The Thames Gateway Planning Framework* (RPG 9a); and Haughton *et al.* (1997).

29   'Super-regional structure to be in place within two years', *Planning* 30 July 1999, p. 1. See also the supplementary evidence presented to the Select Committee, op. cit., vol. 3, pp. 163–7.

30   HC Environment, Transport and Regional Affairs Committee, Environment Subcommittee (1999) HC 232 (Session 1998–99), 3 vols. See, for example, memorandum by the TCPA (vol. 2, p. 45), and memorandum by the DETR (vol. 2, p. 71).

31   How far these developments are to be judged as 'clusters' (or even one 'cluster') is problematic. There has been strong governmental backing for clusters (in the White Paper *Our Competitive Future: Building the Knowledge Driven Economy* (Cm 4176, 1998) – generally referred to as 'the Competitiveness White Paper – and the Sainsbury Report, *Biotechnology Clusters* (DTI 1999). There has been a surprisingly general enthusiasm for clusters, despite the vagueness of the concept. (Note their popularity in the RDA regional strategies.) For a critical view, see Perry (1999). An informative short account of the Cambridge proposals and the DETR decisions is that of Green.

32   Government Office for the South East (1999) *Regional Planning Guidance for the South East of England: Public Examination May–June 1999 – Report of the Panel*, paras 4.8–4.10.

33   A good case in point is a Surrey plan of 1993 in which overriding importance was given to environmental and infrastructure policies. The Report of the Panel on the Examination in Public held that the plan did not 'adequately provide for industry and commerce' and concluded that the employment policies in the plan needed to be changed 'so as to be more responsive to the needs of business'. A similar stance was taken in

relation to Hampshire, where policies were designed to control the rate of growth in order to safeguard the environment, character, and heritage of the county. The Secretary of State backed the view of the Panel that Hampshire could not be regarded as an 'economic island'. In strong terms he denounced the proposed policies as 'a recipe for economic decline' (Read and Wood 1994: 26).

34   Ibid.

35   The Hunt Report (1996) is the *Report of the Lords Select Committee on Relations between Central and Local Government*. The European Charter of Self-Government is reproduced in an appendix to the McIntosh Report.

36   The Executive's response was issued in October 1999: *The Scottish Executive's Response to the Report of the Commission on Local Government and the Scottish Parliament* (<www.scotland.gov.uk/library 2/doc04>).

37   SO (1998) *A New Ethical Framework for Local Government in Scotland: Consultation Paper*. This stems from the Nolan Report, *Standards of Conduct in Local Government* (Cm 3702, 1997).

38   On the basis of research carried out for the RTPI, Higgins and Allmendinger (1999) argue that 'contracting out now appears to be a permanent and growing feature of UK public planning'. Their research is published by the RTPI (1996) *Contracting Out Planning Services*. The business campaign group London First has pointed to the 'obvious' advantage of the Mayor for London contracting out specific research to specialist external agencies. It notes that there are 'especially strong agencies' in the fields of land use planning and economic development (*Planning*, 30 July 1999: 2).

39   The Local Government Act 1999 defines best value as 'securing continuous improvement in the exercise of all functions undertaken by the local authority, whether statutory or not, having regard to a combination of economy, efficiency and effectiveness'. The 'extremely uneven progress' in introducing Best Value into planning is reported by Thomas (1999). On the application of Best

Value to parish councils, see DETR (1999) *The Application of Best Value to Town and Parish Councils*. See also Warwick Business School (1999).

40 Capping as such is abolished, but there are reserve powers of a more discriminating nature.

41 'Development control is not – or should not be – an after-the-fact regulatory process. It is instead the implementation of the plan and, as such continuously feeds back into it, identifying the changes in circumstances which require changes in criteria, revision of policies and supplementary planning guidance. It is vitally important that those who guide the preparation of plans should be, as far as possible, those who work to implement them'; RTPI evidence to the Joint Committee on the Draft Local Government Organisation and Standards Bill see the Committee's report, paras 117–20. But see Hambleton and Sweeting (1999a).

42 See External Enquiry into Issues of Concern about the Administration of the Planning System in Warwick District Council (1994), and Report of an Independent Inquiry into Certain Planning Issues in Bassetlaw (1996). On the Newark case, see *Planning* 29 October 1999: 2, and 5 November 1999: 15. Note also the North Cornwall case referred to earlier (p. 48).

43 See *Planning* 12 November 1999: 1.

44 Allen (1990: 12) has commented that 'central agencies are often at least as incompetent, inefficient or corrupt as local bodies . . . local authorities are perennially in the news for alleged corruption and graft . . . one or two notorious cases can suffice to keep the whole concept of local government in disrepute'. For a catalogue of cases of corruption in public administration, see Doig (1984).

45 The first report was on members of parliament, ministers and civil servants, and executive non-departmental public bodies (Cm 2850, 1995). This was followed by a report on further and higher education bodies, grant-maintained schools, training and enterprise councils, and housing associations (Cm 2170, 1996). The report on local government was published in 1997 (Cm 3702).

46 *Modern Local Government: In Touch with the People* (Cm 4014, 1998: chapter 6); *Modernising Local Government: A New Ethical Framework* (1998); and *Local Leadership, Local Choice* (Cm 4298, 1999: chapter 4).

47 The DETR has published a training syllabus prepared in conjunction with the LGA and the RTPI: *Training in Planning for Councillors* (DETR 1999).

4

THE PLANNING POLICY FRAMEWORK

Spatial planning is now taking on a new importance with the privatization of development, the increased status accorded to the land use development plan . . . and the marketing of areas to lever in government grants and inward investment. Spatial planning, as an integrative mechanism for coordinating diverse strategies, sits uneasily with the present method of forward planning revolving around the statutory land use development plan.

(Hull 1998: 328)

INTRODUCTION

The framework of planning instruments, and the procedures which are followed in their creation and operation, provide an important starting point for understanding how the planning system works. This chapter and chapter 5 discuss the formal system of town and country planning which is embodied in a huge library of statutes, rules, regulations, directions, policy statements, circulars, and other official documents. However, it is important to appreciate at the outset that the formal system is one thing; the way in which matters work in practice may be very different. The informal planning system operates within the formal structure. It may continue with little modification even when major legislative changes are made; alternatively, there may be significant changes in practice within a stable formal system. Political forces, professional attitudes, and management styles will all affect the ways in which the system operates.

It is also necessary to note that much development (in the everyday, rather than the legal, sense of that word) takes place without any help or hindrance from the planning system. Even where the development is clearly related to some action within the statutory

framework for planning, the actual outcome is affected by 'extraneous' factors, and it may not be at all clear what effect planning has had on the outcome.

It is current government policy to bring much more of the informal operation of planning and development within the statutory planning framework. The emphasis is on plan-led development control. This is intended to reduce the amount of *ad hoc* planning control, and thus provide a firmer foundation for the resolution of conflicts and investment decisions. Plans are seen as providing a more efficient means of conflict mediation than decision-making on a project-by-project basis, as well as a measure of certainty and coordination for the promotion of investment (Healey 1990).

The plan-led system is intended to produce a comprehensive and systematic hierarchy of national and regional guidance, and development plans. It is envisaged that once these plans are in place, they will be a more important factor than their predecessors in land use decision-making. However, high expectations about a new plan framework have not been met. Early optimism is waning in the face of continued problems in establishing comprehensive cover of detailed plans. There is also scepticism about any major shift in emphasis or impact, with central government continuing to enforce a common approach, and reinforcing the need for planning policy to support business.

As we shall see, the fundamental principles of the system remain intact, and recent changes, while superficially significant, may have little effect on the patterns of land use and protection, or on who gains and loses from it. Plans may be more significant (and there are now many more plans), but fundamental questions have been raised about the ability of the system to respond to the complex challenges of sustainable development and economic competitiveness. Europe has been a source of many ideas, and especially important in shifting attention to the strategic level of plan-making and the need for greater sectoral policy integration. For planners who experienced the introduction of the 1947 system or the changes made in 1968, there must be a sense of déjà vu. Many of the questions of those times bear a marked resemblance to the current debate:

- What framework will ensure the accountability of decision-makers and safeguard the interests of those affected by planning, yet be expeditious and efficient in operation?
- How can the framework provide a measure of certainty and commitment, yet allow for flexibility to cope with changing circumstances, local conditions, and new opportunities?
- What objectives should plans pursue, and how will these shape their form and content?
- Who should have influence in the planning process, and what should be the respective roles of central and local government and of local communities?

These perplexing questions have no easy answers: indeed, by their nature, they have to be readdressed constantly. Acceptable answers rarely have stability since conditions and attitudes change over time. The biggest change in recent years has been the gradual growth of supranational planning, led by the European Union. It is appropriate that a discussion of the UK planning framework should begin with this.

(A) SUPRANATIONAL PLANNING

THE RATIONALE FOR PLANNING AT THE EUROPEAN SCALE

The EU is driven by the goals of economic competitiveness, social cohesion, and balanced development of economic activities among its regions, and, since the adoption of the Amsterdam Treaty in 1999, sustainable development. These objectives have an obvious spatial dimension. The main obstacles to meeting them are the great disparities in wealth, jobs, investment, and access to services across Europe. Indeed, recent evidence suggests that despite the actions of the EU, some disparities (especially between the north and the south) are widening, and that economic and political forces will ensure that they continue to do so.

The growing economic and social integration of European nations and regions in the context of globalisation is having a profound affect on spatial development patterns. Significant elements of economic activity together with political and cultural relations are effectively becoming globalised and independent of nation-states. Locational decisions are now more likely to ignore regional and national boundaries. The extent and depth of globalisation are disputed, but it is widely accepted that it has specific implications for changing patterns of spatial development. Of particular note in the European context are increased spatial concentration of economic activity and the central role of global and regional cities; intensified competition between cities across national boundaries; the corresponding polarisation of economic prosperity; and the negative environmental consequences (Sassen 1995). Major development schemes have effects which often go well beyond national borders. In sum, the transboundary interdependencies of spatial development are now much stronger than they once were.

These effects are reinforced by Community policies, especially in the fields of regional policy, transport, environment, and agriculture, though their implications for spatial development are not always explicitly considered in the policy-making process. Spatial planning and state regulation in other spheres play a significant role in addressing these trends, by maximising the competitive position and growth potential of major urban areas while attempting to ensure that, at best, patterns of growth are sustainable and, at worst, the negative impacts are ameliorated.

SPATIAL IMPACTS OF COMMUNITY POLICY: THE STRUCTURAL FUNDS AND COMMUNITY INITIATIVES

The EU's regional policy provides Structural Funds to promote 'the harmonious, balanced and sustainable development of economic activities, and in particular the development of competitiveness and economic innovation'.[1] Allocation of the structural funds from 2000 is divided according to three 'objectives', the first two of which have a strong spatial dimension. Objective 1 is to assist designated regions lagging behind in development[2] (with less than 75 per cent of the Community average GDP). Objective 2 is targeted at the economic and social conversion of designated areas facing structural difficulties. Objective 3 is to support the adaptation and modernisation of policies and systems of education, training, and employment, and is available to all regions not designated as Objective 1. The Funds account for more than a third of the total Community budget, and will amount to 195 billion euros (about £117 billion at 2000 exchange rates) in the programming period 2000–06; the UK will receive about 10 billion euros.[3]

About 70 per cent of the structural funds go to Objective 1 areas, with 12 per cent to Objective 2 and 12 per cent to Objective 3; most of the rest are held in reserve or are allocated through Community initiatives. Thus some 82 per cent of structural funds are targeted on specific regions; this is perhaps not surprising since the main intention is to produce a better economic and social balance across the Community. The areas of the UK which fall under these objectives are identified in Figure 4.1. The main beneficiaries in the UK are the Objective 1 regions, which are now Cornwall and the Isles of Scilly,

Merseyside, South Yorkshire, and West Wales and the Valleys. The designations are effective for the seven years to 2006.

A transitional assistance mechanism has been established until 2005 to soften the blow of the loss of Community funding for areas previously designated as Objective 1: the Highlands and Islands, and Northern Ireland.[4]

EU funding is also made available through Community initiatives which are used to tackle specific problems with a European dimension (such as the conversion of defence industries). Previous initiatives have been criticised because of the relatively small funding available relative to the bureaucracy involved in their implementation. In order to concentrate funding and focus on Community objectives more during the 2000–06 programme, the thirteen Community initiatives have been reduced to four. They are INTERREG (5 billion euros (362 million for the UK) on transnational planning; URBAN 700 million euros (117 million) on urban regeneration; LEADER 2 billion euros (106 million) on rural development; and EQUAL 2.8 billion euros (376 million) on employment and training. All are developments of existing initiatives and have already been prominent tools for planners in the UK. They are explained in more detail as relevant in subsequent chapters.

The UK government also has its own measures to promote the development of economically disadvantaged regions, although assistance has been reduced considerably since the 1960s. The main instrument remaining is regional selective assistance (RSA), which supports projects to create jobs and increase regional competitiveness. In 1998 the European Commission published guidelines on regional aid to promote comparable and transparent systems across the EU.[5] The DTI has subsequently reviewed the boundaries of assisted areas in the UK to bring them into line with the new guidelines. Three tiers of assistance are proposed with different rates of assistance. The first two follow the guidelines and are designated by national government.

Tier 1 is designated according to the same criteria as Objective 1 Structural Funds and is thus coterminous

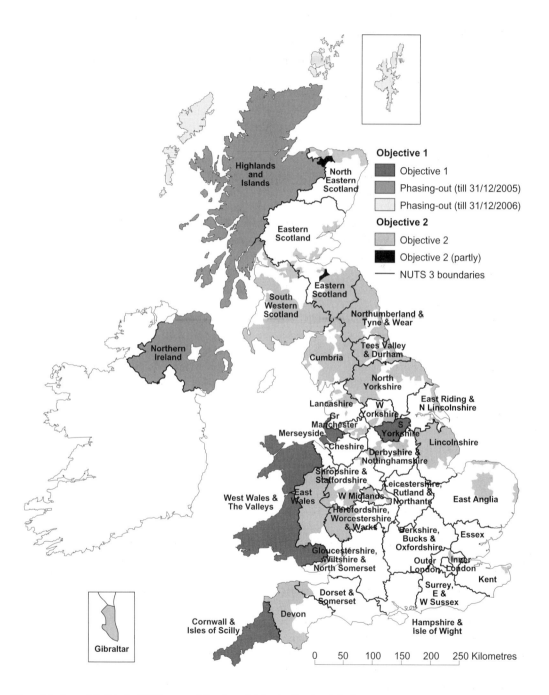

Objective 1

- Objective 1
- Phasing-out (till 31/12/2005)
- Phasing-out (till 31/12/2006)

Objective 2

- Objective 2
- Objective 2 (partly)
- NUTS 3 boundaries

Figure 4.1 Areas Eligible for EU Structural Funds and Selective Regional Assistance

Source: European Commission Directorate General for Regional Policy

with Objective 1. Previously the main recipients of national and European regional assistance were quite different, and the government justified this because of the programmes' differing objectives. Now this applies only to Tier 2 regions, which in the UK have been built up from ward level (in line with the wishes of many consultees) and differ significantly from the Community Objective 2 regions. In England there are also Enterprise Grant Areas (Tier 3), which may also be designated under devolved powers elsewhere in the UK.

Spending of this magnitude has significant impacts on spatial development patterns through investment in infrastructure and changing locational decisions (Williams 1996). This is *de facto* spatial planning. However, coordination of this investment through explicit territorial planning strategy is another matter, and it is only recently that the EU has taken a direct interest in such planning.

All the European institutions have now recognised the importance of European and transboundary spatial development trends. Member states are encouraged to work cooperatively on spatial planning in order to coordinate the spatial impacts of sectoral policies, promote sustainable forms of development, support economic competitiveness, and protect the environment. It is argued that there are important transboundary dimensions to spatial planning which need to be taken up through appropriate institutions and instruments at jurisdictional levels above the nation-state.

EUROPEAN COMPETENCES IN SPATIAL PLANNING

There has been a dramatic increase in spatial planning activities that cut across national borders and involve European institutions, which in turn raises questions about the assignment of competences for spatial planning. Until recently there has been no question that competence over spatial planning (in its various forms) rests solely with the member state governments (which in some countries is devolved to subnational governments). But the growing influence of the supranational and transboundary actions on domestic planning systems and policies challenges that assumption (CEC 1997). Many Community policy sectors impact on town and country planning.

In the transport field there is, in part, an exclusive Community competence. Action has concentrated on harmonisation of national transport policies and the Trans-European Networks (TEN) in the areas of transport, telecommunications, and energy infrastructures. In the environment field, the Directives on Environmental Assessment are well known but the Air Quality, Waste, Birds, and Habitats Directives have significant impacts, along with the Fifth Environmental Action Programme. Community agriculture policies through the Common Agricultural Policy (CAP) and other specific measures such as environmentally sensitive areas (ESAs) have had very significant implications for spatial development patterns and planning in the UK. A recent innovation is Community urban policy, although this is limited so far to promoting pilot projects, studies, and exchange of experience. Coastal management is another important example where the demonstration programme is likely to lead to specific recommendations in relation to planning. New Community legislation and other actions will continue to come forward with impacts on planning in the UK, for example on air quality, landfill, strategic environmental assessment, and renewable energy. In this context it is unsurprising that questions have been raised about the competences of the Community. What should be the role of national, transboundary, and European institutions in spatial planning? Will spatial planning be yet another step in the continuing process of transfer of powers from the member states to the Community? (Tillotson 2000: 52).

The extent to which powers will be transferred, including those relating to spatial planning, will depend to a great extent on the application of the principle of subsidiarity, to which the Community is committed (Nadin and Shaw 1999). In essence, subsidiarity means that competences should be located at the most appropriate level and that they should not be located at a higher jurisdictional level than is necessary.

Aside from formal statements in Community legislation, general concerns about the growing powers of the Community during the 1990s drew attention to the idea of subsidiarity and its use in controlling the extension of Community competences. Since then the making of Community legislation has slowed quite dramatically. Instead there is more emphasis on informal actions which come under less scrutiny. This has been the case for spatial planning, where action is through intergovernmental working led by the member states rather than the Commission (see Chapter 3). The EU is empowered to 'adopt measures concerning town and country planning, land use with the exception of waste management . . . and management of natural resources' (Article 175), but it is agreed that this would support action in these fields only in so far as they help to achieve measures related to environmental protection, and then only when agreed by unanimity (Nadin and Shaw 1999 33; Bastrup-Birk and Doucet 1997).

Nevertheless, ministers responsible for spatial planning in the member states do meet on a six-monthly cycle and have sanctioned increasing attention to supranational planning by the Committee on Spatial Development. The application of subsidiarity has tended to slow progress on European spatial planning, but its continued growth and influence, like the process of European integration itself, seem inevitable.

WHAT IS SPATIAL PLANNING?

It should be emphasised that across Europe spatial planning can be understood in different ways, and the term is often used in a generic sense to describe any land use or physical planning system. *Physical planning* describes government action to regulate development and land uses in pursuit of agreed objectives. This form of planning is one policy sector within government alongside policy sectors such as transport, agriculture, environmental protection, and regional policy. Land use planning may incorporate mechanisms to coordinate other sector policies. *Spatial planning* in the European sense is more than this and

is more centrally concerned with the coordination or integration of the spatial dimension of sectoral policies through a territorially based strategy. The strategy acts as a framework for the formulation and implementation of sectoral policy. One of the sectors will be *land use planning*. In this sense, spatial planning seeks to identify and address the contradictory effects of sectoral policies, and the opportunities for synergy through the territorial strategy. In practice, all planning systems in Europe tend to be land use systems with different degrees of coordination, but mostly weak with considerable sectoral compartmentalisation (Nadin *et al.* 1997). This applies to the UK.

THE EMERGENCE OF SUPRANATIONAL PLANS

The first initiative on systematic planning at the European scale came from the German government, which prompted the establishment of a permanent Conference of European Ministers of *Aménagement du territoire* (CEMAT) through the Council of Europe. The early regional policy of the EU had little spatial content and instead focused on the need to support particular industrial and commercial sectors. Despite a resolution in the European Parliament to draw up a European scheme for spatial planning, progress has been slow. However, the French, German, Danish, and Dutch governments continue to promote supranational planning studies. These have introduced memorable spatial concepts at the European level such as the now infamous 'blue banana' (a concept used by the French to describe the area between south-east England and north Italy where growth and investment have been concentrated).

The Commission's first major contribution to the development of European supranational planning came with the publication of *Europe 2000: Outlook for the Development of the Community's Territory* (European Commission 1991). This document was intended to provide a European reference for planners working on national or regional planning policies. It was effectively a geography text, raising awareness of European-wide spatial development issues. It adopted an approach which cut across country borders to identify seven transnational study areas having shared characteristics. (An eighth area was added with the inclusion of the eastern German *Länder*.) These eight regions, together with adjacent 'external impact areas', became the subject of extensive research studies.

The initial findings for these study areas are reported in *Europe 2000+ Cooperation for European Territorial Development* (European Commission 1994). As its title implies, this publication signalled a change in gear on supranational planning, with a clear call for more cooperation between member states. It charts the trends in the physical development of the European territory and, crucially, makes strong assertions about the preferred development patterns for the future.

The main emphases of the *Europe 2000+* studies are the need to control urban sprawl, a common feature of development predominantly, if not exclusively, in the southern European states; and the strengthening of small and medium-sized towns, especially where this can help in focusing the provision of services in rural areas. Underlying these ideas is the concept of a polycentric urban system, a balanced distribution of urban services across the territory, with more emphasis on development at points along corridors joining the main centres. The rationale is that this will avoid both the congestion problems of very large conurbations and the decline in service provision in rural areas, but it is an idea which largely ignores the dominant centralising tendencies of the market.

Elsewhere, more rigorous protection of areas of environmental importance is promoted, together with further policies aimed at containing the problems of land abandonment in rural areas where traditional agricultural practices are increasingly uneconomic. Urban regeneration and the revitalisation of poor neighbourhoods also has a high priority, but a firm line is being taken on the retention and restoration of the built heritage. Additionally, *Europe 2000+* reports on the Trans-European Networks (TEN), which are explained briefly in Chapter 11.

Whereas the Commission has concentrated on providing a more coherent analysis of European regional geography, the Committee on Spatial Development has taken the lead in producing the *European Spatial Development Perspective* (ESDP) (CSD 1999) This is the most important initiative on spatial planning at this level, and a unique experiment in supranational planning. The ESDP is intended to promote 'coherence and complementarity' of the development strategies of the member states by coordinating the spatial aspects of EC sectoral policies. The task of achieving the necessary consensus among fifteen member states (and many regions within nations) is obviously a difficult one and, needless to say, the statements made are at a very general level. Nevertheless, the final document was endorsed by all governments at their meeting in Potsdam in May 1999. The main contents of the ESDP are shown in Box 4.1.

BOX 4.1 A SUMMARY OF THE EUROPEAN SPATIAL DEVELOPMENT PERSPECTIVE, FINAL POTSDAM VERSION (1999)

Purpose

The main purposes of the ESDP are

- to raise awareness of the significance of spatial development trends for the objectives of the EU;
- to provide a common reference framework to guide action on planning and development decisions;

- to promote integration of policy across different sectors of activity, and coordination and complementarity of member state policies through a common strategy;
- to provide a framework within which sustainable economic development can take place, and enable the EU to meet its international treaty obligations.

Development Trends at the European Scale

The ESDP outlines a wide range of significant spatial development trends for Europe, including:

- little change in population but continuing demand for urbanisation resulting in urban sprawl;
- rapid change in urban structure and increasing competition between urban centres;
- intensification of agriculture, depopulation, and poor service provision in rural areas;
- increasing waste and pollution, the ongoing loss of biodiversity;
- increasing transport flows, congestion, missing links, bottlenecks, poor accessibility in peripheral areas;
- the need to harmonise the infrastructure networks of the accession countries.

It notes the lack of coordination of the spatial effects of EU actions through the Structural Funds, the trans-European networks, the CAP, and environmental policy, and argues that sectoral policy could be much better coordinated.

Policy Options

The ESDP has three general objectives: a more balanced system of towns and cities; parity of access to infrastructure; and prudent management of heritage. The ESDP stops short of promoting specific policies, but makes suggestions of possible approaches – for example:

- improving the balance of urban activity between and across regions by avoiding excessive concentration of activity around a particular metropolitan area, and the creation of alternative economic centres;

- promoting the sustainable development of cities through the control of their physical expansion;
- development of more environmentally friendly means of access, and conservation of the cultural heritage;
- promoting environmental measures in rural areas and the diversification of the rural economy;
- giving special attention to the transport needs of land-locked and remote regions;
- creation of buffer zones and completion of an environmental network of protected areas;
- promoting innovative telecommunications services and applications, particularly in remote regions.

Implementation, Action, and Debate

The implementation of the ESDP will be through the reorientation of member states' own strategic policies, and through the use of existing Community instruments and mainstream spending. The significance of the ESDP is that it is focusing debate on the potential of spatial planning to address significant problems in the coordination of public sector policy and the guidance of private-sector development. The ESDP itself raises a number of key questions:

- How can spatial planning help to co-ordinate sectoral policy, and how can transnational co-ordination be achieved?
- Is there a need for more spatial planning powers for the Commission?
- How can national and regional planning laws 'be adopted to take on board cross-border and transnational planning aspects', and is there a need for transnational planning instruments?

INTERREG IIC

While the member states and the Commission completed the ESDP, the Community initiative Interreg IIc encouraged local authorities and other public bodies to take part in transnational planning projects. Interreg IIc ran from 1996 to 2000 and was the first Community initiative specifically to promote action on spatial planning. Its objectives were to promote strategies for sustainable development, to foster transnational cooperation within a common planning framework, and to improve the impact of Community policies. A major stream in the programme also promotes cooperative approaches to the problems of flood and drought.[6] Interreg IIc has had a major impact in the UK, encouraging local authorities and other bodies to work cooperatively with partners in other countries on planning issues by co-financing projects. Its successor from 2000, Interreg III, will have more than seven times as much funding.

The argument for these initiatives follows the logic of the ESDP: more integration of spatial planning policy between nations will contribute to a better balance of development between regions, and thus increase the social and economic cohesion and economic competitiveness of the Community. Funding was made available over the period 1997–99 amounting to 413 million ecu (about £248 million) – a moderate amount in comparison with other Community initiatives.

Seven transnational regions were defined within which public bodies could bid for funding to support transnational spatial planning. The UK was involved in three Interreg IIc transnational regions: the Atlantic Area, the North Western Metropolitan Area, and the North Sea Region. These were extensions of the transnational study regions first identified in *Europe 2000+*, with an original Centre Capitals Region being extended to cover the whole of the UK and Ireland. This produced three overlapping programme areas in the UK (Nadin and Shaw 1998).

To be eligible for funding, each project had to involve at least two member states and have an impact in at least three. Examples of successful projects include collaboration on planning issues related to the high-speed train network in north-west Europe; sharing of experience and strategies on providing access to urban services in rural areas and increasing the vitality of small towns in the North Sea Region; and devising joint strategies for the conversion of fishing infrastructure to ecotourism in the Atlantic Area.[7]

Three principles are important in the selection of projects. They must have a transnational dimension; there must be a strong multiplying effect giving added value, especially in improving the prospects for achieving planning strategies; and the projects must have the potential to influence other operational programmes so that they make a contribution to transnational planning strategies. In practice, many activities are accepted as 'spatial planning' which are not closely related to town and country planning in the UK, such as support for innovation in industry and research. Also, the transnationality criterion has been given a broad interpretation, and the level of actual joint working across national boundaries on many projects is quite low (Nadin and Brown 1999).

An overriding theme of the initiative is the need for programmes to be consistent with the aims of the ESDP and to contribute to the further elaboration of transnational planning strategies or frameworks. The operational programme measures and specific projects funded within them will need to contribute to a broader transnational strategy or framework. Two such transnational frameworks or 'vision statements' have been produced which affect the UK: the *Norvision* and the *Vision for North West Europe*.[8]

Interreg IIc has dramatically expanded the effort and experience on transnational planning in the UK, and this will be increased further through Interreg III, which is operational until 2006 and places a continued emphasis on balanced development among countries. Interreg III has total funding of nearly 5 billion euros (about £3 billion) and is divided into three strands: (a) promoting cross-border cooperation on joint strategies for sustainable territorial development; (b) transnational cooperation among large groupings of European regions (a development of Interreg IIc); and (c) interregional cooperation and networking. For those authorities and organisations with a history of transnational collaboration the initiative provides a welcome opportunity to contribute in a bottom-up way to the formulation of transnational planning policy and the ESDP. Where there is less experience, Interreg may offer the potential for capacity-building through education, discussion and the sharing of ideas.

CONVERGENCE OF EUROPEAN PLANNING SYSTEMS

Increasing attention to spatial planning policy at the European level and increasing transboundary co-operation will inevitably affect the planning mechanisms in member states. There are large differences in the way that planning operates across Europe: each member state (and in some cases an individual region) has developed its system in response to local economic and physical development problems arising within particular cultural, legal, and social contexts. There is likely to be a trend towards some degree of convergence of these systems as countries work together.

In order to facilitate understanding about the way that spatial planning operates in different member states, and thus to promote more effective cross-border and transnational planning, DG Regio commissioned a *Compendium of EU Spatial Planning Systems and Policies*. This demonstrates the diversity in planning systems and policies (especially in their operation), but also notes similar trends as the different countries respond to the same macroeconomic forces. There is a distinct trend in much of Europe towards greater flexibility in the operation of regulation. New mechanisms are being introduced to establish more strategic planning frameworks and to allow for decisions which are contrary to the characteristic binding zoning plans.

Another common trend is the integration of spatial plans and sectoral spending programmes. The spatial plan is more widely recognised as the coordinative mechanism for sectoral policy and spending. New instruments are being introduced to tackle cross-border issues, and there is increased transnational cooperation between planners dealing with similar issues in different regions. In the UK this is most pronounced in Kent, where the County Council has worked cooperatively with four other 'regional authorities' of France and Belgium in the production of *A Vision for EuroRegion* (Kent County Council 1995).

These impressive developments in supranational planning have not been made without some resistance. The UK government in particular has been less than enthusiastic about 'universal spatial planning policies' and has instead emphasised the usefulness of exchanges of experience. There is certainly room for debate on the implications of seeking dispersed but concentrated development, urban containment, and a focus on corridor development. There will be other important issues that have not yet been considered, such as the availability and price of land for development. The proposals may not fully address powerful market forces, especially at a time when the private sector is taking a greater share of investment in virtually all member states. The limitations of the spatial planning systems in bringing about desired objectives of sustainable and balanced development across the community are recognised, and this may eventually lead the Commission to put more emphasis on other policy options for regulating and promoting development such as taxation measures or development incentives.

However, there is a general assumption in Brussels (and many member states) that more supranational planning is inevitable and that it is to be welcomed. This sentiment is shared by the transition countries which, while not wishing to reinvent centralised state planning, do want to deal in a coordinated way with their massive problems of environmental degradation and economic decline.[9]

(B) NATIONAL AND REGIONAL PLANNING

NATIONAL PLANNING GUIDANCE

There is no national land use planning in the UK in the sense that policies or plans are prepared for the whole country. Indeed, one outcome of devolution is less consistency across the UK. There is, however, a growing amount of 'national' land use *guidance*, and all four systems make use of policy statements at this level. This developed first in Scotland, where it was prompted by the need for a strategy to deal with the unprecedented problems posed by North Sea oil and gas. Since these were considered to be of national importance, the Scottish Office decided to issue guidelines for use by local authorities, especially in relation to coastal development (Gillett 1983). *North Sea Oil and Gas Coastal Planning Guidelines* was published in 1974, and *national planning guidelines* (NPGs) on other topics soon followed. These guidelines were intended to fill a gap between relatively inflexible policy expressed in circulars, and general advice that could be ignored (Raemaekers 1995). They had formal status, but they did not tie ministers or local authorities to particular solutions. They did not go so far as to constitute a 'national plan', and they were not intended to be comprehensive, but they were locationally specific.

The benefits of this system soon became apparent (Diamond 1979). National guidance enabled local authorities to explain the way in which their plans took account of national policies; a higher degree of coordination was possible among the various branches of central government; and national interests in which the Secretary of State needed to be involved could be readily separated from local matters. Later assessments continued to recognise the strengths of the NPGs (Nuffield 1986; Rowan-Robinson and Lloyd 1991). Nevertheless, as the series expanded to cover more topics and non-locationally specific guidance, questions arose about the precise status of NPGs and their overlap with other policy statements (Rowan-Robinson and Lloyd 1991; Planning Exchange 1989).

In response to these concerns, the Scottish Office introduced in 1993 a rationalised structure for national policy and advice through a series of *national planning policy guidelines* (NPPGs), together with continued use of planning advice notes (PANs) and circulars. The role of NPPGs is 'to provide statements of government policy on nationally important land use and other planning matters, supported where appropriate by a locational framework'. NPPGs are broader in scope than their predecessor NPGs, and are intended to provide more comprehensive coverage of topics of national concern. The role of PANs is 'to identify and disseminate good practice and to provide advice and other information' (NPPG 1: para. 13).

Because of the recognised success of the NPG series, the shift to NPPGs has been watched closely. A review of early experience (Raemaekers *et al.* 1994) assessed NPPGs as 'a convincing effort to produce a successor to the pioneering series of the 1970s, fit for changed and complex circumstances'. It also called for greater breadth in topic coverage, and identified particular omissions in the proposed list, especially on major transport infrastructure. Subsequently, guidance and good practice in Scotland has expanded considerably,[10] and a further review was conducted (Land Use Consultants 1999), again with generally positive findings. Many detailed recommendations were made for improving the form and content of national guidance, among them the need for a more open and participatory process in both preparation and review. The Minister's response has been to establish an advisory group made up of a wide range of public-, private-, and community-sector interests to assist in the policy review process.[11]

The national policy guidance review also argued that national policy could be strengthened through the preparation of a general overarching 'vision' for the planning system in Scotland, and the preparation of locationally specific guidance for discrete geographical areas in Scotland. The Executive is now considering, with other interests, the feasibility and value of preparing a national spatial planning perspective for Scotland along the lines of national spatial strategies in other smaller European countries such as Denmark and the Netherlands.

In England and Wales, national guidance did not arrive until 1988, although its form and content have provided a model for more recent changes to the Scottish system. National guidance in England is made up of planning policy guidance notes (PPGs) and minerals planning guidance notes (MPGs). As in Scotland, PPGS have taken on the role of expressing national land use and development policy, leaving circulars to be used mainly for elaboration of procedural matters. They have certainly clarified and extended the national policy framework, but they have tended to be more general than the NPPGs in Scotland (and certainly NPGs), broader in scope, and not at all location specific.

PPGs have had a considerable impact on planning practice. An evaluation of their effectiveness concluded that they had 'assisted greatly in ensuring a more consistent approach to the formulation of development plan policies and the determination of planning applications and appeals' (Land Use Consultants 1995a: 47). This is because PPGs are important material considerations in development control and have a determining influence on the content of development plans. Conformity between national guidance and plans is ensured through regional office scrutiny of development plans, but the study also found that most professional planners have a high regard for national guidance and welcome the order and consistency in policy that it brings. Councillors are generally more sceptical, which is not surprising, because national guidance constrains their discretion to respond according to their interpretation of local needs.

Having successfully introduced more systematic planning policy at national level, the government now has the problem of maintaining consistency and clarity in the series. There have been many calls for more PPGs, while at the same time some concerns have been voiced over perceived contradictions between one PPG and another, and between the series and other government policy statements. One example (from the Land Use Consultants 1995a study) is the concern over the different explanations of the term 'sustainable development' in government guidance. The RTPI (1999) has recommended that the policy content of PPGs be separated from general guidance and advice and replaced by an annual statement of national planning policy. The weight of recently issued guidance notes suggests that there is certainly some scope for rationalisation. At the same time there is also a need for more specific guidance on some topics so as to reduce the effort needed to debate national policy at inquiries on development projects of national significance.[12] This would require some aspects of national policy guidance to be much more locationally specific (currently, only

minerals guidance mentions locations). To some extent the enhanced *regional planning guidance* will meet this need. Another approach, as in Scotland, would be a 'national spatial planning perspective', and the RTPI has investigated the feasibility of such a perspective for the whole of the UK prior to making a proposal to government.[13]

In Wales the publication of national guidance was, until 1995, mostly shared with England through joint publications of the DoE and the Welsh Office. On occasion, separate advice was thought necessary to reflect distinctive Welsh conditions (for example on land for housing and on strategic planning), but these followed the English version quite closely. By 1995 the Welsh Office had decided to go its own way, and published two draft planning policy guidance notes intended to replace PPGs originally shared with the DoE; final versions were published in 1996 (since revised). The first deals solely with the new unitary development plan structure following local government reorganisation. The second summarises policy guidance for a long list of topics covered previously in separate notes.

The effect was to reduce the amount of guidance considerably. This was criticised by the RTPI because it 'significantly dilutes essential policy advice, lacks coherence and comprehensiveness, and will provide Wales with inferior guidance to that available to local planning authorities in England'. Technical Advice Notes (TANs) have been published, filling in much of the detail lost, but concerns centre on the problem of lack of strategic direction, especially in the light of the reforms to the development plan system in response to the reorganisation of local government. However, one interesting innovation is the inclusion of a strategic diagram for the whole of Wales, illustrating areas of growth and protection, and main communication routes (although it effectively states little more than the obvious).

Northern Ireland lagged behind in the production of 'national' guidance until the mid-1990s. Planning officers in the Province make use of PPGs and NPPGs to keep in touch with policy developments. Since 1995 *national planning policy statements* (PPSs) have been published by the Planning Service for Northern

Ireland.[14] The statements are similar in form to guidance in the rest of the UK but reflect the special planning and political circumstances in Northern Ireland – not least the centralisation of planning in the Planning Service. It may be that the strong competence of central government is responsible for the definitive nature of some policies, for example that 'there is no justifiable need for any new regional out-of-town shopping centres' (PPS 1, para. 35). The Planning Service also publishes *development control advice notes* (DCANs) providing more detailed guidance on good practice.

An explicit policy framework from national government is especially important in Northern Ireland because it is national government which makes almost all decisions. The PPSs must conform to the Regional Strategic Planning Framework for Northern Ireland being prepared by the Department of the Environment for Northern Ireland, which is discussed below. However, less progress has been made here in involving a wider range of interests in the production of national planning policy.

In whatever form, guidance issued by national government carries considerable weight and thus the content of that guidance is discussed in some detail in subsequent chapters. But though local planning authorities are required to have regard to guidance, they are not bound by it. Indeed, other material considerations may be of greater importance in particular cases, and planning authorities may wish to take a different line, so long as they can give adequate reasons. Moreover, the advice in one guidance note may contradict that in another, perhaps as a result of piecemeal revisions at different times. Nevertheless, guidance notes command a great deal of respect and are closely followed in development control and development planning. They exert significant influence in planning policy and decisions, and are quoted profusely in decision-making, especially at inquiries. Some government policy is still to be found in circulars and also, from time to time, in ministerial statements. Major changes in policy are often published in White Papers. All these documents can be regarded as material considerations in planning, and thus central government has an array of instruments in which

national policy can be expressed. Indeed, the result can be confusing, if not actually contradictory. The Environment, Planning and Regional Affairs Committee Report on the Planning Inspectorate noted the difficulties this presented for planning inspectors.

There are likely to be important further developments in this field, including more locationally specific 'spatial guidance' both for particular sectors and for the national territories as a whole. A critical task will be the incorporation of other sectoral policy-makers and operators into the policy-making process, such that there is wider ownership of national spatial policy among government departments, agencies, and service providers.

REGIONAL PLANNING GUIDANCE

In addition to the strengthening of national guidance, the government has taken steps in two stages towards expanding the role of regional planning guidance in England. The first set of changes began with the 1986 Consultation Paper *The Future of Development Plans*, which noted the progress that had been made in some regions (such as East Anglia, the West Midlands, and the South-East) by local authorities cooperating voluntarily to produce regional strategies.

Official encouragement was given in this Consultation Paper to the formation of other regional groupings. No precise procedures were suggested, and a warning was given that such arrangements would not represent a formalised regional structure, nor would they be a return to the type of large-scale regional planning which was attempted in the 1960s and 70s. Regional planning conferences of local authorities were invited to prepare draft guidance looking ahead twenty years or more, which the Secretary of State considers before publishing final guidance. (Initially, regional guidance was published in the PPG series, but since 1989 there has been a separate RPG series.) The 1989 White Paper recommended that business organisations and other bodies should be involved as well as local authorities, and also required the joint bodies to detail the results of consultations and the response. Conservation and agricultural interests were later added to this list. Strategic guidance produced by the metropolitan counties as part of the unitary development planning process was gradually merged with the new regional guidance, except in London, where separate strategic planning guidance remains.

The recognition by government of the need for more strategic planning was welcomed by all sides, but the early efforts came in for considerable criticism. An indication of the varying style and quality is given in the *Town Planning Review* symposium (Minay 1992). From this it appears that much of the guidance has been no more than a further detailing of national guidance and restatement of current policies. For example, the Regional Strategic Guidance for East Anglia is described as 'bland and incremental . . . it mostly describes existing situations and trends'. Most of the others are described in a similar tone, though Roberts (1996) gives a more positive appraisal. A later review by Baker (1998) notes increasing specificity of regional guidance through the use of subregional divisions; first signs of attempts to integrate a wider range of sectoral policy interests (particularly transport and economic development); and growing institutional capacity for planning at the regional level.

The second stage of strengthening regional planning came from 1997. The new Labour government's commitment to devolution and regionalisation at last provided the right context for progress to be made on long-standing and widely held views on the need to further strengthen the regional dimension to planning.[15] In February 1998 a ministerial statement was published, entitled *Modernising Planning*, together with a consultation paper, *The Future of Regional Planning Guidance*. One of the main themes of the ministerial statement was the need to strengthen strategic planning capabilities at the regional and subregional levels, reflecting both the need to respond to developments at the European level and the government's desire to decentralise to the regions.

The consultation paper accepted the validity of criticisms of regional planning guidance in not providing a real strategic direction for the regions and not having the confidence of regional stakeholders. Proposals were made for strengthening regional

guidance by extending its scope and specificity beyond land use, by making the process for its production more inclusive and transparent; and by giving the regional bodies the main competence for its production. The proposals stopped short of recommending that regional guidance should become a statutory document on the basis that this would require primary legislation. Local authorities were encouraged to begin to take the principles of the consultation document forward prior to the publication of formal guidance. A *Draft Planning Policy Guidance Note 11: Regional Planning* was published in March 1999 and a final version with much the same content in 2000.

European developments and practice in other countries have had a considerable influence on the new approach to regional planning. The new RPG is intended to provide a *regional spatial strategy* for the region; that is,

> to provide a broad development strategy for the region over a fifteen to twenty year period and identify the scale and distribution of provision for new housing and priorities for the environment, transport, infrastructure, economic development, agriculture, minerals and waste treatment and disposal. . . . By virtue of being a spatial strategy it also informs other strategies and programmes, in particular: . . . the regional context for the preparation of local transport plans; and it should also provide the longer term planning framework of the Regional Development Agencies' regional economic strategies.
> (PPG 11 (2000), para. 1.03)

But the government also urges caution, and necessarily so, since the approach to developing regional spatial planning is effectively to widen the remit of one sectoral policy instrument. The central purpose of RPG is still to provide a regional framework for the preparation of local authority development plans (para. 1.03). The impact of RPG on other sectoral policy will rely heavily on indirect influence. The new approach recognises the need to involve regional stakeholders more fully in the process, notably the regional development agencies (RDAs); other public bodies in such sectors as health and education; the Environment Agency; business and commercial organisations, including transport operators and utility companies; and voluntary bodies. But the extent to which these interests, which have hitherto shown little interest in regional spatial planning, can be success-fully incorporated into the process is yet to be seen. The proposals for wider scope and a more open process may assist. The proposals for the content and procedure are summarised in Figure 4.2.

The principal procedural innovation follows an experiment in the West Midlands where the draft advice to the Secretary of State was subject to public examination in a similar way to that conducted for structure plans. The examination in public now becomes a requirement for all RPGs. A draft strategy will be 'tested' at an examination before an independent panel with invited participants. The panel's report will then form an input to the Secretary of State's consideration of the guidance before a draft RPG is published. After experiencing the examination of the South-East's RPG. Crow (2000) concluded that it can make a positive contribution – although the final decision still rests with the Secretary of State. Two other innovations are the requirement for sustainability appraisal and the identification of clear targets and performance indicators. The DETR commissioned research on both matters which provides the basis for good practice guidance for the RPBs.[16]

The ECOTEC study on targets and indicators provides a fascinating review of the haphazard proliferation of targets and indicators (and might usefully be replicated for development plans). Despite the rapid increase in their use, there is little systematic consideration of indicators or targets in relation to the policy objectives; rather, they mostly represent general aspirations for the region, and the constraints of data availability. Nevertheless, considerable progress is being made, and the government's view is that 'systematic use is made of quantified regional and subregional targets that provide the benchmark for RPG success'.[17]

The main responsibility for preparation shifts from central government to the regional planning bodies (RPBs), although in close liaison with the government offices. Rather than preparing advice for the Secretary of State, the RPBs will be producing a draft of the final guidance. Since there are no regional authorities, the final RPG will still be issued by the Secretary of State, but there is a clear intention to provide as much

discretion to the regional bodies as possible, so long as they work within the parameters of national planning policy. What constitutes the RPB is left to the local authorities within each region. In some cases the emerging regional chambers may take on the regional planning role; in others the lead may remain with the conferences of planning authorities.

The capacity to undertake the regional planning function varies considerably across the country, but after decades of neglect of the regional planning function, it is generally weak and 'dependent upon local government officers coming together and carrying out the necessary studies on the back of their mainstream jobs' (Kitchen 1999a: 12).[18] Not only do technical skills and resources need expanding, but the regional conferences (and RPG itself) have not had a strong influence on other sectors. In some cases, it has had little influence on the actions of its own local planning authority partners, which is the explanation for the weak common denominator style of some documents.

The expectations that an enhanced regional planning guidance system can command the commitment of a whole range of regional and national actors seems ambitious in the face of the dominance of a few key corporate stakeholders (Vigar and Healey 1999). There is a need for a system which obliges all the key regional interests to work cooperatively on strategic spatial planning process and to sign up to its conclusions. Though the new arrangements move in this direction, they are still firmly rooted in the land use planning sector with little formal control over other sectors.

The relationship between the RPG and RDA strategies will be critical. In this approach much emphasis is given to cooperative working between the regional bodies to ensure complementarity of the strategies. The RDA strategies 'will need to operate within and alongside the long term spatial context for the region provided by RPG' (PPG 11: para. 4.5).

The RPB should draw on the RDA's work in identifying the strengths, weaknesses, opportunities and threats to the regional economy, in order to identify the areas where land needs to be released for economic development or where policies targeted at facilitating regeneration need to be focussed.

(PPG 11: para. 4.4)

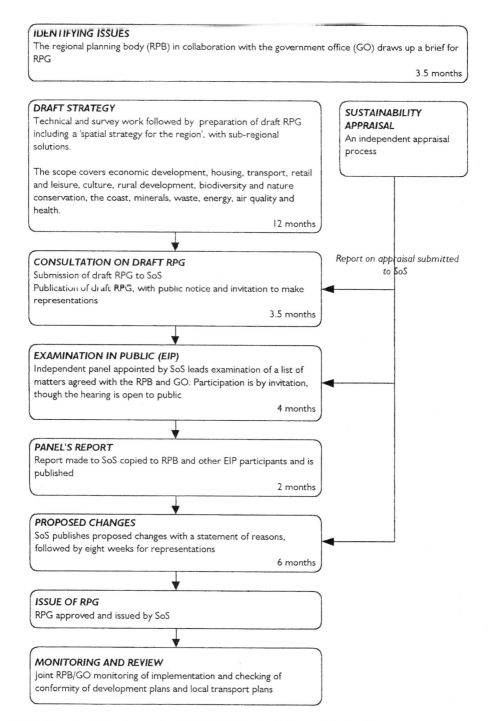

Figure 4.2 The Procedure for the Preparation of Regional Planning Guidance in England

Kitchen asks: what will happen when push comes to shove, as at some time in most regions it will? Will the RPG with its environmental and sustainability appraisals have sufficient teeth to make a real difference to what RDAs actually do?' (1999a: 13). The relationships between the RPBs and RDAs are to be monitored by the government offices, and there is no doubt that there will need to be considerable work to avoid inconsistency and contradiction.

Until the 1980s, Scotland had a tier of regional reports which provided a corporate policy statement for the regions as well as a framework for the preparation of structure plans. There were few formal procedures governing their preparation, and they did not require central government approval, but were simply published with the Secretary of State's observations. They were much admired, but as the statutory development plan framework was put into place, they became regarded as redundant.[19]

In Wales a series of guideline documents was prepared as *Strategic Planning Guidance in Wales* by the Welsh Office. The documents were intended to 'consolidate and re-present the wide range of available strategic guidance material in a consistent and accessible form' and particularly to provide a framework for the preparation of structure plans. Given the changes to a unitary structure in Wales, strategic guidance is perhaps more important than ever, but neither of the two Welsh planning guidance documents refers to it.

Innovation in regional planning is perhaps most evident in Northern Ireland. The first regional plan in Northern Ireland was the Belfast Regional Plan published in 1964 (the Matthew Plan). This proposed the stopline, a system of radial motorways, and a major new town, Craigavon, modelled on the English experience (Hendry 1989). Like its counterparts in England, the plan was overtaken by the effects of dramatic economic recession. The subsequent *Regional Physical Development Strategy 1975–85* sought to concentrate growth in the Province to twenty-six key centres, but the depressing effects on other areas were widely challenged (Blackman 1985). A new rural planning policy published in 1978 took a much more relaxed approach to development in three-quarters of the rural

territory, which led to extensive development of single houses in the countryside and ribbon development.[20] This led to a reappraisal of the need for regional planning and the publication of *A Planning Strategy for Rural Northern Ireland* in 1993. This is unlike any other UK regional planning document in that it includes both strategic objectives for the overall development of the territory and detailed development control policies. This could only be a product of a system where central government has set the strategy, made local plans, and undertaken development control. The Strategy introduces new restrictions on development in the countryside, while introducing the novel designation of 'dispersed rural communities'.

The Belfast Agreement of 1998 gave added impetus to the increasing activity on regional planning in Ulster. In the same year the Secretary of State for Northern Ireland launched *Shaping Our Future: Towards a Strategy for the Development of the Region – Draft Regional Strategic Framework for Northern Ireland*. The strategy is a considerable step forward in strategic planning in Northern Ireland, and, interestingly, bears close comparison with new regional planning guidance in the Republic of Ireland.

(C) DEVELOPMENT PLANS

ESTABLISHING DEVELOPMENT PLANS 1947–68

The main instrument of land use control in Britain during the first half of the twentieth century was the planning scheme. This was, in effect, development control by zoning. As discussed in Chapter 2, zoning was replaced in 1947 by a markedly different system which attempts to strike a distinctive balance between flexibility and commitment. The approach is, in many important ways, the same in the 1990s as it was in the 1950s. It is fundamentally a discretionary system in which decisions on particular development proposals are made as they arise, against the policy background of a generalised plan. The 1947 Act defined a development plan as 'a plan indicating the manner in which a local planning authority propose that land in their area should be used'.

Unlike the prewar operative scheme, the development plan did not of itself imply that permission would be granted for particular developments simply because they appeared to be in conformity with the plan. Though developers were able to find out from the plan where particular uses were likely to be permitted, their specific proposals had to be considered by the local planning authority. When considering applications, the authority was expressly directed to 'have regard to the provisions of the development plan', but the plan was not binding, and indeed, authorities were instructed to have regard not only to the development plan but also to any other material considerations'. Furthermore, in granting permission to develop, local authorities could impose 'such conditions as they think fit'.

However, though the local planning authorities had considerable latitude in deciding whether to approve applications, it was intended that the planning objectives for their areas should be clearly set out in development plans. The development plan consisted of a report of survey, providing background to the plan but having no statutory effect; a written statement, providing a short summary of the main proposals but no explanation or argument to support them; and detailed maps at various scales. The maps indicated development proposals for a twenty-year period and the intended pattern of land use, together with a programme of the stages by which the proposed development would be realised. The plans were approved by the minister (with or without modifications) following a public inquiry. Initially a three-year target was set for submission of the plans, but only twenty-two authorities met this, and it was not until the early 1960s that they were all approved.

By this time the requirement to review plans on a five-yearly cycle had brought forward amendments, many taking the form of more detailed plans for particular areas. These had to follow the same process of inquiry and ministerial approval as the original plans, and many authorities were still engaged on the first review in the mid-1960s. Furthermore, although the system of development control guided by development plans operated fairly well without significant change for two decades, 1947-style plans did not prove flexible in the face of the very different conditions of the 1960s. The statutory requirement for determining and mapping land use led inexorably towards greater detail and precision in the plans and more cumbersome procedures. The quality of planning suffered, and delays were beginning to bring the system into disrepute. As a result, public acceptability, which is the basic foundation of any planning system, was jeopardised.

It was within this context that the Planning Advisory Group (PAG) was set up in May 1964 to review the broad structure of the planning system and, in particular, development plans. In its report, published in 1965, PAG proposed a further fundamental change to the planning system, one which would distinguish between strategic issues and detailed tactical issues. Only plans dealing with the former would be submitted for ministerial approval: the latter would be for local decisions within the framework of the approved policy. Legislative effect to the PAG proposals was given in 1968 (for England and Wales) and 1969 (for Scotland), creating a two-tier system of structure plans and local plans.

SEPARATING STRATEGY AND TACTICS FROM 1968

The essential features of the 1968 system are still in place today, though there have been numerous incremental changes. *Structure plans* provide a strategic tier of development plan and, until 1985, were prepared for the whole of England by county councils (and the two national park boards). They were originally subject to the Secretary of State's approval, but since 1992 have been adopted by the planning authority itself. They consist of a written statement and key diagram setting out the broad land use policies (but not detailed land allocations) for the area, measures for the improvement of the physical environment, and policies for the management of traffic. Accompanying these is an explanatory memorandum in which the authority summarises the reasons which justify the policies and general proposals in the plan.

The central department's view of the functions of the structure plan has been fairly consistent but views about its scope – the range of topics which should be considered – have varied considerably. The initial conception was that they should be wide-ranging, but the government narrowed the range of competence of structure plans over the years, only to widen it again in 1999. The content of plans is discussed further below. The functions, as now set out in PPG 12 (1999), are 'to state in broad terms the general policies and proposals of strategic importance for the development and use of land in the area, taking into account national and regional policies' (para. 3.7). The structure plan should indicate the scale of provision including figures for housing and other land uses,

> and the broad location of major growth areas and preferred locations for specific types of major development . . . [and] the general location of individual major and strategic developments likely to have a significant effect on the plan are; and . . . broad areas of restraint on development.
>
> (ibid., para. 3.8)

The structure plan makes use of a key diagram rather than map, thus avoiding the identification of particular parcels of land. This limits debate to the general questions of *strategic location* rather than the use of specific sites. General land use policies can thus be determined before detailed land use allocations are made, albeit not always to the liking of those affected by later, more detailed plans. In practice, counties formulated their 'policy and general proposals' in greater detail than anticipated by government, including quite detailed land allocations and development control policies in some cases. An argument in favour of more detail was that few local plans were being produced, but more detail also gave the county council more control over the implementation of policy.

Local plans provide detailed guidance on land use. They consist of a written statement, a proposals map, and other appropriate illustrations. The written statement sets out the policies for the control of development, including the allocation of land for specific purposes. The proposals map must be on an Ordnance Survey base, thus showing the effects of the plan to precise and identifiable boundaries.

Under the 1968 system there were three types of local plan: general plans (referred to as 'district plans' before 1982), action area plans, and subject plans. *General local plans* were prepared 'where the strategic policies in the structure plan need to be developed in more detail'.[21] *Action area local plans* dealt with areas intended for comprehensive development; and *subject plans* dealt with specific planning issues over an extensive area, typically minerals and green belt, but many others such as caravans and pig farming.

Local plans have never been subject to approval by the Secretary of State, but are adopted by the planning authority (although the Secretary of State has rarely-used powers to call in plans and to require modifications). The original rationale for this was that a local plan would be prepared within the framework of a structure plan; and since structure plans would be approved by the Secretary of State, local authorities could safely be left to the detailed elaboration of local plans. This went to the very kernel of the philosophy underlying the 1968 legislation, namely that central government should be concerned only with strategic issues, and that local matters should be the clear responsibility of local authorities. The Secretary of State, having approved the structure plan, could safely leave the detailed elaboration of its policies at the local level to the (same) local authority *without the necessity of further approval*.

This division of plan-making functions was predicated on the creation of unitary planning authorities responsible for preparing both the structure and local plan. But the 1972 Local Government Act established two main types of local authority in England and Wales, and divided planning functions between them. The two levels of local government do not share the same views about planning policy across much of the country, which has exacerbated conflict in the system. For some, this has always been a fundamental weakness of the system, leading to calls for the abolition of structure plans, but for others it has been a useful separation of powers, with the conflict usefully exposing critical issues in planning. Two mechanisms were introduced to promote effective cooperation in the planning field and to minimise delay, dispute and duplication: the development plan scheme (later the local plan scheme) and the certificate of conformity.[22]

EVALUATION OF THE 1968 DEVELOPMENT PLANS

There has probably never been a time when development plans, of whatever vintage, did not have their critics – and many of the criticisms have never changed. A decade after the start of the new system, Bruton (1980: 135) summarised the problems as 'delay and lack of flexibility; an over-concentration on detail; [and] ambiguity in regard to wider policy issues'. The same is probably true today.

Plans were very slow in coming forward to statutory approval and adoption.[23] The first structure plan cycle took fourteen years to complete: over the years 1981 to 1985 the time taken from the submission of structure plans to their final approval averaged twenty-eight months. One of the main reasons for this long delay was that many of the written statements and explanatory memoranda were very lengthy: in the first round, several contained more than 100,000 words. They also contained too many policies – typically more than a hundred, many of which the DoE considered to be irrelevant to structure plans: 'building design standards, storage of cycles, the costs of waste collection, the development of cooperatives, racial or sexual disadvantage, standards of highway maintenance, parking charges, the location of picnic sites and so-called nuclear-free zones'.

The disputes and delays over structure plan approval also held back the adoption of local plans; indeed, the first local plan was not adopted until 1975. However, the rate of deposit and adoption increased sharply after the initial round of structure plans was mostly completed and by March 1987, 495 local plans had been adopted in England and Wales (Coon 1988). Unfortunately, many of the plans were out of date by the time their processing was complete. This is not surprising when it is noted that the average time taken to prepare and adopt a local plan was about five years.

During the 1970s and 1980s non-statutory planning documents ('informal policy') proliferated.

At one time these informal policy documents out-numbered statutory plans by about ten to one (Bruton and Nicholson 1985). They took many forms, from single-issue or area policy notes to comprehensive but informal plans. Central government has consistently pressed for the elimination of non-statutory policy, except where it might be legitimately described as supplementary planning guidance (SPG) which is properly related to an adopted plan and subject to consultation. The value of SPG is that it avoids excessive detail in the statutory plan, and for this reason its use is supported by government, but a clear distinction is drawn between SPG and 'bottom-drawer' plans and policies which seek to avoid public scrutiny.

The experience of structure planning was disappointing, and there was at this time growing confusion about its role. Certainly it had not lived up to the expectations of the PAG report. Though it undoubtedly provided a forum for debate about strategy, it did not provide the firm lead that was promised. The uncertainties and complications of structure planning in practice carried over to local planning and contributed, in some areas, to a professional culture that was at best indifferent to statutory plans (Shelton 1991). There were more positive attitudes in other areas. Where the stakes involved in development applications were high, as in London and counties such as Hertfordshire (where full statutory plan cover was completed during the 1980s), statutory plan-making was vigorously pursued. Also, despite turbulent economic conditions, the plans proved to be reasonably robust and effective in implementing policy and defending council decisions at appeal.

Considerable research was carried out during the 1980s and 1990s on the preparation and operation of development plans. The main references are given in the 'further reading' section at the end of this chapter. Healey et al. (1988) concluded that plans had proved to be effective in guiding and supporting decisions, and in providing a framework for the protection of land. They were particularly useful in shaping private-sector decisions, especially in the urban fringe. Conversely, the difficulty encountered in controlling public-sector investment in housing, economic development, inner-city policy, and infrastructure provision was shown to be an impediment to effective implementation of strategy. The research team argued that this criticism was not one that plans alone could address. In similar vein, Carter *et al.* (1991) high-lighted the 'considerable confusion' about the rela-tionship between development plans and other plans for housing (HIPs) and transport (TPPs).

Davies *et al.* (1986a, b, c) concluded that plans might play only a small part in guiding development control decisions overall, but were much more impor-tant when a case went to appeal – what they termed the 'pinch points' of the system. They suggested that this reflected the system's chief virtue: its ability to enable a sensitive response to local conditions. It was recommended that the DoE should encourage local authorities to provide better written policy cover; to reduce its complexity by incorporating as much as possible in statutory plans; and to facilitate more speedy adoption.[24]

A big contribution to the failure to produce plans has been the vacillating and confused attitude of central government. The status of statutory plans reached a low point in 1985 when the White Paper *Lifting the Burden* denigrated both structure and local plans, and criticised the procedures for preparing plans as 'too slow and cumbersome'. More flexibility was also called for – somewhat at odds with previous advice, which had sought to reduce administrative discretion in the system via a planning framework that offered more certainty, clarity, and consistency to private-sector investors (Healey 1986).

PLANNING IN LONDON AND THE METROPOLITAN COUNTIES FROM 1985

The Thatcher government's precipitate decision to abolish the GLC and the MCCs forced hasty action concerning the planning system in these areas. This was simple in the extreme: London boroughs and the metropolitan districts became 'unitary' planning authorities. Thus, in precisely those parts of the country where there is a particular need for a two-tier planning system, it was lost.

Initially the government had proposed that the borough and district authorities should have respon-sibility for both structure and local plans, but later it was decided that this would be too cumbersome. Instead, a new *unitary development plan* was proposed, together with a joint planning committee for Greater London (a role which was undertaken by the London Planning Advisory Committee). The intention was that after consultations, the Secretary of State would provide strategic guidance to assist in the preparation of the unitary development plans. Little advice was given to the districts about their input to the development of strategic guidance except that it was to be produced on a cooperative and voluntary basis by the districts themselves.[25]

Unitary development plans (UDPs) are in two parts:

Part I is analogous to the structure plan in non-metropolitan areas. It consists of a written statement of the authority's general policies for the development and use of land in their area. The broad development and land-use strategy of Part I provides a framework for the authority's detailed proposals in Part II, which is analogous to the local plan in non-metropolitan areas. Part II contains a written statement of the authority's proposals for the development and use of land; a map showing these proposals on an Ordnance Survey base; and a reasoned justification of the general policies in Part II of the plan. The proposals in Part II of a plan must be in general conformity with the policies in Part I. Action areas (see paragraph 6 above) may also be designated in Part II of a UDP (PPG 12 (1999) Annex A para. 7).

The different parts of a UDP should be presented as one document, and they have a ten-year horizon. The UDP is adopted by the district council, and is not subject to the approval of the Secretary of State (although reserve powers of central intervention have been maintained).

There was a good deal of initial scepticism about these new arrangements, though they are more closely allied to the 1965 thinking of PAG than the system that was then put into place. There were particular concerns about the future of strategic thinking in the metropolitan areas, difficulties of cooperation between districts, and problems of participation and coping with the statutory right to objection in plans which embrace such large areas. For the districts themselves, many of these worries have proved unfounded. It has been possible to accommodate policy and political differences among districts, but this has been very much on a lowest common denominator level (Hill 1991; Williams *et al.* 1992). It has also proved possible, perhaps even desirable, to produce the strategy and detail concurrently. However, there remain serious concerns about the extent to which the public, interest groups, and even some professionals can engage effectively in the process. There have also been considerable delays in some metropolitan districts, notably Leeds and Bradford, where very detailed plans were produced in particularly contentious circumstances, generating great conflict and many thousands of objections.

The provisions for the London Mayor implemented in 2000 also include a requirement for the creation of a spatial planning strategy for London which will supersede the strategic guidance for London, although at the time of writing, the arrangements had yet to be finalised.

THE FUTURE OF DEVELOPMENT PLANS

As early as 1977, proposals were being made for a review of the 1968 system but, at this time, review was considered premature since only twenty-four structure plans had been submitted, and only seven had been approved. By 1985, prompted by the concern for 'freeing' enterprise from unnecessary restraints, the White Paper *Lifting the Burden* announced that there were to be changes in the development plan framework. The following year a consultation paper was published proposing the abolition of structure plans in England and Wales (but not in Scotland, where they had 'not in general given rise to the same problems as have been experienced south of the Border') and their replacement by statements of county planning policies on a limited range of issues (to be specified by the Secretary of State) which would not form part of the statutory development plan; a wider coverage of regional and subregional planning guidance to be issued by the Secretary of State after consultations and public comment; and the introduction of single-tier

district development plans covering the whole of each district.

The context for preparation and discussion of these proposals centred on the growing dissatisfaction of many different interests about the making of many *ad hoc* and apparently inconsistent decisions by both the Secretary of State and local authorities. The lobby for change created some unusual bedfellows (both the development and the conservation lobbies). The common concern was for more certainty in the system and a reduction in the growing number of speculative applications. There was also some dissatisfaction among government supporters about decisions taken centrally which went against local (often Conservative) opinion. Local authorities were concerned that more of their decisions were being overruled, and complained at the lack of clarity in central policy. By comparison, matters looked better in Scotland and in the emerging system in the metropolitan counties.

Many of the 500 responses to the Consultation Paper argued very strongly against the proposed abolition of structure plans. In November 1988, PPG 12 was published urging local authorities to extend statutory plan coverage, normally by district-wide plans, and to replace non-statutory policy, which it described as 'insufficient and weak'. Strategic green belt boundaries were singled out as requiring further specification in detailed local plans. In return, the government offered an enhanced status for plans.

Early in 1989, virtually the same set of proposals were published in the White Paper *The Future of Development Plans*, with the addition of a mandatory provision for all counties to prepare minerals development plans. County councils were urged to press ahead with the revision and updating of structure plans and to cooperate on the elaboration of regional guidance. Considerable lobbying by major pressure groups had been successful in bringing about a reprieve for structure plans. Counties were encouraged to review their structure plans and to take them forward to 2006, with the promise that the delays after submission would be reduced. The counties, for their part, were to ensure that plans were less bulky and concentrated on strategic issues. Shortly afterwards the government announced an end to the requirement that the

Secretary of State must themselves approve all structure plans and alterations. During debate on the Planning and Compensation Bill that followed, provisions were added to further increase the status of the statutory plans in development control.

DEVELOPMENT PLANS IN ENGLAND SINCE 1992

The Planning and Compensation Act 1991 made four major changes to the planning framework. The first made the plan the primary consideration in development control. In commending the amendment (now section 54A of the 1990 Act), Sir George Young coined a phrase in saying that 'the approach shall leave no doubt about the importance of *the plan-led system*'.

Two other changes involved making the adoption of *district-wide local plans* mandatory, and abolishing the requirement for central approval of structure plans. Central government has retained its powers of intervention, and the structure plan is still subject to an examination in public. Finally, the Act introduced a mandatory requirement for counties to produce minerals plans and waste plans for the whole of their areas. Small-area local plans and subject plans were abandoned except for minerals and waste, although local authorities may still designate action areas (Figure 4.3).

In the first part of the 1990s it seemed that the framework of local planning policy in England and Wales was to become more coherent. Given that most local authorities had little coverage of statutory plans (but many other informal policy documents) and others had produced a mix of interlinked subject and small-area-based policy documents, those investigating policy did so with not a little uncertainty. Now the prospect was for a much clearer system. Those needing to know about planning policy would make reference to the structure plan, the district-wide local plan, and the minerals and waste plans, with some certainty that they would exist.

Local government reorganisation introduced considerable extra complexity in England, as illustrated in Figure 4.4. Where the two-tier system remains, the planning framework is not affected: counties prepare the structure plan and waste and minerals plans (or one plan for both topics), and districts prepare the district-wide local plan. Where new unitary authorities are created, they are effectively county councils, but in almost all cases prepare joint structure plans with the neighbouring county councils. The joint arrangements are summarised in Table 4.1. The exceptions are Halton, Warrington, Herefordshire, the Isle of Wight, and Thurrock, which are to prepare unitary development plans (as in the metropolitan districts). All the other unitary authorities will prepare their own district wide local plans. The county will continue to prepare waste and minerals plans (or one plan for both topics) for its area. The metropolitan districts are of course, unaffected by local government reorganisation and will continue with their unitary development plans.

DEVELOPMENT PLANS IN NORTHERN IRELAND

The formal change to a discretionary system of development plans and control did not come to Northern Ireland until 1972. Prior to this, the system was much the same as for the rest of Britain before 1947, with local authorities able to prepare planning schemes. Practice was similar also in that very little progress was made on the preparation and approval of such schemes, and a system of interim development control operated. The 1972 Order introduced the development plan, with similar status to those in the rest of the UK. There are three types of development plan (area, local, and subject plans) which are produced and adopted by the DoENI (shown in Figure 4.5). Area plans, which can cover the whole or a substantial part of one or more district council areas, are the main reference for development control, and include both strategic and detailed policies.

The provisions of the 1991 Act, including those on the primacy of the development plan, do not yet apply in Northern Ireland, although in 1999 the Northern Ireland Office consulted on proposals to make the plan the primary consideration. This was in response

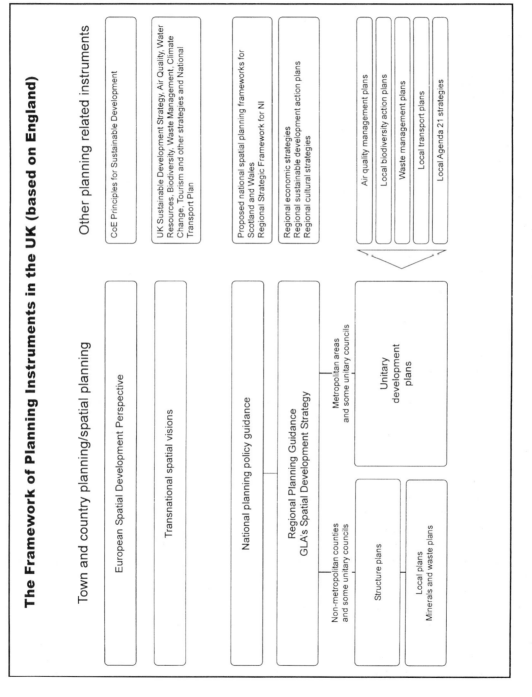

The Framework of Planning Instruments in the UK (based on England)

Town and country planning/spatial planning

- European Spatial Development Perspective
- Transnational spatial visions
- National planning policy guidance
- Regional Planning Guidance
  GLA's Spatial Development Strategy

Non-metropolitan counties and some unitary councils
- Structure plans
- Local plans
  Minerals and waste plans

Metropolitan areas and some unitary councils
- Unitary development plans

Other planning related instruments

- CoE Principles for Sustainable Development
- UK Sustainable Development Strategy, Air Quality, Water Resources, Biodiversity, Waste Management, Climate Change, Tourism and other strategies and National Transport Plan
- Proposed national spatial planning frameworks for Scotland and Wales
  Regional Strategic Framework for NI
- Regional economic strategies
  Regional sustainable development action plans
  Regional cultural strategies
- Air quality management plans
- Local biodiversity action plans
- Waste management plans
- Local transport plans
- Local Agenda 21 strategies

Figure 4.3 The Framework of Planning Instruments in the UK (Based on England)

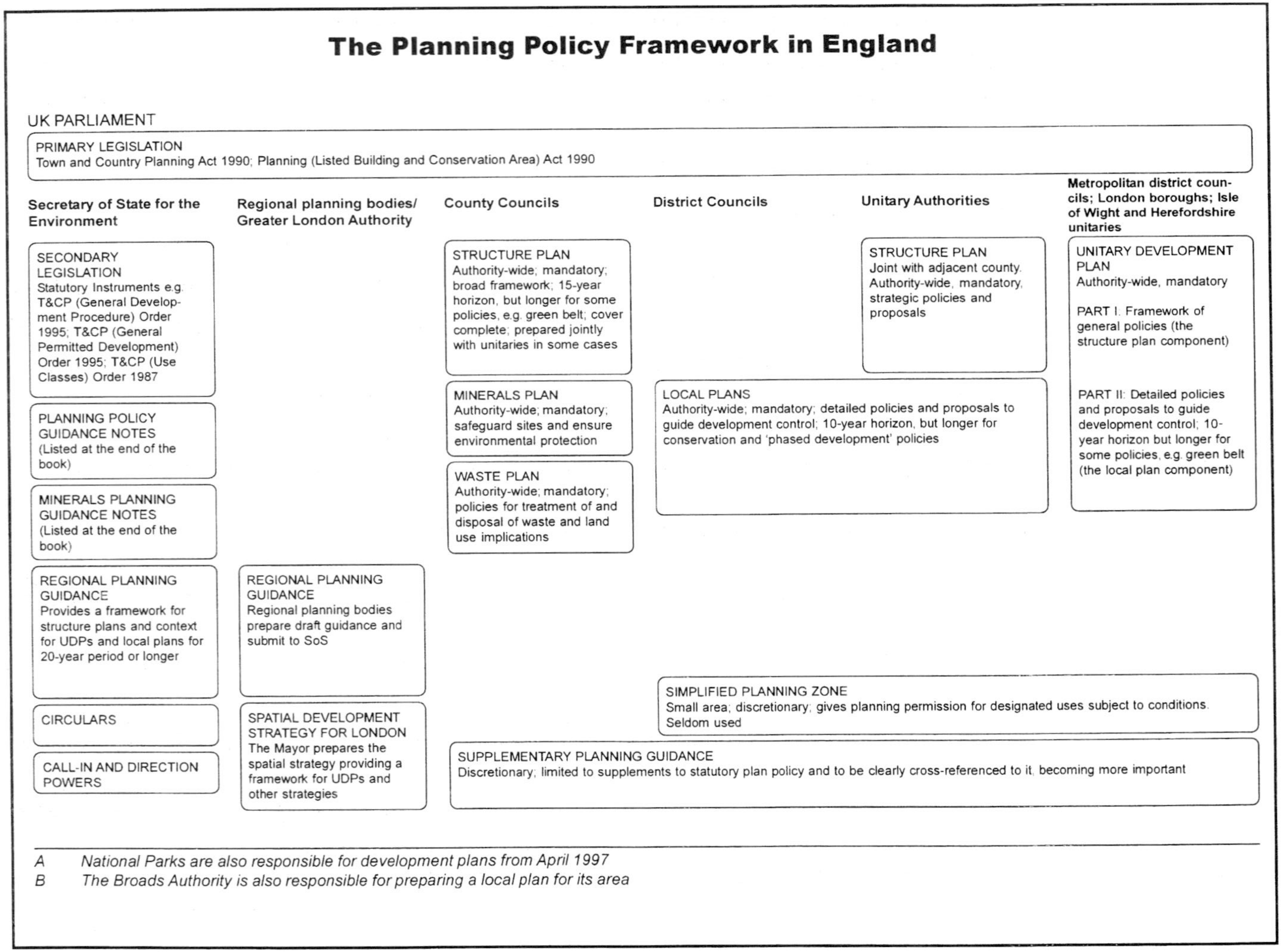

The Planning Policy Framework in England

UK PARLIAMENT

PRIMARY LEGISLATION
Town and Country Planning Act 1990: Planning (Listed Building and Conservation Area) Act 1990

Secretary of State for the Environment

SECONDARY LEGISLATION
Statutory Instruments e.g. T&CP (General Development Procedure) Order 1995, T&CP (General Permitted Development) Order 1995, T&CP (Use Classes) Order 1987

PLANNING POLICY GUIDANCE NOTES
(Listed at the end of the book)

MINERALS PLANNING GUIDANCE NOTES
(Listed at the end of the book)

REGIONAL PLANNING GUIDANCE
Provides a framework for structure plans and context for UDPs and local plans for 20-year period or longer

CIRCULARS

CALL-IN AND DIRECTION POWERS

Regional planning bodies/ Greater London Authority

REGIONAL PLANNING GUIDANCE
Regional planning bodies prepare draft guidance and submit to SoS

SPATIAL DEVELOPMENT STRATEGY FOR LONDON
The Mayor prepares the spatial strategy providing a framework for UDPs and other strategies

County Councils

STRUCTURE PLAN
Authority-wide, mandatory; broad framework. 15-year horizon, but longer for some policies, e.g. green belt; cover complete, prepared jointly with unitaries in some cases

MINERALS PLAN
Authority-wide, mandatory; safeguard sites and ensure environmental protection

WASTE PLAN
Authority-wide, mandatory; policies for treatment of and disposal of waste and land use implications

District Councils

LOCAL PLANS
Authority-wide, mandatory; detailed policies and proposals to guide development control. 10-year horizon, but longer for conservation and 'phased development' policies

SIMPLIFIED PLANNING ZONE
Small area, discretionary; gives planning permission for designated uses subject to conditions
Seldom used

SUPPLEMENTARY PLANNING GUIDANCE
Discretionary; limited to supplements to statutory plan policy and to be clearly cross-referenced to it, becoming more important

Unitary Authorities

STRUCTURE PLAN
Joint with adjacent county. Authority-wide, mandatory, strategic policies and proposals

Metropolitan district councils; London boroughs; Isle of Wight and Herefordshire unitaries

UNITARY DEVELOPMENT PLAN
Authority-wide, mandatory

PART I: Framework of general policies (the structure plan component)

PART II: Detailed policies and proposals to guide development control. 10-year horizon but longer for some policies, e.g. green belt (the local plan component)

A   National Parks are also responsible for development plans from April 1997
B   The Broads Authority is also responsible for preparing a local plan for its area

Figure 4.4 The Planning Policy Framework in England

Table 4.1 Structure Plan Areas in England

Previous structure plan authority	New arrangements
Avon County Council	Joint structure plan: Bristol UA, North Somerset UA (formerly Woodspring), Bath and NE Somerset UA (formerly Wansdyke and Bath), South Gloucestershire UA (formerly Northavon and Kingswood)
Bedfordshire County Council	Joint structure plan: Luton, Bedfordshire County Council
Berkshire County Council	Joint structure plan: Bracknell Forest UA , Newbury UA, Reading UA, Slough UA, Windsor and Maidenhead UA, Wokingham UA
Buckinghamshire County Council	Joint structure plan: Milton Keynes UA, Buckinghamshire County Council
Cambridgeshire County Council	Joint structure plan: Peterborough UA, Cambridgeshire County Council
Cheshire County Council	UDPs for the UAs and a structure plan for the remainder Halton UA , Warrington UA, Cheshire County Council
Cleveland County Council	Joint structure plan also with Darlington UA, Middlesbrough UA, Hartlepool UA, Redcar and Cleveland UA (formerly Langbaurgh-on-Tees), Stockton on Tees UA
Cornwall	Structure plan
Cumbria	Joint structure plan: Cumbria CC, Lake District NPA
Derbyshire County Council	Joint structure plan: Derby City UA, Derbyshire County Council
Devon County Council	Joint structure plan: Plymouth UA, Torbay UA, Devon County Council UA
Dorset County Council	Joint structure plan: Bournemouth UA, Poole UA, Dorset County Council
Durham County Council	Darlington UA – joint structure plan with former Cleveland LAs structure plan for Durham County Council
East Sussex County Council	Joint structure plan: Brighton and Hove UA, East Sussex County Council
Essex County Council	UDP for Thurrock. Joint structure plan; Southend UA, Essex County Council
Hampshire County Council	Joint structure plan: Portsmouth UA, Southampton UA, Hampshire County Council
Hereford and Worcester	Structure plan for Worcestershire County Council UDF for Herefordshire

Table 4.1 continued

Previous structure plan authority	New arrangements
Humberside County Council	Joint structure plan: Kingston upon Hull UA, East Riding UA (formerly East Yorks, Beverley, Holderness and part of Boothferry) Joint structure plan: North East Lincolnshire UA (formerly Cleethorpes and Great Grimsby), North Lincolnshire UA (formerly Glandford, Scunthorpe and part of Boothferry)
Isle of Wight	UDP: Isle of Wight UA
Kent County Council	Joint structure plan: Medway Towns UA (formerly Rochester and Gillingham), Kent County Council
Lancashire County Council	Joint structure plan: Blackburn with Darwen UA, Blackpool UA, Lancashire County Council
Leicestershire	Joint structure plan: Leicester City UA, Rutland UA, Leicestershire County Council
North Yorkshire County Council	Joint structure plan: York UA, North Yorkshire County Council, Yorkshire Dales NPA
Northamptonshire	Structure plan
Nottinghamshire County Council	Joint structure plan: Nottingham City UA , Nottinghamshire County Council
Oxfordshire	Structure plan
Shropshire County Council	Joint structure plan: The Wrekin UA, Shropshire County Council
Somerset	Joint structure plan: Exmoor NPA, Somerset County Council
Staffordshire County Council	Joint structure plan: Stoke-on-Trent City, Staffordshire County Council
Suffolk	Structure plan
Surrey	Structure plan
Warwickshire	Structure plan
West Sussex	Structure plan
Wiltshire County Council	Joint structure plan: Swindon UA, Wiltshire County Council

Note: For clarity, the term 'unitary authority' (UA) is used here rather than district council. The Peak District National Park and Lake District National Park are also structure plan authorities. From April 1997 all national parks became the sole planning authority for their area.

The Planning Policy Framework in Scotland, Northern Ireland and Wales

SCOTLAND

Scottish Parliament **Unitary Councils**

PRIMARY LEGISLATION
Town & Country Planning (Scotland) Act, 1997

SECONDARY LEGISLATION
e.g.: Town & Country Planning (Use Classes) (Scotland) Order 1989; Town & Country Planning (General Permitted Development) (Scotland) Order, 1992

CIRCULARS
Elaboration of procedural matters

NATIONAL PLANNING POLICY GUIDELINES
Statements of government policy on nationally important land use issues

Proposed national spatial planning framework for Scotland

PLANNING ADVICE NOTES
Identify and disseminate good practice

STRUCTURE PLAN
Provides a framework for local plan production and general guidance for development control
6 joint plans
11 single authority plans
new structure plan arrangements introduced 1996

LOCAL PLAN
Complete coverage mandatory but many small area plans; provide detailed guidance for development control

SIMPLIFIED PLANNING ZONE
Grants planning permission in advance. Seldom used

SUPPLEMENTARY PLANNING GUIDANCE

NORTHERN IRELAND

UK Parliament **Northern Ireland Assembly** **The Planning Service Executive Agency**

LEGISLATION
The Planning (Northern Ireland) Order 1991 under the Northern Ireland Act, 1974

SECONDARY LEGISLATION
e.g.: Planning (Use Classes) Order (Northern Ireland) 1989; Planning (General Development) Order (Northern Ireland) 1993

PLANNING POLICY STATEMENTS
To provide guidance for plan preparation and development control

DEVELOPMENT CONTROL ADVICE NOTES
Detailed guidance for development control

REGIONAL STRATEGIC FRAMEWORK FOR NORTHERN IRELAND
Strategic spatial planning guidance to 2025

BELFAST METROPOLITAN AREA PLAN

AREA PLAN
Covers the whole or substantial part of area of a district council; strategic policies providing a framework for local plans and for development control; 10-15 year horizon

LOCAL PLAN
Covers part of the area of one or more district councils; detailed policies to guide development control

SUBJECT PLAN
Covers any area for a particular planning project

WALES

UK Parliament **The Welsh Assembly** **Unitary Councils/National Parks**

PRIMARYLEGISLATION
The Town & Country Planning Act, 1990

Proposed national spatial planning framework for Wales

SECONDARY LEGISLATION
e.g.: Town & Country Planning (General Permitted Development) Order 1996; Town & Country Planning (General Development Procedure) Order 1996; Town & Country Planning (Use Classes) Order 1992

PLANNING GUIDANCE WALES
• Unitary Development Plans
• Planning Policy

TECHNICAL ADVICE NOTES

STRATEGIC PLANNING GUIDANCE IN WALES

UNITARY DEVELOPMENT PLANS
Same instrument as applies in metropolitan districts in England with Part I: Strategy and Part II: Detailed policies.
This system was introduced in 1996 to replace the previous two-tier system of structure plans and local plans.

Figure 4.5 The Planning Policy Framework in Scotland, Northern Ireland, and Wales

Table 4.2 Structure Plan Areas in Scotland

Single authority structure plan areas	Joint structure plan areas
Argyle and Bute	Aberdeen City
Borders	Aberdeenshire
Dumfries and Galloway	East Ayrshire
Falkirk	North Ayrshire
Fife	South Ayrshire
Highland	Angus
Moray	Dundee
Orkney Islands	Stirling
Perthshire and Kinross	Clackmannanshire
Shetland Islands	East Lothian
Western Isles	Edinburgh
	Midlothian
	West Lothian
	Dumbarton and Clydebank
	East Dumbartonshire
	East Renfrewshire
	Glasgow City
	Inverclyde
	North Lanarkshire
	Renfrewshire

to the House of Commons Northern Ireland Affairs Committee's 1996 report *The Planning System in Northern Ireland*. The committee expressed serious concerns about the lack of a clear strategy for the Province as a whole (which is now met by the creation of the *Regional Planning Strategy*) and the inadequacy of the development plans system. Northern Ireland is not affected by changes in local government. Area and local plans will continue to be prepared by the six divisional offices of the DoENI.

DEVELOPMENT PLANS IN SCOTLAND

The Scottish system differs in several significant ways from that in England and Wales, but the two-tier system of development plans and the procedures for the adoption and approval were broadly similar until 1996. Some differences can be attributed to the particular geographical characteristics of Scotland;

others may legitimately be attributed to a desire to avoid some of the difficulties of the English system. Because of the different administrative structure and larger planning areas in Scotland, there is a slightly different emphasis in the functions of structure plans, which are to indicate policies and proposals concerning the scale and general location of new development, and to provide a regional policy framework for accommodating development (PAN 37:7). Progress on the approval of structure plans was a significant problem, with an average of seventeen months needed for Secretary of State approval.

Changes to the development plan system itself have followed closely those introduced south of the Border. For example, the procedure for making alterations to structure plans and local plans has been made simpler. Also, certain adjustments have been made to the division of planning responsibilities between regions and districts.

The 1991 Act brought some of the same changes made in England and Wales to Scotland, notably the enhanced status of development plans in development control and insertion of section 18A into the 1972 Act, with the same effect as s. 54A in England and Wales; calls for more succinct statements of policy; and the emphasis on 'physical land use development' (PAN 37). However, in a number of ways the Scottish development plan system remains distinctive. The structure plan still has to be approved by central government. The survey still plays a part in the approval, and must be put on deposit and accompany the deposited plan in the submission. Local government reorganisation created unitary authorities in Scotland in 1996, and although the two-tier system of structure and local plans was retained, joint working is now necessary for the production of some structure plans, and the arrangements are summarised in Figure 4.5 and Table 4.2. The Scottish Office has designated seventeen structure plan areas, six of which cover more than one unitary authority. Local planning continues unchanged in the new unitary districts.

Evaluation of development planning in Scotland for the Scottish Executive (Hillier Parker *et al.* 1998) also reveals similarities with the situation in England and Wales. Local government reorganisation delayed the production of plans, and it was not until 1989 that full structure plan cover was achieved. However, progress on local plans has been better overall than in England and Wales, mainly because of the mandatory requirement. Recommendations are made for widening the ownership of plans through more effective involvement of other interests and for giving the plan more focus to avoid over-complexity.

DEVELOPMENT PLANS IN WALES

In Wales the system of development plans was virtually the same as that for England until 1996. One important variation was that the responsibility for waste rested with the districts (rather than counties) and thus waste policies were included in local plans rather than separate county-wide subject plans.

Local government reorganisation created unitary councils in 1996, and the plan framework was amended to require each authority (including the national parks) to prepare a unitary development plan. The Welsh UDP will have a similar form to the UDPs in the English metropolitan districts, with a Part I and Part II. The councils may prepare UDPs jointly, and the Part II element may be prepared by area committees. Arrangements for the transition to the new framework have been put in place. Welsh local authorities were able to seek approval from the Secretary of State to continue through to the adoption of plans already in preparation. With changes to local authority boundaries, local plans (including some yet to be adopted) may cover only part of an authority's area or be split between two.

THE CONTENT OF PLANS

The 1947 legislation was largely concerned with land use: 'a development plan means a plan indicating the manner in which a local planning authority propose that land in their area should be used'. The 1968 Act signalled a major shift in focus: emphasis was laid on major economic and social forces and on broad policies or strategies for large areas. It was held that land use planning could not be undertaken satisfactorily in isolation from the social and economic objectives which it served. Thus the plans were to encompass such matters as the distribution of population and employment, housing, education, and leisure.

This broader concept of planning did not survive, and by 1980 central government had moved back to a predominantly land use approach. This radical departure from the ideas of 1968, and the contraction of the scope of structure plans, have been widely documented (Cross and Bristow 1983; Healey 1986). Central government also intervened to restrict plan content significantly. Thornley (1991: 124) provides a useful summary of what he describes as the 'attack on structure plans'.[26] During the 1990s, Departmental advice about plan content has become increasingly specific and restrictive. The impact has been that, while local plans embraced wide-ranging social and economic objectives, their proposals nevertheless are 'primarily about land allocation' (Healey 1983: 189) Moreover, though local plans vary substantially in form, and 'appear local in orientation and specific to particular areas and issues', there is considerable consistency in scope and content. Consistency arises from the need for central government support for policy, and because of the limited planning powers provided by legislation. Perhaps another cause lies in the professional training and culture of planners, which are rooted in land use and physical concerns.

Criticisms of the weaknesses of environmental planning, the sustainable development agenda, and the need for more policy integration have led the government to promote a wider scope for development plans. The 1999 revision of PPG 12 provides some clarification of the government's current position on plan content, as shown in Box 4.2.

BOX 4.2 THE SCOPE OF DEVELOPMENT PLANS IN ENGLAND

The topics that the Act requires local planning authorities to consider:

- the conservation of the natural beauty and amenity of the land;
- the improvement of the physical environment; and
- the management of traffic

In addition subjects that may be considered in structure plans are:

- housing, including figures for additional housing requirements in each district, and targets for development on greenfield and brownfield sites;
- green belts;
- the conservation and improvement of the natural and built environment;
- the economy of the area, including major industrial, business, retail, and other employment-generating and wealth-creating development;
- a transport and land use strategy and the provision of strategic transport facilities including highways, and other infrastructure requirements;
- mineral working (including disposal of mineral waste) and protection of mineral resources; waste treatment and disposal, land reclamation and reuse;
- tourism, leisure, and recreation; and
- energy generation including renewable energy (para. 3.10).

In relation to sustainable development, PPG 12 suggests 'other issues that may be addressed in plans, either as land use policies or as considerations which influence policies in the plan'. These include:

- environmental considerations: energy, air quality, water quality, noise and light pollution, biodiversity, habitats, landscape quality, the character and vitality of town centres, tree and hedgerow protection and planting, revitalisation of urban areas, conservation of the built and archaeological heritage, coastal protection, flood prevention, land drainage, groundwater resources, environmental impacts of waste and minerals operations, unstable land;
- economic growth and employment: revitalisation and broadening of the local economy and employment opportunities, encouraging industrial and commercial development, types of economic development; and generally to take account of the needs of businesses while ensuring that proposals are realistic;
- social progress: impact of planning policies on different groups, social exclusion, affordable housing, crime prevention, sport, leisure and informal recreation, provision for schools and higher education, places of worship, prisons, and other community facilities, accommodation for gypsies; but 'to limit the plan content to social considerations that are relevant to land use policies'.

A consequence of the increasing attention that many organisations now give to plans in the light of the 'plan-led system' has been the production of 'model policies'.[27] Numerous national and local organisations such as the Environment Agency, English Nature, and Friends of the Earth (1994a) have suggested policy wording that can be taken 'off the shelf' rather than written anew for each local plan. General advice is now also given by the POS (1997). The documents provide both general advice about how to write policy and what is admissible content, and a bank of generic policy statements that can be tailored to local circumstances. Other advice is given in individual planning policy and good practice guidance, notably the DETR's *Planning for Sustainable Development: Towards Better Practice* (1998). The amount of advice given to planning authorities about plans is somewhat overwhelming, and the detail suggested often goes beyond what is appropriate in local plans. The pressure to incorporate many additional policies has been a major cause of delay and uncertainty in the system.

THE STATUTORY PROCEDURES AND MANAGEMENT OF THE PLAN PROCESS

A particularly helpful feature of the 1991 Act is that it brought the procedures for the various types of plan (in England and Wales) much more closely into line with each other, but the most recent changes are forcing them apart again. Devolution may also create more variation across the country. The general procedure for the preparation of local plans in England is illustrated in Figure 4.6.

Essentially, the procedures comprise 'safeguards' to ensure the accountability of government in the planning process. This is particularly important in the UK, where there is no constitutional safeguard of private property or other rights (other than that provided by the European Convention on Human Rights) and where there is wide administrative discretion in decision-making. There is no appeal to the courts on the policy content of plans. The procedures also provide for increased involvement of other organisations and the public in policy formulation. The process of open discussion and formal adoption lends authority and standing to plans, and provides an element of legitimacy even though the plans are not subject to direct ministerial approval.

In the following discussion the focus is on the key safeguards, the main criticisms of the procedure, and recent amendments. The knotty questions about the extent to which the public and other objectors are effectively able to make use of the safeguards and how this influences plan content are dealt with in Chapter 12.

The main safeguards in plan preparation and adoption are as follows:

- the opportunity for all interests to be consulted in the formative stages of plan preparation;
- the need for authorities to consider conformity between plans and regional and national guidance;
- the right to make objections to both strategic and detailed plans, and to have objections to the latter heard before an independent further inspector;
- a further right to object to any proposed modifications or where the authority proposes to reject the recommendation of an inspector or a panel;
- the overarching right of the Secretary of State to intervene and to direct modifications; and
- a limited right to challenge the plan in the courts.

The central focus of the formal adoption procedure is the hearing. In the case of a local plan or UDP, this is a public local inquiry; in the case of a structure plan, it is an examination in public. At the inquiry an independent inspector hears 'objections', whereas the examination in public (EIP) is a 'probing discussion' of selected matters which the authority needs to consider before taking the structure plan forward. (The forms of hearing are discussed further in Chapter 12). Anyone can object to a development plan, and the planning authority has a duty to consider all objections. For local plans and UDPs, objectors also have a right to present their case to the inquiry. The EIP deals only with those matters which the authority considers need examination in public, and the planning

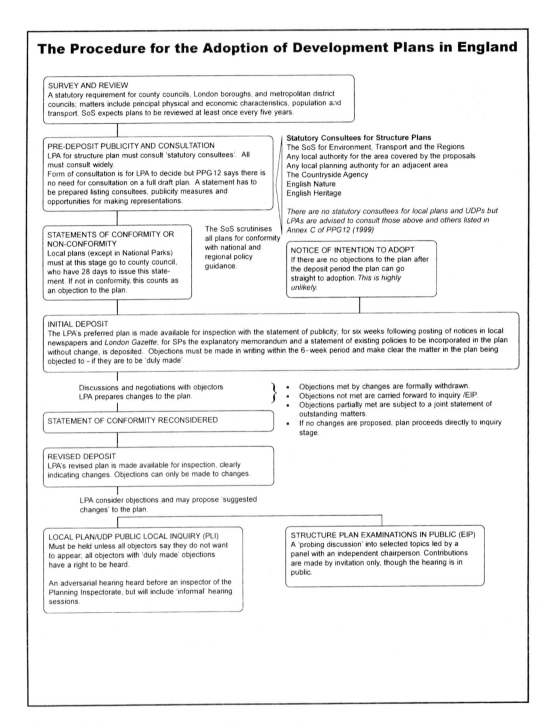

The Procedure for the Adoption of Development Plans in England

SURVEY AND REVIEW
A statutory requirement for county councils, London boroughs, and metropolitan district councils; matters include principal physical and economic characteristics, population and transport. SoS expects plans to be reviewed at least once every five years.

PRE-DEPOSIT PUBLICITY AND CONSULTATION
LPA for structure plan must consult 'statutory consultees'. All must consult widely.
Form of consultation is for LPA to decide but PPG12 says there is no need for consultation on a full draft plan. A statement has to be prepared listing consultees, publicity measures and opportunities for making representations.

Statutory Consultees for Structure Plans
The SoS for Environment, Transport and the Regions
Any local authority for the area covered by the proposals
Any local planning authority for an adjacent area
The Countryside Agency
English Nature
English Heritage

There are no statutory consultees for local plans and UDPs but LPAs are advised to consult those above and others listed in Annex C of PPG12 (1999)

STATEMENTS OF CONFORMITY OR NON-CONFORMITY
Local plans (except in National Parks) must at this stage go to county council, who have 28 days to issue this statement. If not in conformity, this counts as an objection to the plan.

The SoS scrutinises all plans for conformity with national and regional policy guidance.

NOTICE OF INTENTION TO ADOPT
If there are no objections to the plan after the deposit period the plan can go straight to adoption. *This is highly unlikely.*

INITIAL DEPOSIT
The LPA's preferred plan is made available for inspection with the statement of publicity; for six weeks following posting of notices in local newspapers and *London Gazette*; for SPs the explanatory memorandum and a statement of existing policies to be incorporated in the plan without change, is deposited. Objections must be made in writing within the 6-week period and make clear the matter in the plan being objected to – if they are to be 'duly made'.

Discussions and negotiations with objectors
LPA prepares changes to the plan.

}
- Objections met by changes are formally withdrawn.
- Objections not met are carried forward to inquiry /EIP.
- Objections partially met are subject to a joint statement of outstanding matters.
- If no changes are proposed, plan proceeds directly to inquiry stage.

STATEMENT OF CONFORMITY RECONSIDERED

REVISED DEPOSIT
LPA's revised plan is made available for inspection, clearly indicating changes. Objections can only be made to changes.

LPA consider objections and may propose 'suggested changes' to the plan.

LOCAL PLAN/UDP PUBLIC LOCAL INQUIRY (PLI)
Must be held unless all objectors say they do not want to appear; all objectors with 'duly made' objections have a right to be heard.

An adversarial hearing heard before an inspector of the Planning Inspectorate, but will include 'informal' hearing sessions.

STRUCTURE PLAN EXAMINATIONS IN PUBLIC (EIP)
A 'probing discussion' into selected topics led by a panel with an independent chairperson. Contributions are made by invitation only, though the hearing is in public.

Figure 4.6 The Procedure for the Adoption of Development Plans in England

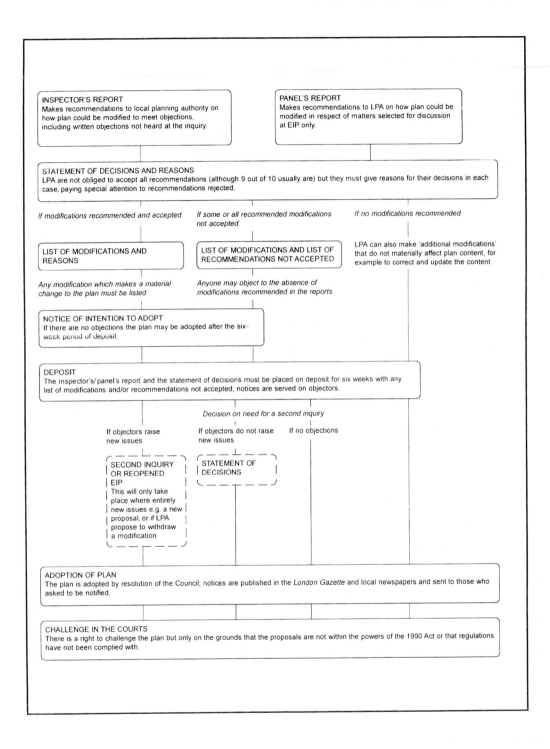

INSPECTOR'S REPORT
Makes recommendations to local planning authority on how plan could be modified to meet objections, including written objections not heard at the inquiry.

PANEL'S REPORT
Makes recommendations to LPA on how plan could be modified in respect of matters selected for discussion at EIP only.

STATEMENT OF DECISIONS AND REASONS
LPA are not obliged to accept all recommendations (although 9 out of 10 usually are) but they must give reasons for their decisions in each case, paying special attention to recommendations rejected.

If modifications recommended and accepted

If some or all recommended modifications not accepted

If no modifications recommended

LIST OF MODIFICATIONS AND REASONS

LIST OF MODIFICATIONS AND LIST OF RECOMMENDATIONS NOT ACCEPTED

LPA can also make 'additional modifications' that do not materially affect plan content, for example to correct and update the content

Any modification which makes a material change to the plan must be listed

Anyone may object to the absence of modifications recommended in the reports

NOTICE OF INTENTION TO ADOPT
If there are no objections the plan may be adopted after the six-week period of deposit.

DEPOSIT
The inspector's/panel's report and the statement of decisions must be placed on deposit for six weeks with any list of modifications and/or recommendations not accepted; notices are served on objectors.

Decision on need for a second inquiry

If objectors raise new issues

If objectors do not raise new issues

If no objections

SECOND INQUIRY OR REOPENED EIP
This will only take place where entirely new issues e.g. a new proposal, or if LPA propose to withdraw a modification

STATEMENT OF DECISIONS

ADOPTION OF PLAN
The plan is adopted by resolution of the Council; notices are published in the *London Gazette* and local newspapers and sent to those who asked to be notified.

CHALLENGE IN THE COURTS
There is a right to challenge the plan but only on the grounds that the proposals are not within the powers of the 1990 Act or that regulations have not been complied with.

authority determines who shall participate in the examination (whether or not they have made objections or representations).

It has been a constant theme that the planning system has performed poorly in preparing plans and keeping them up to date, especially in the 1990s, when there was strong support from central government for the preparation and effective use of plans. The plan-making process takes about five and half years on average (Steel *et al*.1995), and 'in excess of four years in Scotland' (Hillier Parker *et al*. 1998: 11). These averages mask great variation, extending from three to ten years. In England the time taken to adopt plans has tended to get longer because the enhanced status of plans and the district-wide format have led to an increase in the number of objections. However, overall productivity in the system may have improved because the increase in time taken to prepare and adopt is much less than the extra work that is entailed. The greatest proportion of time taken in the process is still in the preparation of a draft plan prior to deposit – a problem confirmed by the recent research on structure plans (Baker and Roberts 1999).

Overall, the generalisations made here need to be treated with care. Some planning authorities are able to cope very well with the procedures and produce plans in good time and keep them up to date.[28] In contrast, the performance of some planning authorities is abysmal, with very slow progress and little information about programmes. There are periodic calls from ministers for improved performance – the latest in 2000, when 20 per cent of planning authorities in England had still failed to adopt a local or unitary development plan.[29] Nevertheless, government initiatives to improve plan-making performance concentrate on amending the procedures, when this is plainly not the main problem.

Various changes have been made to attempt to simplify and to speed up the procedures. The amendments include reducing the requirements for public participation; allowing the local plan to be adopted in advance of a structure plan review; powers for the Secretary of State to request modifications; and providing more opportunity for objections after the inquiry.[30] In the wake of the 1991 Act, new targets were set for complete cover of plans by 1996. It quickly became apparent that the targets were not going to be met. In 1994 the DoE issued a consultation paper setting out proposals for further amendments to the procedure. This was followed in 1996 by revisions to the Code of Conduct for inquiries and EIPs, and regulations governing the preparation of plans. The proposals include removal of excessive detail from plans, more effective consultation early in the process, more emphasis on dealing with objections in writing, and shorter reports from inspectors.

While substantial, the changes have not reduced the work for local authorities, especially in the context of escalating numbers of objections. Criticism continued to be made about the cumbersome and time-consuming procedures. The TCPA, the RTPI, the LGA, and the Planning Officers' Society all published their own recommendations for improving the efficiency of the procedures and effectiveness of the system. All the papers place an emphasis on procedural changes, and relatively little attention is given to the form and content of plans. This is surprising, given the massive changes in the context for development planning since this was last reviewed. It is also disappointing to note the lack of attention to research findings and previous debates.

Recommendations for reform included abolition of the inquiry altogether, and making the inspector's recommendations binding on the local authority (Roberts 1998; RTPI 1999). In Scotland the 1998 Hillier Parker report on development planning came to similar conclusions, arguing that the report of the public inquiry should be binding and that there should be a national timetable for plan production. Research on the efficiency and effectiveness of local plan inquiries (Steel *et al*. 1995) suggests that making the inspector's report binding would be difficult to implement and probably counter-productive. The report pointed up the weakness in the capacity (and sometimes willingness) of many planning authorities to manage the procedure according to the guidance laid down. It also noted other factors which can be critical to the effectiveness of the procedure, including the form and content of the plan, especially the level of detail and the types of policies and proposals.

In 1996 the DoE began a more fundamental review of the development plan process. A 1997 consultation paper set out the options for speeding delivery of plans, and a further paper in 1998, under the *Modernising Planning* initiative, made specific proposals (its title was *Improving Arrangements for the Delivery of Local Plans and Unitary Development Plans*). During 1999 the DETR consulted on a revised PPG 12 along with draft revised regulations, and final versions were published at the end of that year. Also during 1999 the DETR consulted on the structure planning system by circulating the research report (Baker and Roberts 1999).

The 1999 version of PPG 12 rejects calls to make the inspectors' recommendations binding or to limit the rights to object and appear at the inquiry. Instead, the emphasis of the new procedure is on pre-deposit consultation based on issues papers, drawing out objections at an early stage in the process, reducing the length and detail of plans, and improving local authority management of the process, including the requirement for a publicly adopted timetable for plan production. The new regulations require a 'two-stage deposit' for local plans and UDPs, as shown in Figure 4.6. The first *initial deposit* will allow local authorities to gather objections, and this will be followed by a period of negotiation and revision of the plan. A second *revised deposit* stage will gather objections on any of the changes to the plan made by the planning authority. The regulations also reduce the statutory requirement for consultation with consultees prior to deposit, although this is unlikely to have much effect in practice. Of equal note in the PPG are the guidelines on the role of plans in enabling sustainable development, and the integration of transport and land use policies in plans, both of which are explained in the relevant chapters. Here it is sufficient to note that there are inevitable implications for the complexity of plans and the process, and thus the time required for plan preparation.

Despite the constant amendment to the plan-making process, the basic procedure has proved resilient. This is perhaps because it nicely balances the concerns of local authorities (which typically call for fewer procedural requirements in order to speed the process) and the concerns of objectors of all kinds (who naturally desire more influence in the local planning process). But the criticisms have been consistent and certainly a factor in the lack of enthusiasm of planning authorities for statutory plan-making.

EVALUATION OF THE IMPACT OF THE PLAN-LED SYSTEM

There were high hopes for the new development plan regime. It was anticipated that the planning system should become simpler and more responsive, reducing costs for both the private sector and local authorities and making it easier for people to be involved in the planning process. Central government set a target of the end of 1996 for the adoption of the new round of plans, and considerable encouragement was given to local authorities to produce the plans expeditiously. In many parts of the country plans have been put into place or updated. But the great frustration for many supporters of plan-led development is that many authorities have failed to respond to the challenge – hence the unusually frank statements on local authority performance given by ministers. Every opportunity has been given to local authorities to put their plans into place, but in some places lack of political will, professional expertise, or failures in management, together with some confusing advice from central government, have compounded to undermine progress. By the end of 2000 about a fifth of local planning authorities had no adopted statutory plan, and many others had plans that were out of date.

The plan-making process today is for most places much more complex and contentious than it was in the 1980s primarily because of the increasing participation of interests who recognise its potential significance for later development control decisions. Objections to plans are now typically counted in thousands, while few plans would have been subject to this level of objection twenty years earlier. One example (among many) is the case of the East Lothian Local Plan, where in 1998–99 the Scottish Inquiry Reporters Unit engaged three reporters on the inquiry who simultaneously addressed eight associated planning appeals.

The DTLR too has become increasingly active in scrutinising plans prior to the inquiry and has recruited additional staff to check plans for consistency with central government policy and regional guidance. The result has been more departmental objections to plans, and frequent and sometimes lengthy requests for changes. This represents a sharp increase in central government involvement in local planning.

Despite the increasing complexity, government intervention and the effective practice of some planning authorities, it would not be unfair to sum up the record of plan-making in much of the country as poor. Coverage of up-to-date plans is still patchy, it does not reflect the costs of the service, it has suffered major delays, and monitoring and review is weak. The situation is an embarrassment for central government and local authorities alike. The *Planning Concordat* drawn up between the LGA and DETR in 1999 recognises that 'under-performance by some local authorities . . . could undermine the plan-led system'. The issue has also been taken up in a series of ministerial meetings with planning officers in the regions during 1999.

The evidence suggests that the problem of delay relating to the statutory procedure has been overplayed. The most time-consuming part of the plan preparation process is in preparing the draft plan, not in the formal stages. The form of many adopted local plans and UDPs reflects a 'blueprint mentality', where the objective is to produce a blueprint for all development and anticipate all future decisions in the one document. This in part is because of government guidance which still (rather contradictorily) requires local authorities to include all policies that the local authority may use to refuse planning applications or to impose conditions on permissions. There is also an evident lack of experience and skills in many authorities and there are relatively few professionals who have experience of successful plan formulation. The planning schools may also have overlooked this in the curricula. Management, by officers and elected members, has often been weak (even now, some authorities cannot specify the target date for adoption), and often there is little political pressure to get the plans completed.

Overall, the quality or 'fitness for purpose' of plans has been lacking. The Inspectorate has gathered evidence from many hundreds of planning inquiries and its views and advice have been widely circulated.[31] The main problems have been failure to plan ahead from the start; poor attitudes to objectors, who are seen as a nuisance; shifting priorities during the plan-making process, which reflects a lack of commitment of officers and members; unnecessary conflicts with the regional offices and the DETR; provoking conflict through unnecessary plan content that seeks to cover all contingencies; too much prescription and detail, including policies not capable of implementation or monitoring; and failure to identify costs of these problems for senior managers and politicians. It should be emphasised that many successful authorities do not exhibit these problems. Also, external factors have played a part: fluctuating attitudes of central government to plan-making; problems with the two-tier system; political conflicts; and the former Conservative government's negative attitude towards planning and local government.

ZONING INSTRUMENTS

There are two examples of attempts to reintroduce the zoning approach in the UK planning system, in an effort to simplify planning: enterprise zones and simplified planning zones. Both reflect economic rather than land use planning objectives.

A major plank in the Conservative government's response to economic recession in the early 1980s was the proposed reduction in the 'burden' of regulation on business and enterprise'.[32] In enterprise zones (EZs), amendments to the planning regime were part of a much wider range of advantages offered, including exemption from rates on industrial and commercial property. The enterprise zone scheme had the effect of granting planning permissions in advance for such developments as the scheme specifies, and it was up to the planning authority to determine what planning concessions were offered. The scheme was simply proposed by the local authority and approved by the Secretary of State. Thirty-two enterprise zones were designated.[33] The enterprise zone initiative was closely monitored, and findings show in some cases a dramatic increase in development activity.[34] Overall, however, the liberalisation of land use planning controls made only a minor contribution to any success. A considerable amount of negotiation (whether it be termed 'planning' or not) still had to take place, both between the developers and local authorities and also between developers and other agencies.

Whatever the research on enterprise zones might have concluded, the government was so enamoured of the idea that it introduced a new type of *simplified planning zone* (SPZ) based upon it. The general notion of zoning as an alternative to the development plan had been rejected, but the DoE did see a limited role for zoning in particular locations where greater certainty, and some flexibility in the detail of development proposals, would contribute to economic development objectives. An SPZ is the local equivalent to a development order made by the Secretary of State. It replaces the normal discretionary planning system with advance permission for specified types of development.

Two broad types of scheme are possible: the *specific scheme*, which lists certain uses to be permitted, and the *general scheme*, which gives a wide permission but excludes certain uses. Conditions can be made in advance, and certain matters can be reserved for detailed consideration through the normal planning process. SPZs cannot be adopted in national parks, the Norfolk Broads, AONBs, SSSIs, approved green belts, conservation areas, and other protected areas. SPZs were particularly promoted for older industrial sites (especially those in single ownership) where there is a need to promote regeneration (Lloyd 1992).

The introduction of the SPZ provisions has excited very limited interest, and progress has been slow. Such interest as there was tended to come from authorities with experience of (or failure to obtain) enterprise zones: these authorities had fewer fears about the loss of normal development control powers over the quality of development (Arup Economic Consultants 1991). In all important respects the procedures for adoption of SPZs were identical to those of local plan preparation and adoption. The prospect of taking a scheme through these lengthy procedures was daunting, and it rapidly became clear that they were (in the words of the research report) 'undoubtedly cumbersome'.

As with development plans, the procedures have been amended to try to encourage more interest, but with little effect (Blackhall 1993, 1994). Some of the reasons are perhaps obvious. There is little difference between the allocation of land in a development plan and an SPZ: both indicate the type of development that is acceptable. Moreover, the extra 'certainty' provided by an SPZ designation is to some extent illusory since formal relationships are replaced by informal discussions. Additionally, decisions on the fulfilment of conditions and negotiations on reserved matters may still be needed. There are only a very small number of zones, and these operate in a narrow range of circumstances.

Overall, as Allmendinger (1996b) argues, the SPZ concept largely failed because it lacked clear and consistent objectives. It sought to offer deregulation and more certainty for developers, but in fact led to greater uncertainty when put into practice. But, above all, the recession at the end of the 1980s undermined property-led development, on which the idea rested. The reintroduction of zoning into the British planning system through SPZs has thus been unsuccessful.

BEST VALUE AND DEVELOPMENT PLANS

Given the issues raised above, the Best Value initiative is going to be a considerable challenge for development planning. Chapter 3 described how Best Value takes forward the value-for-money theme underpinning CCT. But Best Value 'is not just about economy and efficiency, but also about effectiveness and the quality of services'. The principles of Best Value are as relevant to development planning as any other local authority service: clear service standards – as regards both cost and quality; targets for continuous improvement by the most effective, economic, and efficient means; more say for service users; independent audit and inspection; and new powers for government to act on service failures.

Development planning has tended to escape value-for-money initiatives previously (unlike development control), but Best Value is comprehensive: it applies to all services (whether provided directly or indirectly) and all resources. Also, Best Value is supposed to help in addressing 'cross-cutting issues' such as sustainable development, which is central to development planning. But Best Value is also a very positive opportunity to improve performance in development planning because it focuses on the main management failure of local authorities: 'lack of attention to how resources are used in relation to common objectives'. Henceforth, planning authorities will need to judge their performance in development planning against performance indicators and targets, some set at national level and some by the authority (the national Best Value performance indicators (BVPIs) are listed in Chapter 3). National targets for quality will need to be consistent over five years with performance in the top quartile of all authorities when the targets are set, and for cost and efficiency consistent with the top 25 per cent in the region.

Very few plans have been put out to tender for preparation by the private sector, but under Best Value, local authority work must be subject to real competitive pressure – for example, by commissioning an independent benchmarking report to compare with other authorities or by putting out to tender all or part of the service to the private sector and other local authorities, with or without the in-house team bidding.

Some authorities already have performance plans in place for the planning service, including development planning. Elsewhere, there may be a need for urgent action in relation to plan-making. Among the issues that will need to be addressed are the need to demonstrate the value and quality of the plan; giving more attention to balancing the costs of the planning exercise with the value of the plan product; and extending participation and consultation beyond plan content to evaluating the satisfaction of participants with the process.

FURTHER READING

General

A good starting point for investigation of the policy framework is DETR PPG 1: *General Policy and Principles* (1997) and PPG 12: *Development Plans and Regional Planning Guidance* (1999) (these apply to England; see also their equivalents in the other countries of the UK). There are no textbooks which cover all the material in this chapter, but selected sources on particular aspects are noted below.

Supranational Planning

The literature on European spatial planning is growing rapidly. Williams (1996) *European Union Spatial Policy and Planning* provides an excellent introduction to European planning instruments. A good summary of the emergence of planning at the European level is by Fit and Kragt (1994) 'The long road to European spatial planning'. Other material includes the ESDP itself which is available on the Inforegio web site at <http://inforegio.cec.eu.int/wbdoc/docoffic/official/sdec/som_en.htm>. For interpretation and critique, see Bengs and Böhme (1998) *The Progress of European Spatial Planning*; Böhme and Bengs (1999) *From Trends to Visions: The European Spatial Development Perspective*; the special edition of *Built Environment* 23(4), especially the article by Bastrup-Birk and Doucet 'European spatial planning from the heart'; Faludi (2000) 'The European Spatial Development Perspective: what next?'; the special edition of *Town and Country Planning* 67 (2) (1998); and Zonneveld and Faludi (1996) 'Cohesion versus competitive position: a new direction for European spatial planning'.

On the impact of European spatial planning in the UK, see Nadin (1999) 'British planning in its European context', Wilkinson *et al.* (1998) *The Impact of the EU on the UK Planning System*; and Bishop *et al.* (2000) 'From spatial to local: the impact of the European Union on local authority planning in the UK'. On Interreg IIc, see Nadin and Shaw (1998) 'Transnational spatial planning in Europe: the role of Interreg IIc in the UK'.

There is increasing interest in the different planning systems in Europe. The European Commission's *Compendium of Spatial Planning Systems and Policies* (Nadin *et al.* 1997) is the most comprehensive source, Other recent sources include Newman and Thornley (1996) *Urban Planning in Europe*; Healey *et al.* (1994) *Trends in Development Plan Making in European Planning*; Schmidt-Eichstaedt (1996) *Land Use Planning and Building Permission in the European Union*; and the updated ISOCARP *International Manual of Planning Practice* (Lyddon and Dal Cin 1996). The older Davies *et al.* (1989b) *Planning Control in Western Europe* is also still relevant.

National and Regional Planning

PPG 11 and the regional guidance notes themselves are key references. Beyond that, there are now several critical reviews of national guidance, including Land Use Consultants (1995a) *The Effectiveness of Planning Policy Guidance Notes*. A comprehensive review of regional guidance is given in the symposium edited by Minay (1992), with updates by Roberts and Lloyd (1999) 'Institutional aspects of regional planning, management, and development: models and lessons from the English experience'; Quinn (1996) 'Central government planning policy'; Kitchen (1999a) 'Consultation on government policy initiatives: the case of regional planning guidance'; and Baker (1998) 'Planning for the English regions: a review of the Secretary of State's regional planning guidance'. On Scottish guidance, see Raemaekers *et al.* (1994) *Planning Guidance for Scotland* and Land Use Consultants (1999) *Review of National Planning Policy Guidelines*. On Wales, see Alden and Offord (1996) 'Regional planning guidance'.

There are many references on regional planning, including the standard textbook by Glasson (1992b) *An Introduction to Regional Planning*. Wannop (1995) *The Regional Imperative* provides a review with international comparisons. The resurgence of interest in strategic planning is analysed by Breheny (1991) 'The renaissance of strategic planning' and Roberts and Lloyd (1999) 'Institutional aspects of regional planning, management, and development: models and

lessons from the English experience'. Current practice in Northern Ireland is reflected in the DoENI (1998) *Shaping Our Future: Towards a Strategy for the Development of the Region – Draft Regional Strategic Framework for Northern Ireland*. Proposals for the UK spatial planning framework have been made by Wong *et al.* (2000).

Development Plans

PPG 12 (1999), and its equivalents elsewhere in the UK, are the basic sources. Baker and Roberts' *Examination of the Operation and Effectiveness of the Structure Planning Process* provides a much-needed comprehensive review of current structure planning practice. More detailed procedures are considered by Phelps (1995) 'Structure plans: the conduct and conventions of examinations in public' and Jarvis (1996) 'Structure planning policy and strategic planning guidance in Wales'. The RTPI has published a guide to good practice and a research report both entitled *Fitness for Purpose* (2001a, b).

There is more material on local planning, although much of it concentrates on procedures. The most recent textbook is Adams (1994) *Urban Planning and the Development Process*; and Tewdwr-Jones (1996) *British Planning Policy in Transition: Planning in the 1990s* gives a wide-ranging set of papers. The standard texts are now ageing. Their explanations for success and failure are still pertinent, but the details of practice less so: Healey (1983) *Local Plans in British Land Use Planning*; Bruton and Nicholson (1987) *Local Planning in Practice*; and Fudge *et al.* (1983) *Speed, Economy and Effectiveness in Local Plan Preparation and Adoption*. Other notable sources on local planning include Hull (1998) 'Spatial planning: the development plan as a vehicle to unlock development potential?'; Hull and Vigar (1998) 'The changing role of the development plan in managing spatial change'; Vigar and Healey (1999) 'Territorial integration and "plan-1ed" planning'; Healey (1994) 'Development plans: new approaches to making frameworks for land use regulation'; Healey (1986) 'The role of development plans in the British planning system'; Healey (1990) 'Places, people and politics: plan-making in the

1990s', and several chapters of case studies in Greed (1996a) *Implementing Town Planning*.

An important text summarising a major evaluation of the impact of development plans is Healey *et al.* (1988) *Land Use Planning and the Mediation of Urban Change*. See also Davies *et al.*(1986b) *The Relationship between Development Plans and Development Control*, and MacGregor and Ross (1995) 'Master or servant? The changing role of the development plan in the British planning system'. Kitchen (1997) *People, Politics, Policies and Plans* gives a view of plan making from inside a local authority.

Up-to-date summaries of procedures are given in the latest editions of Duxbury (1999) *Telling and Duxbury's Planning Law and Procedure* and Moore (2000) *A Practical Approach to Planning Law*. Steel *et al.*(1995) *The Efficiency and Effectiveness of Local Plan Inquiries* examines the procedures in practice, as does the RTPI Study by Cardiff University and Buchanan Partnership (1997) *Slimmer and Swifter: A Critical Examination of District Wide Local Plans and UDPs*. For Northern Ireland, see Dowling (1995) *Northern Ireland Planning Law*; and for Scotland, Collar (1994) *Green's Concise Scots Law: Planning* and McAllister and McMaster (1994) *Scottish Planning Law*.

The value of enterprise zones is considered by PA Cambridge Economic Consultants (1995) *Final Evaluation of Enterprise Zones*, while the number of publications on SPZs outnumbers the zones. See Blackhall (1993) *The Performance of Simplified Planning Zones*; Lloyd (1992) 'Simplified planning zones, land development, and planning policy in Scotland'; and Allmendinger (1996a) 'Twilight zones'. Official guidance is given in PPG 5, PAN 31, and TAN (W)3, all entitled *Simplified Planning Zones*.

NOTES

1   Council Regulation (EC) no. 1260/1999 *Official Journal L* 161 21.6.99. For a review of the differences between the 2000–06 and 1995–99 programmes, see the European Commission paper *Reform of the Structural Funds 2000–2006 Comparative Analysis*. June 1999. The full texts of the regulations and explanatory memoranda are available on the Inforegio web site: <http://www.inforegio.cec.eu.int>.

2   Areas are designated according to the Nomenclature of Territorial Units for Statistics (NUTS). In England NUTS 1 equates to the standard regions, NUTS 2 to groups of counties, and NUTS 3 to individual counties or groups of local authorities. The definition of the NUTS regions is very contentious as it can determine whether the area is eligible for Community assistance (Casellas and Galley 1999). The DTI has favoured the use of smaller statistical areas (wards) for the definition of assisted areas Tier 2 in the UK.

3   The Funds are the European Regional Development Fund (ERDF), the European Social Fund (ESF), the Guidance section of the European Agricultural Guidance and Guarantee Fund (EAGGF), and the Financial Instrument for Fisheries Guidance (FIFG). Objective 1 areas make use of all four, Objective 2 areas are eligible for ERDF and ESF, and Objective 3 makes use of the ESF only. There is in addition an 18 billion euro Cohesion Fund available for structural assistance to Greece, Ireland, Portugal, and Spain, and 21 billion euros earmarked for the accession countries. The comparative figure for the five-year programming period to 1999 is 155 billion euros, of which the UK received 13 billion (about £7.8 billion).

4   Northern Ireland also benefits from a special PEACE programme under the Community Initiatives which is worth 500 million euros (100 million of which is to be spent in the Republic of Ireland).

5   *Official Journal* 98/C 74/06.

6   INTERREG IIc was launched in 1996 (OJ 96/C200/07) and builds on previous cross-border cooperation programmes through Interreg I and IIa. It is funded under the ERDF Regulation EEC 4254/88. Competence for the EU to provide funding in this way comes from Article 161 (ex-Article 130d) under Title XVII Economic and Social Cohesion, and the Structural Fund

regulations under this Article give more detailed justification. Article 10 funding was also used to promote transnational planning in some areas. A parallel programme, TERRA, has been launched to promote transnational cooperation on spatial planning in areas which are 'vulnerable'.

7   Details of projects funded through Interreg IIc are available on the programme web sites, which are accessible via <http://www.interregiicuk.org.uk/>.

8   The Vision for North West Europe is available at <www.uwe.ac.uk/fbe/vision>, and the Norvision at <www.mem.dk/lpa/landsplan/international/norvision.thm>.

9   These arguments have been developed and consolidated over time in a succession of documents. Those not mentioned elsewhere in this chapter include the *European Regional and Spatial Planning Charter* (CoE 1983), *Guiding Principles for Sustainable Spatial Development* (CEMAT 1999); and the *Fifth Environmental Action Programme on the Environment: Towards Sustainability* (European Commission 1992) (the sixth is to be published in 2001).

10 The Scottish Executive has also taken a very positive attitude towards the dissemination of national planning guidance. Copies are freely distributed and many are available for reference on the Scottish Executive web site <www.scotland.gov.uk/>.

11 Minister for Transport and the Environment (Scotland) Speech to the RTPI National Conference, 25 November 1999.

12 The DETR consultation paper *Planning for Major Projects* draws attention to the problem of long inquiries debating policy in the absence or more specific national guidance on significant developments such as airports.

13 A series of papers have been prepared on behalf of the RTPI considering the potential of a national spatial planning framework. A collection of papers exploring the subject was published in *Town Planning Review* 70 (3) (1999).

14 The planning policy statements supersede relevant policy advice in the Planning Strategy for Rural Northern Ireland, and the whole strategy is to be replaced by 2002. The first PPSs to be published in 1995 were on transport, retailing, town centres, and industrial development.

15 The Town and Country Planning Association has for many years been at the forefront of the campaign for better strategic planning; see, for example, D. Hall (1991) and TCPA (1993). The RTPI published its case for enhancing regional planning guidance, *Regional Planning in England*, in 1997.

16 The research projects were the *Scoping Study: RPG Targets and Indicators* undertaken by ECOTEC (1999) and *Proposals for a Good Practice Guide on Sustainability Appraisal of Regional Planning Guidance* undertaken by Baker Associates (1999).

17 Examples of targets given in PPG 11 are traffic reduction and modal split; development of town centre versus out-of-town floorspace; rural accessibility to services; enhancement of biodiversity; and the use of recycled materials. The research found that current practice only sets quantified housing targets.

18 The ECOTEC *Scoping Study: RPG Targets and Indicators* (1999) provides some interesting findings on the capacity of the regional bodies to undertake regional planning, and has some positive conclusions too – for example, recent considerable improvements in the information base in some regions.

19 The report of Raemakers *et al.* on national planning guidance in Scotland (1994) recommended a tier of regional planning guidance for Scotland, and a return to regional reports for the city regions in the central belt to provide greater coordination in joint structure plan preparation in the wake of local government reorganisation, but this has not yet been taken up.

20 Twenty-five thousand houses were built in the countryside over ten years, many close to the urban centres, and the annual figure of 2,500–3,000 for Northern Ireland is more than for the rest of the UK put together.

21 Unlike the structure plan, which was prepared by

all relevant authorities for the whole of their area, local authorities were advised that local plans would not be needed in all areas, for example where there was little pressure for development and no need to stimulate growth. This discretion was used: a small number of authorities prepared a single plan for the whole area, others prepared one or more plans for parts of their area, while others prepared none at all.

22 The *development plan scheme* (later replaced by the *local plan scheme*) set out the agreed programme for the preparation and amendment of local plans. With the introduction of a mandatory require-ment to produce district-wide local plans in 1992 such schemes have been made redundant. The 1986 Act required local authorities to keep a register of development plan policies and the new regulations require a similar index of information in respect of the development plan. The *certificate of conformity* is given by the structure planning authority. Before 1992 the lack of a certificate would delay the local plan process, but today any disputes about conformity are taken to the inquiry as an objection.

23 By 1977 only seventeen of the necessary eighty-nine structure plans for England and Wales had been submitted, and seven approved. By 1980, of seventy-nine English structure plans which were expected, sixty-four had been submitted and thirty-eight approved.

24 Further support has been lent to these arguments by subsequent research. Rydin *et al.*(1990) and Collins and McConnell (1988) argued for recog-nition of the value of plans, and for a stronger development plan framework, but also for flexibility for local variation in form and content.

25 Strategic guidance was published for all the metropolitan regions, but except for London (where it has a statutory character) and the Thames Gateway, it has been or is planned to be incorporated into RPG.

26 In the 1980 Manchester Structure Plan, for example, the Secretary of State 'deleted more than 40 per cent of the policies, and a further 20 per cent were substantially modified'. Thornley

argues that such actions reflected the govern-ment's intention to allow market forces to operate at the cost of social and other wider objectives.

27 The idea of model policies for local plans was proposed in the early 1970s (Fudge *et al.*1983) but received little support.

28 Some of the authorities with the best records of plan production are in the areas where the planning process is under most pressure, as in the Home Counties around London.

29 In February 2000 Nick Raynsford, Minister for Planning, said, 'It is not acceptable that nine years after the requirement to produce an area-wide local plan was introduced, many authorities have still not done so. Indeed 5 per cent do not expect to have an adopted plan before 2002 and 2 per cent of authorities have not yet placed a plan on deposit. . . . I am also extremely concerned that there are over 200 local authority adopted development plans which will reach, or will have passed their original end-date [due date?] by 2001' (DETR Press Notice, 14 February 2000). In fact, the mandatory requirement was anticipated at least two years before its implementation. Concerned or not, these regular statements have not led to any intervention by government.

30 The Local Government and Planning Act 1980 introduced an expedited procedure which, in certain circumstances, allowed the local plan to be adopted in advance of a structure plan review. The Housing and Planning Act 1986 gave powers to the Secretary of State to request modifications to plans (in addition to the seldom used powers to call in). Most recently, the 1991 Act abandoned the need for the six-week consultation period prior to the deposit for objections, and an extra opportunity has been provided for objections after the inquiry where the planning authority does not accept the inspector's recommendations.

31 Most of this advice is written up in the Inspectorate's guidance to local authorities. This account also incorporates comments made during the delivery of University of the West of England short courses by the former head of local plan inquiries at the Inspectorate, David John. See also

the RTPI guide to good practice listed under 'Further reading'.

32 Sir Geoffrey Howe credited the notion to Sir Peter Hall, who in turn identified the origins of the concept in a 1969 article (Banham *et al.*1969). Hall (1991) has reviewed the ways in which this notion was transposed and 'sanitised' into the enterprise zone initiative in Britain.

33 The remaining enterprise zones are the East Midlands, Dearne Valley, North East Derbyshire, and Tyne Riverside.

34 The Corby EZ, for example, was virtually fully committed after seven years with 5,600 jobs and 293,000 million sq. ft of new floorspace (PA Cambridge Economic Consultants, 1987).

5

THE CONTROL OF DEVELOPMENT

The extent of vilification to which development control has been subject in Britain over the past 25 years suggests that it may be a rather more important process than its detractors allow. The importance of the process has, at one level, to do with matters of substance. Questions of land-use and urban form affect profoundly the welfare and enjoyment of life of those who live in urbanized societies like ours. Decisions taken in the course of development control have a long-term impact. At another level, however, the development control process serves as a focus for a whole range of questions about how we govern ourselves and on whom we confer power to take decisions on our behalf.

(Booth 1996: 1 and 2)

THE SCOPE OF CONTROL

Most forms of development (as statutorily defined) are subject to the prior approval of the local planning authority, though certain categories are excluded from control because they are thought to be trivial or beneficial. Local planning authorities have considerable discretion in giving approval. The legislation requires them to 'have regard to the provisions of the development plan, so far as material to the application', but also to 'any other material considerations'. Thus the authority can approve a proposal that does not accord with the provisions of the plan. The 1991 Act also increased the significance of plans, such that the plan becomes the first and primary point of reference in decision-making, effectively introducing a presumption in favour of development proposals which are in accordance with the development plan.

Planning decisions of a local planning authority (LPA) are of three kinds: unconditional permission, permission subject to conditions, or refusal. The practical scope of these powers is discussed in a later section; here it is necessary merely to stress that an applicant has the right of appeal to the Secretary of State against conditional permissions and refusals. If the action of the LPA is thought to be *ultra vires* (beyond its legal powers), there is also a right of recourse to the courts. Furthermore, planning applications which raise issues that are of more than local importance, or are of a particular technical nature, can be 'called in' for decision by the Secretary of State. The general development control process in England is illustrated in Figure 5.1. Readers should be aware that this is only a general guide and reference should be made to the reading listed at the end of this chapter on matters of detail.

Development control necessarily involves measures for enforcement. This is provided by procedures which require anyone who carries out development without permission or in breach of conditions to consult with the LPA and, in certain circumstances, to 'undo' the development, even if this involves the demolition of a new building. A *stop notice* can also be used to put a rapid end to the carrying out or continuation of development which is in breach of planning control, when serious environmental problems are being caused by the unauthorised activity.

These are very strong powers, and it is clearly

important to establish the meaning of *development*, particularly since the term has a legal connotation far wider than its use in ordinary language.

THE DEFINITION OF DEVELOPMENT

In brief, development is 'the carrying out of building, engineering, mining or other operations in, on, over or under land, or the making of any material change in the use of any buildings or other land' (and, since the 1991 Act, now covers some categories of demolition and rebuilding). There are many legal niceties attendant upon this definition with which it is fortunately not necessary to deal in the present outline. Some account of the breadth of the definition is, nevertheless, needed. 'Building operations', for instance, include rebuilding, structural alterations of or additions to buildings and, somewhat curiously, 'other operations normally undertaken by a person carrying on business as a builder'; but maintenance, improvement, and alteration works which affect only the interior of the building or which do not materially affect the external appearance of the building are specifically excluded.

The second half of the definition introduces a quite different concept: development here means not a physical operation, but a change in the *use* of a piece of land or a building. To constitute 'development', the change has to be *material*; that is, substantial – a concept which it is clearly difficult to define, and which, indeed, is not defined in the legislation. A change in *kind* (for example from a house to a shop) is material, but a change in *degree* is material only if the change is substantial. For instance, the fact that lodgers are taken privately in a family dwelling-house does not of itself constitute a material change so long as the main use of the house remains that of a private residence. On the other hand, the change from a private residence with lodgers to a declared guest house, boarding house, or private hotel would be material. Difficulties arise with changes of use involving part of a building, with ancillary uses, and with the distinction between a material change of use and a mere interruption.

This is by no means the end of the matter, but enough has been said to show the breadth of the definition of development and the technical complexities to which it can give rise. Reference must, nevertheless, be made to one further matter. Experience has shown that complicated definitions are necessary if adequate development control is to be achieved, but the same tortuous technique can be used to exclude matters over which control is not necessary. First, there are certain matters which are specifically declared not to constitute development (for example, internal alterations to buildings, works of road maintenance, or improvement carried out by a local highway authority within the boundaries of a road). Second, there are others which, though possibly constituting development, are declared not to require planning permission.

There is provision for the Secretary of State to make a *General Permitted Development Order* (GPDO) specifying 'permitted development rights' for matters that constitute development but do not require permission because it is effectively granted by the Order. The *Use Classes Order* (UCO) specifies groups of uses within which a change of use does not constitute development and is therefore permissible. Also, the Secretary of State can make *Special Development Orders* (SDOs) granting planning permission for specific locations or categories of development.

THE USE CLASSES ORDER AND THE GENERAL DEVELOPMENT ORDERS

The Use Classes Order groups all land uses into classes. Table 5.1 shows the use classes in different parts of Britain. Changes within each class do not constitute development and therefore do not need planning permission. Thus, class A1 covers shops used for all or any of a list of ten purposes, including the retail sale of goods (other than hot food); the sale of sandwiches or other cold food for consumption off the premises; for hairdressing; for the direction of funerals; and for the display of goods for sale. Class A3 covers 'use for the sale of food or drink for consumption on the premises or of hot food for consumption off the premises'. As a result of these classes, a shop can be

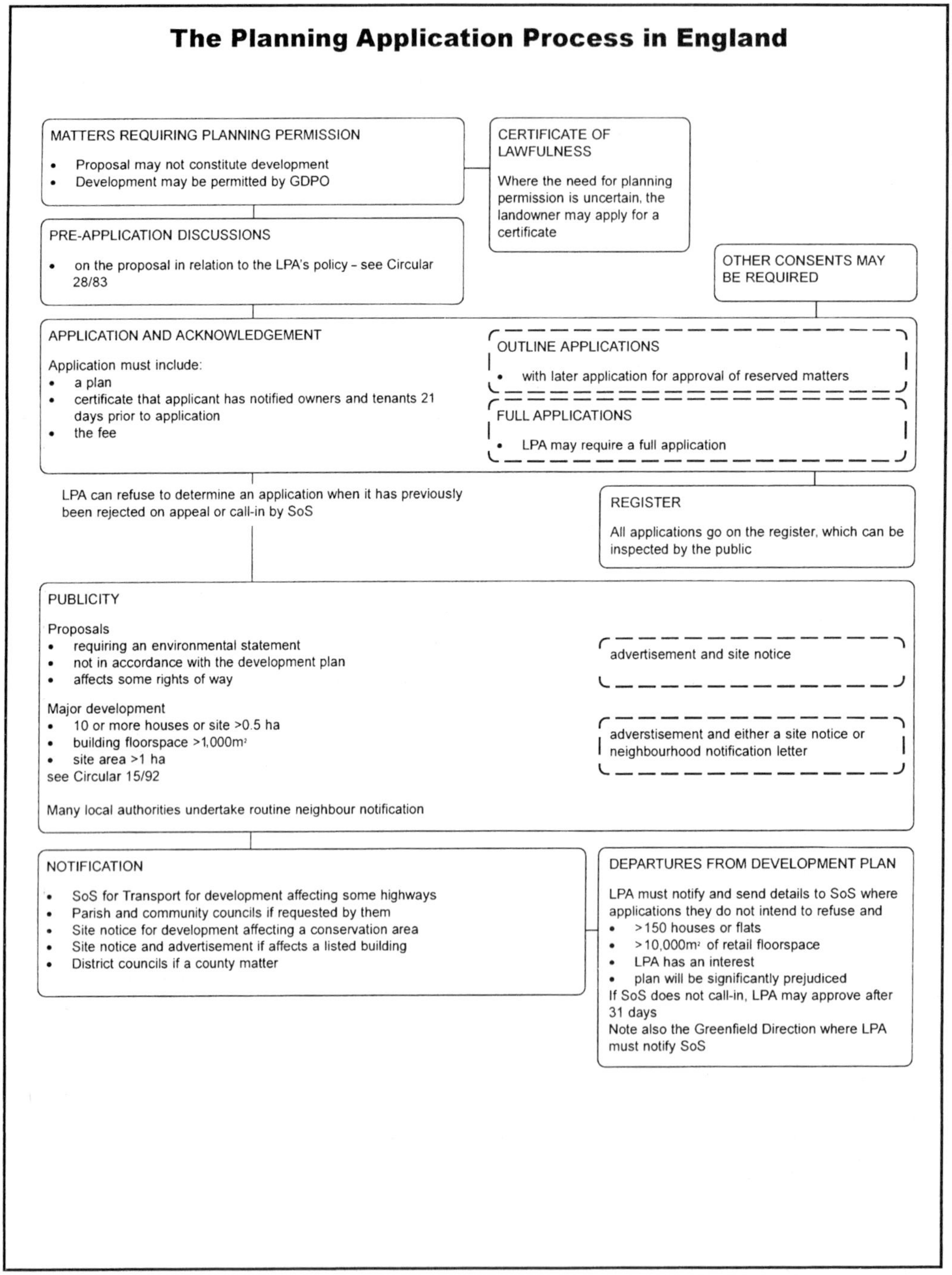

The Planning Application Process in England

MATTERS REQUIRING PLANNING PERMISSION

- Proposal may not constitute development
- Development may be permitted by GDPO

CERTIFICATE OF LAWFULNESS

Where the need for planning permission is uncertain, the landowner may apply for a certificate

PRE-APPLICATION DISCUSSIONS

- on the proposal in relation to the LPA's policy – see Circular 28/83

OTHER CONSENTS MAY BE REQUIRED

APPLICATION AND ACKNOWLEDGEMENT

Application must include:
- a plan
- certificate that applicant has notified owners and tenants 21 days prior to application
- the fee

OUTLINE APPLICATIONS

- with later application for approval of reserved matters

FULL APPLICATIONS

- LPA may require a full application

LPA can refuse to determine an application when it has previously been rejected on appeal or call-in by SoS

REGISTER

All applications go on the register, which can be inspected by the public

PUBLICITY

Proposals
- requiring an environmental statement
- not in accordance with the development plan
- affects some rights of way

advertisement and site notice

Major development
- 10 or more houses or site >0.5 ha
- building floorspace >1,000m²
- site area >1 ha
see Circular 15/92

advertisement and either a site notice or neighbourhood notification letter

Many local authorities undertake routine neighbour notification

NOTIFICATION

- SoS for Transport for development affecting some highways
- Parish and community councils if requested by them
- Site notice for development affecting a conservation area
- Site notice and advertisement if affects a listed building
- District councils if a county matter

DEPARTURES FROM DEVELOPMENT PLAN

LPA must notify and send details to SoS where applications they do not intend to refuse and
- >150 houses or flats
- >10,000m² of retail floorspace
- LPA has an interest
- plan will be significantly prejudiced
If SoS does not call-in, LPA may approve after 31 days
Note also the Greenfield Direction where LPA must notify SoS

Figure 5.1 The Planning Application Process in England

CONSULTATION

In various circumstances consultation is required with:
- Health and Safety Executive
- MAFF/SoS for Transport/SoS Heritage
- Sports Council
- Highway Authority
- The Coal Authority
- Environment Agency (formerly NRA)
- HMBC (English Heritage)
- The Theatre Trust
- Waste regulation authorities

LPA will have long list of local consultees.

There is formally 14 days to respond.

See article 10 of GDPO.

PREPARATION OF REPORT

Planning officers will prepare a report on the application, undertake discussions and negotiations with the applicant and other interests, consider consultation returns and policy context, undertake site visits and request further information or changes to the proposal.
Reports are considered by planning committees and sometimes area committees or parish councils.
Many minor decisions are delegated to planning officers.

DECISION

Application is determined in accordance with the development plan, unless material considerations indicate otherwise, decision will be made within 8 weeks.

REFUSED

LPA must give clear and precise reason.

GRANTED

Development to be begun within a specified period or 5 years.

APPEAL

Made to SoS within 6 months – must include:
- original application
- plans and correspondence
- notices
determined by Inspectorate by
- written representations
- informal hearing (no cross-examination)
- public local inquiry (inquiry procedure rules apply)

SoS 'recovers' some
- appeals for own decision e.g. if >150 houses or of significant controversy.

DECISION

Inspector makes most decisions but may report to SoS if 'recovered'.

CHALLENGE

Appellant can seek 'statutory review' in the High Court within 6 weeks on the grounds that decision
- not within powers of the Act
- procedural requirements not met.
Decision may only be to quash or uphold previous decision.

For a more comprehensive explanation, see Moore (2000).
For variations in Scotland, see McAllister and McMaster (1999) and Collar (1999).
For variations in Northern Ireland, see Dowling (1995).

Table 5.1 Summary and Comparison of the Use Classes Orders

England and Wales (Town and Country Planning (Use Classes) Order 1987) (SI No. 764) (as amended)			Scotland (The Town and Country Planning (Use Classes) (Scotland) Order 1989) (SI N. 147) (as amended)			Northern Ireland (Planning (Use Classes) Order (NI) 1989) (as amended)		
Class	Use	Development permitted by the GDPO (which may be subject to limitations)	Class	Use	Development permitted by the Permitted Development Order	Class	Use	Development permitted
A1	Shops	from A3, A2 if premises have a display window at ground floor level, or for the display or sale of motor vehicles	1	Shops	from sale and display of motor vehicles	1	Shops	from a betting office or from food or drink
A2	Financial and professional	to A1 from A3	2	Financial, professional, and other services	to 1	2	Financial, professional and other services	from a betting office or from food or drink; to 1 if has a display window at ground floor level
A3	Food and drink	to A1 and A2	3	Food and drink	to 1 and 2			
B1	Business	to B8 (max. 235 m²) from B2 and B8 (max. 235 m²)	4	Business	to 11 (max. 235 m²))	3	Business	
						4	Light industrial	to 11
B2	General industrial	to B1 and B8 (max. 235 m²)	5	General industrial	to 4 and 11 (max. 235 m²)	5	General industrial	to 4 or 11 (max. 235 m²)
			7–10	Special industry groups		7–10	Special industrial groups	
B8	Storage or distribution	to B1 (max. 235 m²) from B1 or B2 (max. 235 m²)	11	Storage or distribution	to 4	11	Storage and distribution	to 4 (max. 235 m²)

Table 5.1 continued

England and Wales (Town and Country Planning (Use Classes) Order 1987) (SI No. 764) (as amended)			Scotland (The Town and Country Planning (Use Classes) (Scotland) Order 1989) (SI N. 147) (as amended)			Northern Ireland (Planning (Use Classes) Order (NI) 1989) (as amended)		
Class	Use	Development permitted by the GDPO (which may be subject to limitations)	Class		Development permitted by the Permitted Development Order	Class		Development permitted
C1	Hotels, boarding houses, and guest houses		12	Hotels and hostels (not including public houses)		12	Guest houses and hostels	to 14
C2	Residential institutions		13	Residential institutions		13	Residential institutions	to 14
C3	Dwelling-houses		14	Houses		14	Dwelling houses	
D1	Non-residential institutions		15	Non-residential institutions		15	Non-residential institutions	
D2	Assembly and leisure		16	Assembly and leisure		16	Assembly and leisure	

Note: The sub-division of residential dwellings into two or more separate dwellings is a change of use

changed from a hairdresser to a funeral parlour or a sweetshop (or vice versa), but it cannot be changed (unless planning permission is obtained) to a restaurant or a hot food take-away, which is in a different class. The classes, it should be stressed, refer only to changes of use, not to any building work, and the Order gives no freedom to change from one class to another. Whether such a change constitutes development depends on whether the change is 'material'.

The General Permitted Development Order gives the developer a little more freedom by listing classes of 'permitted development' – or, to be more precise, it gives advance general permission for certain classes of development, typically of a minor character.[1] If a proposed development falls within these classes, no application for planning permission is necessary: the GPDO itself constitutes the permission. The Order includes minor alterations to residential buildings, and the erection of certain agricultural buildings (other than dwelling-houses). A summary of permitted development rights is given in Box 5.1. The GPDO also permits certain changes of use within the UCO, such as a change from an A3 use (the food and drink class) to an A1 use (shop), but not – because of the possible environmental implications – the other way round. While the use changes allowed by the UCO are all 'bilateral' (any change of use within a class is reversible without constituting development), the GPDO builds upon this structure by specifying a number of 'unilateral' changes *between* classes for which permission is not required. The rationale here is that the permitted changes generally constitute an environmental improvement. The rights given by the GPDO can be withdrawn by Article 4 directions and conditions on planning permissions (discussed on p. 128).

The cynic may be forgiven perhaps for commenting that the freedom given by the UCO and the GPDO is so hedged by restrictions, and frequently so difficult to comprehend, that it would be safer to assume that any operation constitutes development and requires planning permission (though it may be noted with relief that painting is not normally subject to control, unless it is 'for purpose of advertisement, announcement or direction'). The legislators have been helpful here. Application can be made to the LPA for a

certificate of lawfulness of a proposed use or development (CLOPUD). This enables a developer to ascertain whether or not planning permission is required.

The Orders were modified by the Conservative government of 1979–97 as part of its policy of 'lifting the burden' on business. Some of the changes have proved to be very controversial. For example, it is now allowable to change a restaurant to a fast hot-food take-away, or to change a public house to more profitable uses such as professional offices and other uses appropriate to a shopping area. The impact of these changes on such matters as local amenity and traffic generation is detailed in a 1992 report by Sandra Bell for the London Boroughs Association (LBA). The government response to the representations of local government is set out in the report: it is maintained that 'the advantages of the present arrangements in terms of the certainty and flexibility they provide for the commercial sector, and the reduction in intervention and bureaucracy, far outweigh the disadvantages'. A clearer illustration of the political nature of planning would be difficult to find (though more persuasive are the photographs in the LBA report depicting the detrimental effects of 'change of use').

VARIATIONS IN NORTHERN IRELAND AND SCOTLAND

Development control operates in a similar way across the whole of the UK, although it is established by separate law and policy in Northern Ireland and Scotland, and there is a separate 'national' policy context in Wales. A comparison of the Use Classes Orders shows minor variations of the kind that exist in other parts of the system. The most important difference is that in Northern Ireland development control is operated by the Planning Service, an executive agency of the Department of Environment for Northern Ireland, which operates through six divisional planning offices. Local authorities in Northern Ireland have only a consultative role, and planning applications are made to the Department (Trimbos 1997). The Planning Service makes recom-

---

BOX 5.1 A SUMMARY OF PERMITTED DEVELOPMENT RIGHTS IN ENGLAND

Permitted development rights are granted by the General Permitted Development Order 1995 (GPDO). The Order grants planning permission for certain minor forms of development which are listed in Schedule 2. The permissions can be withdrawn by Article 4 directions or conditions attached to planning permissions. The application of the Order is complex and this is only a brief summary.

- Development within the cartilage of a dwelling house, limited to 10 per cent of the cubic content of terraced houses and 15 per cent of detached houses, and an overall maximum of 115 m^3
- Minor operations such as painting and erection of walls and fences but not over 2m in height
- Temporary buildings and uses in connection with construction, and temporary mineral exploration works
- Caravan sites for seasonal and agricultural work

- Agricultural and forestry buildings and operations (although the local planning authority must be notified in certain circumstances)
- Extension of industrial and warehouse development up to 25 per cent of the cubic content of the original building
- Repairs to private driveways and services provided by statutory undertakers and local authorities (including sewerage, drainage, postboxes), maintenance and improvement works to highways by the highway authority
- Limited development by the local authority such as bus shelters and street furniture
- Certain telecommunications apparatus not exceeding 15 m height, and closed circuit television cameras, subject to limitations
- Restoration of historic buildings and monuments
- Limited demolition works

---

mendations to the local district councils, which can request the Service to reconsider. The Service may reconsider the application but, if there is no agreement, the matter is referred to the Chief Executive's Office and a decision is made by the Management Board (senior civil servants). Appeals in Northern Ireland are heard by the Planning Appeals Commission; this is an independent body whose members are appointed by the Secretary of State for Northern Ireland. The Commission also hears inquiries into major planning applications and development plans.

The provisions of the 1991 Planning and Compensation Act (which have made important changes to the procedure for enforcement, the control of demolition, and the relationship between plans and decisions) have not been implemented in Northern Ireland. Another variation is in neighbour notification, where Northern Ireland has had a more thorough system. This is guided by a non-statutory notification scheme requiring, for example, advertisement of all applications.

Scotland's legislation also differs in this respect, and requires the applicant (rather than the local authority) to serve notices on neighbours. Neighbour notification is discussed further in Chapter 12.

SPECIAL DEVELOPMENT ORDERS

While the GPDO is applicable generally, special development orders (SDOs) relate to particular areas or particular types of development. SDOs (like other Orders) are subject to parliamentary debate and annulment by resolution of either House. They have provided an opportunity for testing opinion on controversial proposals such as the reprocessing of nuclear fuels at Windscale, but most of the nineteen SDOs made in England and Wales were to facilitate the operation of UDCs. In these cases the order granted permission for development that was proposed by the corporations and approved by the Secretary of State.

The use of the SDO procedure raises considerable controversy since it involves a high degree of central involvement in local planning decisions. One very contentious case was the granting of permission for over a million square feet of offices and homes at the eastern end of Vauxhall Bridge in London. At that time the DoE said that 'the purpose of making fuller use of SDOs would not be to make any general relaxation in development control, but to stimulate planned development in acceptable locations, and speed up the planning process' (Thornley 1993: 163). In practice, central government has not made use of the orders in recent years and has instead opted for other means to shape major decisions.

WITHDRAWAL OF PERMITTED DEVELOPMENT RIGHTS

The development rights that are permitted by the GPDO can be withdrawn by a Direction made under Article 4 of the Order (and hence known as *Article 4 Directions*). The effect of such a direction is not to prohibit development, but to require that a planning application is made for development proposals in a particular location. The direction can apply either to a particular area (such as a conservation area) or, unusually, to a particular type of development (such as caravan sites) throughout a local authority area. Since the direction involves taking away a legal right, compensation may be payable. Article 4 Directions should be made only in exceptional circumstances and where there is 'a real and specific threat'.[2]

The most common use of an Article 4 Direction is in areas where special protection is considered desirable, as with a dwelling-house in a rural area of exceptional beauty, a national park, or a conservation area. Without the direction, an extension of the house would be permitted up to the limits specified in the GPDO. The majority of Article 4 Directions in fact relate to 'householder' rights in conservation areas. They are also used in national parks and other designated areas to control temporary uses of land (such as camping and caravanning) which would otherwise be permitted (Roger Tym & Partners 1995a).

THE PLANNING APPLICATION PROCESS

All planning authorities provide guides on the planning application process and readers should make reference to them for the finer points. For many minor applications it is a straightforward process, but in some cases it can become very complex. Figure 5.1 gives an overview of the process in England, and it is much the same elsewhere. Many applications will begin with pre-application discussions with the local authority. It is especially important for the local authority to ensure that the application is complete and meets its requirements so that there is minimum delay in processing. On receipt of the application and fee the authority will acknowledge receipt and begin publicity, notification, and consultation procedures, all of which will vary depending on the type of application. Many applications will require other consents from the authority and other agencies, notably building regulations approval, and these may be coordinated by the authority in the 'one-stop shop approach', as discussed on p. 152. The extent and type of public involvement in the application and appeal procedure are reviewed in Chapter 12.

On the basis of consultation returns, the relevance of national and local policies, previous decisions, and a site visit, the planning officer will make a report to the planning committee with a recommendation on the decision to be made.[3] This report, along with the committee agenda and minutes and consultation returns, are public documents. The applicant may be able to make a presentation to committee but this is at the discretion of the authority. Decision notices are sent to the applicant, who can appeal against refusal or conditions imposed. Most applications in many authorities will be decided by the planning officer under delegated powers, subject to meeting criteria such as the application being in accordance with development plan policies and below certain thresholds. When elected members consider applications they may not always agree with officers and there are some celebrated cases where members have decided applications against the advice of their officers, such as the cases of Ceredigion and North Cornwall.[4] This is one category of cases that has led to planning authorities' decisions being subject to judicial review (discussed on p. 47).

More complex applications will require negotiations between the applicant or agent and the authority. The officer will be seeking to ensure that the application meets policy and will be working from past experience of committee decisions. The discretionary nature of the British planning system allows for negotiation prior to the final decision. In theory this offers scope to ensure that the final development is closer to meeting the needs of all parties, so long as officers and applicant recognise the benefits of negotiation to achieve better outcomes (Claydon 1998). In practice it appears that local authority officers are less prepared to make good use of the opportunity of negotiation than developers (Claydon and Smith 1997).

In complicated cases it is sometimes convenient for an applicant or the LPA (or both) to deal with an application in outline. *Outline planning permission* gives the applicant permission in principle to carry out development subject to *reserved matters* which will be decided at a later stage. This is a useful device to enable a developer to proceed with the preparation of detailed plans with the security that they will be not be opposed in principle. In a few cases there will need to be an environmental impact assessment – the procedures for which are described in Chapter 7.

THE DEVELOPMENT PLAN IN THE DETERMINATION OF PLANNING APPLICATIONS

Crucial to the development control process is the concept of *material considerations*. These are exactly what the term suggests: considerations that are material to the taking of a development control decision. The primary consideration is the development plan.[5] The 1990 Act states that 'where in making any determination under the planning Acts regard is to be had to the development plan, the determination shall be made in accordance with the plan unless material considerations indicate otherwise' (s. 54A).

The introduction of these words in section 54A of

the Act signalled the government's intention to move to a *plan-led system* of development control.[6] This may sound strange to those new to planning, and it may be appropriate to ask if there can be any other sort of planning system. But the status of the plan in development control prior to 1991 was more ambiguous and most of the country did not have a local plan in place (as explained in Chapter 4). The implications of s. 54A have been the subject of much debate. In the light of experience, government has clarified the official meaning in the revisions of PPG 1 in 1992 and 1997. The current guidance is given in Box 5.2.

Section 54A certainly had a major impact on the planning system. There is much more emphasis on the

---

BOX 5.2 THE MEANING OF SECTION 54A OF THE 1990 ACT

The Government is committed to a plan-led system of development control. This is given statutory force by section 54A of the 1990 Act. Where an adopted or approved development plan contains relevant policies, section 54A requires that an application for planning permission or an appeal shall be determined in accordance with the plan unless material considerations indicate otherwise. Conversely, applications which are not in accordance with relevant policies in the plan should not be allowed unless material considerations justify granting a planning permission. Those deciding such planning applications or appeals should always take into account whether the proposed development would cause demonstrable harm to an interest of acknowledged importance. In all cases where the development plan is relevant, it will be necessary to decide whether the proposal is in accordance with the plan and then to take into account other material considerations.

Source: Planning Policy Guidance Note 1 (1997: para. 40)

development plan, and the parallel requirement for the preparation of statutory local plans as well as UDPs or structure plans for all areas has attempted to ensure that there is an adequate framework of policy against which to test applications. Planning committee reports now tend to emphasise more strongly the relevant policies of the development plan in relation to each application.

The 'presumption in favour of development' dating back to the beginnings of planning control (M. Harrison 1992) has effectively been changed to a presumption in favour of the development plan, or more accurately, in the words of Malcolm Grant (1997),

> it is if anything, a presumption *in favour* of development that accords with the plan; and a presumption *against* development that does not. In each case, the development plan is the starting point, and its provisions prevail until material considerations indicate otherwise.
>
> (*Encyclopedia* P54A.07; emphasis in original)

But in most cases other material considerations will also play a part in the decision, and this has always been the case. Whether or not other material considerations outweigh the development plan is a matter of judgement for the decision-makers.[7] Even with much more comprehensive plan coverage, many issues raised by planning applications will not be addressed in policy, and there is a limit to which governments at any level can, or wish to, commit policies to paper. The more this is done, the more inflexible will planning become, the less will it be able to adapt to changing circumstances, the greater is the likelihood of conflict between policies, and the more confusing the situation will be. The merit of the British development control system is the discretion given to decision-makers at the point of considering a planning application. Thus it has flexibility and adaptability to differing and changing situations.[8]

Legal niceties aside, how do planning authorities actually decide planning applications? What if there is no adopted local plan (which in 1999 was still the case for one-third of all planning authorities)? What if the local plan does not address an issue raised by a planning application? Since the early 1980s there has

been a blossoming of research which considerably increases our understanding of how this part of the machinery of planning works. Findings from research in the 1980s on the role of development plans is still relevant. Davies *et al.* (1986a) found from a major review of decisions that development plans were used in development control, but that many considerations were not covered by the plans. Even when there were relevant policies, they were typically expressed in general terms and needed 'translation' into operational terms for each development application. Non-statutory documents, including development control policy notes, design guides, development briefs, informal local plans, and 'policy frameworks', were also important. With some caveats, notably the more comprehensive nature of many plans in the 1990s, these findings are probably no less valid today.

The same authors also examined the role of plans in appeals. This revealed that 'one third of all appeal decision letters contained no mention of policy, national or local, statutory or non-statutory', but these included a high proportion of the householder appeals and comparatively few major proposals. By contrast, two-thirds of the appeal decision letters referred to some form of policy, and in these cases the policy usually became the main determinant of the outcome of the appeal: a high proportion were rejected. 'Inspectors nearly always dismissed appeals, and supported the local authority, on proposals for which there was relevant cover in the development plan.' On the other hand, they 'more often allowed appeals which turned on practical appeal considerations lacking firm local policy coverage, but in which national policies were invoked in favour of the appellant'. There is an important message here: where there is an articulated policy which can clearly be applied to a case, development control and appeal decisions tend to abide by it – and this was well before the term 'plan-led system' was coined. It suggests that the mandatory requirement for local plan policy is more important than the introduction of s. 54A.

OTHER MATERIAL CONSIDERATIONS

Since planning is concerned with the use of land, purely personal considerations are not generally material (though they might become so in a finely balanced case). The courts have held that a very wide range of matters can be material. Indeed, in the oft-quoted *Stringer* case, it was stated that 'any consideration which relates to the use and development of land is capable of being a planning consideration' (*Stringer v. Minister of Housing and Local Government* 1971). Whether a particular consideration falling within that broad class in any given case is material will depend on the circumstances. In another important case (*Newbury*), the House of Lords formulated a threefold 'planning test': to be valid, a planning decision had (i) to have a planning purpose; (ii) to relate to the permitted development; and (iii) to be reasonable (*Newbury District Council v. Secretary of State for the Environment* 1981).

These and similar judicial statements provide only the flimsiest of guidelines, and a succession of cases have shown that the issues are legion. The list of possible considerations begins with the siting and appearance of the proposed buildings; the suitability of the site and its accessibility; relationship to traffic and infrastructure provision; landscaping; and the impact on neighbouring land and property. But many other matters may be relevant: environmental impacts; the historical and aesthetic nature of the site; economic and social benefits of the development; considerations of energy and 'sustainable development'; impact on small businesses; previous appeal decisions; the loss of an existing use; whether the development is likely to be carried out; and, in a few cases, financial considerations, including the personal circumstances of occupiers. Whether or not any of these considerations is material depends on the circumstances of each case. Very few considerations have been held by the courts to be immaterial but ones that have include the absence of provision for planning gain; and to make lawful something that is unlawful (Moore 2000: 206).[9]

Given this wide range of possibilities, one looks to 'planning guidance' for a lead. Circulars and planning policy guidance notes (and their equivalents) are always material considerations in making planning decisions. Although they have no formal statutory force, the local planning authority must have regard to them. Where the local authority does not follow national guidance it must give 'clear and convincing reasons'.[10] Changes to national policy that post-date the development plan are particularly important, as in the case of the revisions to PPG 3, *Housing*, made in 2000. Government policy is helpful where there are special requirements. For example, for out-of-town shopping centres it is explicitly advised (in PPG 6, para. 1.16) that 'key considerations should be applied', including the likely impact of the development on the vitality and viability of existing town centres . . . their accessibility by a choice of means of transport; and their 'likely effect on overall travel patterns and car use'. But for many topics guidance can be found to justify alternative positions. Also, planning law is not neatly codified: it lies all over the place, and is not limited to circulars and guidance notes. As long as a minister has stated a policy somewhere, it seems that it is acceptable as a material consideration: 'The courts have accepted that policies should take into account white papers, circulars, policy guidance, previous decisions, written parliamentary answers, and even after dinner speeches. There is an increasing tendency to announce policy at professional conferences' (Read and Wood 1994: 13).

In short, there are few limits to the matters which can be regarded as material to a planning decision, and few clear (but many unclear) guidelines. Two considerations warrant further discussion: the design and appearance of development, and amenity.

GOOD DESIGN

Much of the built heritage is worth preserving because it is well designed. It is therefore of more than contemporary concern that new buildings should be well designed. Nevertheless, the extent to which 'good' design can be fostered by the planning system (or any other system) is problematic.

Good design is an elusive quality which cannot easily be defined. In Holford's words (1953),

> design cannot be taught by correspondence; words are inadequate, and being inadequate may then become misleading, or even dangerous. For the competent designer a handbook on design is unnecessary, and for the incompetent it is almost useless as a medium of instruction.

Yet local authorities have to pass judgement on the design merits of thousands of planning proposals each year, and there is continuous pressure from professional bodies for higher design standards to be imposed.

There is a long and inconclusive history to design control (well set out by John Punter, in various publications from 1985). A 1959 statement by the MHLG stressed that it was impossible to lay down rules to define good design. Developers were recommended to seek the advice of an architect (presumably a good one!). The policy should be to avoid stifling initiative or experiment in design, but 'shoddy or badly proportioned or out of place designs' should be rejected – with clear reasons being given.

The reader is referred to Punter's work for the fascinating details of the continuing story, recounting the personal achievements of Duncan Sandys, particularly in founding the Civic Trust in 1957, and later in promoting the Civic Amenities Act; the high buildings controversy ('sunlight equals health'); the problem of protecting views of St Paul's Cathedral; the arguments over the Shell Tower (which prompted the quip that the best view of the Shell Tower was to be obtained from its roof); the publication of Worskett's *The Character of Towns* (1969); the unpublished Matthew-Skillington report, 'Promotion of high standards of architectural design', which led to the appointment of a Chief Architect in the Property Services Agency; the property boom and a spate of books bearing titles such as *The Rape of Britain* (Amery and Cruikshank 1975) and *The Sack of Bath* (Fergusson 1973); the Essex *Design Guide for Residential Areas* (1973) – 'the most influential local planning authority publication ever'; the attempt (in 1978) to prevent the building of the National Westminster Tower; and so on.

New contexts emerged with the return of a Conservative government in 1979, and there was the unprincely attack in 1984 by Prince Charles on the 'monstrous carbuncle' of the proposed extension to the National Gallery and the 'giant glass stump' of the Mies van der Rohe office block adjacent to the Mansion House (described by one architect – who will not be named – as a building which would be 'unsurpassed in elegance and economy of form'). The Prince followed up his criticisms with *A Vision of Britain* (1989): 'a personal view of architecture' spelling out, with telling illustrations, how 'we can do better'.

In his case study of office development control in Reading, Punter demonstrates the interesting point that it is only since the late 1970s that the local authority 'have begun to influence the full aesthetic impact of office buildings, though they have controlled height, floorspace and functional considerations since 1947'. Moreover:

> Aesthetic considerations do not operate in a vacuum: they are merely one set of considerations amongst many in deciding whether a development gets planning permission. In the case of office development, despite its visual impact, the control of floorspace and the provision of associated facilities and land uses have been higher order goals in Reading. . . . Aesthetic considerations are inevitably the first to be sacrificed in the cause of 'speed and efficiency' in decision-making, by clients, developers, architects and planners . . . There is a lack of design and architectural skills within the control section of the planning authority, but while its presence would strengthen the planning effort, its scope would be severely constrained by wider policy constraints, by general manpower shortages and, most of all perhaps, by the relevant circulars and the appeal process.[11]

The Conservative administration of 1979 started off with a strong bias against design controls, and the views of the Secretary of State (Michael Heseltine), which had been expressed at the Town and Country Planning Summer School, were reproduced in (the now withdrawn) DoE Circular 22/80:

> Far too many of those involved in the system – whether the planning officer or the amateur on the planning committee – have tried to impose their standards quite unnecessarily on what individuals want to do. . . . Democracy as a system of government I will defend

against all comers, but as an arbiter of taste or as a judge of aesthetic or artistic standards it falls short of a far less controlled system of individual, corporate or institutional patronage and initiative.

Mr Heseltine may have been expressing a strong personal view here, but the official policy clearly reflected it.[12] The 1992 version of PPG 1 included an annex on design control (based on a draft prepared jointly by the RIBA and RTPI) and tried to square the circle by advising that, on the one hand, 'the appearance of a proposed development and its relationship to its surroundings are material considerations' but, on the other, that good design 'is primarily the responsibility of designers and their client'. 'Planning authorities should reject obviously poor designs' but [they] 'should not impose their taste on applicants for planning permission simply because they believe it to be superior'. Current guidance in the 1997 PPG 1 includes much the same mixture of statements stressing the importance of good design, supporting the role of design professionals, and accepting the need for planning intervention, if only in limited circumstances. But the balance has swung in favour of intervention. More is made of the role of the development plan and supplementary planning guidance (if subject to public consultation) in justifying control 'to promote or reinforce local distinctiveness', but 'local planning authorities should not concern themselves with matters of detailed design, except where such matters have a significant effect on the character or quality of an area' (para. 18). There is more than a hint of the government trying to please everyone here.

Meanwhile, public outcry over the poor quality of much that passes for urban design continues, including the relatively weak contribution to sustainable development. A large number of public, private, and voluntary bodies have joined in the call for improved standards of design. An *Urban Design Campaign*, spearheaded by the enthusiastic John Gummer, Secretary of State for the Environment, had a high profile during his term of office,[13] though it is only one of many attempts to arouse a greater awareness of, and concern for, design issues. Others include design advice issued by government departments and local authorities, the CPRE's *Local Attraction* campaign, and the efforts of

bodies ranging from the Countryside Commission and the Royal Fine Art Commission and Commission for Architecture and the Built Environment, to Common Ground, and the interesting discussion on the concept of 'local distinctiveness'. Some very attractive publications have emerged as a result of this interest. Whether there has been a parallel emergence of better-designed buildings is an open question. The latest attempt to improve practice comes from the DETR's own *By Design: Urban Design in the Planning System – Towards Better Practice*. In view of the mixed messages in PPG 1, how frustrating it must be for local authority planners and design professionals to read in the first paragraph of this report that 'Everything hangs on how well a local authority draws up, and then uses, the tools it has available to foster better urban design.'

Many of the more difficult decisions are made by inspectors. Durrant's (2000) explanation of the reasoning that an inspector makes in cases of dispute over quality of design reveals the very subjective nature of the task – in his case including an example of allowing a twenty-storey 'glass mountain' adjacent to a grade 1 listed parish church on the south bank of the Thames at Battersea. Durrant argues that the reasoning process has two principal components: the context (both aesthetic and functional) and the scale of buildings; but at the appeal stage the options available to the inspector decision are really only yes or no. It must be in the development process that most improvements can be made to the quality of proposals. Recent government statements are therefore welcome in that they are directed equally to developers:

> We are not going to beat about the bush. When applying for planning permission, house builders will have to demonstrate to local planning authorities how they have taken the need for good design into account.
>
> The point is that good urban design is not just about aesthetics. It concerns the quality of life people experience. For example, it can help prevent crime and the fear of crime. It can help create a sense of community. It is not trite to say that good urban design helps make good places and satisfied people. It will help us put land to better use, because wasting land in the towns means more land lost in the countryside.
>
> (The Minister for Housing and Planning, Nick Raynsford, in a speech to the House Builders Federation, 27 January 2000)

Although the statement is welcome, it should be noted that good design is a necessary but not sufficient condition in achieving the minister's wider aspirations. Findings from Alice Coleman's *Design Improvement Controlled Experiment* should temper design ambitions. This experiment began in 1989 and concerned the radical improvement to the design of local authority housing estates. Design improvements were made to seven estates, including replacing open courtyards with more private gardens and removing overhead walkways. The design improvements may have been welcome in themselves, but an evaluation by Price Waterhouse published in 1997 found that the social and sustainability objectives were not met. Indeed, this sort of investment performed less well than Estate Action.[14]

Nevertheless, for Punter, 'design issues occupy a more important position in contemporary planning practice today than at any stage over the last 50 years' (1999: 151). Carmona's survey of residential design guidance (1998 and 1999) shows that most authorities are making efforts to improve the quality of design, although practice 'remains varied in the extreme'. About half of all planning authorities have at least three forms of design guidance, often linked in hierarchical fashion from strategic through local to site specific. 'Together, the evidence illustrates a strong belief in the value of pre-conceived prescription as the basis for controlling residential design, but tremendous variety – and therefore inconsistency – in the chosen approaches used to prescribe that design' (Carmona 1999: 36).

More disappointing are the attempts to achieve more sustainable solutions in residential development (only 4 per cent in Carmona's survey), with most authorities using only easily measurable design criteria, repeating national guidance, or concentrating on the impacts of the car.

AMENITY

'Amenity' is one of the key concepts in British town and country planning: yet nowhere in the legislation is it defined. The legislation merely states that 'if it appears to a local planning authority that it is expedient in the interests of amenity', the authority may take certain action, in relation, for example, to unsightly neglected waste land or to the preservation of trees. It is also one of the factors that may need to be taken into account in controlling advertisements and in determining whether a discontinuance order should be made. It is a term widely used in planning refusals and appeals: indeed, the phrase 'injurious to the interests of amenity' has become part of the stock-in-trade jargon of the planning world. But like the proverbial elephant, amenity is easier to recognise than to define, and there is considerable scope for disagreement on the degree and importance of amenities: which amenities should be preserved, in what way they should be preserved, and how much expense (public or private) is justified.

The problem is relatively straightforward in so far as trees are concerned. It is much more acute, for example, in connection with electricity pylons, yet the electricity generating and supply companies are specifically charged not only with maintaining an efficient and coordinated supply of electricity but also with the preservation of amenity. Here the question is not merely one of sensitivity but also of the additional cost of preserving amenities by placing cables underground.

Apart from problems of cost, there is the problem of determining how much control the public will accept. Poor architecture, ill-conceived schemes, and mock-Tudor frontages may upset the planning officer, but how much regulation of this type of 'amenity-injury' will be publicly acceptable? And how far can negative controls succeed in raising public standards? Here emphasis has been laid on design bulletins, design awards, and such ventures as those of the Civic Trust, a body whose object is 'to promote beauty and fight ugliness in town, village and countryside'. Nevertheless, LPAs have power not only to prevent developments which would clash with amenity (for example, the siting of a repair garage in a residential area) but also to reject badly designed developments which are not intrinsically harmful. Indeed, outline planning permission for a proposal is often given on the condition that detailed plans and appearances meet the approval of the authority.

CONDITIONAL PERMISSIONS

A local planning authority can grant planning permission subject to conditions, and almost all permissions are conditional. This can be a very useful way of permitting development which would otherwise be undesirable. Many conditions are simple, requiring, for example, that materials to be used are agreed with the local authority before development starts. But there are many more complex permutations. Thus a service garage may be approved in a residential area on condition that the hours of business are limited. Residential development may be permitted on condition that landscape works are carried out in accordance with submitted plans and before the houses are occupied. Office development may be permitted subject to means of access for people with disabilities being agreed with the local authority, and so on.

The power to impose conditions is a very wide one. The legislation allows LPAs to grant permission subject to 'such conditions as they think fit'. However, this does not mean 'as they please'. The conditions must be appropriate from a planning point of view:

> the planning authority are not at liberty to use their power for an ulterior object, however desirable that object may seem to them to be in the public interest. If they mistake or misuse their powers, however *bona fide*, the court can interfere by declaration and injunction.
>
> (*Pyx Granite Co Ltd v. Minister of Housing and Local Government* 1981)

DoE Circular 11/95, *The Use of Conditions in Planning Permissions*,[15] stresses (in para. 2) that

> If used properly, conditions can enhance the quality of development and enable many development proposals to proceed where it would otherwise have been necessary to refuse planning permission. The objectives of planning, however, are best served when that power is exercised in such a way that conditions are clearly seen to be fair, reasonable and practicable.

As might be expected, there is considerable debate on the meaning of these terms. Circular 11/95[16] elaborates specifically on the meaning of six tests: conditions should be necessary; relevant to planning; relevant to the development to be permitted; enforceable; precise; and reasonable in all other respects. Numerous court

judgements provide guidance on how the tests should be applied. For a development to meet the test of being necessary, the local authority should ask whether permission would be refused if the condition were not imposed. Relevance to planning and to the development may be difficult to judge. Though planning conditions should not be used where they duplicate other controls such as those of pollution control, they may be needed if the other method of regulation does not secure planning objectives. Conditions should not be imposed on one site to seek to improve conditions on a neighbouring site – for example, where existing car parking is insufficient. But it may be appropriate to impose conditions to address problems elsewhere as a result of the new development – for example, increasing congestion on another part of the site. In any event, it is permissible to impose conditions only on the use of land under the control of the applicant. The enforceability test requires that the local planning authority should be able to monitor and detect whether the applicant is complying with it. Enforceability is also closely related to precision in drafting of the condition. Both the authority and the applicant need to be able to understand exactly what is required by a condition. An appendix to the Circular gives numerous examples of how conditions should be drafted so as to avoid vagueness and ensure clarity.

The reasonableness test requires that the condition is not unduly restrictive. In particular, it should not nullify the benefit of the permission. A condition may also be unreasonable if it is not within the powers of the applicant to implement it, for example where it relates to land in the ownership of a third party. A striking example of a condition which was quite unreasonable was dealt with in the *Newbury* case. There the district council gave permission for the use of two former aircraft hangars for storage, subject to the condition that they be demolished after a period of ten years. The House of Lords held that since there was no connection between the proposed use and the condition, it was *ultra vires*. In granting permission for development at Aberdeen Airport the planning authority sought to impose a number of conditions to minimise the impact on the local area. One condition restricted the direction of take-off and landing of aircraft, but this was found to be both unreasonable and unnecessary since the Civil Aviation Authority (and not the airport) controls flight paths (McAllister and McMaster 1994: 136–7).

Up to 1968, conditions were also imposed to give a time limit within which development had to take place. The 1968 Act, however, made all planning permissions subject to a condition that development is commenced within five years. If the work is not begun within this time limit, the permission lapses. However, the Secretary of State or the local planning authority can vary the period, and there is no bar to the renewal of permission after that period has elapsed (whether it be more or less than five years).

The purpose of this provision is to prevent the accumulation of unused permissions and to discourage the speculative land hoarder. Accumulated unused permissions could constitute a difficult problem for some LPAs: they create uncertainty and could make an authority reluctant to grant further permissions, which might result in, for example, too great a strain on public services.[17] The provision relates, however, only to the beginning of development, and this has in the past been deemed to include digging a trench or putting a peg in the ground.[18]

As well as imposing conditions, the local authority may also reach agreement with the developer about planning obligations or 'planning gain', where the developer pays for related works without which planning permission could not be granted. Agreements typically cover the provision of infrastructure such as traffic management and access, public open spaces, and other improvements as 'compensation' for loss through development, a proportion of affordable housing in residential schemes, and even commuted payments to support public transport serving the development. The government emphasises that planning conditions should be used in preference to planning obligations, but for planning authorities the obligations are more important in larger-scale developments. A fuller explanation of obligations is given in Chapter 6. Here it should be noted that planning conditions should not duplicate obligations, and permission cannot be granted on condition that an obligation is entered into. While a developer can appeal against planning

conditions, there is no such possibility for obligations which are entered into 'voluntarily'.

FEES FOR PLANNING APPLICATIONS

Fees for planning applications were introduced in 1980. This represented a marked break with planning traditions, which had held (at least implicitly) that development control is of general communal benefit and directly analogous to other forms of public control for which no charges are made to individuals. The Thatcher administration had a very different view; to quote from the parliamentary debate:

> We do not believe that the community as a whole should continue to pay for all sorts of things that it has paid for in the past. . . . In the general review that has taken place to see where we can reduce spending from the public purse . . . we came to the conclusion that the cost of development control was an area where some part of the cost should be recovered.

The 1980 Bill provided additionally for fees for appeals but this provision was dropped in the face of widespread objections from both sides of the House.

The current regulations were made in 1989 and amended in 1991 and 1992. The fee structure is subject to change over time, and a detailed schedule is therefore not appropriate.[19] The government's ultimate aim is to recover the full administrative costs of dealing with planning applications. The power to charge fees for planning applications does not extend to the pre-application discussion stage. A decision of the Court of Appeal allowing such charges was reversed by the House of Lords in 1991.

PLANNING APPEALS

An unsuccessful planning applicant can appeal to the Secretary of State, and a large number do so. Appeals are allowed on the refusal of planning permission, against conditions attached to a permission, where a planning authority has failed to give a decision within the prescribed period, on enforcement notices, and on

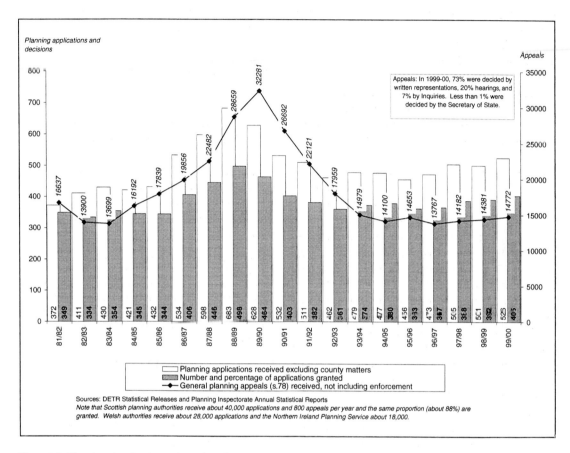

Figure 5.2 Planning Applications, Appeals and Decisions in England 1981–82 to 1999–2000

other matters as discussed below. Appeals decided during 1998/99 (England and Wales) numbered 12,877, of which about a third were allowed. Figure 5.2 illustrates trends in the number of appeals.

Although (in England) the appeal is made to the Secretary of State, the vast majority of appeals are considered by inspectors 'standing in the Secretary of State's shoes'. The same applies in the other countries of the UK, but the Welsh Assembly is so far unique in establishing a cross-party Planning Decision Committee with four members to make the final decision on important appeals and called-in applications (see p. 51).[20] Until 1969 the ministry responsible for planning dealt with all appeals.[21] In view of increasing

delay in reaching decisions and the huge administrative burden, the 1968 Planning Act introduced a system whereby decisions on certain classes of appeal were 'transferred' to professional planning inspectors who had previously only made recommendations to the minister. Over time the range of decisions transferred to inspectors has been extended such that virtually all are now decided by the Planning Inspectorate. Matters of major importance may be 'recovered' for determination by the Secretary of State. In fact, fewer than 1 per cent of all appeals are recovered, although it can be argued that the significance is much greater than the figure suggests. Even where decisions are recovered, it is the senior civil servants in the Department rather than the minister who make most decisions.[22]

Wide powers are available to the Secretary of State and inspectors. These include the reversal of a local authority's decision or the addition, deletion, or modification of conditions. The conditions can be made more onerous or, in an extreme case, the Secretary of State may even go to the extent of refusing planning permission altogether if it is decided that the local authority should not have granted it with the conditions imposed.

Before reaching any decision, the inspector or Secretary of State needs to consider the evidence, and this can be done in three ways: by inquiry, hearing, or written representation. Most appeals are considered by written representation with 73.1 per cent of all planning appeals in England in 1998–99 falling into this category, while hearings account for 19.1 per cent and inquiries 7.8 per cent. The procedures are governed by the rules of natural justice and by inquiry procedure rules which have recently been updated in England.[23] Both the applicant and the planning authority have the right to demand a full inquiry if they so wish, but the emphasis over recent years has been to get as many appeals as possible heard by the other two less expensive and time-consuming methods. The efficiency of procedures leading up to and during inquiries has been strongly criticised (Graves *et al.* 1996; O'Neill 1999).[24] Since 1968 the Secretary of State has had power to refer development proposals of a far-reaching or novel character to an *ad hoc* Planning Inquiry Commission, but this power has never been used. Proposals are in hand for a new approach to dealing with major projects of national significance. These are discussed in Chapter 12 in the context of the history of public involvement in planning through inquiries. In essence, it is proposed to strengthen the national policy context in relation to major projects and to create new parliamentary procedures for approving projects in principle before consideration of the detailed issues through the development control process.

Inquiries are *adversarial* debates conducted through the presentation and questioning (cross-examination) of evidence. The proceedings are managed by inspectors, but advocates, often barristers, play a dominant role in the proceedings, thus lending a courtroom atmosphere. Such an approach has benefits in safeguarding the principles of open, impartial, and fair consideration of the issues. Nevertheless, it is widely acknowledged as unnecessary for certain less complex appeals, especially where one party is not professionally represented. Thus, the hearing procedure has been created, which proceeds in an *inquisitorial* way, with the inspector playing an active role in structuring a round-table discussion and asking questions, but with no formal cross-examination. But the most popular and straightforward procedure is through 'written reps'.

Over the years the mechanisms for considering appeals have been streamlined. The substantial increase in the number of appeals in the late 1980s quickly had the effect of slowing down the procedures, ironically at a time when the government was seeking to 'lift the burden' of planning regulation. Reviews of the system were undertaken to tackle this embarrassment for government, the first of which was completed in 1985 and published together with an 'action plan' in 1986. Rules were introduced to govern the written representation procedure in a similar way to the rules for inquiries, which were also strengthened. Further minor changes were made in 1992 and further substantial revisions in 2000, aimed at speeding up the process and to provide statutory rules for the hearings process.[25]

The rules govern the exchange of information among the parties to the appeal and set a timetable for this to happen. The latest amendments are intended to ensure that the appeal processes follow the predetermined timetable more often than is the case now. Sanctions have been strengthened such that evidence may be disregarded unless it is submitted on time, or in some cases to impose costs on the guilty party (except for written representations). There is a stronger emphasis on the appellant and local authority agreeing the matters in dispute beforehand, keeping evidence concise,[26] and inspectors are encouraged to take more control over unnecessary cross-examination. The revisions also include an amendment to the General Developmental Procedures Order (GDPO) requiring local planning authorities to specify all relevant development plan policies and proposals when refusing permission or imposing conditions, reflecting the increased emphasis placed on development plans providing a firm basis for planning decisions. Evaluations of more informal methods of holding appeals are generally positive (Stubbs 1999, 2000).

The appeals procedure is a microcosm of the whole planning system. It is where the system and its policies are challenged and where the most contentious and difficult issues are addressed. It is a 'pinch point' of the system. What happens here is critically important for the system as a whole, in terms both of planning policy and how the system should be operated. Although each appeal is considered on its own merits, the cumulative effect is to operationalise policy. It is here that the sometimes vague, sometimes contradictory, messages in government policy must be resolved. The wider effect of appeal decisions may be difficult to assess but clearly they have a very real influence on other decisions made by planning authorities, and are a route for the imposition of central government policy on local authorities. Inspectors pay particular attention to national policy, which is the determining factor in many appeals (Rydin et al. 1990; Wood et al. 1998).[27]

On the operation of the system, recent discussions on the appeal system give a pointer to the government's overall philosophy on decision-making in planning, as explained by Shepley (1999: 403). The approach is one which supports the fundamental principles of openness, fairness, and impartiality but also recognises the need to make decisions more quickly, more cheaply, and earlier in the development process. Experiments have been conducted on 'alternative dispute resolution' in planning through mediation. The intention is that the 'win–lose' style of deciding planning appeals may be replaced with a process that seeks a solution which is acceptable to both parties. Pilot studies involved the Planning Inspectorate providing trained mediators where local authorities suggested cases that might benefit. Evaluation of the pilot found that there is a role for mediation generally for householder applications involving disputes over design or layout. But further use of mediation would bring about only a modest reduction in appeal cases, and some incentive would be needed to establish more use (Welbank et al. 2000). Perhaps the most important finding was that the current planning application regime was unsuitable for householder applications and an alternative is needed.

Another idea under consideration is the 'environmental court', which would 'extend public access to environmental justice' but also involve far-reaching changes to the system – making for example the Planning Inspectorate part of the machinery of the courts rather than government departments.[28] Irrespective of the merits of environmental courts, it is likely that there will have to be some changes to the appeal and enforcement procedures in the light of the 1998 Human Rights Act.

CALL-IN OF PLANNING APPLICATIONS

The power to 'call in' a planning application for decision by the Secretary of State is quite separate from that of determining an appeal against an adverse decision by a local planning authority. The power is not circumscribed: the Secretary of State may call in *any* application. During the year 1998–99, 119 applications in England (of 503,000) were called in. There are no statutory criteria or restrictions, and no prescribed procedures for handling representations from the public, although if either the applicant or the local planning authority so desires, the Secretary of State must hold a hearing or public inquiry. The type of issue likely to warrant use of call-in has been clarified in a recent statement to the House of Commons in answer to a written question:

His policy is to be very selective about calling in planning applications. He will in general, only take this step if planning issues of more than local importance are involved. Such cases may include, for example, those which, in his opinion:

- may conflict with national policies on important matters;
- could have significant effects beyond their immediate locality;

- give rise to substantial regional or national controversy;
- raise significant architectural or urban design issues; or
- may involve the interests of national security or of foreign Governments.

(Hansard, 16 June 1999, col. 138)

To assist Secretaries of State in making these decisions, all applications for development involving a substantial departure from the provisions of a development plan which the planning authority intends to permit must be sent to the Secretary of State together with a statement of the reasons why it wishes to grant the permission. This procedure enables the Secretary of State to decide whether the development is sufficiently important to warrant its being called in. The Secretary of State also makes directions requiring local authorities to consult with him or her on certain types of application, so that consideration can be given to the use of call-in powers. For example, directions have been made requiring local authorities to consult the Secretary of State on leisure, retail, and office uses over 5,000 square metres and which do not conform to the development plan (reduced in 1999 from 10,000 square metres). There is also now a requirement to consult where the authority intends to grant permission for development of playing fields against the advice of the Sports Council. Further powers are available for directions that prevent a planning authority granting permission for a particular application or a class of application; again this may be used to give the Secretary of State time to consider whether the application should be called in.

Certain types of development tend to involve the Secretary of State: new settlements and other very large housing developments, applications involving the green belt, large-scale minerals proposals, and development affecting buildings of national significance are most common. Mineral workings often raise problems of more than local importance, and the national need for particular minerals has to be balanced against planning issues. It is argued that such matters cannot be adequately considered by local planning authorities (which will invariably face massive local opposition), and involve technical considerations requiring expert opinion of a character more easily available to central

government. A large proportion of applications for permission to work minerals have been called in. Furthermore, there is a general direction calling in all applications for the winning and working of ironstone in certain counties where there are large-scale ironstone workings. On important questions of design, the Commission for Architecture and the Built Environment[29] has, in its terms of reference, the power 'to call the attention of any of our departments of state . . . to any project or development which [it considers] may appear to affect amenities of a national or public character'.

Inevitably the Secretary of State has the job of balancing local concerns with national policies and priorities. The Paignton Zoo case illustrates this well. The proposed development included the refurbishment of the zoo, the development of a 6,000 sq. m retail food store, parking spaces for 600 cars, and a petrol filling station. The proposals clearly raised major issues of policy, including those set out in PPG 6 *Town Centres and Retail Development* and PPG 21 *Tourism*. There were several conflicting considerations, including the likely effect of the retail development on the town centre, and the precarious economic position of the zoo, which, so it was argued, was 'likely to close unless it receives a capital injection of the size that only this proposal is likely to provide, thereby causing a loss to the local economy of approximately £6 million per annum and a significant loss of jobs'. The Secretary of State decided that these and other benefits to tourism and the local economy (together with highway improvements) more than outweighed any harm which might be done to the vitality of the town centre, and he therefore granted planning permission. In the words of the decision,

the harm likely to arise from the proposals is less clear cut than the effects that would result from the decline and possible closure of the zoo . . .; the balance of advantage lies in favour of allowing the proposal . . .; the zoo's leading role in the local economy places it in a virtually unique position . . .; [but the decision] should not be regarded as a precedent for other businesses seeking to achieve financial stability

(*JPL* 1995: 657)

ENFORCEMENT OF PLANNING CONTROL

If the machinery of planning control is to be effective, some means of enforcement is essential. Under the prewar system of interim development control there were no such effective means. A developer could go ahead without applying for planning permission, or could even ignore a refusal of permission. The developer took the risk of being compelled to 'undo' the development (for example, demolish a newly built house) when, and if, the planning scheme was not approved, but this was a risk that was usually worth taking. And if the development was inexpensive and lucrative (for example, a petrol station or a greyhound racing track), the risk was virtually no deterrent at all. This flaw in the prewar system was remedied by the strengthening of enforcement provisions.

These are required not only for the obvious purpose of implementing planning policy, but also to ensure that there is continuing public support for, and confidence in, the planning system. To quote PPG 18, *Enforcing Planning Control*,

> The integrity of the development control process depends on the LPA's readiness to take effective action when it is essential. Public acceptance of the development control process is quickly undermined if unauthorised development, which is unacceptable on planning merits, is allowed to proceed without any apparent attempt by the LPA to intervene before serious harm results from it.
>
> (1991: para. 4)

Enforcement provisions were radically changed by the 1991 Planning and Compensation Act following a comprehensive review by Robert Carnwath, QC, published in 1989. Current provisions are summarised in DoE Circular 10/97.[30] The 1991 Act provided a range of tools in addition to the long-standing provision for enforcement notices.

Development undertaken without permission is not an offence in itself, but ignoring an *enforcement notice* is, and there is a maximum fine following conviction of £20,000. (In determining the amount of the fine, the court is required to 'have regard to any financial benefit which has accrued'.) There is a right of appeal against an enforcement notice. An appeal also contains a deemed application for development for which a fee is payable to the planning authority. Appeals can be made on several grounds: for example, that permission ought to be granted, that permission has been granted (e.g. by the GPDO), and that no permission is required. There is also a limited right of appeal on a point of law to the high court. The enforcement appeal procedures in England have been reviewed with a view to improving their efficiency and standardising procedures between these and other planning appeals.[31] Enforcement can be a lengthy process. For example, South Hams District Council issued an enforcement notice in January 1990 for the removal of a house built without consent. In 1993 the owner was fined £300 for breaching the enforcement notice. In 1995 he was jailed for three months for contempt of a court order requiring demolition. He had demolished only the upper storey and grassed over the lower half.[32]

Where it is uncertain whether planning permission is required, an LPA has power to issue a *planning contravention notice*. This enables it to obtain information about a suspected breach of planning control and to seek the cooperation of the person thought to be in breach. If agreement is not forthcoming (whether or not a contravention notice is served), an enforcement notice may be issued, but only 'if it is expedient' to do so 'having regard to the provisions of the development plan and to any other material considerations'. In short, the local authority must be satisfied that enforcement is necessary in the interests of good planning.

In view of the government's commitment to fostering business enterprise (discussed further on p. 29), LPAs are advised in PPG 18 to consider the financial impact on small businesses of conforming with planning requirements. 'Nevertheless, effective action is likely to be the only appropriate remedy if the business activity is causing irreparable harm.' Development 'in breach of planning control' (development carried out without planning permission or without compliance with a planning condition) might be undertaken in good faith, or ignorance. In such a case, application can be made for retrospective permission. It is unlikely that a local authority would grant unconditional permission for a development against which it had served a planning contravention notice, but it might be willing to give conditional approval.

The 1991 Act also introduced a *breach of condition* notice as a remedy for contravention of a planning condition. This is a simple procedure against which there is no appeal, though there may be some legal complexities that will prevent its widespread use (Cocks 1991). Further, there is a new provision enabling an LPA to seek an injunction in the high court or county court to restrain 'any actual or apprehended breach of planning control'. In Scotland the provision is for an interdict by the court of session or the sheriff.

Where there is an urgent need to stop activities that are being carried on in breach of planning control, an LPA can serve a *stop notice*. This is an attempt to prevent delays in the other enforcement procedures (and advantage being taken of these delays) resulting in the local authority being faced with a *fait accompli*. Development carried out in contravention of a stop notice constitutes an offence.

The provisions for enforcement are complex, and the reader is referred to the discussion in chapter 5 of the 1989 Carnwath Report and the relevant circulars. The position can be exacerbated by the lowly esteem in which the enforcement system (and those who staff it) are often held. Several commentators have termed enforcement 'the weakest link in the planning chain', both south and north of the Border. A recent study in Scotland found great variation in the use of enforcement powers. For example, one authority had served 156 planning contravention notices between 1992 and 1996 while another had served none (Edinburgh College of Art *et al.* 1997). Although planning authorities are mostly happy with enforcement powers they are sometimes reluctant to use them.[33] Fortunately, the majority of alleged contraventions of planning control are dealt with satisfactorily and without any recourse to legal action, but the minority have a disproportionate effect on the credibility of the planning process as a whole.

REVOCATION, MODIFICATION, AND DISCONTINUANCE

The powers of development control possessed by local authorities go considerably further than the granting or withholding of planning permission. Authorities can interfere with existing uses and revoke a permission already given, even if the development has actually been carried out.

A *revocation order* or *modification order* is made when the development has not been undertaken (or before a change of use has taken place). The local authority must 'have regard to the development plan and to any other material considerations', and an opposed order has to be confirmed by the Secretary of State. Compensation is payable for abortive expenditure and any loss or damage due to the order. Such orders are rarely made.

Quite distinct from these powers is the much wider power to make a *discontinuance order*. This power is expressed in extremely wide language: an order can be made 'if it appears to a local planning authority that it is expedient in the interests of the proper planning of their area (including the interests of amenity)'. Again confirmation by the Secretary of State is required, and compensation is payable for depreciation, disturbance, and expenses incurred in carrying out the works in compliance with the order. An order will be confirmed only if the case is a strong one. Indeed, cases have established the principle that a stronger case is needed to justify action to bring about the discontinuance of a use than would be needed to warrant a refusal of permission in the first instance.

British planning legislation does not assume that existing non-conforming uses must disappear if planning policy is to be made effective. This may be an avowed policy, but the planning Acts explicitly permit the continuance of existing uses.

DEVELOPMENT BY THE CROWN, GOVERNMENT DEPARTMENTS, AND STATUTORY UNDERTAKERS

The Crown is not bound by statute and thus development by government departments does not require planning permission, but there have been special arrangements for consultations since 1950. Increased public and professional concern about the inadequacy of these led to revised, but still non-statutory, arrangements culminating in DoE Circular 18/84. This asserts clearly that before proceeding with development, government departments will consult LPAs when the proposed development is one for which specific planning permission would, in normal circumstances, be required. In effect, local authorities should treat notification of a development proposal from government departments in the same way as any other application. Where the local authority is against the development the matter is referred to the Secretary of State. Development by private persons on 'Crown land' (that is, land in which there is an interest belonging to Her Majesty or government department) does require planning permission in the normal way, although there are limitations on the ability of the planning authority to enforce in these cases. The government announced in 1992 its intention to bring development by government departments under normal planning control, but so far no action has been taken.

Development undertaken by statutory undertakers is subject to planning control but it is also subject to special planning procedures. Where a development requires the authorisation of a government department (as do developments involving compulsory purchase orders, work requiring loan sanction, and developments on which government grants are paid), the authorisation is usually accompanied by *deemed planning permission*. Much of the regular development of statutory undertakers and local authorities (for example, road works and laying of underground mains and cables) is *permitted development* under the GDO. Statutory undertakers wishing to carry out development which is neither permitted development nor authorised by a government department have to apply for planning permission to the local planning authority in the normal way, but special provisions apply to *operational land*. The original justification for this special position of statutory undertakers was that they are under an obligation to provide services to the public and could not, like a private firm in planning difficulties, go elsewhere.

Planning permission for major infrastructure projects such as railways, light rail systems and bridges was until 1992 granted by Act of Parliament. The Transport and Works Act 1992 provided a new procedure whereby the Secretary of State is able to make *works orders* which among other things will normally include deemed planning permission. Orders are subject to objection and public inquiry if necessary. For works of national significance the works must still be approved by Parliament.

DEVELOPMENT BY LOCAL AUTHORITIES

Until 1992, LPAs were also deemed to have permission for any development which they themselves undertook in their area, as long as it accorded with the provisions of the development plan; otherwise they had to advertise their proposals and invite objections. The only requirement was for the local authority to grant itself permission by resolution. These 'self-donated' planning permissions are problematic: though local authorities are guardians of the local public interest, they can face a conflict of interest in dealing with their own proposals for development. Pragmatic consideration of the merits of a case involving their own role as developers can easily distort a planning judgement. Examples include attempts by authorities to dispose of surplus school playing fields with the benefit of permission for development; and competing applications for superstore development when one of the sites is owned by the authority itself. The local authorities' position was not helped by judgements against them that found many irregularities in the necessary procedures (Moore 2000: 311).

Because of these difficulties, new regulations were issued in 1992 which require LPAs to make planning applications in the same way as other applicants, and generally follow the same procedures including publicity and consultation. There must be safeguards to ensure that decisions are not made by members or officers who are involved in the management of the land or property, and the planning permission cannot pass to subsequent land and property owners. Where other interests propose development on local authority-owned land they must apply for permission in the normal way. The new procedures did not go as far as some had hoped and criticism continues, and inevitably so since the accusation of bias is always possible while local authorities are able to grant themselves planning permission. The Scottish Local Government Ombudsman has for long complained about 'the ease with which planning authorities breach their own plans, particularly considering the time, effort, and consultation which goes into them'. One solution would be for the Secretary of State to play a role in all applications in which the local authority has an interest (as proposed by the Nolan Committee on Standards of Conduct in Local Government).

CONTROL OF ADVERTISEMENTS

The need to control advertisements has long been accepted. Indeed, the first Advertisements Regulation Act of 1907 antedated by two years the first Planning Act. But, even when amended and extended (in 1925 and 1932), the control was quite inadequate. Not only were the powers permissive: they were also limited. For instance, under the 1932 Act the right of appeal (on the ground that an advertisement did not injure the amenities of the area) was to the magistrates' court – hardly an appropriate body for such a purpose. The 1947 Act set out to remedy the deficiencies. There are, however, particular difficulties in establishing a legal code for the control of advertisements. Advertisements may range in size from a small window notice to a massive hoarding; they may be in the form of a poster, a balloon, or even lasers; they vary in purpose from a bus stop sign to a demand to buy a certain make of detergent; they could be situated alongside a cathedral, in a busy shopping street, or in a particularly beautiful rural setting; they might be pleasant or obnoxious to look at; they might be temporary or permanent; and so on. The task of devising a code which takes all the relevant factors into account and, at the same time, achieves a balance between the conflicting interests of legitimate advertising and 'amenity' presents real problems. Advertisers themselves frequently complain

that decisions in apparently similar cases have not been consistent with each other. The official departmental view is that no case is exactly like another, and hard and fast rules cannot be applied: each case has to be considered on its individual merits in the light of the tests of amenity and – the other factor to be taken into account – public safety.[34]

The control of advertisements is exercised by regulations[35] which are explained in PPG 19: *Outdoor Advertising Control*. The Secretary of State has very wide powers of making regulations 'in the interests of amenity or public safety'. The question of public safety is rather simpler than that of amenity, though there is ample scope for disagreement: the relevant issue is whether an advertisement is likely to cause danger to road users, and also to 'any person who may use any road, railway, waterway (including coastal waters), docks, harbour or airfield'. In particular, account has to be taken of the likelihood of whether an advertisement 'is likely to obscure, or hinder the ready interpretation of, any road traffic sign, railway signal, or aid to navigation by water or air'. Amenity includes 'the general characteristics of the locality, including the presence of any feature of historic, architectural, cultural or similar interest'. The definition of an advertisement is not quite as complicated as that of development, but it is very wide:

> Advertisement means any word, letter, model, sign, placard, board, notice, awning, blind, device or representation, whether illuminated or not, in the nature of, and employed wholly or partly for the purposes of, advertisement, announcement or direction and . . . includes any hoarding or similar structure used, or designed or adapted for use, and anything else principally used, or designed or adapted principally for use, for the display of advertisements.
> (Town and Country Planning Act 1990 s. 336(1))

It is helpfully added that the definition excludes anything 'employed as a memorial or as a railway signal'. Various classes of advertisement are currently excepted from all control, although the classes are currently under review. They are advertisements displayed on a balloon; on enclosed land; within a building; and on or in a vehicle. Also excepted are traffic signs, election signs, and national flags. As one might expect, there are some interesting refinements of these categories, which can be ignored for present purposes (though we might note, in passing, that a vehicle must be kept moving or, to use the more exact legal language, must be normally employed as a moving vehicle). With these exceptions, no advertisements may be displayed without *consent*. However, certain categories of advertisement can be displayed without *express consent*; so long as the local authority takes no action, they are *deemed* to have received consent. These include bus stop signs and timetables, hotel and inn signs, professional or business plates, 'To Let' and 'For Sale' signs, election notices, statutory advertisements, and traffic signs.

It needs to be stressed that amenity and public safety are the only two criteria for control. The content or subject of an advertisement is not relevant, and a local authority cannot refuse express consent on grounds of morality, offensiveness, or taste. Thus an advertisement which contained the words 'Chish and Fips' was considered by the Secretary of State, on appeal, to be questionable on grounds of taste, but not detrimental to amenity: the appeal was allowed (*JPL* 1959: 736).

If an advertisement displayed with deemed consent becomes unsafe, unsightly or in any way 'a substantial injury to the amenity of the locality or a danger to members of the public', the LPA can serve a *discontinuance order*. There is the normal right of appeal to the Secretary of State. Advertisements displayed with express consent can be subject to revocation or modification, again with the normal rights of appeal.

Complex though this may seem, it is not all that there is to advertisement control. In some areas – for example, conservation areas, national parks, or areas of outstanding natural beauty – it may be desirable to prohibit virtually all advertisements of the poster type and seriously restrict other advertisements, including those normally displayed by the ordinary trader. Accordingly, local planning authorities have power to define *areas of special advertisement control* (ASACs) where very strict controls are operated. Within such areas the general rule is that no advertisement may be displayed; such advertisements as are given express consent are considered as exceptions to this general rule.

These special controls originated primarily from the need to deal with the legacy of advertising hoardings which were such a familiar sight in the 1930s. It is now argued that they are obsolete, and can be replaced by simpler controls. Added justification is given to this argument by the fact that, in 1995, nearly a half of the area of England and Wales had been defined by local planning authorities as being within areas of special control. Consultation papers in 1996 and 1999[36] have argued that many orders were out of date since they no longer corresponded to the current limits of the built environment, while the system was either obscure or widely misunderstood by the public. The 1999 paper proposes to limit ASACs to national parks, AONBs, conservation areas, SSSIs and the Broads. Other changes are proposed in order to update the regulations, to close loopholes, and to reflect developments in the advertising industry: for example, in relation to bringing balloon advertisements under control; to add the flying of the European flag as an exemption from control; and to bring lasers into the meaning of 'advertisement'. Particular attention is being given to fly-posting following research on the subject (Arup Economics and Planning 1999).

CONTROL OF MINERAL WORKING

The reconciliation of economic and amenity interests in mineral working is an obvious matter for planning authorities (MPAs). It would, however, be misleading to give the impression that the function of planning authorities is simply to fight a continual battle for the preservation of amenity. Planning is concerned with competing pressures on land and with the resolution of conflicting demands. Amenity is only one of the factors to be taken into account. The general policy framework for minerals is set out in MPG 1. It is interesting to compare the current policy, as set out in the 1996 MPG, with that of the earlier (1988) version (both of which are illustrated in Box 5.3). The 1996 version places a significantly greater emphasis on conservation and environmental considerations.[37]

Planning powers provide for the making of the essential survey of resources and potentialities, the

---

BOX 5.3 THE OBJECTIVES OF SUSTAINABLE DEVELOPMENT FOR MINERALS PLANNING

The objectives are:

(i)   to conserve minerals as far as possible, whilst ensuring an adequate supply to meet needs;

(ii)   to ensure that the environmental impacts caused by mineral operations and the transport of minerals are kept, as far as possible, to an acceptable minimum;

(iii)   to minimise production of waste and to encourage efficient use of materials, including appropriate use of high quality materials, and recycling of wastes;

(iv)   to encourage sensitive working practices during minerals extraction and to preserve or enhance the overall quality of the environment once extraction has ceased;

(v)   to protect areas of designated landscape or nature conservation from development, other than in exceptional circumstances where it has been demonstrated that development is in the public interest; and

(vi)   to prevent the unnecessary sterilisation of mineral resources.

PREVIOUS STATEMENT OF OBJECTIVES (1988)

(a)   to ensure that the needs of society for minerals are satisfied with due regard to the protection of the environment;

(b)   to ensure that any environmental damage or loss of amenity caused by mineral operations and ancillary operations is kept at an acceptable level;

(c)   to ensure that land taken for mineral operations is reclaimed at the earliest opportunity and is capable of an acceptable use after working has come to an end;

(d)   to prevent the unnecessary sterilisation of mineral resources.

Source: MPG 1 (1996 and 1988)

---

allocation of land in development plans, and the control (by means of planning permission) of mineral workings. The MPA has to assess the amount of land required for mineral working, and this requires an assessment of the future demand likely to be made on production in its area. Obviously, this involves extensive and continuing consultation with mineral operators. All MPAs are now required to prepare *minerals plans* (which may be produced jointly with their waste plan).

Powers to control mineral workings stem from the definition of development, which includes 'the carrying out of . . . mining . . . operations in, on, over or under land'. However, a special form of control is necessary to deal with the unique nature of mineral operations. Unlike other types of development, mining operations are not the means by which a new use comes into being, but a continuing end in themselves, often for a very long time. They do not adapt land for a desired end-use: on the contrary, they are essentially harmful and may make land unfit for any later use. They also have unusual location characteristics: they have to be mined where they exist. For these reasons, the normal planning controls are replaced by a unique set of regulations.

Two major features of the minerals control system are that it takes into account the fact that mineral operations can continue for a long period of time, and the fact that measures are needed to restore that land when operations cease. It is, therefore, necessary for MPAs to have the power to review and modify permissions and to require restoration. Under current legislation, MPAs have a duty to review all mineral sites in their areas. This includes those which were 'grandfathered' in by the 1947 Act. These old sites, of which there may be around a thousand in England and Wales, often lack adequate records. They present the particular problem that they can include large un-worked extensions which are covered by the permission; if worked, these could have serious adverse effects on the environment. The provisions relating to these sites are even more complicated than those pertaining to the generality of mineral operations, and they have been significantly altered by the 1995 Environment Act. Details are set out in MPG 14.

Policies for restoration (and what the Act quaintly calls 'aftercare') have become progressively more stringent, mainly in response to what the Stevens Report (1976) referred to as a great change in standards and attitudes to mineral exploitation. A lengthy guidance note (MPG 7) fully explains restoration policies and options. In view of the ongoing nature of mineral operations, particular importance is attached to schemes of progressive restoration which are phased in with the gradual working out of the site. (A very effective policy is to make new working dependent upon satisfactory restoration of used sites.) A good idea of the current policy is gained from the following quotation from MPG 7:

> The overall standards of reclamation have continued to improve over recent years, and with the development and implementation of appropriate reclamation techniques, there is potential for land to be restored to a high standard suitable for a variety of uses. Consistent and diligent application of the appropriate techniques will ensure that a wide range of sites are restored to appropriate standards. This may lead to the release of some areas of land which would not otherwise be made available for mineral working, for example, the best and most versatile agricultural land. Conversely where there is serious doubt whether satisfactory reclamation can be achieved at a particular site, then there must also be a doubt whether permission for mineral working should be given.
>
> (MPG 7 1996: para. 3)

The extraction of minerals is one of the most obvious examples of a 'locally unwanted land use' (LULU) and one that has a disproportionate effect on particular locations. But minerals extraction may also bring economic benefits, especially in more remote rural locations. Minerals development control seeks to reconcile these conflicting interests, and recent reviews of minerals planning guidance have taken much more account of the need for sustainable development (see Box 5.3). Nevertheless, a major limitation on the control of minerals exploitation is the emphasis on finding suitable locations, albeit in the interests of mitigating environmental impacts.[38] Much less attention is given to managing the demand for these resources, a question which is taken up in Chapter 7.

CARAVANS

During the 1950s the housing shortage led to a boom in unauthorised caravan sites. The controversy and litigation to which this gave rise led to the intro-duction of special controls over caravan sites (by Part I of the Caravan Sites and Control of Development Act 1960).[39] The Act gave local authorities new powers to control caravan sites, including a requirement that all caravan sites had to be licensed before they could start operating (thus partly closing loopholes in the planning and public health legislation). These controls over caravan sites operate in addition to the normal planning system: thus both planning permission and a licence have to be obtained. Most of the Act dealt with control, but local authorities were given wide powers to provide caravan sites.

Holiday caravans are subject to the same planning and licensing controls as residential caravans. To ensure that a site is used only for holidays (and not for 'residential purposes'), planning permission can include a condition limiting the use of a site to the holiday season. Conditions may also be imposed to require the caravans to be removed at the end of each season or to require a number of pitches on a site to be reserved for touring caravans.

One group of caravanners is particularly unpop-ular: gypsies, or, to give them their less romantic description, 'persons of nomadic life, whatever their race or origin' (but excluding 'members of an organised group of travelling showmen, or persons engaged in travelling circuses, travelling together as such'). The basic problem is that no one wants gypsies around: 'all too often the settled community is concerned chiefly to persuade, or even force, the gypsy families to move on'. In an attempt to deal with the problem, the Caravan Sites Act 1968 gives local authorities in England and Wales (but not in Scotland) the duty to provide adequate sites for gypsies 'residing in or resorting to' their areas. In 1979 100 per cent grants were made available for capital works on sites.[40] But the problems persisted; indeed, they got worse with increases in the estimated number of gypsy caravans, although many of these are in fact 'New Age Travellers'. 'The public visibility of gypsies has grown, while the tolerance of the settled community to them has declined' (Home 1993).[41]

The 1992 consultation paper *Reform of the Caravan Sites Act 1968* heralded a marked shift in policy. Under the new regime, the obligation of local authorities to provide gypsy sites has been abolished ('privatised'?), and central government grants for gypsy caravan sites are being phased out. However, local authorities should 'continue to indicate the regard they have had to meeting gypsies' accommodation needs', with 'broad strategic policies' in structure plans and detailed policies in local plans (DoE Circular 1/94). However, gypsy sites will not be appropriate in green belts or other protected areas, which had been previously allowed for by Circular 28/77. The flexibility to provide sites in the green belt and other protected areas was described as a 'special status' and current guidance says decisions on sites should be made on land use criteria only.

Significantly, the legislation implementing the new policy is not of a planning character: it is the Criminal Justice and Public Order Act 1994 (explained in Circular 1/94). In addition to repealing the obligations imposed on local authorities by the Caravan Sites Act, it provides stronger powers to remove 'unauthorised persons', though the DoE Circular espouses a policy of toleration towards gypsies on unauthorised sites. As a result, it can be expected that appeals relating to gypsy

sites will increase. The first appeal decisions suggest that inspectors may pay special attention to personal circumstances (particularly of children and the elderly) in justifying the grant of planning permission for gypsy sites. Also, cases have been taken to the European Court of Human Rights and have been held to be admissible. Clearly, the abolition of 'the privileged position of gypsies' will not end this sorry saga: it merely opens a new chapter.

TELECOMMUNICATIONS

One area of development control work that has expanded rapidly and with some controversy is telecommunications. The expansion of masts to service mobile phone networks has been a particular concern, at first because of their visual impact, but more recently because of potential health effects. In England, PPG 8 (1992) and Circular 4/99 are the main sources for policy on telecommunications. Similar guidance applies in the other parts of the UK. Given the speed at which this technology is advancing it will be no surprise that, at the time of writing, the guidance is under review. The DETR consultation paper *Telecommunications Mast Development* provides a summary of the background. It begins by pointing out the significance of electronic communication in the economy. More than 213,000 people are employed in this sector directly but it also provides a route to improved competitiveness for many businesses. Government policy is strongly behind the expansion of telecommunications; indeed, the aim is to make the UK the best place in the world for e-commerce by 2002.[42] Whether or not this is achieved, electronic communication presents many challenges for the planning system, not least in its effect on spatial development patterns (see, for example, Graham and Marvin 1999). But the erection of masts and related equipment has been the main talking point so far, and government policy is clearly influenced by the economic argument for efficient communications networks and fostering competition among rival networks.

Masts under 15 m in height are permitted development (with some exceptions)[43] but there is a *prior approval procedure* which gives the planning authority forty-two days to say whether it wishes to approve details of the siting and appearance of the development. Since 1999 the developer has also had to erect a site notice and has been encouraged to make sure that people affected will learn about the proposal. Local plans should include general policies for telecommunications-related development and may allocate sites for large masts. The planning authority should also encourage different operators to share facilities, though competition among the networks obviously limits their willingness to cooperate. There is also an obligation on the developer to site the mast so that it has least effect on the external appearance of buildings. Where this is not followed, the planning authority may serve a breach of condition notice on the basis that a condition of the permitted development right has not been complied with. Masts over 15 m in height require planning consent.

The main proposed change to planning policy relates to health considerations (which are barely mentioned in the 1992 PPG 8). The Independent Expert Group on Mobile Phones conducted an assessment of the health effects and concluded that 'there is no general risk' (quoted in the 2000 Consultation Paper). Nevertheless, the Group recommended a precautionary approach, and the proposed revised planning guidance accepts that the perception of risk can be a material consideration in determining applications for telecommunications apparatus. Thus we are likely to see tighter controls where masts are located in built-up areas, especially around schools and other public buildings.

PURCHASE AND BLIGHT NOTICES

A planning refusal does not of itself confer any right to compensation. On the other hand, revocations of planning permission or interference with existing uses do rank for compensation, since they involve a taking away of a legal right. In cases where, as a result of a planning decision, land becomes 'incapable of reasonably beneficial use', the owner can serve a purchase notice upon the local authority requiring it to buy the property. In all cases, ministerial confirmation is required. The circumstances in which a *purchase notice* can be served include:

- refusal or conditional grant of planning permission;
- revocation or modification of planning permission; and
- discontinuance of use.

In considering whether the land has any *beneficial use*,

> relevant factors are the physical state of the land, its size, shape and surroundings, and the general patterns of land-uses in the area; a use of relatively low value may be regarded as reasonably beneficial if such a use is common for similar land in the vicinity.
>
> (DoE Circular 13/83)

A purchase notice is not intended to apply in a case where an owner is simply prevented from realising the full potential value of his or her land. This would imply the acceptance in principle of paying compensation for virtually all refusals and conditional permissions. It is only if the existing and permitted uses of the land are so seriously affected as to render the land incapable of reasonably beneficial use that the owner can take advantage of the purchase notice procedure.

There are circumstances, other than the threat of public acquisition, in which planning controls so affect the value of the land to the owner that some means of reducing the hardship is clearly desirable. For example, the allocation of land in a development plan for a school or for a road will probably reduce the value of houses on the land or even make them completely unsaleable. In such cases, the affected owner can serve a *blight notice* on the local authority requiring the purchase of the property at an 'unblighted' price. These provisions are restricted to owner-occupiers of houses and small businesses who can show that they have made reasonable attempts to sell their property but have found it impossible to do so except at a substantially depreciated price because of certain defined planning actions. These include land designated for compulsory purchase, or allocated or defined by a development plan for any functions of a government department, local authority, or statutory undertaker, and land on which the Secretary of State has given written notice of his or her intention to provide a trunk road or a *special road* (i.e. a motorway).

The subject of planning blight takes us into the much broader area of the law relating to compensation. This is an extremely complex field, and only an indication of three major provisions can be attempted here.

First, there is a statutory right to compensation for a fall in the value of property arising from the use of highways, aerodromes, and other public works which have immunity from actions for *nuisance*. The depreciation has to be caused by physical factors such as noise, fumes, dust, and vibration, and the compensation is payable by the authority responsible for the works. Second, there is a range of powers under the heading 'mitigation of injurious effect of public works'. Examples include sound insulation; the purchase of owner-occupied property which is severely affected by construction work or by the use of a new or improved highway; the erection of physical barriers (such as walls, screens, or mounds of earth) on or alongside roads to reduce the effects of traffic noise on people living nearby; the planting of trees and the grassing of areas; and the development or redevelopment of land for the specific purpose of improving the surroundings of a highway 'in a manner desirable by reason of its construction, improvement, existence or use'. Third, provision is made for *home loss payments* as a mark of recognition of the special hardship created by compulsory dispossession of one's home. Since the payments are for this purpose, they are quite separate from, and are not dependent upon, any right to compensation or the *disturbance payment* which is described below. Logically, they apply to tenants as well as to owner-occupiers, and are given for all displacements whether by compulsory purchase or by any action under the Housing Acts. These provisions were slightly extended in the 1991 Planning and Compensation Act.

Additionally, there is a general entitlement to a *disturbance payment* for persons who are not entitled to compensation. Local authorities have a duty 'to secure the provision of suitable alternative accommodation where this is not otherwise available on reasonable terms, for any person displaced from residential accommodation' by acquisition, redevelopment, demolition, closing orders, and so on.

EFFICIENCY IN DEVELOPMENT CONTROL

There has been a succession of attempts on the part of central government to 'streamline the planning process' and to make it more 'efficient'. The reasons for these have differed. In the early 1970s the concern was with the enormous increase in planning applications and planning appeals which, of course, stemmed from the property boom of the period. By 1981, government concern was with the economic costs of control, with cutting public expenditure, and with 'freeing' private initiative from unnecessary bureaucratic controls. During the 1990s the emphasis has been on speeding and raising standards of the 'planning service' so as to achieve better efficiency and value for money, although the objective of relieving business from the burden of regulation has remained important. Indeed, the main themes of the discussion on making the system more efficient in the 1970s are much the same today. It is therefore useful to look briefly at the analysis of the problem made by an inquiry chaired by George Dobry, QC, in 1975.

The starting point for Dobry's inquiry was the lengthening delay in the processing of planning applications, but he was quick to point out that 'not all delay is unacceptable: it is the price we must pay for the democratic planning of the environment'. Moreover, his review took account of factors which were very different from those relevant to 'streamlining the planning machine': the increasing pressure for public consultation and participation in the planning process; and the 'dissatisfaction on the part of applicants because they often do not understand why particular decisions have been made, or why it is necessary for what may seem small matters to be the concern of the planning machinery at all'. Additionally, he noted that 'many people feel that the system has not done enough to protect what is good in an environment or to ensure that new development is of a sufficiently high quality'.

Dobry therefore had a difficult task of reconciling apparently irreconcilable objectives: to expedite planning procedures while at the same time facilitating greater public participation and devising a system which would produce better environmental results. His solutions attempted to provide more speed for developers, more participation for the public, *and* better-quality development and conservation.

This was to be effected by the division of applications into minor and major. Despite the inherent difficulties of determining in advance the category into which an application falls (at least to the satisfaction of the public and the local amenity societies), it is nevertheless a fact that some 80 per cent of all applications are granted (that is still the case today) and that many of these *are* simple and straightforward. Dobry's proposal, in essence, was that simple applications should be distinguished and treated expeditiously by officials, though with the opportunity for some participation and with a safety channel to allow them to be transferred to the major category if this should prove appropriate.

Dobry's scheme was a heroic attempt to improve the planning control system to everyone's satisfaction (Jowell 1975). Inevitably, therefore, it disappointed everybody. For example, though he made a number of proposals to increase public participation, his overriding concern for expediting procedures forced him to compress 'simple' applications into an impracticable time scale.

The Dobry inquiry was instigated by a Conservative government at a time when the property market was booming. On its completion the boom had collapsed and a Labour government had published its outline proposals for the community land legislation. Thus, the planning scene had changed fundamentally. The government rejected all Dobry's major recommendations for changes in the system, though it was stressed that their objectives could typically be achieved if local authorities adopted efficient working methods. Dobry's view that 'it is not so much the system which is wrong but the way in which it is used' was endorsed, and his *Final Report* was commended 'to students of our planning system as an invaluable compendium of information about the working of the existing development control process, and to local authorities and developers as a source of advice on the best way to operate within it'.

The Conservative government which was elected in 1979 lost no time in preparing a revised development control policy. A draft circular was sent out for comment in mid-1980. This created alarm among the planning profession, partly because of its substantive proposals but also partly because of its abrasive style. 'The Most Savage Attack Yet', expostulated *Municipal Engineering*, while *Planner News* remonstrated that the results of the circular 'could be disastrous'.[44] The revised circular, as published (22/80), was written with a gentler touch, but much of the message was very similar. The emphasis was on securing a 'speeding up of the system' and on ensuring that 'development is only prevented or restricted when this serves a clear planning purpose and the economic effects have been taken into account'.

Regular publication of performance figures (the percentage of planning applications decided within eight weeks) became the standard by which the efficiency of the development control system is measured. Quarterly figures have been published since 1979, and are used by both the government and the development industry to bolster criticisms of the system.

The policy 'to simplify the system and improve its efficiency' (to use the words of the 1985 White Paper *Lifting the Burden*) continued with revised circulars, new White Papers, and the introduction of planning mechanisms which reduced or bypassed local government control such as simplified planning zones and urban development corporations. However, towards the end of the 1980s a greater emphasis on 'quality' emerged, as environmental awareness and concern increased. A change in direction was signalled by the 1992 Audit Commission report on development control, significantly entitled *Building In Quality*. Though the major emphasis was still on the process of planning control rather than its outcome, there was a very clear recognition of the importance of the latter. The report noted that there had been a preoccupation with the speed of processing planning applications, 'ignoring the mix of applications, the variety of development control functions, and the quality of

Strategic objectives	Percentage of new homes built on previously developed land
Cost and efficiency	Planning cost per head of population
Service delivery outcomes	The number of advertised departures from the statutory plan approved by the authority as a percentage of total permissions granted. Percentage of applications determined within eight weeks (excluding applications involving environmental assessment). Average time taken to determine all applications
Quality	Percentage of applicants and those commenting on planning applications satisfied with the service received (based on a list of questions specified by the DETR). Score against a checklist of planning practice; for development control these include: • providing for pre-applications discussions; • having a published charter setting targets for stages of the process; • having fewer than 40% of appeals overturning the council's original decision; • delegation of 70% or more decisions; • no costs or ombudsmen reports finding against the authority; • provision of a one-stop service; and • equal access to the planning service for all groups.

Figure 5.3 Best Value Performance Indicators for Planning

outcomes'. But there had been no 'shared and explicit' concept of quality, yet 'The quality of outcomes is more important than the quality of the process because buildings will be seen long after memories of the decision process have lapsed, but it is far harder to assess'. Quality of development control is seen as involving an 'adding of value' by the local authority. What that 'added value' may be is dependent upon the authority's overall objectives: 'in areas under heavy development pressure or in rural areas, environmental, traffic, or ecological considerations may be paramount'; in Wales 'the impact of the development on the Welsh language can be a consideration'.

The effect of *Building In Quality* has been to redress the balance somewhat from the emphasis on lifting the burden of regulation, but the importance of meeting the eight-week targets remains. There was a considerable overall improvement in performance in the first part of the 1990s: from 46 per cent of applications decided within eight weeks in 1989–90 to 65 per cent in 1993–94, but performance has

remained much the same since then. In 1998–99 the figure was 63 per cent.[45] A review of progress on *Building In Quality* in 1999 points to the value of increasing delegation of decision making to officers for those authorities whose performance has improved. The reduced number of applications from the peak in 1988–89 (illustrated in Figure 5.2) and the increasing coverage of local development plan policy are also significant factors. Nevertheless, the review clearly demonstrates great variation in performance and the need for improvement in matters of customer care and service delivery – a point made by a succession of planning ministers, but apparently with little effect. And the need to make the system more responsive to business and 'customer' needs remains. Particular attention is being given to the idea of creating a 'one-stop shop' approach providing a more user-friendly service for those who will be seeking more complicated consents.[46]

The introduction of Best Value, the new initiative to improve performance in the delivery of local government services across the board, is explained in Chapter 4. Best Value seeks to marry the need for increased efficiency with recognition of the importance of maintaining and improving quality, and has established a wide framework of performance indicators and targets, some dictated 'nationally' and others defined by the local authority itself. Figure 5.3 shows the performance indicators for planning; most relate to the development control function, and specific reference is made to the quality of the service. Local authorities are preparing five-year plans for improving performance under Best Value, and these will be reviewed annually, although planning will not be a priority for some in the first round for 2001. Authorities will supplement the national indicators with their own local indicators in making comparisons with other authorities. Good practice in implementing Best Value in planning is being prepared by the Planning Officers' Society with support from the DETR. Although planning authorities have long made the case for assessing quality as well as efficiency, there is no doubt that more than a few will struggle to meet the criteria here.

FURTHER READING

Legal Texts

The law and procedure of development control are explained fully in several textbooks. For England and Wales, see K. Thomas (1997) *Development Control: Principles and Practice*; Moore (2000) *A Practical Approach to Planning Law*, and Duxbury (1999) *Telling and Duxbury's Planning Law and Procedure*; for Northern Ireland: Dowling (1995) *Northern Ireland Planning Law*; and for Scotland: Collar (1999) *Green's Concise Scots Law: Planning* and McAllister and McMaster (1999) *Scottish Planning Law. The Encyclopedia of Planning Law and Practice* (Grant 1997) provides excellent commentary on planning legislation and policy. Some of the RTPI's practice advice notes (listed at the end of this text) are particularly relevant to development control.

Use Classes and Development Orders

For a study on the use and complexity of the GPDO, see Edinburgh College of Art *et al.* (1997) *Research on the General Permitted Development Order and Related Mechanisms*; and also BDP in association with Berwin Leighton (1998) *The Use of Permitted Development Rights by Statutory Undertakers*. On Article 4 Directions, see Roger Tym & Partners (1995a) *The Use of Article 4 Directions by Local Planning Authorities*, and Larkham and Chapman (1996) 'Article 4 Directions and development control'. The standard legal text is Grant (1996) *Permitted Development*.

The Development Plan as a Consideration

The impact of the introduction of section 54A (s. 25 of the Scottish Act) (note that there is no equivalent in the Northern Ireland legislation) is reviewed by Gatenby and Williams (1996) 'Interpreting planning law', and see also their earlier paper (1992) 'Section 54A: The legal and practical implications'; MacGregor and Ross (1995) 'Master or servant? The changing role of the development plan in the British planning system'; Purdue (1991) 'Green belts and the presumption in favour of development'; M. Harrison (1992) 'A presumption in favour of planning permission?'; and Herbert-Young (1995) 'Reflections on section 54A and *plan-led* decision-making'.

Other Material Considerations

PPG 1 presents a succinct statement concerning material considerations. For a categorisation of considerations, see Davies *et al.* (1986a) *The Relationship between Development Plans, Development Control and Appeals*; this work is also summarised by K. Thomas (1997) *Development Control: Principles and Practice*. See also HC Welsh Affairs Committee (1993) *Rural Housing*, where the issue is examined in depth.

Design

There are two very good starting points for considering the role of design as a consideration in planning: the

DETR and CABE publication *By Design: Urban Design in the Planning System – Towards Better Practice* (2000) includes checklists of design considerations and a list of other references. The reader by Greed and Roberts (1998) *Introducing Urban Design* has a number of relevant papers, but in particular Carmona 'Urban design and planning practice' presents the history of design control and a very thorough review of design considerations in control. See also Carmona's two-part article in *Planning Practice and Research* (1998, 1999) 'Residential design policy and guidance'; and Taylor (1999) 'The elements of townscape and the art of urban design'. Punter's work is notable in this field. See his 1990 book *Design Control in Bristol, 1940–1990*; and (1986–87) 'A history of aesthetic control: the control of the external appearance of development in England and Wales'. Another recent text is Smith Morris (1997) *British Town Planning and Urban Design: Principles and Policies*.

There is a huge library on particular design issues. See, for example, Llewelyn-Davies (1998b) *Planning and Development Briefs: A Guide to Better Practice*; Countryside Commission (1993) *Design in the Countryside*; DoE (1994) *Quality in Town and Country* (Discussion Paper); and (1995) *Quality in Town and Country: Urban Design Campaign*; National Audit Office (1994) *Environmental Factors in Road Planning and Design*; NIDoE (1994) *A Design Guide for Rural Northern Ireland*; Scottish Office (1994) *Fitting New Housing Development into the Landscape* (PAN 44), Barton *et al.* (1995) *Sustainable Settlements*; Bishop (1994) 'Planning for better rural design'; and Owen (1991) *Planning Settlements Naturally*. There are numerous guides to better design in the built environment including English Partnerships' *Urban Design Compendium* (2000); the Scottish Office PAN 59: *Encouraging Higher Standards of Design*; and the DETR's *By Design: Urban Design in the Planning System: Towards Better Practice* (2000) and *Training for Urban Design* (2000).

Amenity

For a discussion of statutory provisions in relation to 'amenity', see Sheail (1992) 'The *amenity* clause', and also an unusual historical study of the development of the notion of amenity in Millichap (1995a) 'Law, myth and community: a reinterpretation of planning's justification and rationale'.

Appeals

The *Planning Inspectorate Journal* has provided numerous perspectives on appeals. The Chief Planning Inspector has given his view of current issues in Shepley (1999) 'Decision-making and the role of the Inspectorate'. An example of the analysis of appeals data is given by Wood *et al.* (1998) 'The character of countryside recreation and leisure appeals'.

Enforcement

Two standard legal texts are Millichap (1995b) *The Effective Enforcement of Planning Controls* and Bourne (1992) *Enforcement of Planning Control*. DoE Circular 10/97 gives a comprehensive explanation of the enforcement provisions of the 1991 Act alongside the DETR *Enforcing Planning Control: A Good Practice Guide*; see also PPG 18. The operation of the procedures in Scotland has been investigated by the Edinburgh College of Art *et al.* (1997) *Research on the General Permitted Development Order and Related Mechanisms*; see also Upton and Harwood (1996), 'The stunning powers of environmental inspectors' which provides a striking contrast to the planning enforcement system. The DETR consultation paper *Improving Enforcement Appeal Procedures* (1999) reviews the finer points of implementing the 'new' system.

Advertisements

The regulations are explained in DoE Circular 19/92; policy guidance is given in PPG 19 and TAN (W) 7. The fullest exposition of the law of advertisement control is given in Mynors (1992) *Planning Control and the Display of Advertisements*, and a useful update is given in the 1999 DETR consultation paper on *Outdoor Advertisement Control*. For Scotland, see SOEnD Circular 31/92.

Minerals

The minerals planning guidance notes provide a rich source of information, in particular MPG 1 on general considerations and MPG 2 on how development control of minerals is undertaken. A bibliography on reclamation for various uses in given in MPG 7: on aggregates, see MPG 6 (revised 1994); DoE. The DETR sponsors extensive research on minerals and the reader is directed to the *Planning and Minerals Research Newsletter*. An excellent review of mineral resource planning and sustainability is given by Owens and Cowell (1996) *Rocks and Hard Places*. Minerals guidance is being reviewed and further consultation papers are expected.

Caravans and Gypsies

The appalling living conditions of gypsies, combined with public attitudes towards them, have presented difficult political problems for governments: an excellent review of the contradictions in government policy is provided by Morris (1998) 'Gypsies and the planning system'; see also Gentleman (1993) *Counting Travellers in Scotland*. The main official reference is DoE Circular 1/94 *Gypsy Sites and Planning*.

Efficiency in Development Control

The 1992 Audit Commission report *Building in Quality: A Study of Development Control* and the subsequent 1998 *Building in Quality: A Review of Progress on Development Control* are the main sources. The various responsible government departments publish quarterly figures on the development control performance which are accompanied by press notices giving the latest ministerial pronouncements on the need for improvement. The Planning Officers' Society has made its good practice guidance on Best Value and planning available via the Web at <www.barnsley.gov.uk/planning/index.html>. Further details on Best Value are available at the DETR web site <www.local-regions.detr.gov.uk/bestvalue/indicators/bvaudit/index.htm>. See also the DETR good practice guide *The One Stop Approach to Development Consents*, which includes discussion on related issues such as pre-application discussions; and the Report of the Scottish Executive *Targets Working Group on Planning Services* (1999), which includes an analysis of what factors delay planning permissions.

NOTES

1   Until 1995 the General Development Order (GDO) contained both permitted development rights and procedural matters (relating to planning applications). In 1995 these were separated (following the Scottish model introduced in 1992). There is therefore now a *General Permitted Development Order* (GPDO) and a *General Development Procedure Order* (GDPO). See Circular 9/95, *General Development Order Consolidation*. Though these new orders are predominantly consolidations, they contain a number of changes.

2   Circular 9/95: *General Development Order Consolidation* 1995 para. 1. The Secretary of State approves Article 4 Directions except those that are specifically related to dwelling-houses in conservation areas. PPG 15 explains the application of Article 4 Directions in conservation areas in England.

3   The changes in local authority management structures mean that many decisions are now made by a cabinet rather than committees, but in the case of planning and other regulatory activities local authorities must retain the committee decision-making procedure (although some decisions will be delegated). Some large authorities will divide up the committee into smaller local area committees.

4   Essex (1996) reviews these two cases and the general issue of relationships between officers and members in decision-making. The issue is also taken up in Chapter 12.

5   The development plan may mean the unitary development plan or the structure and local plan depending on the area in question and includes the minerals and waste plans and old subject plans where these have not yet been replaced with new plans; see Chapter 4.

6 Exactly the same provision is made in the Scottish legislation as s. 25 of the 1997 Act (formerly s. 18A of the 1972 Act). No such provision has been made for Northern Ireland.

7 In his judgement in the case of *The City of Edinburgh v. the Secretary of State for Scotland*, Lord Hope said, 'it requires to be emphasised however, that the matter is nevertheless still one of judgement, and that this judgement is to be exercised by the decision taker. The development plan does not, even with the benefit of section 18A, have absolute authority. The planning authority . . . is at liberty to depart from the development plan if material considerations indicate otherwise' (Encyclopedia P54A.05/2 (Grant 1997)).

8 The discretion to take into account a wide range of considerations in making decisions on planning applications is perhaps where the British planning systems differ most from most systems elsewhere, which tend to rely more on the provisions of a legally binding planning instrument. See Booth (1996) and Nadin *et al.* (1997).

9 The research by Davies *et al.* (1986a) noted above provides long lists of material considerations. See also K. Thomas (1997: 95–103) and Roger Tym & Partners (1989a).

10 PPG 1, *General Policy and Principles* (1997), notes that the government's statements of planning policy 'cannot make irrelevant any matter which is a material consideration in a particular case. But, where such statements indicate the weight that should be given to relevant considerations, decision-makers must have proper regard to them. If they elect not to follow relevant statements of the Government's planning policy they must give clear and convincing reasons (*EC Grandsen and Co Ltd v. SSE and Gillingham BC 1985*). Emerging planning policies, in the form of draft Departmental Circulars and policy guidance, can be regarded as material considerations, depending on the context. Their very existence may indicate that a relevant policy is under review and the circumstances which have led to that review may need to be taken into account' (paras 52 and 53).

11 This dismal conclusion is corroborated by the studies of Booth and Beer (1983). They found that though nearly two-thirds of all permissions granted 'carried conditions that were intended to modify the landscape design, layout and architectural detailing of the developments', the conditions were frequently vague and not site specific. Enforcement was lax. Many applications were considered 'without anybody fully trained in design having been involved in their processing'. The suggestion is made that 'this may well be a factor in the generally poor quality and particularly the monotony of the new environments which can be observed in many parts of Britain'.

12 The role of regulation of design was taken forward in Circular 31/85, which emphasised that 'a large proportion of planning appeals involve detailed design matters' and that 'far too many planning applications are delayed because the planning authority seeks to impose detailed design alterations'.

13 Chapman and Larkham (1999) note the poor level of commentary on and failure to disseminate lessons from this initiative in a paper generally sceptical of its wider impact in the face of lack of interest after a change in government.

14 Price Waterhouse (1997) *The Design Improvement Controlled Experiment: Evaluation of the Impact, Costs and Benefits of Estate Re-modelling* (London: DETR).

15 Circular 11/95 does not deal with conditions in respect of minerals or waste which are dealt with in the minerals planning guidance notes and PPG 23, *Planning and Pollution Control*.

16 In Scotland guidance is given in Circular 4/1998, and in Wales it is Welsh Office Circular 35/95, both entitled *The Use Conditions in Planning Permissions*.

17 PPG 3, *Housing* (2000), notes that it is common practice to renew planning permissions, but encourages local planning authorities to review permissions in the light of current planning policy and if necessary not renew permissions or impose new conditions (para. 40).

18 But the trench-digger may be brought up against a further provision: the serving of a *completion*

notice. Such a notice states that the planning permission lapses after the expiration of a specified period (of not less than one year). Any work carried out after then becomes liable to enforcement procedures.

19 The fees are amended on a regular basis. For illustration, at the time of writing, the fee in England and Wales for residential development is £190 per dwelling (up to a maximum of £9,500 or £4,750 for outline applications); £95 for extensions to dwellings; and for commercial and industrial buildings it varies according to gross floorspace created: £35 for 40 sq. m, and for larger developments £190 for each 75 sq. m up to a maximum of £9,500 (The Town and Country Planning (Fees for Applications and Deemed Applications) (Amendment) Regulations 1997 Statutory Instrument 1997 no. 37). Fees in Scotland have been updated more recently: The Town and Country Planning (Fees for Applications and Deemed Applications) (Scotland) Amendment Regulations 2000, SI 2000 no. 150.

20 In England appeals are made to the Secretary of State for Transport, Local Government and the Regions and in Wales to the Welsh Assembly. In both cases the Planning Inspectorate Executive Agency considers and makes decisions on most (the nature of the Agency is explained in Chapter 3). In Scotland the Inquiry Reporters Unit considers appeals, representing the Scottish Minister for Planning. In Northern Ireland the Planning Appeals Commission has the same role as the Planning Inspectorate.

21 The Franks Committee on Administrative Tribunals and Inquiries argued that it was not satisfactory 'that a government department should be occupied with appeal work of this volume, particularly as many of the appeals relate to minor and purely local matters, in which little or no departmental policy entered' (Franks Report 1957: 85).

22 Cases may be recovered by the Secretary of State where they involve substantial development (over 150 houses or retail development over 100,000 sq. ft), significant proposed development in the green belt; major mineral planning appeals; where other government departments have an interest, or where there is major controversy over the development.

23 Circular 5/2000 explains the procedures and gives references to the inquiry, hearing and written representation rules. At the time of writing, Scotland and Wales had not reviewed their appeal procedures, although devolution is likely to make them more distinctive to the specific needs of these countries.

24 Government targets for the processing of appeals by the Inspectorate in England are eighteen weeks for 80 per cent of written representations, twenty-four weeks for 80 per cent of hearings and thirty-six weeks for 80 per cent of inquiries. Performance has improved dramatically over recent years and is reaching the target for written representations (seventeen weeks) but is below it for hearings (twenty-seven weeks) and inquiries (forty-one weeks) (Planning Inspectorate Executive Agency, *Statistical Report*, 1999). Similar targets are set in Scotland and Wales.

25 See the DETR consultation papers *Modernising Planning: Improving Planning Appeal Procedures* (1998) and *Modernising Planning: The Recovery of Costs of Public Local Inquiries Held into Planning Matters* (1998).

26 The Inspectorate have agreed and published *Better Presentation of Evidence in Chief* with the Local Government Planning and Environment Bar Association (2000). See also the RTPI Practice Advice Note No. 9 *Development Control: Handling Appeals* (1995), although this does not take account of the new procedures in England.

27 The difficulties of the interpretation of aggregate appeals data (since each decision is made on its merits) has been a subject of continuing debate. For example, see Brotherton (1993).

28 See Cambridge University, Department of Land Economy (1999) *Environmental Court Project* (London: DETR).

29 Until September 1999 this role was undertaken by the Royal Fine Arts Commission, which requested intervention by the Secretary of State

on numerous occasions, although not always successfully. Another important body is the Urban Design Alliance, which comprises professional bodies that seek to improve the quality of life through urban design.

30   The Circular should be read in conjunction with 'Planning Policy Guidance Note 18, *Enforcement*, and the DETR', *Enforcing Planning Control: Good Practice Guide for Local Planning Authorities*. In Scotland the key references are Circular 4/1999 and Planning Advice Note 54, *Enforcement*. See also RTPI Practice Advice Note no. 6 (1999) and *Enforcement of Planning Control* (1996).

31   The DETR consultation paper *Modernising Planning: Improving Enforcement Appeal Procedures* (1999) makes numerous recommendations including requirements for a list of relevant development plan policies and time limits for notification and representations to be made.

32   *Independent*, 9 December 1995: 9.

33   The study found that stop notices are not used because of the fear of compensation payments; that breach of condition notices may be difficult to employ because conditions are not worded with sufficient specificity; and that there was some frustration at the difficulty of employing the 'ultimate sanction' through the courts.

34   The government has suggested that a third test should be added: the policies of the development plan.

35   In England, the Town and Country Planning (Control of Advertisements) Regulations 1992 (SI 666) as amended in 1994 (SI 2351) and 1999 (SI 1810).

36   DoE consultation paper *Outdoor Advertisement Control: Areas of Special Control of Advertisements* (1996) and DETR consultation paper *Modernising Planning: Outdoor Advertisement Control* (1999).

37   The first revision of MPG 1 was published in 1994. The second and sixth points in the 1996 list were added in 1996 strengthening policy on both preventing negative environmental impacts and ensuring that mineral resources are kept available.

38   The DETR has funded a series of research projects on the environmental impacts of minerals exploitation that inform national policy on development control. The most recent reports are Arup Environmental and Ove Arup & Partners (1995) *The Environmental Effects of Dust from Surface Mineral Workings*; Vibrock Ltd (1998) *The Environmental Effects of Production Blasting from Surface Mineral Workings*; ENTEC UK Ltd (1998) *The Environmental Effects of Traffic Associated with Mineral Workings*; and University of Newcastle upon Tyne (1999) *Do Particulates from Opencast Coal Mining Impair Children's Respiratory Health?*

39   This legislation has remained as a separate code and is not consolidated in the Town and Country Planning Act of 1990. The Caravan Sites Act 1968, which deals mainly with the protection from eviction of caravan dwellers and gypsies, is similarly separate.

40   The report by John Cripps (1977) *Gypsy Site Policy and Illegal Camping: A Report on the Working of the Caravans Sites Act 1968*, was instrumental in the changes. Circular 28/77 clearly conveyed the government policy of the time to give gypsies special protection in the planning system; it even accepted the necessity of establishing gypsy sites in the protected areas such as green belts and AONBs. It was anticipated that caravan sites would be located in such protected areas, especially when close to the urban fringe.

41   The number of gypsies has been estimated at '9,000 families in 13,500 caravans, 9,000 of which are parked on legal sites' (excluding 'New Age Travellers') (Morris 1998).

42   The UK's e-commerce strategy is being implemented by the Electronic Communications Act 2000 and a host of measures including the requirement for 50 per cent of government services to be delivered through the Internet by 2005 and all services by 2008. Details are available at the DTI web site <www.dti.gov. uk/cii/ecommerce/ukecommercestrategy/index.sh tml>.

43   Some masts or antennas may be so small that they do not constitute development – for example, television aerials have been treated as outside the

definition of development (despite their sometimes significant impact on the external appearance of buildings). The exceptions from permitted development for masts under 13 m in height include proposed masts on listed buildings, scheduled ancient monuments and where the planning authority has made an Article 4 Direction withdrawing permitted development rights.

44 An example of the matters to which objection was taken was the call for relaxation of controls over private-sector housing: 'Local authorities should not lay down requirements on the mix of house types, provision of garages, internal standards, sizes of private gardens, location of houses on plots and in relation to each other, provision of private open space.' The Circular also stated that planning authorities should not attempt to compel developers to adopt designs which were unpopular with customers or clients, 'and they shouldn't attempt to control such details as shapes of windows or doors or the colour of bricks'.

45 Only twenty-seven planning authorities in England met the government's target of deciding 80 per cent of applications within eight weeks; ninety-seven planning authorities decided 70 per cent or more and ten planning authorities decide fewer than 40 per cent within eight weeks.

46 In 1998 the DETR published a good practice guide, *The One-Stop Shop Approach to Planning Consents*. There is also increasing interest in comparisons of practice in the UK with other countries; on development control, see GMA Planning *et al.* (1993) *Integrated Planning and the Granting of Permits in the EC*.

6

LAND POLICIES

It is clear that under a system of well-conceived planning, the resolution of competing claims and the allocation of land for the various requirements must proceed on the basis of selecting the most suitable land for the purpose, irrespective of the existing values which may attach to the individual parcels of land.

(Uthwatt Report 1942)

THE UTHWATT REPORT

It was the task of the Uthwatt Committee, from whose report the above epigraph is taken, to devise a scheme which would make its striking claim possible. Effective planning necessarily controls, limits, or even completely destroys the market value of particular pieces of land. Is the owner to be compensated for this loss in value? If so, how is the compensation to be calculated? And is any 'balancing' payment to be extracted from owners whose land appreciates in value as a result of planning measures?

This problem of compensation and betterment arises fundamentally 'from the existing legal position with regard to the use of land, which attempts largely to preserve, in a highly developed economy, the purely individualistic approach to land ownership'. This 'individualistic approach', however, has been increasingly modified during the past hundred years. The rights of ownership were restricted in the interests of public health: owners had (by law) to ensure, for example, that their properties were in good sanitary condition, that new buildings conformed to certain building standards, that streets were of a minimum width, and so on. It was accepted that these restrictions were necessary in the interests of the community (*salus populi est suprema lex*, the welfare of the people is the supreme

law) and that private owners should be compelled to comply with them even at cost to themselves.

> All these restrictions, whether carrying a right to compensation or not, are imposed in the public interest, and the essence of the compensation problem as regards the imposition of restrictions appears to be this – at what point does the public interest become such that a private individual ought to be compelled to comply, at his own cost, with a restriction or requirement designed to secure that public interest? The history of the imposition of obligations without compensation has been to push that point progressively further on and to add to the list of requirements considered to be essential to the well-being of the community.
>
> (*Uthwatt Report*, para. 33)

But clearly there is a point beyond which restrictions cannot reasonably be imposed on the grounds of good neighbourliness without payment of compensation – and 'general consideration of regional or national policy requires so great a restriction on the landowner's use of his land as to amount to a taking away from him of a proprietary interest in the land'. This, however, is not the end of the matter. Planning sets out to achieve a selection of the most suitable pieces of land for particular uses. Some land will therefore be zoned for a use which is profitable for the owner, whereas other land will be zoned for a use having a low, or even nil, private value. It is this difficulty of *development value* which raises the compensation problem in its most acute form. The expectations (or hopes) of owners extend over a far larger area than is likely to be developed. This *potential* development value is therefore speculative, but until the individual owners are proved to be wrong in their assessments (and how can this be done?), all owners of land having a potential value can make a case for compensation on the assumption that their particular pieces of land would in fact be chosen for development if planning restrictions were not imposed. Yet this *floating value* might never have settled on their land, and obviously the aggregate of the values claimed by the individual owners is likely to be greatly in excess of a total valuation of all pieces of land.

Furthermore, the public control of land use necessarily involves the shifting of land values from certain pieces of land to other pieces: the value of some land is decreased, while that of other land is increased. Planning controls, so it was argued, do not destroy land values: in the words of the Uthwatt Committee, 'neither the total demand for development nor its average annual rate is materially affected, if at all, by planning ordinances'. Nevertheless, the owner of the land on which development is prohibited will claim compensation for the full potential development of his land, irrespective of the fact that the value may shift to another site.

In theory, it is logical to balance the compensation paid to aggrieved owners by collecting a betterment charge on owners who benefit from planning controls (Hagman and Misczynski 1978), but previous experience with the collection of betterment had not been encouraging.[1] The Uthwatt Committee concluded that the solution to these problems lay in changing the system of land ownership under which land had a development value dependent upon the prospects of its profitable use. They maintained that no new code for the assessment of compensation or the collection of betterment would be adequate if this 'individualistic' system remained. The system itself had inherent 'contradictions provoking a conflict between private and public interest and hindering the proper operation of the planning machinery'. A new system was needed which would avoid these contradictions and which so unified existing rights in land as to 'enable shifts of value to operate within the same ownership'. The Uthwatt Committee's solution was the nationalisation of development rights in undeveloped land.

THE 1947 ACT

Essentially, this is what the 1947 Town and Country Planning Act did: development rights and their associated values were nationalised. No development was to take place without permission from the local planning authority. If permission were refused, no compensation would be paid (except in a limited range of special cases). If permission were granted, any resulting increase in land value was to be subject to a development charge. The view was taken that 'owners who lose development value as a result of the passing of the Bill are not on that account entitled to compensation'. This cut through the insoluble problem posed in previous attempts to collect betterment values created by public action. Betterment had been conceived as any increase in the value of land arising from central or local government action. The 1947 Act went further: all betterment was created by the community, and it was unreal and undesirable (as well as virtually impossible) to distinguish between values created, for example, by particular planning schemes, and those due to other factors such as the general activities of the community or the general level of prosperity.

If rigorous logic had been followed, no payment at all would have been made for the transfer of development value to the state, but this, as the Uthwatt Committee had pointed out, would have resulted in considerable hardship in individual cases. A £300 million fund was therefore established for making 'payments' (as distinct from 'compensation') to owners who could successfully claim that their land had some development value on the appointed day – the day on which the provisions of the Bill which prevented landowners from realising development values came into force. Considerable discussion took place during the passage of the Bill through Parliament on the sum fixed for the payments, and it was strongly opposed on the ground that it was too small. The truth of the matter was that in the absence of relevant reliable information, any global sum had to be determined in a somewhat arbitrary way, but in any case it was not intended that everybody should be paid the full value of their claims. Landowners would submit claims to a centralised agency, the Central Land Board, for *loss of development value*: that is, the difference between the *unrestricted value* (the market value without the restrictions introduced by the Act) and the *existing use value* (the value subject to these restrictions). When all the claims had been received and examined, the £300 million would be divided between claimants at whatever proportion of their 1948 value that total would allow. (In the event, the estimate of £300 million was not as far out as critics feared: the total of all claims finally amounted to £380 million.)

These provisions, of which only the barest summary has been given here, were very complex and, together with the inevitable uncertainty as to when compensation would be paid and how much it should be, resulted in a general feeling of uncertainty and discontent which did not augur well for the scheme. The principles, however, were clear. To recapitulate, all development rights and values were vested in the state: no development could take place without permission from the local planning authority and then only on payment of a betterment charge to the Central Land Board. The nationalisation of development rights was effected by the 'promised' payments in lieu of compensation. As a result, landowners owned only the existing use rights of their land and it thus followed, first, that if permission to develop was refused, no compensation was payable, and second, that the price paid to public authorities for the compulsory acquisition of land would be equal to the existing use value: that is, its value excluding any allowance for future development.

The scheme did not work as smoothly as was expected. Land changed hands at prices which included the full development value. This was largely due to the severe restrictions which were imposed on building. Building licences were very scarce, and developers who were able to obtain them were willing to pay a high price for land upon which to build. The Labour government was in process of reviewing the scheme when it lost office.

THE 1954 SCHEME: THE DUAL LAND MARKET

The Conservative government which took office in 1951 was intent on raising the level of construction activity and particularly the rate of private house-building. Though, within the limits of building activity set by the Labour government, it is unlikely that the development charge procedure seriously affected the supply of land, it is probable that the Conservative government's plans for private building would have been jeopardised by it. This was one factor which led the new government to consider repealing development charges. There is no doubt that these charges were unpopular, particularly since they were payable in cash and in full, whereas payments on the claims on the £300 million fund were deferred and uncertain in amount. Given the political and technical problems involved, it was decided that the best solution was the complete abolition of development charges. However, to safeguard the public purse, acquisitions of land by public authorities were to remain at the existing use value.

The effect of the complicated network of legislation which now (1954) operated was basically to create two values for land according to whether it was sold in the open market or acquired by a public authority. This was an untenable position, and as land prices increased, partly owing to planning controls, the gap between existing use and market values widened, particularly in suburban areas near green belt land. The greater the amount of planning control, the greater did the gap become. Thus, owners who were forced to sell their land to public authorities considered themselves to be very badly treated in comparison with those who were able to sell at the enhanced prices resulting in part from planning restrictions on other sites. The inherent uncertainties of future public acquisitions – no plan can be so definite and inflexible as to determine which sites will (or might) be needed in the future for public purposes – made this distinction appear arbitrary and unjust. The abolition of the development charge served to increase the inequity.

The contradictions and anomalies in the 1954 scheme were obvious. It was only a matter of time before public opinion demanded further amending legislation.

THE 1959 ACT: THE RETURN TO MARKET VALUE

Opposition to this state of affairs increased with the growth of private pressures for development following the abolition of building licences. Eventually the government was forced to take action. The resulting legislation (the Town and Country Planning Act 1959) restored *fair market price* as the basis of compensation for compulsory acquisition. Owners now obtained (in theory at least) the same price for their land irrespective of whether they sold it to a private individual or to a public authority.

These provisions thus removed a source of grievance, but they did nothing towards solving the fundamental problems of compensation and betterment, and the result proved extremely costly to public authorities. If this had been a reflection of basic principles of justice there could have been little cause for complaint, but in fact an examination of the position shows clearly that this was not the case.

In the first place, the 1959 Act (like previous legislation) accepted the principle that development rights should be vested in the state. This followed from the fact that no compensation was payable for the loss of development value in cases where planning permission was refused. But if development rights belong to the state, surely so should the associated development values? Consider, for example, the case of two owners of agricultural land on the periphery of a town, both of whom applied for planning permission to develop for housing purposes – the first being given permission and the second refused on the ground that the site in question was to form part of a green belt. The former benefited from the full market value of his or her site in residential use, whereas the latter could benefit only from its existing use value. No question of compensation arose since the development rights already belonged to the state, but the first owner had these given back without payment. There was an obvious injustice here which could have eventually led to a demand that the 'penalised' owner should be compensated.

Second, as has already been stressed, the comprehensive nature of the planning system has a marked effect on values. The use for which planning permission has been, or will be, given is a very important factor in the determination of value. Furthermore, the value of a given site is increased not only by the development permitted on that site, but also by the development not permitted on other sites. In the example given above, for instance, the value of the site for which planning permission for housing development was given might be increased by virtue of the fact that it was refused on the second site.

THE LAND COMMISSION 1967-71

Mounting criticism of the inadequacy of the 1959 Act led to a number of proposals for a tax on betterment. The Labour government which was returned to power in 1964 introduced the Land Commission Act, which provided for a new levy and had two main objectives: 'to secure that the right land is available at the right time for the implementation of national, regional and local plans', and 'to secure that a substantial part of the development value created by the community returns to the community and that the burden of the cost of land for essential purposes is reduced' (White Paper, *The Land Commission*, 1965).

To enable these two objectives to be achieved, a Land Commission was established. The Commission could buy land either by agreement or compulsorily, and it was given very wide powers for this purpose. The second objective was met by the introduction of a betterment levy on development value. This was necessary not only to ensure that a substantial part of the development 'returned to the community', but also to prevent a two-price system as had existed under the 1954 Act. The levy was deducted from the price paid by the Commission on its own purchases and was paid by owners when they sold land privately. Landowners thus theoretically received the same amount for their land whether they sold it privately, to the Land Commission, or to another public authority.

The levy differed from the development charge of the 1947 Act in important ways. Most significantly, it did not take all the development value. Though the Act did not specify what the rate was to be, it was made clear that the initial rate of 40 per cent would be increased to 45 per cent and then to 50 per cent 'at reasonably short intervals'. (It never was.) The Land Commission's first task was to assess the availability of, and demand for, land for housebuilding, particularly in the areas of greatest pressure. In its first annual report it pointed to the difficulties in some areas, particularly in the South-East and the west Midlands, where the available land was limited to only a few years' supply. Most of this land could not, in fact, be made available for early development. Much of it was in small parcels; some was not suitable for development at all because of physical difficulties; and, of the remainder, a great deal was already in the hands of builders. Thus there was little that could be acquired and developed immediately by those other builders who had an urgent need for land. All this highlighted the need for more land to be allocated by planning authorities for development.

The Land Commission had to work within the framework of the planning system, and was subject to the same planning control as private developers. The intention was that the Commission would work harmoniously with local planning authorities and form an important addition to the planning machinery. As the Commission pointed out, despite the sophistication of the British planning system, it was designed to control land use rather than to promote the development of land. The Commission's role was to ensure that land allocated for development was in fact developed, by channelling it to those who would develop it. It could use its powers of compulsory acquisition to amalgamate land which was in separate ownerships and acquire land whose owners could not be traced. It could purchase land from owners who refused to sell for development or from builders who wished to retain it for future development.

In its first report, the Land Commission gently referred to the importance of its role in acting 'as a spur to those local planning authorities whose plans have not kept pace with the demand for various kinds of development'. Though it hoped that planning authorities would allocate sufficient land, it warned that in some cases it might have to take the initiative and, if local authorities refused planning permission, go to appeal. In its second interim report, a much stronger line was taken. It pointed out that in the pressure areas, it had had only modest success in achieving a steady flow of land onto the market. This was largely because these were areas in which planning authorities were aiming to contain urban growth and preserve open country.

It is not easy to appraise what success the Land Commission achieved. It was only beginning to get into its stride in 1970 when a new government was returned which was pledged to its abolition on the grounds that it 'had no place in a free society'. This pledge was fulfilled in 1971 and thus the Land Commission went the same way as its predecessor, the Central Land Board.

THE CONSERVATIVE YEARS 1970–74

Land prices were rising during the late 1960s (with an increase of 55 per cent between 1967 and 1970), but the early 1970s witnessed a veritable price explosion. Using 1967 as a base (100), prices rose to 287 in 1972 and 458 in 1973. Average plot prices rose from 908 in 1970 to 2,676 in 1973 (DoE, *Housing and Construction Statistics 1969–1979*, table 3).

Not surprisingly, considerable pressure was put on the Conservative government to take some action to cope with the problem, though it was neither clear nor agreed what the basic problem was (Hallett 1977: 135). The favourite explanation, however, was 'speculative hoarding', and it was this which became the target for government action (in addition to a series of measures designed to speed up the release and development of land). A White Paper, *Widening the Choice: The Next Steps in Housing*, set out proposals for a *land hoarding charge*. This was to be levied 'for failure to complete development within a specified period from the grant of planning permission'. After this 'completion period' (of four years from the granting of outline planning permission or three years in the case of full planning permission), the charge was to be imposed at an annual rate of 30 per cent of the capital value of the land. The scheme was clearly a long-term one and, to deal with the urgent problem ('urgent' in political if not in any other terms), a *development gains tax* and a *first letting tax* were introduced.

The development gains tax provided for gains from land sales by individuals to be treated not as capital gains, but as income (and thus subject to high marginal rates). The first letting tax, as its name implies, was a tax levied on the first letting of shops, offices, or industrial premises. In concept, it was an equivalent to the capital gains tax which would have been levied had the building been sold.

Both taxes came into operation at the time when the land and property boom turned into a slump. Indeed, it has been suggested that they contributed to it (Hallett 1977: 137).

THE COMMUNITY LAND SCHEME

The Labour government which was returned to power in March 1974 lost little time in producing its proposals for a new scheme for collecting betterment. The objectives of this were 'to enable the community to control the development of land in accordance with its needs and priorities' and 'to restore to the community the increase in value of land arising from its efforts'. The keynote was 'positive planning', which was to be achieved by public ownership of development land. In England and Scotland, the agency for purchasing development land was to be local government (thus avoiding the inter-agency conflict which arose between local authorities and the Land Commission). In Wales, however, with its smaller local authorities, an *ad hoc* agency was to be created (this became the Land Authority for Wales, now part of the Welsh Development Agency).

In order 'to restore to the community the increase in value of land arising from its efforts', it was proposed that 'the ultimate basis on which the community will buy all land will be current use value'. Sale of the land to developers, on the other hand, would be at market value. Thus, all development value would accrue to the

community. Provisionally, however, development values were to be recouped by a development land tax.

The ensuing legislation came in two parts: the 1975 Community Land Act provided wide powers for compulsory land acquisition, while the Development Land Tax Act 1976 provided for the taxation of development values. Thus the twin purposes of 'positive planning' and of 'returning development values to the community' were to be served. The Community Land Scheme was complex, and became increasingly so as regulations, directions, and circulars followed the passing of the two Acts. The intention was for it to be phased in gradually, thus enabling programmes to be developed in line with available resources of finance, human resources, and expertise.

In the first stage, which started on the 'first appointed day' (6 April 1976), local authorities had a general duty 'to have regard to the desirability of bringing development land into public ownership'. In doing so, they had 'to pay particular regard to the location and nature of development necessary to meet the planning needs of their areas'. To assist them in carrying out this role, they had new and wider powers to buy land to make it available for development.

The second stage was to be introduced as authorities built up resources and expertise. The Secretary of State would make orders providing that land for development of the kind designated in the order, and in the area specified by the order, would pass through public ownership before development took place. These *duty orders* were to be brought in to match the varying rates at which authorities became ready to take on such responsibilities.

When duty orders had been made covering the whole of Great Britain, the 'second appointed day' (or SAD Day as critics dubbed it) could be brought in. This would have had the effect of changing the basis of compensation for land publicly acquired from a market value (net of tax) basis to a current use value basis: that is, its value in its existing use, taking no account of any increase in value actually or potentially conferred by the grant of a planning permission for new development.

The scheme, like its two predecessors, had little chance to prove itself before the return of a

Conservative government. The economic climate of the first two years of its operation could hardly have been worse, and the consequent public expenditure crisis resulted in a central control which limited it severely (Grant 1979; Emms 1980).

Thus three attempts to solve the compensation and betterment problem failed, though the problems to which they were directed are still very much with us. Moreover, as the following discussion shows, there are still attempts to secure the recoupment of betterment.

PLANNING AGREEMENTS AND OBLIGATIONS

The failure of comprehensive schemes for the collection of betterment was one of a number of factors which, in the early 1980s, stimulated an already established trend for increasing the levying of charges on developers. Other influences included a general move from a regulatory to a negotiatory style of development control, increased delays in the planning system, and the financial difficulties of local authorities in providing infrastructure (Jowell 1977a; Sheaf Report 1972).

Planning authorities have had power to make 'agreements' since 1932, but it was not until the property boom of the early 1970s that they became widely used – or, as some argue, abused. The term *planning gain* is popularly used, but with two different meanings. The term can denote the provision of facilities which are an integral part of a development, but it can also mean 'benefits' which have little or no relationship to the development, and which the local authority requires as the price of planning permission. There has been very extensive debate on this issue, and the list of relevant publications is very long. Unfortunately, neither publications nor statutory changes and ministerial exhortations have done much to settle the arguments. The extremes range from the Property Advisory Group's (1981) categorical statement that planning gain has no place in the planning control system, to Mather's (1988) proposal that planning gain should be formalised by allowing local authorities to sell or auction planning consents.

Essentially, the issue is the extent to which local authorities can legitimately require developers to shoulder the wider costs of development: the needed infrastructure, schools, and other local services.

The extremes are easy to identify: the cost of local roads in a development is clearly acceptable, while financial contributions to the cost of running a central library are not. But, of course, most items fall well within these extremes. The general view, supported by a number of studies, was that the majority were legitimate (Byrne 1989; Eve 1992; Rowan-Robinson and Durman 1992a). These studies effectively demolish the argument that there was widespread extortion by way of planning gain.

Fewer than 1 per cent of planning decisions involve planning agreements; the largest proportion are concerned with regulatory matters (contracts, plans and drawings, building materials, etc.); and over half deal with occupancy conditions (for example, restrictions required for sheltered housing, agricultural dwellings, and social housing). Agreements serve an important function in securing the provision of infrastructure necessitated by a development (particularly local roads), and in environmental improvement (such as landscaping). Only a very small number of agreements are concerned with wider planning objectives. In Scotland 'most agreements are useful adjuncts to the development control process'; abuse of power does not present a problem; and for the most part, the benefits secured by agreements have been related to the development proposed: where they have not, the benefits have been of a relatively minor order (Rowan-Robinson and Durman 1992a: 73).

The statutory provisions relating to agreements were amended by the Planning and Compensation Act 1991. Agreements were replaced by 'obligations' and can now be unilateral – not involving any 'agreement' between a local authority and a developer at all. Though the wider debate has been on the ethics of planning gain, this provision in fact deals only with a narrow legal difficulty. A DoE Consultation Paper issued in August 1989 explained that a logjam could arise where the Secretary of State decided that a planning appeal should be allowed if a certain condition were met, but there was no legal basis for imposing

BOX 6.1 PLANNING GAIN: THE PAIGNTON ZOO CASE

The Paignton Zoo case is a revealing case of the extent to which planning benefits are acceptable as legitimate. A proposed development included a 65,000 sq. ft retail store, parking spaces for 600 cars, a petrol station, and the refurbishment of the zoo.

The proposals clearly raised major issues of policy including those set out in PPG 6 *Town Centres and Retail Development* and PPG 21 *Tourism*. There were several conflicting considerations, including the likely effect of the retail development on the town centre, and the precarious economic position of the zoo (which was 'likely to close unless it receives a capital injection of the size that only this proposal is likely to provide, thereby causing a loss to the local economy of approximately 6 million per annum and a significant loss of jobs'). The Secretary of State decided that these and other benefits to tourism and the local economy (together with highway improvements) more than outweighed any harm which might be done to the vitality of the town centre, and he therefore granted planning permission. In the words of the decision, 'the harm likely to arise from the proposals is less clear cut than the effects that would result from the decline and possible closure of the zoo; the balance of advantage lies in favour of allowing the proposal; the zoo's leading role in the local economy places it in a virtually unique position'. However, the Secretary of State stressed that the decision 'should not be regarded as a precedent for other businesses seeking to achieve financial stability'.

Source: JPL (1995: 657)

the condition (typically because it involved off-site infrastructure). The new provision allows a developer to make an agreement to provide the necessary off-site works even if the local authority is not prepared to be a party to the agreement. This seems a small point on which to base the change from 'agreements' to 'obligations'; it is possible that a more important function of the provision was to give the appearance of a change in policy which would curb the alleged excesses of planning gain.

In fact, nothing could be further from the reality. DoE Circular 7/91 had already made it clear that local authorities could negotiate with developers for the provision of social housing. This represented a major extension of the arena of planning agreements. But, in the debates on the 1991 Bill, the minister (Sir George Young) went further:

I think we are all agreed that planning gain is a useful part of the planning system and should be preserved and even encouraged. . . . A planning gain would do more than merely provide facilities that would normally have been provided at public expense. It would provide facilities that the public would never have afforded.

Similarly, the RICS, in its response (1991) to the White Paper *This Common Inheritance*, expressed the hope that agreements would be extended: 'It is hoped that consideration can be given to an increased use of agreements where major developments are proposed so that the community can gain some off-setting benefit, particularly when there is a loss of amenity.' As the report on the Scottish study notes (Rowan-Robinson and Durman 1992a), these views (from such eminent sources) amounted to a major change in opinion since the Property Advisory Group (1981) declared the pursuit of planning gain to be unacceptable.

At the root of this is a significant change in the expected roles of the private and public sectors in land development. Whereas it used to be the case that the responsibilities of developers were clearly limited, it has become generally (even if not unanimously) accepted that the public sector is financially unable to meet the associated costs in the traditional way. The move from a regulatory to a negotiatory style of control is another aspect of this, as has been the willingness of developers to shoulder these costs.

Economic and social factors now loom large in planning decisions. The courts have clearly stated that financial issues can be 'material considerations' in planning, as long as they are secondary to planning matters. Thus, in the case of office development granted (contrary to the local plan) to enable the re-development of the Covent Garden Opera House to be financially viable, it was argued:

> Financial constraints on the economic viability of a desirable planning development are unavoidable facts of life in an imperfect world. It would be unreal and contrary to common sense to insist that they must be excluded from the range of considerations which may properly be regarded as material in determining planning applications. . . . Provided that the ultimate determination is based on planning grounds and not on some ulterior motive, and that it is not irrational, there would be no basis for holding it to be invalid in law solely on the ground that it has taken account of, and adjusted itself to, the financial realities of the overall situation.
> (*R. v. Westminster City Council ex. p. Monahan*, *JPL* 1989: 107).

Social factors may present greater difficulties, as when Lord Widgery held that the London Borough of Hillingdon could not impose a condition that the occupants of a private housing development should be people on the council's waiting list (*R. v. London Borough of Hillingdon ex. p. Royco Homes* [1974] 2 All ER 643). Nevertheless, the matter is not settled – as is instanced by the debate on the role of planning policies (as distinct from housing policies) in the provision of affordable housing.

PLANNING AND AFFORDABLE HOUSING

The stance of the central government on the role of planning in relation to affordable housing has been a curious one for some time. On the one hand, 'planning conditions and agreements cannot normally be used to impose restrictions on tenure, price or ownership', but 'they can properly be used to restrict the occupation of property to people falling within particular categories of need'. Both statements are from Circular 7/91, *Planning and Affordable Housing*, which was an early attempt to wrestle with this politically difficult issue. This was replaced by Circular 9/98 (with the same title), which repeats the warning against tenure

---

BOX 6.2   PLANNING OBLIGATIONS: GENERAL POLICY

Properly used, planning obligations may enhance the quality of development and enable proposals to go ahead which might otherwise be refused. They should, however, be relevant to planning and directly related to the proposed development if they are to influence a decision on a planning application. In addition, they should only be sought where they are *necessary* to make a proposal acceptable in land-use planning terms. When used in this way, they can be key elements in the implementation of planning policies in an area. For example, planning obligations may involve transport-related matters (e.g. pedestrianisation, street furniture and lighting, pavement and road surface-design and materials,

and cycle ways). Planning obligations may relate to matters other than those covered by a planning permission, provided that there is a direct relationship between the planning obligation and the planning permission. But they should not be sought where this connection does not exist or is considered too remote. Planning obligations may have a useful role to play in the planning system. *The tests to apply for their use are that they should be necessary, relevant to planning, directly related to the proposed development, fairly and reasonably related in scale and kind to the proposed development and reasonable in all other respects.*

Source: Circular 1/97, *Planning Obligations* (Annex B, para. B2)

conditions, and defines affordable housing in these terms:

> The terms 'affordable housing or 'affordable homes' are used in this Circular to encompass both low-cost market and subsidised housing (irrespective of tenure, ownership whether exclusive or shared or financial arrangements) that will be available to people who cannot afford to rent or buy houses generally available on the open market.

This is criticised as quite inadequate on three grounds. First, it leads to the provision of small houses for sale at full market prices. Second, housing may be less expensive than other housing in a development, but still not 'affordable' to local people. Third, on resale, houses are sold at full market prices, thus losing the benefit of any discount and also control over future occupants (Chartered Institute of Housing *et al.* 1999).

Though the Circular uses the language of voluntary provision and relies upon developers' contributions being secured through negotiation, neither the purpose nor the effect of its requirements is voluntary (Grant 1999b: 71). Developers are expected to provide affordable housing on developments above a certain size (twenty-five dwellings, or more than one hectare, except in Inner London, where the requirement relates to fifteen dwellings or half a hectare). Where a developer is unwilling to accept such a condition, planning permission may be refused. This policy is seen as a means of catering for a range of housing needs and of encouraging the development of mixed and balanced communities in order to avoid areas of social exclusion. It is to be noted that this policy has no specific legislative provision and, though this does not make it illegal, a developer has little chance of successfully opposing it. An appeal is hardly likely to succeed when the principle is set out in a departmental circular. In fact, at the time of writing, increasing use is being made of the requirement for low-cost housing. The Stockport case (see Box 6.3) exemplifies the situation.

Even more curious is the policy of 'exceptional release' of land, outside the provisions of the development plan, for 'local needs' housing. This is an explicit 'use of the planning system to subsidise the provision of low cost housing through containment of land value'.[2] The extent to which authorities can

BOX 6.3 LACK OF LOW-COST HOUSING LEADS TO DISMISSAL OF APPEAL

McCarthy and Stone proposed to develop a brownfield site in Stockport with two blocks of sheltered flats for elderly owner-occupiers. The development would have met a market demand and would have improved the character and appearance of a derelict site. However, it was rejected on appeal because the scheme did not provide any low-cost housing. The inspector said that he considered that the failure to make provision for an element of low-cost housing on what is a suitable site would be so harmful as to amount to a compelling planning objection. He also maintained that the provision of affordable housing would not render the development cost of the flats uneconomic. Moreover, there was no 'convincing evidence that the development's success would be jeopardised because of any incompatibility between affordable housing and sheltered housing for the elderly'.

Source: *Planning* 17 September 1999: 9

achieve planning benefits depends, of course, on their bargaining power, which in turn may be related to current (and local) economic conditions. The situation varies over time and by region. In some circumstances, 'getting a developer to build anything is, in our eyes, a planning gain' (quoted in Jowell 1977a: 428); in others, the local pressures for development are so strong that local authorities can secure considerable benefits. The London situation, however, is unique, with housing costs at record levels, and acute pressures on affordable housing.[3] By 1998 the number of households on local council waiting lists had risen to 178,000, and the average private rent for a three-bedroom dwelling was £349 a week. (By contrast, the rent of a semi-detached house in Leeds was £85 a week.) The problem will worsen as the number of households in London increases (by 600,000 between 1996 and 2021, according to the latest household projections) and new provision remains low (only 11,170 housing association dwellings were built in 1998, while local authority building has ceased). Only a major increase in both the allocation of housing land for affordable housing and the necessary funding will make any improvement.

A (perhaps surprisingly) little-used policy is to reserve new housing for local people. This is particularly appropriate in areas where there is great competition for housing on the part of commuters or holiday home buyers. One authority that has designated areas where new houses are restricted to locals is the North York Moors National Park. This policy was introduced in 1992, and was being extended at the end of 1999. An alternative, of course, is the provision of social rented housing, where occupancy is straightforward. Unfortunately, funding for such housing is scarce.

The growth of planning agreements gives rise to a number of concerns. The ethics of bargaining are debatable; there is scope for unjustifiable coercion; and equal treatment as between applicants can be abandoned in favour of charging what the market will bear at any particular time. Additionally, bargaining is a closed, private activity which sits uneasily astride the current emphasis on open government and public participation.

Much of the difficulty in this area may arise from the discretionary nature of the British planning control system, in which negotiation is an important feature. However, studies of US land use regulation (which supposedly emphasises property rights and reduces development uncertainties) show that negotiation is equally prevalent there (Cullingworth 1993: chapters 6 and 7). An essential issue is that though development rights in land are nationalised, their associated values are privately owned. Much of the case for 'planning gain' is that it is a means of capturing some of this value for the public benefit.

LAND POLICIES IN THE 1980S

Though the Community Land Act was repealed by the Conservative government in 1980, local authorities still retained considerable powers of compulsory acquisition of land. They could acquire, with the consent of the Secretary of State, any land in their area which

(a) is suitable for and required in order to secure the carrying out of development, redevelopment, or improvement; or
(b) is required for a purpose which it is necessary to achieve in the interests of the proper planning of an area in which the land is situated.

These powers (which are still possessed by local authorities under section 226 of the 1990 Act) specifically provided for compulsory acquisition of land for disposal to a private developer. Indeed, the government made it clear that these 'planning purposes' powers (which could be of particular importance in bringing land on to the market) were generally to be used to assist the private sector. The compulsory purchase regime is now under review. The *Interim Report*[4] of the review notes the widespread perception that the process 'is slow in operation, inefficient, and not always fair to those whose property is acquired' (p. 7). As a result, compulsory purchase is now less often used than local authorities would like it to be, and is therefore not the aid to urban regeneration that it might be (Freilich 1999). Such powers have been widely used for land assembly, and the redevelopment of many towns from the 1950s to the 1990s would have been impossible without them, though many of the results are hardly an advertisement for increasing their use now. In the words of the minister, the main objective is to create a system 'which is efficient, effective and fair'. That last criterion becomes even more important in the context of the Human Rights Act (Redman 1999).

Additionally, the Secretary of State has some formidable powers. First, he or she has the reserve power to direct a local authority to make an assessment of land available and suitable for residential development. Second, the powers to acquire any land necessary for the public service include the authorisation of acquisitions 'to meet the interests of proper planning of the area, or to secure the best or most economic development or use of land'. (Ironically these provisions are a modified re-enactment of a section of the repealed Community Land Act.) However, little use has been made of these powers; instead, reliance has been placed on ensuring that local planning authorities planned for sufficient available land.

LAND AVAILABILITY AND URBAN CAPACITY STUDIES

It was a major objective of the postwar planning system to ensure that land required for development would become available – if necessary by the use of compulsory purchase powers. As previous discussion has shown, things did not work out like this despite three attempts (in 1947, 1967, and 1975). Except in special cases, such as new towns and comprehensive development areas, there has been little use of compulsory purchase powers. Thus the land 'allocations' in plans remained just that: allocations on paper. There is no necessary relationship between the allocation of land and its *availability*. It is therefore not surprising that there has been considerable controversy over the extent to which allocated land is in fact available for development. In Hooper's (1980) words,

> The planning system and the house building industry operate not only with a different definition, but with a different conception, of land availability – the former based on public control over land use, the latter on market orientation to the ownership of land.

However, land availability studies were the centre-pin of the planning system[5] until they were supplanted by urban capacity studies.

This new system, introduced by the revised PPG 3, *Housing* (2000), represents a major change in policy. It places emphasis on the reuse of land in urban areas. This is the favoured location for new development in view of its assumed 'sustainability'. This is interpreted in various ways: it is held that urban locations reduce traffic (and emissions) and help to safeguard the countryside; they provide accessibility to goods and services, and allow new energy-saving technologies such as combined heat and power systems; and they provide a more lively and interactive social milieu.[6] But, above all, the policy has widespread popular support, particularly in terms of 'saving the countryside'. Eloquent of this is the Select Committee's forthright declaration that 'the only way that the government's proposals for urban regeneration and for greater use of recycled land can be achieved are by restricting the amount of greenfield land brought forward'.[7] This has been backed up by a new *Greenfield Housing Direction* (2000), which requires local authorities to consult the Secretary of State on planning applications for major housing developments of more than 5 hectares or 150 dwellings.

The same commitment is evident in changes in government policy, above all the commitment to maximising the reuse of previously developed land and the conversions of buildings for housing in order both to promote regeneration and to minimise the amount of greenfield land being taken for development. This policy permeates the revised PPG 3, in which the policy is spelled out in some detail. Potential sites should be assessed against a number of criteria such as the availability and net cost of previously developed sites; their location and accessibility by public transport; the capacity of the infrastructure and services such as schools and hospitals, and the potential for developing and sustaining local services; and physical constraints on development.

A sequential approach to the phasing of sites is introduced under which greenfield sites should not be developed for housing until the following options have been considered:

- using previously developed sites within urban areas;
- exploiting fully the potential for the better use and conversion of existing dwellings and non-residential properties;
- increasing densities of development in existing centres;
- releasing land held for alternatives uses, such as employment; and
- identifying areas where, through land assembly, area-wide redevelopment can be promoted.

Local planning authorities are asked to undertake 'urban capacity' studies. These are to replace the housing land availability studies as the principal means for determining the location of potential housing sites. No guidance has as yet been provided on the methodology of these,[8] other than that they should take account of the National Land Use Database and examine the implications of policies for increasing densities, reducing car parking, and reviewing the potential over allocation of land for employment. Regional planning bodies are to use the capacity studies in proposing land recycling targets and allocating them among planning authorities.

The government's target is for 60 per cent of new housing to be provided on previously developed land or through conversions. This proportion is regarded by some (such as the TCPA) as overambitious, and by others (such as the CPRE) as too low.[9] The actual figure for 1998 (the last year for which statistics were obtained) was 57 per cent or 53 per cent excluding conversions (DETR, *Land Use Change in England Bulletin* 2001 and *Urban Task Force* 1999: 174). The proportion varies considerably among regions, from 35 per cent in the South-West to 82 per cent in London. Reliable statistics are scarce on this issue, and the Urban Task Force considered the evidence carefully. As an illustration of the difficulties, there were two quite different figures for the amount of derelict land: 34,500 hectares according to the Derelict Land Survey, and 17,300 hectares recorded in the National Land Use Database. The Urban Task Force devised its own estimates for the various types of recyclable land and also of the number of dwellings that are likely to be accommodated on this land under current policies. It can be argued that these estimates are heavily influenced by wishful thinking, particularly since they are almost three times the estimate of the National Land Use Database. They are shown in Table 6.1.

Despite the questionable nature of these figures, they imply that the 60 per cent target may not be achieved, particularly in London and the South-East. However, in the view of the Task Force, changes in current policies would make them achievable. The types of policy changes envisaged include increased urban regeneration, increased densities for development, allocation of more urban land to housing, and greater recycling of underused and empty buildings.

Table 6.1 Estimated Number of Houses Likely to be Built on Previously Developed Land, England, 1996–2021

Vacant previously developed land	163,510
Derelict land/buildings	170,210
Vacant buildings	101,800
Projected windfall and other sources (1996–2021)	1,526,000
Total	1,961,520

Source: Urban Task Force (1999a: 305), updated by Government Statistical Bulletin 500: *National Land Use Database*, which gives a much less optimistic forecast for the reuse of existing vacant buildings (101,800 rather than 247,000). The period 1996–2021 is that used for the current household projections.

On a range of assumptions, the Task Force (p. 305) manages to reach an 'attainable target' of 62.2 per cent.

It could equally well be argued that, rather than an increase, it is more likely that there will be a decline. The argument here is that the increase in recycled land for housing has been due solely to the development of vacant brownfield sites which have been relatively easy to deal with (Llewelyn-Davies 1996). There has been little or no increase in the supply of other urban redevelopment sites or unused vacant land (Table 6.2).

Certainly, the actual figures calculated by the Task Force are debatable, though, in a somewhat cavalier manner, it notes that it is 'not making any great claims' for the figures, but nevertheless maintains that they show that 'over a significant period, the cumulative effect of a consistent and continued policy commitment could be considerable'. In one sense, therefore, its detailed calculations are of less import than the message it tried to convey, which is captured by the title of its report: *Towards an Urban Renaissance*.[10]

A major issue in the urban renaissance drive is the fear that it will lead to 'town cramming'. An early expression of this was the response to the 1990 EC *Green Paper on the Urban Environment*. At what point do higher urban densities give rise to cramming? There is, of course, no mathematical answer to the question,

Table 6.2 Previous Use of Land Changing to Residential Use, England, 1985–93

	1985 (%)	1989 (%)	1993 (%)
All rural uses	52	47	39
Urban uses:			
Redeveloped sites	27	29	24
Brownfield sites (vacant and previously developed)	11	14	25
All vacant land previously developed	38	43	49
Vacant: not previously developed	10	10	12
All urban	48	53	61
All uses	100	100	100

Source: Breheny (1997: 212)

though there is an abundant literature on the issue.[11] Aspects of design are often of greater significance, as are even more elusive elements of 'character'. But most important is the very 'richness of cities', so well captured in the 1999 report with this title by Worpole and Greenhalgh.

What is conspicuously missing from much of the debate is the question of the acceptability of increased densities (urban compaction). It is not easy to measure this in any straightforward way since the term is capable of varying interpretations, but a good proxy is provided by the findings of the authoritative DoE-sponsored *Housing Attitudes Survey* (Hedges and Clemens 1994). This showed 'central urban dwellers to be much less satisfied than those in the suburbs, and these again less than those in rural areas'. This finding is reinforced by the analysis of population density, 'which shows a marked inverse relationship between satisfaction and density'.[12] The survey also showed a clear preference for houses rather than flats. This can hardly be surprising, since this has been a consistent finding of housing research, but the issue has gained prominence in view of the very large increase in one-person households shown in the household projections (who make up three-quarters of the total increase). Though it may seem reasonable to assume that many of these will want small dwellings, possibly in flats, the evidence is that the greater part of the demand

is for houses with gardens (Hooper *et al.* 1998). 'A preference for a flat starts at 11 per cent, falls to 1 per cent as the family grows, and then climbs to 31 per cent among single older people' (Hedges and Clemens 1994). The 1996 White Paper *Household Growth: Where Shall We Live?* concluded that despite the increase in small households, 'there is little evidence of any increase in demand for smaller housing units; there has, moreover, been a decline in one bedroom houses and flats completed in the last ten years, and a growth in the number of larger houses (four bedrooms)'.

Even more persuasive is the fact of long-term decentralisation from the cities. This has eased the traditional problems of cities, though it has proved difficult to attune policies to the problems which remain. Movement out of the cities has been a dominant feature of demographic and economic geography for a century. (However, it should be stressed that the arithmetic of this is usually expressed in net terms, ignoring the fact that people are moving *into* as well as out of urban areas.)

Much of the debate on the urban renaissance is couched in terms of redevelopment of the inner city, ignoring the problems and opportunities of the suburbs (where the majority of people live and where much development activity has been concentrated during the 1990s). The suburbs do not typically need large-scale redevelopment plans but, as a Civic Trust study shows, they can be in need of careful improvement to arrest decline and to enable them 'to play a more positive and sustainable role within city regions' (Gwilliam *et al.* 1998).

There are also a variety of measures that can improve both suburban and inner-city environments while, at the same time, providing additional housing. Policies in relation to empty properties can clearly make a modest but useful contribution to both, as the work of the Empty Homes Agency (Plank 1998) demonstrates.[13] The LOTS ('living over the shop') scheme was less successful, largely because of the lengthy and often difficult negotiations required with the owners of the shop! It does, however, have potential when included as an element of wider-based regeneration schemes.[14]

In addition to housing issues, a relatively neglected matter is that of the geography of jobs. Though there are no figures on this for recent years, employment in the 1980s showed an employment exodus from urban areas. Is this continuing? Patterns of commuting have become more complex, and there is now suburb-to-suburb and even city-to-suburb commuting. There are many questions here, and few answers. As more housing is provided in the cities (often involving the replacement of places of employment), will reverse commuting grow? Does this matter? Is there a need for policy intervention, and could this prove practicable? Such questions are being addressed in an ongoing TCPA project (Breheny 1999b). In the meantime, there is a more urgent issue of unemployment among inner-city residents. Continued replacement and conversion of industrial, warehousing, and office buildings will further reduce restricted employment opportunities.

HOUSEHOLD PROJECTIONS

Demographic analysis and forecasting are crucial to any method of determining housing needs and land requirements. Projections of households are made on a periodic basis by the Government Statistical Service. Until quite recently these were widely accepted as a basis for policy. The national figures are used by the central department to determine regional and county housing requirements. Concern about these projections grew in the 1980s, particularly in the South-East, 'where years of continuous housing development have generated a militant resistance to what are seen as excessive impositions of yet further housing development' (Breheny 1997). The household projections for the period 1991–2016, published in 1995, gave rise to an even more vociferous and wider debate, which was kept informed by a series of CPRE publications that strenuously put forward the case both for protecting the countryside against housing development and for disparaging the methodology used in the official household projections.[15]. The press also took up the popular outcry and led a 'greenfield campaign'. There were sufficient legitimate grounds of criticism in their arguments for them to be credible, particularly among those who were convinced of their conclusions. A DETR research project reinforced some of these criticisms: for example, that the projections 'extrapolate forward past trends in a technically complex way, but take limited account of the underlying causal processes or relationships that might affect the rates at which households form' (Bramley *et al.* 1997).

The official household projections are now more than a technical input to the planning system: they are matters of widespread controversy. Particularly attractive to critics are two points: first, that no projections are wholly satisfactory; and second, that housing supply does have *some* effect on household formation. This latter point is popularly viewed in terms which are similar to the now accepted argument that new roads generate traffic. Thus the shortcomings of 'predict and provide' which were seen to be valid in relation to roads were translated into housing terms: more houses lead to more households in the same manner as more roads lead to more traffic. This appealing (and greatly exaggerated) argument has been taken over in the government's redesign of the arrangements for determining housebuilding needs at regional and local authority level (DETR, *Planning for the Communities of the Future*, 1998). In place of 'predict and provide' there is now 'plan, monitor and manage'. This is a neat piece of political semantics, which appears to mean that both the assessment of housing requirements and its distribution within the region should be kept under review, and if there are signs of either under- or over-provision, both RPG and development plans should be reviewed accordingly.[16] It remains to be seen if this makes any real difference. Interestingly, in his evidence to the Select Committee, the Deputy Prime Minister, John Prescott, gave his view that it did not: 'I do not think that planning, predict and provide is contradictory to planning, monitoring and managing; one is a process and the other one is how you achieve it'.[17]

More fundamentally, the arguments about household predictions reflect a widespread opposition to change. This is a compound of a desire to maintain existing amenities, fears of increased traffic and congestion, and the traditionally strong countryside preservation ethic. The battles over the latest regional planning policy statements (RPG) amply illustrate this. They are dressed up in emotive and vague slogans which confuse the issues. Thus the draft RPG for the South-East[18] states that 'the countryside should be more strongly protected from inappropriate development', but, as the Panel report on the public examination pointed out, though this sounds incontrovertible at first blush, the use of the term 'inappropriate' without qualification begs the question of what is inappropriate. The report continues:

> All too often we found that it simply meant any form of urban expansion, particularly for house building. Whilst it must be an objective to minimise the loss of countryside to urban expansion, we do not consider that this one objective should dominate all others. It should not result in denying the opportunity of a decent home for all who desire one in the region, nor should it stand in the way of economic success, nor – and we see this as a particular danger – should it compromise real urban renaissance by providing an excuse for town cramming. . . . If urban concentration is forced upon towns for reason of preserving countryside and without due balance of the other elements of urban renaissance, then the cities and towns will simply become worse places to live in, and the pressures on the countryside will be unnecessarily increased.[19]

This critique of the anti-development stance taken by SERPLAN and the local authorities in the South-East permeates the Panel's report. The housebuilding strategy proposed in the draft PPG interprets 'plan,

monitor and manage' to mean 'short-term incremental decisions of planning to meet need as and when it arises'. (As a result, it proposes a baseline housing provision of 862,000 dwellings between 1991 and 2016, with increases to this figure as needed; the Panel proposed a figure of 1.1 million for this period.[20]) In the Panel's view, this is 'the antithesis of a plan-led system'. The essence of planning lies in taking a view of what is likely to happen in the future and planning to meet it. It continues:

> The SERPLAN approach, in our view, will serve only to perpetuate planning by appeal resulting on the ground in disjointed increments of added on development in apparently random locations with little coherence to the established structures of towns nor genuine opportunities for their development to be accompanied by planning extension of public transport and other infrastructure. This is not a sustainable way to meet development needs, and it is hardly surprising that it attracts so much opposition from local people when it occurs.[21]

The SERPLAN strategy is by no means unusual: indeed, it typifies much of current planning for development needs. It reflects public opinion, and presents a major problem for central government. It is difficult to see how responsible planning at the regional level can be squared with planning which is responsive to public opinion.

Nevertheless, there does seem to be a genuine desire to move to a 'bottom-up approach' to the estimation of housing requirements. One reason for this is that it is difficult to predict migration across regional boundaries, especially in the long term. Migration flows can change with economic conditions, and indeed in some areas there may be cross-regional influences from more than one direction. (The East Midlands, for example, has been simultaneously experiencing development pressures from the South-East, West Midlands, Greater Manchester and South Yorkshire. In the North, pressures have crossed county boundaries from West Yorkshire to North Yorkshire.) As a result,

> All this strongly underlines the need for a flexible two-stage approach: we must plan to meet medium-term (15–20-year) projections, while recognising that in the short term conditions may vary greatly. This means that land allocations need to be established in regional guidance and structure plans to meet longer-term needs,

while releases from those allocations are made locally according to a system of close year-by-year monitoring agreed by the authorities with each regional conference.

(Breheny and Hall 1996: 45)

Much of the opposition to the household projections is concentrated in the South. This is, at least in part, due to the fact that migration from the North to this region has markedly increased housing demand in this part of England. Indeed, 'the speed of migration appears to have significantly increased with the upturn in the economy since 1993'.[22] The government's concern for intra-regional policy is not matched by its action on inter-regional issues.[23] Indeed, the essential remit of the regional development agencies is the fostering of regional economic development, and accordingly this is being fostered in the South-East, as in the other regions.[24]

There is also a marked movement out of the urban areas, which has brought about a repopulation of small towns in the countryside. As Peter Hall has pointed out, 'already by the 1980s, the map of population change was the exact reverse of the equivalent map of the 1890s: the counties and the districts that were then suffering the biggest population losses have become the areas with the biggest gains' (Hall and Ward 1998: 106). What is interesting about this centrifugal movement is what Champion and Atkins have termed 'the counterurbanisation cascade'. As can be seen from Table 6.3, 'at the beginning of the 1990s, migration within Britain was producing a clear redistribution of population down the settlement hierarchy from larger metropolitan areas to medium-sized and smaller cities and towns and more rural areas' (Champion and Atkins 1996: 26).[25] However, the 2000 Urban White Paper notes that during the late 1990s there were indications of a slowing of population decline in the metropolitan areas, and some were even growing, particularly London.

Migration has not, however, been at a level to bring about any general collapse of housing demand in the north of England, but it has led to increased 'departures and higher vacancies in local authority housing and has produced local surpluses in the least popular localities' (Holmans and Simpson 1999). It is also important to note that it is in the older industrial cities of the North that the greatest scope exists for the

Table 6.3 Population Changes Resulting from Within-Britain Migration 1990–91

District type	Population number	Net migration number	Net migration percentage
METROPOLITAN BRITAIN	19,030,230	-85,379	−0.45
Inner London	2,504,451	-31,009	−1.24
Outer London	4,175,248	-21,159	−0.51
Principal metropolitan cities	3,922,670	-26,311	−0.67
Other metropolitan districts	8,427,861	-6,900	−0.08
NON-METROPOLITAN BRITAIN	35,858,614	85,379	0.24
Large non-metropolitan cities	3,493,284	-14,040	−0.40
Small non-metropolitan cities	1,861,351	-7,812	−0.42
Industrial districts	7,475,515	7,194	0.10
Districts with new towns	2,838,258	2,627	0.09
Resort, port and retirement	3,591,972	17,736	0.49
Urban–rural mixed	7,918,701	19,537	0.25
Remote urban–rural	2,302,925	13,665	0.59
Remote rural	1,645,330	10,022	0.61
Most remote rural	4,731,278	36,450	0.77

Source: Champion and Atkins (1996) and Hall and Ward (1998: 106)

Table 6.4 Population of the UK 1981–97, and Projected 2001–2021 (in millions)

	1981	1991	1997	Projected 2001	Projected 2011	Projected 2021
England	46.8	48.2	49.3	49.9	51.2	52.5
Wales	2.8	2.9	2.9	2.9	3.0	3.0
Scotland	5.2	5.1	5.1	5.1	5.0	5.0
N. Ireland	1.5	1.6	1.7	1.7	1.7	1.7
U.K.	56.4	57.8	59.0	59.6	60.9	62.2

Source: Annual Abstract of Statistics 1999 (Table 5.1)

BOX 6.4 HOUSEHOLD PROJECTIONS FOR ENGLAND TO 2016

The number of households in England is projected to grow from 20.2 million in 1996 to about 24.0 million in 2021, an increase of 3.8 million or about 150,000 households per year. Slightly more than three-quarters of the projected increase in the number of households can be attributed to changes in the size and age structure of the adult population.

The South East, East of England and the South West are all projected to have around a quarter more households in 2021 than in 1996. For London and the East Midlands growth is around a fifth, and in other areas projected growth is significantly lower. The North East has the lowest projected growth of just 8 per cent.

If international migration increased or decreased by 40 thousand per annum over the projected period, this could mean a projected change at national level at 2021 of over 0.4 million households. Similarly, if real interest rates throughout the period were one percentage point higher or lower, the projected number of households in 2021 could change by 0.2 million.

Source: DETR (1999) *Projections of Households in England to 2021* (selected passages from pp. 5–6)

---

development of brownfield sites. In the newer, less industrialised locations, the prospects for recycling are the poorest.[26] Geographical factors are not the only important issues: others include the state of the local economy, and also the 'flow' of previously developed sites.[27]

NEW SETTLEMENTS

The conclusion of the new towns programmes, coupled with increasing concern with the 'land for housing' problem, naturally prompted debate on additional new towns. The TCPA had traditionally maintained that this should be a major plank in regional policy but, during the 1980s, against the background of a buoyant housing market, proposals came from the private sector for private enterprise new towns that would fill the gap left by the completion of the existing new towns. The best known of these came from the now disbanded Consortium Developments, which proposed a ring of new villages around the South-East which would form 'balanced communities' developed to high standards of design.

Consortium Developments Ltd, by working on a relatively large scale, can negotiate a keen price that allows investment in a quality product: a high quality infrastructure in the paving and road surfaces, high quality landscaping, sensitive design of public spaces, variety in both form and tenure of housing provision, and a wide range of supporting facilities.

(Roche 1986: 312)

These were words in the direct tradition of the new towns movement, but their spokespersons now had to contend with a sophisticated planning machine. Proposals for Foxley Wood in Hampshire, Stone Bassett in Oxfordshire, Westmere in Cambridgeshire, and Tillingham Hall in Essex were all rejected on appeal. As Hebbert (1992: 178) comments, their experience 'demonstrated that even the presence of the most radical free enterprise British government of recent times is no guarantor of profitable large scale private developments in green field sites'. However, they have not been completely ruled out: the 1992 version of PPG 3 notes (with no conscious irony) that there had now been 'considerable experience of planning proposals [sic] for new settlements' which had 'almost invariably been deeply controversial'. It advised that future proposals should be contemplated only in cases where they represented a clear expression of local preference supported by local planning authorities. Politically, the importance attached to 'local choice' effectively meant that any proposal for a new settlement was likely to be killed, though a study was commissioned of 'alternative development patterns for new settlements.'[28]

The analysis of the differing types of development did not go very far in demonstrating the superiority of any one type of development over another.[29] This is not surprising: *general* issues of urban form are of limited practical value since the real problems are not general but site specific. The advantages and disadvantages of particular development forms vary according to the features of alternative sites and their location in a specific subregion (and, with larger development, perhaps the wider region as well). They will vary also according to the size, character, and purpose of the development, its transport links and potentialities, and its present and future relationships with the surrounding areas. Additionally, there are

issues of finance, administration, politics, and suchlike that can prove to be of decisive importance. All these (and no doubt other) factors combine to make generalisations highly problematic, and thus any major development proposal requires thorough and lengthy study and negotiation. Given the high sensitivity to development almost anywhere (no doubt increased by the generally poor quality of design), it is not surprising that proposers of new settlements have had a very tough time making any progress at all. (An American environmental acronym points to the problem: BANANA – Build Absolutely Nothing Anywhere Near Anything.)

In the meantime, development has proceeded (or did not proceed) on particular sites for which builders sought planning permission. This non-planning approach was checked by a process that contained one or more of the elements of strong local opposition, public inquiries, and ministerial decisions. Somewhere buried in this process was a vestige of planning policy, but it was a hit-or-miss affair. Certainly, it was a far cry from positive planning or ensuring that the right development went ahead at the right place at the right time. But, of course, the basic dominant political philosophy not only was unsympathetic to 'planning policies', but held that market mechanisms were superior. And so little progress was made in fashioning the planning system to the needs of the time. The question now is what difference a new government is making.

It is too soon to attempt to answer this question. Much thought has been given to recasting the planning system, both procedurally and substantively, but the difficulties are all too apparent. Major changes involve either huge public expenditure, particularly on infrastructure, or strong opposition from vested interests. New settlements would encounter both. As a result, there has been little change in the attitude to new settlements on the part of the Blair government, though the revised PPG 3 (2000) is arguably slightly more positive:

The Government is not against new settlements and believes that in the right location and with the right concept, they can make a contribution to meeting the need for housing. However, the cost of developing a new community from scratch, including the full range of new services and infrastructure, means that they will not always be a viable solution. New settlements will not be acceptable if their principal function is as a dormitory of an existing settlement. New settlements, whether large-scale additions to existing settlements or completely new, may under certain circumstances prove to be a sustainable development where

- they are large enough to support a range of local services, including schools, shops and employment;
- they exploit existing or proposed public transport by locating in a good quality public transport corridor;
- they can make use of previously used land; and
- there is no more sustainable alternative.

Proposals for 'larger new settlements' have to be brought forward through the new regional guidance machinery (discussed in Chapter 3). It is warned that 'proposals for new settlements will be controversial and all schemes will need to be agreed between the tiers of plan-making authorities'. That this warning is fully justified is illustrated in the declaration of the Sane Planning in the South-East protest group:

The Sane Planning in the South East protest group have presented to the Secretary of State a declaration to mark the tenth anniversary of the protest against Foxley Wood (when an effigy of the then Secretary of State was burned). The group maintains that new settlements still have no place in the South East.

(*Planning* 30 July 1999)

More favoured are planned extensions to existing urban areas. Indeed, it is even contemplated that green belt boundaries may have to be reviewed where possibilities for development within urban areas are limited. However, the Panel Report on the South East Draft Regional Planning Guidance proposed that 'areas of plan-led expansion should be designated in Ashford, the Milton Keynes/Bedford/Northampton triangle, the Crawley/Gatwick area, and an area close to Stansted.[30] Though the report does not suggest any special mechanism for the last three of these, for Ashford it comments that 'substantial town expansion, possibly up to a population of 150,000, should be assisted by action under new town legislation'.

GREEN BELTS

The policy of maintaining an adequate supply of land for housing can be difficult to reconcile with policies relating to green belts and the safeguarding of agricultural land. Though 1987 saw a major policy shift on the latter (which is discussed later), green belts have, for a variety of reasons, remained a strong policy issue for both central and local government, well supported by public opinion.

Green belt policy emerged in 1955 after the expression of considerable concern at the implications for urban growth of the expanded house building programme. Unusually, the policy can be identified with a particular minister, Duncan Sandys (who later made another contribution to planning with the promotion of the Civic Trust and the Civic Amenities Act). Sandys' personal commitment involved disagreement with his senior civil servants, who advised that it would arouse opposition from the urban local authorities and private developers who would be forced to seek sites beyond the green belt. Experience with the Town Development Act (which provided for negotiated schemes of 'overspill' from congested urban areas to towns wishing to expand) did not suggest that it would be easy to find sufficient sites. Sandys, however, was adamant, and a circular was issued asking local planning authorities to consider the formal designation of clearly defined green belts wherever this was desirable in order to check the physical growth of a large built-up area; to prevent neighbouring towns from merging into one another; or to preserve the special character of a town.

The policy had widespread appeal, not only to county councils, which now had another weapon in their armoury to fight expansionist urban authorities, but also more widely. One planning officer commented that 'probably no planning circular and all that it implies has ever been so popular with the public. The idea has caught on and is supported by people of all shades of interest.' Another noted that 'the very expression *green belt* sounds like something an ordinary man may find it worthwhile to be interested in who may find no appeal whatever in "the distribution of industrial population" or "decentralisation". . . . Green

belt has a natural faculty for engendering support'
(Elson 1986: 269).

The green belt also formed a tangible focal point for
what is now called the environmental lobby. However,
initially its biggest support came from the planning
profession, which in those days still saw planning in
terms of tidy spatial ordering of land uses. Desmond
Heap, in his 1955 presidential address to the (then)
Town Planning Institute, went so far as to declare that
the preservation of green belts was 'the very *raison d'être*
of town and country planning'. Their popularity,
however, has not made it any easier to reconcile
conservation and development.

The green belt policy commands even wider support
today than it did in the 1950s. Elson concluded his
1986 study with a discussion of why this is so:

> It acts to foster rather than hinder the material and non-
> material interests of most groups involved in the planning
> process, although it may be to the short term tactical
> advantage of some not to recognise the fact. To *central
> government* it assists in the essential tasks on interest
> mediation and compromise which planning policy-
> making represents. . . . To *local government* it delivers a
> desirable mix of policy control with discretion. To *local
> residents* of the outer city it remains their best form of
> protection against rapid change. To the *inner city local
> authority* it offers at least the promise of retaining some
> economic activities that would otherwise leave the area;
> and to the *inner city resident* it offers the prospect, as well
> as often the reality, of countryside recreation and relaxa-
> tion. To the *agriculturist* it offers a basic form of protection
> against urban influences, and for the *minerals industry*
> it retains accessible, cheap, and exploitable natural
> resources. *Industrial developers* and *housebuilder* complain
> bitterly about the rate at which land is fed into the
> development pipeline, yet at the same time are dependent
> on planning to provide a degree of certainty and support
> for profitable investment. Planning may be an attempt to
> reconcile the irreconcilable, but green belt is one of the
> most successful all-purpose tools invented with which to
> try.
>
> (Elson 1986: 264)

The latest policy statement on green belts in
England (the revised PPG 2 of 1995) confirms the
validity and permanence of the green belts policy.
Green belts, now cover over one and a half million
hectares (12 per cent) of England. The general location
of the green belts is shown in Figure 6.5. Table 6.5

and Table 6.6 give the area of green belt in England
and Scotland respectively. Until recently there has
been no formal green belt policy in Wales. However,
the Welsh planning guidance is encouraging them in
the most heavily populated areas. A proposal for the
first green belt, between Newport and Cardiff, was
included in the draft UDP for Newport.

The 1993 study by Elson *et al.*, which was
undertaken at a time when the earlier (1988) PPG 2
was operative, concluded that the green belts had been
successful in checking unrestricted sprawl and in
preventing towns from merging. Green belt boundary
alterations in development plans had affected less than
0.3 per cent of green belts in the areas studied over an
eight-year period. Most planning approvals in green
belts had been for small-scale changes which had no
significant effect on the open rural appearance of green
belts. The appeal system had strongly upheld green
belt policy.

The relationship between green belt restraint and
the preservation of the special character of historic
towns was much more difficult to evaluate. Though
the idea had 'a well-established pedigree', and though
the green belt boundaries were particularly tight, there
was little evidence to connect policy and outcomes. It
was difficult also to assess how far green belts had
assisted in urban regeneration. Though the green belts
did 'focus development interest on sites in urban
areas', local authorities tended to regard the creation
of jobs as more important than any land development
objective *per se*. Indeed, urban regeneration was often
seen as requiring the selective release of employment
sites in the green belt. The supply of adequate sites
within urban areas was not sufficient for development
needs (though it might be increased by an expanded
programme of land reclamation). Moreover, refusal
to allow development on the periphery of an urban
area could lead to leap-frogging beyond the green
belt, or development by the intensification of uses in
towns located within the green belt. A note is made
of the suggestion that 'the inner city will rarely be
a substitute location for uses seeking planning
permission on the urban fringe:

> The housing market potential in the two locations is quite
> different (in terms of the size and price range of houses

Key

Green Belt

Urban areas

Aberdeen

Falkirk/
Grangemouth

Greater Glasgow

Edinburgh

Ayr/Prestwick

Newcastle
upon Tyne

York

Blackburn

Leeds

Liverpool

Manchester

Sheffield

Stoke on Trent

Nottingham

Derby

Birmingham

Coventry

Cambridge

Gloucester

Oxford

London

Bristol

Bournemouth

0   50   100   150   200   250 Kilometres

Figure 6.1 Green Belts in the UK

which may be marketed for example), and many of those developing other uses require the better accessibility (normally by private car) which a peripheral or outer location affords.

(Elson *et al.* 1993: para. 2.37)

There was seen to be a clear need for further research here. One piece of DoE-sponsored research on the green belt which is particularly striking was undertaken on the Oxfordshire settlement strategy, which concentrates development in selected country towns beyond the Oxford green belt. It was intended that this would facilitate the provision of public transport. Things worked out very differently. A study of travel

Table 6.5 Green Belts, England, 1997

	Green belt area hectares
Tyne and Wear	52,500
York	25,400
South and West Yorkshire	252,800
North-West	251,700
Stoke-on-Trent	44,100
Nottingham and Derby	62,000
Burton and Swadlincote	700
West Midlands	230,400
Cambridge	26,700
Gloucester and Cheltenham	7,000
Oxford	35,100
London	512,900
Avon	68,500
SW Hampshire and SE Dorset	82,300
TOTAL ENGLAND	1,652,300

Source: DETR Information Bulletin 1183, December 1999, *Green Belt Statistics: England 1997*

Notes: North-West includes Greater Manchester, Merseyside, Cheshire, and Lancashire; London excludes metropolitan open land; SW Hampshire and SE Dorset includes the New Forest area. The 1997 green belt statistics for England cannot be compared with earlier figures since they are based on a new and more accurate methodology. Details are given in *Green Belt Statistics: England 1997*, DETR Information Bulletin 1183, 8 December 1999.

Table 6.6 Green Belts, Scotland, 1999

	Green belt area (hectares)
Aberdeen	23,039
Ayr/Prestwick & Troon	3,024
Clackmannan	981
Edinburgh	15,869
Falkirk/Grangemouth	3,803
Glasgow	109,917
TOTAL SCOTLAND	156,633

Figures supplied by the Scottish Executive Development Department

patterns of the new residents in three of these towns (Bicester, Didcot, and Witney) reveals high travel distances, high levels of car use, little use of public transport, and almost 90 per cent of employed residents travelling to work outside the town. By contrast, a new housing development on the edge of Oxford has far less car travel since the public transport system provides a better alternative. The DoE report laconically comments that 'these conclusions suggest that local authorities will need to consider carefully the regional dimension of location planning, and the transport policies applied in individual settlements' (DoE 1995, *Reducing the Need to Travel through Land Use and Transport Planning*). The same story can be told of many other places beyond the green belt.

In Scotland, green belts have been established around Aberdeen, Ayr/Prestwick, Edinburgh, Falkirk/Grangemouth, and Glasgow. Interestingly, the Dundee green belt has been replaced by a general countryside policy (Regional Studies Association 1990: 22). Scottish green belts have had somewhat wider purposes than those in England: these include maintaining the identity of towns by establishing a clear definition of their physical boundaries and preventing coalescence; providing for countryside recreation and institutional uses of various kinds; and maintaining the landscape setting of towns. There is a greater emphasis on the

---

Box 6.5 GREEN BELT POLICY IN ENGLAND

Purposes:

- to check the unrestricted sprawl of large built-up areas;
- to prevent neighbouring towns from merging into one another;
- to assist in safeguarding the countryside from encroachment;
- to preserve the setting and special character of historic towns; and
- to assist in urban regeneration, by encouraging the recycling of derelict and other urban land.

Use of Land in Green Belts:

- to provide opportunities for access to the open countryside for the urban population;
- to provide opportunities for outdoor sport and outdoor recreation near urban areas;
- to retain attractive landscapes, and enhance landscapes near to where people live;
- to improve damaged and derelict land around towns;
- to secure nature conservation interest; and
- to retain land in agricultural, forestry, and related uses.

Source: PPG 2, *Green Belts*

---

environmental functions of the green belts, and recreation is included as a primary objective. The title of the Scottish circular (24/85) is significant: *Development in the Countryside and Green Belts*, underlining the links between general countryside policies and green belts. 'As a result, a much more integrated approach to the planning of green belt and non-green belt areas is achieved in Scotland.' The Regional Studies Association study commends the Scottish approach, arguing that 'green belts have become an outmoded and largely irrelevant mechanism for handling the complexity of future change in the city's countryside'.

Green belts are the first article of the British planning creed. They are hallowed by use, popular support, and fears of what would happen if they were 'weakened'. Fierce arguments are waged by a wide range of groups from national bodies such as the CPRE to local green belt residents. There are, however, other issues which do not attract the same concern, such as the costs imposed by green belts, and the inadequacy of a planning policy which lays such a great emphasis on protection and a lesser emphasis on instruments for meeting development needs. On this line of argument, green belts should be part of a more comprehensive land use/transport policy.

In this connection the more recent articulation of green belt policy in Wales is noteworthy (Tewdwr-Jones 1997). There has been increasing pressure from environmentalists for the establishment of green belts around the main urban areas which are under development pressure, and the 1999 planning guidance for Wales set out guidelines. Though these echo the points set out in the English PPG, they also emphasise the importance of development land:

> When considering green belt designation, local planning authorities will need to ensure that a sufficient range of development land is available which is suitably located in relation to the existing urban edge and the proposed green belt, bearing in mind the longer term need for development land, the effects of development pressure in areas beyond the green belt and the need to minimise the need to travel.

Moreover, the Welsh policy requires that local authorities must justify the need for such areas – must demonstrate why normal planning and development control policies, including green barrier/green wedge policies, would not be adequate. Thus the rationale for designation is far stricter than in England.

As in England, the debate on Welsh green belts largely ignores the issue of managing the countryside within green belts, though both specifically refer to opportunities for access and for outdoor sport and outdoor recreation. These do not figure significantly in the public debate: the overwhelming concern is with preventing development.[31].

TOWN CENTRES AND SHOPPING

Out-of-town shopping centres have been blamed for weakening or even killing off traditional town centres and for increasing car travel (and its accompanying pollution). On the other hand, they are clearly popular in themselves, and on a market test are successful. However, concern for the decline of the centres of smaller towns has led to initiatives to promote 'vital and viable town centres' and as well as stronger planning controls under PPG 6 (1993, revised 1996). The publication of the revised PPG 6 in 1996, following the 1994 PPG on transport, marked the end of laissez-faire policies (Truelove 1999: 207). Since then, controls over out-of-town shopping centres have become increasingly strict (strongly supported by the House of Commons Environment Committee[32]).

Official policies are designed to serve several objectives:

* to sustain and enhance the vitality and viability of town centres;
* to focus development, especially retail development, in locations where the proximity of businesses facilitates competition from which all consumers are able to benefit and maximises the opportunity to use means of transport other than the car;
* to maintain an efficient, competitive, and innovative retail sector; and
* to ensure the availability of a wide range of shops, employment, services, and facilities to which people have easy access by a choice of means of transport.

As a disclaimer, it is stressed that 'it is not the role of the planning system to restrict competition, preserve existing commercial interests, or to prevent innovation'. But, of course, the impact of policy may have exactly these effects, since new competitors are seldom allowed.

Translated into practical terms, current policies are based on a sequential approach which gives first priority to town centre sites, followed by edge-of-centre sites, district, and local centres, 'and only then out-of-centre sites in locations that are accessible by a choice of means of transport'. To quote PPG 6 (*Town Centres and Retail Development*, para. 3.13):

> In the case of many smaller centres, particularly historic towns, the best solution may be an edge-of-centre foodstore with parking facilities, which enables car-borne shoppers to walk to the centre for their other business in town, and shoppers who arrive in the centre by means of other means of transport to walk to the store. One trip can thus serve several purposes, and the new shop is likely to help the economic strength of the existing town centre, be accessible to people without cars and overall generate less car use.

Students of planning appeals will know what arguments can rage over such an apparently simple statement. First, how is 'edge of town' to be defined? A proposal by Sainsbury's to build a store on a redundant site in Brighton was rejected on appeal on grounds which included its location being not genuinely on the edge of town. It was 145 metres from the primary shopping area (*Planning* 16 October 1998). Second, what evidence is there that proximity to the town centre will have the benefits that are claimed? A study for the DoE suggested (on the rather small sample of two case studies) that 'in terms of linked trips, edge-of-centre stores do not necessarily generate significantly higher degrees of linkage with town/district centres than out-of-centre stores'.[33] There are many issues of this kind which beset the seeker after the truth of PPGs.

The latest in what is a long saga of court cases revolves around the concept of 'need'. The situation became so confused that the Planning Minister (Richard Caborn at the time) issued a 'clarification'. It is worth looking at this:

> [T]he requirement to demonstrate 'need' should not be regarded as being fulfilled simply by showing that there is capacity (in physical terms) or demand (in terms of available expenditure within the proposals catchment area) for the proposed development. Whilst the existence

of capacity or demand may form part of the demonstration of need, the significance in any particular case of the factors which may show need will be a matter for the decision-taker.

This requires several readings before one realises that it means (in David Lock's words) that 'you must prove "need", but the Government will not tell you whether you have succeeded until you have succeeded' (Lock 1999). As a manager from Sainsbury's has pointed out, the one thing that is clear, however, is that the Minister has certainly provided the lawyers with potentially rich pickings when arguing about how need is defined and by whom (H. Williams 1999).

These issues have been set out here at some length, not because they are exceptional, but because they are the very stuff of planning arguments. As with the debates on 'vital and viable' town centres, these can mask secular social and economic trends such as changes in retail trading patterns and distribution, changes in trading laws (as with the relaxation of the Sunday trading laws, which affect small retailers much more than the superstores), changes in branch banking, and (still unclear) the effect of the Internet on buying patterns.

VACANT AND DERELICT LAND

Some of the estimates for vacant and derelict land have been discussed above in the context of urban capacity studies. Here the focus is on the nature of this land, and the policies that have evolved to deal with it.

Much land that was once useful and productive has become waste land, particularly in the inner cities and in mining areas. It is unsightly, unwanted, and, at worst, derelict and dangerous. The planning system is not designed to deal with such land easily: its essential characteristic is to allocate land between competing uses. Where there are no pressures for development, there is a severe limit to what can be done, especially when the amount of waste land is large, as it is in older industrial areas. Major efforts have been made to deal with the problems. Between 1988 and 1993 some 9,500 hectares of derelict land was reclaimed. Unfortunately, a large amount of new dereliction is

continually being created, and the result is that the total amount remains high. Though the amount of derelict land in England decreased by 2 per cent between 1988 and 1993, the 1993 Derelict Land Survey showed a total of 39,600 hectares of derelict land at the latter date. Unfortunately, this was the last year in which the survey was undertaken. The latest figure, from the National Land Use Database in 1999, is 17,000 hectares of derelict land and 16,000 hectares of previously developed vacant land (though this is not completely comparable and possibly is an under-estimate).[34] This is an area about twice the size of the whole of the city of Glasgow.

Reclamation policies have changed over the years. Originally the objective was to remove eyesores and potential dangers caused by spoil heaps and other waste. Much of this was located in rural areas, and the policy was to return the land to agriculture or forestry, or to make it available for public open space (known in the jargon as a 'soft end-use'). Since the 1980s, emphasis has shifted to 'hard end-uses' such as industrial, commercial, or residential development, particularly in older urban areas. Increasingly, the focus has been on brownfield sites for housing development as a favoured alternative to greenfield sites. A range of policy instruments to deal with derelict land have been developed. Some of these have been part of broader policies in relation to urban regeneration (through urban development corporations, enterprise zones, and the Urban Programme).

Vacant land is conceptually different from derelict land, though the two categories can overlap.[35] Research shows that land vacancy is typically a transient feature of the environment. Though much of it has been vacant for a long time (two-thirds of a sample had been vacant for more than twelve years), some of this idle land – perhaps a third – can be used when subsidies are paid to overcome physical constraints. However, some vacant land – perhaps two-thirds – is so because of institutional factors, owners' intentions, or poor demand. As the evaluation study explained,

> Many sites remain vacant for non-physical reasons. Some are delayed by the legitimate workings of the planning system, and by legal and other institutional difficulties. Existing policy instruments can do little to overcome

these difficulties. Others are delayed by owners', particularly private sector owners', intentions that they should remain vacant for various (largely obscure) reasons.
(Whitbread *et al.* 1991, para. 3.147).

An earlier report suggested a long list of reasons why vacant land is not put to temporary uses: expenditure by the owner would be needed to meet fire, safety, and insurance requirements, in providing access, and in site clearance; temporary tenants tend to be unreliable and to cause environmental problems; demand from temporary users is deficient and uncertain, and often provides landowners with a very low financial return; there are often problems in securing vacant possession; landowners may be unaware of the potential of temporary uses; or they may think that keeping sites vacant preserves existing use rights, or puts pressure on local authorities to grant planning consent for development (Cameron *et al.* 1988). Much of this land is in private ownership and, short of compulsory acquisition, there is little that 'policy' can do to speed up the reuse of the land.[36]

But these are the failures; more striking are the undoubted successes of other policy instruments, of which the evaluation study counted thirteen (ranging from grant-aid to planning and promotional action by local authorities). Foremost among these was the Derelict Land Grant programme, which was replaced by the English Partnerships Land Reclamation Programme, and which in turn has been transferred to the new Regional Development Agencies. Before discussing this, it is relevant to note another policy initiative which failed and which has been overtaken by this same programme. This concerns contaminated land (see below).

A new category of vacant land is redundant military land. The 'peace dividend' following the demise of communism in Russia and certain European states has generated significant amounts of land in the Ministry of Defence Estate, and has closed some US Air Force bases in the UK.[37] Much of this land is located in rural areas or economically depressed urban areas (including former naval dockyards). Disposal of MoD land is subject to the Crichel Down rules requiring surplus land to be offered back to the original owners at current market values. Such values will take into

account the conditions of the site, which may be contaminated. Many sites also contain listed buildings, monuments, and environmental and landscape designations that will affect reuse. The MoD is required to maximise income from disposals, which typically leads to proposals for new housing.

CONTAMINATED LAND

There is no clear line between vacant, derelict, and contaminated land (or neglected, underused, waste, and despoiled land). The terms are used in different ways, sometimes for different purposes, sometimes with the same or similar meanings (and new terms arise from time to time, such as 'previously used land' and 'brownfield sites'). Contaminated land is particularly difficult to define, though the term is commonly used to imply the existence of a hazard to public health. The 1995 Environment Act introduced a new definition which incorporates this long-standing idea. Though there is an overlap with 'derelict' land, there are important differences. A chemical waste tip may be both derelict and contaminated; a disused chalk quarry may be derelict but not contaminated; an active chemical factory may be contaminated but not derelict. It is the additional health danger which is the characteristic feature of contaminated land, and this also implies a severe degree of pollution and, typically, an increased difficulty in abating it. However, the health risk arises only in relation to the use to which the land is to be put. A piece of land may pose no risk if used for one purpose, but a severe risk if it is used for another. The site of an oil refinery may be contaminated, but that is of no consequence if no other use is intended (and assuming that there are no effects beyond the site). 'A scrapyard contaminated by metal traces would constitute a hazard for subsequent agricultural use, but the contamination would be of no account in the construction of an office block'.[38]

Partly because of a characteristically pragmatic approach, there has never been an attempt to quantify the amount of contaminated land in Britain. Instead of identifying contaminated land and then determining appropriate policies for dealing with it, the

British approach has been to regard contamination as a general concept which is given substance only in relation to particular sites and particular end-uses. The nature of policy flows from this: 'Policy is to ensure that the quality of land is fit for the purpose to which it is being or will be used.' There is no requirement for land to be brought up to a minimum quality standard regardless of use, unless that land poses a threat to the public health or the environment.

The House of Commons Environment Committee considered this approach to be inadequate since (in its judgement) there is land which is so contaminated that it is 'a threat to health and the environment both on site and in the surrounding area'. The Committee also recommended that local authorities should be given a duty 'to seek out and compile registers of contaminated land'.

There was a remarkably swift response to this: the Environment Protection Bill was amended to provide such a duty. The implementation of this, however, rapidly ran into severe difficulties, and the initial proposals had to be drastically changed.[39] The problem underlying all this is that it is relatively simple to register land that is *possibly* contaminated, but extremely labourious and costly to identify land that is in fact contaminated. Even at the low rate of £15,000 per hectare, it would cost around £600 million merely to investigate the 40,000 hectares of land identified in the 1988 Derelict Land Survey (Thompson 1992: 22). To cover all relevant land would cost many times this amount, and would take many years to complete.[40]

It was because of difficulties such as these that the government was eventually forced to abandon the scheme as originally envisaged. The difficulties of changing from the traditional British reactive approach to a genuinely proactive approach are manifest (A. Harrison 1992: 809). However, a renewed attempt was made in 2000. The revised provisions are set out in the long and complex Circular 02/2000.[41] It will take some time for the implications of the new regime to be understood.

> Suffice it to say at this stage that they will create a regime to enforce remediation on certain contaminated sites where there is a serious degree of health or environmental

risk and where there is a justification for requiring compulsory remediation. Compulsory remediation is a drastic remedy, and these powers are likely to be used only in limited circumstances where voluntary remediation, e.g. in the course of redevelopment (or otherwise) is unlikely to occur. It is therefore a legislative supplement to the planning and development control process which is likely to continue to govern the overwhelming majority of remediation of contaminated sites.
>
> (Winter 1998: 10)

Local authorities are required to prepare and implement a strategy for identifying land falling within the statutory definition of 'contaminated' and to require its remediation. This means inspection of sites and identification of responsibilities for remediation and monitoring – although the Environment Agency will also monitor implementation and provide advice on specific problems. The statutory definition now concentrates on the notion of 'where significant harm is being caused or where there is a significant possibility of such harm being caused. . . [or where] pollution of controlled waters is being, or is likely to be caused'. In order to avoid blighting land, only sites that are identified as contaminated land and where the local authority is taking action will be listed on registers.

In 1994 a new agency came into operation to implement 'a new approach to vacant land' which includes unused, underused or ineffectively used urban land, land which is contaminated, derelict, neglected, or unsightly, or land which is likely to be affected by subsidence. The agency was statutorily termed the Urban Regeneration Agency but, to underline the nature of its role, it took the non-statutory name of English Partnerships. (Its remit applies only to England: similar functions are undertaken in Scotland by Scottish Enterprise, and in Wales by the Welsh Development Agency.) These functions have now been taken over by the new Regional Development Agencies, though English Partnerships will (appropriately enough) have a partnership role with the RDAs. English Partnerships (which now has this as its statutory as well as its informal name) is continuing with a focus on national and cross-regional coordination. (Its main areas of operation are developing assets, creating partnerships, improving the environment, and finding new sources of funding in the field of regeneration).[42] In its new form, English Partnerships will continue its programme for the regeneration of the Greenwich Peninsula, the 'national coalfields portfolio, and millennium communities Competitions. (See also the discussion in Chapter 10.)

SCOTTISH LAND REFORM

There are numerous policies relating to land, but rarely is there anything which might be termed a 'land policy'. Scotland presents a fascinating exception. Following a very long history of attempts to reform the Scottish feudal land system, the Scottish Parliament is embarking on 'an integrated programme of action and legislation' over a four- to five-year period starting in 1999. This was summarised in the very first White Paper to be published by the Scottish Executive (*Land Reform: Proposals for Legislation*, 1999).

The Scottish system is extraordinarily complex, and only a short indication of its character can be given here.[43] Essentially, feudal land ownership is a hierarchical system in which land rights derive from the highest authority, theoretically God, but in practice the Crown. The Crown is known as the Paramount Superior, and all other landowners are known as vassals of the Crown. The relationship need not be direct, however, and a vassal can convey land (to a new vassal), retaining interests which are set out in the title deed. There is no limit to the number of times this 'feuing' can take place. Each superior can reserve rights and impose additional 'burdens' (such as a restriction on building on the land or carrying on a business). Nearly all privately owned land in Scotland is held under feudal tenure, and the survival of such characteristically feudal elements as superiorities and feu duties is indicative of the extraordinarily archaic and complex nature of Scotland's current system of land ownership.

Previous reforms have attempted to simplify this system, but have not tackled the more political issue of landownership. It is claimed that Scotland has the most concentrated pattern of private landownership in the world: '343 landowners own over half of the entire privately owned rural land in the country'. In the Highlands and Islands, half of all the private land (about 1.5 million hectares) is owned by fewer than a hundred landowners.[44]

Proposals for reform were issued by a Land Reform Policy Group appointed by the government, and a White Paper, *Land Reform: Proposals for Legislation*, was issued in 1999. This first instalment of reform is limited in its scope to giving 'community bodies the right to buy rural land which is to be sold', and to creating a right of 'responsible access to land'. The latter is outlined in Chapter 9. Here a brief summary is given of the proposals relating to the former.

The intention is to create new opportunities for 'community ownership'. This is to be done by providing for the registration of community bodies (set up for the purpose and incorporated) that are interested in acquiring land when it comes to be sold in their area. Registered bodies will have the right to buy such land (whether privately or publicly owned). The price will be assessed by a government-appointed valuer, with disputes being settled by the Lands Tribunal for Scotland. A minimum percentage of those aged 18 or over and who live and/or work on the land in question must support the proposed purchase. To deter evasion, Scottish ministers will be able to exercise a new compulsory purchase power where this is in the public interest.

The proposals have been characterised by Wightman as 'based on a flawed, shallow and partial analysis of the problem [and revealing] a timidity and poverty of imagination when it comes to tackling landed power' (*Guardian* 30 August 1999). Others might argue that it is perhaps early days to judge. Some progress has already been made (in advance of general legislation) in the Highlands and Islands, where a Community Land Unit has been established and is operating schemes of both technical and financial assistance in its region.[45]

FURTHER READING

Good introductions to some of the major areas covered in this chapter are Hall and Ward (1998) *Sociable Cities: The Legacy of Ebenezer Howard*; and Bramley *et al.* (1995) *Planning, the Market and Private House-Building*.

Land Values and Prices

There has been surprisingly little study of the operation of the various experiments in capturing land values for the public benefit. A long and detailed account of the legislative history is given in Cullingworth (1980) *Environmental Planning 1939–1969*, vol. 4: *Land Values, Compensation and Betterment*. More digestible accounts are provided by McKay and Cox *(1979) The Politics of Urban Change* (chapter 3) and Cox (1984) *Adversary Politics and Land*.

The effect of planning on the land market has been the subject of a long-standing debate both in theoretical terms, as in Evans (1983) 'The determination of the price of land', and in the context of British planning, as in the same author's (1988) *No Room! No Room! The Costs of the British Town and Country Planning System* and (1991) 'Rabbit hutches on postage stamps'. Less tendentious are the study commissioned by DoE on *The Relationship between House Prices and Land Supply* (Eve and Department of Land Economy, University of Cambridge 1992); Monk *et al.* (1996) 'Land-use planning, land supply and house prices'; and Bramley and Watkins (1996) *Steering the Housing Market: New Building and the Changing Planning System*. There is a convenient brief summary, 'The planning system and house prices', in Annex E of the White Paper *Household Growth: Where Shall We Live?* (Cm 3471, 1996).

Planning Gain

By contrast there have been a large number of studies of planning agreements, planning obligations, and planning gain. Selected titles (listed by date of publication) are Rowan-Robinson and Young (1989) *Planning by Agreement in Scotland*; Callies and Grant (1991) 'Paying for growth and planning gain: an Anglo-American comparison'; Eve (1992) *Use of Planning Agreements*; Fordham (1993) 'Planning gain in ten dimensions'; Healey *et al.* (1992a) 'Rationales for planning gain'; and Bunnell (1995) 'Planning gain in theory and practice: negotiation of agreements in Cambridgeshire'. A comprehensive study is Healey *et al.* (1995a) *Negotiating Development: Rationales and Practice for Development Obligations and Planning Gain*.

The December 1997 special issue of *Urban Studies* is devoted to 'developer contributions: the bargaining process', and includes papers on the USA, Canada, and the Netherlands. Of particular interest are Ennis 'Infrastructure provision, the negotiating process and the planner's role', and Claydon and Smith 'Negotiating planning gains through the British development control system'. DoE Circular 1/97, *Planning Obligations*, is, of course, of major importance. See also Cornford (1998) 'The control of planning gain', and Planning and Environment Law Reform Working Group (1999) 'Planning obligations'. A succinct overview is given in Wenban-Smith and Pearce (1998) *Planning Gains: Negotiating with Planning Authorities*. The currently relevant Circular is 1/97 *Planning Obligations*.

The importance of the increased cost of infrastructure in negotiations for contributions from developers was discussed in the 1972 Sheaf Report (*Local Authority/Private Enterprise Partnerships Schemes*). A contemporary analysis is Marvin and Guy (1997) 'Infrastructure provision, development processes and the co-production of environmental value'. For a broader discussion, see Ward (1999) 'Public–private partnerships'.

Planning and Affordable Housing

The use of planning powers to require the provision of affordable housing has attracted much debate. See, for example, Kirkwood and Edwards (1993) 'Affordable housing policy: desirable but unlawful?'; Barlow *et al.* (1994a) *Planning for Affordable Housing*; Elson *et al.* (1996) *Green Belts and Affordable Housing: Can We Have Both?* and 'Planning mechanisms to secure affordable housing' in Joseph Rowntree Foundation (1994) *Inquiry into Planning for Housing*; and Gallent (2000) 'Planning and affordable housing: from old values to New Labour'. The DETR has produced *Guidance for Local Authorities on Housing Strategies* (available direct from DETR). The Shelter report by Holmans *et al.* (1998) *How Many Homes Will We Need* (1998) is an assessment of the need for affordable housing in England. (The production of affordable houses greatly falls short of the need).

On the strange 'exceptions policy' (the exceptional release of land for local needs housing), see Annex A to PPG 3 and Circular 6/98 *Planning and Affordable Housing*; Bishop and Hooper (1991) *Planning for Social Housing* (which contains a good bibliography of publications up to this date); Williams *et al.* (1991) *Evaluating the Low Cost Rural Housing Initiative*; Elson *et al.* (1996) *Green Belts and Affordable Housing: Can We Have Both?*; and Gallent and Bell (2000) 'Planning exceptions in rural England: past, present and future'.

More generally on housing need issues, see HC Environment Committee (1996) *Housing Need*. A comprehensive analysis of English housing conditions is given in Green *et al.* (1996) *Housing in England 1994/95*.

Land Availability and Urban Capacity Studies

Central government policy in relation to land availability studies was set out in the 1988 and 1992 versions of PPG 3, *Housing* (with an annex on the organisation and methodology). The equivalent for Wales was TAN 1, *Joint Housing Land Availability Studies*. There was an extensive literature on the subject, for example Bramley (1989) *Land Supply, Planning, and Private Housebuilding*; Jackson *et al.* (1994) *The Supply of Land for Housing: Changing Local Authority Mechanisms*; Bramley *et al.* (1995) *Planning, the Market and Private House-Building*; and Bramley and Watkins (1996) *Steering the Housing Market: New Building and the Changing Planning System*.

High-quality housing capacity studies are few in number but this will no doubt change as they become the successor to the land availability studies. In the meantime the sole title is UK Round Table on Sustainable Development (1997) *Housing and Urban Capacity*, which contains a review of studies (it is available from DETR). See also Llewelyn-Davies (1994) *Providing More Homes in Urban Areas* and Llewelyn-Davies Planning (1997) *Sustainable Residential Quality: New Approaches to Urban Living*. Lord Rogers' Task Force report (1999) *Toward an Urban Renaissance* makes an unconvincing assessment of the potentialities of brownfield sites. A useful collection of essays is edited

by Jenks *et al.* (1996) *The Compact City: A Sustainable Urban Form?*.

On brownfield sites, Llewelyn-Davies (1996) *The Re-use of Brownfield Land for Housing* deals with the difficulty of the remaining brownfield sites and the need for substantial government subsidies. The difficulties are illustrated in a short report by the Civic Trust, *Brownfield Housing 12 Years On* (1999). See also Breheny and Hall (1996) *The People: Where Will They Go?* and Alker *et al.* (2000) 'The definition of brownfield'. Bibby and Shepherd (1999), in 'Refocusing national brownfield housing targets', discuss some important neglected issues relating to brownfield housing targets.

Household Projections

The latest (1996-based) projections are published in DETR (1999) *Projections of Households in England to 2021*. The most accessible discussions of household projections are given in the Conservative government's White Paper *Household Growth: Where Shall We Live?* (Cm 3471, 1996) and in Breheny and Hall (1996) *The People: Where Will They Go?* More technical is Bramley *et al.* (1997) *The Economic Determinants of Household Formation: A Literature Review*. See also Allinson (1999) 'The 4.4 million households: do we really need them anyway?'. For rural areas, see Rural Development Commission (1998) *Household Growth in Rural Areas: The Household Projections and Policy Implications*.

New Settlements

An account of the long-standing British campaign for new settlements is discussed by Ward (1992) *The Garden City: Past, Present and Future*; and by Hardy (1991a) *From Garden Cities to New Towns* and (1991b) *From New Towns to Green Politics* (which is a two-volume history of the TCPA). A volume in the official history *Environmental Planning 1939–1969* by Cullingworth (1979) provides a detailed deadpan record of government policy over this thirty-year period. Breheny *et al.* (1993) *Alternative Development Patterns: New Settlements* provide a more up-to-date picture. Invaluable as a succinct historical account, as well as an overview and analysis of the current policy issues, is Hall and Ward (1998) *Sociable Cities: The Legacy of Ebenezer Howard*.

Green Belts

Two major publications on green belts are Elson (1986) *Green Belts: Conflict Mediation in the Urban Fringe*, and the report of a study for DoE by Elson *et al.* (1993) *The Effectiveness of Green Belts*. Broader in scope is the classic study by Peter Hall *et al.* (1973) *The Containment of Urban England*. On green belts in Scotland, see Regional Studies Association (1990) *Beyond Green Belts* and Pacione (1991) 'Development pressure and the production of the built environment in the urban fringe'. On Welsh policy in relation to green belts, see *Planning Guidance (Wales): Planning Policy First Revision* (1999) and Tewdwr-Jones (1997) 'Green belts or green wedges for Wales? A flexible approach to planning in the urban periphery'. A short critical appraisal of green belt policy is Cherry (1992) 'Green belt and the emergent city'.

Town Centres And Shopping

Obviously, PPG 6 is crucial on the subject of out-of-town and edge-of-town shopping centres in England, even though it is somewhat opaque in parts. The equivalent Planning Policy Statement for Northern Ireland is PPS 5 *Retailing and Town Centres* (1996); for Scotland it is NPPG 8 *Town Centres and Retailing* (1998). Good reviews of the issues involved arising with out-of-town shopping centres include BDP Planning and Oxford Institute of Retail Management (1994) *The Effects of Major Out-of-Town Retail Developments* and CB Hillier Parker and Saxell Bird Axon (1998) *The Impact of Large Foodstores on Market Towns and District Centres*. See also Sparks (1998) *Town Centre Uses in Scotland*; URBED (1994) *Vital and Viable Town Centres: Meeting the Challenge*; HC Environment Committee (1997) *Shopping Centres*; Ravenscroft (2000) 'The vitality and viability of town centres'; and National Retail Planning Forum (1999) *A Bibliography of Retail Planning*.

An analysis of the changing economics of superstore development is given by Wrigley (1998) 'Understanding store development programmes in post-property-crisis UK food retailing'. For a discussion of retail parks, see Guy (1998) 'High Street retailing in off-centre retail parks', and 'Alternative-use valuation, open A1 planning consent, and the development of retail parks'. More generally, see Guy (1994) *The Retail Development Process*.

Vacant, Derelict, and Contaminated Land

The research on vacant land includes Cameron *et al.* (1988) *Vacant Urban Land: A Literature Review* and Whitbread *et al.* (1991) *Tackling Vacant Land: An Evaluation of Policy Instruments*. The latter provides a review of previous research. For a broader overview of urban land policies, see Chubb (1988) *Urban Land Markets in the United Kingdom*.

Derelict land is dealt with in a number of government reports including *The Strategic Approach to Derelict Land Reclamation* (1992); *Assessment of the Effectiveness of Derelict Land in Reclaiming Land for Development* (1994); *Derelict Land Survey 1993* (1995); and *Derelict Land Prevention and the Planning System* (1995).

Policy on contaminated land is succinctly set out in PPG 23 *Planning and Pollution Control*. A commentary on this is given by Graham (1996) 'Contaminated land investigations: how will they work under PPG 23?'. An exhaustive legal guide is Tromans and Turrall-Clarke (1994 with 1996 supplement) *Contaminated Land*. In September 1999 DETR issued *Draft Circular on Contaminated Land*. The final version of this is to be published in early 2000.

On redundant military land, see Bateman and Riley (1987) *The Geography of Defence*; National Audit Office (1992) *Ministry of Defence: Management and Control of Army Training Land*; Farrington (1995) 'Military land in Britain after the Cold War'; Fuller Peiser and Reading University (1999) *Development of the Redundant Defence Estate*; and Fyson (1999b) 'Iron out defence land policy to get the full benefits'.

Scottish Land Reform

The main book used in the text is Callander (1998) *How Scotland is Owned*. He also wrote *A Pattern of Landownership in Scotland* (1987). Another author in this field is Wightman (1996) *Who Owns Scotland* and (1999) *Scotland: Land and Power*. See also Ogilvie (1997) *Birthright in Land*, and McCrone (1997) *Land, Democracy and Culture in Scotland*.

NOTES

1   The principle had been first established in an Act of 1662 which authorised the levying of a capital sum or an annual rent in respect of the 'melioration' of properties following street widenings in London. There were similar provisions in Acts providing for the rebuilding of London after the Great Fire. The principle was revived and extended in the Planning Acts of 1909 and 1932. These allowed a local authority to claim first 50 per cent and then (in the later Act) 75 per cent of the amount by which any property increased in value as the result of the operation of a planning scheme. In fact, these provisions were largely ineffective since it proved extremely difficult to determine with any certainty which properties had increased in value as a result of a scheme or, where there was a reasonable degree of certainty, how much of the increase in value was directly attributable to the scheme and how much to other factors. The Uthwatt Committee noted that there were only three cases in which betterment had actually been paid under the planning acts.

2   *Planning Policy and Social Housing* (RTPI 1992: 5). Grant (1999b) discusses this policy explicitly as a form of betterment recoupment. He adds that 'the tenuous link drawn in the circular between private and affordable housing is demonstrated by the government's willingness for the obligation to be commuted to a financial contribution by the developer towards the provision of affordable housing elsewhere in the local authority's area'.

3   See Chartered Institute of Housing *et al.* (1999);

National Housing Federation (1999); and Whitehead *et al.* (1999).

4   *Fundamental Review of the Laws and Procedures Relating to Compulsory Purchase and Compensation: Interim Report* (London: DETR, January 1999); see also the reports of the Symposium on Compulsory Purchase: An Appropriate Power for the 21st Century, available at <www.detr.gov.uk/planning/cp/01.htm>.

5   There is a more detailed account in the previous edition of this book.

6   The claimed benefits of living in 'compact cities' vary greatly. Arguments in favour include Jacobs (1961); Elkin (1991), Sherlock (1991) ECOTEC (1993); and various official publications on sustainable development. Arguments suggesting that the benefits are illusory, infeasible, or overstated include Breheny (1997); P. Hall (1999c); and K. Williams (1999).

7   See the *Government Response to the Environment, Transport and Regional Affairs Committee: Housing* (Cm 4080, 1998: para. 145).

8   A review by Llewelyn-Davies revealed that few local authorities had undertaken such studies. Where they had been carried out, they seriously underestimated the amount of land available for housing. They recommended that studies should (1) be based on original site work, (2) include a significant physical design element, and (3) not be constrained by existing policies and standards. None of the studies reviewed met these criteria (UK Round Table on Sustainable Development 1997).

9   Evidence to the HC Select Committee on *Housing PPG 3*, HC 490-I. See also the Friends of the Earth report by Rudlin (1998).

10   There will no doubt be lengthy arguments about the figures. Following the publications of the Urban Task Force report, a Statistical Service Information Bulletin (no. 500) was released giving provisional results from the National Land Use Database (web site www.nlud.org.uk>). (The National Land Use Database is a partnership between DETR, English Partnerships, the Improvement and Development Agency, and the

Ordnance Survey.) This showed that in 1998 'there were some 33,000 hectares of previously vacant and derelict land in England, of which just over 12,000 hectares either had planning permission for housing or were judged suitable for housing. At current densities, this land could accommodate over 325,000 new dwellings. An estimated additional 385,000 dwellings could be provided through the reuse of vacant commercial buildings, land currently in use but allocated in plans for redevelopment or with planning permission for housing, and other sites expected to become available for redevelopment.' Whatever assumptions are made about density these figures are far short of the up to four million dwellings required. Fyson (1999b: 15) comments that 'the crucial fallacy exposed by this report is the supposed connection between urban revival and the concentration of most new housing in existing urban areas'.

11 See, for instance, Jenkins *et al.* (1996); Breheny (1997); and P. Hall (1999c) *Sustainable Cities and Town Cramming*.

12 Hedges and Clemens (1994): tables 6.17 and 7.17 and commentary pp. 132 and 158). Breheny (1997) discusses these and other relevant issues. See also Todorovic and Wellington (2000).

13 There is also the issue of empty properties owned by government departments: see DETR (1999) *Revised Guidance on Securing the Better Use of Empty Homes*. From April 2000, council tax will be payable at the rate of 50 per cent on dwellings that have been vacant for a year or more.

14 See DETR (1997) *Evaluation of Flats over Shops*, London Planning Advisory Committee (1998b), and Urban Task Force (1999a: 253–4).

15 See, for example, CPRE (1994c); Bramley and Watkins (1995); Bramley (1996b); and Green Balance (1999).

16 DETR (1999) *Planning Policy Guidance Note 11: Regional Planning Public Consultation Draft*, para. 5.4. See also HC Select Committee on the Environment, Transport and Regional Affairs, 10th Report (session 1997/98) *Housing*, vol. 1, para. 2.11, and *The Government's Response* (Cm

4080), paras 127–36. Stephen Crow, in his evidence to the HC Environment Subcommittee, argued that both the expressions 'predict and provide and 'plan, monitor and manage were 'slogans which can mean all things to all men'.

17 HC Select Committee, op. cit. para. 210. For the CPRE view on this, see Wenban-Smith (1999) and *Sprawl Patrol* campaign and briefing sheet *Plan, Monitor and Manage* (details at <www.cpre.org.uk>). The Select Committee's report *Housing PPG 3* (1999) criticised the draft PPG for its lack of clear and specific guidance; op. cit. para. 14.

18 SERPLAN *A Sustainable Development Strategy for the South East* (SERP 500, 1998). Accompanying documents are listed in appendix 2 of *RPG for the South East, Public Examination: Report of the Panel*.

19 *RPG for the South East*, Public Examination: *Report of the Panel*, paras 4.54 and 4.63.

20 See also Whitehead *et al.* (1999). (This study includes the Eastern region.)

21 *RPG for the South East*, op. cit., para. 4.67.

22 Evidence of Professor Tony Crook and Dr Christine Whitehead to the Select Committee, op. cit., p. 74.

23 It is curious and unfortunate that 'there is no mechanism in England whereby the desirability of inter-regional migration can be debated, (Breheny and Hall 1996).

24 See the discussion and references on RDAs in Chapter 3.

25 The major factor, of course, is the changing pattern of employment. Though there has been a general loss of manufacturing jobs, the loss has been most dramatic in the conurbations. These losses have not been offset by a corresponding growth in alternative employment in the affected areas. The expansion in service jobs has been located almost entirely mainly in towns and rural areas where there are attractive and cheap sites. See Turok and Edge (1999), Turok and Webster (1998), and Rowthorn (1999).

26 See Breheny (1997): 213) and Breheny and Hall (1996).

27 'The capacity of urban areas to absorb new

housing depends not only on the *stock* of brownfield land that is currently available, but also on the *flow* of previously developed sites that will become available over the period to which the target relates' (Bibby and Shepherd 1999).

28 Breheny *et al.* (1993). Much of this is of a technical nature, comparing the costs and benefits of different forms of development. This is a difficult and complex matter, since so much depends on site-specific issues. The authors neatly point up the difficulties by stressing that their analysis is 'intended to focus discussion rather than present a definitive assessment'. But central government is urged to come off the fence, and to give a clear statement on the management of urban growth. It is unequivocally stated that 'unless much tougher containment policies are introduced – at the very time when concerns are being expressed over urban intensification – it is inevitable that significant greenfield/village development will take place in the UK'.

29 It did, however, carefully avoid making the mass of assumptions which flawed an earlier study by the National Institute of Economic and Social Research (Stone 1973; see also Cullingworth (1979: 473).

30 See also Hall and Ward (1998); particularly chapter 9 on 'sustainable social cities of tomorrow.

31 But see the study commissioned by the Sports Council for Wales (Elson 1991). Tewdwr-Jones (1997) suggests that the alternative policy of green wedges in areas of possible development pressures could provide a flexible way of meeting both current recreation needs and future development needs.

32 There have been two inquiries by the HC Environment Committee (1994): *Shopping Centres and Their Future* (and the *Government Response* 1995) and *Shopping Centres* (1997). The *Government Response* to this was published later in the same year.

33 CB Hillier Parker and Saxell Bird Avon (1998: para. 10.12). See also the series of reports on the employment impact of out-of-town superstores published by the National Retail Planning Forum.

34 The figures are from the Government Statistical Service *Information Bulletin* 500 (20 May 1999), and details of the NLUD findings are available at <www.nlud.org.uk>.

35 For Scotland, see *Scottish Vacant and Derelict Land Survey 1998* (Edinburgh, TSO: 1999).

36 Policies can be founded on myths as well as on adequate understanding of problems. So it was with the land registers established by the 1980 Local Government, Planning and Land Act. The myth was that one of the major causes of urban dereliction was the hoarding of land by public authorities. By requiring local authorities and other public bodies to 'register' their land, it was expected that it would find its way into the development process. In fact, with the reality being much more complicated than the perception, the registers were of little effect. (See the evaluation undertaken for the DoE by Whitbread *et al.* (1991).)

37 This section draws on the DETR study *Development of the Redundant Defence Estate* (Fuller Peiser and University of Reading 1999), which notes that the Ministry of Defence is the second largest estate in single ownership in the UK, with about 226,000 hectares of land. (Only the Forestry Commission has more land.)

38 This and the following quotations are from the HC Environment Committee report *Contaminated Land* (1990).

39 The crux of the problem lay in the concept of 'contamination'. Instead of referring to land that is contaminated, the Act relates to 'land which is being or has been put to any use which may cause that land to become contaminated with noxious substances'. This very inclusive definition was made particularly onerous in the initial draft regulations because of the very large number of contaminative uses which were specified. There was strong criticism that the registers would create widespread blight and, in an attempt to pacify objectors, the number of specified uses was greatly reduced.

40 Another objection to the initial regulations was that they prohibited the deregistration of sites.

This was defended on two grounds. One is that factual information on the site's history (which cannot by definition change) will be necessary when any future change of use is proposed. The other is that contamination from the site may have migrated to adjacent sites; owners, regulatory authorities, and developers are expected to use registers to identify such sources of contamination.

41 See also the DETR web pages on contaminated land at <www.environment.detr.gov.uk/ landliability/ index.htm>, which provide a summary of the current regime.

42 The Commission for the New Towns has also combined with English Partnerships.

43 This account leans heavily on Callander (1998), from which extensive quotations are taken.

44 A. Wightman, 'A land (un)divided: land reform proposals for Scotland fall far short of what is needed for the redistribution of power', (*Guardian* 30 August 1999). See also Wightman (1996, 1999).

45 Highlands and Islands Enterprise Community Land Unit *Action Framework 1998–2001*. In its first year the Unit's achievements included financial assistance to Abriachan Forest Trust towards the purchase of 50 hectares of woodland on the side of Loch Ness, and assistance to some twenty smaller community land initiatives.

7

PLANNING, THE ENVIRONMENT, AND SUSTAINABLE DEVELOPMENT

In the last few decades, much has been achieved in reversing the environmental damage of previous centuries. Few people, for example, would have foreseen, even fifty years ago, that a river like the Don, despoiled by the filth of two centuries of industrial intensification and decline, would flow clean enough to support thriving fish populations by the dawn of the new Millennium. Few probably even spared a thought for whether such a turn-around in environmental fortunes might be desirable, let alone achievable.

(Sir John Harman, Chairman of the Environment Agency for England and Wales in the Foreword of *Creating an Environmental Vision Consultation Draft*, 2000)

THE ENVIRONMENT

In one sense, all town and country planning is concerned with the environment; but the reverse is not true, and it is difficult to decide where to draw the boundaries. The difficulty is increased by the rate of organisational change over recent years including the shifting of responsibilities from local government to *ad hoc* bodies, and by the flood of new legislation, prompted in part by the EU. Further complications arise because of the increased concern for the environment and the rise of sustainable development as a political slogan and goal.

The implications for 'town and country planning' are still working themselves out, and not always easily, as some of the implications touch at the heart of the planning system. Thus, it has been a long-standing feature of planning control that permission is given unless there are good reasons for refusal. It is for the local planning authority to demonstrate (to the Secretary of State if necessary) that an application should be refused. With 'environmental' procedures, however, the onus shifts somewhat: the developer's proposals have to be demonstrably acceptable, and permission can be refused if they are not. Though official pronouncements and advice are coy in acknowledging this, it is clear that environmental factors can be decisive in a planning decision and that applicants may even be required to discuss the merits of alternative sites. In the words of PPG 23, environmental statements, which must accompany particular applications 'may – and as a matter of practice should – include an outline discussion of the main alternatives studied by the developer and an indication of the reasons for choosing the development proposed, taking account of environmental effects' (PPG 23: para. 3.16).

Local authorities have specific powers in relation to some environmental issues such as certain aspects of pollution, waste, and noise, but they are not environmental planning authorities. Other, specific 'pollution control regimes' exist for this purpose, but there is no clear dividing line. A related issue here is that of *sustainability* – a concept around which much environmental policy revolves.

BOX 7.1 PLANNING AND POLLUTION CONTROL

The dividing line between planning and pollution controls is not always clear cut. Both seek to protect the environment. Matters which will be relevant to a pollution control authorisation or licence may also be material considerations to be taken into account in a planning decision. The weight to be attached to such matters will depend on the scope of the pollution control system in each particular case. . . .

In deciding whether to grant planning permission, planning authorities must be satisfied that planning permission can be granted on land-use grounds, and that concerns about potential releases can be left for the pollution control authority to take into account in considering the application for the authorisation or licence. Alternatively, they may conclude that the wider impact of potential releases on the development and use of land is unacceptable in all the circumstances on planning grounds, despite the grant, or potential grant, of a pollution control authorisation or licence.

Source: PG 23, *Planning and Pollution Control* (paras 1.34–36)

BOX 7.2 DEFINITIONS OF SUSTAINABILITY

Sustainable development: development that meets the needs of the present without compromising the ability of future generations to meet their own needs. It contains within it two key concepts:

- the concept of 'needs', in particular the essential needs of the world's poor, to which overriding priority should be given;
- the idea of limitations imposed by the state of technology and social organisation on the environment's ability to meet present and future needs.

(World Commission on Environment and Development: *Our Common Future* (Brundtland Report) 1987)

To promote development that enhances the natural and built environment in ways that are compatible with

- the requirement to conserve the stock of natural assets, wherever possible offsetting any avoidable reduction by a compensating increase so that the total is left undiminished;

- the need to avoid damaging the regenerative capacity of the world's ecosystems;
- the need to achieve greater social equality;
- the avoidance of the imposition of added costs or risks on future generations.

(Blowers 1993)

Sustainability means making sure that substitute resources are made available as non-renewable resources become physically scarce, and it means ensuring that the environmental impacts of using those resources are kept within the Earth's carrying capacity to assimilate those impacts.

(Pearce 1993)

Sustainable development is not simply about creating wealth and protecting the environment. It is also about caring for people and their quality of life. It is about ensuring that the quality of life of future generations will be as good as, or better than, it is for us.

(Environment Agency, *Creating an Environmental Vision*, 2000)

SUSTAINABILITY

Words cast a spell which can, at one and the same time, command respect and create great confusion. No word illustrates this better than the ubiquitous 'sustainability'. There is a view that the word has been so badly abused and misused that it has lost any useful meaning; it now serves to obscure rather than reveal the real issues. General public awareness and understanding of the concept remains low.[1] That there is a broad political consensus on the importance of the general idea of sustainability is surely an indicator of how widely it can be interpreted. Thus sustainability and sustainable development are not capable of precise scientific definition.[2] They are instead social and political constructs used as a call to action but with little in the way of practical guidance (O'Riordan 1985; Baker *et al.* 1997).[3] Indeed, the ambiguity inherent in the terms can be seen as a positive as it presents an opportunity for local political debate on sustainability issues among competing positions. Debate around the sustainability concept ensures that some of the key conflicts and contradictions in public policy (and planning practice) are at least exposed and perhaps addressed (Myerson and Rydin 1996).

But acceptance of the political, vague, and uncertain meaning of the sustainability concept is not an excuse for inaction (any more so than the contested nature of the term 'democracy' is an excuse not to improve our democratic processes). Many academics, environmental groups, and government officials are devoting earnest effort to establishing what sustainability means – or what it should mean – for public policy. There are, without question, important implications for town and country planning arising from the fundamental principles of sustainability – but the nature of these principles can be confusing because of the great variety of definitions. One famous poetic rendering is by Chief Seattle: 'We do not inherit the world from our ancestors: we borrow it from our children.' This encapsulates the essential idea which is more prosaically expressed in the well-known formulation of the 1987 Brundtland Report: 'Sustainable development is development that meets the needs of the present without compromising the ability of future generations to meet their own needs' (World Commission on Environment and Development 1987).

Shiva (1992: 192) has pointed to two very different uses of the concept. One ('the real meaning') relates to the primacy of nature: 'sustaining nature implies the integrity of nature's processes, cycles and rhythms'. This is to be contrasted with 'market sustainability', which is concerned with conserving resources for development purposes, and, if they become depleted, finding substitutes. On this latter approach, sustainability is convertible into substitutability and hence a cash nexus. The distinction is given eloquent expression in the words of a Native American elder who, in epitomising the non-convertibility of money into life, said: 'Only when you have felled the last tree, caught the last fish, and polluted the last river, will you realize that you can't eat money' (Shiva 1992: 193).[4]

This distinction between fundamental (or strong) definitions of sustainability and superficial (or weak) definitions has been made in numerous ways.[5] Owens (1994a) explains that the strong definition places fixed and inviolable constraints on economic activity, whereas the weak definition simply gives environmental capacities greater weight in the decision process. Broadly speaking, the first formulation challenges whether it is right to continue to meet various demands and needs if this cannot be accomplished without reducing current levels and quality of environmental stock. Thus demand management of resource use should be the central policy response. Needless to say, it is generally the weaker formulations that actually dominate. The policy response at this level has been described as *ecological or environmental modernisation* (Jacobs 1999). Here there is an emphasis on meeting sustainability through securing greater eco-efficiency by reducing waste, conserving energy, and reducing pollution, while, to put it crudely, the economy continues to function as before. The objective is to influence market forces rather than regulate or replace them – for example, with devices such as environmental designations or financial mechanisms, though the limitations in providing adequate rewards to the market are well understood (Milton 1991). A key role of planning here is in finding appropriate locations to meet resources demands where environmental costs are lower or where the trade-off of environmental loss against economic gains is more acceptable.[6] Cowell and Owens (1998) have shown how the planning system mediates the questions of demand management and spatial location in a case study of aggregates planning – though the general argument can be applied more widely.

Those who advocate that sustainability is familiar in the history of planning (Hall *et al*. 1993) are in effect presenting the weak interpretation – the planning system's traditional role has been to deal with the locational issues so as to reduce environmental damage and achieve some sort of balance between new urban development and environmental protection. But strong (or even moderate) interpretations of sustainability raise questions about the capability of the planning system to deal with the structural questions of the relationship between social justice (the distribution of costs and benefits), economic demands, and environmental capacities. This is not to say that the spatial or territorial questions are unimportant – they are – but that additional dimensions should also be considered, not least in demand management. Recent changes to the planning system such as in relation to meeting housing land requirements (discussed in Chapter 6) hint at changes in this direction. They also reflect growing consensus about the fundamental and very challenging principles which should govern public policy for sustainability.

In this respect it should be noted that the UK approach has traditionally differed from that in other European countries, particularly Germany. An important difference in principle (differences in practice may be less marked) is that of 'anticipation' as distinct from reaction. Whereas the UK has taken the view that environmental problems should be defined in terms of their measurable impacts, other countries have gone beyond this, and anticipated problems before the degree of environmental damage can be ascertained – this is related to the *precautionary principle*.

In Germany the concept of *vorsorgeprinzip* is applied, meaning broadly the principle of 'prevention' or 'anticipation' (but this fails to capture its full meaning). The German word connotes a 'notion of good husbandry which represents what one might also call best practice'. Möltke (1988) comments that '*Vorsorgeprinzip* is more than just prevention as an efficient means to an end but rather prevention as an end of itself.' The aim is, therefore, to establish pollution control policy, not merely as a means of reducing economic or social cost but also as a means of preserving wider ecosystems. Typically, the European approach involves the avoidance of 'excessive cost'. This, of course, is no easier to define than concepts such as 'reasonable cost', but it is clearly intended to be more demanding. Shed of its more philosophical overtones, the issue is fundamentally 'whether to protect environmental systems before science can determine whether damage will result, or whether to apply controls only with respect to a known likelihood of environmental disturbance' (O'Riordan and Weale 1989: 290).

The principles for sustainability for territorial development and land use planning have been explained in many different ways but are summarised in the EU Sustainable Cities Report (1996), Blowers (1993), Selman (1992) and others. They are:

- to develop within *environmental capacities* and apply the *precautionary principle* where these are uncertain;
- to protect and enhance the stock of *natural capital*, ensuring that it is passed on in good condition to future generations (*intergenerational equity* and *futurity*);
- to ensure that most human benefit is obtained from economic activity; and that there is a fair distribution of the benefit from the use of resources (*intragenerational equity*);
- not to export the costs of economic growth and environmental quality to other places (however distant) and promote *local self-sufficiency*;
- to close *resource loops* through reuse and recycling and the active management of resource flows;
- to ensure that the costs of environmental damage are borne by those who cause them (*the polluter pays principle*); and
- to ensure active *involvement* of local communities in decisions that affect them.

The shortlist has been developed into a more comprehensive framework of sustainability principles as shown in Table 7.1. This conceptualisation of sustainability was developed specifically for territorial development with the aim of assisting in transposing the very general notions of sustainability into planning and development practice and also for appraising existing planning policies and actions (Brown *et al.* forthcoming). The conclusions from the research indicate how sustainability has been 'operationalised' or put into practice by identifying which sustainability principles are actually used in plan- and decision-making and how.

Assessments of the take-up of sustainability principles into aspects of town and country planning are now coming forward.[7] In sum, there has been only partial and fragmented conversion of the principles into planning policies and actions. Policies tend to follow well-worn formulae or 'checklists' and are seldom ambitious in addressing the strong definition of sustainability through, for example, demand management. Even where there is a positive attitude to the notion of sustainable development, the planning response is understood in relatively narrow terms – predominantly the organisation of land uses and transport links – and because of institutional fragmentation rarely addresses the wider impacts in fields such as energy, waste air, noise, and water, or even quality of life. Policy compartmentalisation and departmentalism are strong barriers to effective integrative approaches to sustainable development. The positive results derive largely from linkages between the planning process and Local Agenda 21 and the application of environmental appraisal (discussed on p. 221).

Owens (1994a) suggested there was lots of 'sustainability rhetoric' but in practice it was business as usual. By 1998 Counsell reported that translation of sustainability principles into operational policies in structure plans was still 'proving difficult', though there is great variation in performance – perhaps as much related to local short-term self-interest as to concerns about long-term intergenerational equity. In the meantime, the stock of advice to planning authorities about how to incorporate sustainability into plans and decisions has increased sharply.[8] But aspirations still outstrip achievements. Even the most ambitious experimental projects such as the government's 'Millennium Villages' have 'not yet delivered the order of magnitude of improvement needed to demonstrate true sustainability' (Llewelyn-Davies 2000: 3). The explanation is of course complex and the references noted here point to many factors, but planners will often cite the contradictory and unhelpful nature of national policy and actions (especially outside the planning system) and the limited scope of planning. Recent changes in PPG 3, *Housing* (2000), suggest that significant efforts are being made to provide a stronger framework, but considerable ambiguity remains at the national level.

AGENDA 21 IN THE UK

The UK has made a very positive response to the commitments of Agenda 21. The 1992 Rio Earth Summit gave a major impetus to the elaboration of 'sustainable' policies. Agreement was given to *Agenda 21*: a comprehensive world-wide programme for sustainable development in the twenty-first century. In formulating this programme, major emphasis was placed on a very wide degree of participation. In the UK this is organised at central and local government levels.

Two years after the Rio Summit the government published *This Common Inheritance: Britain's Environmental Strategy*, which was followed by annual monitoring reports. In 1994 this was effectively replaced by *Sustainable Development: The UK Strategy*. The 1997 Labour administration undertook to revise the strategy and published numerous consultation documents during 1998. The revised strategy, *A Better Quality of Life*, was published in May 1999. Scotland published its own sustainable strategy consultation document, *Down to Earth*, in the same year.

The 1999 Strategy promotes four main objectives: social progress (the main addition from the previous strategy), protection of the environment, prudent use of natural resources, and maintenance of high levels of economic growth (see Box 7.3). The strategy identifies 147 sustainable development indicators, including fourteen headline indicators, and makes a fairly frank

Table 7.1 Sustainability Principles for Territorial Development

Principles	*Criteria*
Overarching	
Futurity and intergenerational equity	Precautionary principle (no irreversible decisions)
	Include cumulative and long-term impacts in decision-making
	Commitment to equity at local, national, and international levels
Intersocietal equity	Ensure commitment to equity so environmental impacts and the costs of protecting the environment do not unfairly burden any one geographic or socio-economic sector
Local and regional self-sufficiency	Reduce externality effects so that environmental impacts and costs do not unfairly burden any one geographic group or socio-economic sector
	Use close in preference to distant resources
	Natural disasters
Risk prevention and reduction	Human-made disasters
Environmental	
Maintain the capacity of natural systems	Absolute protection of critical natural capital
	Defence of improvement of soil quality and stability
	Defence and improvement of key habitats and biodiversity
	Respect absorption and assimilation capacities of natural systems
	Efficient use of renewable resources
Minimise resource consumption	Minimum depletion of renewable resources
	Minimum depletion of non-renewable resources
	Energy efficiency
	Minimise waste, recycling and reuse
Environmental quality	Reduce pollution emissions; protect air and water quality and minimise noise
	Protect and enhance environmental amenity and aesthetics
	Protect natural and cultural heritage
Economic and societal	
Protect and develop the economic system	Encourage and develop connections between environmental quality and economic vitality
Develop the human social system (education, democracy, human rights)	Satisfy and protect basic needs (shelter, food, clean water, etc.)
	Provide entrepreneurial and employment opportunities
	Protect basic human rights
Develop the capacity of the political system	Ensure health and safety
	Improve local living conditions
	Satisfy the economic and living standards to which people aspire
	Ensure transparent decision-making processes
	Develop open, inclusive and participatory governance
	Apply subsidiarity and ensure that competences are exercised at the most appropriate level

Source: Adapted from Brown *et al.* (forthcoming)

Table 7.2 Main Events in the Growth of the Sustainable Development Agenda

Year	Major events in environmental planning
1972	UN Conference on the Human Environment, Stockholm
1973	First EC Action Programme on the Environment
1985	First EC Directive on Environmental Assessment
1987	World Commission on Environment and Development (WCED): Brundtland Report – *Our Common Future*
1990	*This Common Inheritance: Britain's Environmental Strategy*
1992	UN Conference on Environment and Development (UNCED or the Earth Summit), Rio and creation of the UN Commission on Sustainable Development (CSD)
	Agenda 21: a comprehensive world-wide programme for sustainable development in the 21st century
	Climate Change Convention: international agreement to establish a framework for reducing risks of global warming by limiting 'greenhouse gases'
	Biodiversity Convention: international agreement to protect diversity of species and habitats
	Statement of Forest Principles for management, conservation, and sustainable development of the world's forests
1994	*Sustainable Development: The UK Strategy*
1996	UN Habitat II Conference, Istanbul
	EU Expert Group on the Urban Environment Report on European Sustainable Cities
	The Aalborg Charter on Local Agenda 21 and the setting up of the European Sustainable Cities and Towns Campaign
1997	Earth Summit +5: five-year review and adoption of Programme for the Further Implementation of Agenda 21 by UN General Assembly
	EU Amsterdam Treaty incorporates sustainable development as a fundamental objective of the EU
1998	Consultation on draft revised UK Strategy on Sustainable Development *Opportunities for Change* and supplementary strategies on business, tourism, biodiversity, forests, construction, and sustainability indicators
	EU Communication *Sustainable Urban Development: A Framework for Action*
1999	*A Better Quality of Life: A Strategy for Sustainable Development for the United Kingdom* published (and additional special papers)
	Down to Earth: A Scottish Perspective on Sustainable Development published
2000	EU Global Assessment of the Fifth Action Programme on the Environment
2001	EU Sixth Framework Programme on the Environment
	OECD Analytical Paper on Sustainable Development
2001	UN Habitat III Conference

BOX 7.3 THE FOUR OBJECTIVES OF A BETTER QUALITY OF LIFE: A STRATEGY FOR SUSTAINABLE DEVELOPMENT FOR THE UNITED KINGDOM

What is sustainable development? At its heart is the simple idea of ensuring a better quality of life for everyone, now and for generations to come.

Social progress which recognises the needs of everyone. Everyone should share in the benefits of increased prosperity and a clean and safe environment. We have to improve access to services, tackle social exclusion, and reduce the harm to health caused by poverty, poor housing, unemployment and pollution. Our needs must not be met by treating others, including future generations and people elsewhere in the world, unfairly.

Effective protection of the environment. We must act to limit global environmental threats, such as climate change; to protect human health and safety from hazards such as poor air quality and toxic chemicals; and to protect things which people need or value, such as wildlife, landscapes and historic buildings.

Prudent use of natural resources. This does not mean denying ourselves the use of non-renewable resources like oil and gas, but we do need to make sure that we use them efficiently and that alternatives are developed to replace them in due course. Renewable resources, such as water, should be used in ways that do not endanger the resource or cause serious damage or pollution.

Maintenance of high and stable levels of economic growth and employment, so that everyone can share in high living standards and greater job opportunities. The UK is a trading nation in a rapidly changing world. For our country to prosper, our businesses must produce the high quality goods and services that consumers throughout the world want, at prices they are prepared to pay. To achieve that, we need a workforce that is equipped with the education and skills for the 21st century. And we need businesses ready to invest, and an infrastructure to support them.

---

assessment of the baseline position and trends for each.[9] These are summarised in Table 7.3. The range of indicators, including, for example, levels of crime, gives a clear indication of the very broad definition that the government has given to sustainable development.[10]

The strategy has found favour with many interests, not least because all areas of public policy are given some prominence in the objectives of sustainable development. There is a strong theoretical argument for a holistic perspective that recognises the part that must be played by all sectors of government in achieving social, economic, and environmental sustainability objectives. But clarity of purpose is sorely compromised, especially in comparison with approaches elsewhere, and there is little doubt that the economic imperative still holds sway. The strategy is in the ecological modernisation approach with a concentration

on increasing economic growth but to be achieved while reducing pollution and the use of natural resources. Thus some indicators for UK sustainability have more than a passing resemblance to the OECD's indicators for economic competitiveness.

Levett (2000) describes the list of indicators as 'a towering achievement', especially in their breadth, but notes that many are concerned with inputs as proxies for ends or measuring actual progress towards greater sustainability — as for example in measuring the existence of Agenda 21 strategies rather than their impacts. Such criticisms of indicators are well known. Selection is intensely political because the indicators are in effect the definition of sustainability, and they may reveal great shortcomings — as illustrated in Box 7.3. Above all, as the strategy itself accepts, increasing eco-efficiency will not be able to keep pace with 'business-as-usual economic growth'. As Levett

Table 7.3 The UK's Strategic Objectives and Headline Indicators for Sustainable Development

Headline indicator	Baseline assessment	Overall performance 1990–98
Maintaining high and stable levels of economic growth and employment		
H1 Total output of the economy (GDP and GDP per head)	Between 1970 and 1998 the output of the economy has grown 86% in real terms.	positive
H2 Total and social investment as a percentage of GDP	Total investment has declined from 20% of GDP in 1970 to 17% of GDP in 1998 – and business has invested consistently less per head than in other G7 countries.	negative
H3 Proportion of people of working age who are in work	In May/July 1999 the employment rate was 74% of those of working age, about the same as 1970, though it has increased for women and decreased for men.	no change
Social progress which recognises the needs of everyone		
H4 Success in tackling poverty and social exclusion (children in low-income households, adults without qualifications and in workless households, elderly in fuel poverty)	Little change over the past ten years with 19% of children in low-income households; 17% of working-age people with no qualifications; 13% people in workless households, and about 60% of single elderly households in fuel poverty.	negative
H5 Qualifications at age 19	The proportion of 19-year-olds with NVQ level 2 (or 5 GCSEs grade C or above) was 45% in 1984 and 74% in 1999.	positive
H6 Expected years of healthy life	Life expectancy has increased (74 years for men and 79 years for women in 1995), but more years are spent in poor health.	no change
H7 Homes judged unfit to live in	Improvement from 8.8% unfit homes in 1986 to 7.2% in 1996 (1.5m homes).	no change
H8 Level of crime	Recorded crime of all types has increased substantially since 1970. Burglary and theft from cars has decreased since 1993 but violent crime continues to rise.	positive and negative

Effective protection of the environment

H9 Emissions of greenhouse gases	UK emissions of greenhouse gases fell by 9% between 1990 and 1997 mainly because of the switch from coal to gas and nuclear power electricity generation. Transport emissions are becoming more significant.	positive
H10 Days when air pollution is moderate or higher	The average number of days recorded as moderate or higher per recording site fell from 60 days in 1993 to 25 days in 1998.	positive
H11 Road traffic	Over the past twenty years, the amount of car mileage per head has increased by 65%; road traffic is now eight times that in 1950 (car traffic fourteen times), and it is forecast to grow by a third over the next twenty years.	no change
H12 Rivers of good or fair quality	Nearly 95% of the river network is of good or fair quality. River lengths that are of good chemical quality rose from 48% in 1990 to 59% in 1998.	positive
H13 Populations of wild birds	Populations of some farmland and woodland birds has fallen by more than half since the mid-1970s, though populations of others, including open-water birds, has been fairly stable.	negative
H14 New homes built on previously developed land	The proportion of new homes built on previously developed land has been much the same since 1989 (though it increased from 1985) and in 1997 was 55%.	no change

Prudent use of natural resources

H15 Waste arisings and management	Household waste has increased by 26% from 1983/84 to 1997/98 and now stands at between 170 and 210m tonnes, 60% of which is disposed of by landfill.	negative

Source: *Quality of Life Counts: Indicators for a Strategy for Sustainable Development for the United Kingdom: A Baseline Assessment*, Government Statistical Service, 1999

Note: In the context of devolution, the Northern Ireland Executive, Scottish Executive, and Welsh Assembly are responsible for elaborating on these national goals and indicators. In total there are 147 indicators

explains, 'eco-efficiency may have a useful contribution to make, but it is fanciful to the point of irresponsibility to expect it to be the main means of reconciling economic and environmental aims' (p. 6). Thus ecological modernisation is not a long-term solution. Nevertheless, improvement in the sustainability indicators is fast becoming an end in itself, though the political significance and impact of the strategy are questionable.[11]

REGIONAL SUSTAINABILITY FRAMEWORKS

In February 2000 the DETR published guidance for the preparation of regional sustainable development frameworks (RSDFs), with a requirement for them to be in place by the end of 2000 (although this has proved to be optimistic).[12] The regional bodies are responsible for adopting the frameworks, and they will need to be coordinated with the sustainable development work of the regional development agencies (RDAs) and regional planning guidance (RPG). The agencies had previously been issued guidance on incorporating the principles of sustainable development into their economic strategies, and some have set up extensive sustainability issue networks or 'round tables'.

The RSDFs are non-statutory guidance but it is widely recognised that the regional and subregional levels are crucial for many sustainable issues such as waste, water management, renewable energy, agriculture, tourism, and urban–rural interdependencies (McLaren 1998). Progress on regional sustainable development frameworks was underway in some regions, not least because of concerns that neither the regional economic strategies nor the regional guidance has fully addressed Agenda 21. To counter this, the RSDFs are to propose a long-term and high-level vision and establish regional indicators and targets. The objective is to join up social, economic, environmental, and resource considerations, which should certainly provide a common context for the preparation of both RDA strategies and RPG. Wales, Scotland, and Northern Ireland are all in the process of preparing sustainable development strategies.

LOCAL AGENDA 21

At the local level, *Local Agenda 21* calls for each local authority to prepare and adopt a local sustainable development strategy. These local efforts have been aided by the work and publications of the Improvement and Development Agency (IDeA) (formerly the Local Government Management Board).[13] A major feature of the consultation programme at the local level is that it involves much more than the term 'consultation' often means. Groups have been established in local areas to debate the meaning of sustainability and to determine how progress towards it can be achieved and assessed ('you can only manage what you can measure'). These local endeavours are designed to produce policies and indicators which are locally appropriate. The research has underlined the importance of this local 'ownership'. There is a positive and a negative aspect to this. Positively, 'Agenda 21 is as much concerned with the *process* of sustainable development – participative, empowering, consensus-seeking, and democratic – as it is with *content*; and 'social processes of securing agreement on and commitment to sustainability aims are indispensable' even where the requirements for sustainability are determined externally (LGMB 1995b). Also, sustainable development strategies draw together many actors into an inclusive network, but 'this, paradoxically, is potentially its greatest weakness, as excessive inclusivity may lead to a lack of clear purpose, direction and commitment' (Selman and Wragg 1999).

In short, the changes in attitudes and behaviour which will be required by policies of sustainability will come about only if they are acceptable. The negative side to this is the widespread distrust of both local and central government which research has uncovered (Macnaghten *et al*. 1995). Agenda 21 emphasises equality and economic, social, and political rights. Among the top concerns are poverty, unemployment, and deterioration in the quality of life and the health of local communities. These are reflected to some extent in the local sustainability indicators chosen.[14] But similar to practice at the national level, the indicators generally reflect the data that are routinely collected and readily available, and there is limited opportunity for comparison from one authority to another (Cartwright 2000). The process can 'easily become cosmetic and bogged down in group dynamics and inertia' (Scott 1999). In addition, although almost all local authorities are working on a Local Agenda 21 Strategy, their commitment varies considerably (Cartwright 1997). Local Agenda 21 has certainly contributed to the growing awareness of environmental and sustainability issues in local politics, but the sum of evaluations (and a review of examples of strategies) suggests that they have succeeded simply in presenting the agenda, with limited impact on mainstream policy. The question now is how the Local Agenda 21 process proceeds. The likely direction is integration with community planning and Best Value (LGA and IDeA 1998; Hams 2000; Christie 2000).

ENVIRONMENTAL POLITICS AND INSTITUTIONS

Environmental politics has become an energetic force on the British scene over the past thirty years, and this is reflected in the growth of environmentally related government units and agencies, advisory panels, and interest groups. Its rise has been prompted by a miscellany of matters, with the most significant first step being prompted by the oil crises of the 1970s, which prompted a new look at resource depletion. Fear of environmental disasters has also played a part, and these seemed more credible after catastrophes such as Seveso, Bhopal, Chernobyl, and, at home, Windscale and Flixborough. The impact of development on natural resources has become clearer with the swing from widespread drought in the late 1990s to even more devastating floods in 2000. Radical campaigners, especially the anti-road tree-dwellers, have also played their part. Thus the environment has become part of the political coinage, and the parties vie with each other in producing convincing statements not only of their concern but also of their workable programmes of action.

Curiously, the growth of environmental consciousness was in part due initially to the lack of government concern. The environment was rarely the subject of political battles. Yet England has been a world pioneer on a number of environmental issues. The Alkali Inspectorate, which was established in 1863, was the world's first environmental agency. Some of the earliest voluntary organisations had their origin in England: for example, the Commons, Open Spaces and Foot-paths Preservation Society in 1865, and the National Trust in 1895 – an organisation that (with over two million members) has grown to be the largest conservation organisation in Europe. The 1947 Town and Country Planning Act introduced a remarkably comprehensive land use planning system (even though, in the circumstances of the time, much of rural land use was purposely omitted). Legislation on clean air has a long history, with its major landmark being the 1956 Act, passed following the killer smog of 1952. The UK also had the first cabinet-level environment department (the Department of the Environment was established in 1970), though its name was, for many years, more impressive than its achievements. Yet these historical events stand as lonely peaks in an otherwise flat plain: until the 1980s, the environment was not a salient political issue (McCormick 1991; Robinson 1992). Part of the reason for this has been the idiosyncratic nature of British pollution control: instead of the formal, legalistic, and adversarial styles common elsewhere, Britain has traditionally operated a system of comfortable negotiation between government technicians and industry. This curiously informal and secretive system avoids confrontation and legalistic procedures (McAuslan 1991).[15]

All recent administrations have had strong advocates of the environmental cause at Cabinet level, but the topic has not quite made it to the premier-division issues in the UK as it has in some other countries. Margaret Thatcher, for example, was initially averse to environmental concerns, which she viewed as a brake on enterprise. Her administrations followed traditional British practice in responding 'pragmatically and flexibly, even opportunistically, when environmental issues have threatened to become too contentious' (Lowe and Flynn 1989: 273).[16] Tony Blair was in government for more than three years before making a speech on environment policy, though in 1997 he made a call for all local authorities to complete Agenda

21 strategies by the end of 2000.[17] His contribution was very much about environmental problems and there was little acknowledgement of the need for more fundamental changes to achieve more sustainable development, but rather an explicit commitment to continue with the tradition of voluntary agreements with business (Warburton 2000).

Even though, global summits aside, prime ministers have not prioritised the environment, certain ministers,[18] parliamentary select committees, agencies, advisory bodies, and interest groups have continued to raise the profile of environmental issues. Parliamentary committees are often regarded as ineffectual, but they have been of great value to environmental groups by providing a new public platform and a route for exerting pressure on Parliament. In particular, the reports of the Environment Select Committee have become a respected source of alternative wisdom and relatively accessible information.

In 2000 the government established the Sustainable Development Commission (SDC),[19] which subsumed the UK Round Table on Sustainable Development and the British Government Panel on Sustainable Development. Its purpose, like that of its predecessors, is to review the extent to which sustainable development is being achieved, identify trends in unsustainability, and deepen understanding of the concept. The two previous organisations made a considerable contribution to government policy with annual and *ad hoc* reports. The Royal Commission on Environmental Pollution has also been an important advocate of improved environmental policy through such reports as *Transport and the Environment* (1994) and *Energy: The Changing Climate* (1999). Thus its current review of environmental planning, to be published early in 2001, is of particular interest.[20]

Another feature of British environmental politics is the active character of some of the important interest groups. Some of these are not merely interest groups: they own and manage extensive areas of land, and they fulfil a range of executive responsibilities. The National Trust and the Royal Society for the Protection of Birds, for instance, own and manage large areas of protected land. Such bodies are also characteristically charities and therefore debarred from overt political activity. Lobbying is thus not only well mannered, but also discreet. The emphasis may be more on education than propaganda, though the distinction can be a fine one.

Governments may try to outflank environmental groups, but increasingly they cannot ignore them, particularly with their new access to power via the EU. Some thirty British groups, together with eighty from other countries, are members of the European Environmental Bureau, which gives them access to the European Commission and the Council of Ministers (Deimann 1994). The British groups have been able to make good use of their experience in lobbying. According to Lowe and Flynn (1989: 272), they 'have adapted more easily than many of their counterparts to the successive rounds of consultation and detailed redrafting of directives and regulations that characterise Community decision-making'.

THE IMPACT OF THE EU

There can be no doubt that the EU has had a major impact on British environmental policy. Indeed, it is not much of an exaggeration to say that much of the government's policy has been dictated by its directives (Milton 1991: 11; Wilkinson *et al.* 1998). This is so despite the fact that the Treaty of Rome imposed no environmental obligations on member states, and the Community initially had no environmental competences. Indeed, Article 2 of the Treaty provided that sustained rather than sustainable growth was the aim: 'a continuous and balanced expansion'. The international scene changed in the late 1960s and early 1970s, with a significant influence being the UN Conference on the Human Environment, which was held in Stockholm in 1972. In the same year, the EC determined that economic expansion should not be 'an end in itself', and that 'special attention will be paid to protection of the environment' (Robins 1991: 7).

In 1973 the first EC *Action Programme on the Environment* was agreed, covering the period 1973–76. Further programmes followed: the fifth covers the period 1993–2000 and at the time of writing a sixth is in preparation. The Single European Act of 1987 gave added legitimacy by including environmental goals in the Treaty and, significantly, added the important provision that 'environmental protection requirements shall be a component of the Community's other policies' (Haigh 1990: 11). Since then the European Environment Agency (EEA) has been established, with headquarters in Copenhagen, providing a monitoring service for the European institutions.[21]

The environmental action programmes have had increasing impact on policy and practice in member states. They are 'forward planning' documents for emerging policies to be implemented by the EU and followed by national, regional, and local governments. Though they have no binding status, many of the proposals result in directives and other action. The Fifth Action Programme has brought a more comprehensive and long-term approach. The overriding aim of the programme is to ensure that all EU policies have an explicit environmental dimension. It stresses the potential of spatial planning instruments. EU documents are not noted for their brevity, and the programme documents are far too wordy to reproduce, but the following gives some flavour of their character. It also illustrates the importance attached to spatial planning instruments:

> The community will further encourage activities at local and regional level on issues vital to attain sustainable development, in particular to territorial approaches addressing the urban environment, the rural environment, coastal and island zones, cultural heritage and nature conservation areas. To this purpose, particular attention will be given to: further promoting the potential of spatial planning as an instrument to facilitate sustainable development . . .; developing a comprehensive approach to urban issues . . . [and] developing a demonstration programme on integrated management of coastal zones . . . to facilitate sustainable development, and promote the potential of the European Spatial Development Perspective in addressing the territorial impact of sectoral policy, and the need for greater integration of land use and transport planning.
>
> (EC 1992: *Towards Sustainability*)

But these are only objectives; they need to be transposed into agreed Community law and action. The great majority of EU environmental laws are in the form of directives (see Chapter 3) which give member states some freedom to choose the manner in which they are transposed into national law. It is unusual for directives to be transposed into national legislation by the due date – which is typically two months after adoption by the Council of Ministers. Nevertheless, they must be implemented 'in a way which fully meets the requirements of clarity and certainty in legal situations'. States cannot rely on administrative practices carried out under existing legislation (Wägenbaur 1991). Moreover, if a directive is not implemented by national law, it is possible for legal action to be taken by private parties to seek enforcement.

In spite of the seemingly very strong powers already possessed by the EU, there is more to come, although the use of Community legislation will give way in some areas to more general agreements and guidelines. The 1987 Amsterdam Treaty incorporated sustainable development as a fundamental objective of the EU and since then there have been commitments to ensure environmental appraisal of all Community policies and actions. In 1999 the Commission undertook an evaluation of the Fifth Action Programme and reported in the *Global Assessment*, which recognised that though some environmental improvements have been made, 'less progress has been made overall in changing economic and societal trends which are harmful to the environment'. The report notes that existing environmental policy 'will not be able to keep pace with . . . demand for road transport, house or road building, etc. . . . growth in these areas simply outweighs the improvements attained by stricter environmental controls'. The paper sets out some ideas for possible future action for discussion and gives a pointer to what might be expected in the sixth action programme.[22] This includes:

- full implementation of the Habitats and Birds Directives, integration of biodiversity and environmental policy into other Community policies;
- implementation of the Communication *Sustainable Urban Development: A Framework for Action*;
- urgent action to protect coastal areas at risk from urbanisation (reckoned to be 85 per cent of the coast);
- action to prevent further degradation and loss of soil;

- improved implementation and enforcement of Community environmental legislation;
- more attention to market-based instruments (taxes, subsidies, etc.);
- a more integrated approach through spatial instruments (with reference to the ESDP); and
- stronger involvement of citizens and other stakeholders.

THE ENVIRONMENT AGENCIES

The 1995 Environment Act provided for the establishment of an Environment Agency for England and Wales and an equivalent Scottish Environment Protection Agency. The idea of such an agency had been resisted by the government for a number of years, and the change of heart was primarily in response to demands from industry for a one-stop shop for environmental regulation.[23] Another factor in the debate was the importance of having an agency that was able to negotiate with the EU from a position of strength.

Against this background, the Environment Agency has taken over the responsibilities of bodies which had been established by a reorganisation only a few years earlier. In England and Wales these were the National Rivers Authority, Her Majesty's Inspectorate of Pollution, and the local waste regulation authorities. In Scotland they were the river purification authorities, HM Industrial Pollution Inspectorate, and the waste regulation and local air pollution responsibilities of the district and islands councils. In Northern Ireland the DoENI has all the responsibilities for environmental protection except waste disposal which lies with the local authorities.

The agencies are non-departmental public bodies, operating on Next Steps lines: the management has a large degree of freedom within the framework of ministerial guidance and its management framework. The framework is based on the government's overall environmental strategy set out now in *A Better Quality of Life: A Strategy for Sustainable Development for the UK* (1999). The agencies are in the process of reviewing their policies in the light of the new strategy, but an important implication of sustainable development is

that they take an integrated approach to their responsibilities: this, indeed, is its essential *raison d'être*. Sustainable development is also leading the agencies to reflect on the traditional reactive and regulatory approach and to add a 'more forceful dimension'. Part of this for the Environment Agency includes a commitment to creating 'a single regulatory system that covers the environmental impact of processes and their resource use, products and their effects, and their impact on land use'. And it foresees an increasing role for local authorities and development agencies such that land use planning and development control are 'more closely aligned to environmental risks and steps necessary to avoid them'.[24]

Establishing more integration and prevention rather than regulation is a considerable challenge. The functions of the agencies are already very wide, including industrial pollution; aspects of waste, including radioactive substances, water resources, and quality; the implementation of a number of EU Directives; and, in Scotland, local air pollution control. Further links with land use planning are also anticipated. In their long-term strategies the environment agencies have adopted a thematic approach for its objectives as shown in Box 7.4. The main tools by which these objectives are addressed by the agencies and other bodies are discussed below.

BATNEEC, BPEO, AND BPM

In their regulation-of-pollution role, environmental bodies have generally sought to achieve the *best practicable means* (affectionately known as BPM) of dealing with problems – 'means' that will go as far as seems reasonable towards meeting desirable standards but which do not involve too great a strain on the polluter's resources. This approach has a long history: indeed, it has been the cornerstone of industrial air pollution control since the Alkali Act of 1874. Its modern version has been expanded to BPEO: *best practicable environmental option*, which retains the element of negotiation but involves a wider consideration of environmental factors and an openness which was foreign to its predecessor (RCEP 1988: para. 1.3).[25] Central to this principle is the recognition of the need for a coordinated approach to pollution control, taking into account the danger of the transfer of pollutants from one medium to another, as well as the need for prevention.

The current regime under the Environment Protection Act 1990 requires the regulating authority to ensure that the *best available techniques not entailing excessive cost* (BATNEEC) are being used:

(1) for preventing the release of prescribed substances into an environmental medium, or, where that is not practicable, for reducing the release to a minimum; and

(2) for rendering harmless any other substance which could cause harm if released into any environmental medium.

BATNEEC is the concept favoured by and introduced in EU Directives, and has now been adopted in English environmental law. It is the responsibility of the operator to demonstrate that the requirements of BATNEEC are met and also to demonstrate their competence and experience, and that effective environmental management controls are in place. Additionally, certain statutory environmental standards ('quality objectives'), specified emission limits, or national quotas have to be met.

Where a process involves the release of harmful substances to more than one medium, BPEO must also be demonstrated – thus there may be trade-offs among the effects in one environmental medium against another. In order to judge the effects of different emissions in different media an integrated permit process has been adopted.

INTEGRATED POLLUTION PREVENTION AND CONTROL

Among the innovations made in environmental policy a particularly interesting one is that of *integrated pollution control* (IPC).[26] This is the administrative apparatus for implementing the *best practicable environmental option* (BPEO). It contrasts with the customary British method of operating different controls in isolation, with separate approaches to individual forms of pollution. The crucial problem with this is that pollution does not abide by the boundaries of air, land, and water: pollution is mobile. In the jargon, it is a 'cross-media' problem.

A 1996 EU Directive (96/61) extended the regulatory regime and controls when implemented by the Pollution, Prevention and Control Act and Regulations in 2000. The new regime, *integrated pollution, prevention, and control* (IPPC) will cover 6,500 industrial processes with integrated controls over pollution, noise, waste reduction, energy efficiency, and site restoration. All operators of installations covering any of the listed processes require a permit which will cover all controls where the applicant will need to demonstrate that best available techniques (BAT) are being used.

PENALTIES FOR POLLUTION

A striking feature of the recent environmental legislation is the severity of the penalties for polluting (Harris 1992a). One feature in particular is noteworthy: the use of 'strict liability'. Generally, under English law, the prosecution has the burden of proving that a defendant is guilty beyond reasonable doubt. The 1990 Act, however, provides that where it is alleged that BATNEEC has not been used in a prescribed operation, 'it shall be for the accused to prove that there was no better available technique not entailing excessive cost than was in fact used'. This makes an offence one of 'strict liability', in contrast to the traditional one of 'fault-based'.[27] Though its use is likely to be rare, it is indicative of the change in official attitudes to pollution (documented in Rowan-Robinson and Ross 1994). It will also involve highly technical matters, which may present severe difficulties for the existing courts. Indeed, some have argued that there is a need for a specialised court (Carnwath 1992; Cambridge University 1999).

ECONOMIC INSTRUMENTS OF ENVIRONMENTAL POLICY

Public opinion is in favour of regulatory standards because of their apparent fairness: all are required to

---

BOX 7.4 THE LONG-TERM OBJECTIVES OF THE ENVIRONMENT AGENCIES

- People will have peace of mind from knowing that they live in a clean, safe, and diverse environment that they can use, appreciate, and enjoy.
- Both urban and rural areas will have an obvious and overall improvement in the extent and quality of their habitats and the wildlife that they support.
- Industry and businesses generally will be managed in a way that fully protects human health and the environment.
- Waste and wasteful behaviour will no longer be a major environmental threat because of the re-use of resources and the adoption of sustainable waste management practices.
- Neither human health nor the natural and man-made environments will be damaged by emissions to the atmosphere.
- There will be sufficient clean and healthy waters to support people's needs and those of wildlife.
- The natural resources provided by the land will be enhanced, harm to people and wildlife will be avoided, and a wide range of land uses will be supportable.
- Flood warnings and sustainable defences will continue to prevent deaths from flooding; property damage and distress will have been minimised; and all the benefits to be derived from natural floods will be exploited.
- Greenhouse gas emissions will have been greatly reduced and society will have adapted efficiently to climatic change and be prepared for further changes.

(Environment Agency, *Creating an Environmental Vision*, 2000)

- Scotland will have an economic base of sustainable industries providing the wealth creation thus enabling the population to enjoy and appreciate the high quality of their environment.
- The way in which society is organised and managed will have changed in response to pressure for sustainable development.
- Emissions of contaminants will be licensed such that they do not exceed the capacity of the environment to assimilate them.
- Pollution from diffuse sources will be coming under control through the wide adoption of good practice.
- Waste will be effectively minimised, reused or recycled.
- The amenity value of rural and urban environments will be significantly improved, with reductions in litter and other unacceptable impacts.
- The rate of loss of greenfield land will be reduced, with development taking place more often on brownfield sites.
- There will be increased energy supply from renewable and low impact sources.
- Decisions which affect people's quality of life will be taken according to local circumstances.

(Scottish Environment Protection Agency, *Environmental Strategy*, 1998)

---

meet the same target. Polluters may also like them because of the certainty which they give to the market. In fact, the fairness is illusory. Fixed standards impose quite different costs on different firms, depending for example on the state of their machinery and processes. More importantly in terms of effective environmental improvement, firms will tend not to seek anything beyond the regulatory standard even if they can achieve a higher standard at relatively low cost. They have no incentive to do so, unless they thereby obtain other benefits.

There are considerable advantages to be derived from designing pollution controls in a way that gives firms economic incentives to reduce pollution to the maximum extent. If, for example, a tax is levied for every tonne of waste produced, a firm will be motivated to review its processes to reduce its waste to the minimum. Positive market incentives may also overcome the reluctance of some firms to meet regulatory standards and reduce the costs of regulation. Since administrative resources are typically inadequate, this is a significant issue. Overstretched agencies may well know that some firms are in default, but they may have some difficulty proving it, or they may have to accept a firm's assurance that it is doing the best it can. Particularly bad cases may be prosecuted, but this takes even more time and resources, and the courts can be unpredictable. In all, as the UK Round Table on Sustainable Development (2000) and others have pointed out, there is a strong case for further developing the use of economic instruments for implementing environmental policy and sustainability in the UK, especially when used as part of complementary packages including regulation, negotiated agreements, and changes to mainstream spending programmes.

Economic instruments can take many forms. The simplest economic instrument is a tax – either to deter negative actions (waste) or to promote positive ones (technological developments). For example, a tax may be levied on pollution at a rate determined in relation to the damage caused and the costs of clean-up. Such a tax could be levied on lead or carbon content. (Several European countries have such a carbon tax.) The tax provides an immediate incentive to firms to reduce their use of the pollutant – and it is a continuing incentive. The difficulty arises in setting an equitable rate – a problem which also arises with marketable pollution permits, which the government is intending to introduce.

Economic incentives can be applied to some types of waste with a deposit refund system. This is essentially the same as the charges on returnable bottles, though rather more complicated. The producer of something which would become a waste after it has been used in a manufacturing process (a solvent for instance) would be required to pay a charge for each unit produced. This would increase its price (thereby introducing an incentive for reduction in its use). A refund of the charge would be payable to anyone who returned the solvent after its use. This system has the advantage of providing a disincentive to illegal tipping. The same system can be applied to motor vehicles.

All the advisory bodies have spent considerable time in debating and recommending the use of economic instruments and they recognise progress made, while pointing out that there is undoubtedly much more that can be done.[28] Some possible innovations, such as road pricing, have been debated for many years, but the technical and political difficulties constitute a major obstacle. Implementation of congestion charging is expected in London during the current term of the GLA. Progress has been made in the fields of landfill tax implemented in 1996 (following the EU Directive) and the creation of environmental trusts (including one that supports cars with alternative fuels). Satellite national accounts have been prepared that address economic, social, and environmental costs but so far are separate from the main national accounts. The Budget statement now includes a note on its environmental impacts. Proposals have been mooted in consultation papers,[29] but there is also lots of scope to make improvements by amending existing mainstream spending, especially in relation to procurement and subsidies. Environmentally damaging subsidies have been estimated at £20,000 million per year (Government Panel on Sustainable Development Third Annual Report 1997). Some, including company car tax benefits and road fund licensing, are now being amended to reflect environmental costs.

LOCAL ENVIRONMENT AGENCY PLANS

The requirement for the production of catchment management plans previously held by the NRA was transferred on their creation to the Environment Agencies. In England and Wales these have been supplemented by local environmental agency plans (LEAPs), which are of smaller scale, covering a small or sub-catchment area and cover the full range of topics for which the Environment Agency is responsible – primarily pollution, waste, water, and air quality. LEAPs are non-statutory documents and progress in their preparation has been slow but they may be a material consideration in development control. Local planning authorities are encouraged to take them into account in the review of development plans. There is wide consultation with local authorities, other bodies, and the public during the preparation of LEAPs. Note that although the catchment type of boundary is particularly useful for water management, it is less relevant for air quality and waste management, which are traditionally much closer to local authority boundaries. Nevertheless, they offer possibilities for the better integration of environmental policy in a territorial plan and in this sense perhaps may develop in a similar way to the German landscape plans.

The proposed EU Water Framework Directive will require the preparation of river basin management plans (RBMPs) for geographical areas around catchments, coasts, and estuaries. The plan will need to link water management to other environmental and economic activities in the area, for example in relation to impacts on the demand for water and the water environment. Almost certainly RBMPs will be developed from existing catchment plans (and in England LEAPs). They will have statutory force and may be binding, with yet further implications for town and country planning.

CLEAN AIR

Concern about air pollution is not new: it was as early as 1273 that action in Britain was first taken to protect the environment from polluted air. A royal proclamation of that year prohibited the use of coal in London and one man was sent to the scaffold in 1306 for burning coal instead of charcoal. Those who pollute

the air are no longer sent to the gallows, but though gentler methods are now preferred, it was not until the disastrous London smog of 1952 (resulting in 4,000 deaths) that really effective action was taken. The Clean Air Acts of 1956 and 1968 introduced regulation of emissions of dark smoke, grit, and dust from furnaces, chimney heights, and domestic smoke. Local authorities were empowered to establish *smoke control areas*, which were very effective (coupled with the switch from coal fires to central heating).

Air quality has improved considerably over the past three to four decades: smoke emissions have fallen by 85 per cent since 1960; the notorious big-city smogs are a thing of the past; and hours of winter sunshine in central London have increased by 70 per cent. In matters of the environment, however, problems are never 'solved': they are merely replaced by new ones – and there is now a long list of damaging air-borne pollutants that are the subject of new research, policies, and actions.[30] Current trends show that the improvements made in respect of industrial and domestic sources of air pollution are being eroded by the damaging effect of increased traffic sources (Banister 1999; Stead and Nadin 2000). Moreover, severe problems in the shorter run can be expected in 'hotspots', particularly in congested urban centres.

Government has been very active on air quality, and in this field the UK is a leader. The 1994 and 1999 UK sustainable development strategies both give prominence to improving air quality. A UK National Air Quality Strategy (NAQS) (1997) has been agreed with national standards and targets, and is already under review in the light of new research findings. A comprehensive network of air quality monitoring stations is in place.[31] Local authorities have to undertake periodic reviews of air quality and identify the areas where national targets are not likely to be met. They then produce *local air quality management plans* for the specified areas.[32] Air quality management plans will seek to reduce emissions through addressing the sources and distribution, especially traffic (which is discussed further in Chapter 11). They may, in principle, designate areas which should be closed to traffic or be restricted to low-emission vehicles – although care will be needed to avoid displacement effects, and some 'local' pollution will have a non-local source. In practice they are not proving to be so radical (Miller 2000). Recent advice to planning authorities suggests that the long-held separation of the role of land use planning from pollution control is being eroded. The 1997 paper *Air Quality and Land Use Planning* says that

> Where the impact of development is likely to be significant in air quality terms, then, provided the impact relates to the use and amenity of land, the planning application may be refused or the impact mitigated by the imposing of conditions.
>
> (DETR 1997: para. 370)

Although it is restricted to questions of the use and amenity of land (as opposed to health) there is a clear signal to authorities to use planning powers to improve air quality conditions. But air quality standards are not easy to determine: the scientific base is inadequate, and a great deal of judgement is necessary. The governmental response to this has been to work towards two measures: a long-term goal, and an operational threshold which indicates when conditions are of so low a quality as to require an immediate response. (Confusingly, these are both termed 'standards'.) Local planning authorities are expected to have regard to the local air quality management plans and to the national standards in preparing land use development plans, and in carrying out other duties such as transport planning.

The EU has played a critical role in bringing about air quality initiatives, largely supported by the UK. The Air Quality Framework Directive of 1996 sets target values for twelve air pollutants which are to be elaborated and revised under 'daughter directives' (and provided a stimulus for the publication of the NAQS. The first daughter directive was agreed under the UK Presidency in 1998 and adopted in 1999. It covers sulphur dioxide, nitrogen oxides, particles, and lead. The second addresses carbon monoxide and benzene, and at the time of writing is moving towards adoption. A third on ozone is in preparation.

THE WATER ENVIRONMENT

The Environment Agencies hold the main regulatory powers over the water environment, although they have no operational responsibilities (these are carried out by the water service companies, or in some cases local authorities). The Agencies have statutory functions in relation to water resources, and the control of pollution in inland, underground, and coastal waters. Their powers are wide, but there are three critical issues in relation to planning: water quality and pollution; the maintenance of water supplies; and flooding.[33]

On water quality the Agencies can take preventive action to stop pollution, take remedial steps where pollution has already occurred, and recover from a polluter the reasonable costs of doing so. The Agencies have inherited and continued to develop a sophisticated and relatively public regulatory system which involves the setting of water quality objectives and a requirement that consent is obtained for discharges of trade and sewage effluent to controlled waters. Extensive monitoring programmes include surveys of the quality of rivers, estuaries, and coastal waters. The highly detailed figures produced from these surveys are not easy to summarise or to interpret, and performance is mixed. River quality is improving and there was a net upgrading of 26 per cent of total monitored length of rivers and canals in England and Wales between 1990 and 1994, but pollution incidents rose over the past decade. Nitrate levels in rivers have remained constant but in some groundwater sites are still rising (Stead and Nadin 2000: 348). Mostly as a result of investment in sewerage works, bathing water quality is improving. In 2000, 211 (44.8 per cent) of the 471 beaches tested in England and Wales met the 'blue diamond' standard[34] of the EU Bathing Waters Directive, and 95 per cent passed the mandatory tests. In Scotland 26 out of 60 got the blue flag and 53 passed the mandatory standard.

As a country surrounded by water and with an annual rainfall of around 1,100 millimetres, one might expect that there would be no question about adequate supply of water. However, rain falls unevenly over both area and time. In the mountainous areas of the Lake District, Scotland, and Wales, average annual rainfall exceeds 2,400 mm, and for most of the country there is a significant margin between effective rainfall and abstraction. But in the Thames estuary rainfall is less than 500 mm, and for much of the Thames and Anglian regions licensed abstractions are more than two-thirds effective annual average rainfall. This is of great concern, even given the high level of reuse, because these are also the regions with the highest demands for new development. The drought of 1988–92 and long, hot, dry summer of 1995 raised awareness about the impact of demand, with unacceptably low levels in some rivers, and supply constraints. As a result, there has been a stream of official reports and consultation papers, and development of academic studies in this area, which had received remarkably little attention previously (at least in the UK).

The need for a major programme of new investment is now widely recognised, not only to replace outworn facilities, but also to meet new demands for water, for environmental protection, and for sustainability. At the same time, increased concerns about water supply have come from developers and the public. The result is a renewed awareness of the importance of the relationship between water and land use planning (Slater *et al*. 1994: 376; see also PPG 12). In addition, government has made a requirement for twenty-five-year resource plans from water companies and targets for reduction of leakage of 25 per cent over three years (in 1997 about 25 per cent was lost through leakage). Demand management is certainly coming to the fore in relation to water supplies, but some areas have more water than they can cope with.

Flooding has moved up the priority list of critical issues during the 1990s, and especially since the floods of Easter 1998. This prompted a review of planning guidance on flooding and development, but no sooner was this completed than the worst floods since records began (400 years ago) devastated much of the country in 2000. The wettest autumn on record (457 mm in three months) resulted in floods in England and Wales affecting 7,406 homes at an estimated cost of £500 million and two deaths (but let's not forget that at the same time, defences protected more than 400,000 homes). The severity of the floods is reflected in one report that 'RNLI lifeboats operated on the High Street to rescue residents trapped in the upper floors of buildings'.[35] Flooding is inevitable, of course; it is part of a natural environmental cycle and cannot be

prevented, but there is much that can be done to prevent flood risk and mitigate the consequences. The environment agencies predict that flooding events will become more numerous and severe.[36]

The reasons for flooding are complex and very much dependent on the conditions of particular catchments and 'coastal cells'. Global warming and associated sea level rise (and land movements) with greater and more intense periods of rainfall play a part. So do engineering works to drainage systems, rivers, and coastlines (flood defences in one location can cause problems elsewhere), and agricultural practices that increase the rate of run-off.

The type and location of new development are the major issues for the planning system, both within flood plains and adjacent rivers and the coast, but also in more distant locations, where it can increase the amount and speed of run-off. The planning system has come in for considerable criticism during the latest round of floods. There is no doubt that building on flood plains in particular has had an impact, though the environment agencies and their predecessors will have been consulted on these developments. The Select Committee on Environment, Transport and Regional Affairs undertook a review of development on, or affecting, the flood plain in the aftermath of the 2000 floods and prior to the finalisation of new government guidance. Its conclusions were clear on the critical effect of increased run-off caused by new development; development in flood plains; and particularly development in functional flood plains or washlands that are used for storage during floods. There are now more than 1.8 million homes in flood risk areas. The Environment Agency estimates that 8 per cent of the land area of England and Wales is at risk from flooding and that if current development patterns persist, a further 342,000 homes may be added to those at risk by 2021 (*Creating an Environmental Vision: Progressing the Environmental Agency's Contribution to Sustainable Development*, 2000).

Planning guidance on flood risk has for some years emphasised that it is a material consideration and that it is appropriate to refuse permission in cases where risk is unacceptable. Recent revisions have strengthened this advice with reference to the sustainable development and the precautionary principle.[37] In the words of the Scottish guidance, 'planning authorities should first, seek to avoid increasing the flood risk by refusing permission where appropriate, and secondly, seek to manage the threat of flooding only in cases where other reasons for granting permission take precedence over flood risk' (NPPG 7, 1995: para. 42). Particular care is promoted in dealing with development proposals that lie just beyond existing flood defences, where a breach may involve a high risk of life. In these cases the advice is even stronger: 'development should not be permitted where the existing flood defences would not provide an acceptable level of safety'. The same applies to caravans in areas of high risk.

Responsibility for determining the extent of risk formally lies with the landowner, although all planning authorities have been issued with flood risk maps by the environment agencies. A particular problem applies where intensification may result from development that does not require planning permission, in which case planning authorities should consider Article 4 directions to remove permitted development rights. Development plans should take into account flood risk, especially where there is a history of flooding, and the environment agencies are important consultees on this matter. Policies and decisions need to be consistent with shoreline management plans and LEAPs.

Although the current and proposed guidance is firm, it did not satisfy the Select Committee, which made clear recommendations for stronger national guidance, and they are likely to be accepted by government. The Committee wants to see in effect a presumption against development in the flood plain and the adoption of a sequential approach – and this will apply to brownfield sites too. Land already allocated for development which does not pass stringent new tests should be deallocated in plans, and MAFF has been told to implement the environmental impact assessment to agricultural practices. In conclusion, the Committee points to the enormous costs of improving flood defences over coming years, with the Thames Barrier for one becoming redundant before the middle of the century.

WASTE PLANNING

The UK produces over 400 million tonnes of waste each year. Details of the recycling and disposal of the 116 million tonnes of industrial, commercial and domestic waste are given in Table 7.4. The rest is mostly agricultural, mining, and quarrying waste and sewage sludge. The legislation covering waste management is immense, with twenty-eight relevant EU directives alone. Essentially, the 1990 Act imposes a *duty of care* on all who are concerned with controlled waste. This duty, similar to that imposed on employers by the Health and Safety at Work Act 1974, is designed to ensure that waste is properly managed. It should be collected, transported, stored, recovered, and disposed of without harm to human health or the environment. The law also ensures that the responsible authorities develop plans for managing and disposing of waste.[38]

Waste regulation functions are the responsibility of the environment agencies. Waste collection remains with local government. Waste planning is the responsibility of local planning authorities (in two-tier areas it is the county), and the regional bodies also have responsibility for waste planning policy for the region. Waste planning authorities must identify suitable sites for the disposal and handling of waste in the context of BPEO, the integrated approach to environmental management, and the government's national objectives for waste. Guidance for waste planning authorities in England is provided in PPG 10, *Planning and Waste Management*.[39] National waste policy was initially set out in the 1995 strategy for sustainable waste management, *Making Waste Work*, which has subsequently been superseded by the *Waste Strategy 2000* for England and Wales. The strategy and the targets and indicators it promotes are material considerations in planning.

Given the nature of waste policy, regional bodies are specially encouraged to address waste policy in revised regional planning guidance. Local authorities will rarely be able to address waste issues independently and there is much to be gained from cooperation at the regional level, although self-sufficiency within regions is encouraged. The waste planning authority (the

Table 7.4 Estimated Waste Production Recycling and Disposal 1998–99

Waste type	Generation (m tonnes)	Recovered and recycled	Disposal	
			landfill	Other
Inert, in-house construction	2	39%	56%	5%
Paper and card	7	77%	22%	1%
Food	3	80%	7%	13%
Other general and biodegradable	9	63%	26%	11%
Metals and scrap equipment	6	89%	10%	1%
Contaminated and health care	5	36%	42%	22%
Mineral waste and residues	6	38%	62%	0%
Chemicals	4	28%	45%	27%
General commercial	23	22%	78%	0%
General industrial	13	13%	86%	1%
Municipal (household) waste	28	9%	83%	8%
Total	106			

Source: Waste Strategy 2000, England and Wales (Part 2: 13–14)

authority that deals with waste planning applications) will be the county in two-tier areas and the unitary council, national park, or London borough elsewhere. Where there is a structure plan, this will also include policies on waste. There will be a specific waste local plan also at the county level. Most unitary authorities include waste policies within their unitary development plans (see Chapter 4). Minerals come under a different provision, but, since a significant proportion of waste arises from mineral workings, waste and mineral plans can be combined.

Waste policies deal with all types of waste, including scrapyards, clinical and other types of waste incinerator, landfill sites, waste storage facilities, recycling and waste reception centres, concrete crushing and blacktop reprocessing facilities, and bottle banks. National policy and targets are playing an increasingly important role. National policy is now very comprehensive (if not always very ambitious) and includes the general principles of moving away from landfill towards recycling, composting, and recovering energy from waste. Nevertheless, there is still a requirement for making a realistic assessment of the need for waste facilities and 'ensuring that there is adequate scope for the provision of the right facilities in the right places'. Planning authorities have the responsibility of ensuring that waste facilities are not developed in locations where they would be harmful or otherwise unacceptable for land use reasons. In this they need to work closely with the environment agencies to ensure that planning and pollution regulation are consistent. A closely integrated 'twin-track' approach is being promoted by the agencies.

Planning authorities also have an important positive planning role in waste management through promoting 'the proximity principle' and the 'regional self-sufficiency principle'. These stem from the desirability of waste recovery or disposal being close to the place where it is produced. This 'encourages communities to take more responsibility for the waste which they – either themselves as householders, or their local industry – produce. It is their problem, not someone else's.' It also limits environmental damage due to the transportation of waste.

The potential for recycling slipped down the policy agenda for some years after peaking in the 1960s, but is now being renewed, presenting more challenges to the planning system. The provision of waste disposal sites was relatively problem free for the system, which relied on the availability of mineral workings to provide suitable sites. The big issues were the responsibility of pollution control (Davoudi 1999). But as stricter controls, EU policy, and rising development pressures to be accommodated in a plan-led process, together with more demand for waste sorting and bulking depots and recovery facilities, the planning issues became more complex. Waste generates considerable public concern, and waste plans and policies are among the most contentious. As a result, planning is playing a more central role in the waste management process.

Again the EU has played an important part in stimulating action in the UK, including the preparation of the national strategies. The EU Landfill Directive (99/31) requires ambitious national targets to be set for the reduction of biodegradable municipal waste sent to landfill, banning the disposal of hazardous and non-hazardous wastes together, and banning the landfill of tyres and liquid wastes. Other objectives for waste management that the planning system needs to take into account are summarised in Box 7.5.

BOX 7.5 SOME OF THE KEY TARGETS FOR WASTE MANAGEMENT

Landfill of municipal waste
By 2010 to reduce biogradable municipal waste landfilled to 75% of that produced in 1995
By 2013 to reduce biogradable municipal waste landfilled to 50% of that produced in 1995
By 2020 to reduce biogradable municipal waste landfilled to 35% of that produced in 1995

Landfill of industrial and commercial waste
By 2005 to reduce the amount of industrial and commercial waste sent to landfill to 85% of that landfilled in 1998

Recovery of municipal waste
To recycle or compost at least 25% of household waste by 2005
To recycle or compost at least 30% of household waste by 2010
To recycle or compost at least 33% of household waste by 2015

To recover value from 40% of municipal waste by 2005
To recover value from 45% of municipal waste by 2010
To recover value from 67% of municipal waste by 2015
(Recover means recycling, composting, and energy recovery)

Source: Waste Strategy 2000

NOISE

'Quiet costs money . . . a machine manufacturer will try to make a quieter product only if he is forced to, either by legislation or because customers want quiet machines and will choose a rival product for a lower noise level.' So stated the Wilson Committee in 1963. This, in one sense, is the crux of the problem of noise. More, and more powerful, cars, aircraft, portable radios and the like must receive strong public opprobrium before manufacturers – and users – will be concerned with their noise level. Similarly, legislative measures and their implementation require public support before effective action can be taken.

As with other aspects of environmental quality, attitudes to noise and its control have changed in recent years, partly as a result of the advent of new sources of noise such as portable music centres, personal stereos, and electric DIY and garden equipment, as well as greatly increased traffic. (Developments in electronics have also provided easier methods of obtaining data on noise.) The increased concern about noise is reflected in a succession of inquiries and new planning guidance (PPG 24: *Planning and Noise*). More substantively, two Acts have been passed to provide stronger measures for dealing with the problems. The Noise and Statutory Nuisance Act, which was passed in 1993, strengthened local authority powers to deal with burglar alarms, noisy vehicles and equipment, and various other noise nuisances. Second, the Noise Act of 1996 provided a summary procedure for dealing with noise at night (11 p.m. to 7 a.m.). This includes powers for local authorities to serve a warning notice, and to seize equipment which is the source of offending noise. The 1996 Act does not require local authorities to use its provisions, but the situation is to be reviewed in the light of experience.

There are three ways in which noise is regulated: by setting limits to noise at source (as with aircraft, motorcycles, and lawnmowers); separating noise from people (as with subsidised double glazing in houses affected by serious noise from aircraft or from new roads); and exercising controls over noise nuisance. Where intolerable noise cannot be reduced and reduces property values, an action can be pursued at common law or, in the case of certain public works, compensation can be obtained under the Land Compensation Act 1973.

Noise from neighbours is the most common source of noise nuisance and complaints. This is a difficult problem to deal with, and official encouragement is being given to various types of neighbourhood action, such as 'quiet neighbourhood', 'neighbourhood noise watch', noise mediation, and similar schemes (Oliver and Waite 1989). There is provision under the Control of Pollution Act 1974 for the designation by local authorities of *noise abatement zones*, though the statutory procedures for these are cumbersome and, in any case, they are not well suited to dealing with neighbourhood noise in residential areas (though they are useful for regulating industrial and commercial areas).

Traffic noise takes many forms and is being tackled in various ways (conveniently summarised in chapter 4 of the Royal Commission on the Environment 1994 report *Transport and the Environment*). Road traffic noise is the most serious in the sense that it affects the most people. Here emphasis is being put on the development of quieter road surfaces and vehicles. Aircraft noise has long been subject to controls both nationally and (with the UK in the lead) internationally. The principal London airports are required by statute to provide sound insulation to homes seriously affected by aircraft noise, and similar non-statutory schemes apply to major airports in the provinces.

Noise is a material consideration in planning decisions, and development plans may contain policies on noise particularly where there are major noise generators such as airports (although the reproduction of detailed noise contours in plans is not recommended). PPG 24 sets out four noise exposure categories (NECs) and in the worst case (category D) permission should normally be refused. The definition of the boundaries between categories is difficult for non-experts, but they are clearly insufficient to prevent the building of houses adjacent to motorways. Such decisions aside, local authorities are taking more interest in noise and one – Birmingham City Council – with the support of the DETR (and building on practice in other European countries) has produced a noise map of the whole of the city, including the impact of road, rail, and air traffic and ambient noise levels during both the day and night. The exercise anticipates legislation that may require such noise mapping for all urban areas. The CPRE has already produced a map of tranquil areas for the whole of England comparing the 1960s with the 1990s, which demonstrates the extensive intrusion of noise.[40]

It will not come as a surprise that the EU has a noise directive in preparation which includes a requirement for noise mapping together with action plans to address identified problems and reduce the number of people exposed to excessive noise, and for the provision of information on noise levels to the public.

ENVIRONMENTAL IMPACT ASSESSMENT AND APPRAISAL

As environmental issues have become more complex, ways have been sought to measure the impacts of development. Cost–benefit analysis was at one time seen as a good guide to action. By taking into account non-priced benefits such as the saving of time and the reduction in accidents, it can 'prove' that developments such as the Victoria Underground line in London are justified. Useful though this technique is for incorporating certain non-market issues into the decision-making process, it has serious limitations. In particular (quite apart from the problems of valuing 'time'), some things are beyond price, while others have quite different 'values' for different groups of the population. Reducing everything to a monetary price ignores factors such as these. Alternatives such as Lichfield's *planning balance sheet* and Hill's *goals achievement matrix* attempt to take a much wider range of factors into account.

Environmental impact assessment (EIA) is a procedure introduced into the British planning system as a result of an EC Directive.[41] Though it might appear that environmental assessment is nothing new on the British planning scene (hasn't this always been done with important projects?), it is in fact conceptually different in that it involves in theory a highly systematic quantitative and qualitative review of proposed projects – though practice is somewhat different (Wood and Jones 1991). Nevertheless, unlike some European countries, Britain has had, since the 1947 Act, a relatively sophisticated system which involves a case-by-case review of development proposals. Indeed, the UK government resisted the imposition of this scheme through the Directive. A summary of the procedure is given in Figure 7.1.

It is important to appreciate that EIA is a *process*. The production of an *environmental statement* (ES) is one part of this. The process involves the gathering of information on the environmental effects of a development. This information comes from a variety of sources: the developer, the local planning authority, statutory consultees (such as the Countryside Agency and environment agencies), and third parties (including environmental groups). There are now many evaluations of practice both in the UK and elsewhere.

For some types of development an EIA is mandatory. These are listed in schedule 1 of the regulations (and are therefore inevitably known as 'Schedule 1 projects'). They include large developments such as power stations, airports, installations for the storage of radioactive waste, motorways, ports, and suchlike. Projects for which EIA *may* be required ('Schedule 2 projects') are those which have *significant* environmental impacts. There are three main types of development where it is considered that an EIA is needed:

1 for major projects which are of more than local importance, principally in terms of physical size;
2 'occasionally' for projects proposed for particularly sensitive or vulnerable locations, for example a national park or a SSSI; and
3 'in a small number of cases' for projects with unusually complex or potentially adverse effects, where expert analysis is desirable – for example, with the discharge of pollutants.

There is a marked resemblance between this and the circumstances in which the Secretary of State may exercise the powers of 'call-in' – they both relate to developments of particular importance which require more than a normal scrutiny for planning and environmental purposes. In 1995 *permitted development rights* were withdrawn from projects listed in Schedule 1, and also for projects having likely significant environmental effects (DoE Circular 3/95). Good practice guidance based on evaluation of the implementation of the EIA process is extensive (see the list of DETR publications) with useful comparisons among countries (European Commission, *Evaluation of the Performance of the EIA Process Final Report*, 1998).

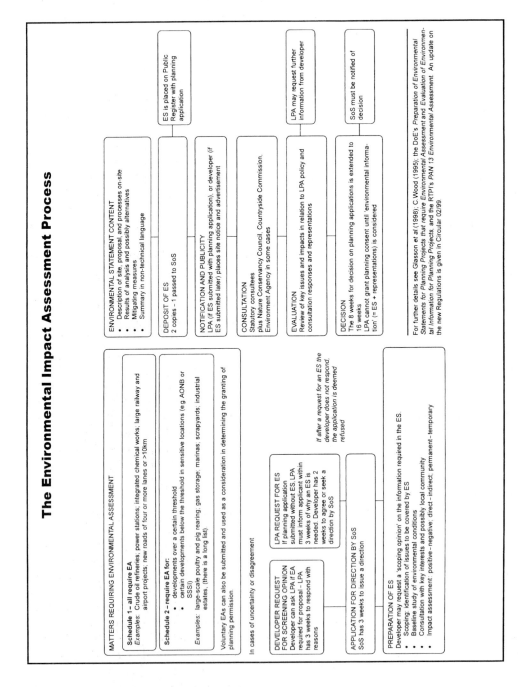

The Environmental Impact Assessment Process

MATTERS REQUIRING ENVIRONMENTAL ASSESSMENT

Schedule 1 – all require EA
Examples: Crude oil refineries, power stations, integrated chemical works; large railway and airport projects; new roads of four or more lanes or >10km

Schedule 2 – require EA for:
• developments over a certain threshold
• certain developments below the threshold in sensitive locations (eg AONB or SSSI)
Examples: large-scale poultry and pig rearing, gas storage, marinas, scrapyards, industrial estates. (there is a long list)

Voluntary EAs can also be submitted and used as a consideration in determining the granting of planning permission.

In cases of uncertainty or disagreement

DEVELOPER REQUEST FOR SCREENING OPINION
Developer can ask LPA if EA required for proposal – LPA has 3 weeks to respond with reasons

LPA REQUEST FOR ES
If planning application submitted without ES, LPA must inform applicant within 3 weeks of why an ES is needed. Developer has 2 weeks to agree or seek a direction by SoS

If after a request for an ES the developer does not respond, the application is deemed refused

APPLICATION FOR DIRECTION BY SoS
SoS has 3 weeks to issue a direction

PREPARATION OF ES
Developer may request a 'scoping opinion' on the information required in the ES.
• Scoping: identification of issues to be covered by ES
• Baseline study of environmental conditions
• Consultation with key interests and possibly local community
• Impact assessment: positive–negative; direct–indirect; permanent–temporary

ENVIRONMENTAL STATEMENT CONTENT
• Description of site, proposal, and processes on-site
• Results of analysis and possibly alternatives
• Mitigating measures
• Summary in non-technical language

DEPOSIT OF ES
2 copies – 1 passed to SoS

ES is placed on Public Register with planning application

NOTIFICATION AND PUBLICITY
LPA (if ES submitted with planning application), or developer (if ES submitted later) places site notice and advertisement

CONSULTATION
Statutory consultees
plus Nature Conservancy Council, Countryside Commission, Environment Agency in some cases

EVALUATION
Review of key issues and impacts in relation to LPA policy and consultation responses and representations

LPA may request further information from developer

DECISION
The 8 weeks for decision on planning applications is extended to 16 weeks
LPA cannot grant planning consent until 'environmental information' (= ES + representations) is considered

SoS must be notified of decision

For further details see Glasson *et al* (1998); C.Wood (1995); the DoE's *Preparation of Environmental Statements for Planning Projects that require Environmental Assessment* and *Evaluation of Environmental Information for Planning Projects*; and the RTPI's *PAN 1/3 Environmental Assessment*. An update on the new Regulations is given in Circular 02/99.

Figure 7.1 The Environmental Impact Assessment Process

Wood and Bellinger (1999) record that in the first ten years of the implementation of the Directive in the UK (from 1988), 3,000 environmental statements had been prepared. Drawing on other evaluations, Glasson (1999: 367) notes that

> EIA is a more structured approach to handling planning applications . . . projects and the environment benefit greatly from EIA . . . and consultants feel that EIA has brought about at least some improvements in environmental protection, in project design and the higher regard given to environmental issues.

Against this there are problems connected to the 'dual consent procedure': EIA in the planning process takes place alongside IPPC and leads to duplication of effort; the lack of attention to alternative options (although the revised Regulations address this); and the blueprint 'build it and forget approach' that tends not to consider the environmental effects of development over its full lifetime, or cumulative impacts. Other evaluations have also shown that EIA may have only a marginal effect on some projects (Blackmore *et al.* 1997).

STRATEGIC ENVIRONMENTAL ASSESSMENT

After many years of debate a political agreement was reached at the end of 1999 on the proposed *Directive on the Assessment of the Effects of Certain Plans and Programmes on the Environment* (the SEA Directive).[42] The Directive will require EU member states to establish procedures to ensure that environmental consequences of plans and programmes are identified before they are adopted, and that effective consultation is undertaken on the environmental implications. The main requirements in the amended proposed Directive are:

- assessment of the environmental effects of 'all plans and programmes';[43]
- preparation of an environmental report identifying and evaluating the environmental effects of implementing the plan or programme;
- consultation with relevant authorities, NGOs, and the public, and with other member states if there are transboundary impacts; and

- statements summarising how the environmental considerations have been integrated into the plan or programme alongside the plan or programme itself.

The main change from the previous draft Directive is the link made to projects which require environmental assessment under the EIA and Habitats Directives. Plans that do not deal with this relatively significant scale of project will not be subject to SEA. In effect, provisions for SEA are already in place in the UK for development plans through the requirement for environmental assessment, and considerable expertise has been developed since publication of the DoE *Good Practice Guide*. The scope of appraisal has been widened to become, in some cases, 'sustainability appraisal' and its use extended to the regional level. Nevertheless, it has been argued that the appraisal process is not contributing to policy-making; that it is highly subjective; and that it leads to inconsistent conclusions (Russell 2000).[44] It is unlikely, therefore, that the Directive in its current form will make a big impression on town and country planning. The next stage may be of more consequence as experience from other European countries becomes better known in the UK. For example, the DETR is proposing research on the idea of 'territorial impact assessment', and others are proposing more radical steps in relation to environmental or ecological compensation schemes (Wilding and Raemaekers 2000). Both are already employed in Germany.

FURTHER READING

For a history of pollution control and much else on the origins of 'environmental policy', see Ashby and Anderson (1981) *The Politics of Clean Air*. Also strongly recommended is Ashby's reflective *Reconciling Man with the Environment* (1978). A detailed legal sourcebook is the *Encyclopaedia of Environmental Law* edited by Tromans *et al.* (looseleaf; updated regularly). Less daunting is Hughes (1996) *Environmental Law*. Miller's background paper for the RCEP, *Planning and Pollution Revisited* (2000) and Wood (1999) 'Environmental planning' both trace the history of the relationship. PPG 24 provides a summary of government policy, though this is now a little dated.

Sustainability

Only the briefest indication of the mass of publications on sustainability can be given here. The Brundtland Report (Report of the World Commission on Environment and Development, 1987) *Our Common Future* is perhaps the most quoted and misquoted source on sustainability; though its interest is increasingly historical, it is still an important original source. Major UK official references on sustainability are cited in the text, and it is certainly worthwhile to start with *A Better Quality of Life: A Strategy for Sustainable Development for the United Kingdom*; the related national strategies and *Sustainability Counts* on indicators (bearing in mind that this is a government interpretation of sustainability). For a more critical review, see Jacobs (1999) *Environmental Modernisation*; Owens (1994a) 'Land, limits and sustainability'; Khan (1995) 'Sustainable development: the key concepts, issues and implications'; and Church and McHarry (1999) *One Small Step: A Guide to Action on Sustainable Development in the UK*. For US perspective, see Board on Sustainable Development, Policy Division (1999) *Our Common Journey*. There is an extremely long list of Web resources on sustainability at the World Wide Web Virtual Library<www.ulb.ac.be/ceese/meta/sustvl. html>.

References that specifically address planning's contribution to sustainability are Blowers (1993) *Planning for a Sustainable Environment* and Breheny (1992) *Sustainable Development and Urban Form*; Williams *et al.* (2000) *Achieving Sustainable Urban Form*; Buckingham-Hatfield and Evans (1996) *Environmental Planning and Sustainability*, Selman (1996) *Local Sustainability*; Selman (1999) *Environmental Planning*; IDeA (1998) *Sustainability in Development Control*; World Health Organisation (1997) *City Planning for Health and Sustainable Development*; and Kenny and Meadowcroft (1999) *Planning Sustainability*. For European comparisons, see Brown *et al.* (forthcoming) *Sustainability, Development and Spatial Planning in Europe*.

On urban and regional sustainability, see Elkin *et al.* (1991) *Reviving the City: Towards Sustainable Development*; Gibbs (1994) 'Towards the sustainable city'; Haughton and Hunter (1994) *Sustainable Cities*; the EU Expert Group on the Urban Environment (1996) *The European Sustainable Cities Report*; Barton (2000) *Sustainable Communities*; and Ravetz (2000) *City Region 2020*.

Local Agenda 21

Recent publications and many cases studies of Agenda 21 are available free on the LGA and IDeA web site <www.la21-uk.org.uk>. For international comparisons, see ICLEI's Agenda 21 web site <www. iclei.org/la21.htm> and the UN sustainable development web site <www.un.org/esa/sustdev/ csdgen.htm>, which also has a surprisingly useful summary of action on sustainability in the UK. For government publications, see LGA *et al.* (1998) *Sustainable Local Communities for the 21st Century: Why and How to Prepare an Effective Local Agenda 21 Strategy*; and LGA and IDeA (1998) *Integrating Sustainable Development into Best Value*. Other main sources include Wilkes and Peter (1995) 'Think globally, act locally: implementing Agenda 21 in Britain' and IDeA (1997) 'Local Agenda 21 in the UK – The First 5 Years'. *EG Magazine* is a monthly publication concerned with practice on LA21 and is published by Westminster University.

Environmental Politics and the Impact of the EU

The subject of environmental politics is also well covered by many text books, including Fischer and Black (1995) *Greening Environmental Policy: The Politics of a Sustainable Future*; Lowe and Goyder (1983) *Environmental Groups in Politics*; Worpole (1999) *Richer Futures*; and Doyle and McEachern (1997) *Environment and Politics*. See also Newby (1990) 'Ecology, amenity, and society', which shows that environmental politics is not simply a modern fad. Beckerman (1995) gives an iconoclastic appraisal of 'environmental alarmism' in *Small is Stupid: Blowing the Whistle on the Greens*. On

Europe, see Vogel (1995) 'The making of EC environmental policy'; and Shaw *et al.* (2000) *Regional Planning and Development in Europe.*

Economic Instruments of Environmental Policy

Cairncross (1993) *Costing the Earth* (chapter 4) is a good non-technical discussion and there is a more recent book by O'Riordan (1997) entitled *Ecotaxation.* There is a discussion of economic instruments in chapter 16 of Cullingworth (1997a) *Planning in the USA*, on which parts of the text are based.

Air, Water and Waste

There are a number of general sources for information and statistics, and of particular interest is McLaren *et al.* (1998) *Tomorrow's World: Britain's Share in a Sustainable Future*, which sets out the theory of 'environmental space' and explains how, realistically, the UK could drastically cut its use of resources. See also the DETR *Digest of Environmental Statistics* (published annually), the Environment Agency's Strategy and State of the Environment Report at <www.environment-agency.gov.uk/strategy/strategy. html>, and Stead and Nadin (1999) 'Environmental resources and energy in the United Kingdom'.

Air pollution policy and an explanation of trends are set out in the *National Air Quality Strategy* (1999) and DETR (1999) *Economic Analysis of the National Air Quality Strategy Objectives.* In Scotland, see SEPA (2000) *Air Report* and Scottish Executive (2000) *Local Air Quality Management General Guidance Series.* See also Elsom (1996) *Smog Alert: Managing Urban Air Quality* and Colls (1996) *Fundamentals of Air Pollution.*

The environment agencies' web sites are probably the best starting point for policy on the water environment, and a set of case study materials for higher education is in preparation and will be at <www. environment-agency.gov.uk/education/studies/ case.htm>. A number of official publications have arisen from both the drought and flooding crises, although they have little to say about planning. Of interest are DETR (1998) *Water Resources and Supply:*

Agenda for Action; DETR (2000) Water Quality in England: A Guide to Water Protection in England and Wales at <www.environment.detr.gov.uk/ wqd/guide/ water.htm>, DETR (2000) *Code of Practice on Conservation Access and Recreation*; and in Scotland, SEPA (2000) *Improving Scotland's Water.* See also Slater *et al.* (1994) 'Land use planning and the water sector'. On flooding, the relevant policy guidance notes are comprehensive and are cited in the text. They should be read in conjunction with the Select Committee Report (2000) *Development Affecting the Flood Plain.*

On waste, both volumes of the *Waste Strategy 2000* (or the Scottish Executive's *National Waste Strategy*) together with PPG 10 (1999) *Planning and Waste Management* (or the equivalents) provide a very comprehensive source. On Scottish policies, see NPPG 10 (1996) *Planning and Waste Management.* For a critical review of the policy of encouraging the recycling of paper products, see Collins (1996) 'Recycling and the environmental debate'.

Noise

PPG 24 (1994) deals with planning and noise. The Batho Report (noise review working party report, 1990) examined a wide range of issues concerned with noise. Later reports have dealt with particular aspects such as the Mitchell Report (1991) *Railway Noise and the Insulation of Dwellings* and the Building Research Establishment report *The Noise Climate around Our Homes* (1994).

Environmental Assessment and Appraisal

The principal texts on environmental impact assessment are Glasson *et al.* (1998) *Introduction to Environmental Impact Assessment* and Wood (1995) *Environmental Impact Assessment: A Comparative Review*, See also DoE (1993) *Environmental Appraisal of Development Plans: A Good Practice Guide*; Elvin and Robinson (2000) 'Environmental impact assessment'; Jones *et al.* (1998) 'Environmental assessment in the UK planning process: a review of practice'; Glasson (1999) 'The first 10 years of the UK EIA system:

strengths, weaknesses, opportunities and threats'; Weston (2000) 'Reviewing environmental statements: new demands for the UK's EIA procedures'; and Wood (2000) 'Ten years on: an empirical assessment of UK environmental statement submissions'. On appraisals, see Russell (2000) 'Environmental appraisal of development plans'.

NOTES

1 In a public survey in Scotland in 1995 only 12 per cent of respondents could define sustainable development and only 2 per cent could explain Agenda 21. However, most respondents (64 per cent) thought that protecting the environment is more important than economic growth; that technological development is not a solution to resource depletion; and that government intervention to improve sustainability is welcome (McCaig *et al.* 1995).

2 See, for example, the monumental 1995 report of the Select Committee of the House of Lords on Sustainable development for the range of definitions.

3 Interestingly, the British Government Panel on Sustainable Development, in its first report (1995), commented that the term was 'not so much an idea as a convoy of ideas'. It is a rallying cry, a demand that environmental issues need to be taken into account; but it provides little guide to action.

4 On the other hand, there are some formidable (if not popular) economic arguments which more prosaically point to the differences between notions of sustainability, optimality, and ethical superiority. The fact that a particular path of development is unsustainable does not necessarily mean that it is undesirable or sub-optimal. In the words of Beckerman (1995: 126), 'most definitions of sustainable development tend to incorporate some ethical injunction without apparently any recognition of the need to demonstrate why that particular ethical injunction is better than many others that one could think up'.

5 See for example Baker *et al.* (1997).

6 It has been argued that there is a process of 'peripheralisation' as locally unwanted land uses (LULUs) (in their case hazardous and polluting industries) are exported to areas beyond the main metropolitan centres to peripheral areas that have less power or will to resist them.

7 See for example, Bruff and Wood (1995), Counsell (1998), and Hales (2000) in the UK, Burke and Manta in the USA, and the findings of the EU-funded SPECTRA project, which are available at www.uwe.ac.uk/fbe/spectra.

8 Of particular note here are LGMB (1995a), Barton *et al.* (1995), Barton (2000), the DETR good practice guide *Planning for Sustainable Development: Towards Better Practice* (1998), and the DETR research report on the Millennium Villages and sustainable communities (Llewelyn-Davies 2000). See also Friends of the Earth (1994a), Ravetz (2000), Levett *et al.* (1999), and the Town and Country Planning Association's *Tomorrow Series* of booklets on environmental planning issues including D. Hall (1999), Hooper (1999), Marsden (1999), O'Riordan (1999), and Winter (1998). See also relevant research findings on urban intensification, particularly Breheny (1992), Rudlin (1998), and K. Williams (1999).

9 The indicators were first published as *Indicators of Sustainable Development for the United Kingdom* but were revised for the 1999 Strategy after consultation on proposed headline indicators through the document *Sustainability Counts* (1998).

10 The broad definition of sustainable development taken in the UK is in distinct contrast to the approach in some other countries such as Sweden or New Zealand, where the ecological dimension is given much more prominence. In Sweden the objective of sustainability is defined in national legislation as being to protect the environment, to conserve the supply of environmental resources, and to make most efficient use of natural resources (see Seaton and Nadin 2000).

11 When faced with public protest in 2000, the government did not hesitate to withdraw from

one commitment: 'to increase duty on petrol and diesel each year by 6 per cent above inflation to reduce carbon dioxide emissions from road transport, 1 per cent higher than the previous Government's commitment' (para. 5.8).

12 *Guidance on Preparing Regional Sustainable Development Frameworks* (London: DETR, 2000). See also the UK Round Table on Sustainable Development report *Sustainable Development Opportunities for Devolved and Regional Bodies* (1999), and *Building Partnerships for Prosperity: Sustainable Growth, Competitiveness and Employment in the English Regions* (London: DETR, 2000).

13 See, for example, the LGMB reports (1993a, 1993b, 1995a, 1995b).

14 For local indicators of sustainable development see the DETR report *Local Sustainability Counts*, which is a handbook of twenty-nine indicators for LA21 and local community planning. Local authority Agenda 21 strategies are also a good source.

15 The system of voluntary compliance is a striking feature of other British regulatory systems. For example, it also appears in a different guise in countryside policy, where, to quote from *This Common Inheritance* (1990: para. 7.3), 'the government works in partnership with [countryside] owners and managers to protect it through voluntary effort'.

16 Although there was her remarkable conversion to the environmental cause in 1988 when she surprised everybody by testifying her personal 'commitment to science and the environment'. With resounding words she rallied her followers to environmentalism, declaring that Conservatives were 'not merely friends of the earth' but also 'its guardians and trustees for generations to come'.

17 The text of the speech is available at <www.number10.gov.uk/news>.

18 The government has appointed 'green ministers' to review the environmental and sustainability impacts of sectoral policies and to establish a programme of action to deliver sustainable development through mainstream policies. The green

ministers meet in a Cabinet Committee on the Environment which is chaired by the Deputy Prime Minister.

19 The SDC is chaired by Jonathon Porritt, covers the whole of the UK and is sponsored by the Cabinet Office. It reports to the Prime Minister, the First Ministers in Scotland and Northern Ireland, and the First Secretary in Wales.

20 At the time of writing, background papers prepared for the Review of Environmental Planning was addressing five main themes: the extent to which planning supports environmental sustainability; the barrier effect of administrative boundaries; the extent of integration and coordination of environmental policy and action; subsidiarity and democracy in environmental policy; and assessment approaches. Background papers are available on the RCEP web site, <www.rcep.org.uk>.

21 The EEA was established in 1993 with the objective of providing 'a seamless information system' on the environment for policy-makers. It does this by collecting and presenting in compatible format existing information through the European Environment Information and Observation Network (EIONET), which comprises 600 environmental bodies and agencies across Europe. Its membership includes EFTA countries as well as EU ones. A major achievement was the preparation of the Dobris Assessment of Europe's Environment (1995) and the Second Assessment (1998), which cover forty-six countries and are the principal sources for state of the environment information in Europe. The EEA web site is <www.eea.eu.int>.

22 Further details are available at <europa.eu.int/comm/environment/newprg/ fromrio.htm>.

23 For an explanation of the creation of the Agencies, see HC Environment Committee (1992) *The Government's Proposals for an Environment Agency*, which points to the important comments made by the Advisory Committee on Business and the Environment and the Institute of Directors. The Environment Committee argued that the Agency

should have more functions than it was actually given by the government. The TCPA went further and argued the need for integration between environmental planning and land use planning, a point now taken up in the review of environmental planning by the RCEP.

24 These quotations are taken from the draft *Environmental Vision* of the Environment Agency for England and Wales (2000: 38–9). A similar (though only five-year) Vision document, *State of the Environment Report*, was published by the Scottish Environment Protection Agency in 1996.

25 ALARA should also be mentioned – the principle of 'as low as reasonably achievable', which applies in the regulation of emissions from radioactive sources.

26 In fact IPC was first recommended by the Royal Commission on Environmental Pollution but 'was largely ignored by government until 1987' (Miller 2000). The EU has been partly responsible for the creation of a more integrated pollution control regime.

27 From 1997 to 1999 the Environment Agency successfully prosecuted 1,700 people for pollution offences resulting (among other penalties) in fifteen prison sentences and a £4 million fine in the case of the *Sea Empress* (reduced to £750,000 on appeal) (Environment Agency; *Creating and Environmental Vision*, 2000).

28 The Advisory Committee on Business and the Environment reports annually (to the president of the Board of Trade and the Secretary of State for the Environment) on economic instruments. Its recent reports have dealt with tradeable permits for water pollution, the landfill tax, and the promotion of alternative fuels. The Government Panel on Sustainable Development's First Annual Report (1995) considered economic instruments, and all subsequent annual reports have noted progress. Proposals for alternative fuels, as well as various economic instruments, were discussed by the Royal Commission on Environmental Pollution in its reports *Transport and the Environment* and *Energy*. The Round Table on Sustainable

Development has recently reported on the use of Economic Instruments (2000).

29 For example in *Economic Instruments and the Business Use of Energy* (1998), *Economic Instruments for Water Pollution* (1997), and *Economic Instruments in Relation to Water Abstraction* (2000).

30 For a summary of the main air pollutants and recent trends, see Stead and Nadin (2000), the UK National Air Quality Strategy, and the DETR *Digest of Environmental Statistics* (published annually). The main air pollutants are carbon dioxide (CO_2), a 'global pollutant' thought to be mainly though not solely responsible for global warming; sulphur dioxide (SO_2), which contributes to acidification of soil and water; nitrogen oxides (NO_X), which also contribute to acid deposition and with other pollutants give rise to smog and poor air quality; ozone, which is created in the atmosphere by chemical reactions involving sunlight, NO_X, and volatile organic compounds (VOCs) and has impacts on the health of the lungs; particles or particulates (PM_{10}) of many types, which contribute to respiratory and cardio-vascular health problems; carbon monoxide (CO); benzene and 1,3-butadiene, which are human carcinogens, the former associated particularly with leukaemia; and lead, which has many negative health effects. Proposed EU Directives will also introduce dioxins into this list.

31 At the end of 2000 the monitoring network comprised eighty-four urban, twenty rural and thirteen specialised hydrocarbon stations, with more related and non-automated sites, which together makes up the most sophisticated monitoring system of any EU state. Considerable information is available via the Internet: <www.aeat.co.uk/netcen/airqual/bulletins>.

32 Monitoring of the production of air quality management plans is under way at the Centre for Air Quality Management at the University of the West of England. It was estimated that by the end of 2000 only three local authorities would have designated areas, although forty more are near completion and 94 per cent of the forty-three local authorities (not including Northern

Ireland) would have begun the process of designation.

33   The environment agencies also have certain powers to prevent flooding, as well as responsibilities for the licensing of salmon and freshwater fisheries, for navigation, and for conservancy and harbour authority functions.

34   The blue diamond is awarded to beaches where between May and September seawater contains less than 10,000 coliforms (tiny living organisms) per 100 ml. A green circle is awarded where quality is moderate and a red square where it fails the test.

35   Select Committee on Environment, Transport and Regional Affairs Second Report 2000: *Development Affecting the Flood Plain*, para. 1.

36   Draft PPG 25 (see note 37) suggests that the 1-in-100 year high-water level on the east coast may be exceeded every twenty years on average by 2050, and that rainfall will increase by 0–10 per cent by 2050. These changes add up to increases in peak flow of up to 20 per cent in the Thames and Severn catchments within fifty years, although the uncertainty of forecasting is acknowledged.

37   In England planning guidance is to be found in the joint DoE and MAFF Circular 30/92 *Development and Flood Risk*. In Wales this has been superseded by Technical Advice Note 15: *Development and Flood Risk* (1998). Scotland has taken the lead in providing a planning policy guidance note on the subject: NPPG 7, *Planning and Flooding* (1995), while a new PPG 25 is imminent in England following consultation on the draft: *Development and Flood Risk* (2000). The planning policy guidance notes are particularly useful, with explanations of the causes of flooding and bibliographies of research and other guidance.

38   The definition of waste gives rise to problems of a byzantine character: the lengthy DoE Circular 11/94 explains all. The legal definitions in the UK now follow that in the EU Waste Framework Directive, which describes waste as 'any substance or object [.] which the holder discards or intends or is required to discard'. Sixteen categories are listed and summarised in Annexe B to the *Waste Strategy 2000* for England and Wales. An alternative definition by Mary Douglas is 'waste is matter in the wrong place (quoted in Worpole 1999: 24). Worpole goes on to say that 'a newspaper on the café table is a highly esteemed cultural artifact; blowing around the street an hour later, it becomes a threat to our very sense of meaning and belonging. Ten newspapers scattered on the pavement and there goes our neighbourhood.'

39   In Scotland there is a parallel NPPG 10: *Planning and Waste Management* (1996); and in Wales a Technical Advice Note: *Planning Pollution Control and Waste Management* (which is due to be revised). See also *Re-inventing Waste: Towards a London Waste Strategy* (London: LPAC, 1998), *The Landfill Campaign Guide* (London: Friends of the Earth, 1997), and the Select Committee on Environment, Transport and Regional Affairs report *Sustainable Waste Management* (1998). Park (2000) has found that the landfill tax is being implemented but that little funding is finding its way to local clean-up projects as intended.

40   The map of tranquil areas of England (1995) was produced in cooperation with the Countryside Agency and is available at <www.greenchannel.com/cpre>.

41   Environmental assessment was introduced by the 1985 EC Directive on the Assessment and Effects of Certain Public and Private Projects on the Environment (85/337). The Directive was amended in 1997 by the *Amending Directive* (97/11). The amendment extended the range of projects that are subject to EIA and made other requirements in relation to the need for the planning authority to tell the developer what should be included in the EIA (scoping), the provision of information on alternative options, and other procedural matters. The Directives are implemented through regulations in the UK – for England and Wales the Environmental Impact Assessment Regulations 1999 SI no. 293, and in Scotland by the Environmental Impact Assessment Regulations 1999 SSI no. 1, which

are explained in Scottish Executive Development Department's PAN 58 and Circular 15/1999.

42 An amended text was formally adopted in March 2000, and at the time of writing adoption of the final directive was imminent. It is available at <http://europa.eu.int/comm/environment/eia/sea-legalcontext.htm>. Member states will need to implement the Directive within three years.

43 Plan and programmes covered include those prepared for 'agriculture, forestry, fisheries, energy, industry, transport, waste management, water management, telecommunications, tourism, town and country planning or land use *and* which set a framework for future development consent of projects' and which fall under the Environmental Assessment and Habitats Directives. Plans and programmes dealing only with finance and budgets, or serving national defence, or which determine the use of small areas at the local level and minor modifications, will be exempt, unless the member state determines that they are likely to have significant environmental effects. Plans and programmes for the Structural Funds are also exempt, although there are other provisions in their regulations which require that environmental considerations are taken into account.

44 The DETR has published *Guidance for the Sustainability Appraisal of Regional Planning Guidance* (2000).

8

HERITAGE PLANNING

It is time to build a new future from England's past. Conservation is not backward looking. It offers sustainable solutions to the social and economic problems afflicting our towns and cities. It stands in the vanguard of social and economic policy, capable of reversing decades of decay by injecting new life into familiar areas.

(Conservation-led Regeneration: The Work of English Heritage, 1998)

EARLY ACTIONS TO PRESERVE

Britain has a remarkable wealth of historic buildings, but changing economic and social conditions often turn this legacy into a liability. The cost of maintenance, the financial attractions of redevelopment, the need for urban renewal, the roads programme, and similar factors often threaten buildings which are of architectural or historic interest.

The first state action came in 1882 with the Ancient Monuments Act, but this was important chiefly because it acknowledged the interest of the state in the preservation of ancient monuments. Such preservation as was achieved under this Act (and similar Acts passed in the following thirty years) resulted from the goodwill and cooperation of private owners. A major landmark in the evolution of policy in this area was the establishment, in 1908, of the three Royal Commissions on the Historical Monuments (of England, Scotland, and Wales). They had (and still have) the same purpose, exemplified by the original terms of reference of the English Commission:

> to make an inventory of the Ancient and Historical Monuments and constructions connected with or illustrative of the contemporary culture, civilisation and conditions of life of the people of England, from the

earliest times to the year 1700 and to specify those that seem most worthy of preservation.

The quotation is instructive: the emphasis is on preservation and on 'ancient'. There was no concern for anything built after 1700 – a prejudice which Ross (1996: 14) notes was typical of the time. Slowly changing attitudes were reflected in 1921 when the year 1714 was substituted for 1700! The date was advanced to 1850 after the end of the Second World War, and in 1963 an end-date was abolished.

The Commissions were established to record monuments, not to safeguard them. It was not until 1913 that general powers were provided to enable local authorities or the Commissioners of Works to purchase an ancient monument or (a surprising innovation in an era of sacrosanct property rights) to assume 'guardianship' of a monument, thereby preventing destruction or damage while leaving 'ownership' in private hands. Major legislative changes were made in the 1940s, though in practice the most important innovation was the establishment of a national survey of historic buildings. This was a huge job (quite beyond the capabilities of the slow-moving Royal Commissions). It was undertaken, county by county, by so-called 'investigators', and by 1969 gave statutory protection to almost 120,000 buildings, and non-statutory recognition (but not protection) to a further 137,000 buildings.[1]

Statutory protection, however, is not sufficient by itself: the owners of historic buildings often need financial assistance if the cost of maintaining old structures is to be met.[2] Grants were introduced in 1953 for preserving houses which were inhabited or 'capable of occupation'. Further big changes were made in 1983, and later most of the provisions relating to heritage properties were consolidated in England and Wales in the Planning (Listed Buildings and Conservation Areas) Act 1990.[3]

CONSERVATION AND HERITAGE

In considering the role of this regulatory system, it is important to appreciate what is meant by the term 'conservation'. Though it is often used synonymously with 'preservation', there is an important difference. Preservation implies maintaining the original in an unchanged state, but conservation embraces elements of change and even enhancement. To provide an economic base for the conservation of an old building, new uses often have to be sought. It is quite impossible to conserve all buildings in their original state irrespective of cost, and there frequently has to be a compromise between 'the value of the old and the needs of the new' (Ross 1991: 92). Thus 'new uses for old buildings' is a major factor in conservation, and it necessarily implies a degree of change, even if this is restricted to the interior.[4] Again, for conservation purposes it may be necessary to enhance a site to cater for public enjoyment. The difference is more than one of name.

'Heritage' is the fashionable term, although its use is not always welcomed.[5] Heritage takes the conservation idea further and embraces consideration of the use of what is conserved. It includes 'the process of evaluation, selection and interpretation – perhaps even exploitation – of things of the past' (Larkham 1999a: 105). For some, heritage presents historical buildings and places as commodities to be traded, packaged,

and marketed. And much of the UK is now neatly packaged into heritage products, carefully denoted by the brown signs marking entrances to 'Shakespeare's County', Brontë Country, 'Lawrence Country', and many more. On the positive side, the notion of heritage draws attention to the economic potential of conservation, but it has been argued that the commodification process pays much less attention to authenticity and accuracy. Use of the term, and indeed action on conservation, was given a boost by European Architectural Heritage Year in 1974 and since then has been used widely, although government policy documents have stuck to the more appropriate 'historic environment'.[6]

Delafons (1997: 168–71) reviews current conservation policy in England (as set out in PPG 15) in view of conservation doctrine built up since the nineteenth century, and in comparison with its predecessor, Circular 8/87. Despite views to the contrary, the presumption in favour of preservation remains in place but a more realistic and flexible approach is given to alternative uses for historic buildings. Even though the PPG was prepared during the time of deregulation and emphasis on economic development, the emphasis of PPG 15 is tipped heavily in the direction of conservation – so much so that it tends to downgrade the potential of well-designed replacement buildings. It says that claims about their architectural merits cannot justify the demolition of any listed building.[7] Since the PPG (and in a different political context), English Heritage and the other agencies have put more emphasis on the regeneration potential of conservation and the concept of 'conservation for everyone'. However, it will take some time for conservation to rid itself of the well-deserved criticism of elitism, if indeed it can. The whole ethos of conservation policy in the UK has been about selection.

The latest development in terminology (if not necessarily in action) is the idea of 'sustainable conservation' or, to put it more accurately, conservation for sustainability. It has been argued that the concepts are two sides of the same coin. The historic environment is a finite resource that should not be depleted. Conservation encourages the recycling of existing buildings and materials, the use of local resources, and diversity in the environment. It can be argued that the historic city in many ways is a model for a sustainable city (Manley and Guise 1998: 86). But there is still much to be debated on the relationship between sustainability and conservation, not least in the widening gap between the quality of 'sheltered' historic areas and the rest of the public realm.

HERITAGE RESPONSIBILITIES

There has been much reorganisation of responsibilities for managing the historic environment in the UK over recent years. In England, responsibility for heritage was transferred from the Department of the Environment to the newly created Department of National Heritage, which in 1997 became the Department of Culture, Media and Sport. The newly devolved administrations in Scotland, Wales, and Northern Ireland have taken on responsibility for heritage matters; and executive agencies have been created to manage the heritage – English Heritage, Historic Scotland, Cadw (Welsh Historic Monuments) in Wales, and the Environment and Heritage Service in Northern Ireland.[8]

Major institutional change is often a politically adept technique of seeming to be doing something substantive while only giving the appearance of so doing. In this case, however, the institutional change has been part of a new commitment to preserving and enhancing the historic legacy. This has been greatly facilitated by the advent of funds from the National Lottery as discussed below.

Many governmental and voluntary organisations play a role in heritage planning and the main ones are shown in Table 8.1. The executive agencies manage most government funding for the heritage (except lottery funding), maintain historic buildings and sites in government ownership, and advise government on heritage matters, including planning decisions. The Royal Commissions survey and compile the historic monuments records (in England the Royal Commission has been merged with English Heritage). The historic buildings councils advise government in heritage matters, notably regarding the listing of

buildings. The advisory body, the Commission for Architecture and the Built Environment (CABE), has a wide remit 'to inject architecture into the bloodstream of the nation' (England), and in this role will often advise about the impact of new development on the heritage.

This is a field in which voluntary organisations have been particularly active. The first of these dates back to 1877 when William Morris (horrified at the proposed 'restoration' of Tewkesbury Abbey) inspired the founding of the Society for the Protection of Ancient Buildings (Ross 1996). Many others have followed; the National Trust with 2.5 million members is the largest. Others with more specialist concerns include the Georgian Group, the Victorian Society, and the Twentieth Century Society.[9] Other organisations have a wider remit: the Civic Trust champions improvement in all places where people work but supports heritage conservation and interpretation through, for example, heritage open days, when buildings normally closed are opened for visitors. The main organisation for planners (both officers and members) is the English Historic Towns Forum, formed in 1987, and there is also a Conservation Officers' Society.

ARCHAEOLOGY

One indicator of the increase in the public popularity of archaeology is the number of television programmes now devoted to the subject. Planning policy has contributed to this through changes over the past fifteen years that have strengthened protection and provided funding from developers for archaeological works. *Rescue archaeology* has been widely publicised through such finds as the streets from Saxon Lundenwic at the Covent Garden Opera House site; the Dover Boat – 'the Bronze Age cross-channel ferry' unearthed during road works in Dover; and the Roman Lady burial at Spitalfields in London. Planning provisions have provided an opportunity for investigation, recording, and removal of these archaeological remains prior to proposed development. But planning has also reduced such situations by recording remains and anticipating problems in local plans.

Table 8.1 Government Departments, Agencies, and Advisory Bodies for Heritage in the UK

	England	Northern Ireland	Scotland	Wales
Government department	Department of Culture, Media and Sport	Northern Ireland Executive	Scottish Executive	Welsh Assembly
Executive agencies	English Heritage; Royal Parks Agency	Environment and Heritage Service	Historic Scotland	Cadw (Welsh Historic Monuments)
Other advisory bodies	Commission for Architecture and the Built Environment	Historic Buildings Council for Northern Ireland; Historic Monuments Council for Northern Ireland	Historic Buildings Council for Scotland	Historic Buildings Council for Wales
Royal commissions	(Royal Commission on the Historical Monuments of England was merged with English Heritage in 2000)		Royal Commission on the Ancient and Historical Monuments of Scotland	Royal Commission on the Ancient and Historical Monuments of Wales
Other	Historic royal palaces			
Other funding bodies	National Heritage Fund and Heritage Lottery Fund (with separate committees in Northern Ireland, Scotland, Wales, and the English regions.			

PPG 16 *Archaeology and Planning*, and similar policy statements outside England,[10] make it clear that there is a presumption in favour of the preservation of important remains, whether or not they are scheduled. There is thus a measure of protection over the large number of unscheduled sites that are on the lists maintained by county archaeological officers. (These are known as SMRs: county *sites and monuments records*.) Such sites are a 'material consideration' in dealing with planning applications.

Planning authorities make provision in their development plans for the protection of archaeological interests, often with good cooperation from large developers. What is perhaps surprising is the extent to which some developers are prepared to go to assist rescue archaeology, and even to fund it. Funding from developers for archaeological work is now four times that available from other sources. Such funding is generally welcomed, but there are criticisms that this gives the developer rather than the archaeologists control over the work. Also, developers do not have to pay for the production or dissemination of reports. Much archive material is being produced but is not widely available. A useful mechanism for liaison is provided by the Code of Practice of the British Archaeologists and Developers Liaison Group.[11]

However, rescue archaeology is, at best, of limited benefit: it is certainly far inferior to preservation *in situ*. The cost of this can be enormous and, given the incredible range of archaeological remains in Britain, some selection is inevitable.

The Secretary of State can also designate *areas of archaeological importance*. In these areas, developers are required to give six weeks' notice (an *operations notice*) of any works affecting the area, and the 'investigating authority' (e.g. the local authority or a university) can hold up operations for a total period of up to six months. The powers have been used very sparingly, and only five areas have been designated, comprising the historic centres of Canterbury, Chester, Exeter, Hereford, and York. A 1996 consultation paper *Protecting Our Heritage* argued that the powers are now redundant, but this recommendation (like all the others in the report) has not yet been acted upon (but see the discussion on the heritage review, below).

ANCIENT MONUMENTS

The term 'ancient monument' is defined very widely: it is 'any scheduled monument' and 'any other monument which in the opinion of the Secretary of State is of public interest by reason of the historic, architectural, traditional, artistic or archaeological interest attaching to it'. This is so broad a definition that it could include almost any building, structure, or site of archaeological interest made or occupied by humans at any time. It includes, for instance, a preserved Second World War airfield complex at East Fortune (near Haddington, in Lothian).

The legislation requires the Secretary of State to prepare a schedule of monuments 'of national importance', which are then given special protection through the planning system. This 'scheduling' is a selective and continuing process. It has been under way for over a century and, for many years, proceeded at a very slow rate.[12] The pace of recording monuments has quickened over recent years following initiatives by the Royal Commissions but it will still be many years before the schedule could be described as near completion. At the end of 1999 there were 33,900 scheduled sites in the UK (see Table 8.2). Estimates of the total number of archaeological sites in Britain vary but it is in the region of one million. Since there are such a huge number of known archaeological sites and monuments, it is not surprising that estimates differ.[13] The number is in decline, with one estimate suggesting that one site has been lost every day since 1945 (Bryant 1999). During 1998 and 1999 a Monuments at Risk Survey (MARS) revealed that at least 70,000 monuments are at risk, the main culprit being damage by ploughing.[14]

In England the aim is to review 70–80 per cent of the total by 2003. In Scotland it is considered that the most outstanding monuments have been scheduled; there are over 10,000 other monuments that might be scheduled but have yet to be assessed. At the current rate of progress (according to the 1995 NAO report *Protecting and Presenting Scotland's Heritage Properties*) the schedule will not be complete for at least a further twenty-five years.

It is recognised that the present schedule is not only very incomplete, but also an inadequate and

Table 8.2 Numbers of Listed Buildings, Scheduled Monuments, Conservation Areas, and World Heritage Sites in the UK, 1999

	Listed buildings	Scheduled monuments (in care)		Conservation areas	World heritage sites	Historic parks and gardens
England	453,011	17,759	(400)	8,819	10	1,335
Scotland	44,462	7,035	(330)	674	3	275
Wales	22,308	3,000	(129)	400	1	n.a.
Northern Ireland	8,563	1,365	(181)	40	1	n.a.
Total UK	572,225	33,900		9,114	15	

Note: There are also three world heritage sites in overseas dependent territories. The nature of 'historic parks and gardens' varies.

unrepresentative sample of the archaeological heritage. In PPG 16 the DETR advises that 'where nationally important archaeological remains, whether scheduled or not, and their settings, are affected by proposed development there should be a presumption in favour of their physical preservation'. The Acts also provide further protection from damage by users of metal detectors, although they also require monuments to be open to the public.

The fact that a monument is scheduled does not mean that it will automatically be preserved under all circumstances. It simply ensures that full consideration is given to the case for preservation if any proposal is made which will affect it. The need to preserve is a material consideration in development control and whether a monument is scheduled or not, and planning authorities may seek Article 4 directions to remove permitted development rights.

Any works have to be approved by the Secretary of State (who receives advice from the agencies, commissions, and other advisory bodies). Such approval is known as *scheduled monument consent*. Where consent is refused, compensation is payable (under certain limited circumstances) if the owner thereby suffers loss.[15] In practice, the great majority of applications for consent are approved, often with conditions attached. The issue here is seen as one of balancing the need to protect the heritage with the rights and responsibilities of farmers, developers, statutory undertakers, and other landowners. The legislation also

empowers the Secretary of State to acquire (if necessary by compulsion) an ancient monument 'for the purpose of securing its preservation' – a power which applies to any ancient monument, not solely those which have been scheduled.

Though most heritage properties remain in private ownership, a small number are managed by the Heritage Departments – officially known as being 'in care'. These are generally of important historical, archaeological, and architectural significance. A very high proportion are of great antiquity, including prehistoric field monuments such as Maiden Castle; prehistoric structures such as Stonehenge; Roman monuments such as Wroxeter and parts of Hadrian's Wall; and a large number of medieval buildings.[16]

LISTED BUILDINGS

Under planning legislation, and quite separate from the provisions relating to monuments, the central departments maintain lists of buildings of *special architectural or historic interest*.[17] The preparation of these lists has been a mammoth task which has progressed slowly because of inadequate funding and (the underlying factor) a lack of public interest.

Although the national listing survey is now substantially complete, listing is a continuing process, not only for additional buildings but also for updating information on existing listed buildings, particularly in terms of their condition. Existing listed buildings can be up- or downgraded. In addition, individual buildings can be spot-listed. This arises because of individual requests, often precipitated by the threat of alteration or demolition. The majority of these requests are made by local authorities (ideally well in advance of development proposals being submitted for planning permission) so that 'any application for listed building consent can be considered in tandem with the planning application for the new development' (Ross 1996: 81).

At one time, listing often came as a surprise to owners who were not aware that their property was under consideration for listing. Since 1995 the departments have consulted on listing, though there is still no duty to consult anyone (including owners) and no right of appeal. Listing has been described as 'a fearful prospect' for owners of younger commercial and industrial properties because of the costs and delays in making changes to what are often obsolete and inefficient buildings (Derbyshire *et al.* 1999). The costs of retaining the building or any financial consequences are not considered in the listing process. An owner can apply for a certificate of immunity from listing which lasts for five years, but this may simply raise government's awareness of the need to list.

There are two objectives in listing. First, it is intended to provide guidance to local planning authorities in carrying out their planning functions. For example, in planning for redevelopment, local authorities will take into account listed buildings in the area. Second, and more directly effective, when a building is listed, no demolition or alteration that would materially alter it can be undertaken by the owner without the approval of the local authority.[18] This is *listed building consent* and is separate to planning permission, but there is no fee. There have been numerous celebrated cases where people have been caught out because it was not recognised that the works require listed building consent. This arises because it is not 'development' (as defined for planning permission) that is controlled, but any works to a listed building *that affect its character as a building of special architectural or historic interest*. Thus painting a building (or even a door) may need consent if it affects archi-

tectural or historic character. Furthermore, the definition of what is listed is very wide and includes certain fixtures and fittings.[19]

The procedure for obtaining listed building consent is summarised in Figure 8.2. Applications have to be advertised, and any representation must be taken into account by the local authority before it reaches its decision. Where demolition is involved, English local authorities have to notify English Heritage, the appropriate local amenity society, and a number of other bodies.[20] If, after all this, the local authority intends to grant consent for the demolition (or, in certain cases, the alteration) of a listed building, it has to refer the application to the DETR so that it can be considered for 'call-in' and decision by the Secretary of State. English Heritage advises on this, and in most cases the DETR accept its advice. It is the Department's policy that consent applications should generally be decided by local authorities, and only a very small number are referred.

Conditions can be imposed on a listed building consent in the same way as is done with planning permissions. The type of conditions that can be imposed are set out in DoE Circular 8/87 and include the preservation of particular features, the making good of damage caused by works of alteration, and the granting of access (before work commences) to a named body to enable a photographic record or measured drawings to be made.

All these provisions apply to listed buildings, but local authorities can serve a *building preservation notice* on an unlisted building. This has the effect of protecting the building for six months, thus giving time for the DETR to consider (on the advice of English Heritage) whether or not it should be listed. As mentioned above, owners and developers who wish to be assured that they will not be unexpectedly made subject to listing can apply to the LPA for a *certificate of immunity from listing*.

With a listed building, the presumption is in favour of preservation. It is an offence to demolish or to alter a listed building unless listed building consent has been obtained. This is different from the general position in relation to planning permission, where an offence arises only after the enforcement procedure has been invoked. Fines for illegal works to listed build-ings are related to the financial benefit expected by the offender.

The legislation also provides a deterrent against deliberate neglect of historic buildings. This was one way in which astute owners could circumvent the earlier statutory provisions: a building could be neglected to such an extent that demolition was un-avoidable, thus giving the owner the possibility of reaping the development value of the site. In such cases the local authority can now compulsorily acquire the building at a restricted price, technically known as *minimum compensation*. If the Secretary of State approves, the compensation is assessed on the assumption that neither planning permission nor listed building consent would be given for any works to the building except those for restoring it to, and maintaining it in, a proper state of repair; in short, all development value is excluded.

The strength of these powers (and others not detailed here) reflects the concern which is felt at the loss of historic buildings. However, they are not all of this penal nature. Indeed, ministerial guidance has emphasised the need for a positive and comprehensive approach. Grants are available towards the cost of repair and maintenance. Furthermore, an owner of a building who is refused listed building consent can, in certain circumstances, serve a notice on the local authority requiring it to purchase the property. This is known as a *listed building purchase notice*. The issue to be decided here is whether the land has become 'incapable of reasonably beneficial use'. It is not sufficient to show that it is of less use to the owner in its present state than if developed. Local authorities can also purchase properties by agreement, possibly with Exchequer aid. Exceptionally, a neglected building can be compulsorily acquired. There is only one case of this: the St Ann's Hotel building in Buxton, which is part of a late eighteenth-century crescent. It had had a long history of neglect which continued through various ownerships. After all alternatives had been exhausted, the Secretary of State served a compulsory purchase order in 1993.

In spite of all these (and other) provisions, many listed buildings are at risk. In Scotland (according to

the 1995 National Audit Office report *Protecting and Presenting Scotland's Heritage Properties*) the ongoing *Buildings at Risk Register* contained, in 1994, 860 listed properties which were unoccupied or derelict and which had a dubious future (over 2 per cent of the total). The position is relatively worse in England: an English Heritage report showed that 36,700 listed buildings (7 per cent of the total) are at risk from neglect; twice as many are in a vulnerable condition and need repair if they are not to fall into the 'at risk' category. Of course, most listed buildings are in private ownership, and the owners may well not feel the respect for their buildings which preservationists do; or they simply may be unable to afford to maintain them adequately. Advice, grants, and default measures cannot achieve all that might be hoped and, though a precious building can be taken into public ownership, this is essentially a matter of last resort. Sharland (2000) has put the case for more careful scrutiny of how preservation can be put into effect so that we list buildings that can be preserved, with a statutory duty on owners to keep those that are listed in good repair.

CRITERIA FOR LISTING HISTORIC BUILDINGS

Criteria for listing historic buildings are divided into four groups according to the date of building:

Before 1700 All buildings which survive in anything like their original condition are listed.

1700–1840 Most buildings are listed, though selection is necessary.

1840–1914 Only buildings of definite quality and character are listed and the selection is designed to include the best works of the principal architects.

Post-1914 Selected buildings of high quality are listed.

In Scotland, the grouping is: prior to 1840; 1840–1914; 1914–45; and post-1945.

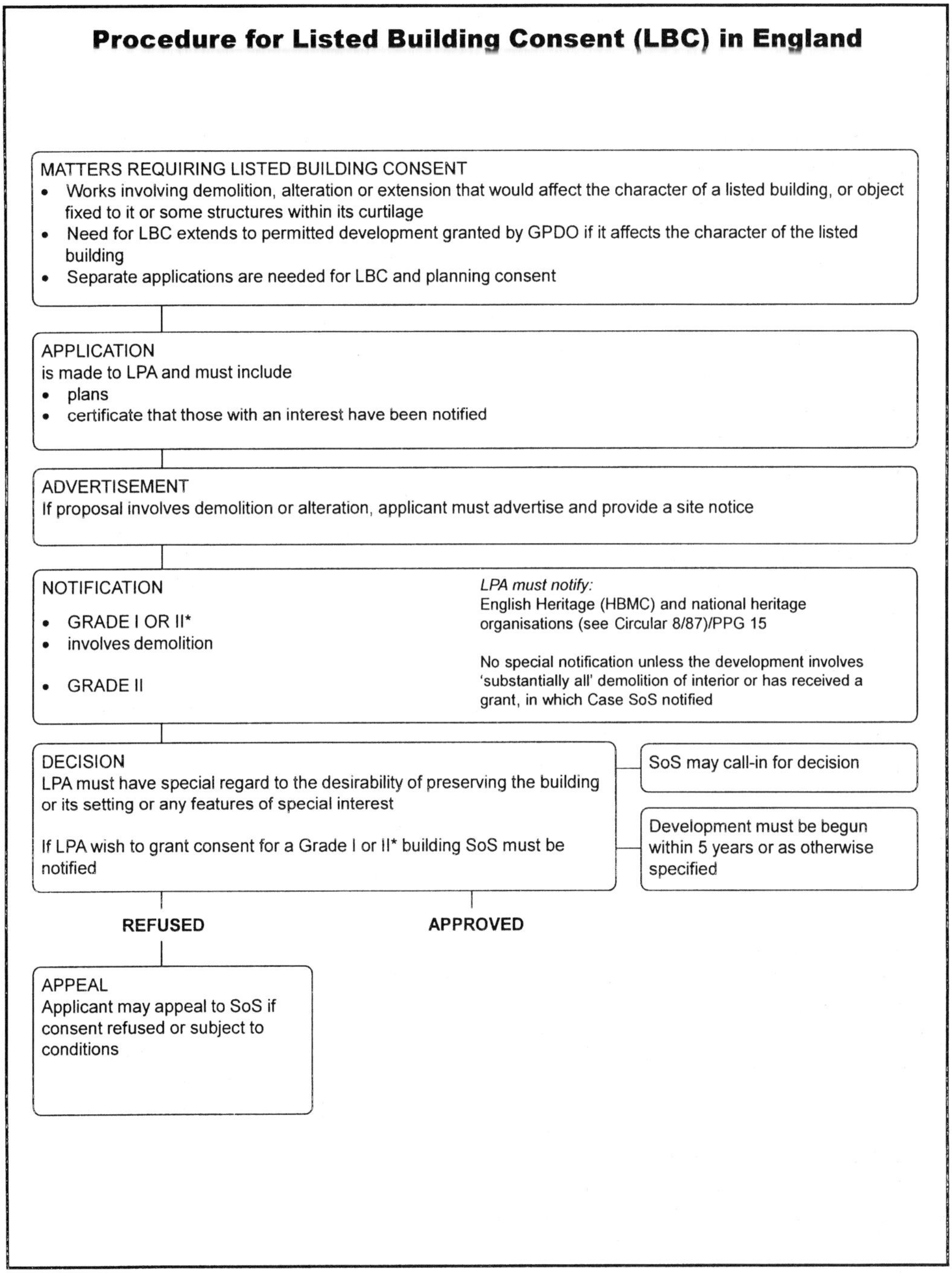

Procedure for Listed Building Consent (LBC) in England

MATTERS REQUIRING LISTED BUILDING CONSENT
- Works involving demolition, alteration or extension that would affect the character of a listed building, or object fixed to it or some structures within its curtilage
- Need for LBC extends to permitted development granted by GPDO if it affects the character of the listed building
- Separate applications are needed for LBC and planning consent

APPLICATION
is made to LPA and must include
- plans
- certificate that those with an interest have been notified

ADVERTISEMENT
If proposal involves demolition or alteration, applicant must advertise and provide a site notice

NOTIFICATION

- GRADE I OR II*
- involves demolition

- GRADE II

LPA must notify:
English Heritage (HBMC) and national heritage organisations (see Circular 8/87)/PPG 15

No special notification unless the development involves 'substantially all' demolition of interior or has received a grant, in which Case SoS notified

DECISION
LPA must have special regard to the desirability of preserving the building or its setting or any features of special interest

If LPA wish to grant consent for a Grade I or II* building SoS must be notified

SoS may call-in for decision

Development must be begun within 5 years or as otherwise specified

REFUSED   **APPROVED**

APPEAL
Applicant may appeal to SoS if consent refused or subject to conditions

Figure 8.1 The Procedure for Listed Building Consent in England

BOX 8.1 A SAMPLE OF THE HALF-MILLION LISTED BUILDINGS IN THE UK

1000 old red telephone boxes

123 cinemas

The Essex County Cricket Club pavilion

City Hall, Cardiff

Jodrell Bank Radio Telescope

Alexandra Palace, London

12,000 churches

A petrol pump at Oxton, Nottinghamshire

Ribblehead Viaduct, North Yorkshire

Coventry Cathedral

Carrickfergus Castle, Co. Antrim

The Rotunda, Birmingham

In choosing buildings, particular attention is paid to 'special value within certain types, either for architectural or planning reasons or as illustrating social and economic history'; to technological innovation or virtuosity (for instance, cast-iron prefabrication or the early use of concrete); to any association with well-known characters or events; and to 'group value', especially as examples of town planning such as squares, terraces, or model villages (DoE Circular 8/87). Buildings are graded according to their relative importance. The grading systems are set out in Table 8.2.

Scotland has for long had a rolling thirty-year rule under which any building of that age could be considered for listing. This was initially thought to be too problematic in relation to the much larger number of buildings that would be covered by such a rule in England. How was the quality of buildings to be assessed over such a short time period? Would apparent 'successes' soon be seen as 'failures' – and vice

versa?[21] Many buildings have been demolished which would today attract vociferous defence. On the other hand, some more recent architecture would have difficulty in finding a place in the hearts of those who support the protection of good interwar buildings. Clearly, this is an area where attitudes differ and firm guidelines are far from easy to determine – as is also the case with contemporary design and amenity guidelines.

Matters were suddenly accelerated when Sir Albert Richardson's Bracken House in the City of London was threatened with demolition. The Secretary of State decided to list this grade II*, thus copying the Scottish principle that buildings under thirty years old could be listed. At the same time, going one better than the Scots, it was decided that outstanding buildings that were only ten years old could be listed if there was an immediate threat to them.

PUBLIC PARTICIPATION IN LISTING

Theoretically, anyone can propose that a building be listed, but in practice, buildings have usually been proposed by the Commissions and decided by the central departments. There was no advance publicity in this system. It was therefore a surprise when the Secretary of State in England announced that the system was to be opened up, not only for proposals from the public, but also for comments on proposals for listing. In 1995, following an earlier initiative, the public were invited to comment on proposals from English Heritage for the listing of forty modern buildings and thirty-seven textile mills in the Manchester area. Among the former were the Centre Point office block in central London, Millbank Tower, the John Lewis warehouse at Stevenage, and the signal box at Birmingham New Street station.

There was concern that such a highly publicised process of listing might incite owners to demolish their earmarked buildings at speed (as happened to the Firestone building. Spot-listing is one answer to this (if it is done quickly enough), or the imposition of a building preservation order (though this renders the local authority liable to compensation if the building

Table 8.3 Listed Building Categories in the UK

England and Wales		Northern Ireland		Scotland	
Grade	Criteria	Category	Criteria	Category	Criteria
I	Buildings of outstanding or exceptional interest	A	Of national importance	A	Buildings of national or international importance, either architectural or historic, or fine and little-altered examples of some particular period, style or building type
II*	Particularly important buildings of more than special interest but not in the outstanding class	B+	Of national importance but with minor detracting features or of national importance with some exceptional features	B	Buildings of regional or more than local importance, or major examples of some period, style, or building type which may have been somewhat altered
		B1	Of national or local importance, or good examples of some period		
		B2	or style.		
II	Buildings of special interest which warrant every effort being made to preserve them	C	Of positive architectural interest or historic interest but are not 'special' and including those that contribute to the value of groups of buildings	C(S)	Buildings of local importance; lesser examples of any period, style, or building type, whether as originally constructed or as the result of subsequent alteration; simple, well-proportioned traditional buildings often forming part of a group

is not in fact eventually listed). A better solution would be a new power for an instant listing which carried no compensation penalties for the local authority.[22]

As these examples show, public opinion (when aroused) can play an important part in this planning field. The same is true with listed buildings under threat, as is well exemplified by the successful campaign to save St Pancras Station in London. There is a wide degree of public support for conservation, which has been heightened by the increased concern for environmental issues (Lane and Vaughan 1992).

Some questions remain, however, and they are likely to come to the fore as local planning authorities continue to develop their competences in the area. One question is the justification for the existence of two regimes: one for the listing of historic buildings and the other for the scheduling of ancient monuments and archaeological remains. Other questions relate to the division of responsibilities between planning authorities and central government and, in particular, the degree of the integration between heritage planning and the other functions of local planning authorities (Redman 1990; Scrase 1991). Some of these issues have been taken up in 'the heritage review' described below.

CONSERVATION, MARKET VALUES, AND REGENERATION

A major barrier to the conservation of some listed buildings is finding a contemporary use for them that is compatible with the character which it is desired to preserve. Research shows that listed office buildings have a 'market performance' which is generally as good as that of other buildings, and sometimes better. On the other hand, listing can reduce market value, particularly of small buildings in areas of high development outside conservation areas. However, the reduction is a one-time cost which is borne by the owner at the time of listing: future 'market performance' is not affected. But listing can also increase values because of the 'prestige' thereby accorded; and this can also raise neighbouring values. As with all such issues, much depends on local factors (Scanlon *et al*. 1994).

In deciding whether or not to list a building, the Secretary of State is required to have regard only to the special architectural or historic interest. No account can be taken of economic issues (such as the condition of the building and the cost of conserving it, or the possibilities of finding a viable use for the building). Nor can the personal circumstances of the owner be considered. (Such issues become relevant only when an application is made for listed building consent to demolish or alter a listed building.)[23]

In view of the renewed accent on urban policy and the work of the Urban Task Force, English Heritage has strongly promoted the regeneration potential of conservation. The epigraph at the head of this chapter gives a very positive message about the economic value of conservation, which is justified in the 1998 English Heritage report *Conservation-Led Regeneration*, which includes numerous successful and inspiring urban

regeneration schemes across the country. Major projects such as the Albert Dock, Liverpool; Saltaire, West Yorkshire; and Dean Clough Mills, Halifax, are well known, but there are very many smaller schemes that are equally impressive.[24]

Looking at this issue the other way around, development and regeneration can often enable restoration and conservation of the historic environment. But 'enabling development' often calls for considerable adaptation of historical assets. English Heritage defines enabling development as that 'which, whilst it would achieve significant benefit to a heritage asset, would normally be rejected as clearly contrary to other objectives'. The argument is that the benefits of safeguarding the heritage asset – a country house, for instance – offset the negative impact (and detriment to the asset itself) of, say, new housing development within the grounds of the house.[25] The development makes up the 'conservation deficit': the difference between the cost of repair and renovation to bring it into viable use and the resulting value of the property on the market. While English Heritage agrees that enabling development can be a useful planning tool, it has concluded that too often schemes 'destroy more than they save'. Therefore in 1999 English Heritage adopted a presumption against enabling development unless it meets strict criteria, including the requirements that the development must not detract from the heritage asset or its setting, and that it should be demonstrated that it is the minimum necessary to secure benefits for the asset.

In circumstances where 'enabling' is resisted, and thus private investment deterred, the only answer is subsidy from public or charitable sources. English Heritage has begun to target grant assistance on more deprived areas, and where investment in existing buildings will contribute to economic and social regeneration. Other funding bodies (described below) are doing the same. Nevertheless, there are still questions about the extent to which the government's urban renaissance policy has taken conservation fully on board. Conservation has not been given a prominent position in the Urban Task Force report, or funding priorities (see Chapter 10).

CONSERVATION AREAS

Of particular importance in heritage planning is the emphasis on areas, as distinct from individual buildings, of architectural or historic interest. Statutory recognition of the area concept was introduced by the Civic Amenities Act 1967.[26] Local planning authorities have a duty 'to determine which parts of their area are areas of special architectural or historic interest, the character of which it is desirable to preserve or enhance', and to designate such areas as *conservation areas*. When a conservation area has been designated, special attention has to be paid in all planning decisions to the preservation or enhancement of its character and appearance. Demolition of all buildings (unlisted as well as listed) is controlled.[27] There are also special provisions for preserving trees.

But owners of unlisted buildings have 'permitted development rights': they are not subject to the restrictions applied to owners of listed buildings. However, LPAs can withdraw these permitted development rights by use of an *Article 4 direction* (discussed in Chapter 5). Indeed, this is the common use of such directions (Roger Tym & Partners 1995a). They are typically intended to prevent piecemeal erosion of the character of an area through the cumulative effects of numerous small changes. Local planning authorities also have a duty to seek 'the preservation and enhancement' of conservation areas. Though some authorities take this duty seriously, it is generally poorly implemented, often on the grounds of inadequate resources.

The statutory provisions relating to the establishment of conservation areas are remarkably loose: there is no formal designation procedure, there is no requirement for a formal public inquiry (though proposals have to be put before a public meeting), and there is no specification of what qualifies for conservation area status. Circular 8/87 notes that 'these areas will naturally be of many different kinds':

> They may be large or small, from whole town areas to squares, terraces and smaller groups of buildings. They will often be centred on listed buildings, but not always. Pleasant groups of other buildings, open spaces, trees, and historic street patterns, a village green or features of historic or archaeological interest may also contribute to the special character of an area. Areas appropriate for designation as conservation areas will be found in almost every town and many villages. It is the character of areas, rather than individual buildings, that [section 60 of the Planning (Listed Buildings and Conservation Areas) Act 1990] seeks to preserve or enhance.

Larkham (1999a:113) points out that in practice the scope of designations has widened over recent years and there is also a trend to designating much larger areas.[28] The 1996 English Heritage consultation paper proposed that designations should 'include a statement identifying the specific features of the area that it is considered desirable to preserve or enhance'. In 2000 a similar call was made for better assessment of the qualities of conservation areas, both existing and proposed.

The number of conservation areas has grown dramatically, and by the end of 1999 there were more than 9,000 in the UK (Table 8.2).[29] Over a million buildings are in these areas. Indeed, it was suggested some time ago that perhaps a 'saturation point' had been reached in that the resources are simply not available for 'enhancing' such a large number of areas (Morton 1991; Suddards and Morton 1991). There continues to be a widespread view that more attention should be given to managing existing conservation areas and less to designating additional ones (Larkham and Jones 1993), but there has been little government action on this so far. Townshend and Pendlebury (1999) point to the continuing poor performance of professionals in involving residents in designation and management of conservation areas. Though an expert-led approach may be required for conservation areas of national significance, it may be that a more community-led approach, facilitated by the expert, is more appropriate in many thousands of conservation areas. This leads to the tentative suggestion that there may be a need for a grading scheme for conservation areas, in a similar way to listed buildings, so as to allow for different forms of control.

English Heritage has targeted its resources in conservation on priority areas through *town schemes* and *conservation area partnerships*, where it jointly funded works with planning authorities, and others. The latest in this form of initiative is the Heritage Economic

Regeneration Scheme (HERS), and this succeeds the area partnerships. The schemes will give more emphasis to economic and community regeneration as well as physical improvements and provide £15 million over three years from 1998. The idea is to concentrate on neighbourhood businesses, high streets, and corner shops, and 'where areas based assistance through building repair and enhancement will tip the balance in favour of continued local employment, new homes and inward investment'.

WORLD HERITAGE SITES

The UNESCO World Heritage Convention established a World Heritage List of sites that UN member states are pledged to protect. The eighteen sites within UK jurisdiction (of the 582 world-wide) are listed in Box 8.2. In 2000 the UK updated its 'tentative list' of sites that may be nominated for the world heritage status over the next five to ten years.[30] The tentative list is a requirement of the committee that oversees designation. The committee had expressed the wish to consider further natural and industrial sites to provide a better balance with the large number of architectural sites. Thus the current UK tentative list of twenty-five sites includes, for example, the Lake District, the New Forest, Shakespeare's Stratford, and the Mount Stewart Gardens, Northern Ireland. Six of the twenty-five have already been nominated.

The inclusion of a site on the World Heritage List carries no additional statutory controls, though of course it underlines its outstanding importance. This is a relevant material consideration in planning control. Local planning policies should, in the words of PPG 15, 'reflect the fact that all these sites have been designated for their outstanding universal value, and they should place great weight on the need to protect them for the benefit of future generations as well as our own'. Significant development proposals affecting a world heritage site generally require an environmental assessment. This has not protected sites in the UK from the pressure of new development – even Stonehenge. The world heritage committee now requires a management plan for all listed sites, and English Nature published in 2000 *The Stonehenge World Heritage Site Management Plan*.

HISTORIC PARKS AND GARDENS[31]

The 5,000 or more public parks in the UK have played an important role in quality of life in towns and cities, and many have great historical and landscape significance. But their quality is generally deteriorating – lamentably so. The neglect of public parks is one manifestation of the general decline in local authority services, but it also reflects changing social needs and behaviour. There is no statutory duty on local authorities to provide or maintain parks and open spaces. Indeed, there seem to be no clear responsibilities in relation to parks. The Urban Parks Forum claims that neither the DETR nor DCMS has recognised the problem, and the lack of even basic statistics on the amount of parkland and its quality would seem to bear that out. But Britain boasts some of the finest historical parks in the world. Most, other than the royal parks, are in private ownership or under the management of the National Trust and other charitable organisations.[32] The eight royal parks are managed by an executive agency of the DCMS, the *Royal Parks Agency*, with a budget of £20 million.

A *Register of Parks and Gardens of Special Historic Interest in England* is compiled by English Heritage, in county volumes.[33] The Register uses a grading system similar to that for listed buildings with grades I, II*, and II, but unlike for listed buildings, there are no additional consents required. Given that the Register is still a recent innovation, it may be that statutory controls will be imposed in the future (Pendlebury 1999). In the meantime, the Register is a material consideration in development control, and planning authorities must consult English Heritage on applications likely to affect Grade I or II* parks and gardens; and consult the Garden History Society on development that may affect any site on the register. There are about 1,200 sites on the Register, 120 of which are urban parks.[34] A review is under way, to be completed by 2002, but English Heritage expects this to result in only an additional 100 urban park designations.

BOX 8.2 DESIGNATION OF WORLD HERITAGE SITES UNDER THE JURISDICTION OF THE UK

1986	Giant's Causeway and Causeway Coast
	Ironbridge Gorge
	Stonehenge, Avebury and associated sites
	Durham Castle and Cathedral
	Fountains Abbey and St Mary's, Studley Royal
	The castles and town walls of Edward I in Gwynedd
	St Kilda
1987	Blenheim Palace
	City of Bath
	Hadrian's Wall Military Zone
	Palace of Westminster and Westminster Abbey and St Margaret's Church
1988	Henderson Island (Pitcairn Islands)
	Tower of London
	Canterbury Cathedral, St Augustine's Abbey, and St Martin's Church
1995	Edinburgh Old and New Towns
	Gough Island Wildlife Reserve (St Helena Islands)
1997	Maritime Greenwich
1999	The heart of Neolithic Orkney

Current nominations:

Blaenavon industrial landscape in south Wales
Town of St George, Bermuda
Derwent Valley Mills, Derbyshire
Saltaire, West Yorkshire
Dorset and east Devon coast
New Lanark, Scotland

Ironically, the poor condition of many parks may be a reason why they do not appear on the register and thus miss out on the benefits that the register might give.

CHURCHES

The situation regarding 'ecclesiastical buildings' is exceptional and also complicated. In essence, there is what is technically termed 'the ecclesiastical exemption' from listed building and conservation area controls. The exemption may apply to many buildings: the Church of England alone has more than 16,700 churches.[35] The Church introduced measures to control demolition more than seven hundred years ago, and has been regularly inspecting churches for three hundred years. It spends a large amount each year on the upkeep and maintenance of its buildings (mainly funded by its congregations). The result is that 'a listed Church of England church has a chance of avoiding demolition nearly three times better than a listed secular building'.

There are two parallel statutory systems of control over Church of England churches: the Church's system and the secular system. The Church's system is much stricter and more comprehensive. It involves regular inspection of every church, and embraces not only the fabric of the buildings, but their contents and churchyards. There are two separate statutory procedures applying to parish churches (whether listed or unlisted), according to whether they are in use or redundant. Churches in use are subject to a system of inspection and reporting at the local level, and to monitoring at higher levels: by Diocesan Advisory Committees at diocesan level, and by the Council for the Care of Churches at the national level.

Redundant churches are safeguarded by the Pastoral Measure 1983, which provides procedures for deciding whether a church is still required for worship, and, if not, what the future of the building should be. The *Churches Conservation Trust* (formerly the Redundant Churches Fund) finances the management, maintenance, and repair of churches judged of sufficient architectural or historic importance. The fund receives 70 per cent of its funds from the DCMS and the remainder from the Church Commissioners, and in 1999 had 300 redundant churches in its care.

Until recently, cathedrals were outside any planning procedure and, despite their huge popularity with visitors (and contribution to tourism), were not eligible for grant aid. A separate system of controls over building works was introduced by the Care of Churches Measure 1990. This is administered by the individual cathedrals jointly with a Cathedrals Fabric Commission in consultation with English Heritage, which also provides grant aid.

All church buildings are subject to normal planning control over, for example, changes of use and significant alterations. They are also listed in exactly the same way as other buildings of special historic or architectural interest. However, because of the Church's separate statutory procedure, listed building consent is not required for churches where the primary use is as a place of worship. Such consent is required, however, for alterations to redundant churches, though not if demolition is carried out pursuant to a scheme under the Pastoral Measure 1983.

A government review of the ecclesiastical exemption completed at the beginning of 1993 led to a decision to extend it to churches of all denominations where an acceptable system of control operates on principles set out in a code of practice.[36] The 1994 Ecclesiastical Exemption Order revoked the exemption for religious bodies that had not adopted their own regulation systems. For those that have, it temporarily extended exemption to other buildings within the Churches' estates, but this too will be revoked unless the Churches introduce their own controls over such buildings. In 1999 the Church of England adopted the Care of Places of Worship Measure, which empowers the Council for the Care of Churches to compile a list of the other buildings used for purposes of the Church (such as school and college chapels) that it wishes to fall within the protection of the Church. At some point the ecclesiastical exemption will be removed for those religious bodies that have not been included within a 'self-regulatory regime'.

FUNDING FOR CONSERVATION OF THE HISTORIC ENVIRONMENT

Following the outcry over the controversial sale of the assets of the Mentmore estate in 1977, a *National Heritage Memorial Fund* (NHMF) was established by the National Heritage Act of 1980. This is dedicated as 'a memorial to those who have died for the United Kingdom'.[37] The Fund gives financial assistance 'towards the cost of acquiring, maintaining or preserving land, buildings, works of art and other objects of outstanding interest which are also of importance to the national heritage'. It is doing an important job in preventing heritage assets from being exported. But given the value of some assets it is a relatively small fund. During 1999-2000 it made grants worth £2.7 million in respect of eleven items.[38] In addition to normal Exchequer payments into the fund, further payments can be made in relation to property accepted in satisfaction of tax debts.

The NHMF now also distributes the *Heritage Lottery Fund* (HLF). The scale of conservation work made possible by the Lottery was probably not imagined ten years ago. The HLF allocated £148 million in 1,872 grants in 1999–2000.[39] As the chairman of the Fund has said, 'every age needs its patrons and the private patronage on which we have largely relied in this country is now vigorously supported (but happily not supplanted) by lottery money' (*HLF and NHMF Annual Report* 1999–2000: 2). The HLF now has four general priorities: conservation; national heritage; local heritage; and access and education. Its formal aim reflects current concerns with equal access and sustainability. It is:

> to improve quality of life by safeguarding and enhancing the heritage of buildings, objects and the environment, whether man-made or natural, which have been important in formation of the character and identity of the United Kingdom, in a way which will encourage more sections of society to appreciate and enjoy their heritage and enable them to hand it on in good heart to future generations.
>
> (*HLF and NHMF Corporate Plan*, 2000)

It is difficult to do justice here to the range of projects that have received a contribution from the HLF, since its impact is so pervasive. The better-known projects include the National Maritime Museum at Greenwich, the American Air Museum at Duxford, St George's Market in Belfast, Robert Owen's school in New Lanark (part of the world heritage site), and the Big Pit Mining Museum, Blaenafon (the National Mining Museum for Wales). Planning authorities have been important players in both large and small projects as initiator and sometimes co-funder. Of particular interest is the townscape heritage initiative. This provides small grants towards heritage scheme feasibility studies, and more substantial funding to improve the vitality of many towns, some of which are not traditionally seen as heritage centres.

After early criticism about elitism and unequal distribution of funding across the country, the HLF now strongly emphasises the need for funded projects to deliver wider public benefits, for example through economic regeneration and social inclusion. There are more small-scale 'community grants', providing a simpler application process. They have contributed £3.9 million in 1,284 awards in 1999–2000. Regional

offices have been set up in Northern Ireland, Scotland, Wales, and the English regions. A special study is under way to examine how the regeneration of coalfield communities may be supported (they have done badly in the distribution of funds so far). Another study is looking at the needs of urban parks (discussed on p. 224.[40]

Lottery funding aside, it should not be forgotten that some local authorities and many voluntary organisations have a long-standing record of funding conservation. Also, in England there is a relatively small (£546,000 in 1999–2000) fund now administered by English Heritage for innovatory or experimental projects that contribute to government's objectives for the heritage. Projects have been funded involving new records of the historic environment, promoting access, and improving management practice. The fund supported the Civic Trust's Heritage Open Days initiative.

PRESERVATION OF TREES AND WOODLANDS

Trees are a delight in themselves; they also have the remarkable quality of hiding developments which are best out of sight. Trees are clearly, so far as town and country planning is concerned, a matter of amenity. Indeed, the powers which local authorities have with regard to trees can be exercised only if it is 'expedient in the interests of amenity'. Where a local authority is satisfied that it is expedient, it can make a *tree preservation order* (TPO) applicable to trees, groups of trees, woodlands, and trees planted as a result of a planning condition. Such an order can prohibit the cutting down, topping, or lopping of trees except with the consent of the local planning authority.[41] Orders are made according to a model given in the Regulations. People affected must be consulted and the planning authority must consider all objections and other comments before confirming the TPO. Subsequently any proposals to cut or lop the protected trees need consent from the planning authority. In conservation areas, trees otherwise not protected by TPOs are also subject to a special regime. The planning authority must be given six weeks' notice of any works, during which it can consider the need for a TPO.

Mere preservation, however, can lead eventually to decay and thus defeats its object. To prevent this, a local authority can make replanting obligatory when it gives permission for trees to be felled. The aim is to avoid any clash between good forestry and the claims of amenity. But the timber of woodlands or orchards always has a claim to be treated as a commercial crop, and though the making of a tree preservation order does not necessarily involve the owner in any financial loss (isolated trees or groups of trees are usually planted expressly as an amenity), there are occasions when it does.

Yet though woodlands are primarily a timber crop from which the owner is entitled to benefit, two principles have been laid down which qualify this. First, the national interest demands that woodlands should be managed in accordance with the principles of good forestry; and second, where they are of amenity value, the owner has a public duty to act with reasonable regard for amenity aspects. It follows that a refusal to permit felling or the imposition of conditions on operations which are either contrary to the principles of good forestry or destructive of amenity ought not to carry any compensation rights. But where there is a clash between these two principles, compensation is payable. Thus, in a case where the principles of good forestry dictate that felling should take place, but this would result in too great a sacrifice of amenity, the owner can claim compensation for the loss which he or she suffers. Normally, a compromise is reached whereby the felling is deferred or phased. The commercial felling of timber is subject to a licence by the Forestry Commission.

Planning powers go considerably further than simply enabling local authorities to preserve trees. Planning permission can be made subject to the condition that trees are planted, and local authorities themselves have power to plant trees on any land in their area. With the increasing vulnerability of trees and woodlands to urban development and the needs of modern farming, wider powers and more Exchequer aid have been provided by successive statutes. Local planning authorities are now *required* to ensure that conditions (preferably reinforced by tree preservation orders) are imposed for the protection of existing trees and for the planting of new ones.

In 1994 the DoE reviewed the TPO system and then consulted on new regulations to overcome certain anomalies. Following the change in administration in 1997, the proposals were put on hold until 1998, when a further consultation paper was published. New regulations were made in 1999.[42] The changes are generally quite minor and relate to the simplification of the model order by which TPOs are made: maps, inspection, and clarification of the exemptions afforded to statutory undertakers following the wave of privatisation. Guidelines were published in 1995 on consultation between statutory undertakers and planning authorities, and a code of practice is proposed. Local authorities can undertake work on trees on any land, which obviously puts considerable onus on their internal consultation processes between, say, the planning officers and those who undertake the work. New legislation is proposed to amend the provisions further, mainly in relation to offences committed in damaging protected trees.

A different approach has been taken to the protection of 'important hedgerows', which, because of devastating losses, have been given protection through Regulations made in 1997 (see also Chapter 9).[43] The Regulations apply only to hedgerows with a continuous length of 20 m or more and which meet another hedgerow at either end. They need to be growing on, or adjacent to, common land, protected land, or land used for agriculture, forestry, or for keeping horses, etc. The Regulations do not apply to hedgerows around houses. If all these conditions are satisfied, before removing a hedgerow the owner must notify the planning authority, which, if it wants the hedge retained, has forty-two days to serve a 'hedgerow retention notice'. But the planning authority can do this only for an 'important hedgerow', which means it must have been in existence for thirty years or more and have some archaeological, historical, wildlife, or landscape qualities.

TOURISM

Heritage is an important factor in tourism. Together with 'culture' and the countryside, heritage stimulates about two-thirds of the visits made by foreign tourists. (They certainly do not come to enjoy the weather!) Heritage attractions are also the most important reason that domestic tourists give for having made their visit within the UK.[44] Since tourism plays a very significant role in economic prosperity, it follows that heritage is very important in Britain's economy, and, of course, in the perception of Britain that many visitors gain. But over the past twenty years, the UK's share of world tourism has generally declined (although there has been some improvement since 1994).[45]

There is an abundance of figures that demonstrate the impact of tourism.[46] There is, however, a big downside: tourism can lead to excessive wear and tear on the fabric of buildings, to congestion, to litter, and even to open hostility by residents to visitors. The generally accepted implication is that tourism has to be 'managed'. Several organisations are now devoted to this: the Historic Towns Tourism Management Group, the Heritage Cities Association (a marketing consortium), and the English Historic Towns Forum. A concern for 'vital and viable town centres' has grown, and other specialist bodies have been established, such as the Association of Town Centre Management.

That tourism, as well as heritage, is a matter of importance for local planning authorities is self-evident. However, it is not a policy area which can be isolated from related ones. It is interesting to note that the PPG on tourism (PPG 21) refers to a long list of other relevant PPGs. This list is an eloquent testimony to the interconnectedness of planning issues; and it also points to the inherent difficulty of reconciling numerous considerations – or even giving adequate consideration to all of them.

Responsibilities for tourism, like most of government, have shifted over recent years with a view to addressing the 'untidy structure' for tourism and its linkages with the heritage.[47] The creation of the DNH, now DCMS, brought together tourism and heritage from six departments. The Scottish Executive and the Welsh and Northern Ireland Assemblies have responsibilities for tourism and fund 'national tourist boards'. The British Tourist Authority (BTA) advises government on issues that affect the UK as a whole. The English Tourism Council (ETC) was created in 1999 (as successor to the English Tourist Board) in an attempt to provide a more strategic body to take forward a national tourism strategy with less direct involvement in implementation, which is now concentrated in the English regional tourist boards (RTBs). The RTBs are limited companies funded by ETC and other sponsors and commercial sources.[48] Local authorities also promote and manage tourism, and this is particularly important in the heritage towns.

The 1997 Labour government established a Tourism Forum which contributed to a major review of government policy for tourism. The results were published in the 1999 DCMS Report *Tomorrow's Tourism*, and similar reviews are under way elsewhere in the UK.[49] The reports have the same emphasis on promoting tourism for its economic and regeneration potential. In particular, there is a need to ensure that the tourism industry in the UK performs at a similar rate to its competitors, since it is losing ground. An action programme was produced for England, and later discussed at a national tourism summit. Of particular relevance for planning are action points to provide a sustainable blueprint for tourism to safeguard the countryside and heritage; to encourage more integrated promotion of the heritage; to develop niche markets in such areas as 'film tourism' so as to 'unlock the potential of Britain's unique cultural and natural heritage'; and to encourage regeneration of traditional tourist resorts.

As with all matters concerning sustainability there is more than a hint of contradiction in the action points. On the one hand, business and economic priorities mean that the heritage has to earn a return on the public funds invested. On the other hand, there is a need to preserve the integrity of the properties and places visited. The relevant documents all note this problem, and much needs to be addressed in tourism management at the local level. It is well recognised that the presentation and management of historic properties cannot be a purely commercial operation.

In any case, too great a success in attracting custom could place unsustainable pressures on the very experience which the customers are seeking. For example, marketing efforts have helped to increase visitor numbers at historic houses from 3 million in 1970 to nearer 11 million at the end of the 1990s. But 'it has been estimated that the wear and tear due to visitors in one year exceeds the previous domestic wear and tear of two decades and in some cases up to a whole century' (Lloyd 1999: 1).[50]

The ETC has taken a lead role on the idea of 'sustainable tourism' and has created a Sustainable Task Force (*sic*). A report, *Sustainable Tourism: An Action Plan*, is proposed for 2001. Unfortunately, the ETC's 'vision for sustainable tourism' hardly touches on the concept of sustainability. It reads: 'England will promote and develop tourism that exceeds visitor expectations, ensures the long term viability of the industry, benefits local communities and helps to protect and enhance the places in which it takes place.'[51]

Some critics have been very outspoken about so-called 'sustainable tourism'. Croall (1995) presents strong arguments for greater protection of the environment against the effects of tourism. An even stronger case is made by Minhinnick (1993), who argues that 'the idea of making tourism an environmentally sustainable activity is at best an exciting pipe-dream and at worst a deceit'. This elegantly written essay is merciless in its criticisms: 'the trouble with tourism is that moderation is not part of its language'; 'local distinctiveness is erased and replaced by mediocre uniformity'. Though these fundamental conflicts barely figure in official reports, there are lots of practical suggestions, but these are predominantly concerned with minimising environmental impacts.

The 1998 consultation on the national sustainability strategy included a paper entitled *Tourism: Towards Sustainability* which considers some of the issues. These include the potential of tourism to benefit local communities, the need to manage visitor flows, the transport impacts, and planning. Indeed, planning figures very prominently in the responses to this consultation paper, particularly the need to amend planning guidance to ensure that tourism development meets sustainability criteria; and to strengthen plans and development control powers to ensure that tourism investment is concentrated where it can do most benefit for regeneration and least damage. But much of the agenda for tourism is no more than the agenda for everyone else: better strategic coordination of policy through planning, closer integration of transport modes, quality public transport, opportunities for cycling and walking, effective reuse and renewal of the heritage, and the need to find a way to spread tourism and its benefits beyond the mainstream 'honeypots' and out to the regions.

BOX 8.3 THE SIGNIFICANCE OF TOURISM, 1999

25.5 million overseas visitors came to the UK

UK residents took 122 million trips of one night or more in the UK

Overseas tourists spent about £13 billion in the UK

Expenditure on travel to British operators was £3.3 billion

Total expenditure on tourism is £61 billion

Employment in tourism is about 18 million

There are 125,000 tourism businesses

The share of GDP attributable to tourism is 6 per cent

Over fifty historic towns attract over 20,000 overseas overnight visitors

Six historic towns receive over 150,000 overseas overnight visitors

The government provides almost £100 million grant to the tourist boards

UK visitors abroad spent £8.2 billion more than overseas visitors spent in the UK

THE HERITAGE REVIEW

In November 1999 the government announced its intention to carry out a systematic review of policy for the historic environment in England, led by the DETR and DCMS. Early in 2000, English Heritage was instructed to undertake stage 1: a review of current policies in consultation with all interests. The terms of reference for the review were restrictive: the principles of PPG 15 were to stay in place, resources would remain much the same, and there would be no major structural reorganisation of responsibilities. The review was to consider in particular the relationship between heritage and tourism and the roles of the numerous bodies involved in conservation. In addition, English Heritage was asked to review the condition of the historic environment, the need for dealing with heritage at risk, possible simplification of the procedures, and connections with urban regeneration and emerging policy on sustainable development.

The report, *Power of Place*, was published at the end of 2000, as a step towards a strategy for the historic environment in England. The report's many recommendations are mostly precise and send a strong message about how to strengthen conservation in all areas of public policy. This is to be done by, for example, ensuring that conservation is reflected in all government sectoral policy; ensuring that tax and funding regimes support conservation at least as well as new-build;[52] strengthening regulation and powers for designated areas; and improving information and skills. Despite the limitations imposed by the terms of reference, the report makes a strong case for organisational rationalisation:

> DETR lacks an effective historic environment dimension to wider policy objectives and DCMS has never given the issue the attention or priority it deserves. These problems are in addition to the split responsibility for planning and listed building and other consent procedures.
>
> (p. 43)

The theme of getting a more consistent approach to conservation across all arms of government appears also in relation to ensuring that conservation figures in the policies of RDAs and local strategic partnerships.[53] Another strong theme is the promotion of the economic value of the historic environment in terms of investment returns (at least when it can be used as offices), job creation, and tourism ('pound for pound, repair and maintenance create more employment than new-build').

On the role of the planning system, the report notes that designations have been more successful in relation to buildings and monuments than with areas of land: conservation areas, parks and gardens, and battlefields. One area that is lagging behind badly is marine heritage. Of the 34,000 known marine archaeological sites in English territorial waters, only thirty-eight are afforded statutory protection – thus the report argues that the marine heritage should be brought within the remit of English Heritage.

The report argues for more systematic evaluation of all buildings and sites that are identified for conservation; and that character appraisal or assessment of historical assets should be normal practice and could form the basis of 'spatial masterplans'[54] for their future development. Similarly, conservation plans should be prepared for historic sites and can provide the basis for management agreements with owners. Capacity studies are also mentioned, especially in relation to tourism impacts. But for all this work, there is a considerable shortfall of relevant skills and qualified staff. Many planning authorities (22 per cent in England) have no staff in the conservation field (which says a lot about some local authority attitudes to the historic environment). Therefore, English Heritage recommends that appropriate performance criteria on heritage management should be included in the Best Value regime (see Chapter 3). The need to improve the information about historic environments, complete records, and provide easier access to them is also given attention.

Some recommendations are more challenging, notably that permitted development rights should be withdrawn in all conservation areas. This recommendation and the effective withdrawal of rights of owners is unlikely to be acceptable without significant tightening of the criteria and procedure for designation.[55] It would also have the effect of increasing the difference between quality of the built environment in conservation areas and elsewhere, when it might be argued that the deteriorating quality of all environments is the more significant issue.

One significant message for government is that it needs to put its own house in order. It applies to a number of public bodies such as the Ministry of Defence whose actions in some cases suggest that they see the heritage as an obstacle to realising a quick return rather than an asset. It is stated unequivocally that 'examples of best practice in estate management are found in the large private estates and not the public sector.'[56]

FURTHER READING

History and General Context

Excellent, sympathetic, and informative accounts of the history of historic preservation are given by Ross (1996) *Heritage and Planning* and Delafons (1997) *Politics and Preservation*. (Ross was former head of the Listing Branch of the DoE and Delafons was for twelve years Deputy Secretary responsible for land use planning at the Department of the Environment, and so they give an insider's account, and inevitably concentrate on the government view.) Pickard (1996) *Conservation in the Built Environment* is a standard text. Larkham (1996) *Conservation and the City* sets conservation in the context of urban morphology and international comparisons; and his paper 'Preservation, conservation and heritage' (1999a) is a most useful summary and contains advice on further reading. There is an extensive list of sources at the Historic Environment Information Resources Network: <http://www.britarch.ac.uk/HEIRNET/>. The DCMS and Royal Commissions' annual reports are useful sources on current activities. The *Heritage Review* (noted on p. 251) will be an important source.

Conservation Law

General texts covering many points in this chapter are Mynors (1998) *Listed Buildings, Conservation Areas and Monuments* and Suddards and Hargreaves (1996) *Listed Buildings*. An excellent summary of the legal provisions covering heritage is given in Moore (2000) *A Practical Approach to Planning Law*.

Policy for the Historic Environment

In England, policy guidance is given in PPG 15 (*Planning and the Historic Environment*), PPG 16 (*Archaeology and Planning*), and DoE Circular 8/87 *Historic Buildings and Conservation Areas: Policy and Procedures*. See also the DETR paper *Contemporary Issues in Heritage and Environment Interpretation*. English Heritage publishes a *Conservation Bulletin* three times a year.

For Northern Ireland, see PPS 6, *Planning, Archaeology and Built Heritage* (1999). A useful reference is Hendry (1993) 'Conservation in Northern Ireland'.

For Scotland, see NPPG 18, *Planning and the Historic Environment* (1999); PAN 42, *Archaeology*; Scottish Office (1998) *Planning and the Historic Environment*; Historic Scotland Circular 1/1998: *The Memorandum of Guidance on Listed Buildings and Conservation Areas*; National Audit Office (1995) *Protecting and Presenting Scotland's Heritage Properties*; and the Scottish Executive's *Report of the Conservation Control Working Group*.

For Wales, see *Planning Guidance Wales* (first revision 1999) Section 5; WO Circular 60/96 *Planning and the Historic Environment: Archaeology* and WO Circular 61/96 *Planning and the Historic Environment: Historic Buildings and Conservation Areas*; and the National Audit Office, Wales, *Protecting and Conserving the Built Heritage in Wales*.

Archaeology and Scheduled Monuments

In addition to PPG 16 (and variants outside England), see Roger & Partners Tym (1995b) *Review of the Implementation of PPG 16: Archaeology and Planning* and English Heritage (1992) *Development Plan Policies for Archaeology*. A legal text is Pugh-Smith and Samuels (1996a) *Archaeology in Law*. Many examples are given on the CD-ROM of the Royal Commission on Historical Monuments in England (1998) *Monuments on Record: Celebrating 90 Years of the Royal Commissions on Historical Monuments* (Swindon: RCHME (now available from English Heritage in England and the Royal Commissions elsewhere in the UK).

Conservation Areas

The Character of Conservation Areas (RTPI 1993) is the most recent comprehensive survey and still provides a very useful overview and (in the second volume) some useful supplementary material, including a bibliography.

Historic Parks and Gardens and Trees

Pendlebury (1999) traces the history of controls over historic parks and gardens; and see also Pendlebury (1997) 'The statutory protection of historic parks and gardens: an exploration and analysis of "structure", "decoration" and "character".' There is a growing literature on parks: see the Comedia report *Park Life: Urban Parks and Social Renewal* (1995); DETR (1996) *People Parks and Cities: A Guide to Current Good Practice in Urban Parks*; and the extensive evidence in the House of Commons Environment, Transport and Regional Affairs Committee's *Twentieth Report: Town and Country Parks*. On trees, see DETR: *Protected Trees: A Guide to Tree Preservation Procedures*.

Economic Aspects and Regeneration

There have been a number of studies of the consequences of listing and conservation for returns on property. The most recent is the Investment Property Databank report for English Heritage (1999) *Investment Performance of Listed Buildings*. See also Scanlon *et al.* (1994) *The Economics of Listed Buildings* and Drury (1995) 'The value of conservation'. On regeneration, see the English Heritage reports *Heritage Dividend: Measuring the Results of English Heritage Regeneration* (1999) and *Conservation-Led Regeneration: The Work of English Heritage* (London: English Heritage, 1998).

Tourism

The government reviews of tourism policy consider the link with heritage. See DCMS *Tomorrow's Tourism: A Growth Industry for the New Millennium* (2000) and *Tourism: Towards Sustainability* (1998); Scottish Executive *A New Strategy for Scottish Tourism* (2000);

and the Welsh Assembly *Achieving Our Potential* (2000). Croall (1995) *Preserve or Destroy: Tourism and the Environment* presents a well-argued case for greater protection of the environment against the effects of tourism. Some of these issues are taken up in a special edition of *Built Environment* volume 26 (1) (2000) edited by MacDonald. Also invaluable is the annual *English Heritage Monitor*, published by the English Tourist Council.

NOTES

1 Ross gives an interesting account of how this mammoth job was done, often on a voluntary or near-voluntary basis. The survey took twenty-two years and even then it was incomplete. Because of the attitudes of the time, Victorian architecture was almost totally neglected.

2 The issue was highlighted by the 1950 Gowers Report *Houses of Outstanding Historic and Architectural Interest*. The Historic Buildings and Ancient Monuments Act of 1953 followed, which established the Historic Building Councils for England, Scotland, and Wales and the first system of grants.

3 Curiously, those relating to ancient monuments are still separate in the Ancient Monuments and Archaeological Areas Act 1979, and both are separate from the planning legislation, and the need for consolidation is widely recognised. The separation of the conservation and planning legislation is reflected in that the primacy of development plans does apply in the case of listed building consent.

4 It was no doubt with this necessity for change in mind that the CPRE altered the second word in its name from 'preservation' to 'protection'.

5 The phrase 'national heritage' was in fact used in arguing for a statutory duty to list buildings of merit in the 1947 Act (Delafons 1997: 60, quoted in Larkham 1999a).

6 Policy guidance for the different countries of the UK is listed at the end under 'Further reading'.

7 Delafons gives a review of the case of No. 1

Poultry, an application for demolition of grade II buildings that found its way to the House of Lords before consent was granted. While the Lords' decision supported the discretion of the Secretary of State to allow demolition in certain cases, paradoxically the outcome was a further limitation of discretion by reinforcing the preservation doctrine.

8 English Heritage was created in 1983. Its formal name is the Historic Buildings and Monuments Commission.

9 It was Octavia Hill who founded the National Trust in 1895. The Georgian Group was founded in 1937, after the Commissioners for Crown Lands demolished Nash's Regent Street and threatened to do the same with Carlton House Terrace. The widespread destruction of Victorian and Edwardian buildings, particularly the Euston Arch, led to the creation of the Victorian Society in 1957. The Twentieth Century Society, which was set up in 1979 (originally as the Thirties Society) to safeguard interwar architecture, nearly saved the Firestone factory on the Great West Road, but was thwarted by the developers (Trafalgar House), who moved the bulldozers in over the August 1980 holiday weekend before the procedure for 'spot-listing' had been completed. SAVE Britain's Heritage was formed in 1975, the European Architectural Heritage Year, with a special emphasis on finding alternative uses for historic buildings. There are now many heritage organisations, several of which have statutory consultee status on proposals to demolish listed buildings.

10 Policy guidance for archaeology is given at the end of the chapter under 'Further reading'.

11 This is an organisation promoted by the British Property Federation and the Standing Conference of Archaeological Unit Managers (itself a representative body of some seventy-five professional archaeological units).

12 The term 'schedule' originates from the Ancient Monument Protection Act 1882, which provided for the protection of twenty-nine monuments which were set out in a *schedule* to the Act; the term has persisted. The Royal Commissions and English Heritage undertake many surveys on specific types of buildings such as churches, hospitals, and even monuments of the Cold War such as the 368m wide 'listening post' antenna at Cruicksands in Bedfordshire, which was demolished in 1996. A large selection of examples of monuments for the whole of Britain is available on the CD-ROM *Monuments on Record: Celebrating 90 Years of the Royal Commissions on Historical Monuments* (1998).

13 The problem of estimating is further complicated by the fact that some figures refer to register entries (which cover more than one site) and others to individual sites.

14 MARS was established by English Heritage and was based on a sample of 1,300 1 km × 5 km transects of land and examination of photographic records from the past fifty years (Bryant 1999).

15 A proposed development at Abbey Mead near Swindon revealed a Roman water temple and gardens. The developers stopped work and have subsequently been paid £1 million in compensation by English Heritage.

16 Properties in the care of Historic Scotland include Edinburgh Castle, Stirling Castle, Fort George, and Urquhart Castle (near Loch Ness). Welsh Historic Monuments manages Chepstow Castle, the Blaenavon Ironworks, the Welsh Slate Museum (Llanberis), Neath Abbey, and Tintern Abbey. In Northern Ireland, 166 monuments are in the care of the DoENI, including Londonderry's city walls, Newtownards Priory, Enniskillen Castle, Tully Castle, Carrickfergus Castle, and, perhaps surprisingly, the Carrickfergus Gas Works.

17 In England this is the Department of Culture, Media and Sport. Consideration was given recently to devolving control over listing to English Heritage but it was thought that this would lead to extra costs and it was not taken up.

18 In 1997 the House of Lords issued the 'Shimizu judgement', which affected the meanings of demolition and alteration (*Shimizu (UK) Limited v. Westminster City Council* ([1997] 1 All ER 481).

DETR Circular 14/97 (DCMS Circular 1/97) summarised the implications: 'Demolition refers to pulling down a building so that it is destroyed completely or at least to a very significant extent'. Demolition of part of a building (previously treated as demolition) 'falling short of the whole demolition will be works for the alteration of the listed building'. In listed building practice this had little effect – since the alterations continue to require consent – but see the note on conservation areas on p. 256. (See the DETR's June 2000 Consultation Paper, *The Impact of the Shimizu Judgement*.)

19 In March 2000 Ian Hislop, editor of *Private Eye*, found himself in the news because he had to apply retrospectively for listed building consent after planning officers (visiting in respect of a major proposal to extend the building) noticed that recent repairs had been made to the wooden frame of his sixteenth-century cottage. Not long before, Conservative MP Teresa Gorman was convicted of twenty-nine breaches of the listed building consent regulations in alterations made to her grade II listed farmhouse. Scrase (2000) refers to various cases where the removal of fixtures and fittings has been determined to constitute alterations to the listed building, including in one case the removal of paintings.

20 Consultees may include the Ancient Monuments Society, the Council for British Archaeology, the Georgian Group, the Society for the Protection of Ancient Buildings, the Victorian Society, and the Royal Commission on Historical Monuments. Scottish and Welsh authorities are required to notify their respective Royal Commissions on Ancient and Historical Monuments. Planning authorities must consult the Secretary of State and English Heritage on all applications affecting Grades I and II* buildings; and all buildings if demolition is involved. In Greater London the authorisation of English Heritage is required for all listed buildings before consent can be granted.

21 There have been marked changes in attitudes to different architectural styles. Perhaps the most famous comes from Paris, where the Eiffel Tower was once described in terms of 'the grotesque mercantile imaginings of a constructor of machines', but is now 'the beloved signature of the Parisian skyline and an officially designated monument to boot' (Costonis 1989: 64). Perhaps the Millennium Dome may follow the same path?

22 The 1996 consultation paper discusses a proposal for a new power for the Secretary of State to provisionally list an endangered building to allow consultation to take place. Such a procedure (as with the current listing regime) would not involve any compensation.

23 The issue is raised in the 1996 consultation paper *Protecting Our Heritage* and, though views were sought, the government view is clearly that no change is appropriate: 'In the government's view, the logic of the current regime is clear and defensible – the purpose of listing a building is to register its quality. The fact of listing does not rule out any future change to the building. Rather, it ensures that any proposals affecting it can be fully assessed, having regard to all the circumstances prevailing at the time, including economic and financial considerations.'

24 The Albert Dock, Liverpool, comprises five grade I dock buildings now converted into 93,000 sq. m of television studios, galleries, offices, and shops. The Salt Mills and Saltaire Village projects have created 1,800 jobs from mixed public–private investment of £50 million. Dean Clough Mills, Halifax, was a private-sector scheme where under the direction of Sir Ernest Hall 3,500 jobs have been housed in redundant carpet mills. The SAVE Britain's Heritage paper *Catalytic Conversion* (1998) makes a similar argument with a set of examples of good practice.

25 An excellent (or perhaps notorious) example is the estate of Hardwick Hall on the Nottinghamshire–Derbyshire border, where the National Trust has sold estate properties and land for housing in order, it argues, to maintain the Hall. The quality of the development leaves much to be desired, with bright red factory-made pantile roofs of new bungalows clearly visible from the Hall.

26 The Act was promoted as a private member's bill

by Duncan Sandys, president of the Civic Trust, and passed with government backing.

27 The Shimizu judgement, referred to earlier, has had a particular impact on conservation areas. Previously, the understanding was that the Act provided for all forms of demolition to all buildings in conservation areas to be subject to control, including the demolition of any part of a building (which, for example, could include boundary walls). The Shimizu judgement confirmed that demolition 'refers to pulling down a building so that it is destroyed completely or at least to a very significant extent' (DETR/DCMS *Consultation Paper on the Impact of the Shimuzu Judgement*, 2000). In practice, therefore, partial demolition no longer required conservation area consent. Thus, substantial control is lost over piecemeal changes to conservation areas, especially to dwelling-houses, where householders' permitted development rights allow them to undertake partial demolitions (which would allow, for example, demolition of boundary walls to create a parking space). A number of solutions are being considered in England at the time of writing, including withdrawal of permitted development rights in conservation areas. In Scotland the matter has been dealt with through Circular 1/200.

28 Larkham (1999a: 113) suggests that perhaps the largest conservation area is the Yorkshire Barns and Walls area, which covers tens of square kilometres; but the most complex must be Bath, where the single conservation area now covers 1,914 hectares (66 per cent of the city's area). In another article Larkham (1999b) considers the recent trend for the conservation of residential suburbs.

29 The lower figure for Northern Ireland is probably a reflection of the fact that only central government can make designations. Five new designations in the Edwardian suburbs of Belfast were made by the Minister in 2000.

30 See DCMS (1999) *World Heritage Sites: The Tentative List of the United Kingdom of Great Britain and Northern Ireland.*

31 We are grateful to the Garden and Landscape team at English Heritage for pointing out the omission of historic parks and gardens in the previous edition.

32 The National Trust boast that its 161 gardens comprise the largest and most diverse group of private gardens in the world.

33 Similar lists are being compiled in the other countries of the UK. In Scotland it is known as the Register of Historic Gardens and Designed Landscapes and in Wales the Register of Landscapes, Parks and Gardens of Special Historic Interest in Wales and the Register of Landscapes of Outstanding Historic Interest in Wales.

34 The figures are taken from English Heritage's evidence to the Select Committee on Environment, Transport and the Regional Affairs inquiry in *Town and Country Parks* (1999) <www.publications.parliament.uk/pa/cm/cmselect.htm>. The numerous memoranda of evidence provide interesting reading on the demise of urban parks generally, as well the ones of historic value. The comments from the Urban Parks Forum are taken from its submission to the enquiry.

35 The Environment Committee found during a 1986–7 inquiry that of the Church of England's 16,700 churches, 8,500 were pre-Reformation, and 12,000 were statutorily listed (2,675 in the highest grade).

36 The code includes the requirement that all works to a listed church building which would affect its character are submitted for the approval of an independent body, and that there is consultation with the LPA, English Heritage, and national amenity societies. The religious bodies to which the exemption now applies because they have an agreed self-regulatory system are the Church of England, the Church in Wales, the Roman Catholic Church, the Methodist Church, the Baptist Union of Great Britain, the Baptist Union of Wales, and the United Reformed Church.

37 A National Land Fund had been established with similar intent in 1946 with funding of £50 million, part of which was used for the purchase by the Secretary of State of buildings of

outstanding architectural or historic interest, together with their contents. The fund was raided by the Exchequer in 1957 and became moribund.

38 The largest grant was £1.1 million to the British Library (the second of two instalments) towards the acquisition of the Sherborne Missal (one of the finest illuminated Gothic manuscripts in the world, with 694 pages of information on medieval life). Grants relating to the built environment have previously included £4.9 million for the purchase of Croome Park, Worcester, and support to the Barlands Farm Romano-Celtic Boat (found during development of a supermarket).

39 In total 1,200 grant applications were made, requesting £620 million and 450 projects were completed. There are forty-four projects worth more than £5 million, and about a quarter of the funding goes to 'conservation and restoration' and a third to 'historic building repair or refurbishment'. In 1999–2000 in Northern Ireland, eighty-seven grants were awarded, worth £4.7 million; in Scotland 167 grants worth £19.8 million; and in Wales 129 grants worth £20.2 million.

40 £16.3 million is already directed to the *Urban Parks Programme*, and another £4.9 million to the *Places of Worships Scheme*.

41 This discussion is based on the system in England. The arrangements are similar elsewhere, but there are significant differences for Northern Ireland. Note that the cutting down of a tree is not 'development' and thus not subject to the normal controls – hence the need for special provisions. There is an 'informal' tree register (a registered charity) which is compiling a list of notable trees: <www.tree-register.org>.

42 *Tree Preservation Orders Draft Regulations: A Consultation Paper* (1998).

43 The Hedgerows Regulations 1997 (SI no. 1160).

44 In a 1993 survey of overseas visitors to London undertaken by the British Tourist Authority and the London Tourist Boards, 69 per cent of holidaymakers gave heritage as the main reason for visiting London (the figure for visitors from North America was 80 per cent). With domestic tourists, the most frequently mentioned main purpose of a holiday (after walking holidays) was to visit heritage attractions and sites.

45 In 1997 Britain was the fifth most visited country in the world, after France, the United States, Spain, and Italy.

46 The statistics in the figure are taken from a House of Commons research paper (Bardgett 2000) (which gives a very useful account of recent trends and policies) and the ONS datasets on tourism.

47 The quote comes from the House of Commons National Heritage Committee (1994: para. 54), which complained that 'there is a serious lack of coherence about policy for the preservation of our heritage and its very important links with the tourist industry'. It involved thirty-four quangos.

48 Each RTB takes the lead on a specific issue: sustainability and rural tourism is led by the Cumbria RTB, resorts by the North-West RTB, and walking by the South-West RTB.

49 See the Scottish Executive Report *A New Strategy for Scottish Tourism* (2000) and National Assembly for Wales: *Achieving Our Potential*, In Northern Ireland an 'Action for Tourism Task Force' has been created.

50 Environmental capacity studies have been undertaken at numerous visitor attractions. The best-known is the Chester Study, undertaken by the local authority and English Heritage. The studies can go into great detail. At Stonehenge the study has involved mathematical modelling of visitor movements and studies on the best type of grass.

51 The quote is taken from the ETC's web site, <www.wisegrowth.org.uk>, which has been set up to provide a national source of expertise, advice and information on sustainable tourism. The English Historic Towns Forum is investigating best-practice advice for historic towns on sustainable tourism.

52 The issue raised most by consultees was VAT. Conservation and repair is subject to VAT, but replacement is not. Thus, for example, it may pay to replace windows rather than repair them. The main recommendation in relation to tax is to

equalise VAT at 5 per cent for all building work.

53 Perhaps because of the restrictive brief, another very long-standing issue does not appear to be addressed: the separate systems and lists for buildings, monuments, gardens, and battlefields, first mentioned in 1950 (Larkham 1999a: 108).

54 Parish or village appraisals are well established (see Moseley 1997). The Heritage Review draws attention to the work of the Urban Design Alliance, which is carrying out ten pilot 'place-check' projects that have similar objectives.

55 One in ten planning authorities have used Article 4 directions to withdraw permitted development rights in conservation areas. This recommendation is justified with reference to the many changes that can take place in conservation areas with no control at all. For example, 'the local authority can resurface streets and pavements in new materials and patterns, install traffic signs, and together with over 30 private utility companies install cabinets, kiosks and other street furniture without any control or co-ordination' (*Power of Place*: 35).

56 In relation to the management of historic public buildings, see the English Heritage report *Better Public Buildings*.

9

PLANNING AND THE COUNTRYSIDE

The challenge for rural communities is clear. Basic services in rural areas are overstretched. Farming has been hit hard by change. Development pressures are considerable. The environment has suffered.

(White Paper, *Our Countryside: The Future: A Fair Deal for Rural England*, 2000)

THE CHANGING COUNTRYSIDE

There have been great changes in the British country-side since the early postwar policies were forged. Suburban commuter residential development, roads and transport, people seeking recreation, the changing economy, forestry, conservation, and a host of other pressures have grown beyond any expectation. The changes show no sign of abating: they never have. The British countryside has been subject to continual change: the 'natural' scenery which is now the concern of conservationists is the human-made result of earlier economic change. The changes continue: the most recent are those which come with the crisis in the agricultural industry, and with the growth of both population and economic activity.

A major plank of postwar policy was that a prosperous agriculture would not only be of strategic economic value but also provide the best means of preserving the countryside. Aided by European policies, and by technological advances, the promotion of agricultural production has been a huge success. Unfortunately, as so often happens with policy successes, the solution of one problem gives rise to another. In place of the need for increased agricultural production is the problem of dealing with surpluses and finding ways of reducing output. Matters are further complicated (throughout Europe) by more productivity

increases resulting from a number of factors including continuing technological advances (notably biotech-nological developments) and agricultural development in eastern Europe (historically a major food-producing region). The pressures for change in agricultural policy have been increased by mounting concern over the rural landscapes, which have changed in response to newer production methods, and by growing demands for conservation and recreation.

A reversal of long-established, popular policies does not come easily; and the difficulties are increased when so many interests benefit from the subsidised regime. Above all is the overwhelming impact of the EU Common Agricultural Policy. The acronym (CAP) is oddly inappropriate since the one thing that is proving extremely difficult to do is to 'cap' it. Reform is a long and difficult process. Current proposals are a continuation of earlier attempts to move support for commodity production towards direct payments to farmers, particularly for environmental benefits. The issues are complex both in economic terms (reform of the CAP could have a major impact on agricultural land values) and politically (the problems – and the political muscle – of farmers vary across the EU). UK policies are severely constrained and the scale and speed of change are highly uncertain.

As a result, change is taking place more slowly than is required. Nevertheless, changes in policy have been made. A major milestone was the 1986 Agriculture Act, which required agriculture ministers to maintain a balance between the interests of agriculture and wider rural and environmental interests. In particular, attention had to be given to 'the conservation and enhancement of the natural beauty and amenity of the countryside', and to 'the promotion of the enjoyment of the countryside by the public'. (More tangibly, the legislation provided for the establishment of *environmentally sensitive areas*, which are discussed later.)

A subsequent DoE Circular (16/87) noted that 'the need now was to foster the diversification of the rural economy so as to open up wider and more varied employment opportunities'. The circular continued:

> The agricultural quality of the land and the need to control the rate at which land is taken for development are among the factors to be considered [in assessing planning applications affecting agricultural land], together with the need to facilitate development and economic activity that provides jobs, and the continuing need to protect the countryside for its own sake rather than primarily for the productive value of the land.

The circular was a mere twelve paragraphs long, but it represented a dramatic change in policy.

The new dimensions of policy involve issues of land management for which the postwar planning system is inadequate. This system was designed to deal with land use, not its management; and it was largely restricted to urban land, not the countryside. It was assumed that a prosperous agriculture would by itself deal with any problems of the rural economy. One implication of this was that there appeared to be no problem about the division of central government countryside responsibilities between departments concerned with planning and those concerned with agriculture. Other divided responsibilities, particularly between countryside conservation and nature conservation, seemed more problematic: these later gave rise to several organisational changes. Current policies for the countryside are based on an 'overarching vision', set out in Box 9.1.

A first reaction to a 'vision' of this character is to question why separate White Papers are deemed necessary or appropriate for rural and urban areas. (Some wags such as Butt (1999) have speculated that

---

BOX 9.1 COUNTRYSIDE POLICY FRAMEWORK

Our vision is of:

- a *living* countryside, with thriving rural communities and access to high quality public services;
- a *working* countryside, with a diverse economy giving high and stable levels of employment;
- a *protected* countryside in which the environment is sustained and enhanced, and which all can enjoy;
- a *vibrant* countryside which can shape its own future and with its voice heard by Government at all levels.

Our aim is to sustain and enhance the distinctive environment, economy and social fabric of the English countryside for the benefit of all.

Source: DETR and MAFF White Paper *Our Countryside: The Future: A Fair Deal for Rural England* (2000)

---

there might also be a White Paper on the suburbs.) Given the contemporary emphasis on 'joined-up' thinking, and the complex interrelationship between town and country, together with the number of households in rural areas projected to increase by a million by 2011 (RDC (1998) *Household Growth in Rural Areas*), half of which will be from migration, the question is of some significance. Dividing the two is unhelpful and may increase the urban versus rural controversy.

Be that as it may, there is no doubt that some of the 20 per cent of people who live in the countryside[1] face problems of a particular nature. Public transport is limited (75 per cent of rural parishes have no daily bus service); there is a shortage of shops (two-fifths have no shop or post office, and 70 per cent have no general store); most (four-fifths) have no GP based in the parish; a half have no school, and there is a shortage of affordable housing for local people.[2] On the other hand, unemployment in rural districts is lower than the national average, employment growth is higher (with a fall in land-based industries, and a growth in service sectors), and the rural economy is becoming increasingly diversified; housing conditions in rural areas are generally better than in urban areas; and car ownership is higher in rural areas.

Of course, one can play around with such figures to 'prove' different points, but there is clearly a range of conditions in rural areas, and some of the problems require specifically rural policies. Housing and transport in particular require distinctly rural programmes. Examples of these are the rural exceptions policy, which permits rural sites to be released for affordable housing as an exception to normal development plan policies (see Annex A to PPG 3, *Housing*) and rural traffic strategies along the lines suggested by the Countryside Commission's *Rural Traffic: Getting it Right* (1998). Following the Comprehensive Spending Review, planned expenditure on DETR rural programmes was increased from £128 million in 1998–99 to £174 million in 2001–2. Countryside programmes are to be jointly undertaken by the DETR and MAFF. Also following the Comprehensive Spending Review, the Countryside Commission and most of the functions of the Rural Development Commission were merged in a new Countryside Agency.[3] The rationale of this was that it would allow the development of a more strategic and integrated approach to rural policy.

Planning policies in rural areas have been subjected to criticism by the (former) Rural Development Commission (*Rural Development and Land Use Policies* 1998). Many planning authorities, in embracing a policy of 'sustainable development', have interpreted this in environmental terms:

> It is often taken simply to mean environmental protection and reducing travel needs by concentrating development into larger settlements. Those strands of sustainable development relating to economic development and social equity tend to be overlooked. . . . Emphasis should be given to the social and economic implications of not providing for development.

The new Countryside Agency has set out a positive and challenging statement of its conception of future planning policy in its *Planning for Quality of Life in Rural England* (1999). This stresses 'the essential interdependence between a thriving rural economy, sustainable communities and the proper care and enjoyment of the countryside'. It argues that in judging development proposals, 'the new philosophy should be "is it good enough to approve", not "is it bad enough to refuse"'. This implies a proactive approach by local planning authorities: their policies need to be 'criteria based' and clearly state 'the qualities the plan wishes to pursue, and which development proposals should reflect'.

The Cabinet Office Performance and Innovation Unit's *Rural Economies* (1999) report also addressed the role of the planning system. This report was offered as a contribution to the debate on 'modernising the policy framework for rural economies'. (The plural is used because 'the diversity across the country makes it inappropriate to think of a single "rural economy", separate from and different to economic activity in urban areas'.) Among the ideas put forward for discussion are the extension of planning controls to agricultural development, the relaxation of development control in rural areas to allow 'sympathetic and appropriate' economic activity, and a change in the presumption against development of the best and most versatile agricultural land (but with appropriate protection for areas of high environmental value).

The 2000 White Paper issued jointly by the DETR and MAFF, *Our Countryside: The Future: A Fair Deal for Rural England* (2000), is a long and repetitive document, in different parts presenting a celebration of past achievements, a review of existing sectoral policies that affect the countryside, good practice examples, and (only in part) government policy. The views expressed about planning continue in the same vein as previous statements. Planning has protected the countryside, important landscapes, and the environment, but in future there needs to be a more flexible and positive attitude to development in the countryside, especially where this supports the provision of services and affordable homes and diversification of the rural economy. Such attitudes are based on the (largely erroneous) assumption that planning is the problem. The most recent research shows that planning has not generally blocked rural diversification (although there is room for improvement in positive action), and the problems of providing for affordable housing have long been recognised in the system (as discussed in Chapter 6).[4] Nevertheless, serious questions are raised about the implications of current planning policies for the countryside. These are all matters where urban and rural interdependence is critical (ESPRIN 2000; Stead and Nadin 2000).

The White Paper ranges over many policy fields and there are explicit commitments to change – notably in measures to help retain services in rural areas with proposals for a 50 per cent rate relief for services that offer community benefit, and more money for rural transport and market town regeneration. The planning system figures prominently. National policy is promised to assist the planning system to secure more affordable homes through planning obliga-tions; landscape and biodiversity protection is to be strengthened; and access to the countryside is to be made easier. Some of the proposals are discussed at the relevant points below.

A proposal of particular interest is the programme of town and village plans. This takes forward existing best practice on parish appraisals and village design statements (Osborne and Tricker 2000), and will provide £5 million for 1,000 rural communities to prepare a town or village plan. The plan will be non-statutory and will have to be in conformity with national and regional guidance and the development plan, but it may be endorsed by the local planning authority as supplementary planning guidance. The plan will also tackle issues of service provision and community needs. More guidance is promised.

NATIONAL PARKS

The demand for public access to the countryside has a long history (Eversley 1910), stretching from the early nineteenth-century fight against enclosures, James Bryce's abortive 1884 Access to Mountains Bill and the attenuated Access to Mountains Act of 1939, to the promise of the National Parks and Access to the Countryside Act of 1949: an Act which, among other things, poetically provided powers for 'preserving and

enhancing natural beauty'. Many battles have been fought by voluntary bodies such as the Commons, Open Spaces and Footpaths Preservation Society and the Council for the Protection of Rural England, but they worked largely in a legislative vacuum until the Second World War. The mood engendered by the war augured a better reception for the Scott Committee's (1942: para 178) emphatic statement that 'the estab-lishment of national parks is long overdue'. The Scott Committee had very wide terms of reference, and for the first time an overall view was taken of questions of public rights of access to the open country, and the establishment of national parks and nature reserves within the context of a national policy for the preservation and planning of the countryside.

Government acceptance of the necessity for establishing national parks was announced in the series of debates on postwar reconstruction which took place during 1941 and 1943, and the White Paper *The Control of Land Use* referred to the establishment of national parks as part of a comprehensive programme of postwar reconstruction and land use planning. Not only was the principle accepted but, probably of equal importance, there was now a central govern-ment department with clear responsibility for such matters as national parks. There followed the Dower (1945) and Hobhouse (1947) Reports on national parks, nature conservation, footpaths, and access to the countryside, and, in 1949, the National Parks and Access to the Countryside Act, which established the National Parks Commission and gave the main responsibility for the parks to local planning authorities.

The administration of the national parks has been a matter of controversy throughout their history. Dower had envisaged that there would be *ad hoc* committees with members appointed in equal numbers by the Commission and the relevant local authorities. Local representation was necessary since the well-being of the local people was to be the first consideration; but the parks were also to be *national*, and thus wider representation was essential. The lengthy arguments on this issue were eventually resolved by the 1949 Act in favour of a local authority majority, with only one-third of the members being appointed by the Secretary of State. (In line with his conception of truly national parks, Dower had proposed that the whole cost of administering them should be met by the Exchequer – an idea which was never accepted.)

Increasing pressures on the countryside have led to a succession of policy and legislative changes. In 1968 the Countryside Act replaced the National Parks Commission with a more powerful Countryside Commission. A Countryside Commission for Scotland was established under the Countryside (Scotland) Act 1967. The Wildlife and Countryside Act 1981 strengthened the provisions for management agreements and introduced compensation for farmers whose rights were restricted (a major change in principle). Later there were major structural changes in the organisation of agencies responsible for countryside matters, including the establishment of a separate Countryside Council for Wales and the merging of the Countryside Commission for Scotland with the Nature Conservancy for Scotland as Scottish Natural Heritage. The Environment Act 1995 established independent national park authorities, which took over the responsibilities previously exercised by local government. These are now the sole local planning authority for a national park area. In addition to the normal plans, a national park authority is required to prepare a *national park management plan* (Countryside Commission 1997). This goes further than the scope of development plans: in addition to establishing policies, it is intended to spell out how the park is to be managed.

From their inception, the national parks have had two purposes: 'the preservation and enhancement of natural beauty', and 'encouraging the provision or improvement, for persons resorting to national parks, of facilities for the enjoyment thereof and for the enjoyment of the opportunities for open air recreation and the study of nature afforded thereby'. There is inevitably some conflict between these twin purposes, and the National Parks Review Panel (Edwards Report 1991), set up by the Countryside Commission, recommended that they be reformulated to give added weight to conservation – as did the earlier Sandford Report (1974).[5] This argument, which continues, was a major issue in the debates on the sections of the Environment

Act 1995 which established independent national park authorities. Controversy centred on the need 'to promote the *quiet* enjoyment and understanding' in national parks. The final outcome is set out in Box 9.2.

There are ten national parks in England and Wales: the Brecon Beacons, Dartmoor, Exmoor, the Lake District, Northumberland, the North York Moors, the Peak District, the Pembrokeshire Coast, Snowdonia, and the Yorkshire Dales (two more are proposed – see below). Most of the land in the national parks is in private ownership (74 per cent), and only about 2 per cent is owned by the park authorities. The National Trust and the Forestry Commission each own 7 per cent. Three-quarters of the funding for the parks comes from the Exchequer: in 1998–99 this amounted to £31 million.

THE BROADS, THE NEW FOREST, AND THE SOUTH DOWNS

Though proposed by the Dower Report as a national park, the Broads was rejected as such because of its deteriorated state and the anticipated cost of its management (Cherry 1975: 54). However, the need for some type of special protection continued to be debated, and the proposal surfaced again in the late 1970s. The local reaction was against this and instead a voluntary consortium, called the Broads Authority, was formed by the relevant public authorities (with powers and financial resources under the provisions of the Local Government Act 1972, and with 75 per cent Exchequer funding).

Discussions continued over several years among the large number of interested bodies, and in 1984 the Countryside Commission reviewed the problems of the area and the progress that had been made by the Broads Authority. Its conclusion was that despite some achievements, the authority had not made significant improvements in water quality. Moreover, an effective framework for the integrated management of water-based and land-based recreation had not been established; and the loss of traditional grazing marsh was continuing. The outcome was the designation of the area as a body of equivalent status to a national

BOX 9.2 NATIONAL PARKS: PURPOSES

The Environment Act 1995 provides that the purposes of national parks shall be 'conserving and enhancing the natural beauty, wildlife and cultural heritage of the areas', and 'promoting opportunities for the understanding and enjoyment of the special qualities of those areas by the public'. If there is a conflict between these purposes, any relevant authority 'shall attach greater weight to the purpose of conserving and enhancing the natural beauty, wildlife and cultural heritage of the area'.

A national park authority, in pursuing these purposes, 'shall seek to foster the economic and social well-being of local communities within the national park, but without incurring significant expenditure in doing so, and shall for that purpose cooperate with local authorities and public bodies whose functions include the promotion of economic and social development within the area of the national park'.

Source: Environment Act 1995

---

park, but with a constitution, powers, and funding designed to be appropriate to the local circumstances. A new Broads Authority (with the same name as its predecessor) was established by the Norfolk and Suffolk Broads Act 1988. The duties of the authority are extensive. It is the local planning authority and the principal unit of local government for the area. It has strong environmental responsibilities, and is required by the Act to produce a plan which has a wider remit than those required under the planning acts: it is more akin to a national park management plan.

While it took many years to devise an adequate system for administering the Broads, it is taking even longer to do so for the New Forest. The New Forest was 'new' in 1079 when William the Conqueror appropriated it as his new royal hunting ground. Situated in south Hampshire, close to a large urban population, it has unique qualities of landscape and habitats, and a singular set of administrative arrangements. The Crown land, of some 27,000 hectares, is managed by the Forestry Commission. The so-called Perambulation is a wider area of about 38,000 hectares which is defined in the New Forest Act of 1964 (one of several Acts relating to the area). A larger area is the New Forest Heritage Area, which, though lacking statutory designation, was adopted in 1985 by the New Forest District Council, and is identified in local development plans.

In addition to the public authorities in the area, there is a corporate body of Verderers which is responsible for managing the grazing and commoning within the forest. As well as the protection provided in these ways, much of the Crown land in the New Forest is designated as *sites of special scientific interest*. The southern fringe of the forest is within the South Hampshire Coast AONB, and the whole of the heritage area is within the South-West Hampshire Green Belt. It might thus appear that the New Forest is adequately (even if confusingly) protected. However, quite apart from questions of coordinating all these protectors of the forest, there is a further need to safeguard the surrounding grazing lands, which are under pressure for development, and also to ensure that adequate provision is made for recreation in a manner which is in harmony with conservation requirements.

The New Forest has for long been a candidate for designation as a national park. In 1998 the Countryside Commission urged that special legislation be passed to deal with the particular needs of the New Forest, but in 1999 the government announced that it wished to create two new national parks: for the New Forest and for the South Downs (Dewar 1999).

The South Downs is currently designated an AONB (there are two areas designated) across fifteen local authorities. The Downs were among the list of proposed designations back in 1947 but the perceived lack of recreational potential justified a decision not to designate. The decision on the two proposed parks is expected in late 2001 and at the time of writing consultation is under way. If a positive decision is made, it may be that the new parks will not necessarily have the same planning powers as the existing parks, at least in the immediate future (*Planning* 24 November 2000: 13).

LANDSCAPE DESIGNATIONS

Both the Dower and Hobhouse Reports proposed that, in addition to national parks, certain areas of high landscape quality, scientific interest, and recreational value should be subject to special protection. These areas were not considered, at that time, to require the positive management which it was assumed would characterise national parks, but 'their contribution to the wider enjoyment of the countryside is so important that special measures should be taken to preserve their natural beauty and interest'. The Hobhouse Committee proposed that such areas should be the responsibility of local planning authorities, but would receive expert assistance and financial aid from the National Parks Commission. A total of fifty-two areas covering some 26,000 sq. km were recommended, including, for example, the Breckland and much of central Wales, long stretches of the coast, the Cotswolds, most of the Downland, the Chilterns, and Bodmin Moor (Cherry 1975: 55). More modestly, ten national parks were designated.

The 1949 Act did not contain any special provisions for the care of such areas, the power under the Planning Acts being considered adequate for the purpose. It did, however, give the Commission power to designate *areas of outstanding natural beauty* (AONBs), and provided for Exchequer grants on the same basis as for national parks. Forty-one areas have been designated, covering some 21,000 sq. km (over 15 per cent of the area of England, and 4 per cent of Wales). In Northern Ireland there are nine AONBs, covering 2,800 sq. km. (The Scottish equivalents of AONBs are the forty *national scenic areas*, which cover 10,000 sq. km.)

Areas of outstanding natural beauty are, with some notable exceptions, generally smaller than national parks. They are the responsibility of local planning authorities, which have powers for the 'preservation and enhancement of natural beauty' similar to those of park planning authorities. There has been continuing debate on the question as to whether the designation of areas of outstanding natural beauty serves any useful function.[6] The latest initiatives from the Countryside and Rights of Way Act 2000 are the requirement for management plans for all AONBs and allowing the creation of 'conservation boards' that would take over the management function (including the preparation of the management plan) from the local authorities.

In addition to AONBs, there are many local authority designations designed to assist in safeguarding areas of the countryside from inappropriate development. Some of these have been given additional status through inclusion in structure and local plans. Though these are like AONBs in that they involve the application of special criteria for control in sensitive areas, they do not imply any special procedures for development control. The DoE Consultation Paper *The Future of Development Plans* (1986) made reference to *areas of landscape quality*, *areas of great landscape value*, *landscape conservation areas*, *coastal preservation areas*, and *areas of semi-national importance*. The same paper proposed a new statutory designation, the *rural conservation area*, which, it was suggested, would provide a more coherent framework. This idea, however, found little favour during the consultation process and it was dropped. It was felt that the desired objectives could be achieved through statements of policy in development plans.

The current advice (in PPG 7) takes a strong line against local designation, which 'may unduly restrict development and economic activity without identifying the particular features of the local countryside which need to be respected or enhanced'.[7]

A significant feature of the countryside landscape is the hedgerows. Mainly because of agricultural economics, there has been a dramatic loss of hedgerows: since 1947 over half of them have disappeared. Though hedges are protected (under section 97 of the Environment Act 1995 and the Hedgerow

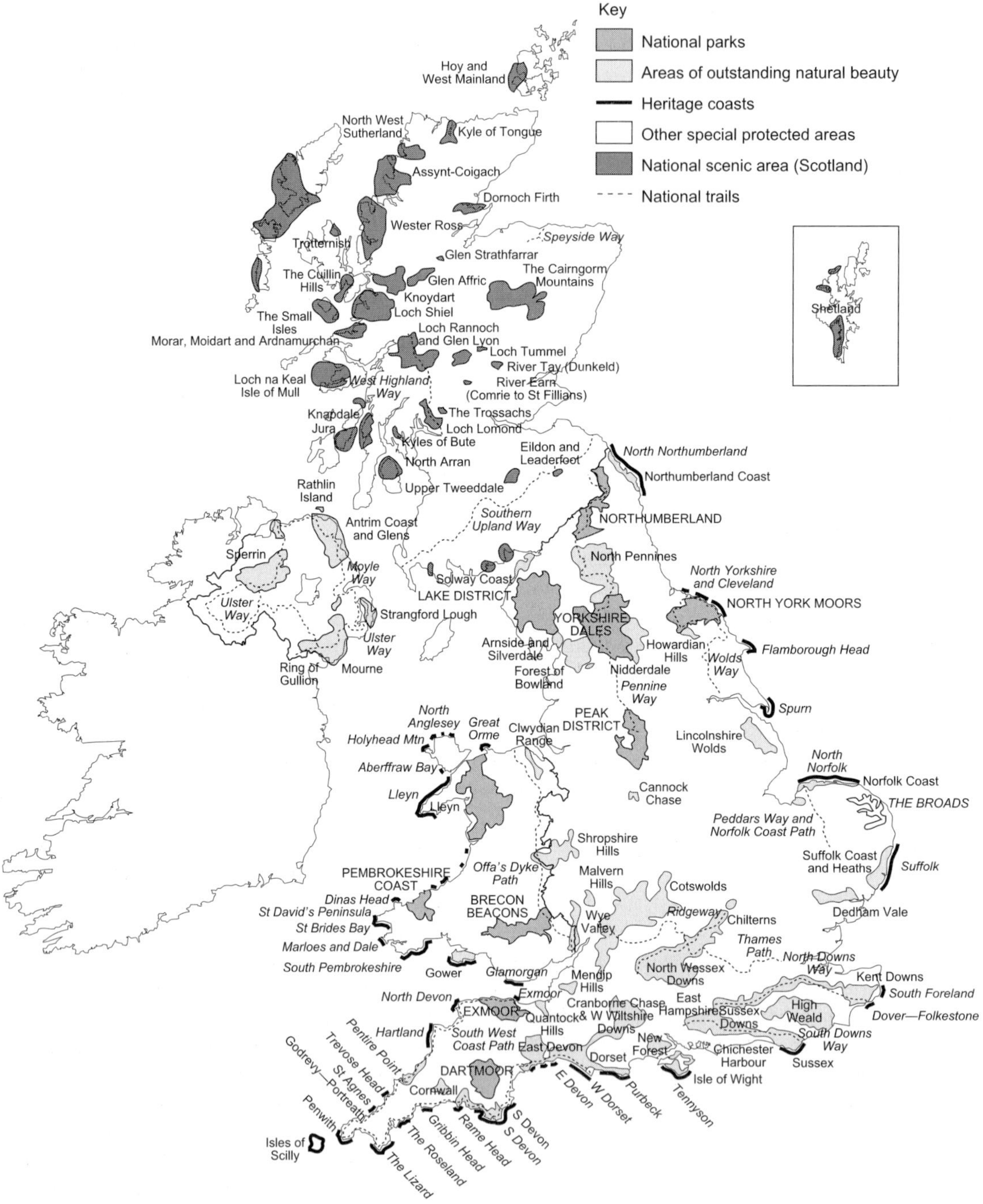

Figure 9.1 Selected Protected Areas in the UK

Table 9.1 National Parks, Areas of Outstanding Natural Beauty, and National Scenic Areas

	Number	Area (square km)	Percentage of total land area
England			
National parks	7	9,936	8
Areas of outstanding natural beauty	37	20,510	16
Northern Ireland			
Areas of outstanding natural beauty	9	2,849	20
Scotland			
National scenic areas	40	10,018	13
Proposed national parks	2	n.a.	n.a.
Wales			
National parks	3	4,129	20
Areas of outstanding natural beauty	5	727	4

Source: Office of National Statistics, *Britain 2001: An Official Handbook* p. 316

Note: The area of national parks in England includes the Broads and the area of AONB in Wales includes a cross-border area with England.

Regulations 1997), this protection is limited. A Hedgerow Review Group recommended in 1998 that, in the long term, the statutory provisions should be amended. The practical effect of the current legislation is to centralise the decision about what are important hedgerows. Moreover, the criteria are complex and difficult to operate.[8] As a result, local authorities are severely constrained. Unfortunately, to deal with this adequately would require primary legislation; and, given the current legislative programme, this is unlikely in the short term. In the meantime, however, revisions could be made to the regulations. (This would require less parliamentary time.) The review therefore concentrated on how the regulations and the associated guidance might be changed to deliver stronger hedgerow protection.[9] The detailed recommendations provide an alternative set of criteria. These are currently being considered by the government.

SCOTTISH DESIGNATIONS[10]

Scotland contains large areas of beautiful unspoiled countryside and wild landscape. It has the majority of Britain's highest mountains, with nearly 300 peaks of over 900 m; it has the great majority of the UK islands; and its coast is over 10,000 km in length.

Despite expectations to the contrary, there are (as yet) no national parks in Scotland. Though a Scottish committee (the Ramsay Committee) recommended, in 1945, the establishment of five Scottish national parks, no action followed. The reasons for this inaction were partly political and partly pragmatic (Cherry 1975: chapter 8). A major factor was that (with the exception of the area around Clydeside and, in particular, Loch Lomond) the pressures which were so apparent south of the Border were absent.

Nevertheless, the Secretary of State used the powers of the 1947 Planning Act to issue *National Parks Direction Orders*. These required the relevant local planning authorities to submit to the Secretary of State all planning applications in the designated areas (which included Loch Lomond/Trossachs, the Cairngorms, and Ben Nevis/Glencoe). In effect, therefore, in an almost Gilbertian manner, while Scotland at this time did not have any national parks, it had an administrative system which enabled controls to be operated as if it did! But, of course, this approach was

Table 9.2 Scottish Designations

	Number	Area (hectares)
Management agreements	1,137	514,000
National nature reserves	71	114,000
Local nature reserves	29	9,000
SSSIs	1,448	920,000
Ramsar wetlands sites	48	234,000
Special protection areas (ECC Birds Directive)	103	372,000
Special areas of conservation (EEC Habitats Directive)	128	643,000
National scenic areas	40	1,000,000
Regional parks	4	86,000
Country parks	36	6,000
Historic gardens and designed landscapes	275	n.a.
Long-distance routes	4	(604 km)

Source: SNH, *Facts and Figures 1998–99*

inherently negative, and it was not until the Countryside (Scotland) Act of 1967 that positive measures could be taken on a significant scale.

This Act provided for the establishment of the Countryside Commission for Scotland – later joined with the Nature Conservancy Council for Scotland in Scottish Natural Heritage (SNH). It also enabled the establishment of regional parks and country parks. A policy framework for these was set out in the Commission's 1974 report *A Park System for Scotland*. The report also recommended the designation of national parks in Scotland, though the term *special park* was used. Until recently, this has not been accepted, though objectives similar to those of national parks have been achieved under other designations. Despite the reluctance to establish national parks in Scotland, pressure for them has mounted (Rice 1998), and in 1997 the Secretary of State for Scotland announced that

> national parks would be the correct way forward for Loch Lomond and the Trossachs, quite probably in the Cairngorms, and possibly in a few other areas as well . . . I see national parks in Scotland as integrating economic development with proper protection of the natural heritage.

Scottish Natural Heritage was asked to provide advice for action by the Scottish Parliament. This it did

by undertaking wide consultations, inviting views, and commissioning research.[11] In 1998 it produced a consultation paper, *National Parks for Scotland*, outlining its initial proposals.

The proposals emphasise the importance of an integration of social and economic purposes along with protection and enhancement of the natural and cultural heritage, and the enjoyment, understanding, and sustainable use of natural resources. Also stressed are local community involvement in national parks, and a strong Park Plan 'prepared through consensus with a zoning system to help reconcile differing needs'. These and other proposals make this document an outstanding statement on participatory planning.[12] At the time of writing, national parks have been proposed for Loch Lomond and the Trossachs.

Areas proposed for national parks are large in area and small in number. There are, however, other designations which are more numerous. There are forty national scenic areas, four regional parks, and thirty-six country parks. The *national scenic areas* are of similar status to the AONBs. They extend over an area of a million hectares, and include such marvellous sites as Ben Nevis and Glencoe, Loch Lomond, the Cairngorms, and the World Heritage Site of the islands of St Kilda. Development control in these areas is the responsibility of the local planning authorities, which are required to consult with SNH for certain categories of development.[13] As in England and Wales, there is an increasing concern for 'positive action to improve planning and land use management' in these areas, and for dealing with the erosion of footpaths. There is also a similar complaint about the lack of resources.

A *regional park* is statutorily defined simply as 'an extensive area of land, part of which is devoted to the recreational needs of the public'. The four parks are Clyde-Muirshiel, Loch Lomond, the Pentland Hills, and Fife. These parks, which cover 86,000 hectares, are primarily recreational areas, and each has a local plan which sets out management policies. Emphasis is laid on *integrated land management* schemes to ensure that public access is in harmony with other land uses. In this, they give effect to Abercrombie's green belt philosophy, articulated in the Clyde Valley Regional Plan. He conceived these *outer scenic areas* not only as recreational areas but also as a means of protecting the rural setting of the conurbations (Smith and Wannop 1985). Since the passing of the 1967 Act, Scottish local authorities have provided thirty-six *country parks* spread across the central belt and north-east. The parks are 'registered' with SNH, which makes grants for capital development expenditure and also towards the cost of a ranger service. Country parks not only are of direct benefit to their 11 million annual visitors, but also have a conservation objective of 'drawing off areas that are sensitive due to productive land uses and fragile wildlife habitats'.

NORTHERN IRELAND DESIGNATIONS

Northern Ireland boasts some of the finest countryside in the UK, often with special value as wildlife habitat. One reason for this is that farm and field sizes are smaller than on the mainland, and almost all the farms are owner-occupied (Glass 1994). Much of the countryside remains unspoiled, but development pressures are increasing, and there has been extensive building of isolated houses in the countryside. Progress with planning for landscape and nature conservation

has been slower than in the rest of the UK. Legislation is far less developed, and there has been much criticism about the delays in designating areas needing protection and management (Dodd and Pritchard 1993).

Criticisms of the backwardness of countryside and nature conservation led to a review on behalf of the Secretary of State by Dr Jean Balfour, whose 1983 report, *A New Look at the Northern Ireland Countryside*, confirmed the low priority given to conservation and its lack of status in the work of the Northern Ireland Department of the Environment (DoENI). The result was the setting up of a unit within the Department and an advisory Council for Nature Conservation and the Countryside. Legislation extending nature conservation powers soon followed. Nevertheless, designations remained 'pitifully slow'. A new initiative was taken in the early 1990s, culminating in the publication in 1993 of a comprehensive *Planning Strategy for Northern Ireland*.

Statutory designations are much the same as in England and Wales, and the DoENI has all powers in respect of designation and management of special areas (including those exercised by English Nature and the Countryside Agency). It can designate national parks (though it has not done so), areas of outstanding natural beauty (of which there are nine, covering 285,000 hectares), nature reserves (of which there are forty-five, covering 4,300 hectares), and *areas of special scientific interest* (the equivalent of the SSSI), of which there are 161, covering 83,000 hectares (*Whitaker's Almanac 2000*: 576–7). An additional non-statutory designation is the *countryside policy area*, which is employed to restrict building in the countryside.

THE COAST

A few figures underline the particular significance of the coast, and therefore of coastal planning: nowhere in the UK is more than 135 km from the sea; the coastline is 18,600 km in length, and the territorial waters extend over about a third of a million square kilometres. About a third of the coast of England

and Wales is included in national parks and AONBs. Large areas of the coast are owned or protected by the National Trust. Following the *Enterprise Neptune* fund-raising appeal, the Trust protects 920 km of the coastline in England, Wales, and Northern Ireland (mainly by ownership; the remainder by covenant). North of the border, the National Trust for Scotland owns large stretches of the coastline and protects further parts by way of conservation agreements. In addition, there are *marine nature reserves* (discussed later).

In spite of all this protection, the pressures on the coastline are proving increasingly difficult to cope with. Between a quarter and a third of the coastline of England and Wales is developed and this proportion has been increasing.[14] Growing numbers of people are attracted to the coast for holidays, for recreation, and for retirement. There are also economic pressures for major industrial development in certain parts, particularly on some estuaries (which have international importance for nature conservation).

The problem is a difficult one which cannot be satisfactorily met simply by restrictive measures: it requires a positive policy of planning for leisure provision. This has long been accepted, and the *heritage coast* designation, introduced in 1972, implies recreational provision as well as conservation. The Countryside Commission (now Agency) has urged that every heritage coast should have a management plan. It has also established the *Heritage Coast Forum* as 'a national body to promote the heritage coast concept and to act as a focus and liaison point for all heritage coast organisations'. This is seen as a needed addition to the activities of the Commission, whose capacity to promote all the initiatives that are necessary is limited.

There are forty-five heritage coasts in England and Wales, protecting some 1,500 km, about a third of the total length of the coastline (*Britain 2000*: 323). In Scotland twenty-six *preferred coastal conservation zones* have been defined with a total length of 7,546 km, covering three-quarters of Scotland's mainland and island coastlines.

The Environment Committee, in its 1992 report *Coastal Zone Protection and Planning*, complained of the lack of coordination among the host of bodies concerned with coastal protection, planning and management. (There are over eighty Acts which deal with the regulations of activities in the coastal zone, and as many as 240 government departments and public agencies involved in some way.) Not surprisingly, there have been suggestions that action is required to simplify, rationalise, coordinate, or consolidate matters. Though this is an apparently obvious and sensible idea, it is remarkably difficult to see how the situation can be significantly changed; and the Environment Committee contented itself by asking for a review of legislation and responsibilities. The government response was negative. It was pointed out that though there were many Acts relating to the coast, the same could be said about the land! Indeed, it was neither possible nor desirable to treat the coast separately from the adjoining land or from the territorial and international waters. Moreover, the suggestion that the town and country planning system might be extended seaward was not persuasive, though it was agreed that 'it is now time to take this debate further'.

The EU is also active in coastal management, through the *Demonstration Programme on Integrated Coastal Zone Management* (ICZM). The Programme involves thirty-five projects throughout the EU, including several in the United Kingdom. It is questionable whether the Community has competence in this area, but there is no doubt that this is an issue that requires coordination at the transnational as well as national and regional levels. The Demonstration Programme reported in 1999 with proposals for a Europe-wide strategy, and support for rethinking the institutional arrangements for coastal zone management in the member states.[15]

Two discussion papers were issued by the DoE in 1993 (*Managing the Coast* and *Development below Low Water Mark*). These drew upon the Countryside Commission's experience with heritage coasts and underlined the usefulness of management plans drawn up by local authorities in liaison with the relevant bodies concerned. The Commission responded that there was a need for guidance on the form and content of such plans, and for integration with the shoreline management plans for flood and coastal defence which are being promoted by MAFF. In 1994 an organisation of maritime local authorities, the National Coasts and Estuaries Advisory Group, published a *Directory of Coastal Planning and Management Initiatives in England* (1994). This was followed, in 1995, by a DoE guide to government policies affecting the coastal zones (*Policy Guidelines for the Coast*). The latest proposal is for the designation of *marine environmental high risk areas* covering no more than 10 per cent of the coastline. They are intended to address the potential pollution from marine shipping. The debate continues.[16]

WATERWAYS

The advent of a Labour government in 1997 proved to be a turning point for the waterways. After years of inadequate funding, a 1999 DETR paper *Unlocking the Potential: A New Future for British Waterways* announced precisely what its title promised. This was one sector of public expenditure which benefited from the government's Comprehensive Spending Review. Additional funding was provided both for current expenditure and to deal with outstanding debt in relation to uneconomic expenditure on uneconomic activities which British Waterways is no longer carrying out (such as freight carrying). At the same time, its status as a nationalised industry was considered to be no longer appropriate since its trading activities had become a small and declining part of its operations. On the other hand, its leisure, amenity, and property activities have expanded rapidly. It has therefore become a public corporation. The government is to publish a Framework Document which will set new aims and objectives for British Waterways. With changes as significant as these, it is not surprising that its 1998–99 annual report enthuses that it had been 'a splendid year for the waterways'.

There are some 3,700 miles of waterways in Britain, of which British Waterways is responsible for about 2,000. Most of the latter are canals; the others are stretches of river navigation. British Waterways is responsible for maintaining the safety and operational quality of the waterways, and safeguarding and

improving the canals and the wider environment of which they are a part. This includes more than one might expect. In the words of British Waterways' 1999 *Plan for the Future 2000–2004*,

> the waterways offer an outstanding historical, environmental and ecological resource; a rare example of eighteenth century technology still working today to perform its original purpose; and a focus for communities to build their shared understanding of the past, whilst at the same time working to secure a future.

An important source of funding for improvements to the waterways is the Board's property portfolio. In conjunction with other organisations, and with extensive use of partnerships, the Board has undertaken several major urban renewal developments, such as in the development of Sheffield Basin, Paddington Basin, and the comprehensive development of the Gas Street Basin in Birmingham (which gained the city the Excellence on the Waterfront award – jointly with Boston and New York!). Since about a quarter of the total length of the waterways falls within the boundaries of the (former) metropolitan counties there is considerable potential for waterside development (though it should be noted that most waterside property is not owned by British Waterways).

Some development has attracted criticism from the pleasure craft operators because it has been seen as destroying the ambience of the canals and thus making them less attractive for cruising. As with many leisure pursuits, there is a problem of satisfying conflicting interests. Those who love the often closed and secretive world of much of the urban canal system may not welcome the new focal point developments, but these have proved to be extremely popular with others. Another conflict arises between the use of the waterways for leisure and their function as an aquatic habitat. Such conflicts cannot be prevented, but it is an explicit policy of British Waterways to achieve an appropriate balance among differing interests and uses.

The canal network has been described as a country park which is two thousand miles long by ten yards wide; but it is much more than that: the British Waterways owns over 2,000 listed structures and ancient monuments, and sixty-four SSSIs. It has a major programme of management and conservation in relation to these and other heritage features. Over 8 million people use the waterways in the course of a year, mainly for informal recreation rather than boating. In its 1993 report *The Waterway Environment and Development Plans*, British Waterways comments that this informal recreation 'offers the greatest prospect for increased use of the waterways', and this is recognised in its initiative in seeking joint study and action with local authorities and a host of other relevant agencies.

Though British Waterways cannot impose direct charges, related leisure facilities are income-generating: for example, shops, public houses, hotels, restaurants, and museums. A good example is the National Waterways Museum at Gloucester, which has over 100,000 visitors a year. In 1999 an initiative was taken to affiliate three waterway museums (Ellesmere Port, Stoke Bruerne, and Gloucester) to a new charitable Waterways Trust. This has gained national designation from the Museums and Galleries Commission, and also Heritage Lottery Fund grant.

PUBLIC RIGHTS OF WAY

The origin of a large number of public rights of way is obscure. As a result, innumerable disputes have arisen over them. Before the 1949 Act, these disputes could be settled only on a case-by-case basis, often with the evidence of 'eldest inhabitants' playing a leading role. The unsatisfactory nature of the situation was underlined by the Scott, Dower, and Hobhouse Reports, as well as by the 1948 report of the Special Committee on Footpaths and Access to the Countryside. All were agreed that a complete survey of rights of way was essential, together with the introduction of a simple procedure for resolving the legal status of rights of way which were in dispute. The 1949 Act attempted to provide for both.

This Act has been amended several times; under the current provisions, county councils have the responsibility for surveying rights of way (footpaths, bridleways, and 'byways open to all traffic') and preparing and keeping up to date what is misleadingly called a *definitive map*. The maps are supposedly conclusive

evidence of the existence of rights of way, but in fact they are not necessarily either complete or conclusive. They are incomplete because inadequate resources have been devoted to undertaking the necessary surveys; and they can be inconclusive because the map may wrongly identify a right of way. The latter is a legal matter which is not discussed here (see Chesman 1991), but the former is a continuing problem of planning policy and administration.

The definitive maps show some 225,000 km of rights of way in England and Wales. Most of these are 'existing': few new recreational paths have been designated, though they are certainly needed in some parts of the country. Moreover, there has been a loss of access by both neglect and deliberate obstruction. Each year some 1,500 formal proposals, affecting 500 km of the network, are made to change rights of way (by creation, diversion, or extinguishment). Of these, about three-quarters are unopposed. The net change is negligible. It is difficult to establish what the overall effect is, though the Ramblers' Association has maintained that over a half of the public rights of way 'are unavailable to all but the most determined and agile person' (Blunden and Curry 1989).

Another footpath problem arises from their popularity: this is the wear and tear caused by a great intensity of use. The Pennine Way in particular has suffered from this, and 'damage limitation' experiments are under way. Recent measures have included laying flagstones which are delivered by helicopter. The Pennine Way is one of the long-distance routes, which now stretch over some 2,700 km. The designation of these hikers' highways has been laborious, but they have had the attention and backing of the Countryside Agency, which has official responsibility for their establishment.

Rights of way provide a structured framework for public access to the countryside, and they are of particular value to the energetic walker. However, they meet only part of the need: the number of people who enjoy a wander in the countryside is far greater than the number who hike long distances. The 1947 Hobhouse Report had argued for a public right of access to all open country: among the many benefits the Committee foresaw was the freedom to ramble across the wilder parts of the country. Its idyllic view was very much in line with the long-standing arguments for a 'right to roam' over all open country. The 1949 Act was much more circumspect, and provided for a right of access only where an access agreement was made with the owner.

The essential arguments on this issue have not changed; but circumstances have. There is now a much wider demand for access to the countryside, fostered both by increasing leisure pursuits and by a huge increase in the ease of travel. But this very increase has strengthened the arguments of landowners about 'inappropriate use' of the countryside. The Labour Party manifesto stated that 'our policies include greater freedom for people to explore our open countryside. We will not, however, permit any abuse of a right to greater access.' This was hardly a full-blooded commitment to legislate for a 'right to roam', and a succession of proposals and consultation papers led to more modest ideas of 'improving rights of way'.

The government has acted on this commitment through the Countryside and Rights of Way Act 2000.[17] The rights of way provisions start to come into effect from April 2002, although it may be 2005 before they are implemented in full. The Act gives walkers the right to walk across mountain, moor, heath and down, and registered common land. The main exemptions are access by cycle, horse, or vehicle; and to gardens, parks, and cultivated land. The Countryside Agency and Countryside Council for Wales have begun preparation of maps showing where the new right of access will apply, and these will become definitive, with no additions allowed after a certain date yet to be determined. New powers for highway authorities and magistrates' courts will allow for the removal of obstructions, although landowners will be able to restrict access for twenty-eight days each year. Local authorities will have to publish plans for improving rights of way, included for those with impaired mobility.

In Scotland the position in relation to rights of way and access to the countryside is different from that south of the Border. The legal system is distinct, and the pressures on the countryside until recently have generally been fewer. There is relatively freer access to the Scottish countryside: there is 'a well-established system of mutual respect between walker and landowner' (Blunden and Curry 1989: 152). Nevertheless, this is not so in areas close to the towns, where access is severely restricted, and as pressures have mounted, the inadequacy of the legal situation has become apparent. In 1997 the government invited Scottish Natural Heritage to review the situation and to make recommendations to the Scottish Parliament.

Given the wider concerns with land reform (discussed in Chapter 6), SNH decided that the scope of the review should be extended beyond the manifesto commitment regarding access to open country to embrace access over all land and inland water. The review was initially undertaken by an existing Access Forum, which includes representatives of the major interests such as landowners, local authorities, conservationists, and recreation organisations, as well as SNH. Their proposals achieved a remarkable degree of consensus, and thus they have a legitimacy which augurs well for parliamentary action. The main proposal is that there should be a non-motorised right of access to land and water for informal recreation and passage, subject to the responsible exercise of that right, to protection of the privacy of individuals, to safeguards for the operational needs of land managers, and any necessary constraints for conservation needs. In supporting this, SNH believed that there was a compelling case for modernising the current arrangements.

Greater access is consistent with a number of public policies, such as those for greater social inclusion and equity, the improvement of health and education, and the achievement of more sustainable development. The absence of a clear duty on local authorities to promote access (as distinct from discretionary powers), together with the current pressures on their resources, have led to a low priority for access needs. A significant increase in resources is required for local authorities and other public bodies involved in open-air recreation and tourism, as well as for additional incentives to landowners to improve the provision and management of access over their land. A new countryside code of practice should be accompanied by a concerted effort to promote good behaviour and to improve visitor management. Government support for the proposals was announced in early 1999 (*SO Press Release*, 2 February 1999).

PROVISION FOR RECREATION

In the early postwar years, national recreation policy was largely concerned with national parks (and their Scottish shadow equivalents), areas of outstanding natural beauty, and the coast. Increasingly, however, there has developed a concern for positive policy in relation to metropolitan, regional, and country parks. The Countryside Act 1968 (which followed a White Paper with the significant title *Leisure in the Country-side*) gave additional powers to the new Countryside Commission for 'the provision and improvement of facilities for the enjoyment of the countryside', including experimental schemes to promote country-side enjoyment. At the same time, local authorities were empowered to provide *country parks*, including facilities for sailing, boating, bathing, and fishing. These country parks are not for those who are seeking the solitude and grandeur of the mountains, but for the large urban populations who are 'looking for a change of environment within easy reach'. There is now a wide range of country parks, picnic sites, visitor-interpretative sites, recreation paths, interpretative trails, cycleways, and similar facilities provided by local authorities and the Countryside Agency. There are over 250 country parks ranging in size from 11 to 1,875 hectares, and more than half of them attract at least 100,000 visits a year, totalling 57 million visits. Indeed, it has been suggested that the growth in this provision has led to, or at least affected, the decline of urban parks (which receive no similar funding).[18] Certainly the deterioration of the quality and attractiveness of urban parks has been rapid and disastrous. A select committee has proposed the establishment of an Urban Parks and Greenspaces Agency as an urban equivalent of the Countryside Agency, though there are wider issues here which the committee did not address.[19]

English Heritage maintains a Register of Parks and Gardens of Special Historic Interest in England. The main purpose of this is to draw attention to those which constitute an important part of the cultural heritage, and also to encourage and advise local authorities to provide adequate protection for these sites through the development control system.[20] More than 1,300 million day visits are made to the English countryside each year.[21] Recreation and tourism make a substantial contribution to the rural economy: total spending by all visitors was £9 billion in 1994; and total employment supported by visitor activity has been estimated at a third of a million jobs (RDC, *The Economic Impact of Recreation and Tourism in the English Countryside*, 1997).

For the most part, the impact on the environment is of manageable proportions,[22] though there are conflicts in specific areas, and traffic problems certainly can considerably reduce the 'quality of the recreational experience' (the extreme sometimes being reached in the Peak District, where on occasion a condition of gridlock is created by the sheer volume of traffic). Though there are some areas where recreational pressures have undesirable impacts, generally leisure and tourism are less of a threat than industrial and agricultural activities. However, as the Environment Committee pointed out in its 1995 report *The Environmental Impact of Leisure Activities*, there are difficult problems of crowding, overuse, and conflict of activities in certain areas. The favourite answer is 'good management', and there is no doubt that this can help in preventing visitors 'loving to death' the beauty spots they wish to visit (to use the apt phrase adapted as the title of a report on sustainable tourism in Europe by the Federation of Nature and National Parks). So can 'countryside codes', 'visitor awareness' campaigns, and suchlike. But more drastic measures are inevitable in the most popular locations: Dovedale attracts two million visitors a year, of whom 750,000 use the main footpath. On busy Sundays no fewer than two thousand people *an hour* can be crossing the river by the stepping stones. (The photograph on the cover of Jonathan Croall's 1995 book *Preserve or Destroy: Tourism and the Environment* is the most eloquent statement of the problem of which this is only one illustration.) In Northern Ireland there were indications that the 'peace dividend' was having an effect, with tourist pressures on attractive areas such as County Fermanagh.

Various schemes are being tried: from strict control of cars and the provision of public transport to the expansion of facilities in new areas. The Countryside Agency is placing increased importance on funding recreational facilities close to where people live, which has the additional advantage of helping 'to reduce the number of countryside trips made by car and provides opportunities for countryside recreation for people without cars, for the young, and for those with special needs'.

One particularly interesting initiative which started in the early 1980s and is now well established is the work of the *Groundwork Trusts*. Conceived as an *additional* resource for converting waste land to productive uses, particularly in urban fringe areas, it facilitates cooperative efforts by voluntary organisations and business, as well as public authorities. There is now a large network of trusts in the UK, involving 120 organisations in partnership. Their enterprise is wide-ranging and ranges from land reclamation, landscaping, environmental appreciation to provision for recreation, and many other activities seen as desirable and worthwhile in local communities. Examples include the development of the Taff Trail, which is a long-distance footpath and cycleway linking Cardiff and Brecon; re-creating wildlife sanctuaries and access around mining villages in east Durham; the development of the Middleton Riverside Park on a totally derelict site a few miles from Manchester; and a programme which encourages owners of industrial and commercial premises 'to stand back and take a look at the external image of their premises and then to make practical landscape improvements'. The scheme is clearly highly adaptable to local conditions and aspirations (Jones 1998, 1999).

Some activities are simply impossible to accommodate in popular locations: high-powered recreational vehicles and boats are not only environmentally damaging, but also destroy the pleasure that others are seeking. The Environment Committee in its 1995 report *The Environmental Impact of Leisure Activities* argued that 'the principle of sustainability in leisure and recreation involves the provision of facilities for all activities, not only for the aesthetically pleasing and non-intrusive ones'. This is an interpretation of the concept of 'sustainability' which would not be universally accepted; but there would be more support for the Committee's proposal that sites should be selected which are 'suitable for noisy and obtrusive activities'. Whether such sites could be readily found is more problematic, even on 'derelict land or land of low amenity'. An economist would point out that this is a classic case for a charging mechanism which would enable compensation to be paid to those adversely affected.

Much of the greater willingness to provide additional opportunities and facilities for recreation in the countryside has emanated from the changed economics of agriculture. Less agricultural production and more diversified activities are desirable. Indeed, without changes in the pattern of economic activity, many rural areas will be adversely affected by the changes in agriculture. The provision of recreational facilities is a potentially lucrative business, as has been dramatically illustrated in those areas where it has been realised.[23] A study of Center Parcs holiday villages demonstrates the substantial benefits which these had brought to the local areas: £4.5 million around Sherwood (Nottinghamshire) and £5.3 million around Elveden (Norfolk). The employment created might eventually amount to well over 1,000 jobs. Tourism also brings more indirect benefits to rural areas such as the survival of bus services and village shops. There is, of course, a cost to be borne for these advantages – in terms of changed character, and conflicts between the interests of visitors and residents (particularly in areas which have attracted new residents). Such conflicts can be reduced by 'good management', but they are inherent in the dynamics of social and economic change.

COUNTRYSIDE GRANT PROGRAMMES

Changes in policies relating to farming have had a more tangible effect than the heated arguments of contenders for and against freer countryside access. The lower priority for food production has led to

---

BOX 9.3 SOME AGRI-ENVIRONMENT SCHEMES

Countryside Access Scheme
Aims to increase the benefits from land which is set aside by offering incentives to farmers to increase public access on the best-located sites.

Countryside Stewardship Scheme
Aims to protect and restore targeted landscapes, their wildlife habitat, and historical features, and to improve opportunities for public access. Currently being developed as the government's main incentive scheme for the wider countryside.

Environmentally Sensitive Areas
Incentives are provided to farmers to safeguard the area and to improve public access in areas of particularly high value because of landscape,

wildlife, or history which are threatened by changes in farming practices.

Farm Woodland Premium Scheme
Grants are made towards cost of planting trees on agricultural land.

Habitat Scheme
Incentives are provided to create or improve wildlife habitats.

Moorland Scheme
Aims to protect or improve moorland by reduction of grazing.

Nitrate Sensitive Areas
For reduction of nitrate leaching.

Source: *MAFF Departmental Report 1999*

---

attempts to broaden the role of landowners as 'managers' of the countryside. Within this changing framework, 'access' becomes a means of diversifying the agricultural economy. Successive measures have reflected the changing priorities, and there has been increased emphasis on the role of farmers as 'stewards of the countryside' which has led to a greater concentration of funds on environmental schemes.[24]

The Countryside Commission has been in the lead in promoting conservation and recreation as explicit objectives of agricultural policy. Its 1989 policy statement *Incentives for a New Direction in Farming* argued that the diminishing need for agricultural production provided an opportunity for 'environmentally friendly' farming. It presented a menu of incentives for farmers and landowners to provide environmental and recreational benefits. These ideas were translated into the *countryside premium*, an experimental scheme which gave incentives for land to be set aside for recreation. It was followed by *countryside stewardship*, which provides incentives for the protection and enhancement of valued and threatened landscapes. This scheme proved to be a successful one: in its first four years,

some 5,000 agreements were concluded, covering 91,000 hectares. It has now been transferred (together with the hedgerow incentive scheme, which provides financial incentives for the restoration of hedgerows) to the Ministry of Agriculture, which operates other, similar land management schemes. The Countryside Commission for Wales has a parallel scheme (*Tir Cymen*). Other 'countryside' bodies also have schemes: the Forestry Commission operates a *woodland grant scheme* which includes payments for the management of woods to which the public have access; and under its *countryside access scheme* the Ministry of Agriculture makes payments for land which is 'set aside' for public access.

The 1986 Agriculture Act made provision for *environmentally sensitive areas* where annual grants are given by MAFF to enable farmers to follow farming practices which will achieve conservation objectives. The introduction of ESAs marked a fundamental policy change (Bishop and Phillips 1993: 325). They provide financial support for practices which result in environmental benefits, in contrast to earlier schemes which gave compensation for forgone profits. ESAs have developed into the main plank of the Ministry's countryside protection policy. There were nineteen ESAs in the original scheme introduced in 1987. By 1999 the number (in the UK) had increased to forty-three.

Under an EU Directive to prevent nitrate pollution in sensitive areas, MAFF administers two schemes (covering nitrate-sensitive areas and nitrate-vulnerable zones) which involve payments to farmers to compensate them for making changes to farming practices which reduce nitrate leaching, thus protecting public drinking water sources.

The Ministry of Agriculture also operates grant schemes for farm diversification and for farm woodlands. The former (now partly amalgamated with the EU-funded *farm and conservation grant scheme*) is mainly aimed at encouraging environmentally beneficial investments. The latter (the *farm woodland scheme*) began as an experiment in 1991, but the results were disappointing and a comprehensive review led to the introduction of a new *farm woodland premium scheme* in 1992. The objectives are to enhance the farmed landscape and environment and to encourage a productive land use alternative to agriculture.

New schemes are being introduced at a bewildering rate. In addition to those already mentioned, there are the *wildlife enhancement scheme*, *parish paths partnership*, and the Development Commission's *countryside employment programme* and *redundant building grant*. Indeed, it is sometimes difficult to be clear where one scheme ends and another begins. Box 9.3 lists most of the MAFF 'agri-environment' schemes 'designed to encourage environmentally friendly farming and public enjoyment of the countryside'. Nevertheless, it is apparent that a determined effort is being made to offset some of the effects of the stimulation to excess production and degradation of the countryside. The effectiveness of these various schemes is a matter of controversy (W. M. Adams 1996).

NATURE CONSERVATION

The concept of wildlife sanctuaries or nature reserves is one of long standing and, indeed, antedates the modern idea of national parks. In other countries some national parks are in fact primarily sanctuaries for the preservation of big game and other wildlife, as well as for the protection of outstanding physiological features and areas of outstanding geological interest. British national parks were somewhat different in origin, with an emphasis on the preservation of amenity and providing facilities for public access and enjoyment (though, as noted earlier, the trend has been to give increasing priority to conservation). The concept of nature conservation is primarily a scientific one concerned particularly with the management of natural sites and of vegetation and animal populations.

The Huxley Committee argued in 1947 that there was no fundamental conflict between these two areas of interest:

> their special requirements may differ, and the case for each may be presented with too limited a vision: but since both have the same fundamental idea of conserving the rich variety of our countryside and sea-coasts and of increasing the general enjoyment and understanding of nature, their ultimate objectives are not divergent, still less antagonistic.

However, to ensure that recreational, economic, and scientific interests are all fairly met presents some difficulties. Several reports dealing with the various problems were published shortly after the war (Dower Report 1945; Huxley Report 1947; Hobhouse Report 1947). The outcome was the establishment of the Nature Conservancy, which was later replaced by English Nature, Scottish Natural Heritage, and the Countryside Council for Wales.[25] In Northern Ireland, Scotland, and Wales, central responsibility for nature conservation and access to the countryside rests with a single body, but in England they remain separate.

Legislation often emerges as a response to new perceptions of problems; but sometimes legislation itself fosters such perceptions. So it was with the Wildlife and Countryside Act 1981. Introduced as a mild alternative to the Labour government's aborted Countryside Bill (and stimulated by the need to take action on several international conservation agreements), the Conservative government expected no serious trouble over the Bill. It was very mistaken: the Bill acted as a lightning rod for a host of countryside concerns that had been building up over the previous decade or so: moorland reclamation, afforestation and 'new agricultural landscapes', loss of hedgerows, damage to SSSIs, and suchlike. The Bill had a stormy passage through Parliament, with an incredible 2,300 proposed amendments. Though most of these failed, the Bill was considerably amended during the process. The major focus of argument (with the strong National Farmers Union and the Country Landowners Association holding the line against a large but diffuse environmental lobby) was the extent to which voluntary management agreements could be sufficient to resolve conflicts of interest in the countryside. The government steadfastly maintained that neither positive inducements nor negative controls were necessary. Indeed, it was held that controls would be counter-productive in that they would arouse intense opposition from country landowners.

Of particular concern was the rate at which SSSIs were being seriously damaged; the speed at which moorland in national parks was being converted to agricultural use or afforestation; and the adverse impact of agricultural capital grants schemes both on landscape and on the social and economic well-being of upland communities. On the first issue, the government finally made a concession and provided for a system of 'reciprocal notification'. This required the Nature Conservancy Council (NCC) to notify all landowners, the local planning authority and the Secretary of State of any land which, in their opinion, 'is of special interest by reason for any of its flora, fauna, or geological or physiological features', and 'any operations appearing to the [NCC] to be likely to damage the flora or fauna or those features'. Landowners were required to give three months' notice of intentions to carry out any operation listed in the SSSI notification. This was intended to provide the NCC with an opportunity 'to discuss modifications or the possibility of entering into a management agreement'. This much-vaunted 'voluntary principle' did not work: sites were damaged while consultations were under way. Amending legislation, passed in 1985, was designed to prevent this.

Another issue of contention arose when, during the debates, attempts to extend grants from 'agricultural business' to countryside conservation were defeated, and a host of amendments divided the Opposition and confused the issues. As passed, the amendments did little more than exhort the Minister of Agriculture, when considering grants in areas of special scientific interest, to provide advice on 'the conservation and enhancement of natural beauty and amenities of the countryside' and suchlike, 'free of charge'. There is, however, power to refuse an application for an agricultural grant on various 'countryside' grounds, but such a refusal renders the objecting authority (the county planning authority in national parks and the NCC in SSSIs) liable to pay compensation. This is a return to the pre-1947 planning system (even though it applies to only a small part of the country), and it has, not surprisingly, given rise to a considerable amount of debate.

William Waldegrave (when Minister for the Environment, Countryside and Local Government) has argued:

> I think that the moral and logical position of the farmer who finds that his particular bit of flora or fauna is now rare is such that there should be no hesitation in saying that he deserves public money if he is asked to do better than those who have been allowed to extinguish their bits, and if it is expensive for him to do so. I believe such flows of money, from taxpayer to land-user, for conservation expenses, are thoroughly justified and should become a useful and permanent adjunct to farm incomes for quite a considerable number of farmers, often in the rather more marginal farming areas where the inherent difficulty of farming has prevented our predecessors from extirpating species which may have gone for good elsewhere.
>
> (Waldegrave *et al.* 1986)

The issue here goes much further than appears at first sight since it raises questions about the ownership of development rights. Though the postwar planning legislation nationalised rights of development in land, it effectively excluded agriculture and forestry. The owner of a listed building receives no compensation for the restrictions which are imposed, and may even be charged for repairs deemed necessary by the local authority in default. The farmer, on the other hand, expects – and obtains – payment for 'profits forgone' in 'desisting from socially undesirable activity, or merely for departing from what is conventionally regarded as good agricultural practice'.

Thus, to quote Hodge (1999), there was 'the irony of one UK government agency being obliged to buy out the subsidies being offered by another'. Moreover, there was concern that some landowners threatened changes merely as a means of extracting compensation: a few very large payments had much publicity. Not surprisingly, the Labour government that came to power in 1997 objected in principle to this system and proposed that payments should be made to landowners only where this was in furtherance of conservation management.

> Public support for the proper management of SSSIs is essential and appropriate . . . but it should be given where positive management prescriptions are required . . . The Government is not prepared, in future, to pay out public money simply to dissuade operations which could destroy or damage these national assets.[26]

The main element in the new policy approach is more effective protection and better management. The Countryside and Rights of Way Act 2000 imposes on public bodies a statutory duty 'to secure the positive management' of SSSIs; the powers of conservation agencies have been strengthened; and the courts have been given powers in relation to penalties for damaging an SSSI. However, experience shows that good management is more generally achieved by negotiation and agreement than by legislative fiat (though this can be helpful as a reserve power). A good recent example is the working partnership that has been agreed with farmers in Bowland, where there has been damage to SSSIs through overgrazing. Overgrazing is encouraged by the CAP, which pays farmers according to the numbers of livestock on the land. Though this policy maximises production, it does so at the expense of wildlife. The problem can be met by reducing grazing and channelling agricultural support in the uplands to make hill farming both economically viable and environmentally efficient. The Bowland initiative involves a partnership between government departments and agencies, local authorities, voluntary conservation organisations, and farmers. This partnership includes a complex funding package which enables farmers to address a number of conservation issues.[27]

BIODIVERSITY

Nature conservation has risen on the political agenda with world-wide concern for biodiversity. Though the terms are different, they mean very much the same thing: the variety of life on the earth. Official biodiversity policy was set out in *Biodiversity: The UK Action Plan* and *Sustainable Development: The UK Strategy*, both published in 1994.

Table 9.3 Protected Wildlife Areas in the UK, March 2000

	Number	Area (sq. km)
National nature reserves	383	2,198
Local nature reserves	718	435
Sites of special scientific interest (GB)	6,545	22,682
Areas of special scientific interest (NI)	177	866
Marine nature reserves	3	194
Areas protected by international agreement		
Special areas of conservation (SACs)	340	17,659
Special protection areas (SPAs)	216	9,748
Ramsar Wetland Sites	147	6,702

Source: Office for National Statistics *Britain 2001: Official Handbook* (p. 318)

Policies take two main forms: the protection of particular species of flora and fauna, and the designation and protection of conservation sites. It is the latter which are of particular relevance to this book. Statutory designations have multiplied in recent years, particularly with the impact of European directives. The indigenous designations include nature reserves, SSSIs, marine nature reserves, and sites of importance for nature conservation. Protection is provided by various obligations, agreements with owners and occupiers, and through acquisition by local authorities, English Nature, or other bodies.

There are 376 *national nature reserves* in the UK, and 689 *local nature reserves*.[28] The former are, by definition, sites of national importance. Nature reserves are not sanctuaries: they are preserves where the conditions provide 'special opportunities for the study of, and research into, matters relating to the fauna and flora'. Most of these are in private ownership subject to a management agreement, but voluntary organisations such as the Royal Society for the Protection of Birds and county wildlife trusts own and manage their own non-statutory reserves.

SSSIs number 6,461 and cover 21,788 sq. km (about 8 per cent of the land area of Britain). In these protected areas, occupiers must obtain permission before certain listed activities (known as 'potentially damaging operations' or PDOs) can be carried out; there is also stricter planning control. Draft new guidelines on protecting, managing, and conserving SSSIs in England were published in 2000 to minimise widespread damage.

The designation of *marine nature reserves* was introduced by the Wildlife and Countryside Act 1981. There are three statutory reserves: Lundy Island (off the coast of Devon), Skomer (Dyfed), and Strangford Lough (N. Ireland). There are also a number of non-statutory marine reserves which have been established by voluntary conservation groups.

As required by EC Directives on Conservation, special measures are to be taken to conserve certain habitats (see Table 9.3). These include a number of classifications: *special protection areas* (SPAs) under the Birds Directive and *special areas of conservation* (SACs) under the Habitats Directive. The European network

of these SPAs and SACs is known as the Natura 2000 Network. The number of SACs in England is set to rise from 147 to 228 following proposals announced in August 2000. SACs automatically become SSSIs subject to consultation. *Ramsar sites* are protected wetlands (so called after the town in which the convention was signed).

Provisions relating to *limestone pavement areas* (which is really a geological designation) were introduced by the 1981 Wildlife and Countryside Act. These cover no more than 2,000 hectares in England and Wales, but are 'of great natural beauty and scientific interest'. They are popular with gardeners looking for stone for rockeries: hence the need for protection.[29]

This bewildering (though incomplete) recital of conservation instruments suggests that the time may well be near when some rationalisation may be considered appropriate. However, they are testimony to a heightened regard for 'this common inheritance'. This has been evolving for some time. It was in 1968 that the Countryside Act provided that 'in the exercise of their functions relating to land under any enactment, every minister, government department and public body shall have regard to the desirability of conserving the natural beauty and amenity of the countryside'. Since 1986 there has been a statutory duty to balance the interests of agriculture with rural and environmental interests, and 'agri-environmental regulation' has entered the planning lexicon. A wide range of countryside initiatives are being introduced. Indeed, as the following pages illustrate, it is difficult to keep pace with the changing policies and programmes.

There are several aspects of conservation policy which are worth noting. The first is that it is not restricted to designated sites (despite their large number): in addition to these nationally (and internationally) important sites, there are many more which are of local importance. (In the words of PPG 9, 'these sites are important to local communities, often affording people the only opportunity of direct contact with nature, especially in urban areas'.) A second, related, point is that (since wildlife does not respect human-made boundaries) it is important to safeguard 'wildlife corridors, links or stepping stones from one habitat to another'. The Habitats Directive specifically requires member states of the EU

> to encourage the management of features of the landscape which . . . by virtue of their linear and continuous structure (such as rivers with their banks or the traditional systems for marking field boundaries) or their function as stepping stones (such as ponds or small woods), are essential for the migration, dispersal and genetic exchange of wild species.

What is particularly significant about this approach (which is only briefly illustrated here) is that it is mandatory on local authorities, and that in addition to land use designations actual *management* is involved. (The mandate comes from a combination of two legal instruments: the Planning and Compensation Act's requirement for plans to include policies in respect of 'the conservation of natural beauty and amenity of the land'; and the provision in the Conservation (Natural Habitats etc.) Regulations 1994 that these policies 'shall be taken to include policies encouraging the management of features of the landscape which are of major importance for wild flora and fauna'.

UK work on biodiversity is coordinated by the UK Biodiversity Group, which is chaired by the DETR. There are also country groups for each of the four constituent countries. In 1998, fifty-six Species Action Plans and ten Habitat Action Plans were published (*DETR Annual Report 1999*).

FORESTRY

Forests cover some 2.5 million hectares in the UK: about 8 per cent of England, 16 per cent of Scotland, 12 per cent of Wales, and 6 per cent of Northern Ireland.[30] There has been a steady increase in the forest area; during the 1980s the increase was of some 300,000 hectares. About two-fifths of productive forestry is managed by Forest Enterprise, the development and management arm of the Forestry Commission.

A substantial reorientation of forestry policy has emerged over the past quarter of a century. In particular, there has been a major move away from a preoccupation with production, to a more balanced approach which places importance on amenity and environmental factors. This has come about after a lengthy period of debate and scrutiny. There was, for example, much argument both before and after the 1981 Countryside and Wildlife Act which centred on the effects of hill farming and forest policies. A succession of reports concluded that these policies, far from sustaining the economies and landscapes of the uplands of England and Wales, were major factors in their decline (MacEwen and MacEwen 1982; Sinclair 1992).

The predominant concern for production was badly affecting the vitality of rural communities and the conservation of the countryside. It was argued that much more employment could be created by coordinated policies sensitively directed to the problems of the uplands as a whole, rather than to separate aspects of them. This was essentially a call for 'integrated' policies, which began to emerge later. The immediate result, however, was a provision in the 1985 Wildlife and Countryside (Amendment) Act requiring the Forestry Commission to attempt a reasonable balance between the interests of forestry and of conservation and enhancement of the countryside.

Forestry policy in the UK has been increasingly influenced by international commitments such as the 1992 Rio Earth Summit, which led to a statement on forestry principles, the Rio Declaration, and Agenda 21, which provided an agenda for sustainable development.[31] A review of forestry policy in 1994 (*Our Forests: The Way Ahead*) proclaimed that 'UK forestry policy is based on the fundamental tenet that forest resources and forest lands should be sustainably managed to meet the social, economic, ecological, cultural and spiritual human needs of present and future generations'. In making changes to existing land use policies, however, there are constraints imposed by the Common Agricultural Policy, which is (as always?) under negotiation.[32]

In addition to the broader economic implications of forestry policy for rural areas, there are several issues which are of particular relevance to countryside policy: amenity, wildlife, access, and recreation. Problems arise because, though forest production is essentially a very long-term enterprise, there is need, in the words of the 1986 National Audit Office *Review of Forestry Commission Objectives and Achievements*, 'to have regard

to a number of broadly drawn secondary objectives
[which] can produce conflicts with and constraints
upon the Commission's primary aim of increasing the
supply of timber'. One of these secondary objectives is
the preservation (and enhancement) of the landscape
and of the wildlife it sustains.

Concern for such wider issues has increased as
environmental awareness has grown. Current policy
is one of 'multiple-purpose forestry'. This embraces a
wide range of approaches. Planting of broadleaves is
being expanded. Access to forests is being extended,
with improved arrangements for access agreements.
New *national forests* are being established in the
Midlands and in central Scotland.[33] A new national
forest park (in addition to the sixteen previously
established by the Forestry Commission) has been
opened at Gwydir. A *Community Forests* programme
(operated jointly by the Forestry Commission and the
Countryside Agency, with local authority support) is
under way which aims 'to create attractive green
settings (rich in wildlife and easily accessible) for the
enjoyment and health of residents, and to encourage
economic regeneration'.[34] Additionally there is a
Community Woodlands scheme which promotes new
woodlands near centres of population (with assistance
from the *Woodlands Grant Scheme*).

In 1999 a wide-ranging review was published under
the title of *A New Focus for England's Woodlands*. This
describes the two main aims of forestry policy in
England as being the sustainable management of existing woods and forests, and a continued steady expansion of the woodland area to provide more benefits
for society and for the environment.[35] The strategy
has four components: rural development; economic
regeneration; recreation, access, and tourism; and the
environment and conservation.

The rural development policy area lays stress on
the creation of a higher proportion of well-designed
larger woodland planting. Larger woodlands are
usually capable of providing more public benefits
than is possible with smaller woods, though the
latter 'can make a significant contribution of local
biodiversity, amenity, environmental health and
sustainable development'. However, any major change
from agricultural land to woodland is dependent on a

reform of the Common Agricultural Policy, and thus
the scale, location, and timescale are uncertain.

Economic regeneration involves promoting forestry
in the restoration of former industrial land (the amount
of which in England is some 175,000 hectares). The
Forestry Commission's Land Regeneration Unit,
which was established in 1997, has demonstrated the
potential, as has the National Urban Forestry Unit,
which works in partnership with local authorities, the
private sector, and non-governmental organisations.[36]
There are also the community forests and woodlands
schemes mentioned above, as well as the Woodlands
by the Motorways project (partly sponsored by the
Highways Agency and Esso).

The recreational aspects of forestry are well documented: some 3 million visits a year are made to the
woods and forests of England. The strategy includes
better information about the opportunities, improving
facilities, and developing the Forestry Commission's
Woodland Park network. There are many positive
environmental benefits of forestry including the
improvement of air quality and the promotion of
biodiversity. Among the actions on this front are better
management of woods and forests, more research on
the environmental benefits of forestry, and the promotion of greater appreciation of the value of trees,
woodlands, and forests.

Much of this, of course, is not new, but the
document presents a very positive approach to the
expansion and management of forestry.

INTEGRATED COUNTRYSIDE PLANNING

The proliferation of initiatives in the countryside
underlines the need for some integration of policies.
A remarkable indication of what this might involve
is provided by Scottish Natural Heritage (which, it
will be remembered, combines responsibilities for
conservation, amenity, and recreation). In *An Agenda
for Investment in Scotland's Natural Heritage* (1992), SNH
pointed out that each economic activity related to the
countryside is dealt with independently: agriculture,
fisheries, forestry, mineral extraction, recreation and

tourism, country sports, rural industries. Yet 'all of these activities are based on use, in one way or another, of the natural heritage: *the natural heritage is the common resource*'. It is also a declining resource, since many of these uses contribute to 'a draw-down of Scotland's natural capital'. The deterioration is substantial and 'calls into question the capacity of the natural heritage to sustain the range of uses to which it is subjected'. All this (and more) clearly indicates the need for an integrated approach to the rural environment.

Encouragingly, the Scottish Office followed the SNH lead with a *Rural Framework* strategy, which unequivocally states that 'tackling rural issues in a sectoral manner does not work'. The keynote of the preferred approach is partnership. An early example is to be found in the Cairngorms and the Trossachs, where joint machinery is being established to deal in a comprehensive way with the complex problems of these famous areas.

This theme was taken up in the three rural White Papers *Rural England: A Nation Committed to a Living Countryside* (1995), *Rural Scotland: People, Prosperity and Partnership* (1995), and *A Working Countryside for Wales* (1996). These were preceded by wide public discussion and input, and they stress that there is 'a widespread recognition that we should design policies to meet our varied objectives for the countryside in an integrated way, and that local responses to rural needs are often the most appropriate'. A common theme is the importance of cooperation between various agencies and community organisations. The Scottish paper outlines a rural partnership process, on which details are to be worked out. It is easy to regard much of this as pious hopes which lack substance. Both White Papers contain more than a fair share of motherhood statements, but they also point to a more grass-roots involvement in social and economic development which is untypical of British administrative style. It could prove to be more sensitive and effective than more ambitious top-down plans. Experience with urban planning, however, warns that good intentions and a belief in the efficacy of fine-sounding ideas such as coordination, integration, partnership, and public participation may not be enough.

However, in Scotland the importance of participation has assumed great significance following devolution:

> The process of developing a new approach to policy-making will involve engaging with the policy-making community across Scotland and reaching out beyond those already well represented in established channels of power, to include smaller organisations at national, regional and local level.
>
> (Hassan 1999: 14)

This appears to be widely accepted, as the discussion of Scottish local government in Chapter 3 shows. The latest policy statement, *Towards a Development Strategy for Rural Scotland* (1998), reiterates the philosophy:

> The consultation exercise revealed the need to develop more refined typologies for rural Scotland and indicators to measure progress of sustainable development. . . . The responses identified four key principles which should underlie the over-arching objective of sustainable rural development. They are that rural development must:
>
> • not set rural Scotland apart
> • reflect the diversity of rural Scotland
> • work through an integrated approach
> • facilitate community involvement.

BOX 9.4 THE SCOTTISH 'RURAL FRAMEWORK'

Tackling rural issues in a sectoral manner does not work. For example, if the pattern of agriculture changes in the light of the Common Agricultural Policy of the European Community, then the support it currently gives to rural communities will also change. But new employment for the agricultural community will come from expanding other traditional activities like tourism or developing new markets, and an approach confined to an agricultural viewpoint may not exploit all the opportunities which become available. Those new opportunities lie within the responsibilities of many departments and agencies, the local authorities, and the private and voluntary sectors.

Source: Scottish Office (1992) *Rural Framework*

Getting the national context right is vital for sustainable development. But, in view of the responses which emphasised the diversity of rural Scotland, it is clearly not at the national level that development strategies responsive to local needs can be best formulated and delivered. Nor can delivery at national level meet that other tenet of the discussion paper: the proposition that rural development should be driven by the priorities of local people to a much larger extent than in the past.[37]

FURTHER READING

Note: With the merger of the Rural Development Commission and the Countryside Commission, all their publications are now obtainable from the new Countryside Agency. The address for publications is Countryside Agency Sales, PO Box 124, Walgrave, Northampton NN6 9TL. A free catalogue of publications is available from this address and at <www.countryside.gov.uk/>.

General

Important books dealing with countryside issues include Champion and Watkins (1991) *People in the Countryside: Studies of Social Change in Rural Britain*; Bunce (1994) *The Countryside Ideal: Anglo-American Images of Landscape*; Cherry (1994a) *Rural Change and Planning: England and Wales in the Twentieth Century*; Cloke *et al* (1994) *Lifestyles in Rural England*; and Newby (1985) *Green and Pleasant Land: Social Change in Rural England*. There are many popular books bemoaning the fate of the countryside: a highly readable and informative one, by an agricultural journalist, is Harvey (1997) *The Killing of the Countryside*; in this he makes reference to Edith Holden's once famous *The Country Diary of an Edwardian Lady* (facsimile 1977), which in the later 1970s became a best-seller, particularly among those nostalgic for a half-forgotten countryside. An up-to-date review of countryside planning is Hodge (1999) 'Countryside planning: from urban containment to sustainable development'. A detailed critical examination of countryside policies is Winter (1996) *Rural Politics: Policies for Agriculture, Forestry and the Environment*. See also the essays edited by Allanson and Whitby (1996) *The Rural Economy and the British Countryside*, and Adams (1996) *Future Nature: A Vision for Conservation*. The new Countryside Agency's approach to its role is set out in two important statements: *Planning for Quality of Life in Rural England: The Interim Planning Policy of the Countryside Agency* and *Tomorrow's Countryside: 2020 Vision* (1999). Detailed accounts of change can be found in Gilg's five-volume series *Progress in Rural Policy and Planning* (1991–95), *Countryside Planning: The First Half Century* (1996), and other titles by the same author.

The White Paper *Our Countryside: The Future: A Fair Deal for Rural England* is the main official source for policy. Three rural White Papers published previously by the Conservative government are *Rural Scotland: People, Prosperity and Partnership* (1995); *Rural England: A Nation Committed to a Living Countryside* (1995); and *A Working Countryside for Wales* (1996). See also PPG 7 *The Countryside: Environmental Quality and Economic and Social Development*: Scottish Office (1992) *Rural Framework*; Scottish Office (1998) *Towards a Development Strategy for Rural Scotland*; and Scottish Office (1999) *Rural Development* (NPPG 15). For Wales, see Welsh Assembly (1999) *Rural Wales: A Statement by the Rural Partnership*. A short account of countryside issues in Northern Ireland is given by Lipman (1999) 'Difficult decisions in a rural balancing act'.

A good statement of the problems of the rural economy and of government policies in relation to this is set out in DETR (1998) *Guidance to the Regional Development Agencies on Rural Policy*. There are numerous studies by the former Rural Development Commission: *Disadvantage in Rural Areas* (1997), *Rural Disadvantage: Understanding the Processes* (1998a), *Rural Development and Land Use Planning Policies* (1998b), and *Household Growth in Rural Areas* (1998c). See also A. Rogers (1999) *The Most Revolutionary Measure: A History of the Rural Development Commission*.

On rural housing issues, see Rural Development Commission (1998) *A Home in the Country? Affordable Housing in Rural England*; Cloke (1996) 'Housing in the open countryside'; and Scottish Office (1998) *Investing in Quality: Improving the Design of New Housing in the Scottish Countryside: A Consultation Paper*.

The damage done to the countryside by modern methods of farming is analysed in Shoard (1980) *The Theft of the Countryside*. 'Interpretations of sustainable agriculture in the UK' by Cobb *et al.* (1999) demonstrates how elusive the concept is, and the large gaps in scientific knowledge on the basic issues involved. See also O'Riordan and Voisey (1997) 'The political economy of sustainable development' and Owens, S. (1994a) 'Land limits and sustainability: a conceptual framework and some dilemmas for the planning system'.

Landscapes

Illustrated, measured studies of landscape change are provided by the Westmacott and Worthington studies, of which the latest (1997) is *Agricultural Landscapes: A Third Look*. Rackham's *Illustrated History of the Countryside* (1997) is a highly enjoyable coffee-table type book which presents a very clear picture of aspects of landscape change. Two Scottish Office publications are *Natural Heritage Designations in Scotland* (Scottish Office 1998) and *National Scenic Areas: A Consultation Paper* (SNH 1998).

National Parks, Access to the Countryside and Public Rights of Way

For accounts of the background to and the implementation of the 1949 Act, see Cherry (1975) *National Parks and Recreation in the Countryside* (volume 2 of *Environmental Planning 1939–1969*) and Blunden and Curry (1989) *A People's Charter? Forty Years of the National Parks and Access to the Countryside Act 1949*. Two major reviews of national parks policies are the Sandford Report (1974) and the Edwards Report (1991). A passionate critique of the restrictions on access (and much else) is given by Shoard (1987) *This Land is Our Land* and (1999) *A Right to Roam*.

Official publications on access to the countryside proliferated in 1998 and 1999: *Access to the Open Countryside: Consultation Paper* (1998); *Options on Access* (1999); and finally *Improving Rights of Way in England and Wales* (1999) and *Access to Open Countryside of England and Wales: The Government's Framework for Action*. Sandwiched between all these DETR publications was a report commissioned from the Countryside Commission, *Rights of Way in the 21st Century* (1998). See also the Rights of Way Act Fact Sheets at <www.wildlife-countryside.detr.gov.uk/cl/bill/factsheet/index.htm>. For a foreign comparison, see Peter Scott Planning Services (1998) Access to the Countryside in Selected European Countries: A Review of *Access Rights, Legislation and Associated Arrangements in Denmark, Germany, Norway and Sweden*. For Scotland, the most important document is *Access to the Countryside for Open-air Recreation* (SHN, 1998).

Coastal Issues

On coastal issues, see DoE (1995) *Policy Guidelines for the Coast*; PPG 20 (1992) *Coastal Planning*; HC Environment Committee (1992) *Coastal Zone Protection and Planning*; SODD *Coastal Planning* (NPPG 13); and Cleator (1995) *Review of Legislation Relating to the Coastal and Marine Environment of Scotland*. Two EC reports were issued in 1999: *Lessons from the European Commission's Demonstration Programme on Integrated Coastal Zone Management* and *Towards a European Integrated Coastal Management Strategy: General Principles and Policy Options*; for details, see *Planning* 27 August 1999: 13.

Waterways

The relationship between planning and waterways is dealt with in BWB (1993) *The Waterway Environment and Development Plans*. More recent are British Waterways (1999) *Our Plan for the Future 2000–2004* and DETR (1999) *Unlocking the Potential: A New Future for British Waterways*.

Recreation and Tourism

Recreation in the countryside is discussed at length in HC Environment Committee (1995) *The Environmental Impact of Leisure Activities*. For a review of the impact of different recreation activities on the environment, see Sidaway (1994) *Recreation and the Natural Heritage*. A 1995 report *Sustainable Rural Tourism* by the

Countryside Commission and others is devoted to exploring the concept referred to in its title, and the ways in which it can be translated into practice. See also PPG 21 *Tourism* (1992); Segal Quince Wicksteed (1996) *The Impact of Tourism on Rural Settlements*; Curry (1997) 'Enhancing countryside recreation benefits through the rights of way system in England and Wales'; and Bell (1997) *Design for Outdoor Recreation*. A recent brief statement is Countryside Commission (1999) *Countryside Recreation: Enjoying the Living Countryside* (1999). Further references on tourism are given in Chapter 8.

Nature Conservation and Biodiversity

Current official guidance on nature conservation is given in PPG 9. Additionally, see NPPG 14 *Natural Heritage*; DoENI PPS *Planning and Nature Conservation*; and Ratcliffe (1994) *Conservation in Europe: Will Britain Make the Grade?*. Two publications of the Countryside Council for Wales are *The Welsh Landscape: A Policy Document* (1996) and *Protecting Our Natural Heritage: A Guide to the Designated Sites and Landscapes of Wales* (1997). Despite the special controls, many SSSIs have been damaged: see, for example, two reports of the National Audit Office: *Protecting and Managing Sites of Special Scientific Interest in England* (1994) and *Protecting Environmentally Sensitive Areas* (1997). See also *SSSIs: Better Protection and Management: The Government's Framework for Action* (1999). On local nature reserves, see Barker and Box (1998) 'Statutory local nature reserves in the United Kingdom'.

On the implementation of biodiversity policies, see *Biodiversity: The UK Action Plan* (Cm 2428, 1994); *Sustainable Development: The UK Strategy* (Cm 2426, 1994); Biodiversity Group and Local Government Management Board (1997) *Guidance for Local Biodiversity Action Plans*; Scottish Biodiversity Group (1998) *Biodiversity in Scotland: The Way Forward* and (also 1998) *Local Biodiversity Action Plans: A Manual*. A useful overview is given in RTPI (1999b) *Planning for Biodiversity*.

On the issue of compensation for 'profits forgone' by farmers, see Jenkins (1990) *Future Harvest*. See also Hodge (1989) 'Compensation for nature conservation' and Hodge (1991) 'Incentive policies and the rural environment'; both these articles give useful references to other literature.

Issues relating to soil are not discussed in this chapter: useful references are Royal Commission on Environmental Pollution (1996) *Sustainable Use of Soil*; Gordon (1994) *Scotland's Soils: Research Issues in Developing a Soil Sustainability Strategy*; and Moore (1995) *Soil Diversity: A Literature Review*.

Forestry and Woodlands

Our Forests: The Way Ahead (1994) presents the conclusions from the Forestry Review. See also HC Environment Committee report *Forestry and the Environment* (1993) and the *Government Response* (1993); and HC Welsh Affairs Committee report *Forestry and Woodlands* (1994), and the *Government Response* (1994). There was also the major policy statement issued after the Rio Summit: *Sustainable Forestry: The UK Programme* (1994). See also NAO (1986) *Review of Forestry Commission Objectives and Achievements*; DoE (1992) *Indicative Forestry Strategies* (DoE Circular 29/92); and DETR *A New Focus on England's Forests*. For Scotland, see SODD (1999) *Indicative Forest Strategies* (Circular 9/1999).

Integrated Countryside Planning

The fullest development of the concept of integrated countryside planning comes from Scotland, starting with SNH (1992) *An Agenda for Investment in Scotland's Natural Heritage* and Scottish Office (1992) *Rural Framework* strategy. All three of the rural White Papers *Rural England* (1995), *Rural Scotland* (1995), and *A Working Countryside for Wales* (1996) further promoted the ideas in each of the three countries. Among later statements are *Towards a Development Strategy for Rural Scotland: The Framework* (1998) and the Welsh Rural Partnership statement *Rural Wales* (1999). On Local Agenda 21, see Bell and Evans (1998) 'The national forest and Local Agenda 21: an experiment in integrated landscape planning', and Selman (1998) 'Local Agenda 21: substance or spin?'. See also Countryside Agency (1999) *Planning for Quality of Life in Rural*

England: The Interim Planning Policy of the Countryside Agency.

NOTES

1 This and following figures relate to England. They are from the Countryside Agency's *The State of the Countryside 1999*. See also Countryside Commission (1997) *Disadvantage in Rural Areas*; (1998) *Rural Disadvantage: Understanding the Processes; Developing Indicators of Rural Disadvantage*; and Rural Development Commission (1997) *Survey of Rural Services*.

2 'The combination of low income, high development costs, falling grant rates, restrictions on planning and funding has resulted in much new rural social housing being unaffordable even to working households' (Rural Development Commission (1998) *A Home in the Country? Affordable Housing Needs in Rural England*).

3 The establishment of the Countryside Agency emerged out of two initiatives of the 1997 Labour government. The first was the setting up of the Regional Development Agencies which took over the rural regeneration functions of the RDC, thereby raising the issue of what was to be done with its national policy advisory work and its countryside activities. The second was the Comprehensive Spending Review of countryside and rural policy, which involved the DETR and MAFF. A number of options were considered and rejected, such as the merger of the Countryside Commission, English Nature, the RDC, and the Farming and Conservation Agency of MAFF. See HC Environment, Transport and Regional Affairs Committee, Environment Subcommittee, Fourth Report 1998–99 (HC 6), pp. 18 (evidence of Michael Meacher) and 42 (evidence of Dr Neil Ward).

4 Both the DETR and Welsh Assembly are undertaking research on the planning system and rural diversification which will be published during 2001. This follows previous research sponsored by the Planning Officers' Society (Shorten and Daniels 2000).

5 Countryside Commission (1991) *Fit for the Future: Report of the National Parks Review Panel* (Edwards Report); *Report of the National Parks Policies Review Committee* (1974).

6 In 1999 the government announced new steps to improve measures for conserving and enhancing AONBs, together with proposals to create new national parks in the New Forest and the South Downs. These are currently under discussion.

7 This is the 1997 version of PPG 7 (para 4.16). The earlier, 1992, version simply makes reference to 'locally devised' designations, and notes that, though they have no statutory role, they 'nevertheless serve to highlight particularly important features of the countryside that should be taken into account in planning decisions'.

8 See the list of criteria set out in CPRE (1999b) *Hedging Your Bets*, p. 4. See also DoE/MAFF (1997) *The Hedgerow Regulations 1997: A Guide to the Law and Good Practice*.

9 Some members of the Group favoured extending statutory protection to all traditional field boundaries, and possibly other countryside features such as ponds. However, it was decided that the work needed to assess the problems, and the benefits and costs of regulation, was beyond the scope of the review. On this, see HC Environment Committee report *Field Boundaries* (HCP 969, 1997–98).

10 There is a useful guide *National Heritage Designations in Scotland*, published by the Scottish Office in 1998. The SNH has published twenty-nine reports entitled *Landscape Character Assessment of Scotland*. It has also built up a national landscape character database (SNH *Annual Report 1997/98*, p. 2).

11 Research reviews were commissioned on national parks structures and powers in the UK and elsewhere; the socio-economic impacts of national parks in the UK and elsewhere; best practice in community participation; and the current statutory powers which may be relevant for national parks in Scotland. Free summaries and the full reports are available from SNH, Battleby, Redgorton, Perth PH1 3EW.

12   The Natural Heritage (Scotland) Act 1991, which provided for the establishment of Scottish Natural Heritage, also introduced *natural heritage areas*. It was intended that they would be designated for a wide range of situations in both upland and lowland Scotland where there is both a landscape and a nature conservation interest, and where there is therefore a need for integrated management. However, since 'the Government now supports the introduction of national parks, it is unlikely that any NHAs will be designated' (NPPG 14, *Natural Heritage*, 1998). However, the SNH is developing a Natural Heritage Zones Programme which will identify twenty-one zones 'reflecting the diversity of Scotland's unique natural heritage'. For each zone a statement is being prepared reviewing trends, opportunities, and pressures. Finally, in conjunction with partner organisations, future action will be determined' (SNH *Annual Report 1997/98*, p. 1).

13   A 1998 SNH consultation paper *National Scenic Areas* proposes that local authorities should have a much stronger role through a new duty to prepare and implement a management strategy for the protection of the scenic qualities of each NSA.

14   As part of the Neptune Coastline Campaign, an assessment was undertaken by Reading University of changes to developed and undeveloped coastline. This showed that the length of coastline of high landscape quality in each region had decreased by from 5 to 9 per cent between 1965 and 1995. However, a few counties had increased lengths of high landscape coastline, probably arising from the demise of industrial activities (Burgon 2000).

15   *Lessons from the European Commission's Demonstration Programme on Integrated Coastal Zone Management* (Luxembourg: OOPEC) and *Towards a European Integrated Coastal Zone Management (ICZM) Strategy: General Principles and Policy Options* (Luxembourg: OOPEC).

16   Dorset County Council has pioneered a planning and management coastal strategy which has been commended by the EU, and is expected to lead to a proposal for World Heritage Site status. See *Planning*, 5 November 1999: 6. Information can be obtained from the Dorset Coast Forum at the Dorset CC web site <www.dorset-cc.gov.uk>. For general information about the coastal zone, see the web site <www.theukcoastalzone.com>.

17   See the DETR's Countryside and Rights of Way Act 2000: Fact Sheets at <www.wildlife-countryside.detr.gov.uk/cl/bill/factsheet/index.htm>.

18   See the discussion in the HC Environment Subcommittee on *Town and Country Parks* (1999: xi). There is no proof for this, but it is pointed out that the deterioration of urban parks coincided with the establishment of funded countryside parks. Of course, there was also an accompanying growth in car ownership and thus increased opportunities for countryside visits. The Heritage Lottery Fund has an Urban Parks Programme which has been very popular: by April 1999 it had received 462 applications and award £117 million in grants (ibid. vol. 2: 137).

19   In evidence to the committee, Ken Worpole argued against a dedicated parks agency: 'my worry is if there is a specialist parks agency . . . it will be dominated by professionals who often have vested interests. . . . My version of the good park is very much about popular use' (p. xxxvii, para 170). See also Greenhalgh and Worpole (1995), Greenhalgh *et al.* (1996), and Worpole and Greenhalgh (1999).

20   Further details are given in the HC Environment Subcommittee report *Town and Country Parks* (1999 vol. 2: 50–3).

21   Countryside Commission (1999) *Countryside Recreation: Enjoying the Living Countryside*. The various figures quoted relate to 1996.

22   'Leisure and tourism do not cause significant widespread ecological damage to the countryside' – HC Environment Committee (1995) *The Environmental Impact of Leisure Activities*.

23   See various publications of the Rural Development Commission, e.g. *The Economic Impact of Holiday Villages* (1991) and *Rural Economic Activity* (1995).

24 These schemes have a positive economic benefit, even though it is not easily measured. See Hanley *et al.* (1999).

25 There is also a body called the Joint Nature Conservation Committee (JNCC), which is a statutory committee of English Nature, Scottish Natural Heritage, and the Countryside Council for Wales and is responsible for advice and research on national and international nature conservation matters.

26 DETR (1999) *Sites of Special Scientific Interest: Better Protection and Management*, p. 4. See also the consultation document with the same title (DETR 1998).

27 The issues are the loss of herb-rich hay meadows and rushy grazing pastures which are important to wading birds. The project also promotes the natural regeneration of woodlands by fencing to prevent sheep grazing as well as improving heather regeneration (English Nature press release *Bowland Farmers Take Action to Work for Wildlife* 2 July 1999).

28 Different sources give slightly different figures: those quoted are taken from ONS (1999) *Britain 2000*, p. 323.

29 Limestone Pavement Orders are designated for the most valuable parts of limestone pavements. There is, however, still an illegal trade on these pavements. The Limestone Pavement Working Group is examining the problems. See HC Select Committee of the Environment, Transport and Regional Affairs Committee, 15th Report (1998–99), p. xii.

30 ONS (1999) *Britain 2000*, p. 463.

31 This led to the production of *Sustainable Forestry: The UK Programme* (Cm 2429, 1994). More recently, the social and cultural importance of forestry was given prominence by the Lisbon declaration *Pan European Ministerial Conference on the Protection of Forests in Europe: Declaration and Resolutions L1 and L2* (Lisbon, 1998).

32 Reform is a long and difficult process. The EU's proposals of 1998 are a continuation of reforms to move support for commodity production towards direct payments to farmers, particularly for environmental benefits. The *Agenda 2000* proposals include a new Rural Development Regulation which will seek a better integration of environmental forestry and rural support measures.

33 The English 'national forest' is being planted by the National Forest Company over some twenty years in about 520 sq. km of Derbyshire, Leicestershire, and Staffordshire. It is funded by DETR. By 1999 nearly three million trees had been planted. A brief account is given in *Whitaker's Almanack 2000*: 577.

34 See Countryside Commission (1996) *Community Forests in the Town and Country Planning System*: the quotation is from the DETR *Annual Report 1999*.

35 That there is widespread support for a significant increase in woodland cover across England was shown by the response to *Woodland Creation: Needs and Opportunities in the English Countryside* (Countryside Commission and Forestry Commission 1996). The *Responses* were published in 1997.

36 See the memorandum by the National Urban Forestry Unit in HC Environment Subcommittee Report *Town and Country Parks*, vol. 2: *Memoranda*: 151–3.

37 Scottish Office (1998) *Towards a Development Strategy for Rural Scotland: The Framework*, paras 30 and 60. For an analysis of the diversity of rural areas in Scotland, see Williams *et al.* (1998).

10

URBAN POLICIES

The problem is that for too long urban policy has acted as a 'filler in of gaps', mopping up the worst cases of fallout produced by wider economic and policy changes. It has functioned as both a form of symbolism and crisis management.

(Atkinson 1999: 84)

INTRODUCTION

The breadth of public intervention embraced by the term 'urban policies' denies any simple summary: indeed, the enduring feature of urban policy has been the endless experimentation with new and often disconnected initiatives. What is consistent is the fragmentation of effort, lack of a strategy, weak involvement of local communities; the marginal impacts (in contrast to mainstream public spending and private investment); the heavy bias to property development and economic development; and the persistent criticism of the inadequacies of urban policy. Since the early 1990s there has been a welcome increase in the influence of local government and local communities, though spending remains relatively low and is falling. The evaluation of urban policies that underpins these critical comments is taken up at the end of the chapter. Here, two preliminary points need to be made.

First, the title may be somewhat misleading in that a number of the policies discussed extend to rural areas (housing for example) or may have non-spatial dimensions (e.g. economic development). Nevertheless, the focus of the discussion is on urban areas in general and inner cities in particular. Second, the title is in the plural since there is no such thing as a single urban policy or a set of policies; rather, urban policy

has been described as 'incoherent' and not deserving of the appellation 'policy' (Atkinson 1999: 84). Moreover, it is not a simple matter to define what are, and what are not, urban policies. National economic, welfare, and housing policies may play a more significant role than urban aid or urban regeneration policies. Nevertheless, there are a group of policies which are officially labelled urban (or, increasingly, 'regeneration') and have sufficient urban identity to justify discussing them together.

The chapter opens with a discussion of inadequate housing: the starting point of urban policy in the nineteenth century, and the major concern of policy until at least the end of the 1960s. The focus then moved from physical conditions to the social aspects of housing and, later, to areas of social need. A further shift took place in the late 1970s, when economic issues were seen as being the key to urban regeneration. By the mid-1980s this had become the conventional wisdom, with an accent on large-scale property development undertaken in partnership with the private sector. By the early 1990s the value of these large projects was increasingly questioned. The past decade has seen attempts to create a more coordinated and consistent set of programmes, a greater emphasis on partnership, and some recognition of the need to invest in people as well as places.

INADEQUATE HOUSING

Britain has a very large legacy of old housing which is inadequate by modern standards. This results from the relatively early start of the industrial revolution in Britain and the rapid, unplanned, and speculative urban development which took place in the nineteenth century. (The contrast with, for example, the Scandinavian countries, whose industrial revolutions came later when wealth was greater and standards higher, is marked.) As a result, British policies in relation to clearance and redevelopment are of long standing, though it was the Greenwood Housing Act of 1930 which heralded the start of the modern slum clearance programme. Over a third of a million houses were demolished before the Second World War brought the programme to an abrupt halt.

By 1938, demolitions had reached the rate of 90,000 a year: had it not been for the Second World War, over a million older houses would (at this rate) have been demolished by 1951. The war, however, not only delayed clearance programmes, but resulted in enforced neglect and deterioration. War damage, shortage of building resources, and (of increasing importance in the period of postwar inflation) crude rent restriction policies increased the problem of old and inadequate housing.

It was not until the mid-1950s that clearance could generally be resumed, and well over two million slum houses have been demolished since then. But the problem is still one of large dimensions. The *English House Condition Survey*[1] carried out in 1996 showed a total of more than 1.5 million dwellings (7.5 per cent of the stock) which were defined as unfit for human habitation.[2] In addition, 3.8 million were in need of urgent repairs (80 per cent of all houses are in need of some repair). Little progress was made during the 1980s and 1990s in improving the overall quality of the housing stock, despite the building of some 650,000 new homes between 1991 and 1996. More than 14 per cent of households live in poor housing, with much higher proportions for ethnic minority, young, unemployed, and lone-parent households.[3] There has been a fall in the number of dwellings lacking basic amenities; but there has been a small increase in the number of dwellings below the fitness standard or in serious disrepair. The number of unfit dwellings made fit between 1991 and 1996 (about 500,000) is equalled by the number of fit dwellings becoming unfit. There is a link here with the marked decline in slum clearance since the mid-1970s. This continued into the 1990s, and in 1996–97 the number of houses demolished or closed in Britain was around 10,000 compared with 54,000 in 1974–75, and 90,000 in 1969.

PRIVATE-SECTOR HOMES: FROM CLEARANCE TO RENEWAL

Since the 1950s, wholesale clearance of housing has given way to 'renewal'. The emphasis gradually shifted from individual house improvements, first to the improvement of streets or areas of sub-standard housing, and later to the improvement of the total environment.

Initially, it was assumed that houses could be neatly divided into two groups: according to the 1953 White Paper *Houses: The Next Step*, there were those which were unfit for human habitation and those which were essentially sound. As experience was gained, the improvement philosophy broadened, and it came to be realised that there was a very wide range of housing situations related not only to the presence or otherwise of plumbing facilities and the state of repair of individual houses, but also to location, the varying socio-economic character of different neighbourhoods, and the nature of the local housing market. A house lacking amenities in Chelsea was, in important ways, different from an identical house in Rochdale: the appropriate action was similarly different. Later, it was better understood that appropriate action defined in housing market terms was not necessarily equally appropriate in social terms. A middle-class invasion might restore the physical fabric and raise the quality and character of a neighbourhood, but the social costs of this were borne largely by displaced low-income families. The problem thus became redefined.

Growing concern for the environment also led to an increased awareness of the importance of the factors causing deterioration. It became clear that these are more numerous and complex than housing legislation had recognised. Through traffic and inadequate parking provision were quickly recognised as being of physical importance. The answer, in appropriately physical terms, was the re-routing of traffic, the closure of streets, and the provision of parking spaces (together with floorscape treatments and the planting of trees). Most difficult of all is to assess the social function of an area, the needs it meets, and the ways in which conditions can be improved for (and in accordance with the wishes of) the inhabitants.

For a considerable time, this issue of the social function of areas was largely ignored, although a strong shift towards improvement rather than clearance was heralded by the 1969 Housing Act. This increased grants for improvement, and introduced *general improvement areas* (GIAs), which were envisaged as being areas of between 300 and 800 'fundamentally sound houses capable of providing good living conditions for many years to come and unlikely to be affected by known redevelopment or major planning proposals. The enhanced grants and the GIAs made a significant contribution to the reduction in the number of unfit properties, though in some areas gentrification unexpectedly took place, reducing the amount of privately rented housing and affecting the existing communities (Wood 1991: 52).

The 1974 Housing Act made a major reorientation of policy and brought social considerations to the fore. There was a new emphasis on comprehensive area-based strategies implementing a policy of 'gradual renewal'. The powers (and duties) conferred by the Act focused upon areas of particular housing stress. Local housing authorities were required to consider the need for dealing with these as *housing action areas* (HAAs). Though these were conceived in terms of housing conditions, particular importance was attached to 'the concentration in the area of households likely to have special housing problems – for instance, old-age pensioners, large families, single-parent families, of families whose head is unemployed or in a low income group' (DoE Circular 13/75).

The intention was that intense activity in HAAs would significantly improve housing conditions and the well-being of the communities within a period of about five years. In the event, HAA designation lasted much longer in many cities. Various additional powers were made available to local authorities within housing action areas, for compulsory purchase, renewal, and environmental improvement; and grant aid for renewal was targeted to these areas.

HOUSING RENEWAL AREAS

The area-based approach to private-sector housing renewal was retained, albeit substantially altered, by the Local Government and Housing Act 1989.[4] In addition to individual income-related house renovation grants, the Act introduced *renewal areas* (RAs), which replaced GIAs and HAAs. There are also powers for local authority support for *group repair schemes* to renovate the exteriors of blocks of houses.

The thrust of the changes reflects a concern for a broader strategic approach including economic and social regeneration as well as housing renewal over a longer ten-year period of designation. It involved the resumption of clearance; the use of partnerships to bring together the initiatives of local authorities, housing associations, property owners, and residents; and a system of grants which were mainly both mandatory and income related.

A *neighbourhood renewal assessment* is a central part of the renewal area concept, and has to precede designation. This is, in effect, a plan-making and implementation programme combined, including an assessment of conditions, an estimate of the resources available, and the selection of the preferred options. The procedure goes much further than the typical land use planning process to incorporate a cost–benefit analysis of different alternative policies over a thirty-year period, including the qualitative social and environmental implications. Consultation is a requirement both during and after the declaration process, with twenty-eight days given for responses to an explanatory summary of the proposals which must be delivered to every address in the area. An interesting requirement is that all who make representations which are not accepted must be provided with a written explanation.

Renewal areas are larger than the former HAAs and GIA, with at least 300 properties as a minimum. More than 75 per cent of the dwellings must be privately owned, and at least 75 per cent must be considered unfit or qualify for grants. At least 30 per cent of the households must be in receipt of specified state benefits, thus ensuring that 'a significant proportion of residents in an RA should not be able to afford the cost of the works to their properties'.

By 1999 more than a hundred neighbourhood renewal assessments had been undertaken and renewal areas designated. C. Wood (1996) provides an overview of the programme, and a further evaluation was published by the DoE in 1997 (Austin Mayhead & Co.).[5] The average size of RA was 1,526, or about twice the size of GIAs and HAAs. Wood notes that this is slower progress than might have been expected and points to the variation in enthusiasm across the country. He provides three case studies of RA designation in Birmingham, a city that at that time had 47,000 unfit privately owned dwellings and another 80,000 considered borderline. The city council has estimated that the total bill for the repair and improvement of the private housing stock in the city would be £750 million. It is perhaps not surprising therefore that Wood concludes that 6,400 properties designated in RAs in Birmingham will remain in the same condition as they started (or have further deteriorated) at the end of the ten-year plan period.

The difference between the scale of the problem and funding available was one reason why authorities were not generally enthusiastic about the RA concept, as the 1997 evaluation points out. This was more than just a negative perception. The evaluation considered ten case study RAs in detail and only two had made a significant impact on the full range of objectives, especially in gaining private-sector investment. Nevertheless, the conclusion was that the neighbourhood renewal assessment process is sound, and the partnership elements in particular had an impact on generating successful solutions. Subsequently, the DETR published *Running and Sustaining Renewal Areas: Good Practice Guide* (1999). This puts more emphasis on the partnership dimension to renewal areas, the need for sustainability assessment, and forward strategies to

ensure that improvement is sustained after the designation ends.

Dwellings which fail to meet the statutory fitness standard qualified for mandatory grants (subject to a means test) until 1996. Discretionary grants were available for minor works such as adapting facilities to enable the elderly to remain in their homes, or to adapt houses for the use of people with disabilities. Since the renovation grants scheme was (originally) largely mandatory and therefore demand led, some local authorities experienced growing financial difficulties. Wood (1994) quotes a senior officer in Birmingham: 'In 1993 the city council was faced with a backlog of 8,000 applications for mandatory grants from private owners which, if paid, would have used up its entire renewal budget for many years to come.' This problem was one of a number which were dealt with in a 1993 consultation paper, *The Future of Private Housing Renewal Programmes*. Changes were proposed to target resources on the worst housing; to focus grant-aid on the poorest households; to give local authorities more flexible powers; and to address the problems of the most vulnerable groups. These proposals found general support, and most were enacted in the Housing, Grants, Construction and Regeneration Act 1996. This replaced parts of the 1989 Act, and grants for repairs to housing are now given at the discretion of the local authority. The rationale for this is that it enables expenditure to match the local strategy, while allowing priority to be given to those who are most in need of financial assistance. The Act also provides for grants to be made in excess of the normal maxima for altering property for the needs of people with disabilities, and also for relocation grants to enable those displaced by clearance to buy a house in the same local area.

At the same time as local authorities were being given more discretion, new requirements were introduced to promote a more strategic approach to private housing renewal, linking the area-based approaches with the individual renovation grants, vacancy measures, energy efficiency, and the potential for private investment. Subsequent evaluation of the strategies indicated considerable variation in approaches, with more than half of them only meeting the basic requirements or less, although there was also evidence of fundamental changes in attitudes towards working with the private sector.[6]

PUBLIC-SECTOR HOMES: ESTATE ACTION

In 1979 the DoE set up a *priority estates project* to explore ways in which problem council estates could be improved. The problems of these estates varied, but all had become neglected and run-down; some had been vandalised. A 1981 report, *Priority Estates Project 1981: Improving Problem Council Estates*, considered three experiments in the improvement of such estates, and concluded that the task of improvement involved a great deal more than mere physical renovation: social and economic problems needed to be addressed at the same time. In 1985 the DoE established an *Urban Renewal Unit* (now called *Estate Action*) to encourage and assist local authorities to develop a range of measures to revitalise run-down estates. These measures include transfers of ownership and/or management to tenants' cooperatives or *management trusts* involving tenants; sales of tenanted estates to private trusts or developers; and sales of empty property to developers for refurbishment for sale or rent.

Estate Action funding is allocated on a competitive basis. A wide range of factors are taken into account: the provisions for tenant participation; the devolution of management to tenants; increases in 'tenant mix', and right-to-buy activity; and the role of the private sector.

A research report evaluating six early Estate Action Schemes was carried out by Capita Management Consultancy (1996). In addition to the general objective of improving the quality of life of residents living in deteriorated local authority estates, there were several specific policy objectives: to bring empty properties quickly back into use; to improve residential quality; to reduce levels of crime and 'incivility'; to encourage participative estate-based management; to diversify tenure; and to attract private investment. Despite a large expenditure amounting (for the whole programme) to nearly £2 billion over eight years, the reported results for the estates studied were dis-

appointing. Indeed, the report makes very depressing reading, even though its conclusions are expressed in the guarded language which often characterises such government-sponsored research reports. A particular shortcoming was a lack of success in involving residents in the development and implementation of the schemes. More generally, the report demonstrates the difficulty of using physical plans for achieving social and economic goals. Estate Action is now winding down, with the last schemes approved in 1995.[7]

HOUSING ACTION TRUSTS

The 1987 White Paper *Housing: The Government's Proposals* announced the creation of *housing action trusts* (HATs) to tackle the management and renewal of badly run-down housing estates. HATs are the housing equivalent of the urban development corporations, but they can be introduced only with the consent of a majority of the tenants. Like the urban development corporations, HATs are non-departmental public bodies responsible directly to the Secretary of State, thus usurping the powers of the local authority. But they also bring substantial additional funding for improving the physical conditions of estates by renovations and improvements to the houses and the environment. Their objectives are to secure more diverse tenure (essentially by involving private and voluntary housing agencies) and to improve economic and social conditions.

Although funding was allocated for HATs as early as 1988, there was considerable delay in getting the first ones started because of fierce opposition by affected local authorities and tenants. Rao (1990) has documented two areas in Lambeth and Sunderland that were initially identified for designation. The dilemma facing local authorities is that HATs involve a loss of control to the private sector, but are a potential source of very substantial extra funding. Tenants fear increased rents and reduced availability of housing if it is sold to private owners after improvement.

Following the embarrassing rejection of HAT designation in the first ballots, much greater care was taken with subsequent proposals to involve tenants in early discussions. By 1999 six had been designated and had received £566 million up to 1998–99, together with substantial private finance through the Private Finance Initiative. That funding provided 2,322 new or renovated homes and training for more than 8,000 residents.

The first phase of an evaluation of the six HATs (Capita Management Consultancy 1997) reports that they present 'typical pictures of run down estates: mostly postwar system built, high-rise blocks with complex interconnections and open space of indeterminate purpose . . . design faults, inadequate maintenance, and misuse by some tenants'. It also notes that 'an estate's reputation could blight tenants' day to day lives'. The problems of negative stereotyping and consequent low demand for housing have become the current focus for research and policy for problem estates (Niner 1999). Low demand resulting in high vacancies and rapid turnover affects all parts of the country and all tenures but is particularly acute in council estates in the North, where local authorities are demolishing homes that are impossible to rent or sell. Low demand arising from unpopularity is more important than need in determining the use of housing (Power and Mumford 1999; Burrows and Rhodes 1998). How an estate gets into this position is explained more by severe poverty and unemployment than the quality of housing, although the original break-up of stable communities through the slum clearance programme may be the origin of the problem.

The 2000 Green Paper *Quality and Choice: A Decent Home for All* promises more of the same in housing policy, with extra money for repairs and modernisation; more local authority stock transfers to social landlords; and a requirement for better linkage between housing and planning policies.

SCOTTISH HOUSING

Scottish housing is different from that south of the Border in significant ways. There is a high proportion of tenement properties; dwellings tend to be smaller; rents are lower; and a higher proportion of the housing stock is owned by public authorities. These and other differences reflect history, economic growth and decline, local building materials, and climate. Above all, Scotland has for long faced a major problem of poor-quality tenement housing. Despite the large amount of clearance in the postwar years, there still remains much poor-quality housing in both the private and the public sectors.

The *1991 Scottish House Condition Survey*, published in 1992, suggests that because of the dominance and relatively young age of much of the public housing stock, a high proportion of housing has the basic amenities, but the condition is often poor, and requires high levels of expenditure.

Scottish housing is distinctive in important ways. First, there is the scale and character of public housing. The low council rents of the past contributed to the relatively low demand for private housing, which, coupled with massive public housebuilding programmes, gave Scotland the highest proportion of public-sector housing in western Europe – and made Glasgow City Council the largest public-sector landlord (McCrone 1991). (It should be noted that the Scots use the term 'house' in the English sense of 'dwelling', i.e. it embraces a flat or tenement.) The right to buy has shifted the balance somewhat between the owner-occupied and public housing sectors: owner occupation rose from 35 per cent in 1979 to 57 per cent in 1995, while public-sector renting fell from 54 per cent to 30 per cent. Nevertheless, in spite of the sale of nearly a quarter of the public housing stock, the level of public renting in Scotland is still much higher than in England and Wales.

The Conservative government of 1979–97 believed that the large-scale public ownership of housing was at the root of much of the Scottish housing problem. Its strong desire to reduce the public housing sector (and particularly to break up the public ownership of the large peripheral estates) was an important background issue in Scottish housing policy for many years.

Equally distinctive has been the institutional context of Scottish housing policy. The relationship between the Scottish Office and Scottish local authorities has always been much closer than the corresponding relationship in England, and there has been much easier coordination. This has resulted in what Carley (1990a: 51) describes as 'a much more clearly defined and integrated housing-neighbourhood renewal policy, which covers housing and planning issues together'. The Scottish Office has tended to work with local authorities (which have a single local authority association), rather than exerting central control through such mechanisms as UDCs and HATs. Instead, there have been centrally sponsored bodies which have worked in cooperation with local government. The main housing body is Scottish Homes. Established by the Housing (Scotland) Act 1988, this incorporates the former Scottish Special Housing Association, which dates from 1937, and the Housing Corporation in Scotland.

As in England and Wales during the 1980s, policy in relation to the older private housing stock placed more emphasis on rehabilitation than on clearance. But with much of the older tenement properties the scope for improvement is severely restricted by the decayed fabric of the buildings, their internal layout, and the high cost of alteration, as well as the practical problems of multiple ownership. Some of these difficulties have been met by the use of powers of compulsory improvement of a whole tenement structure, and by the establishment of *ad hoc* housing associations. However, the Scottish legislation has long provided for more flexibility than the English. *Housing action areas* (which have not been superseded by renewal areas) are of three types: for demolition, for improvement, and for a combination of the two. This enables the most appropriate action to be taken according to area conditions.

Scottish practice has emphasized the involvement of residents in housing renewal. There is a requirement for a two-month period of consultation with local residents before an HAA is declared, and small community-based housing associations are established to lead renewal. Given the distinctive Scottish style of implementing HAAs, there has not been the need for the new legislation equivalent to renewal areas in England and Wales, and they have continued in use.

THE EMPHASIS ON AREA POLICIES

Several major elements can be identified in the development of thinking on deprived areas in Britain:

inadequate physical conditions; the perception of 'large' numbers of 'immigrants' (many of whom were in fact born in Britain); educational disadvantage; and a multiplicity of less easily measurable social problems.

For a very long time there was a preoccupation with inadequate physical conditions (particularly in relation to plumbing). Indeed, British housing policy developed from sanitary policy (Bowley 1945), and it still remains a significant feature of it. Area policy in relation to housing was almost entirely restricted to slum clearance until the late 1960s, when concepts of housing improvement widened, first to the improvement of areas of housing and then to environmental improvement. Despite a number of social surveys, the policy was unashamedly physical – so much so that increasing powers were provided to *compel* reluctant owners and tenants to have improvements carried out. Not until the 1970s was attention focused on the social character and function of areas of old housing.

A further area of policy in the field of housing focused on areas of housing stress. The Milner Holland Committee (1965) looked favourably on the idea of designating the worst areas as *areas of special control* in which there would be wide powers to control sales and lettings, to acquire, demolish, and rebuild property, to require and undertake improvements, and to make grants. The National Committee for Commonwealth Immigrants argued for the designation of *areas of special housing need* (1967). Again, the crucial issues were the control of overcrowding, exorbitant rents, insanitary conditions, disrepair, and the risk of fire. These proposals were not accepted by the government, though increased powers to control multi-occupation and abuses were provided. It was left to the Plowden, Seebohm, and Skeffington reports to probe more deeply into this area.

The Plowden Committee was appointed in 1963, to consider primary education *in all its aspects*. It reported in very broad terms, but underlined the complex web of factors which produced seriously disadvantaged areas (1967). It recommended a national policy of positive discrimination, the aim of which would be to make schools in the most deprived areas as good as the best in the country.

The Seebohm Committee had even wider terms

of reference: to review personal social services in England and Wales, and to consider what changes were desirable to secure an effective family service. Of relevance to the present discussion are the Committee's recommendations in relation to *areas of special need*. Unfortunately, its report (1968) did not suggest how these should be identified, in spite of a recommendation that the areas should be accorded priority in the allocation of resources.

More helpful was its reference to citizen participation, which underlined a point hardly recognised by the Skeffington Committee even though it was specifically concerned with it. It was Seebohm, not Skeffington, who clearly saw that if area action was to be based on the wishes of the inhabitants and carried out with their participation, 'the participants may wish to pursue policies directly at variance with the ideas of the local authorities. . . . Participation provides a means by which further consumer control may be exercised over professional and bureaucratic power'.

This is an issue which will be discussed in a broader context in the final chapter. Here we briefly survey some of the ways in which the development of thinking on deprived areas has been translated into policy from the early 1970s to the Blair government.

THE URBAN PROGRAMME

Area policies in relation to housing improvement, however inadequate they may have been, were based on long experience of dealing with slum clearance and redevelopment. With other area policies there was no such base upon which to build, and both legislation and practice were hesitant and experimental. The approach, however, has remained consistently a spatial one, focusing on particular cities and areas within cities.

The *educational priority areas programme* was established in 1966 (Halsey 1972), and the *urban aid programme* in 1972. The *urban aid programme* (later recast as the *urban programme*) funded mainly social schemes, but it was progressively widened in scope to embrace voluntary organisations, and to cover industrial, environmental, and recreational provision.

The 1977 White Paper *Policy for the Inner Cities* and the Inner Urban Areas Act 1978 brought about significant changes. The new policy was 'to give additional powers to local authorities with severe inner area problems so that they may participate more effectively in the economic development of their areas'

It is difficult to give a coherent account of this programme since its objectives were never clearly spelled out, and its extreme flexibility gave rise to a great deal of confusion (McBride 1973). With the development of inner-city partnerships and programme authorities, the position became even more confused. By 1990 the number of individual programmes had swollen to thirty-four (NAO, *Regenerating the Inner Cities*, 1990).

The *urban programme* was the major plank of the 'deprived area policy' stage. Additionally there were the *community development projects*. These produced a veritable spate of publications ranging from carefully researched analyses to neo-Marxist denunciations of the basic structural weaknesses of capitalist society, though the original aim was 'to overcome the sense of disintegration and depersonalisation felt by residents of deprived areas'.[8]

The urban programme continued, with an increasing emphasis on economic development. It became a valuable source of funding for the many thousands of projects and organisations that have been supported. At its height, about 10,000 projects were funded each year in the fifty-seven programme areas, costing £236 million in 1992–93. In its later years, almost half the expenditure was devoted to economic objectives, and the rest was shared roughly equally between social and environmental objectives. Urban programme funding was largely taken over by the Single Regeneration Budget operated from the Government Offices for the Regions from 1995.

POLICY FOR THE INNER CITIES AND ACTION FOR CITIES

The increasing emphasis on economic regeneration objectives came to dominate urban policy during the 1980s and 1990s. The 1977 White Paper *Policy for the*

Inner Cities spoke of local authorities needing 'to stimulate investment by the private sector, by firms and by individuals, in industry, in commerce, and in housing'. The return of a Conservative government led to a review of inner-city policy which concluded that a much greater emphasis needed to be placed on the potential contribution of the private sector.

Ten years later the rhetoric was much the same. After the 1987 election Margaret Thatcher, then Prime Minister, announced her intention to 'do something about those inner cities'. The immediate result was the publication of a glossy brochure entitled *Action for Cities* (Cabinet Office 1988). This maintained that though the UK had benefited during the 1980s by embracing the ethic of enterprise, this change in attitude had not reached into the inner city. The aim of urban policy therefore was to establish 'a permanent climate of enterprise in the inner cities, led by industry and commerce'. *Action for Cities* programmes were claimed to have involved central government expenditure of £3 billion in 1988–99 and £4 billion in 1990–91, but in fact little additional money was involved. Critics argued that the package merely gave the appearance that a determined effort was being made to get to grips with the problems. However debatable this might be (Lawless 1989: 155), it did indicate some reorientation of thinking on urban policy.

URBAN DEVELOPMENT CORPORATIONS

The 1977 White Paper on inner cities considered the idea of using new-town-style development corporations to tackle inner areas, but the then Labour government concluded that this was inappropriate for the inner cities, and that local government should be the prime agency of regeneration. The Conservative government from 1979 thought differently, mainly because it had little faith in the capabilities of local government. The manifest argument, however, was that the regeneration of major areas of Britain's cities was 'in the national interest, effectively defining a broader community who would benefit from the regeneration' (Oc and Tiesdell 1991: 313). The effect of imposing centrally directed agencies into the hearts of major cities created considerable conflict (although not all the local authorities were against the designations) and long-lasting bad feeling about the role of central government in urban regeneration.

The 1980 Local Government, Planning and Land Act made the necessary legislative provisions, and defined the role of an urban development corporation (UDC) as being

> to secure the regeneration of its area . . . by bringing land and buildings into effective use, encouraging the development of existing and new industry and commerce, creating an attractive environment, and ensuring that housing and social facilities are available to encourage people to live and work in the area.

Though their structure and powers were based on the experience of the new town development corporations, the UDCs were different in several important respects. Their task was a limited one and they had relatively short lives, often years or so. The first were designated in 1981 and the last in 1993. (All had wound up operations by 1998.) The designation procedure was rapid, and did not provide for a public inquiry: instead, their areas were designated by statutory instrument.

Fourteen UDCs were created, twelve in England, one in Northern Ireland (Laganside), and one in Wales (Cardiff Bay).[9] (Except for the London Docklands, their areas were not large, though they suffered especially severe derelict land or plant closure problems (Bovaird 1992). The UDCs had extraordinary powers of land acquisition and vesting of public-sector land. They also (unlike the new town development corporations) usurped the local authority's development control functions (except in Wales) for determining planning applications (including their own proposals), enforcement, and other matters. In short, they had very wide planning responsibilities and freedom from local authority controls. This was not accidental or incidental: it was an essential feature of their conception. Furthermore, they were run by boards of directors drawn primarily from business, and they were accountable only to the central government.

Expenditure on UDCs rose to a peak of over £600 million in 1990–91, though in most years stood

Table 10.1 Selected Regeneration and Inner-City Expenditure and Plans 1987–88 to 2001–02

	1987–88	1988–89	1989–90	1990–91	1991–92	1992–93	1993–94	1994–95	1995–96	1996–97	1997–98	1998–99	1999–2000	2000–01	2001–02
UDCs and DLR	160.2	255	476.7	607.2	601.8	515	341.2	287.1	217.4	196.1	168.8	0	0.2		
Estates Action	75	140	190	180	267.5	348	357.4	372.6	315.9	251.6	173.5	95.7	66.9	63.9	39.4
Housing action trusts					10.1	26.5	78.1	92	92.5	89.7	88.3	90.2	86.4	88.4	88.4
City Challenge						72.6	240	209.0	204.9	207.2	142	9.8	1.7		
Challenge Fund									125	265.1	483.1	0.2			
New Deal for Communities													48.5	120.7	450.0
City Grant/Derelict Land Grant[a]	103.5	95.7	73.5	100.6	113.8	145.0	128.7								
English Partnerships (URA)[b]			16.8	–16.4	6.5	24.2	191.7	211.1	224.0	258.8	294.2	225.5	212.2	78.2	
Urban Programme[c]	245.7	224.3	222.7	225.8	237.5	236.2	166.5	67.8							
City Action Teams			4	7.7	8.4	4.6	3.4	0.2							
Inner City Task Forces	5.2	22.9	19.9	20.9	20.5	23.6	18	15.4	11.9	8.7	6.0	1.6	1.2		
Single Regeneration Budget[d]									136.4	277.5	458.8	560.9	181.4		
Regional development agencies[e]												12.2	593.0	628.5	820.0
London Development Agency														241.8	228.5
Manchester Regeneration Fund/Olympic bid/city centre					0.8	12.2	26.8	30.2	2	0.3	1.1	5.4	1.7		
Coalfield Areas Fund							2.3	2	0.4	0	0	0	10.0	15.0	10.0

Sources: DoE Annual Report 1993 (figure 45); DoE Annual Report 1995 (figure 43); DoE Annual Report 1996 (figure 31, p. 50); and DETR Annual Report 2000 (table 10a).

Notes:

a   City Grant and Derelict Land Grant became part of English Partnerships funding from November 1993 and April 1994 respectively.

b   Regional funding for English Partnerships transferred to the regional development agencies from April 2000.

c   Urban Programme funding became Single Regeneration Budget from 1995–96.

d   Single Regeneration Budget transferred to regional development agencies from April 1999; and to the London Development Agency from April 2000.

e   Regional development agencies funding also includes rural development programmes funding from April 1999.

around £200 million. At one time this was by far the largest share of spending on regeneration: in 1992–93 it amounted to a half of all inner-city spending, reflecting the significant shift away from local authority-directed expenditure. The major share of public funding, and similarly high levels of private investment, went to London Docklands, and within that area to major projects. Just one project, the Limehouse Link road, required the demolition of over 450 dwellings and cost £150,000 per metre, making it the most expensive stretch of road in the country (Brownill 1990: 139).

Final evaluation of the urban development corporations confirms that property development, land reclamation, and physical improvement dominated their objectives and outputs. They made a significant impact on the geography of Britain's cities through the transformation of largely derelict and degraded land created primarily through massive industrial restructuring. Powers of land acquisition, site assembly, reclamation, and financial incentives were important tools. But they have not met any wider objectives. Furthermore, their attention rarely strayed from the designated areas to considerations about the wider economies in which they were situated – a problem exacerbated by narrow performance indicators. Most UDCs could quote impressive figures (Roger Tym & Partners 1998), but it has been argued that most of the new activity that was generated would otherwise have occurred elsewhere. Overall, the most important legacy of the UDCs is perhaps the lesson that sustained regeneration can be achieved only by the combined efforts of all agencies and stakeholders working to the same objectives.

INNER-CITY INITIATIVES

Following Michael Heseltine's 1981 visit to Merseyside in the wake of the Toxteth riots, the Merseyside Task Force Initiative was created. Initially this was a task force of officials from the DoE and the then Department of Industry and Employment, established to work with local government and the private sector to find ways of strengthening the economy and improving the environment in Merseyside. This proved exceptionally difficult, partly because of the multiplicity of agencies involved.

As a result, an attempt was made in later initiatives to obtain a greater degree of coordination through City Action Teams and Inner City Task Forces. First set up in 1985, City Action Teams (CATs) were to take a broader, even regional, view of the coordination of government programmes. Each team was chaired by the regional director of one of the main departments involved: the DoE, the DTI, and the Training Agency.[10] Their funding was limited, reflecting their role as coordinators rather than direct providers. Lawless argues that they were 'unable to devise anything that might be termed a corporate central-government strategy towards inner-city areas' (1989: 61). They were disbanded at the time of the setting up of the Government Offices for the Regions, which took on the coordinating role.

A total of sixteen Inner City Task Forces were established in 1986–87. Their role was essentially one of trying to bend existing programmes and private-sector investment and priorities to the inner city. They were initially expected to have a life of two years, but in the event their lifespan has been variable. Their general objective was to increase the effectiveness of central government programmes in meeting the needs of the local communities.

One of Michael Heseltine's early initiatives was the establishment of the *Financial Institutions Group* under his leadership and staffed by twenty-five secondees from the private sector. Their most important proposal was for an *urban development grant* (UDG) on the lines of the US *urban development action grant*. This was introduced in 1982 'to promote the economic and physical regeneration of inner urban areas by levering private sector investment into such areas'. It was flexible in terms of the area covered. Local authorities contributed 25 per cent of grant-aid, and the private-sector contribution to a project had to be significant. It was replaced by the *urban regeneration grant* in 1986. In its lifetime it supported 296 projects at a cost of £136 million with a corresponding private-sector investment of £555 million. This represented a leverage ratio of about 1:4.

Table 10.2 Urban Development Corporations in England: Designation, Expenditure, and Outputs

			Annual gross expenditure (£m)		Cumulative outputs		
			1992–93	1995–96	Jobs created	Land reclaimed (ha)	Private investment (£m)
First generation	London Docklands	1981	293.9	129.9	63,025	709	6,084
	Merseyside	1981	42.1	34	14,458	342	394
Second generation	Trafford Park	1987	61.3	29.8	16,197	142	915
	Black Country	1987	68.0	43.5	13,357	256	690
	Teesside	1987	34.5	52.8	7,682	356	837
	Tyne and Wear	1987	50.2	44.9	19,649	456	758
Third generation	Central Manchester	1988	20.5	15.5	4,909	33	345
	Leeds	1988	9.6	0	9,066	68	357
	Sheffield	1988	15.9	16.4	11,342	235	553
	Bristol	1989	20.4	8.7	4,250	56	200
Fourth generation	Birmingham Heartlands	1992	5	12.2	1,773	54	107
	Plymouth	1993	0	10.5	8	6	0
Total			621.4	398.2	165,716	2,713	11,240

Source: DoE *Annual Reports 1993, 1995* and *1996*

Notes: Leeds UDC was wound up on 31 March 1995. Bristol UDC was wound up on 31 December 1995. The figures for outputs are cumulative over the lifetime of the UDCs and include all activity with the urban development area, not just those of the UDC.

The urban regeneration grant supported ten schemes at a cost of £46.5 million, with private-sector investment of £208 million. Together they provided an estimated 31,966 jobs, 6,750 new homes, and 589 hectares of land brought back into use (Brunivels and Rodrigues 1989: 66). *City grant* was launched by the *Action for Cities* initiative in 1988 and replaced several existing grants, including the UDG. Its aims were similar but it was paid directly to the private-sector developers to support the provision of new or converted property for industrial, commercial, or housing development. The final evaluation of these grant regimes noted that they were successful in assisting private-sector investment but the focus on job creation in evaluation did not 'address the basic causes of poor regional or local growth' (Price Waterhouse 1993: 63).

ENGLISH PARTNERSHIPS

City grant no longer exists, but its objectives, together with those of derelict land grant and of English Estates, are now addressed by English Partnerships (the operating name of the Urban Regeneration Agency, a non-departmental public body). English Partnerships was launched in November 1993 with the objective of promoting the regeneration of areas of need through the reclamation or redevelopment of land and buildings; the reuse of vacant, derelict, and contaminated land; and the provision of floorspace for industry, commercial and leisure activities, and housing. Its main job is to fund the gap between the costs of undertaking development and the end value. In this it takes on the role of the previous land grants, but with a stronger remit to place individual projects within a strategic framework. It operated through six regional offices until April 1999 when its regional operations were separated out and transferred to the RDAs. From May 1999 the national operations combined with the Commission for New Towns and retained the name English Partnerships.

English Partnerships is funded through a ring-fenced element of the Single Regeneration Budget. An investment fund has been created which can be used with more flexibility than previous grant regimes to provide advice; to take a stake in joint ventures with the private sector and others; to provide loans or loan guarantees; and to support direct development. It is intended to contribute to wider strategies, and to work within the planning framework established by regional guidance and development plans. Thus, collaboration, especially with local authorities, is an essential feature of its work. The partnership element of English Partnerships has been taken seriously, not least because the agency can realise its regeneration objectives only by coordinating contributions from a variety of sources. Comprehensive information has been provided for potential partners on the qualities that English Partnerships want to see in project proposals, notably on design and the creation of mixed-use developments.

English Partnerships now concentrates on four objectives: developing its land assets (which originate mostly from the new towns programme); creating development partnerships (mostly with RDAs and local authorities); improving the environment through land renewal and development; and finding new sources of funding to match public resources. Many projects promote housing and commercial development in the new towns. Others include some very high-profile projects, notably the regeneration of the Greenwich peninsula (£180 million) with the Millennium Dome and the Millennium Village, an experimental development incorporating many energy-saving and lifestyle innovations. Less well known is the English Coalfields regeneration programme, which includes eighty-two sites and unenviably boasts one of the largest portfolios of contaminated land in Europe: 3,400 hectares or the equivalent space taken by some 95,000 houses, three times the annual total land reclamation undertaken by English Partnerships.[11]

In order to promote more brownfield development, English Partnerships is working with the DETR and the Improvement Agency to prepare the first part of the National Land Use Database, which will provide a detailed schedule of all land previously in urban use in England (Wrigley and Seaton 2000). The national functions of English Partnerships are not tied to designated areas, although some priorities are identified such as the areas receiving EU Regional Policy funding, the coalfield closure areas, inner cities (especially City Challenge areas), and other assisted areas.

An interim evaluation of English Partnerships (PA Consulting Group 1999) reports that it has made an effective contribution to land-based regeneration. Over the years it has become more positive in involving local communities in its work as well as the big players, and has created innovative ways of maximising the potential to claw back surpluses into the public purse where developments are successful. Additional functions have been taken on, especially in relation to the coordination of foreign inward investment at the national level. It scores less well on the environmental dimension, and there are strong recommendations about the need to consider more fully the environmental impacts of development which the RDAs will need to take on board. The strategic dimension to land reclamation has improved, but continued difficulties of coordination are recognised. The limited role of English Partnerships in social regeneration issues and other aspects of market failure outside the land question has also worked against strategic integration. This is one area that the RDAs, with their much wider remit to create a strategic focus for regeneration within the regions, should be able to improve upon.

CITY CHALLENGE

A major switch in regeneration funding mechanisms was announced in May 1991, in the form of *City Challenge*. This marked a significant change in policy, and one that later evaluations regarded as 'the most promising regeneration scheme so far attempted' (Russell *et al.* 1996). Thus it has provided the model for its successor: the Single Regeneration Budget Challenge Fund.

Though the emphasis on land and property development remained, City Challenge recognised that this should be more closely linked to the needs of local communities and the provision of opportunities for disadvantaged residents. It was also intended to encourage a long-term perspective on change, and to integrate the work of different programmes and agencies.

Only fifteen authorities were invited to bid for funds in the first round, eleven of which were selected to start on their five-year programmes in 1992.[12] Subsequently, a second round was opened to all urban programme authorities and a further sixteen five-year programmes were approved in 1993. Most City Challenge partnership programmes are in inner-city locations, but a few are on the urban fringe. The Dearn Valley is unique in being 52 sq. km in area, covering a number of smaller settlements in the South Yorkshire conurbation. City Challenge encouraged an integrated approach, with a focus on property development but cutting across a range of topic areas, including economic development, housing, training, environmental improvements, and social programmes relating to such matters as crime and equal opportunities. The competition prize was substantial: £7.5 million for each area for each of five years' funding which, together with levered-in funding, meant that partnerships have spent on average £240 million. The total central government expenditure amounted to £1.15 billion over eight years. This was not new money: City Challenge was a different approach to spending rather than an allocation of new funds – it was to be large scale, holistic, strategic, and based on partnership.

Successful authorities prepared an action programme setting out projects which were funded through existing programmes, with similar, if simplified, procedures for making and considering applications. The private sector was expected to play a significant role, and its involvement had to be demonstrated before projects were agreed.

Local communities were seen to be important partners, but their place in the management structures was variable. Sometimes they are represented at the policy-making level, while in others they are involved only as consultees or in the detailed implementation of projects. The great speed needed to respond to invitations to bid (only six weeks in the first round) worked against much meaningful community participation in the early stages (Bell 1993). Nevertheless, City Challenge gave impetus to bring together different sectors and the creation of new and positive relationships between them, shifting attitudes and mainstream investment (de Groot 1992).

City Challenge was an undoubted success and provided a model for subsequent regeneration efforts. Evaluations (Russell *et al.* 1996; Oatley and Lambert 1995; KPMG Consulting 1999a) have emphasised the value of partnership involving local communities and business: where there were low levels of participation, performance of the partnership was poor. Strategic thinking was also important in providing an overall plan for regeneration, linking specific projects and outcomes. It has also been relatively good value, with leverage ratios of challenge funding to private investment and to other public funding of 1:3.78 and 1:1.45. Needless to say, not all partnerships have been equally successful and there have been particular problems of project integration and sustaining improvement after the end of the programme. The problems of deprivation, educational opportunity, crime, and unemployment are deep-seated and, as the KPMG report argues, warrant sustained long-term action. The challenge has now been taken up by the Single Regeneration Budget.

THE SINGLE REGENERATION BUDGET

In response to criticisms of the fragmented nature of programmes for regeneration, and building on the competitive and partnership approach of City Challenge, came the Single Regeneration Budget (SRB). This was introduced in 1994 with the intention of promoting integrated economic, social, and physical regeneration through a more flexible funding mechanism. It is now administered by the Regional Development Agencies, though central government maintains considerable influence in that it both writes the rules and makes the final decisions on funding allocations. Bids from 'local regeneration partnerships' (reflecting 'the content of the bid and characteristics of the area or groups at which it is aimed') are required to address a mix of objectives.[13]

Four principles underpin its design and execution: the need for a strategic approach; partnership among the public, private, community, and voluntary sectors; competitive bidding for available funds; and payment by results. The SRB addresses criticisms of short-termism and narrow approaches directly. Schemes can include projects needing funding for up to seven years. Bids must be 'strategic' and holistic, and they must establish links to other investment plans such as those required for European Structural Funds and economic development strategies. However, there is no mention of the relationship between the SRB and development plans or regional guidance, which is a serious omission. Bids must consider the various dimensions of disadvantage simultaneously rather than focusing on particular facets of disadvantage (as has been the case with earlier programmes). The 1999–2000 guidance identified five strategic objectives which are listed in Box 10.1 (and linked them to the *National Strategy for Neighbourhood Renewal* and the findings from the Policy Action Teams that were established to contribute to the creation of the strategy, discussed on p. 306).

The SRB brought together twenty previously separate funding programmes. The value of the funds incorporated into the SRB was £1.6 billion in 1993–94 but it fell to £1.36 billion in 1995–96. Funding has increased under the Labour administration, but nevertheless the planned expenditure in 2001–02 is £1.7 billion, or in real terms a 30 per cent fall from the 1993–94 figure (Hogwood 1995; Hall and Nevin 1999).

BOX 10.1 OBJECTIVES OF URBAN AND NEIGHBOURHOOD RENEWAL POLICY

Objectives of the Single Regeneration Budget Challenge Fund (Round 6 December 1999)
The overall priority is to improve the quality of life of local people in areas of need by reducing the gap between deprived and other areas, and between different groups. The objectives are:

- improving employment prospects, education and skills;
- addressing social exclusion and improving opportunities for the disadvantaged;
- promoting sustainable regeneration, improving and protecting the environment and infrastructure, including housing;
- supporting and promoting growth in local economies and businesses; and
- reducing crime and drug abuse and improving community safety.

Objectives from *Bringing Britain Together: A National Strategy for Neighbourhood Renewal* (September 1998)

- tackling worklessness;
- reducing crime;
- improving health;
- raising educational achievements.

The bidding process is conducted through two stages, the first requiring the submission of outline plans describing the partners, principal outputs, the funding sought, and relationship to other funding sources. The bids must include a long list of anticipated outputs in terms of jobs created or safeguarded, number of people trained, new business start-ups, and the area of the land to be improved (Foley *et al.* 1998). The government's own evaluation of the SRB is upbeat, with claims that the 750 schemes approved in rounds 1 to 5 worth £4.4 billion will attract £8.6 billion worth of private-sector investment, and it repeats the partnerships' forecasts that they will create or safeguard some 790,000 jobs, complete or improve 296,000 homes, and support 103 community organisations and 94,000 businesses.[14]

The first bidding round was evaluated by the House of Commons Select Committee on the Environment in its 1995 report *Single Regeneration Budget*, where the concept was generally supported subject to a number of recommendations to reduce bureaucracy, improve consistency, and increase the involvement of voluntary and community groups. The Committee also noted the need for the regional offices to take account of existing regional guidance and development plans. Later evaluations of the SRB have noted the domination of employment- and economic development-related output measures (though this is largely in the form of human resource development); the centralised and opaque decision-making on bids; the limited resources in comparison with the scale of the problem, which leads to a thin distribution of funding; and the need for more effort to involve local communities (Hall 1995; Brennan *et al.* 1998; Hall and Nevin 1999). Nevertheless, they also agree that the Challenge Fund has had a positive impact on promoting more strategic thinking and partnership working in regeneration.[15]

From round 5 (the first under the Labour administration) there has been an attempt to refocus and target the funding more directly to places in most need. A two-tier funding approach has been adopted which requires 80 per cent of funding to be channelled to large, comprehensive schemes in the fifty most deprived local authority areas defined by four measures

on the Index of Local Deprivation. The remaining 20 per cent is targeted at pockets of deprivation in the coalfields, rural areas, and coastal towns. A further innovation is the provision for financial support for capacity-building; indeed, from round 6 it is envisaged that this will be a component of most bids and that much of the first year of operation should be devoted to capacity-building so that local communities can play an active and effective role in the creation and management of schemes. Even so, the SRB investments are still relatively minor compared with the main programmes of participating departments (Hill and Barlow 1995). Only 20 per cent of the resources required for each project come from the SRB, which puts considerable pressure on other public and private sources, and they may not be able to meet the requirements (Hall and Nevin 1999). On the positive side, there is no doubt that the SRB has promoted more strategic thinking and an increase in operational partnerships. Indeed, the new ways of working across public, private, and community sectors may be its most important outcome (Fordham *et al.* 1998).

THE NEW DEAL FOR COMMUNITIES AND NEIGHBOURHOOD RENEWAL

The incoming Labour administration of 1997 maintained the approach to urban policy that it inherited, including the competitive elements in the SRB and the emphasis on partnership with the private and community sectors. But it has also brought forward a raft of new initiatives with the intention of making a concerted attack on social exclusion in the worst parts of cities and of redirecting funding through local authorities. In 1998 the Secretary of State announced extra funding for housing renewal and urban regeneration of £5 billion over three years,[16] most of which was to be at the disposal of local authorities. An extra £3.6 billion over three years was allocated to local authorities through their housing investment programmes (HIPs) to start tackling the estimated £10 billion backlog in housing renovation, with the priority to improve the quality of the public housing stock. The enhanced urban regeneration programme has two main elements: the refocused SRB (explained on p. 304) and the *New Deal for Communities* (NDC).

The initiative for the latter came from the Cabinet Office's Social Exclusion Unit report *Bringing Britain Together: A National Strategy for Neighbourhood Renewal* (1998).[17] The report sets out the intention to concentrate regeneration efforts on the most deprived neigbourhoods. It provides £800 million for seventeen first-round (pathfinder) partnerships and twenty-two secondround partnerships.[18] Given this rather limited funding, it is perhaps obvious that the major objective is as much to bend and improve coordination of existing spending programmes as to provide new money.

Eligible local authority areas were invited to establish a partnership and prepare a delivery plan for neighbourhoods of between 1,000 and 14,000 households. The principal objectives of the scheme are to tackle 'worklessness' (poor job prospects), high levels of crime, educational underachievement, and poor health, but the pathfinder partnerships have in fact addressed a much wider range of issues. The revised guidance includes supplementary objectives including a better physical environment, improved sports and leisure opportunities, and better facilities for access to the arts. The programme gives complete flexibility to the local partnership to define its objectives (within the priorities listed), its ways of working, and its actions, though its plan requires approval by central government.

The first evaluations[19] reveal in part just what little progress has been made on establishing local capacity to address regeneration in a holistic way. The pathfinder partnerships have reported delays in community involvement because no structures were in place, a lack of trust among stakeholders and agencies, and difficulties in even understanding what is already spent in the neighbourhood under other mainstream programmes. Nevertheless, the NDC has generated considerable interest and enthusiasm in communities, not least for the freedom it gives to the neighbourhoods to define their own approach. But it needs to be remembered that the opportunities and resources it provides are limited to areas with a population of less

than 60,000. One way of looking at this is to say that it is a targeting of effort on the most deserving cases. Another is to say that it is no more than an experiment.

After publication of the 1998 report *Bringing Britain Together*, eighteen policy action teams (PATs) were created, bringing together practitioners, academics, and residents from deprived neighbourhoods. The policy action teams made almost 600 recommendations and were used in the preparation of a consultation document, *National Strategy for Neighbourhood Renewal*, published in April 2000. Considering the 'completely new approach to policy making' (as the government described the PAT approach), there is little new thinking in the proposals. Indeed, Oatley (2000) points out the strong resemblance to the 1977 White Paper *Policy for the Inner Cities*. Oatley makes a damning critique of the strategy's reliance on area-based policy (discussed on p. 315) and concludes that there is little hope for a significant impact on urban poverty, though PATs may 'soften the damaging consequences' of economic and social change.

DELIVERING AN URBAN RENAISSANCE

The Labour government delivered its long-awaited urban White Paper at the end of 2000: *Our Towns and Cities: The Future: Delivering an Urban Renaissance* (2000). The paper rambles across most areas of government policy and their impacts on urban areas, though a significant part is related to the planning system and explaining the need for a 'complete physical transformation' of our towns and cities. The main challenges are:

- to accommodate the new homes Britain will need by 2021, making best use of brownfield land;
- to encourage people to remain and move back into urban areas;
- to tackle the poor quality of life and lack of opportunity in certain urban areas;
- to strengthen the factors in all urban areas which will enhance their economic success; and

- to make sustainable urban living practical, affordable, and attractive.

On the physical transformation, much of the content reflects the findings of the Urban Task Force: *Delivering an Urban Renaissance: The Report of the Urban Task Force Chaired by Lord Rogers of Riverside* (Lord Rogers 1999).[20] This report presented 105 recommendations to government, many of which, as the White Paper explains, have found a positive response. Recommendations of particular significance for the planning system are shown in Box 10.2.

The most significant planning related-action on the Task Force report is the revision in 2000 of PPG 3, *Housing* (discussed in Chapter 6). Many other actions are under way or promised: a consultation paper on planning obligations to widen the range of community benefits that can be supported and the possible introduction of impact fees; a revision of PPG 1; provisions for the use of masterplans in regeneration; the creation of 'home zones'; initiatives on the training of planners, designers, and developers; the creation of regional centres of excellence to address skills improvement in each region; and the creation of urban regeneration companies.[21] There is also a raft of proposed fiscal measures such as capital allowances for creating flats over shops.

The importance of strategy and cross-cutting or 'joined-up' action is stressed throughout the White Paper. So too is the variation in experience across England, and thus the importance of regional bodies. These are issues raised in the accompanying report, *The State of English Cities*.[22] This report notes that the principles of policy integration, partnership, and local authority leadership that have been developed within urban policy over the past twenty years are also right for the future. However, it also stresses the need for 'rethinking scales of intervention' so as to 'tie city policies to the broader frameworks of regional and subregional strategy' (p. 46). The subregional or city region scale is promoted as the most appropriate for the economic regeneration strategies. Urban policy should be based on relatively large areas – perhaps with specific planning and fiscal regimes, as also promoted by the Urban Task Force. In this the report also

reinforces messages from European cooperation on urban policy about the significance of urban–rural interdependencies (see below).

The White Paper responds with proposals for (limited) extra funding for the regional development agencies and more flexibility in their operation. It also recognises the role of regional planning strategies, but very little is said about the relationship between planning and economic strategies on urban policy. The *State of English Cities* report points to the weakness of regional economic development strategies so far in identifying the distinctively regional aspects of the strategy. The same could be said for regional planning guidance; and both still tend to treat urban and rural issues separately.

Despite the liberal use of the word 'strategy' and the obvious intention to link many disparate themes of public policy and private action to bring about improvement, the long lists of actions taken (and to be taken) in many different sectors reveal just how difficult this job is. This point is emphasised by the recognition that many of the 'urban problems and policies' discussed here are not peculiar to urban areas at all. In recognition of this, a new Cabinet Committee on Urban Affairs is proposed to take an overview of the impact of sectoral policies on urban areas. But this is unlikely to meet the concerns of the Environment, Transport and Regional Affairs Committee and others that intervention in urban areas is 'confused and badly coordinated' and 'that there should be fewer initiatives and they should be better coordinated locally'.[23]

Delivering an Urban Renaissance is less of a presentation of government policy for discussion prior to bringing forward legislation, and more of a stock-take. The paper itself notes that it 'completes the first phase of the Government's long term programme'. Proposals for substantial new legislation or action are thin on the ground. It has much of the feel of a textbook, with, for example, passages explaining the performance of the UK economy over the past fifty years, and numerous vignettes illustrating examples of good practice across the UK and further afield. It is therefore useful reading for students, although they should beware the obvious gloss.

BOX 10.2 SELECTED RECOMMENDATIONS OF THE URBAN TASK FORCE (1999)

- Require local authorities to prepare a single strategy for the public realm and open space, dealing with provision, design, management, funding, and maintenance.
- Revise planning and funding guidance to discourage local authorities from using 'density' and 'overdevelopment' as reasons for refusing planning permission; and to create a planning presumption against excessively low-density urban development.
- Make public funding and planning permissions for areas regeneration schemes conditional upon the production of an integrated spatial master-plan.
- Develop and implement a national urban design framework, disseminating key design principles through land use planning.
- Place local transport plans on a statutory footing and include explicit targets for reducing car journeys.
- Introduce home zones using tested street designs, reduced speed limits and traffic calming.
- Set a maximum standard of one car parking space per dwelling for all new urban residential development.
- Develop a network of regional resource centres for urban development, coordinating training and encouraging community involvement.
- Produce detailed planning policy guidance to support the drive for an urban renaissance.
- Strengthen regional planning: provide an integrated spatial framework for planning, economic development, housing, and transport policies; steer development to locations accessible by public transport; and encourage the use of subregional plans.
- Simplify local development plans and avoid detailed site-level policies.
- Devolve detailed planning policies for neigh-bourhood regeneration into more flexible and targeted area plans, based upon the production of a spatial masterplan.
- Review employment land designations and avoid overprovision.
- Reduce the negotiation of planning gain for smaller developments with a standardised system of impact fees.
- Review planning gain to ensure developers have less scope to buy their way out of providing mixed-tenured neighbourhoods.
- Oblige local authorities to carry out regular urban capacity studies as part of the development plan-making process.
- Adopt a sequential approach to the release of land and buildings for housing.
- Require local authorities to remove allocations of greenfield land for housing from development plans, where they are no longer consistent with planning objectives.
- Retain the presumption against development in the green belt and review the need for designated urban green space in a similar way.
- Prepare a scheme for taxing vacant land.
- Modify the General Development Order so that advertising, car parking, and other low-grade uses no longer have deemed planning consent.
- Streamline the compulsory purchase order legislation and allow an additional 10 per cent payment above market value to encourage early settlement.
- Launch a national campaign to 'clean up our land' with targets for the reduction of derelict land over five, ten, and fifteen years.
- Introduce new measures to encourage the use of historic buildings left vacant.
- Facilitate the conversion of empty space over shops into flats.

Source: *Delivering an Urban Renaissance: The Report of the Urban Task Force Chaired by Lord Rogers of Riverside* (1999). The Task Force Report makes 105 recommendations in total, and many not listed here are of interest to planning. The full list of recommendations together with explanations of the government's response are listed in an annexe to *Our Towns and Cities: The Future: Delivering an Urban Renaissance* (2000).

OTHER REGENERATION EXPERIMENTS AND INITIATIVES

The most consistent feature of urban policy is the never-ending stream of policies and initiatives, and the current phase probably offers more than any other. Since coming to power in 1997, the Labour administration has taken many new initiatives, often with an area or spatial focus, including health action zones, employment zones, education action zones, and crime reduction programmes (zones). At the end of 1999 there were twelve area-based regeneration schemes (not including the specifically housing-related actions noted above) and thirty other general initiatives.[24] In addition, the Local Government Association has established partnerships in twenty-two pathfinder authorities to prepare comprehensive five-year programmes for regeneration. Many of these initiatives have a bearing on town and country planning – although it is difficult to find any mention in their supporting documentation of the existing statutory planning framework for guiding the spatial distribution of investment and activities.

It is hard to accept the claim that Labour's new deal promotes more strategy in the face of such a complex web of initiatives and competences. The reasoning is that the best way to identify problems and tackle them is through partnership, and this should start at the lowest level. The differing conditions of each area require variety and flexibility in the policy response. But emphasis is also placed on the importance of strategy, and there is a requirement that all partnership bids demonstrate how they relate to other area-based initiatives. Moreover, they must work within the regional regeneration strategy of the RDAs, and meet the most challenging of all cross-sectoral objectives: sustainable development. New guidance has been issued on these matters, including *Sustainable Regeneration: Good Practice Guide* (Rogers and Stewart (1999).[25]

The Labour government is making a brave attempt to address the deep-seated problems of urban regeneration, but getting some sensible coordination of actions will be a considerable task. The government recognises this and is promoting the notion of local strategic partnership (LSP) to provide 'a single overarching local co-ordination framework within which other, more specific local partnerships can operate'. A consultation document was issued in 2000[26] recognising the growing concerns about the number of new partnership arrangements and promoting an additional strategic tier of partnership as the solution. This apparently contrary idea is in fact developing existing good practice, and may well lead to the reduction of other partnerships, which will be subsumed under the new arrangements. The local strategic partnership may cover any area (the consultation paper gives numerous examples) and may take a lead role in the new powers for local authorities to promote or improve the economic, social, or environmental well-being of their area; in the new *community strategies* (both provided by the Local Government Act 2000); and in coordinating neighbourhood renewal.

The government has also set up an interdepartmental support unit to help in the exchange of information on regeneration initiatives, to monitor, and to evaluate. Case studies are being analysed in areas where there are many cross-cutting initiatives to provide further guidance on coordinating initiatives. There is clearly a role for the planning system here which has not yet been seized.

EMPLOYMENT, TRAINING, AND ENTERPRISE AGENCIES

Employment is, of course, one of the principal economic considerations in town and country planning, yet it cannot be said that employment policies have ever been successfully integrated with physical planning policies. At least in part, this is due to organisational separatism and the fact that local authorities have little responsibility for employment and training. Under the Conservative administration even this small responsibility waned.

Initiatives in these areas have been a function largely of central government. These grew in number and administrative complexity until a major reorganisation was made following the 1990 White Paper *Employment in the 1990s*. There are now networks of seventy-two *training and enterprise councils* (TECs) in England and Wales, and twenty-two *local enterprise companies* (LECs) in Scotland. This new system is aimed at providing value for money through a privatised organisation linked, in this case, to the additional objectives of localism and decentralisation (Bennett 1990).

Copied from the USA (where they are known as *private industry councils*), TECs and LECs have as their underlying rationale the idea that if local businesses take a central place in guiding the support programmes for employment and training, the result will be a better response to local employment needs, a more business-like mode of operation, and increased leverage of private-sector support for training (N. Lewis 1992).

The TECs have taken over youth and employment training, and business support and growth initiatives, again with the objective of tailoring these to local conditions. Their activities include maintaining a knowledge base of local labour markets and training needs and provision; providing information to employers, employees, and the unemployed; encouraging employers to meet training needs; and directly providing or commissioning training for employees or the self-employed. All this (to the extent that it is actually done) is decided, in the manner of private board meetings, behind closed doors, although there is evidence of increasing attention to public relations if not true accountability (Hart *et al.* 1996), and the legacy of mistrust between TECs and local authorities is breaking down as partnership arrangements have become more widespread (Plummer and Zipfel 1998). As with other current urban policies, accountability has been sacrificed for the elusive goal of efficiency.

The role of TECS is reviewed in the 1999 White Paper *Learning to Succeed*, which proposed a new framework for the planning and funding of post-16 education and training.

THE EUROPEAN DIMENSION TO URBAN POLICY

The EU does not have a mandate from the Treaties to develop an urban policy.[27] Nevertheless, it has been argued that since more than 80 per cent of the EU's population live in urban areas, and since cities and towns are the motors of economic growth, it is sensible for the Community to take a view on the impact and potential of its actions in them.

In 1990 the Commission, led by the Directorate General for the Environment, published a *Green Paper on the Urban Environment* as 'a first step towards debate and reflection, and attempts to identify possible lines of action' (CEC 1990). One of the first actions was to set up an Expert Group on the Urban Environment made up of national representatives and independent experts to advise the Commission. The Expert Group has taken forward many initiatives and played a central part in setting up the Sustainable Cities and Towns Campaign.

The initiative for urban policy has slowly shifted to the Directorate for Regional Policy because of the considerable impact it has on urban areas through structural funding. Community co-financing has played an important role in many urban regeneration initiatives in the UK, primarily in areas covered by Objectives 1 and 2 (see Chapter 4). In addition, the *Urban* initiative provided funding for coordinated action for 'neighbourhoods in crisis'; and *urban pilot actions* promoted innovation in tackling urban regeneration problems. *Urban* is one of only four initiatives that are to be continued under the 2000–06 round of structural funding. The Community has also funded networks of cities and towns which concentrate on exchanging experience – notably *Eurocities* (a network of more than 100 non-capital cities) (Griffiths 1995) and *Metrex* – the network of metropolitan planning authorities.

Urban areas have been recipients of considerable Community funding, which has helped to fill the gap as national funding has declined (Chapman 1995), but doubts have been raised about the coordination of that investment. In 1997 the Commission published *Towards an Urban Agenda in the European Union*. This received very positive support and encouragement from the Committee of the Regions and English Partnerships, and it was followed in 1998 by *Sustainable Urban Development in the European Union: A Framework for Action*. The main objective of this paper is to stimulate better coordination of existing Community actions that affect urban areas. Its impact was considerable. Many important but 'unanticipated' Community initiatives in the environmental field were announced in the Fifth Framework on the Environment document, which provides a template for the urban paper. The Framework sets an unambiguous agenda for Community action on urban matters, so the actions proposed are listed in full in Box 10.3. Of particular note is the emphasis on area-based initiatives, which will need to be 'essential constituents' of the plans and programming documents that guide the spending of Community funds. The UK approach has been to promote further intergovernmental co-operation rather than formal Community action.

An Urban Exchange Initiative was started under the UK Presidency as an intergovernmental initiative with the aim of preparing a non-binding informal framework for national approaches to urban regeneration by 2000. The initiative brings together case studies of good practice in urban regeneration under seven themes, including town centre management, economic development, sustainable land use, and community involvement. Further proposals for taking this initiative forward are expected under future Presidencies.[28]

SCOTTISH URBAN POLICIES

The evolution of urban policy in Scotland has taken a similar path to that in England and Wales (often in the lead), but with some important differences. No urban development corporations were established in Scotland, since Scottish Enterprise fulfils a similar function and can operate anywhere in the country, unrestrained by the boundaries of a designated area. The Urban Programme has continued to be a key instrument for funding regeneration, although it has suffered the same criticisms of lack of strategic direction, and from 1995 has not been open to individual project bids.

There is a longer history of central–local partnerships in urban regeneration (Boyle 1993). There are several reasons for this, including the different urban history of Scotland, its distinctive governmental organisation, and the close relationship that exists between local and central government. Glasgow and its extreme urban problems have been highly significant in determining the course of urban policy. Of particular note was the establishment of the Scottish Development Agency (SDA) in 1975, and its role in the *Glasgow Eastern Area Renewal* (GEAR) project

The rationale for the SDA was that economic regeneration required measures to promote local growth and overseas investment, and that local authorities did not have the managerial expertise and dynamism needed for the task. But, unlike the case with the later urban development corporations in England, importance was attached to retaining the role of local authorities as well as the many other agencies involved (McCrone 1991: 926).

The GEAR project had economic, environmental, and social objectives. Despite initial teething problems associated with a lack of clarity about the scope of the project and consequent delays, the physical impact on the area was massive. Economic revitalisation, however, proved much more elusive, as was underlined in an unpublished 1988 report by PIEDA (McCrone 1991: 927). Though up to 2,000 jobs were retained or created, the benefits to the resident disadvantaged population were limited.

The experience of the GEAR project was used subsequently in tackling disadvantaged areas facing the additional problem of a severe local employment crisis resulting from the closure of a dominant firm: the steel-making plant at Glengarnock in North Ayrshire, the Singer sewing machine factory in Clydebank, and the Leyland plant at Bathgate. The accomplishments were impressive, in terms of job creation, the provision of new infrastructure, and widespread environmental improvement. This was largely attributed to the establishment of a clear strategy; the nature of the approach, with environmental recovery, job creation, and retraining all working together; and the active participation of local authorities in the scheme (in the knowledge that if they did not participate, the scheme would not go ahead) (McCrone 1991: 929).

Also benefiting from experience, the SDA assumed a coordination role in later projects in Leith, Dundee, Motherwell, Monklands, and Inverclyde. Most of these

> **BOX 10.3 SUSTAINABLE URBAN DEVELOPMENT IN THE EUROPEAN UNION: A FRAMEWORK FOR ACTION (1999)**
>
> 1 Explicit urban programming for Structural Fund support
>
> 2 A stronger urban dimension in employment policies
>
> 3 Support for European 'knowledge centres'
>
> 4 Promotion of inter-urban cooperation
>
> 5 Promotion of attractive urban transport
>
> 6 Development of know-how and exchange of experience on urban economic performance
>
> 7 Cooperation against discrimination and social exclusion
>
> 8 Structural fund support to area-based action for urban regeneration
>
> 9 Second-chance schools
>
> 10 Development of know-how and exchange of experience on discrimination, exclusion, and urban regeneration
>
> 11 Better implementation of existing environmental legislation at urban level
>
> 12 Further legislation concerning waste, air quality, water, and noise
>
> 13 Strengthening pollution control and clean-up in towns and cities
>
> 14 Contributing to a reduction of the environmental impact of urban transport
>
> 15 Sustainable urban energy management
>
> 16 Climate protection (moving towards a Directive on the taxation of energy products)
>
> 17 Extending eco-labelling and the eco-management and audit scheme (EMAS)
>
> 18 EU Structural Fund support for protecting and improving the urban environment
>
> 19 Development of know-how and exchange of experience on the urban environment
>
> 20 Awareness-raising, exchange of experience and capacity-building for sustainable urban development
>
> 21 Innovative urban development strategies
>
> 22 Increasing safety by promoting prevention in the field of urban crime
>
> 23 Improving comparative information on urban conditions
>
> 24 Contribution to the member states' Urban Exchange Initiative

projects involved 'issues that were socially related in that the areas suffered from decay and deprivation of long standing', but they all included an objective of economic regeneration. Moreover, the SDA, the local authorities, and other bodies committed themselves in advance to a project agreement which spelled out who was to do what, where, and when. A more ambitious scheme was undertaken jointly by SDA and the City of Glasgow in the old Merchant City area, with the aid of housing improvement grants and the SDA's *local enterprise grants for urban projects* (LEGUP). The outcome has been judged a success, with a transformation of a large area. Together with other initiatives in the city, there has been an about-turn in both the city's morale and outside perceptions of Glasgow as an attractive place for investment.

The Scottish statements *New Life in Urban Scotland* (1988) and *Urban Scotland into the 1990s: New Life Two Years On* (1990) reviewed the achievements and the priorities for further urban action. Much attention has been devoted to the large housing estates on the periphery of Scottish cities. Those of Paisley (Ferguslie Park), Glasgow (Castlemilk), Edinburgh (Wester Hailes), and Dundee (Whitfield) have undergone regeneration through *urban partnerships* established by the Scottish Office. These involve the local authorities, the local community organisations, the Scottish Office, Scottish Enterprise, and other agencies operating in the areas, but were again criticised for being 'inward-looking', with little strategic context (Hall 1997b). The final evaluation of the urban partnerships (Tarling *et al.* 1999) reveals that the £485 million invested over ten years has been cost-effective: 3,726 new homes have been built and 9,253 improved; employment has improved in two of the partnerships (but fallen in one); and, crucially, the partnerships have improved the image of the estates. But the fundamental problems of poverty and disadvantage remain, and a key reason is the 'churning' of the population as newly employed residents move on to be replaced by unemployed households.

Scotland has led the way in involving local communities (McCrone 1991: 936) and linking physical regeneration with training. The latter was facilitated by the merger in 1991 of the SDA with the Training Agency to create Scottish Enterprise and Highlands and Islands Enterprise. The legislation lists the functions of Scottish Enterprise as being to develop the Scottish economy; to enhance the skills of the workforce; to promote the efficiency and competitiveness of Scottish industry; and to improve the environment. Merger brought these activities under one umbrella. But the evaluation of the urban partnerships concludes that not enough is being done to strengthen the capacity of local people to address the problems or removing the barriers to social inclusion.

Further developments in Scottish urban policy included the publication of *Progress in Partnership* (1993); proposals for change set out in *Programme for Partnership* (1995); and a report on *Partnership in the Regeneration of Urban Scotland* (1996). As in England

and Wales, a competitive system of allocating funds to partnership projects was established. The main share of urban regeneration resources was ring-fenced for twelve *priority partnership areas*.[29] Questions have been raised subsequently about insufficient funding given the tasks, the difficulty of demonstrating social and economic disadvantage in the relatively sparsely populated areas; and the negative effects of the pressure of competitive bidding (McCarthy 1999). The Labour government's emphasis on social exclusion in the most deprived areas is also taking hold, with the creation of social inclusion partnerships in 1999 with £1.3 million of extra funding and extra money for the priority partnership areas to develop much-needed dedicated support units.

NORTHERN IRELAND

Urban policy in Northern Ireland has followed a similar pattern to that in the rest of the UK, with early emphasis on social problems giving way to a concentration on property-led regeneration and, in recent years, a shift of attention to place marketing, public–private partnerships, and community development (Berry and McGreal 1995b). However, the special circumstances, notably the violence and subsequent central government control of policy and implementation, have been important in shaping the problem and responses. Urban conditions and unfairness in employment and housing allocations were important factors in the start of 'the troubles' from 1969, and terrorism has subsequently accentuated the difficulties of tackling them.

Urban problems are concentrated in Belfast and Londonderry, although there is a separate *Community Regeneration and Improvement Programme* for smaller towns. The urban policy budget in Northern Ireland amounts to more than £48 million (1990–2000), but to this must be added the considerable funding (£63 million p.a.) that comes by virtue of the designation of the whole of the Province as an ERDF Objective 1 region until 1999 (with transitional funding beyond), and the significant inward cash flows for employment and housing investment through mainstream funding programmes. There is also a Northern Ireland *Urban Initiative for Peace and Reconciliation* which provides over £10 million for urban regeneration to improve the quality of life, enhance the environs of sectarian interface areas, and support a wide range of regeneration projects.

Urban renewal and all other housing functions are undertaken by a single authority, the Northern Ireland Housing Executive. It has followed similar policies to those in England with the sale of public housing to sitting tenants, the progressive renewal of existing stock (where great improvements have been made), and only very limited new building. Urban policy in the most deprived areas of Belfast has been implemented through nine Belfast Action Teams which coordinate the activities of different departments as did the CATs in England (PA Cambridge Economic Consultants 1992). A separate Londonderry Initiative was established in 1988. The Belfast Regeneration Office manages an urban regeneration programme for the thirty-two most deprived wards in Belfast and promotes business growth and training. A third programme, the Community Economic Regeneration Scheme, enables communities in the most severely depressed areas to bid for substantial capital investment (property development) to promote business and employment. There is one urban development corporation, Laganside, which was established in 1989 following closure of shipbuilding yards; this is still in operation.

All initiatives must now fall within the overall *targeting social need* strategy, which follows a similar approach to that in England. It is proposed that by 2001, zones for intensive action will be identified for education, health, social well-being, physical infrastructure, and other initiatives; and area-based partnerships will be set up in Londonderry and Belfast to target resources on the most disadvantaged neighbourhoods.[30]

EVALUATION OF URBAN POLICY

One remarkable change in government policy over the past decade or so has been the embracing of research as an essential part of the policy-making process. This is, at first sight, surprising since governments typically find it easier to express their goals in broad terms which encompass the widest possible range of outcomes. Modern management techniques, however, have raised to the fore the formulation of specific objectives which can be monitored and evaluated. This has made the easy option of vagueness unacceptable in principle; but practice is a different matter, as illustrated by the emphasis laid on undefined 'regeneration'. The intermediate measures or outputs such as training places, jobs, visitors, roads, and reclaimed land can be counted but are often disputed. Worse still, they do not give a real assessment of outcomes. To what extent have the fundamental objectives of regeneration been realised?

In answering this question, there is an extraordinary difficulty at the outset: how to define clearly what the objectives of policy are. The problem is very familiar to policy analysts (Rittel and Webber 1973), but it has to be constantly tackled anew by policy-makers. A good example (and this discussion has to be illustrative rather than comprehensive) is the apparently simple matter of increasing employment. The problem is one of finding a satisfactory definition (or even concept) of the term 'new employment'. Other complications arise when account is taken of the 'life' of new jobs created: many of the jobs created in the course of regional development programmes later disappeared (Hughes 1991). Moreover, there may be a lag in the growth of jobs which could be difficult to take into account. Again, policy may be directed not to the objective of short-term job creation, but towards increasing the long-term competitiveness of the area in a changing national and world economy. This poses obvious difficulties of evaluation. Going further, if the aim of policy is wealth creation (a term that was popular for a short time in the mid-1980s), any thought of evaluation becomes mind-boggling. How is wealth to be defined (particularly in these environmentally conscious days)? Is the object to raise the average level of wealth, or the level of those who are the poorest? Such questions quickly banished the term from general use.

Despite all the conceptual and practical difficulties, researchers have been able to draw some important conclusions from their evaluations of urban policy. For example, many of the jobs that have been 'created' would have arisen without any intervention. Indeed, if the projects that have been evaluated are typical, 'such programmes are unlikely to make more than a modest contribution to the economic regeneration of the inner cities' (S. Martin 1989: 638) On reflection, this is perhaps unsurprising. In a complex interdependent society (and, increasingly, an interdependent world), 'local' issues are elusive. In Kirby's words (1985: 216), 'we cannot attempt to understand the complexities of local economic affairs *in situ*'. Much research corroborates this view. The experience of the SDA in Glasgow showed that though the provision of premises attracted some firms to the area, this was 'at the expense of other parts of the city, and most jobs were filled by inward commuters anyway' (Turok 1992: 372).

Another issue in which difficulties of assessment abound (and in which myths live on) is that of the impact of property-led development. Much urban policy has been based on the assumption that property development will somehow or other stimulate economic growth and social improvements. How this is to happen has not been articulated, and there has been little detailed research on the subject. Such research as has been undertaken offers no clear conclusions, though studies 'suggest that access to markets, management abilities, and the availability of finance are more important than buildings' and 'levels of investment in product development and production technology, together with differences in the way human resources are managed, are most significant' (Turok 1992).

Many factors other than property will play a part in successful regeneration, such as the availability of a skilled labour force, ready finance, and an attractive environment. Moreover, property development can present its own problems: this became clear in 1990, when rental values fell as the economy dipped. The slowing down of property investment in 1990 turned into a spectacular collapse, with catastrophic effects on the construction industry and local economies.

Urban development corporations above all have failed to consider the relationship between the local economy and property development in adopting a policy of 'privatism' – 'the attracting into the inner city of private developers whose activities can in turn demonstrate that regeneration is taking place' (Edwards and Deakin 1992: 362). We have learned that this type of urban policy can have a detrimental effect on local economic activity as, for example, when the precarious position of small local firms is challenged with competition from outside the locality. Urban policy has been characterised by short-term thinking centred on getting the best return from particular sites (CLES 1992b). There is little to support the view that property-led urban regeneration produces a trickle-down of benefits for the local disadvantaged community as the local economy improves. Conclusions from a comprehensive evaluation of urban policy suggest that other policy priorities would have greater impact (Robson *et al.* 1994).

These observations on property-led urban regeneration are not to suggest that physical improvements are not needed. They will need to remain a central part of urban regeneration if only because of the deteriorating state of the physical fabric of urban areas. The point is to understand the ways in which physical regneration opportunities contribute to social and economic development. This entails ensuring that 'non-physical policy interventions . . . keep the momentum of regeneration rolling forward once physical rebuilding is complete' (Carley and Kirk 1998a).

Much policy takes the form of targeting resources on the most deprived areas. In practice this has had only marginal impact, because of the cuts in mainstream public spending (particularly funding to local authorities and the housing investment programme) that have fallen more heavily on some of the most deprived areas, leaving them even worse off per capita (Robson *et al.* 1994). Not surprisingly, therefore, there is increasing concentration of the most disadvantaged in the urban programme areas in comparison with other authorities. Within the areas themselves there is an increasing concentration of unemployment in the worst areas. Thus there is growing polarisation within the conurbations, with the benefits of targeting being felt most by the surrounding areas that are better placed to take advantage.

Given the difficulties of evaluation, it is not surprising that the researchers describe the task of providing a sweeping overview of urban policy impact as daunting. They underline the interlocking nature of urban policy, continual change in programmes, and the difficulty of identifying a single unambiguous set of objectives against which to measure progress, as the main difficulties. On the positive side, Robson *et al.* found that regeneration funding has had a positive impact on residents' perceptions of their area. There has also been some general 'limited success for government policy', particularly in smaller cities and the outer districts of conurbations. Where well-coordinated multi-agency approaches have been taken, some policy instruments have worked well. But 'the amount of money going into urban policy is minuscule compared to the size of the problems which are being tackled . . . many of the poorer areas are not improving or at least not nearly as much as the better-off areas within the districts'. The five main conclusions of Robson's research are listed in Figure 10.4, together with three principal conclusions of a more recent review by Carley and Kirk (1998a).

In a 1996 review of the prospects for change, Lawless points to the need for a context which encourages local strategy-building rather than centrally directed *ad hoc* responses. 'Extraordinary as it may seem, and admittedly with the important exception of jobs and employment, problems which most affect the urban disadvantaged have received minimal attention.' But this in turn calls for a new form of urban governance which would be stronger at both the regional and local community levels, and a more politically mature programme. Bailey *et al.* (1995: 229) conclude more radically that

> cities need to be seen as an important national economy and that the growth, redevelopment and improvement of these assets can and should be linked with redistributive welfare policies as part of a strategic and comprehensive national economic policy driven by the public sector.

There is some indication that the breadth of opinion on the need for change has had some effect over recent years. The introduction of the Single Regeneration Budget and integrated government offices for the regions is about providing more coordination and consistency from different departments. City Challenge has taken a longer-term view, gives a leading role to local authorities, and seeks to incorporate (although not without some difficulty) local communities. The experience of promoting partnerships in City Challenge has been taken forward such that partnership working, first introduced in urban policy in 1995, and now managed competition have become requirements for almost all urban initiatives in England (Oatley 1995; Atkinson 1999a). In Scotland the competitive element is not so fierce, but the process of negotiation between partnerships and central government is still competitive and suffers from lack of transparency. It is agreed that much more change is needed, in both policy themes and the context in which policy is formulated and implemented.

Recent developments under the Labour administration continue in the same way, but with increased concentration of effort on the worst cases of deprivation, an equal emphasis on investment in people as well as the physical environment, moves towards more strategic thinking and joined-up policy on regeneration, and, to some extent, increased resources.

Underlying the transition from a physical policy concerned with deteriorating housing to a social policy in aid of deprived areas and then to an economic policy for strengthening the base for local growth, and most recently to policy addressing social exclusion, is the dramatic change which has taken place in the character of cities. There has been a relentless decline of population and employment in the urban areas generally, and particularly in the inner areas. (The other side of the coin is the growth in smaller towns and rural areas.) The economy of the older industrial British cities has been transformed: their manufacturing base has been eroded, and there has been little of the expanding tertiary industries to take its place. Cameron (1990: 486) suggests that 'the aggregate decline of many major British cities is inevitable and indeed desirable. Probably a growing percentage of British consumers and producers will seek locations for living and producing outside such cities.'

The notion of area-based solutions has never been stronger, partly because it is thought to be the best way to solve the problems, partly for financial reasons

BOX 10.4 ASSESSING THE IMPACT OF URBAN POLICY: CONCLUSIONS FOR FUTURE URBAN POLICY

1   There are clear indications of the importance of creating effective coalitions of 'actors' within localities and that these are most likely to result from . . . mechanisms which encourage or require long-term collaborative partnerships.
2   Local authorities – in their newly emergent roles as enablers and facilitators – need to be given greater opportunities to play a significant part in such coalitions.
3   Local communities equally need to be given opportunities to play roles in such coalitions. The evidence of increasing polarisation suggests the need for specific resources to address the scope for community capacity-building within deprived areas.
4   There remains a need to improve the coherence of programmes both across and within government departments. This requires a greater emphasis on the identification of strategic objectives which can guide departmental priorities. Area targeting has played an important part in those cases where separate programmes have been successfully linked so as to create additionality, thereby suggesting the value of giving even greater emphasis to area based approaches. . . .
5   An important part of such coherence must derive from less ambiguity in the targeting of resources. There is a strong argument for the development of an urban budget which might be administered at regional level so as to reflect the varying constraints and opportunities across different regions, and to improve coordination across programmes and departments.

(Robson *et al.* 1994)

Towards a long-term strategic approach to urban regeneration: key problems

1   Lack of linkage between physical development and economic regeneration which benefits the socially excluded.
2   Lack of regional frameworks to support city and local regeneration and link key policies (such as transport, and education and vocational training), to derive maximum benefit to regeneration.
3   Lack of integration of short-term initiatives within a long-term vision on the future role of cities and their hinterlands, and of an investment framework to support that vision.

(Carley and Kirk 1998b)

(there is not enough money to go round). Britain is a European leader in the design of area-based programmes. (Parkinson 1998). However, a continuing deficiency of urban policies is that they assume the issue to be tackled lies in particular arts of particular cities: but 'inner-city' problems is a misleading abstraction. To adapt a passage from Marc Fried (1969) (who was commenting on the concept of poverty), 'the inner city [is] an empirical category, not a conceptual entity, and it represents congeries of unrelated problems'. The problems posed in the inner city 'are not readily accessible for study or resolution in the name' of the inner city.

Progress will be made not by 'comprehensive action' but by identifying priority fields in which effort should be concentrated. (This, of course, is precisely what the Conservative government did, even if its choice of focus was debatable.) Most of the problems identified in inner cities are matters of national policy relating to all areas. Thus, though poverty is undoubtedly a problem which arises in inner cities, most of the residents are not in poverty: and most poverty is not in inner areas. The arguments *against* any area-based policy are strong (Townsend 1976). Oatley (2000: 89) makes the point most strongly in his critique of the *Neighbourhood Renewal Strategy*:

> Area-based policies are notoriously unsuccessful in addressing 'people poverty'. Concentrating resources on a small number of neighbourhoods is both administratively and politically convenient, masking the widespread nature of deprivation within society and allowing us to feel that the problem is being dealt with. These responses may at best concentrate resources in areas with high need for the wrong reasons, and at worst, seriously mislead us into thinking that we are tackling the problems when in fact we are only producing palliatives to alleviate the worst symptoms.

To the extent that the problems relate to the deprived, it makes more sense to channel assistance to them directly, irrespective of where they live. Only to the extent that the problems are locationally concentrated should remedies focus on specific locations – as in the case of renewal areas. Oatley goes on to call for radical alternative solutions that might 'break out of the dismal cycle of unfulfilled promises' (2000: 94). None of this is to deny the importance of directly tackling those problems of decay and disadvantage which are all too apparent in many inner areas. Nor is there any argument against the desirability of attempting better organisation of services at local levels, or improved coordination both within and between agencies. What is crucial is to identify the forces which have created the problems and to establish means of stemming or redirecting them.

Though the current rhetoric of urban policy is about partnership and strategy, the reality is an agglomeration of initiatives and agencies which even the professional is hard pressed to comprehend. Initiatives continue to give the impression that they are no more than short-life laboratory experiments. The institutional arrangements being put into place in England are of particular concern, with regional development agencies, the government offices, sectoral government departments, the emerging regional chambers, and the planning conferences all having a role in developing strategy and implementation, while local communities are being given the flexibility to invent their own responses. Whether these will be moulded into a responsive, integrated, and efficient governmental machine remains to be seen.

FURTHER READING

There is a wide and expanding literature on urban policies. General textbooks include the earlier reviews by Lawless (1989) *Britain's Inner Cities*; Robson (1988) *Those Inner Cities*; and MacGregor and Pimlott (1990) *Tackling the Inner Cities*; Blackman (1995) *Urban Policy in Practice*; and Atkinson and Moon (1994) *Urban Policy in Britain*, which provides a historical review and evaluation. See also Turok and Shutt (1994), *Urban Policy into the 21st Century* (special issue of *Local Economy*); Lawless (1996) 'The inner cities: towards a new agenda'; Oatley (1998) *Cities, Economic Competition and Urban Policy*; Atkinson (1999b) 'Urban crisis: new policies for the next century'; and Carley and Kirk (1998b) 'Towards a long-term strategic approach to urban regeneration'.

Brief summaries of current programmes, their objectives, and expenditure are given in the annual reports of the relevant government ministries. The DETR report is particularly useful.

Housing Renewal

There are only a few recent texts on housing renewal: Balchin (1995) *Housing Policy* places renewal within its wider context and the Joseph Rowntree Foundation sponsors a considerable body of research on urban renewal and housing improvement with summaries of findings placed on its web site at <www.jrf.org.uk>, or from the Foundation at The Homestead, 40 Water End, York YO30 6WP. See also Balchin and Rhoden (1998) *Housing: The Essential Foundations*, and Wood (1991) 'Urban renewal: the British experience'.

History and Theory of Urban Policies

Early approaches to the urban problem (before the focus on economic development) are summarised in Edwards and Batley (1978) *The Politics of Positive*

Discrimination and Hambleton *et al*. (1980) *Inner Cities: Management and Resources*. There are many commentaries on urban policy during the Thatcher years, notably Lawless (1991) 'Urban policy in the Thatcher decade'; Parkinson (1989) 'The Thatcher Government's urban policy'; Haughton and Lawless (1992) *Policies and Potential: Recasting British Urban and Regional Policies*; Hambleton and Thomas (1995) *Urban Policy Evaluation: Challenge and Change*; and Hambleton (1993) 'Issues for urban policy in the 1990s'. A special edition of *Planning Practice and Research* (10 (3/4) is devoted to urban policy. Healey *et al*. (1995) *Managing Cities* explore the links between local problems and global economic forces.

Urban Development Corporations

There are a large number of writings on urban development corporations, predominantly of a highly critical nature. These include Brownill (1990) *Developing London Docklands*; N. Lewis (1992) *Inner City Regeneration*; National Audit Office (1988) *Urban Development Corporations* and (1993) *Regenerating the Inner Cities*; and Thornley (1993) *Urban Planning under Thatcherism* (Chapter 8). Imrie and Thomas (1993) *British Urban Policy and the Urban Development Corporations* contains case studies from eight UDCs. The final evaluation was undertaken in three projects reporting in 1998: Roger Tym & Partners (1998) *Urban Development Corporations: Performance and Good Practice*; Centre for Urban Policy Studies (1998) *The Impact of Urban Development Corporations in Leeds, Bristol and Central Manchester*; and Cambridge Policy Consultants (1998) *Regenerating London Docklands*.

Land Reclamation Grants and English Partnerships

A DoE-sponsored study is Price Waterhouse (1993) *Evaluation of Urban Development Grant, Urban Regeneration Grant and City Grant*. Earlier evaluations were made by Jacobs (1985) 'Urban development grant'; S. Martin (1989) 'New jobs in the inner city'; and Public Sector Management Research Unit (1988a) *An Evaluation of the Urban Development Grant Programme*.

PA Consulting Group (1999) has undertaken an interim evaluation of English Partnerships for the DETR.

City Challenge and Competitive Bidding

Among the many useful studies are Bailey and Barker (1992) *City Challenge and Local Regeneration Partnerships*; de Groot (1992) 'City Challenge: competing in the urban regeneration game'; Davoudi and Healey (1994) *Perceptions of City Challenge Processes*; Oatley (1995) 'Competitive urban policy and the regeneration game'; and Oatley and Lambert (1995) 'Evaluating competitive urban policy'. On the community view of City Challenge, see Bell (1993) *Key Trends in Communities and Community Development* and McFarlane (1993a, b) *Community Involvement in City Challenge* (2 vols). Numerous case studies of particular city challenge partnerships have been written, including Davoudi (1995a) 'City Challenge: a sustainable mechanism or temporary gesture'; Oc *et al*. (1997) 'The death and life of City Challenge: the potential for lasting impacts in a limited life urban regeneration initiative'; and Taussik and Smalley (1998) 'Partnerships in the 1990s: Derby's successful City Challenge bid'.

Single Regeneration Budget

The bidding guidance documents available on the DETR web site provide a full explanation of how the SRB operates, together with information on the progress of schemes agreed during previous rounds. Early evaluation of SRB is provided by the HC Environment Committee report (1995) on the Single Regeneration Budget, and Mawson (1995) *The Single Regeneration Budget*. Recent evaluations include those of Foley *et al*. (1998) 'Managing the challenge: winning and implementing the Single Regeneration Budget Challenge Fund'; Hall and Nevin (1999) 'Continuity and change: a review of English regeneration policy in the 1990s'; and Brennan *et al*. (1998) *Evaluation of the Single Regeneration Challenge Fund Budget: A Partnership for Regeneration: An Interim Evaluation*.

Training and TECs

Accounts of the performance of the TECs vary dramatically. A critical review of their performance was published in 1994 by Coopers & Lybrand in their final report on the *Employment Department Baseline Follow-Up Studies*. The findings are assessed by M. Jones (1996) 'TEC policy failure'. In contrast, see the glowing account of the success of one TEC by Groves (1992) 'A chairman's view of a TEC'. A different view is expressed in CLES (1992a) *Reforming the TECs: Towards a Strategy*. See also Haughton (1990) 'Targeting jobs to local people: the British urban policy experience'; Bovaird (1992) 'Local economic development and the city'; M. R. Jones (1995) 'Training and enterprise councils: a continued search for local flexibility'; and Plummer and Zipfel (1998) *Regeneration and Employment: A New Agenda for TECs, Communities and Partnerships*.

Partnerships

The concept of partnership and its progress in urban policy is recorded in Bailey *et al.* (1995) *Partnership Agencies in British Urban Policy*, which also contains a number of case studies. See also Lawless (1994) 'Partnership in urban regeneration in the UK'; Hastings and McArthur (1995) 'A comparative assessment of government approaches to partnership with the local community'; Nevin and Shiner (1995) 'Community regeneration and empowerment'; Littlewood and Whitney (1996) 'Re-forming urban regeneration?'; the DETR (2000) *Local Strategic Partnerships: Consultation Document*; and Carley *et al.* (2000) *Urban Regeneration through Partnership: A Study of Nine Regions*. For a theoretical perspective on the value of partnerships, see Mackintosh (1992) 'Partnership: issues of policy and negotiation' and Hastings (1996) 'Unravelling the process of "partnership" in urban regeneration policy'. A 1999 report of the National Audit Office is *English Partnerships: Assisting Local Regeneration*.

Scotland, Wales, and Northern Ireland

Many of the references cited above cover practice across different parts of the UK. A summary review of Scottish Urban Policy and ideas for its further development is given by McCarthy (1999) 'Urban regeneration in Scotland: an agenda for the Scottish Parliament'; and earlier reviews are by McCrone (1991) 'Urban renewal: the Scottish experience' and Boyle (1993) 'Changing partners: the experience of urban economic policy in west central Scotland'. For a critical review of the setting up of Scottish Enterprise and Highlands and Islands Enterprise, see Danson *et al.* (1989) 'Rural Scotland and the rise of Scottish Enterprise' and Hayton (1992) 'Scottish Enterprise: a challenge to local land use planning' (the annual reports of these organisations give a different perspective). A research report by McAllister (1996) reviews partnership in the regeneration of urban Scotland, and Tarling *et al.* (1999) provide an evaluation of the New Life for Urban Scotland initiative. See also Scottish Office: *Programme for Partnership: Urban Regeneration Policy*, and McCarthy and Newlands (1999) *Governing Scotland: Problems and Prospects – The Economic Impact of the Scottish Parliament*.

For reviews of urban policy in Wales, see Alden and Romaya (1994) 'The challenge of urban regeneration in Wales'. See also the Welsh Office paper *Programme for the Valleys: Building on Success* (1993), and a more critical account by Morgan (1995) 'Reviving the valleys? Urban renewal and governance structures in south Wales'.

Berry and McGreal (1995b) 'Community and inter-agency structures in the regeneration of inner-city Belfast' give a perspective of urban policy in Northern Ireland.

Evaluation

Many of the references mentioned so far include some evaluation of particular programmes or projects but there is a body of literature which takes evaluation of urban policy as its central theme. The principal source is Robson *et al.* (1994) *Assessing the Impact of Urban*

Policy, and see also the DETR report (2000) by the same authors, *The State of English Cities*. The National Audit Office reviewed urban policy in the 1980s in *Regenerating the Inner Cities* (1990). A broader perspective, including consideration of the methodological problems of evaluation, is given in the edited text by Hambleton and Thomas (1995) *Urban Policy Evaluation*. See also Pacione (1997) 'The urban challenge: how to bridge the great divide'; Pantazis and Gordon (eds) *Tackling Inequalities: Where are We Now and What Can be Done?*; Edwards (1997) 'Urban policy: the victory of form over substance'; and Shaw and Robinson (1999) 'Learning from experience? Reflections on two decades of British urban policy'.

On the contribution of property development to urban regeneration, see Healey *et al.* (1992b) *Rebuilding the City: Property-Led Urban Regeneration* and Turok (1992) 'Property-led urban regeneration'. Evaluations of recent initiatives are made by Oatley (2000) 'New Labour's approach to age-old problems' (and other papers in the same volume: *Local Economy* 15 (2)); and Lawless and Robinson (2000) 'Inclusive regeneration? Integrating social and economic regeneration in English local authorities'.

NOTES

1 Similar house condition surveys are conducted in Scotland, Wales, and Northern Ireland. A further English house condition survey is planned for 2001 and will involve inspection of a sample of 25,000 properties.

2 Unfitness in England and Wales is defined by the 1989 Local Government and Housing Act in terms of serious disrepair, structural stability, dampness prejudicial to health, and the availability of basic services and facilities. In 1998 the DETR issued a consultation paper on the *Housing Fitness Standard*, proposing changes to its definition to incorporate a wider range of health and safety criteria. Previously, the statutory definition (in England and Wales) was related to structure, physical condition, and plumbing only. The approach was altered in Scotland following

the 1967 Cullingworth Report. The 1969 Housing (Scotland) Act introduced the concept of a *tolerable standard*, which differs in detail from the fitness standard in England and Wales.

3 Poor housing is defined as that which is unfit, in substantial disrepair, or requires essential modernisation. See Leather and Morrison (1997) for a review of the state of housing in the UK.

4 The provisions for renewal areas and neighbourhood renewal assessments have been taken forward into the 1996 Housing Grants, Construction and Regeneration Act largely unchanged.

5 The evaluations of renewal areas make use of surveys undertaken by Couch and Gill (1993) and Austin (1995). The 2000 Green Paper on Housing, *Quality and Choice: A Decent Home for All*, notes that 132 renewal areas have been designated, with proposed investment of £1.75 billion.

6 The 1996 the DoE published Circular 17/96, *Private Sector Renewal: A Strategic Approach*, requiring local authorities to prepare a private-sector housing strategy as an input to the overall housing strategy required as part of the Housing Investment Programme Process. Evaluation was undertaken and reported in the DETR research report *Private Sector Housing Renewal Strategies* (1998). The DETR also published *Private Sector Housing Renewal Strategies: A Good Practice Guide* (1997), based on the research findings.

7 The DETR has published *Sustainable Estate Regeneration: A Good Practice Guide*, which compares success through Estate Action, the SRB, and mainstream funding.

8 There were twelve community development projects in all: in Birmingham, Coventry, Cumbria, Glamorgan, Liverpool, Newham, Newcastle, Oldham, Paisley, Southwark, Tynemouth, and West Yorkshire. Additionally, 1974 saw the introduction of a small number of *comprehensive community programmes* in areas of 'intense urban deprivation'. In the wake of these, large numbers of studies were undertaken.

9 The twelve English UDCs were (with their date of designation) London Docklands (1981),

Merseyside (1981), Trafford Park (1987), the Black Country (1987), Teesside (1987), Tyne and Wear (1987), Central Manchester (1988), Leeds (1988), Sheffield (1988), Bristol (1989), Birmingham Heartlands (1992), and Plymouth (1993). Details of their expenditure and outputs are given in Table 10.2.

10 There were eight CATs: Birmingham, Cleveland, Liverpool, London, Manchester/Salford, Tyne and Wear, Leeds/Bradford, and Nottingham/Leicester/Derby.

11 This comparison is calculated at twenty-eight houses to the hectare, which is the average rate for development on previously developed land according to the DETR's 1999 report *Land Use Change in England*.

12 The first-round authorities completed their five-year programmes in 1997 and were Bradford, the Dearne Valley Partnership (Barnsley, Doncaster, and Rotherham), Lewisham, Liverpool, Manchester, Middlesbrough, Newcastle, Nottingham, Tower Hamlets, Wirral, and Wolverhampton. The second-round five-year programmes (which were open to all fifty-seven urban programme areas) ended in 1998. The authorities were Barnsley – the only authority to win in both rounds, Birmingham, Blackburn, Bolton, Brent, Derby, Hackney, Hartlepool, Kensington and Chelsea, Kirklees, Lambeth, Leicester, Newham, North Tyneside (ended in 1999), Sandwell, Sefton, Stockton-on-Tees, Sunderland, Walsall, and Wigan.

13 This quotation is taken from the SRB Bidding Guidance Round 6 (para. 1.3.1). The guidance is extensive and can be found on the DETR web site at <www.regeneration.dtlr.gov.uk/srb/index.htm>.

14 DETR (1999) *SRB Bidding Guidance Round 6*. See DETR *Annual Report* 1999 for a more detailed breakdown of anticipated outputs.

15 The Department of Land Economy at the University of Cambridge is studying the SRB Challenge Fund for the DETR by monitoring twenty case study partnerships over an eight-year period.

16 Eight hundred million pounds of this was to come from the freed-up capital receipts earned by local government but frozen under the Conservative administration. The extent to which the 'extra' £5 billion is new money rather than transfers is of course contested.

17 The Social Exclusion Unit has established a programme to produce a *National Strategy for Neighbourhood Renewal* that will be based on the findings of eighteen Policy Action Teams which are undertaking 'an intensive programme of policy development'.

18 The seventeen first-round pathfinder partnerships are in the local authority areas of Birmingham, Bradford, Brighton and Hove, Bristol, Hackney, Hull, Liverpool, Leicester, Manchester, Middlesbrough, Newcastle upon Tyne, Newham, Norwich, Nottingham, Sandwell, Southwark, and Tower Hamlets. The twenty-two second-round partnerships are in Knowsley, Satford, Rochdale, Oldham, Hartlepool, Sunderland, Sheffield, Doncaster, Birmingham, Wolverhampton, Walsall, Coventry, Plymouth, Derby, Southampton, Luton, Islington, Lambeth, Haringay, Lewisham, Hammersmith and Fulham, and Brent.

19 Findings from the pathfinder authorities are given in the DETR report *New Deal for Communities: Learning Lessons: Pathfinders' Experiences of NDC Phase 1* (1999). Other material can be found at the web site <www.regen.net>.

20 Another important source for the White Paper is the *New Commitment to Regeneration* approach developed by the Local Government Association and the Cabinet Office. Details are at <http://www.cabinet-office.gov.uk/seu/index/national_strategy.htm[more].

21 See Parkinson and Robson (2000). Three pilot companies have been created to redevelop and bring investment to the most deprived areas of towns and cities in Liverpool, East Manchester, and Sheffield.

22 See Robson *et al.* (2000).

23 Select Committee on Environment, Transport and Regional Affairs Eleventh Report on the

Proposed Urban White Paper (1999). The Select Committee report explains what other bodies have made such criticisms.

24 The DTLR has helpfully provided a summary listing of regeneration initiatives, and this is available on the web site at <www. regeneration. dtlr.gov.uk/policies/area/summaries/index.htm>. In 1999 the DETR commissioned action research on the better coordination of area-based strategies (details at <www.regeneration. dtlr.gov.uk/policies/area/action/index.htm>).

25 See also the review by Campbell *et al.* (2000) of the Joseph Rowntree Foundation Research Programme.

26 *Local Strategic Partnerships: Consultation Document* (London: DETR, 2000). See also Carley *et al.* (2000).

27 The extent of Community competence in urban policy is considered by Nadin and Seaton (2000).

28 Three reports have been prepared under the Urban Exchange Initiative. These are available at various web sites: *Integrated Approach to Urban Regeneration: Town Centre Management and Community Involvement* <www.regeneration.dtlr. gov.uk/uei.index.htm>; *Sustainable Land Use and City Friendly Transport* <www.baunetz.de/ lm.bw/f publick htm>; and *Economic Development and Urban Research Information Systems* <www. intermin.fi/suom/alue/uei.uei.pdf>.

29 The Priority Partnership Areas were designated by the Scottish Office in response to proposals from local authority-led partnerships, covering the whole of their area. The twelve areas are Great Northern (Aberdeen); Ardler (Dundee); Craigmillar and North (Edinburgh); East End, North and Easterhouse (Glasgow); Inverclyde; Motherwell North (North Lanarkshire); Paisley (Renfrewshire) and North Ayr (South Ayrshire).

30 DoENI (1999) *Department of the Environment for Northern Ireland Draft New Targeting Social Needs Action Plan* (Belfast: DoENI <www.dfpni.gov.uk/ ccru/annex4.htm>).

11

TRANSPORT PLANNING

What nobler agent has culture or civilisation than the great open road made beautiful and safe for continually flowing traffic, a harmonious part of a great whole life?

(Frank Lloyd Wright 1963)

the way we travel, and the continued growth in road traffic, is damaging our towns, harming our countryside and changing the climate of our planet.

(Draft Revised PPG 13 1999: 7)

MOBILITY AND ACCESSIBILITY

Transport is many things: it is a means of getting from one place to another. It includes a range of very different forms of travel: walking, cycling, travelling by car, bus, or train, or flying. A journey to work on the London Underground is very different from a country holiday tour. Except perhaps for the latter type of journey, transport is unlike other goods in that it is a means to an end: it is not an end in itself. Indeed, much transport is an impediment to the enjoyment of something else. It is a means of providing access. Mobility is not important in itself: its importance is in providing access. Yet the debate on transport often forgets this elementary point, and focuses on mobility: faster roads, faster trains, and more frequent buses.

The advantage of focusing on accessibility rather than mobility is that it opens up the possibility of alternative means: changing land use relationships, for example. As an advertisement on a condominium tower above a Toronto metro station neatly pointed out, 'If you lived here, you would be home now.' The greater the accessibility, the lower the need for 'transport'. Thus, transport planning is much more

than the building of roads, even though this has not always appeared to be the case. It should involve a consideration of the relationship between different land uses, and between land uses and transport feasibilities, as well as the relationships between different transport modes and their relative effectiveness in meeting economic, financial, social, and environmental goals.

There is nothing profound in these observations, but, until recently, transport policy appeared to deny their validity. Roads have formed the major focus of policy. It is therefore fitting that we start by considering road traffic.

THE GROWTH OF TRAFFIC

Between 1950 and 1960 the number of vehicles on the roads of Britain more than doubled, from 4.0 million to 8.5 million. The number more than doubled again by the end of 1980, to 19 million. By 1999 the number had risen to 28 million. The most dramatic increase was in cars, from around 2 million in 1950 to 22.8 million in 1999 (Table 11.1).

The proportion of households owning a car has grown from a mere 14 per cent in 1951 to around 70 per cent in the late 1990s (Table 11.2). In terms of total road traffic (measured in billion vehicle-kilometres) the increase has been from 53.1 in 1950 to 467 in 1999 (Table 11.3). Despite a massive road-building programme, including some 3,295 km of motorway, the increase in the length of the road network has been far less than the increase in traffic. There were 372,000 km of road in the UK in 1999 – an increase of less than a quarter over the 1951 total of 297,000 km (which was only fractionally greater than the 1909 figure of 282,000). The consequence, of course, has been that roads have become far more crowded.

The increase in traffic shows little sign of abating, and a major increase is currently forecast. Traffic forecasts are, of course, only estimates, and no more reliable than weather forecasts – less so in fact. (Forecasts based on other forecasts are particularly suspect: the traffic forecast is based mainly on assumed economic growth and, of course, on an absence of serious impediments to car ownership and use.) The 1997 forecast, which is the latest, is for an increase in total traffic (from a 1996 base) of between 3 per cent and 15 per cent by 2001, and between 36 per cent and 84 per cent by 2031. The increase for car traffic is forecast at between 3 per cent and 14 per cent by 2001, and between 30 per cent and 75 per cent by 2031 (Table 11.4). But the recent public transport invest-ment proposals (discussed on p. 336) are forecast to slow the growth of car traffic.

With the emphasis which is so often placed on increases in cars and traffic, it is easy to forget that about a third of households do not have a car. The proportion without a car is higher in the North and in Scotland. It is also higher for the economically inactive and for unskilled manual workers. Generally, car ownership increases in inverse proportion to size of town, and is highest in rural areas, where 84 per cent of households have a car.

The car ownership forecasts have two components: car ownership and car use. Car ownership is still increasing. Indeed, the 'saturation' level has not yet

Table 11.1 Number of Vehicles, Great Britain, 1950–99

	1950	1960	1970	1980	1990	1999
Cars ('000)	1,979	4,900	9,971	14,660	19,742	22,785
All vehicles ('000)	3,970	8,521	13,548	19,199	24,673	28,368

Source: Transport Trends 1999 (table 3.3) and *Transport Statistics Great Britain 2000* (table 3.1)

Table 11.2 Proportion of Households with Cars, Great Britain, 1951–99

	1951	1970	1980	1990	1999
No car	86	48	42	33	28
1 car	13	45	44	44	44
2 cars	1	6	13	19	22
3+ cars	—	1	2	4	5

Source: Transport Statistics Great Britain 2000 (table 3.14)

Table 11.3 Road Traffic, Great Britain, 1950–1999 (billion vehicle-kilometres)[1]

	1950	1960	1970	1980	1990	1999
Cars and taxis	25.6	68.0	155.0	215.0	335.9	380.1
Light vans	7.8	14.7	18.9	23.1	35.7	49.2
Goods vehicles	11.2	15.7	19.0	22.6	24.9	28.1
All motor vehicles	53.1	112.3	200.5	271.9	396.5	467.0

Source: *Transport Trends 1999* (table 3.1) and *Transport Statistics 2000* (table 4.7)

Table 11.4 National Road Traffic Forecasts by Vehicle Type 1996–2031 (1996 = 100)

	Cars			Total traffic		
	Low	Central	High	Low	Central	High
2001	103	109	114	103	109	115
2011	116	127	137	117	128	139
2021	126	143	159	129	146	163
2031	130	153	175	136	160	184

Source: DETR National Road Traffic Forecasts (Great Britain) 1997. Note that the Ten-Year Transport Plan, 2000, predicts greater reductions in the growth of car and lorry traffic.

been reached in any country (not even in the USA). It is therefore not easy to guess what this level may be, and, of course, it may well differ among countries. The saturation level is assumed to be the level observed in the highest-income households of each type in recent years. (It can therefore change.)

Car use is also difficult to predict. It fell during the period 1973–76 when GDP fell and real fuel prices rose, but there was no fall when similar conditions applied during 1979–82. Use of second and third cars is not lower than the use of first cars: in fact it is higher. Of the vehicles on the road, four-fifths are cars. Most of the remainder are goods vehicles. The forecasts for these are calculated separately for light vans and heavy goods vehicles. Light goods traffic is forecast to increase by between 85 per cent and 251 per cent by 2031. Heavy goods traffic is more problematic. Previous forecasts proved to be far too low (the high forecast for the period 1982 to 1987 was 4 per cent; the actual was 22 per cent). The latest forecast gives an increase for rigid heavy goods vehicles of between 15 per cent and 56 per cent by 2031. For articulated heavy goods vehicles the range is from 96 per cent to 165 per cent.

Since the forecasts related to road use, they did not deal with rail traffic. Total passenger mileage has fluctuated, but recent years have seen an increase to a level slightly above the rate in the 1950s (Table 11.5). However, in terms of journeys by the constituent systems a different picture emerges. The national rail network has experienced a large, though erratic fall since 1950, but there has been a significant increase in recent years, with 846 million journeys (a 6 per cent increase over the previous year).

There have been recent increases in journeys on the London Underground, and on new systems such as London Docklands Light Railway, Manchester

Metrolink (Altram), and Sheffield Supertram (Stage-coach). The Glasgow Underground has seen no change, while Tyne and Wear (Nexus) has experienced a decline (Table 11.6). Freight traffic by rail has declined steadily, until very recently when there has been a small increase. By contrast, well over half of freight (measured in billion tonne-kilometres) went by road (Table 11.7). It has been held for some time that since road and rail serve mainly different markets, there is little scope for transferring road freight to the railways. As is discussed on p. 343, in the context of the integrated transport policies this view has now changed.

The huge increase in car traffic and the relative decline in bus and rail travel do not signify a massive transfer from public to private transport. On the contrary, the figures show that most of the increase in car usage is newly generated traffic. Indeed, total passenger travel has increased enormously: people are travelling much more than they used to. Though the issue has not been subject to research (incredible though this seems), some of the increase must have resulted from the dispersed pattern of activities and the increased separation of home and work. This itself

has been facilitated by road improvements, thus illustrating the impact of 'transport supply' on demand. Moreover, since this new traffic is based on dispersal, it may be very difficult to change it to a public transport mode. Though Britain is far from being as car-dependent as the USA, much new development continues to be of this American character.

TRANSPORT POLICIES

Public policy on transport has a long history (Barker and Savage 1974), but post-Second World War policy began with a plan for a network of new trunk roads (which was not implemented) and a plan for the nationalisation of road haulage and the railways (which was). Much energy was dissipated in the nationalisation and denationalisation processes, and more attention was paid to ownership and control than to transport policy. Experience with the centralised and, later, the decentralised British Railways left a legacy of unease about railway spending in the Transport Department which persisted (Truelove 1992; Kay and Evans 1992). With road haulage, the role of

Table 11.5 Passenger Travel by Mode, Great Britain, 1952–99 (in billion passenger-kilometres and as a percentage of total)

	1952		1970		1990		1999	
	bpk	*%*	*bpk*	*%*	*bpk*	*%*	*bpk*	*%*
Bus	92	42	60	15	46	7	45	6
Car	58	26	297	74	588	85	621	85
Motorcycle	7	3	4	1	6	1	5	1
Pedal cycle	23	11	4	1	5	1	4	1
ALL ROAD[a]	180	82	365	91	645	94	675	93
Rail	39	18	36	9	39	6	44	6
Air	0.2	0.1	2.0	0.5	5.2	0.8	7.0	1.0
ALL MODES	219	100	403	100	689	100	728	100

Source: Transport Statistics 2000 (table 3.1)

Note: [a] Including vans and taxis

Table 11.6 Rail Systems in Great Britain, 1990–91 to 1997–98

	Passenger journeys (millions)			
	1990/91	*1995/96*	*1996/97*	*1997/98*
National rail network	809	761	801	846
London Underground	775	784	772	832
Docklands Light Railway	8	14	17	21
Glasgow Underground	14	14	14	14
Tyne and Wear Metro	44	36	35	35
Altram Manchester Metrolink		13	13	14
Sheffield Supertram		5	8	9

Source: Focus on Public Transport 1999 (table 5)

Table 11.7 Domestic Freight Transport by Mode, Great Britain, 1953–99

	Goods moved in billion tonne kilometres				
	1953	*1970*	*1980*	*1990*	*1999*
Road	32	85	93	136	157
Rail	37	25	18	16	18
Water	0	*	54	56	53
Pipeline	0.2	4	10	11	12
All modes	89	136	175	219	240

Source: Transport Trends 2000 (table 1.14)

Note: *Water transport figures not comparable with later ones, but are included in the total

government since denationalisation has been largely restricted to safety controls, though there has been acrimonious argument over axle weights (or, in popular parlance, juggernauts). Bus services have been particularly affected by conflicting political philosophies. Indeed, fights over fares policy were a significant factor in the Conservative government's decision to abolish the GLC and the MCCs.

Until recently, cycling and walking have been given relatively little attention. The major focus of transport policy has been on roads, and only in recent years has it become generally accepted that they have to be considered within a wider 'integrated transport' framework. The starting point for any discussion of this must be the Buchanan Report, *Traffic in Towns*.

THE BUCHANAN REPORT (1963)

It is traffic in towns which forcibly demonstrates that the motor car is a 'mixed blessing', to borrow the title of an earlier book by Buchanan (1958). As a highly convenient means of personal transport it cannot, other things being equal, be bettered. But its mass use restricts its benefits to car users, imposes severe penalties (in congestion, pollution, and reduction of public transport) on non-motorists, involves huge expenditure on roads, and at worst plays havoc with the urban environment.

A major landmark in the development of thought in this field was the 1963 Buchanan Report. This eloquent survey surmounted the administrative separatism which had prevented the comprehensive coordination of the planning and location of buildings on the one hand, and the planning and management of traffic on the other. With due acknowledgement to the necessarily crude nature of the methods and assumptions used, the report proposed, as a basic principle, the canalisation of larger traffic movements onto properly designed networks that would service areas within which environments suitable for a civilised urban life could be developed. The two main ideas here were for primary road networks and environmental areas:

> There must be areas of good environment – urban rooms – where people can live, work, shop, look about and move around on foot in reasonable freedom from the hazards of motor traffic, and there must be a complementary network of roads – urban corridors – for effecting the primary distribution of traffic to the environmental areas.

The simplicity of this concept is in stark contrast to the complexity and huge cost of its application. But what of the alternatives? Buchanan stressed that the general lesson was unavoidable: 'if the scale of road works and reconstruction seems frightening, then a lesser scale will suffice provided there is less traffic'. The great danger, in Buchanan's view, lay in the temptation to seek a middle course between a massive investment in replanning and a curtailing of the use of vehicles 'by trying to cope with a steadily increasing volume of traffic by means of minor alterations resulting in the end in the worst of both worlds: poor traffic access and a grievously eroded environment'. (This, of course, is precisely what has happened.)

An improvement of public transport is no answer to these problems, though it must be an essential part of an overall plan. The implication is that there must be a planned coordination between transport systems, particularly with regard to journeys to work in concentrated centres. On this, Buchanan recommended that transport plans should be included as part of the statutory development plans. This was accepted and passed into legislation by the 1968 Town and Country Planning Act.

ROAD POLICIES IN THE 1980S

Despite the fact that the Thatcher government had no doubts concerning the economic and social value of roads, its concern for reducing public expenditure took priority during most of the 1980s. White Papers on the trunk road system stressed the importance of roads for economic growth, but 'national economic recovery' demanded a close rein on public expenditure. The top priority within a restricted road-building programme was for 'roads which aid economic recovery and development' (foremost among which was the M25). Other priorities were for environmental improvement (by the building of bypasses), road maintenance ('preserving the investment already made'), and improved road safety. It was expected that the balance of the programme was likely to change as the major inter-urban routes were completed. Increasingly (so it was thought) the emphasis would shift to schemes which were required to deal with specific local problems. This perception was dramatically altered by the 1989 traffic forecasts, and a White Paper of that year, *Roads for Prosperity*, announced a massive increase in road-building.

Between 1980 and 1990 the road network increased by 18,400 km. By 1989, investment in trunk roads was nearly 60 per cent higher in real terms than ten years earlier.[1] The policies continued to follow earlier ones in emphasising the importance of roads to economic growth (despite little evidence on the matter). This 'massive' investment (as it was advertised by the government) was of course a response to the 1989 traffic forecasts, although it was stressed that the forecasts were not really forecasts at all:

> They are in no sense a target or an option; they are an estimate of the increase in demand as increased prosperity

brings more commercial activity and gives more people the opportunity to travel, and to travel more frequently and for longer distances.

(1990 White Paper *Trunk Roads*, England)

But if the official forecasts are not to be used as a basis for policy, what is the alternative? The forecasts were based on the assumption that the demand for roads would be met with an appropriate supply, that there would be no significant policy of traffic restraint, and that attitudes towards motoring (and its cost) would not change. Though the official stance on such issues was a coy one, they raise important questions which are now at the forefront of debate (Goodwin *et al.* 1991).

Increasing concern about the impact of specific road construction schemes, coupled with a more general concern about traffic congestion, resulted in a near-paralysis of policy. This had the tangible advantage of assisting in the restraint of public expenditure, which was further helped by the 1994 roads review. Before discussing this, however, it is useful to examine how the need for roads was approached by the Department of Transport. Successive governments have discovered that 'need' (whether for roads, health services, or houses) is an elusive concept, and many have tried to find ways of giving it an objective basis. Not only is this appealingly rational, but it also changes the nature of the debate. Argument can be settled by recourse to the 'facts'. In a democracy, such nonsense encounters stiff resistance. The history of assessing the need for roads is a good illustration of this.

ASSESSING THE NEED FOR ROADS

Trunk Road Assessment (1977)

An independent assessment by the newly established Advisory Committee on Trunk Road Assessment of the methods used for assessing the need for roads was highly critical. The conventional methodologies were judged to be essentially 'extrapolatory', 'insensitive to policy changes', and partly self-fulfilling. Public concern about road planning was shown to be well founded.[2] There were, however, no easy solutions:

indeed, the issues were inherently complex. The way forward lay in a more balanced appraisal process, 'ongoing monitoring arrangements', and more openness – with no attempt 'to disguise the uncertainties inherent in the whole process'.

The Labour government's response was positive, and its 1978 White Paper *Report on the Review of Highway Inquiry Procedures* represented a marked change in approach. National policies were to be set out for parliamentary debate in White Papers: these would 'also serve as an authoritative background against which local issues can be examined at public inquiries into particular road schemes'. It was hoped that this would avoid the confusion at local inquiries between national policies and their application in specific areas. It was pointed out, however, that this would work only if the methods of assessing national needs (what the Committee termed 'a highly esoteric evaluation process') were acceptable. Since these methods could not be properly examined at local inquiries (or, indeed, by Parliament), they were to be subject to 'rigorous examination' by the Committee (now elevated to a Standing Advisory Committee on Trunk Road Assessment (SACTRA). The Committee's report was published in 1979, under the title *Trunk Road Proposals: A Comprehensive Framework for Appraisal*.

This report examined the techniques used to evaluate the economic value of proposed road schemes. The Department's system of cost–benefit analysis (the COBA programme) was criticised for the narrowness of its approach, and certain changes followed. For example, instead of using only one traffic forecast, high and low levels were introduced.

Urban Road Appraisal (1986)

The 1979 SACTRA report dealt with inter-urban roads; in 1984 it was given the task of assessing the traffic, environmental, economic, and other effects of road improvements within urban areas. Some urgency for a review was added by the abolition of the GLC and the metropolitan county councils: this gave the Secretary of State, the London boroughs, and the metropolitan district councils new responsibilities for tackling the transport problems of major urban areas.

The Committee's report, *Urban Road Appraisal*, together with the government response, was published in 1986.

Much of the report sets out recommendations as principles rather than as detailed prescriptions, and it thus seems to have more than a fair share of platitudes. As a result, though the government accepted many of these, they 'can only be applied once detailed guidance on their application has been prepared'. Moreover, 'where further development or research is needed, the detailed guidance required to implement them will take some time to prepare, and progress must be subject to the availability of resources'. The Committee could hardly have found this a very encouraging response!

There were four broad areas which the Committee saw as contributing to the nature and extent of change needed:

- concern about the way in which the Department develops, assesses, and justifies major schemes, particularly in urban areas;
- expectations about public involvement;
- the importance of integration between transport proposals and broader land use planning and environmental considerations; and
- the complexity of the planning, assessment and decision-making processes, especially in their application to urban areas.

Interestingly (in view of the common contention that public participation adds to delay), the Committee maintained that one reason why many road schemes took so long to bring to completion was that 'the opportunity to debate their justification comes too late in the procedural chain'. This was accepted, as was a recommendation that national and local objectives should be treated separately 'so that conflicts and common denominators can be readily seen'. However, a recommendation that there should be a two-stage assessment and public inquiry process for the larger and more complex schemes 'presented difficulties'. Though it was regarded as 'a constructive proposal', the government was 'not convinced of the practicality of separating the examination of policy options from consideration of detailed design and local issues'. The

Urban Road Appraisal report also echoed the widespread unease about the methods used to assess the economic value of road schemes, but its recommendation was couched in such broad terms that the government had no difficulty in side-stepping it by maintaining that it reflected existing practice.

Environmental Impact and NATA

The issues, which successive governments might have hoped would be settled by the various inquiries, refused to disappear: in fact, they became more problematic as public attention widened to encompass more and more matters which had not traditionally been regarded as pertaining to roads. Above all, concern has grown enormously about the environmental effects of roads and, indeed, of all forms of traffic. Inevitably, further inquiries were commissioned, including another SACTRA report (*Assessing the Environmental Impact of Road Schemes*), which was published in 1992.

By this date, the arguments about the limitations of COBA had intensified. Not only were environmental considerations now at the forefront of the debate (particularly after the shock of the 1989 traffic forecasts), but it was being argued that any sharp distinction between economic and environmental impacts was false (Pearce *et al.* 1989). Earlier reports had advised that environmental benefits and costs should not be evaluated in money terms but should be subject to 'professional judgement'. The rationale for this is a simple one: there is no acceptable way to estimate the 'value' of a cathedral, a marvellous view, or other such 'non-economic goods' – though this has not stopped economists from trying (Schofield 1987). But this leads to a host of mind-boggling questions. What is the value of land which is safeguarded from development? Is it the 'economic' value for development, or the lower 'social' value which is determined by planning controls? Which value should be used in evaluating alternative routes for a road? Other questions are equally baffling: if environmental factors are important, which should be taken into account and which should be ignored? (The SACTRA report has a long list of local, regional, national, and global factors.)

How are the cumulative effects of a multiplicity of apparently unimportant decisions to be dealt with? Will future increases in traffic increase environmental damage, or will technological innovations more than offset these? The range and number of questions seem endless. No wonder that cost–benefit analysis is having a hard time![3]

By 1998 the DETR had formulated a New Approach to Appraisal (NATA), which was used in the Roads Review of that year. The new approach assesses transport investment projects against the government's five objectives for transport: environmental impact, safety, economy, accessibility, and integration. The emphasis is on transparency in the presentation of the evaluation and requires a one-page summary table drawing together economic environmental and social factors. COBA is still present in the analysis, and monetary values are assigned where they can be, but presented alongside other quantitative and qualitative analysis 'without giving prominence to any one type of effect or to benefits expressed in monetary terms'. It is accepted that different effects cannot be aggregated or compared directly, so consequently the analysis does not make judgements but only provides information for decision-makers. Thus it is now clearer that it is the politicians and their advisers, and not the appraisal technique, who make the decision. A NATA is now required of all transport authorities, and extensive guidance on how it should be applied has been made available.[4]

Do New Roads Generate Traffic? (1994)

A major issue in the debate on forecasting traffic needs is the extent to which new roads actually generate extra traffic. Certainly, their immediate effect on pre-existing roads can be dramatic, but may be short-lived. Traffic seems to increase faster than new roads can be built. US studies have argued that, typically, the creation of new road space is eventually (and it may be sooner rather than later) taken up by increased traffic. Where does this traffic come from? Downs (1992) has put forward an elegant explanation in his theory of 'triple convergence'. This is based on the simple fact that since every driver seeks the easiest route, the cumulative result is a convergence on that route. If it then becomes overcrowded, some drivers will switch to an alternative route which has become relatively less crowded. These switches continue until there is an equilibrium situation (which like any human equilibrium is not stable – conditions constantly change). On this theory, building a new road, or expanding an existing one, will have a 'triple convergence'. First, motorists will switch from other routes to the new one ('spatial convergence'); second, some motorists who avoided the peak hours will travel at the more convenient peak hour ('time convergence'); third, travellers who had used public transit will switch to driving since the new road now makes the journey faster ('modal convergence').

The eventual outcome depends upon the total amount of traffic (actual and potential) in relation to the available roads. If the increase in traffic stimulated by the new road is modest, there will be an observable benefit for all. Though peak-hour traffic may be congested, this is simply because so many drivers are travelling at the time which is most convenient to them. (There may, however, be a loss to transit passengers if the 'modal convergence' leads to a reduction in service.)

On this argument, new roads can generate traffic by diversion from public transport. Are there other ways in which extra traffic can be generated? Intuitively, it would seem obvious that there are: the more congested and difficult a road journey is, the more likely it is that a potential traveller will seek an alternative. Conversely, ease of road journeys must generate increased trips. Of course, if this were self-evident, it would not have taken SACTRA over two hundred pages to discuss it; nor would the Department of Transport have been so resistant to it. Indeed, the matter is a complex one, mainly because it is difficult to establish cause and effect over time on matters where there are many variables. However, the Department accepted the thrust of the SACTRA report, even though it held that 'clear evidence' was lacking. A Guidance Note on Induced Traffic was issued (DoT 1/95), and the Department's ongoing research programme was augmented.

At the risk of overstretching the discussion, a note on the effects of the Newbury Bypass indicates that the argument is still very much alive. This bypass aroused an extremely bitter controversy and violent opposition, resulting in a huge expenditure on security and ejection of protesters. (The cost of building was £74 million; the cost of policing was an extra £26 million). A report by the local authority (West Berkshire) less than a year after its opening in November 1998 states that reduction of peak traffic has been only 25 per cent. (The prediction was 40 per cent.) Three factors are suggested for the limited impact of the bypass:

> People have retimed their journeys through the town; local people who had suppressed car use altogether may have taken to their cars again in the belief that congestion had been reduced; and more traffic has been pushed on to the ring road by the pedestrianisation scheme [in the town centre].
>
> (*Guardian*, 12 July 1999)

This is a telling example of the acute problems raised by road-building; and there is a wider point: new roads not only induce traffic, but also encourage car owner-ship and use. As car use increases, other methods of transport are used less and, as a result, standards of service fall (thereby further increasing the attraction of car use). Moreover, road-building and ease of car use have major impacts on the location of new devel-opments (of all kinds – housing, employment, shopping, and leisure). Many new locations are car dependent, and therefore may increase the demand for road travel, and hence the need for more road construction. In this cumulative way roads certainly generate more traffic. However, it is extremely difficult to forecast patterns of land use, travel, and the inter-actions between land use and transport. Nevertheless, by the early 1990s it was clear that some fundamental changes in transport policy were needed.

Against this background it is not surprising the Major government attempted to fashion a number of new policy instruments. Among these were a review of trunk road building, traffic management in London, traffic calming, a cycling strategy, and a step towards the integration of transport and land use planning with the publication of PPG 13.

TRUNK ROADS REVIEW OF 1994

The early 1990s witnessed increasing recognition of the impossibility of catering for a continuation in the growth of road traffic, though acceptable alternative policies seemed elusive. A bumper crop of reports in 1994 (including the SACTRA report discussed in the previous section) provided conflicting advice and a massive excuse for further delays in taking positive decisions. However, the *Trunk Roads in England: 1994 Review* (and its 1995 successor *Managing the Trunk Road Programme*) did signal a significant shift in road-building policy. It detailed a reduced road programme (announced in the previous year): a total of forty-nine road schemes were withdrawn completely, and many others were postponed.[5]

This was one of the first steps in a major reorientation of transport policy. Though

> it is no part of this Government's policies to tell people when and how to travel . . . we must be aware of the consequences if people continue to exercise their choices as they are at present. There is no realistic possibility of simply halting traffic growth. . . . The Government's policy for sustainable development is to strike the right balance between securing economic development, protecting the environment, and sustaining future quality of life.

Resources were now to be devoted to the improvement of sections of existing key routes which were likely to experience congestion in the near future (primarily by adding lanes to existing motorways) and to providing urgently needed bypasses. Existing proposals for new trunk routes were to be reduced still further, and the programme of major urban road improvements would be a very limited one. Clearly, this represented a sea change in transport policy.

ROADS POLICY SINCE 1997

The 1998 integrated transport White Paper (discussed on p. 336) apparently put the seal on the shift in emphasis away from roads, saying that 'the priority will be maintaining existing roads rather than building new ones. . . . Simply building more roads is not the answer to traffic growth.' The essence of the new roads policy (*A New Deal for Trunk Roads*, 1998) is essentially to build as few as possible. 'Since new roads can lead to more traffic, adding to the problem, not reducing it, all plausible options need to be considered before a new road is built.' The Highways Agency has been given a new strategic aim of giving priority to better maintenance[6] and making better use of existing roads; and putting greater emphasis to environmental and safety objectives. Responsibility for some 40 per cent of existing trunk roads will be transferred to local authorities (that is, they will be 'de-trunked'). The other 60 per cent have been identified as the nationally most important routes (the 'core network'), and will remain the responsibility of the Highways Authority. In future, strategically important improvements will be planned through the

regional planning guidance system 'to ensure integration across all forms of transport with land use planning'. A wide range of policies are being developed to deal with such issues as better safety, better driver information, tackling noise, protecting the environment, and environmental protection. The most radical proposal is the tentative introduction of tolling on trunk roads. As with the road user and workplace charges which are proposed for local authorities (outlined on p. 343), a start will be made with small-scale pilot charging schemes. Technical trials of electronic systems and their impacts are under way.

But intentions are one thing, actions are another. The Environment Transport and Regional Affairs Committee has questioned the government's commitment to the shift in transport policy away from roads, pointing to the plan to invest £21 billion in new roads over ten years.[7] Whilst there have been shifts in

---

BOX 11.1　THE BYPASS DEMONSTRATION PROJECT

Traffic calming is particularly appropriate after the completion of a bypass. Though the town may feel that the bypass has solved their local problems, in fact it can bring new problems in its wake. The old route will have all the features of a heavily trafficked route: it will bear all the marks of a road which has been adapted (and perhaps mutilated) to accommodate high levels of traffic. Quite apart from the poor appearance of the place, the reduced traffic will facilitate higher speeds, and paradoxically, new traffic hazards may appear. And (the final irony) though traffic will initially decrease significantly, it can soon build up again.

To explore how these problems can be dealt with, the DoT in conjunction with local authorities mounted a Bypass Demonstration Project (announced in the White Paper *This Common Inheritance*, and completed in 1995). A major object of this was to demonstrate how the benefits of a

bypass can be enhanced by an overall improvement scheme. Six towns were selected and, with some financial support from the DoT, major traffic calming and other 'town enhancement' works were undertaken.

The report on the study revealed the problems and opportunities. The removal of through traffic allows a radical change in the street space. Inevitably the benefits are not equally shared, but the six project towns have shown how pedestrians, visitors, cyclists, disabled people, and civic uses in general can benefit substantially, whilst still maintaining vehicular access in a traffic-calmed environment. Such schemes demand a great deal of professional input, coordination of effort, public involvement, and cost: the expenditure in the six towns ranged between £1m and £2m. The benefits are striking, and can amount to a transformation of the area.

Source: DoT (1995) *Better Places through Bypasses: Report of the Bypass Demonstration Project*. (The six towns were Berkhamsted, Dalton in Furness, Market Harborough, Petersfield, Wadebridge, and Whitchurch.)

government attitudes on road-building, the road lobby 'has always found a willing ear at the Ministry' and continues to do so, primarily because of the continuing influence of the outdated idea that new roads stimulate economic growth (Woolmar 1997).

INTEGRATED TRANSPORT PLANNING: PPG 13, *TRANSPORT*

The most significant event in the evolution of a transport policy under the Major government was the publication of planning policy guidance on transport. The importance of this was that it provided guidance on the integration of transport and land use planning (replacing an earlier *Highway Considerations in Development Control*). There had been many calls for such an integration, and these were given weighty support in the 1994 report of the Royal Commission on Environmental Pollution (*Transport and the Environment*). The key aim of PPG 13 was to ensure that local authorities carry out their land use policies and transport programmes in ways which help to 'reduce growth in the length and number of motorised journeys; encourage alternative means of travel which have less environmental impact; and hence reduce reliance on the private car'.

Thus transport planning explicitly became a major component not only of land use planning, but also of environmental policy and of the UK Sustainable Development Strategy. Indeed, the PPG underlined the government's commitment 'to providing a policy framework which will help to ensure that people's transport decisions are compatible with environmental goals'. This was to be facilitated by the policy of increasing the real level of fuel duty by at least 5 per cent each year (a pledge that was dropped in 2000 in the face of protests on the increasing costs of fuel). Electronic tolling on motorways was also envisaged when the appropriate technology became available.

It was stressed throughout PPG 13 that the relationships between transport and land use planning had to be carefully examined at all levels of planning, and that integration and coordination had to be promoted by regional planning guidance (through the regional conferences of local authorities) and in structure plans and local plans. Strategies were required that would reduce the need to travel and maximise the opportunities for travel by public transport. Car parking was also a strategic matter (with policies to be set out in regional guidance and structure plans) 'to avoid the destructive potential for competitive provision of parking by neighbouring authorities'. Other matters dealt with included plans for safe and attractive areas for pedestrians; provision for cyclists; traffic management; provision of park and ride schemes; and 'accessibility profiles' for public transport in order to determine locational policies designed to reduce the need for travel by car.

The proposed revision of PPG 13, published in 1999, continues in the same vein, though with stronger emphasis on guaranteeing access by public transport to new developments; ensuring forms of development that encourage non-motorised transport; and implementation through green transport plans, transport assessments, and national car parking standards. The guidance also includes preferred locations for particular types of development (housing, shopping, leisure, and services), but as the Civic Trust has noted, the guidance is very general, mostly in the form of 'put new development where it is accessible', and adds little to what is said in other PPGs.[8]

The PPG is great for ideas, but lacking in advice on how they might be implemented. Further advice is given in the *Guide to Better Practice* (PPG 13). The guidance has quickly found its way into development plans (indeed, some plans provided the ideas) and the policies have became influential in the outcome of appeals. The PPG had an immediate and strong influence on retail development proposals, and later on office and other major developments.[9]

In rural areas, however, there were seen to be fewer opportunities for achieving public transport improvements or control of parking, and there are even greater concerns about the economic impacts of accessibility. There are other difficulties because of potential conflicts between policy objectives (for example between encouraging development in town centres and controlling air pollution). Deciding among different types of non-car transport is problematic, as is the task of obtaining adequate information from developers. There is a need for better information on travel behaviour and modal split. This will be addressed in part by a new system of transport assessments to replace the system of traffic impact assessments. At the time of writing, further guidance is promised on how these will be conducted, but they are likely to be applied to developments over the thresholds that are proposed for the maximum parking standards. As is common with new policy developments, practicalities can be more limited in practice than anticipated, particularly until some experience has been gained.

THE NATIONAL TRANSPORT DEBATE AND THE 'NEW DEAL'

Though the first PPG 13, and similar policy statements for other parts of the UK, attempted to settle some major policy issues, they merely provided a guide to the issues which needed to be debated. The Conservative government was baffled by the complexities, by the need to reduce pollution, by the difficulties of persuading a largely car-owning electorate to use far less convenient forms of transport, and by its preoccupation with privatising British Rail. The Secretary of State for Transport, Brian Mawhinney, made a valiant attempt 'to reclaim the initiative' (Glaister *et al.* 1998: 6) by launching a series of speeches on salient transport issues.

The threads were drawn together in a Green Paper, *Transport: The Way Forward*.[10] This was a well-documented paper, but quite indecisive. Many of the ideas in the paper appear in some form or another in the Blair government's Consultation Paper *Integrated Transport Policy*, and subsequently found their way to the 1998 White Paper, *A New Deal for Transport: Better for Everyone*.[11] The White Paper is an introduction and summary of a series of 'daughter' documents on such issues as roads, buses, freight, and parking charges. The range of issues covered is unprecedentedly wide – from wheel clamping to railways, from safety to public transport links to airports, from vehicle emissions to measures for a more inclusive society. The *New Deal for Transport* promises to improve the urban

environment by creating the conditions for people to move around more easily. More road space and priority are to be given to pedestrians, cyclists, and public transport. The meaning of 'integrated transport policy' is described at the outset (Figure 11.2). Clearly, this is a very broad conception, and the details are spelled out over many pages. Some of the major features are summarised below.[12]

BOX 11.2 INTEGRATION OF TRANSPORT POLICY

An integrated transport policy means:

- integration within and between different types of transport – so that each contributes its full potential and people can move easily between them;
- integration with the environment – so that our transport choices support a better environment;
- integration with land use planning – at national, regional and local level, so that transport and planning work together to support more sustainable travel choices and reduce the need to travel;
- integration with our policies for education, health and wealth creation – so that transport helps to make a fairer, more inclusive society

Source: White Paper, *A New Deal for Transport*, 1998

Separate papers were published for Northern Ireland, Wales, and Scotland, where transport problems are rather different.[13] For instance, the pattern of car ownership is significantly different in Scotland: in 1997 there were 30 motor vehicles for 100 people, compared to 48 in England and Wales. There is therefore 'a stronger need in general to cater for people who do not have access to a car' (though, of course, there is a 'greater potential for further expansion in car ownership'). There are also marked differences in car ownership between different parts of Scotland, reflecting geography, wealth, and economic activity. Rural Scotland, however, has a relatively high rate of car ownership – more a matter of necessity than wealth.[14] In Northern Ireland 'current levels of congestion, even in Belfast, are not at levels that justify dramatic action to actively restrain the use of the private car'. However, 'this provides greater opportunity to . . . ensure that the growth of car dependence is limited', though restraining future traffic growth 'will take a long time and it will raise many complex issues'. For the immediate future, the priority will be for measures designed to encourage a change in travel behaviour away from car dependence, rather than restrictions on choice.[15]

IMPLEMENTATION OF THE INTEGRATED TRANSPORT POLICY

The White Paper provides an indication of the ways in which integrative policies can be implemented. These include the establishment of a new Independent Commission for Integrated Transport; the introduction of local transport plans and revised regional planning guidance; extensive partnerships and cooperation between transport providers; and, 'where necessary', strengthened local authority powers to secure integration. Implementation of ideas in the Papers is under way through the government's £180 billion transport investment programme set out in *Transport 2010: The 10 Year Plan* (DETR 2000), and through the Transport Act 2000. Planning policy guidance on transport (PPG 13) is being revised, and impacts on other policy for development plans housing and other uses are under consideration.

Extra funding was planned for transport following the Comprehensive Spending Review (announced in July 1998). This was for a total of £1.8 billion over the three-year period 1999–2000 to 2001–02. The main elements of this are £700 million more for local transport, £300 million for local bus services, more than £300 million for the rail industry, and over £400 million for the trunk road and motorway network.[16] Since then the government has published the Ten-Year Plan, which promises £60 billion each for rail, roads, and local transport. The objectives in the plan are ambitious: a 50 per cent increase in passenger use on the railway, twenty-five new light rail projects, and 10 per cent increase in bus use, among others.

The Commission for Integrated Transport (CIT) 'is to provide independent advice to government on the implementation of integrated transport policy, to monitor developments across transport, environment, health and other sectors, and to review progress towards meeting our objectives'. Among the specific issues to be addressed by the CIT are the setting of national road traffic and public transport targets, revisions to the National Road Traffic Forecast; lorry weights; the development of rail freight; review of transport safety arrangements; progress with green transport plans; the new rural bus partnership fund in England; and research needs.

TRAFFIC PLANNING MACHINERY

Traffic policies and planning have evolved (and continue to evolve) over a long period of time. Initially, the main, if not the only, relevant matter was a road network plan. There have always been differing views on this. During the 1930s there was rivalry between the highway engineers, championing a 2,800-mile motorway system, and the county surveyors, favouring a more realistic 1,000-mile network – which became the basis for the motorway building programme (Kay and Evans 1992: 18). But there was little interest in plans for transport as a whole.

A major change came with the Labour government's 1967 White Paper *Public Transport and Traffic*. This heralded a new approach to transport planning: 'our major towns and cities can only be made to work effectively and to provide a decent environment for living by giving a new dynamic role to public transport as well as expanding facilities for private cars'. Since local authorities were responsible for 'planning', they were obviously the appropriate authorities for transport. All forms of transport, it was argued, needed to be planned together in a coordinated way. However, in major urban areas, existing local governments were too numerous and too small. Thus some kind of *ad hoc* system was necessary.

This led to the establishment of passenger transport authorities (PTAs), under powers provided by the 1968 Transport Act, in Greater Manchester, Merseyside, the West Midlands, Tyneside, and Greater Glasgow. Subsequent local government reorganisation gave PTA status to all the English metropolitan county councils and the Strathclyde Regional Council. (Following the abolition of the metropolitan county councils, the PTAs were resuscitated as *ad hoc* authorities.) County councils in the non-metropolitan counties were given parallel duties in relation to 'a coordinated and efficient system of public transport'. There was corresponding provision in the Scottish Local Government Act for the regional and islands authorities. In London the Greater London Council was the PTA (until 1984, when London Regional Transport was established).

A major feature of this new organisation was that it facilitated the preparation of comprehensive transport policies and programmes (TPPs), as well as providing a mechanism for channelling financial support not only to roads but also to public transport services. This was particularly attractive to a number of Labour councils which wished to strengthen, and indeed favour, public transport. Unfortunately, a period of financial stringency followed, and this, together with the fragmented nature of the grant system (despite original aims of 'integration'), killed the rational basis of the new system (Skelcher 1985). Nevertheless, TPPs survived in an attenuated form, and they continued to operate until overtaken by local transport plans in 1999. The main concern of TPPs was with roads, but in the early 1990s there was a noticeable shift in emphasis to a more balanced approach which enabled local authorities to submit package bids covering investment proposals for both roads and public transport, together with a supporting comprehensive transport strategy (Cook and Davis 1993).

REGIONAL TRANSPORT STRATEGIES

Planning for transport is being strengthened through integrated strategies at the regional and local levels. In England, a regional transport strategy (RTS) now has to be prepared as an integral, but clearly identifiable, part of regional planning guidance. The RTS must set out regional priorities for all forms of transport, guidance on the integration of different services, accessibility criteria for regionally significant forms of development, and the strategic context for demand management such as road pricing. The regions are encouraged to undertake 'multi-modal studies' (MMSs) which examine the role of different transport modes within an area or corridor. The government is undertaking a series of MMSs for the major corridors around the country, the results of which should be reflected in regional guidance.

A doubt remains: given the current inadequacy of means for central government to ensure that local authorities comply with regional guidance,[17] there must be major concerns about implementation, particularly in view of the great unpopularity of some of the measures proposed, and fears of competitive 'inaction' by neighbouring authorities. It remains to be seen whether the new regional planning system will be sufficiently effective to overcome such problems.

LOCAL TRANSPORT PLANS

Local transport plans (LTPs) (local transport strategies in Scotland) replace TPPs. Initial guidance on LTPs was published by the DETR in 1998 and provisional one-year plans were published in 1999. Revised guidance and a good practice guide were published in 2000, and the Transport Act 2000 made five-year LTPs a mandatory requirement.[18] The emphasis of the LTP should be on integration through involvement of all relevant interests. The guidance is lengthy and detailed, but a selective summary is given in the Box 11.3. LTPs should be consistent with appropriate development plans, and eventually the two should be integrated with each other. They should also be consistent with regional transport strategies.[19]

BOX 11.3 CONTENT OF LOCAL TRANSPORT PLANS: SUMMARY OF SELECTED TOPICS

LTPs must demonstrate consistency with the Government's transport objectives, and cover all travel modes, including:

- voluntary or community transport (or its potential), particularly in rural areas;
- local strategy for cycling and walking targets;
- traffic management and demand restraint;
- enforcement of emission standards;
- proposals for pilot schemes for road user charging and taxation of workplace parking;
- cooperation with major retailers and leisure operators on car access and alternative means of access;
- integrated strategy on parking, planning policies, and transport powers;
- local road casualty reduction target;
- interchange improvements;
- bus-based park and ride schemes and related reduction in town centre parking or pedestrianisation;

- proposals for capital expenditure on public transport information schemes;
- promotion of green transport plans by employers;
- integrated strategy for travel to school;
- planning and management of the highway network;
- strategy for rural transport, and for countryside traffic management schemes;
- issues connected with freight distribution;
- promotion of social inclusion, including disability issues and extension of bus access for welfare to work;
- action on climate change, air quality, and noise.

Source: DETR (1998) *Guidance on Local Transport Plans*

BOX 11.4 IMPACT OF TRANSPORT POLICIES IN FIVE CITIES

There is a wealth of experience on the impact of different types of transport policy on road traffic, but there is a bewildering range of possibilities. An indication is given by a 1994 report from the Transport Research Laboratory (Dasgupta et al.). This investigated the effects of various policies on urban congestion in five cities (Leeds, Sheffield, Derby, Bristol, and Reading). It was found that halving public transport fares increased bus use by between 7 per cent and 20 per cent, but the proportionate effect on car use was slight: only 1 to 2 per cent. Other options examined included raising fuel costs by 50 per cent, doubling parking charges,

halving the number of parking places, and applying a cordon charge of £2 in the peak and £1 in the off-peak period. The latter two measures had the greatest effect: they reduced car use in the central areas by about a fifth (and increased it in the outer area by between 3 and 5 per cent). Different types of policies have different effects, but they also vary among cities, and between peak and off-peak periods. The study concludes that, when interpreting the results, it is important to take into account the complicated interrelationships among modal transfer, redistribution, changes in vehicle-kilometres, and changes in trends.

Source: Dasgupta *et al.* (1994)

TRAFFIC MANAGEMENT IN LONDON

The Road Traffic Act 1991 provided a new legislative framework for traffic management for London. The Act empowered the Secretary of State to designate a network of priority routes (commonly known as red routes because of their distinctive red road markings and signs) which are subject to special parking and other traffic controls. They are aimed at reducing traffic congestion and improving traffic conditions on main routes, particularly for buses, without encouraging additional car commuting into central London. A pilot scheme in 1991 proved successful: overall journey times improved by 25 per cent, bus journey times were reduced by more than 10 per cent, and reliability increased by a third. More people used buses, and road casualties fell significantly. As a result, a permanent scheme was introduced in 1992. This network covers all trunk roads in London as well as local roads which are of strategic importance.

The planning, coordination and implementation, maintenance, and monitoring of traffic management on the network are the responsibility of an *ad hoc* (non-departmental) body, the Traffic Director for London (also established by the 1991 Act). The network plan forms the framework for detailed local plans, which are the responsibility of the London local authorities. Central to this new system was a package of traffic management schemes and a reform of on-street parking in Greater London. Traffic management measures include increased priority for buses, improved pedestrian crossings, enforcement of parking regulations, and encouragement to cyclists to use alternative roads to red routes except where separate cycle tracks can be provided. A range of traffic-calming measures were implemented on side roads which might be affected by the red route traffic. These regulate speed and deter motorists from using side roads as 'rat runs'. In 20 mph zones there is the additional advantage that the road hump regulations are far more relaxed: for example, warning signs are not required. Doubts about the legality of traffic calming measures were settled by the Traffic Calming Act 1992, which provided for the making of regulations governing them.

In addition to other parking restrictions, the 1991 Act provided for a new system of special parking areas. In these areas, which are designated by London local authorities, parking contraventions are no longer criminal offences, and traffic wardens are replaced by local authority parking attendants. The powers of these attendants include issuing penalty charge notices and authorising wheel clamping and the removal of vehicles. A Joint Parking Committee, appointed by the London local authorities, has the duty of setting certain parking charges, and appointing parking adjudicators, who have a comparable role to magistrates' courts under the former system of criminal parking controls. (The new system is technically known as 'decriminalised enforcement'.) These and similar provisions amounted to an elaborate new system for traffic management in London. (PPG 13 promised that these powers would be extended to local authorities outside London, and this has now been done.) Under the new government arrangements for London, the GLA has to prepare a transport strategy covering infrastructure needs, parking, and access.

WALKING AND CYCLING

Interestingly, the first issue discussed in the *New Deal* White Paper is 'making it easier to walk'. This nicely emphasises the priority for pedestrians. Measures include more pedestrian crossings, more direct and convenient routes for walking, and increased pedestrianisation (illustrated by a striking photograph of a pedestrianised Trafalgar Square).[20] Speed limits and 20 mph zones are already being introduced, 'with markedly beneficial effects'.[21] The Transport Act 2000 gives local authorities the opportunity to make orders to create 'home zones' or quiet lanes to govern traffic and reduce speeds. A strategic view of how walking and cycling can be encouraged will be set out in regional planning guidance. Implementation at the local level will be set out in local transport plans and development plans.

Almost one in five car trips on the urban network at 8.50 in the morning are for the purpose of taking children to school. Walking and cycling to school are to be encouraged by planning safer routes. School travel plans are to be produced by local authorities and schools. A new School Travel Advisory Group has produced a best practice guide and a volume of thirty case studies.[22] In Scotland a Scottish Walking Strategy Forum has been established to consider how walking could be made more popular. There is also a Scottish Cycle Challenge Initiative which promotes projects that encourage cycling to work and to school.

Britain lags behind many other countries in the use of cycles.[23] The world has twice as many bicycles (around 800 million) as cars, and bicycle production outnumbers cars by three to one. There are over thirteen million cycles, and over a third of all British households have at least one. About eleven million people use their cycles at least once a year; in an average week about 3.6 million are used. Over a million people use a bicycle as their main means of transport to work. But only 2 per cent of total trips are made by cycle. The total distance travelled by cycles is between five and six million kilometres a year, compared with around 350 million kilometres for cars. On the other hand, a greater distance is travelled by cycle than by bus and train combined. Despite this apparent abundance of statistics, it is difficult to obtain an accurate picture of cycle use (and still less of any potential increase). Cycling rose higher on the political agenda in 1994 when the DoT published its June 1994 *Cycling Statement*. However, this rather unconvincingly extolled the virtues of cycling, and exhorted local authorities to do more to facilitate and encourage it as a healthy, environmentally friendly, economical, and efficient means of transport for local journeys.

A more tangible boost came with the Millennium Commission's £43 million grant towards the 6,000-mile National Cycle Network organised by the charity Sustrans. The benefits of the scheme have been prominently advertised: '6,000 miles of high quality cycle and pathways within two miles of 21 million people will be provided, linking towns and cities from Dover to Inverness to Belfast' (Sustrans 1996). Half of the network is entirely free of vehicular traffic. Sustrans estimates that there will be more than 100 million journeys a year on the network. Perhaps the greater long-term benefit, however, will be an improvement in the image of the cycle as a means of transport.

It was, however, the 1996 *National Cycling Strategy* which marked a genuine change in governmental attitudes to cycling. This claims to represent 'a major breakthrough in transport thinking in the UK'. The target, also present in the *New Deal* White Paper, was to double the number of trips by cycle by 2002, and quadruple the number by 2012. (The *10-Year Transport Plan* changed the target to trebling cycling trips from 2000 to 2010. It is not clear where these figures come from, but they exhibit an eagerness which to date has been restricted to cycling enthusiasts. They are ambitious, and this may be the reason that neither target has been included in the formal list of DETR objectives.) The *Strategy* covers a wide range of relevant issues, including appropriate planning measures, safety, provision of parking for cycles (and accommodation for them on public transport), integrating cycling with traffic management, cycle security, and the 'communication programme' needed to change attitudes to cycling.

BUSES AND RAPID TRANSIT

Buses are seen as 'the workhorses of the public transport system', but for them to assume that role requires improvements to their design and comfort, cleanliness, regularity, reliability, and interconnection (with both other buses and other forms of public transport), greater priority on the road, and improved information for passengers. Experience to date (given the background of privatisation, deregulation, and competition) points to the practicalities of partnerships with the industry. The *New Deal* White Paper says that 'quality partnerships' (already in operation in cities such as Aberdeen, Brighton, Leeds, and Swansea) will be developed by local authorities against the background of their local transport plans. The details of these partnerships are spelled out in a separate document (*From Workhorse to Thoroughbred: A Better Role for Bus Travel*, 1999). Quality partnerships will be given a statutory basis 'so that all concerned can have the confidence to invest'. Local authorities will have a range of powers to ensure service stability, good timetable information, and systems of flexible joint ticketing.

At the national level, a Rural Bus Partnership fund of £45 million a year is already operating to support bus services in rural areas. This is to be supplemented by a new *Rural Transport Partnership* to enable parish councils and local groups to work in partnership with local authorities. The aim here is 'to support schemes which reduce rural isolation and social exclusion through enhanced access to jobs and services'. Local transport plans deal with the operation of public transport through partnerships, but development plans have a role in identifying the key routes for bus improvements, priority measures, and interchange facilities. The planning authority also has a key role in negotiating gains from development that can provide improvements to public transport. Access to public transport is an important material consideration in the control of new development.

Rapid transit systems tend to be very expensive (particularly if they use a fixed rail), and they were out of favour for many years. Schemes had been approved earlier for several cities, including Glasgow, Tyne and Wear, Merseyside, and London (the Jubilee Line), but many more were shelved. Increased road congestion (and prospects of much more in the future), the model of the London Docklands Light Railway (promoted as part of the Docklands renewal strategy), and increasing experience of foreign systems reawakened political interest in rapid transit.

In the early 1990s there were forty urban areas with proposals for rapid transit,[24] but despite government predictions few have been realised. Britain does not compare well with other European countries on rapid transit, and this may in part be due to the fact that public transport generally is expected to cover a large proportion of its operating costs.[25] This makes the outlook for rapid transit in Britain less certain, though there is some comfort in the fact that most foreign public transport networks have improved their revenue–operating cost ratio. There is, however, the added difficulty in Britain that any rapid transit system would find itself in competition with deregulated bus services.

Current government policy promises that the potential for rapid transit will improve for some selected cities, but the overall position is that bus priority measures offer more cost-effective alternatives. (But will they tempt drivers to give up their cars?) The Ten-Year Plan provides for 'up to 25 new rapid transit lines in major cities and conurbations', still well below early proposals, but it will speed implementation of systems that have been growing only very slowly.[26] In the meantime, the government has been wrestling with the problem of modernising the London Underground. Proposals for splitting up the operation among various public–private partnerships are being strongly contested by the Greater London Authority.

RAILWAYS

The government's view, as expressed in the *New Deal* White Paper, that there is the potential for a 'railway renaissance' pre-dated the unprecedented chaos that enveloped the railway system in the winter of 2000-01. Readers will, no doubt, have their own nightmare stories of travelling by rail during that winter, when the simplest of journeys could become a major challenge. The parlous state of British railways was not, of course, created overnight. It is the product of many years of neglect, weak investment, and poor management, and was well known to many rail commuters, especially in the south-east of England. But it became a national scandal almost overnight following the Hatfield rail crash.[27] In the aftermath of the crash Railtrack, which is responsible for the track infrastructure, was forced to bring forward its programme of repair and maintenance and imposed hundreds of speed restrictions. The effects on timetables was exacerbated by severe flooding. The ensuing chaos revealed the depth of the demise of the railway system.

John Major's Conservative government had privatised and broken up British Rail into many separate companies so as to stimulate competition and investment. The current government claims that investment actually fell until 1997. Such claims are contested, but it is a fact that standards have fallen. Indeed, between 1999–2000 and 2000–01, one train operator improved its performance while twenty-two declined. After Hatfield, the collapse of the rail performance was so great as to render 'any meaningful comparison impossible'.[28] There is no question of renationalising the railways, but a national *Strategic Rail Authority* for Great Britain has been established which is intended to provide 'a clear, coherent and strategic programme' for development. It has taken over the functions of the Franchising Director, the British Railways Boards, and some of the responsibilities of the Rail Regulator and DETR. The Authority will be subject to instructions and guidance laid down by ministers in accordance with the new integrated transport policy, but fragmentation, which is at the heart of much of the current problem, will remain for the foreseeable future.

Investments are needed in infrastructure works and rolling-stock improvements. Some lines can be substantially improved at relatively short notice and at moderate cost, using more and longer trains, extended platforms, and improved signalling. Others face 'pinch points' that restrict capacity, such as the Glasgow Central approaches and the East Coast main line between Finsbury Park and Peterborough. In all, there are fifteen key bottlenecks that have been identified by Railtrack: a programme for solving these could be completed by 2006.[29] The Ten-Year Plan proposes £49 billion worth of investment in the railways, £26 billion of which will come from the public sector. These figures look impressive but are not so different from what has gone before. Much reliance is placed on levering huge sums of private-sector investment.

FREIGHT TRAFFIC

Freight traffic is considered at length in *Sustainable Distribution: A Strategy*, published a year after the 1998 White Paper.[30] As the title suggests, great importance is attached to environmental aspects of freight movement, and many of its proposals are concerned with such matters as pollution, pressures on the landscape, noise and disturbance, and accidents. Stress is laid on the importance of 'the entire supply chain' and its management ('logistics'). The analysis is interesting

(not a common notable feature of such documents) and well worth studying. The package of measures, however, contains little that is new. The annual increases in fuel duty were intended to continue, though this has now been significantly amended through measures in response to protests over fuel costs. Vehicle excise duty rates for lorries are under review ('to reflect the environmental damage they cause').

In preparing development plans, local authorities are to consider and, where appropriate, protect sites that provide opportunities for the transfer of freight from road to rail and to consider opportunities for new developments which are served by waterways. The Highway Agency and Railtrack are undertaking a joint assessment of the potential for new intermodal freight facilities; and 'quality partnerships' are to be developed 'between local authorities, the freight industry, business communities, residents and environmental groups to encourage consolidation of deliveries within urban areas'. A lengthy list ends with a commitment to promote research that will improve understanding of freight transport and distribution.[31]

The Strategic Rail Authority will not be subject to the Franchising Director's current narrow focus on passenger travel: 'it will support integrated transport initiatives, and provide for the first time, a clear focus for the promotion of rail freight'.

ROAD USER AND WORKPLACE PARKING CHARGES

Governments are very reluctant to levy charges on motoring (as distinct from taxes on vehicles and on petrol.) One may wonder why charges on car parking, and tolls on roads which go over or under a river, are acceptable, but congestion charges and road charges are not. The issue is, of course, a political one, and there is no simple rationale for these distinctions. The political nature of the issue is apparent in the extreme caution with which it is being broached. The 1998 White Paper laid the path for the Transport Act 2000, which provides for 'local authorities to charge road users so as to reduce congestion, as part of a package of measures in a local transport plan that would include improving public transport'. This neatly delegates responsibility and political risk to local authorities. Central government will be responsible only for introducing pilot schemes on motorways and trunk roads to assess what lessons they provide.[32] The first major road charging scheme will be implemented by the Greater London Authority in an effort to improve conditions in central London.

A consultation document, *Breaking the Logjam*, details the proposals for local congestion and workplace charges. It is stressed that the charges will be optional ('it will be up to those local councils which think it would help in their area to put up well reasoned proposals'), and they will initially be of a pilot nature (to encourage fresh thinking and to enable learning from practical experience). Contrary to the usual dictates of public finance, the income from these charges will be retained by local authorities to be spent on local transport improvements. This is a good way of making charges more acceptable to both local authorities and motorists.[33] Road user charges will have to be 'in keeping with' the local transport plan 'as drawn up to reflect the regional transport strategy' and the National Air Quality Strategy. Levies on private non-residential parking at the workplace face the danger of simply displacing parking onto adjacent streets. Consequently, the introduction of the levy will require the enforcement of on-street parking controls.

GREEN TRANSPORT PLANS

Bringing about major changes in travel behaviour is 'a shared responsibility', requiring cooperation on the part of travellers, employers, hospitals, and educational establishments, and any organisation or company that can have an impact on travel patterns. Major employers are urged to consider preparing green transport plans which will integrate 'the various ways in which an organisation uses transport to ensure that they complement each other and benefit the strategic business objectives' (DETR, *The Benefits of Green Transport Plans*, 1999). These can include measures that encourage travel to work by public transport, cycling, or walking;

a flexible benefits package to provide attractive alternatives to a company car; a review of standard working hours; a car-sharing scheme; using videoconferencing and other IT equipment to reduce business travel;[34] and enhancing the fuel efficiency of the vehicle fleet.

The Advisory Committee on Business and the Environment has recommended that companies seek to reduce by 10 per cent the total number of people commuting to work, alone, by car.[35] Whether there are adequate incentives for the development of green transport plans is, however, in some doubt (Potter 1999). However, all government departments are expected to have green transport plans in operation by March 2000. Hospitals are singled out for special mention in the White Paper because of their significance in travel generation.

The draft revision of PPG 13 notes that some or all of a green transport plan may be made binding, either through conditions attached to a planning permission or through a related planning obligation. Planning applications must be accompanied by a green travel plan for all major developments over certain thresholds (the ones used for parking noted above), for smaller developments that will generate significant amounts of travel in areas where traffic reduction and alternative modes are priorities, and where it will help to address a particular local problem. The objective in all cases is to deliver more sustainable transport. School travel plans promoting safe non-car routes to school are also required where they are to be expanded.

TRAFFIC CALMING

Traffic calming is an expressive term which, though used in different ways, essentially refers to measures for reducing the harmful effects of motor traffic. In its limited sense it refers to speed reductions, parking restrictions, pedestrianisation schemes, and suchlike. In a wider sense it is synonymous with overall traffic policy, including car taxation and land use measures designed to reduce the need for car journeys. Advocates of traffic calming can make some telling points in its support. For instance, a 50 kph speed limit (about 30 mph) in residential areas is 'acknowledged in many European countries' to be 'far too high'; at speeds of 30 kph or below, additional road space is created since cars need less space. If traffic calming is restricted to a few streets, its benefits are reduced: traffic simply redistributes itself to neighbouring streets. Complete exclusion of traffic can have a dramatic impact on town centres, and has been widely adopted on grounds of safety, amenity, and increased turnover for shops in pedestrianised streets – though the economic benefits are far from certain.[36]

The term 'traffic calming' was introduced by Dr Hass-Klau as the translation of the German term *verkehrsberuhigung*. Her book *Civilised Streets* (Hass-Klau *et al.* 1992) contains detailed technical descriptions of well-established methods such as speed bumps, chicanes (kinks in a road to slow down traffic), and pinch points, as well as some less well-known techniques. It also presents an assessment of traffic-calming experience in Germany, the Netherlands, Denmark, and Sweden. It describes and comments on some forty British traffic calming schemes. The traditional approach has been to segregate traffic and pedestrians: with reductions in traffic speed, they can both be accommodated, but with the pedestrian instead of the car being master.

A major shortcoming of many traffic-calming schemes is that they are essentially local in concept and operation. Rather than being parts of a comprehensive transport policy, they are typically reactions to vocal residents. As a result, the effect of calming in some areas is to move the problem elsewhere. Indeed, Banister (1994a: 212) has suggested that 'positive responses from those living in the traffic-calmed area are more than outweighed by anger from those living in adjacent areas where traffic levels (and accidents) have increased'. It is unfortunate that calming does not simply reduce the total amount of traffic; perhaps with proper planning, it can? The same point arises in relation to parking policy, which is arguably the simplest and most effective method of reducing private car use. A further problem is the damage that some traffic-calming measures have on the appearance of attractive towns and villages, with a clutter of intrusive signs and roadworks.

Many of the measures discussed under the heading of traffic calming have more traditionally been known as traffic management, though the concern is now with wide environmental and amenity issues as well as with traffic flow. This is becoming an increasingly sophisticated area of policy. Additional legislation is a testament to the importance now attached to it: the Traffic Calming Act 1992 extends the statutory provisions for 'the carrying out on highways of works affecting the movement of vehicular and other traffic for the purposes of promoting safety and of preserving or improving the environment'.

PARKING RESTRICTIONS AND STANDARDS

Parking restrictions are the simplest and the most acceptable of traffic controls – which certainly was not true when parking meters were first introduced (Plowden 1971). Until recently, parking restrictions were largely confined to cars entering congested areas, and charges have been raised as demand exceeded capacity. A favoured measure is to escalate the charging rate for long-stayers.[37]

National policy on parking in England was set out in the 1994 PPG 13, and extended and made more stringent in the regional policy guidance (RPG 3) for London. However, studies have shown that compliance with the guidance is poor, particularly in the outer London boroughs. Moreover, though it is advised that strategies for parking should be developed in conjunction with neighbouring authorities, there is 'little evidence' of this. Indeed, cooperation 'may even be restricted in order to preserve the use of parking as an independent counter in order to attract development in competition with other authorities'. Planning authorities have allowed parking provision 'well in excess even of peak time demand'. The quotations are from a report commissioned by DETR which concluded that in the South-East, 'Government policy guidance relating to the use of parking standards as a demand management tool is not reflected in the majority of standards adopted by local authorities' (Llewelyn-Davies and JMP Consultants 1998). Other studies have reached the same conclusion.[38]

Implementation is a neglected area of planning policy, but this report forcibly shows that the neglect can jeopardise policies or even render them ineffective. As policy has become increasingly concerned to restrict cars (rather than their use in particular areas), there will need to be a dramatic change in the implementation of parking standards. Of course, these can only apply in the future. Thus, 'considerable reduction compared with present norms of provision will therefore be needed to prompt any modal shift away from the car' (Llewelyn-Davies and JMP Consultants 1998).

It is perhaps not surprising, therefore, that the revised PPG 13 for England should introduce national maximum standards for car parking. The draft PPG 13 (1999) notes that 'levels of parking can be more significant than levels of public transport provision in determining means of travel, even for locations very well served by public transport'. It requires development plans to set maximum levels of parking for broad classes of development, and to ensure consistency in implementation it provides standards for developments over certain size thresholds.[39]

SCOTTISH GUIDANCE ON TRANSPORT AND PLANNING

The Scottish guidance on transport and planning was the first to be published. The Planning Advice Note *Transport and Planning* sets out 'good practice advice' on measures which local planning authorities may consider in fulfilling their integrated land use and transport planning responsibilities in a sustainable manner. Developers will be required to produce a transport assessment for significant travel-generating developments. Though this is to be distinguished from a formal environmental assessment, it may form part of it. The coverage of the transport assessment will be dependent on the scale, travel intensity, and travel characteristics of the proposal. Essentially, it will provide information to enable the local planning authority to determine the suitability of the location for the proposed use. This is to be assessed 'in terms of both the potential and likely accessibility for people

and freight by all modes'. This will enable the local planning authority 'to determine whether the location has the potential to minimise travel, particularly by private car'. (Much of the relevant information will eventually be set out in an up-to-date development plan.) Where a transport assessment is required, the developer will have to demonstrate that:

- The site, as existing or as a result of the development works, is physically accessible by a network of footpaths and cycle routes, and public transport will deposit passengers within a short and easy walk of the development.
- For non-residential developments, the network of public transport, walking, and cycle routes serving the site links with the majority of the forecast catchment population, with public transport being regular and frequent throughout the opening hours of the development.
- For residential developments, there is a high degree of accessibility to local day-to-day services such as convenience shops, schools, clinics, libraries, and community centres, particularly by walking and cycling, and accessibility to significant urban centres providing a range of services and employment, by walking, cycling, and public transport.

Transport assessments should set out the likely effect of the developer's proposals, particularly on reducing the level of car use, and should indicate how these measures relate to any specific targets in the development plan, or in the local transport strategy. The development plan will outline the transport priorities for particular parts of the local authority's area and the likely nature and scope of contributions which would be expected as part of development on a key site in the plan. Development proposals, related to levels of travel demand or to thresholds stated in the plan, will be expected to help deliver the transport objectives of the plan.

As in England and Wales, interim local transport strategies are to be produced by July 1999, and full strategies by July 2000. Further guidance is to be issued 'in due course'.

EUROPEAN UNION TRANSPORT POLICY

A common transport policy, furthering the free movement of people and goods, has been an objective of the EU since the Treaty of Rome. However, little progress was made on this until the mid-1980s, when the European Parliament challenged the Council of Ministers in the European Court for failing to meet its transport obligations. The Commission's 1992 White Paper *The Future Development of the Common Transport Policy* emphasised the positive role that a coordinated transport policy could play in promoting economic growth through the creation of the single market and what is described as 'sustainable mobility'.

Until the 1990s, policy has been directed primarily at measures to deregulate cross-frontier movements and to increase competition in the transport sectors. A more explicit spatial dimension has since been added with the identification of the Trans-European Networks, explained in Box 11.5. There has also been a shift in emphasis on the contribution that policy can make to reducing the impact of pollution on the environment. Implementation of policy relating to new infrastructure is the responsibility of member states (though the EU has made contributions to major projects, for example in the improvement of links between the UK and Ireland and the West Coast main rail line).

The EU is also concerned with *developing the citizens' network* – that is, promoting alternative transport modes in areas that are dominated by car use – and clean urban transport, which it supports through the CIVITAS initiative. CIVITAS seeks to support radical integrated policies and modal shift.

PUBLIC ATTITUDES AND THE FUTURE

Measures such as those outlined above would have been inconceivable without public support for stronger controls; and it is by no means certain that all of them will prove to be acceptable now. This question of public acceptability is a crucial factor in transport policy. It is also one that changes over time, particularly as the impacts of increased traffic – or increased restrictions – are experienced. There is a very real problem in reconciling private and public interests. Each car owner regards congestion problems as being created by other motorists; the individual's contribution to the total is negligible. This zero marginal cost for the individual imposes high costs on the collectivity of users, but car users have no incentive to economise in their use of road space: to them it is a free good.

The car can be more than a means of transport. It can be an extension of a driver's personality, a symbol

---

BOX 11.5 THE TRANS-EUROPEAN TRANSPORT NETWORKS

The objective is to increase the integration of existing networks for transport, telecommunications, and energy as a means of improving the competitiveness of the European economy. Priority projects were identified, including the high-speed rail links Paris–Brussels–Cologne–Amsterdam–London and Cork–Dublin–Belfast–Larne–Stranraer, and the Channel Tunnel Rail Link. Other priority routes extend the TGV from France into Italy and Germany, and across southern Europe, and also the Øresund fixed link between Denmark and Sweden. Some 30,000 km of new and upgraded high-speed rail track and 12,000 km of motorways are planned by 2020. The UK was awarded £28m in 2000 to support the TENs, £18.4m of which is going to the Channel Tunnel Rail Link. Concern has been voiced about the effects of this major investment on increasing the amount of travel, on the attractiveness of the metropolitan nodes, and on the relative disadvantage to peripheral areas.

of affluence or power, an object to be loved as well as used. The 'love affair' with the car is, however, under strain: mass ownership and use have made it less pleasant (though not necessarily less appealing) than it was (Goodwin *et al*. 1991: 144). Whether the disenchantment has gone far enough to warrant more penal methods of controlling its use is the basic political question. Recent surveys are helpful in showing the nature of public opinion and the scope that might exist for radical changes in policy.

A 'poll of polls' (Jones 1991b) summarised the major findings of ten surveys carried out between 1988 and 1990. First, there is no doubt about the general realisation of the seriousness of the problem of traffic congestion, but most people would be reluctant to reduce significantly their dependence on car use: despite the congestion, it still remains the favoured form of transport among car owners. Nevertheless, there seems to be an increasing willingness to consider switching to public transport for some types of journey if public transport were better. In looking at public transport as an alternative to the car, quality of service is much more important than the level of fares. (But the converse seems to be the case for car use: higher petrol prices are more significant than a doubling of journey times.) There is support for an increase in the range and quality of alternatives (or supplements) to car use, and for more effective parking restrictions, but little backing for road pricing except among Londoners. Attitude surveys, of course, are not necessarily a good guide for policy-makers. A measure may be popular but ineffective:

> indeed its popularity may lie precisely in its ineffectiveness and lack of impact on car-based life styles! Some drivers may support better public transport, for example, because they believe that other drivers will use it, and so clear the roads for them.
>
> (Jones 1991b)

In another study, commissioned by the Oxford Transport Studies Unit (not covered by Jones's review), Cullinane noted the extent of car dependence: about a half of households in the survey perceived a car to be essential to their lifestyle and a further 13 per cent would not want to be without one. However, a quarter of households did not have a car and had no intention of getting one. The overall conclusion was that car dependence was increasing, and that if things are allowed to continue as they are, it will become increasingly difficult to persuade owners to reduce their car usage (Cullinane 1992).

Reflecting on the survey as a whole, Cullinane concludes that congestion seems likely to increase,

> and that there will be some voluntary reduction in traffic as the problems intensify. However, the level of attachment of most people to their car is such that it will take some positive action from outside to force any real reduction in traffic, and this positive action will have the most impact if it hits people's purses.

A number of studies have pointed in the same direction. Cars are highly valued by those who can afford them, but the problems of congestion are becoming increasingly burdensome. There is support for better public transport and, though no massive changeover by car users is to be expected, 'all changes take place at the margin'. Though it is very apparent that motorists do not like the idea of road pricing, there is good evidence that they would accept it (reluctantly) if it were part of a package which provided them with some offsetting benefits, particularly in the form of good public transport. This is perhaps the most important finding of research both in Britain and elsewhere. Cervero (1990) has reviewed North American studies of transit pricing and concludes:

> For the most part, riders are insensitive to changes in either fare levels, structures, or forms of payments, though this varies considerably among user groups and operating environments. Since riders are approximately twice as sensitive to changes in travel time as they are to changes in fares, a compelling argument can be made for operating more premium quality transit services at higher prices. Such programs could be supplemented by vouchers and concessionary programs to reduce the burden on low-income users.

Even the most ardent supporter of road pricing admits that there are many unknowns in the matter. Theoretical studies may be suggestive, but many of the issues are empirical – or at least need empirical testing. Unfortunately, it is one of the dilemmas of research on traffic restraint that though empirical evidence is needed, this is difficult to obtain since governmental authorities are unwilling to experiment without adequate predictions (May 1986: 120). This is a case where both academic and political considerations call for more research. The bibliography of research reports seems set to expand.

A major missing element in the analysis of the latest White Paper is that of the car as much more than a means of mobility. In Susan Owens' words,

> The most casual review of car advertising reveals a parallel universe to that of the White Paper: it is a world in which cars confer identity and status, speed thrills, drivers enjoy increasing levels of protection and creature comforts, and, in a parody of the re-allocation of road space, the boundaries of where vehicles may go are seen as more and more permeable. None of the relevant measures – differential taxation, better enforcement of traffic laws, or 'education' to improve driver attitudes – seem any more than Canute-like in the face of this tide of material and the very considerable interests behind it.
>
> (1998: 331)

One final point is that adjustments in one direction can be difficult to reverse because of the many consequential changes which have taken place. Life has adapted to the incredible flexibility and freedom afforded by the motor car. Jobs, shops, schools, leisure facilities – indeed, a wide range of activities – have dispersed: the car has made this possible and even necessary. It is significant that a quarter of journeys and travelling time are for leisure activities (Reid and Margatroyd 1999: 9). Much development is still in the pipeline (and has been given planning permission), and forces for dispersal will continue to operate. Two topical illustrations can be cited. First, in the health service there is a trend towards large multi-purpose hospitals, and the closure of small ones such as accident and emergency units: 'although these may mean longer journey times, care will be at the highest level on arrival, with no need for a transfer' (*The Times*, 30 September 1999). Second, Boots the Chemists are developing two hundred edge-of-town stores, of which forty have already been built (*Planning*, 1 October 1999).

Other social forces are strong: the growth of two-income households can lead to compromise decisions on housing, necessitating two cars and two journeys in different directions. Affluent teenagers demand a car

as soon as they are legally allowed to drive. To many the car has greatly widened opportunities. But the forces that have widened choices for many have reduced them for those without a car. The government has an almost missionary-like desire to enable the carless to have at least some of the freedom enjoyed by car owners. By reducing car traffic, travel is made easier; or so the philosophy maintains: 'better for everyone', as the title of the 1998 White Paper has it. Yet there are doubts: congestion is democratic; rationing by price benefits the richer and penalises the poorer. Car pollution could be attacked by promoting the development of emission-free electric cars.[40] Rationing space by congestion controls is, on one argument, no different from rationing education or health care: in Graham Searjent's words, 'it is about providing for the few or not providing for all' (*The Times* 30 September 1999). This line of argument is unappealing to some, including the present government, but it does give pause for thought, particularly if a major wider policy objective is social inclusion.[41]

FURTHER READING

Transport Statistics

Publications of statistics and statistical commentaries on transport have mushroomed in recent years. In addition to the annual *Transport Statistics for Great Britain*, and (much broader in scope) *Social Trends*, several new series have been launched, including *Focus on Personal Travel* (1998), *Focus on Public Transport* (1999), and *Transport Trends* (annual). Further titles are given in the list of official publications at the end of the book.

Transport Policy and Planning

The Buchanan Report (*Traffic in Towns*, 1963) is historically a landmark, but see also his earlier *Mixed Blessing: The Motor Car in Britain* (1958). Truelove's *Decision Making in Transport Planning* (1992) provides an excellent overview of transport policies and politics.

Banister's *Transport Planning in the UK, USA, and Europe* (1994a) has a broader canvas, with some comparative analysis; it also has a useful bibliography. See also *Transport, the Environment and Sustainable Development*, jointly edited by Banister and Button (1993). Another general work on transport is Glaister *et al.* (1998) *Transport Policy in Britain*.

There is a large library of US books, of which two are particularly recommended as offering an approach which is refreshing to British readers: Downs (1992) *Stuck in Traffic: Coping with Peak-Hour Traffic Congestion*, and Dunn (1998) *Driving Forces: The Automobile, Its Enemies, and the Politics of Mobility* (1998). An influential report by Goodwin *et al.* was published in 1991: *Transport: The New Realism*. See also Goodwin's *The End of Hierarchy? A New Perspective on Managing the Road Network* (1995), and an introduction to the recent literature on the car is provided by David Banister's review essay 'The car is the solution, not the problem?'.

The report of the Royal Commission on Environmental Pollution, Transport and the Environment (1990), and the various reports of the Standing Advisory Committee on Trunk Road Assessment (SACTRA) and the Select Committee on the Environment, Transport and Regional Affairs are referred to in the chapter. A broader economic assessment of the true costs of road transport is given by Maddison *et al.* (1996) SACTRA's latest report is a highly technical economic analysis: *Transport and the Economy* (1999). Quite different is Whitelegg's wide discussion of transport issues in *Critical Mass: Transport, Environment and Society in the Twenty-first Century* (1997).

Road and Congestion Charges

The amount of writing on the charging of road users is in striking contrast to the amount of action. The number and character of government and parliamentary reports on the issues are testament to the political difficulties involved. There is now some likelihood of action following the publication of *Breaking the Logjam: The Government's Consultation Paper on Fighting Traffic Congestion and Pollution through Road User and Workplace Parking Charges* (DETR 1998).

The Smeed Report (1964) is an early milestone, but it was not until the end of the 1980s that it became clear that government had to take some hard decisions (or avoid them, and take a decision by default). At first, attention was more focused on attracting private-sector involvement in the provision of roads (*New Roads by New Means: Bringing in Private Finance*, 1989). Attention soon switched to more pressing matters of congestion with the 1993 Green Paper *Paying for Better Motorways*, which was followed by an HC Transport Committee report, *Charging for the Use of Motorways* (1994), and the governmental response: *Government Observations* (1994). The Transport Committee launched a wider inquiry later that year: *Urban Road Pricing* (1995).

The government has commissioned a number of studies in this and related fields: see, for instance, MVA Consultancy (1995) *The London Congestion Charging Research Programme: Principal Findings*. Outside Parliament, a succession of reports emerged from various interest groups such as the Chartered Institute of Transport and the London Boroughs Association (see below). A succinct survey of the field is provided by Lewis (1994) *Road Pricing: Theory and Practice*; Grieco and Jones (1994) 'A change in the policy climate? Current European perspectives on road pricing'; Jones (1991b) 'UK public attitudes to urban traffic problems and possible countermeasures: a poll of polls'; London Boroughs Association (1990) *Road Pricing for London*; Nevin and Abbie (1993) 'What price roads? Practical issues in the introduction of road-user charges in historic cities in the UK'; and Newberry (1990) 'Pricing and congestion: economic principles relevant to pricing roads'.

Freight Transport

Dealing specifically with freight transport are DETR (1999) *Sustainable Distribution: A Strategy*; National Audit Office (1997) *Regulation of Heavy Lorries*; Plowden and Buchan (1995) *A New Framework for Freight Transport*; Royal Commission on Environmental Pollution (1994) *Transport and the Environment* (chapter 10); DoT (1996) *Transport: The Way Forward* (chapter 15).

Traffic Calming and Management

DoE (1995) PPG 13: *A Guide to Better Practice: Reducing the Need to Travel through Land Use and Transport Planning*; DoT (1995) *Better Places through Bypasses: The Report of the Bypass Demonstration Project*; Hass-Klau (1990) *The Pedestrian and City Traffic*, and Hass-Klau et al. (1992) *Civilised Streets: A Guide to Traffic Calming*.

Details of the traffic management system for London are set out in DoT (1992) *Traffic in London: Traffic Management and Parking Guidance* (DoT Local Authority Circular 5/92). On the extension of special parking areas to local authorities outside London, see DoT (1995) *Guidance on Decriminalised Parking Enforcement outside London* (DoT Local Authority Circular 1/95). The government's *Transport Strategy for London* was published in 1996. In *Speed Control and Transport Policy* (1996) Plowden and Hillman explore the potential for speed limits as a policy instrument.

A rare comparative study of different approaches to transport planning in large cities is London Research Centre (1992) *Paris, London: A Comparison of Transport Systems*. The latest research report on traffic calming is Ross Silcock Ltd and Social Research Associates (1999) *Community Impact of Traffic Calming Schemes*. A special edition of *Built Environment* (25 (2) (1999)), edited by Stephen Marshall, deals with travel reduction.

Walking and Cycling

The DETR's Traffic Advisory Leaflet 3/00 *Walking Bibliography* is useful, though it is mostly limited to official publications. See also the background papers to *Developing a Strategy for Walking* (DETR 1997). The Scottish Executive Central Research Unit has published a report by System Three, *Research on Walking* (1999). The Environment, Transport and Regional Affairs Committee began an inquiry entitled *Walking in Towns and Cities* at the end of 2000, and evidence presented will become an important source.

The *National Cycling Strategy* (DoT 1996) is the most important official document published in this field: it is obtainable free from the DoT. See also McClintock (1992) *The Bicycle and City Traffic*, and the HC Transport Committees report *Cycling*, 1991.

NOTES

1 The M25 is now the most heavily used road in Britain. Far more traffic uses it than was forecast (though at the planning stage, the Transport Department's forecast was fiercely attacked as being too high). This 'immediate success' (to use the Department's eccentric description) led to a major *M25 Action Plan* (1990) which included improved lighting and signalling, and widening the road to dual four lanes, with a later possibility of five lanes in certain sections. Standards of road maintenance have been falling seriously for some years, and there is a major backlog to be made up. The Scottish *Strategic Roads Review* (1999: 9) has some telling figures on this. For example, the proportion of the Scottish trunk road network with a 'residual life' of less than five years almost doubled to over a fifth between 1994-95 and 1997-98. (Residual life refers to the time before reconstruction works become necessary.)

2 Though it was the opposition to specific roads that received most publicity (as at Westway, Airedale, Twyford Down, Archway), there was a broadly based lack of confidence in the system by which roads needs and routes were addressed. Though some activists took an extreme stance, an eloquent justification for this was provided by Tyme (1978).

3 The report emphasises the importance of clear policy objectives and 'strategic' and long-term effects. These, of course, are precisely the policy issues with which the political process has difficulty. The problem falls into that class of which Rittel and Webber (1973) have neatly termed 'wicked problems': problems that are unique to a specific place and time, which defy definitive formulation, and which can be 'resolved' only by political judgement.

4 DETR *Guidance on the New Approach to Appraisal* (2000) and DETR *Understanding the New Approach to Appraisal* (2000).

5 Four programmed new routes were abandoned: M12–M25 (Chelmsford); A5–M11 (Stansted); a new motorway to the south and west of Preston;

and the M55–A585 (near Blackpool). These were in addition to the motorway links between the M56 and M62 (Manchester) and the M1 and M62 (Yorkshire) which had been abandoned earlier because of the difficulty in finding an environmentally acceptable route.

6   See the review article by McKinnon (1999).

7   The Environment, Transport and Regional Affairs Committee's Seventeenth Report on the Departmental *Annual Report 2000* (2000). The government's ten-year plan for transport includes £21 billion worth of investment in strategic roads including thirty trunk road bypasses, the widening of 5 per cent of the strategic road network, and eighty more schemes to tackle 'bottlenecks'. In addition, privately financed schemes are in preparation, and the Birmingham Northern Relief Road (a major new motorway) is under construction.

8   In the Netherlands this general idea has been elaborated much more fully in the 'ABC policy' – but it may not have produced all the desired results (Priemus 1999; Priemus and Konings 2000). The Civic Trust's comments on the draft PPG reflect wider views and are available at <www.civictrust.org.uk/ppg13.shtml>.

9   Ove Arup and University of Reading (1999) *Planning Policy Guidance on Transport (PPG 13): Implementation 1994–96*.

10 *Transport: The Way Forward* (Cm 3234, 1996). An extract from this is reproduced in the Blair government's White Paper *A New Deal for Transport: Better for Everyone* (Cm 3950, 1998).

11 Tewdwr-Jones (1998: 523) suggests that 'The enthusiasm with which the Labour Government has retained in place a set of policy documents released by Conservative governments undoubtedly indicates the non-political nature of their contents.'

12 An interesting overview is given by Brearley (1998). Brearley is Director General of the Planning, Roads and Local Transport Group, DETR.

13 *A Transport Statement for Northern Ireland* (1998); *Travel Choices for Scotland* (Cm 4010, 1998); and

Transporting Wales into the Future (Welsh Office 1998).

14 *Travel Choices for Scotland*, op. cit., para. 2.1.18. See also Farrington *et al.* (1998).

15 *A Transport Strategy for Northern Ireland*, op. cit., pp. 6, 13 and 39. See also *Moving Forward: Northern Ireland Transport Policy Statement* (1998).

16 Supplementary memorandum by the Minister of Transport (Dr John Reid) to the HC Environment, Transport and Regional Affairs Committee, *Integrated Transport White Paper* (20 January 1999, HC 32: 246).

17 See the discussion on p. 343 on parking standards. Some proposals for 'ensuring compliance' with government policy are given in Llewelyn-Davies and JMP Consultants (1998: 30). These include more active involvement of government offices in the development plan process, and clearer direction to the Planning Inspectorate to ensure compliance of plans with government guidance. The position has now changed with the new machinery for regional planning guidance.

18 *Guidance on Full Local Transport Plans*, (DETR 2000) and *A Good Practice Guide for the Development of Local Transport Plans* (DETR 2000).

19 Two Acts preceded this: the Road Traffic Act 1997 and the Road Traffic Reduction (National Targets) Act 1998. These required the publication of reports on current and future levels of road traffic. At the local level, these will now become part of local transport plans. The national reports will be published by the central government.

20 Policies in the White Paper draw on earlier discussions in *Developing a Strategy for Walking* (DETR 1997). A national walking strategy has been in promised for some time but seems to have been overtaken by the other initiatives recorded here. The DETR has published *Encouraging Walking: Advice to Local Authorities* (2000).

21 See Transport Research Laboratory reports *Review of Traffic Calming Schemes in 20mph Zones* and *Urban Speed Management Methods*. (Both of these are summarised in DETR Traffic Advisory Leaflet 9/99, *20mph Speed Limits and Zones*, available from

the DETR Local Transport Division.)

22 DETR (1999) *School Travel: Strategies and Plans. A Best Practice Guide for Local Authorities*. See also Cairns (1999).

23 While cycling in the UK accounts for less than 2 per cent of trips (and is declining), the proportions are 10 per cent in Sweden, 11 per cent in Germany, 15 per cent in Switzerland, and 18 per cent in Denmark (DoT 1996: 7).

24 The HC Transport Committee's 1991 report (*Urban Public Transport: The Light Rail Option*); the report listed twenty-two of the schemes which were well advanced: their total cost was estimated at £2,700 million.

25 In a useful 1992 'state-of-the-art review', Walmsley and Perrett note that 'even before deregulation, bus services in the major cities typically achieved revenue–cost ratios of 70 per cent, and only in the most highly subsidised were the ratios as low as the systems abroad'.

26 Recent announcements have confirmed that extensions will be made to the Tyne and Wear Metro and Manchester Metrolink, and a new scheme will go ahead in Nottingham. The Environment, Transport and Regional Affairs Committee Eighth Report, *Light Rapid Transit Systems*, provides a useful summary of the different forms and how they might contribute to integrated transport policy.

27 The crash was caused by one rail breaking into 300 pieces with the train travelling at 115 mph; four passengers were killed and 70 injured.

28 *On Track 2* (2000) (the newsletter of the Strategic Rail Authority).

29 During 2000, 930 sites on the 20,000 miles of track had speeds lowered temporarily for safety. Investment will bring services up to normal standard and offer improvements, for example the £11.2 billion investment programme for the West Coast Main Line will allow for 140 mph services by 2005. See <www.railtrack.co.uk>.

30 It was also announced (in the November 1999 Pre-Budget Report) that above-inflation increases will go into a ring-fenced fund for transport improvements.

31 The analysis, however, does not challenge the taken-for-granted assumption of continuing growth in freight traffic: 'Reducing impacts is couched in terms of efficiency gains and limited modal split, while fundamental questions about the need to move ever increasing quantities of goods from place to place remain unasked' (Owens, 1998: 330). A very difference stance is taken by the Institute of Logistics and Transport, whose discussion paper *Sustainable Distribution* (1999) argues that 'we must plan for a continuing increase in demand for efficient freight distribution services'.

32 Opponents will no doubt take note of the decision to review the Dartford Crossing charges, which are time-limited (White Paper, para. 4.102).

33 However, the proposals envisage that local authorities will be able to retain net income only if there are 'worthwhile transport-related projects to be funded'. Moreover, central government will have the power to require a portion of the revenue to be paid to the Treasury.

34 See Solesbury (1999) and Worpole and Greenhalgh (1999) (and also the working paper *Good Connections: Helping People to Communicate in Cities*).

35 See also *Changing Journeys to Work: An Employers Guide to Green Commuter Plans* (Transport 2000 (supported by London First)).

36 For a study of the community impact of traffic calming schemes in Scotland, see Ross Silcock Ltd and Social Research Associates (1999).

37 Thus, the central Cambridge car park has a rate of 90 pence for the first hour, £1.80 for the second hour, £2.70 for three hours, and £4.30 for four hours. It then jumps to £7.50 for the next hour, £10 for the following hour, with a top rate of £15 for more than six hours. There are many permutations of this technique.

38 Llewelyn-Davies and JMP Consultants (1998). See also Ove Arup and University of Reading (1997). This showed that nationally the parking policies of local authorities did not reflect PPG 13 guidance. Other studies demonstrating the inadequacy of parking controls include London Transport

Planning (1997). 'This showed that neither provision on parking at typical office developments, nor the standards included in current UDPs in outer London comply with PPG 3 guidance, with parking in some Boroughs exceeding the RPG3 maximum level by a factor of ten or more'. (The quotation is from the Llewelyn-Davies study, p. 16, fn 4.) In relation to housing developments, the 1999 draft PPG 3 *Housing* advises that 'local planning authorities should revise their standards to allow significantly lower standards of parking provision in all housing developments, including less off-street parking. Car parking provision in any development should not exceed 1.5–2 car parking spaces per dwelling, and should normally be less, often significantly so.'

39 The standards are set out in an annexe to the PPG; for example, food retail developments over 1,000 m² should have a maximum of 1 space per 18–20 m²; higher and further education developments over 2,500 m² should have no more than 1 space per 2 staff. Local authorities are encouraged to apply stricter standards where possible and there are indications that this new approach may encourage development on existing car parking spaces.

40 California, which has acute air quality problems, requires car manufacturers to produce specified numbers of 'clean' (i.e. electric) cars. These have to meet severe standards: how they are met will depend on the results of a research and development effort (which so far has been disappointing). A number of other states have adopted similar policies, which are sometimes described as technology-forcing: the standards involved are stricter than can be met with existing technology. Evidence that such an approach can be effective (even if not as quickly as its protagonists would wish) is suggested by the successful development of catalytic converters. But it may seem dangerous to rely on a technological quick fix to environmental problems (based on Cullingworth (1997a: 219).

41 The Transport Foresight Panel for the White Paper on integrated transport gave a 'vision of what lies ahead'. This includes a zero-emission car: 'By 2020 many urban buses and minibuses will be powered by "clean diesels" using re-designed fuels. . . . The ECO-CAR will have arrived. Many inner-city and city-centre businesses and residents will hire, for periods as short as an hour, specialised zero-emission hybrid-powered city cars.' *The Role of Technology in Implementing an Integrated Transport Policy* (Office of Science and Technology, DTI, 1998).

12

PLANNING, THE PROFESSION, AND THE PUBLIC

Planning proposals are generally presented to the public as a *fait accompli*, and only rarely are they given a thorough *public* discussion.

(Cullingworth 1964: 273)

Plans are policies and policies, in a democracy at any rate, spell politics. The question is not whether planning will reflect politics but whose politics it will reflect.

(Long 1959 cited in Taylor 1998: 83)

INTRODUCTION

The right of the public to have a direct say in planning decisions and the inherently political nature of planning are now taken for granted. The formal machinery for objections and appeals, initially devised only for specified uses by a restricted range of interests, is now employed much more widely. Many informal mechanisms have been created by planning authorities and others to improve the capacity of and opportunity for local communities and interest groups to play a part in formulating and implementing planning policy. Even so, many questions remain about the effectiveness of public participation, whose interests are served by planning, and the relationship between professional and political decisions. This chapter explains the history of public participation in planning; some key 'interests' in planning; the mechanisms that enable them to influence the planning process; and the role of the planning profession.

PARTICIPATION IN PLANNING

The lack of concern for public participation indicated in the first quotation (which is taken from the first edition of this book) was a result in part of the political consensus of the postwar period, and in part of the trust that was accorded to 'experts' – which, by definition, included professionals. The time was perceived to be one of rapidly expanding scientific achievement, and the methods that had made such progress in the physical sciences were thought to be transferable to the problems of social and political organisation (Hague 1984). This, together with the advent of new social security, health, and other social and public services, led to a rapid growth in professions and the bureaucracies in which they worked. Town planners, though having identity problems which took many years to settle, had a good public image: they were to be the builders of the Better Britain which was to be won now that the military battle was over. In the same spirit as established professions, they sought to establish a strong, scientific, and objective knowledge base. Armed with the right techniques in manipulating the environment, they were to address the physical spatial development problems of the nation and, at least by implication, the underlying social and economic forces which drive physical development.

In retrospect, the approach implied a depoliticising of issues which were later appreciated to be of intense public concern. This was further obscured by professional techniques and language which the public could not be expected to understand (Glass 1959). Of course, planners were not alone in this: on the contrary, they simply took the same stance as other 'disabling professions' – to use Illich's term (1977). At the time, however, the lack of political debate and participation was not widely recognised as a problem. Professionals were perceived as acting in everyone's interest – the general public interest.

It was in the 1960s that these ideas were effectively challenged in the UK, closely following experience in the USA. (See, for example, Broady (1968) on the UK and Gans (1968, 1991) on the USA.) By this time the political consensus had broken down, and there was widespread dissatisfaction both with the lack of access to decision-making within government, and with the way in which benefits were being distributed. Though it claimed to serve the public interest, the planning system began to be seen as an important agent in the distribution of resources – frequently with regressive effects (Pickvance 1982). The idea of an objective, neutral planning system was increasingly recognised to be false.

In particular, the physical bias of the planning system had failed to address social and economic problems: perhaps it sometimes even made them worse. There was growing concern for a new type of 'social planning' which would seek to redress the imbalance in the access to goods, services, opportunities, and power. To achieve this, some saw the need for 'advocacy planning' (Goodman 1972) which would provide experts to work directly with disadvantaged groups. This critique has had consequences for planning practice of greater permanence than that achieved by the intellectual arguments themselves. Changes were made in the statutory planning procedures, and consultation and participation gradually became an important feature of the planning process.

The Skeffington Committee reported in 1969 on 'the best methods, including publicity, of securing the participation of the public at the formative stage in the making of development plans for their area'. Its report is sometimes celebrated as the turning point in attitudes to public participation in planning, though its recommendations are mundane and rather obvious – for example, on keeping people informed throughout the preparation of plans, and asking them to make comments. This is testimony to the distance which British local government had to go in making citizen participation a reality.

Unfortunately, the report did not discuss many of the crucial issues, though passing references suggest that the Committee was aware of some of them. For instance, it is rightly stated that 'planning' is only one service, and 'public participation would be little more than an artificial abstraction if it becomes identified solely with planning procedures rather than with the broadest interests of people'. The implications for the organisation of local authorities were not discussed. The proposals for the appointment of 'community development officers to secure the involvement of those people who do not join organisations' and for 'community forums' had little impact at the time, although much later variants of these were to be created by numerous authorities.

What was conspicuously lacking in the debate on public participation was an awareness of its implications for local politics. Public participation implies a transfer of some power from local councils to groups of electors, but it was not Skeffington but the Seebohm Committee (1968) which highlighted the tension between participation and traditional representative democracy.

An essential ingredient of effective public participation is a concern on the part of elected members and professional staff to make participation a reality. This cannot be effective unless it is organised, but this, of course, is one of the fundamental difficulties. Though a large number of people may feel vaguely disturbed in general about the operation of the planning machine, and particularly upset when they are individually affected, it is only a minority who are prepared to do anything other than grumble. The minority may be growing, but as far as can be seen, public participation will always be restricted: 'the activity of responsible social criticism is not congenial to more than a minority' (Broady:1968). Despite the increasing numbers actively participating in plan-making in the 1980s and 1990s, the general point still holds (Edmundson 1993) and furthermore, the situation in other countries is the same (Barlow 1995). The 'public' are much more likely to be engaged with the system on site-specific issues which affect them.

Despite its failure to address more fundamental questions, the reforms introduced as a result of Skeffington were generally acclaimed by the profession. It was the academic commentators who first questioned the underlying assumptions. Neo-Marxists drew attention to the more fundamental divisions of power in the political and economic structure of capitalist society, and how these continued to be evident in the outcomes from the planning system. Practitioners too began to see that extensive partici-pation exercises produced only limited gains, and some advocate planners were among the first to reject the approach for its weaknesses. The critics argued that, like all 'agents of the state', planners operate within a structural 'straightjacket' and, irrespective of their own values, will inevitably serve the very interests which they are supposed to control (Ambrose 1986). This was supported by research findings demonstrating that planning had operated systematically in the interests of property owners. There was also substantial theo-retical work concerned with the role played by the planning system in the interests of capital (Paris 1982).

The critiques were powerful but, by their very nature, they could offer little guidance to planners working in a professional, politically controlled system. Indeed, how could they respond to allegations that a fundamental purpose of planning in society is the legitimation of the existing order? If participation merely supports a charade of power-sharing, leaving entrenched interests secure, what alternatives do planners have? Planning and planners became the primary explanation for the failures of urban and rural development: the postwar housing estates that were built as quickly as possible (and with few resources left over for 'amenities'); the motorways that were belatedly built to cope with the great increase in traffic congestion, but which destroyed the social and phys-ical fabric of towns; the demise of village amenities in areas of development restraint; and the participation processes which raised hopes that were dashed by the outcomes. Certainly, planners played a part in all these, but were they the determining factor or were they, in Ambrose's words (1986), the scapegoat?

In response to these failures, some planners and community activists tried more radical approaches. Popular planning aims

> to democratise decision making away from the state bureaucrats or company managers to include the work-force as a whole or people who live in a particular area . . . empowering groups and individuals to take control over decisions which affect their lives, and therefore to become active agents of change.
>
> (Montgomery and Thornley 1990: 5)

Not surprisingly, the few examples of such practice are to be found in the left-wing strongholds of London and some metropolitan district councils, but have had only limited success.[1] Decentralisation of decision-making has been effected (and perhaps only *can* be effected) only on a very limited basis.

Interestingly, popular planning approaches have generally not been innovative in their methods of establishing community needs. Though the language is one of empowerment, they have tended to involve the usual mixture of meetings, publicity, and leaflets. Few community development projects have managed to achieve a real strengthening of a community's social resources and management capabilities to allow for devolution of real power to local people; rather, they have generally lent support to improve the resource base of communities – an admirable but different objective (Thomas 1995). But the interest in popular planning persists. The Labour administration's will-ingness to experiment with new 'bottom-up' approaches to policy development, and the emphasis on strength-ening local governance capacity in communities, provide further incentive.

Though some planning authorities have been able to demonstrate a consistent commitment to enabling community participation in planning from the 1970s on, the dominant influence during the 1980s was 'business'. Housing, industrial, commercial, and minerals interests all effectively enjoyed special treatment through the planning system. Not only were they able to voice concerns directly through such mechanisms as the Property Advisory Group, but the system overall was skewed to their interests. Many decisions were taken out of the hands of local planning authorities (never mind communities) so as to allow for major development of all forms, the most contentious being out-of-town shopping developments and major new housing areas, including the infamous Foxley Wood new settlement proposal in Hampshire.

The then Secretary of State, Nicholas Ridley, argued that in the interests of the country as a whole, local concerns needed to be set aside in favour of a presumption in favour of new development. There was 'a consistent diminution of the significance accorded to general public participation in policy formulation, as part of an effort to "streamline" the system and reduce delays' (H. Thomas 1996: 177).

LOCAL CHOICE

By 1989, in response to fierce criticism from its own party members in the shires, the Conservative government made an about-turn on community involvement under the slogan 'local choice'. At this time, a clearer national planning framework was instituted through national planning guidance. Local authorities were strongly encouraged to produce more plans, and warnings were given that developers would have to bear costs if they pursued applications contrary to up-to-date statutory plans. The incoming Secretary of State, Chris Patten, rejected the proposal for a new settlement at Foxley Wood, and also confirmed the commitment to 'local choice', noting that in planning, 'many of the important choices are decisions which can and should be made locally, to reflect the values which local communities place on their surroundings' (1989: 18–19).

This was not a signal to local authorities that they could respond as they wished to local demands. Local autonomy was to be exercised only where it was within parameters laid down by the centre in national and regional policy statements. Moreover, plans and decisions needed to be 'realistic about the overall level of provision'. In this, the decentralisation of decision-making is all about offloading responsibility without power. Nevertheless, the statement marks an important milestone in the relationship between central and local decision-making, and a clear move away from the previous line of argument with its presumption in favour of development interests.

The 1992 version of PPG 1 confirmed this change by noting the 'presumption in favour of proposals which are in accordance with the development plan'; and the 1997 version eschews any presumption and instead notes simply that 'an application for planning permission or an appeal shall be determined in accordance with the plan, unless material considerations indicate otherwise'. So today, influence in planning depends much more on the extent to which interests are reflected in the plan, a question which is taken up later in this chapter.

In fact, planning legislation now requires only 'publicity' for planning proposals, although government policy also talks of both public 'participation' and 'involvement'. In practice, community involvement varies considerably from one planning authority to another. DoE research set out in *Community Involvement in Planning and Development Processes* (1995) has illustrated the spectrum of relationships between planner and planned from the provision of information to delegation of control. Like many other studies, this research used the classic 'ladder of participation' (Arnstein 1969) to measure community involvement.[2] It concluded that real involvement was the exception rather than the rule but it was able to point to some innovative good practice, especially in the effective use of neutral intermediaries between community and authority (as foreseen by Skeffington). The report called for a greater awareness of the benefits that can accrue from effective community involvement.

Soon after 'local choice' came John Major's Citizen's Charter initiative. Building on pioneering work in some local authorities, the Citizen's Charter applied six principles: monitoring performance standards; the provision of full and accurate information about how services are run; the creation of greater choice between service providers; courtesy and helpfulness in service provision; making sure things are put right where they go wrong; and value for money. A central aim was to furnish more information to the public about the performance of service providers. The general approach is about quality review, a steadily growing component of public-, private-, and voluntary-sector management practice, closely linked to the use of targets and performance indicators. In general, planning from the mid-1980s, like other local public services, was influenced by the shift to the community-oriented enabling role of local government (Higgins and Allmendinger 1999).

The general concept was taken up with enthusiasm by many organisations, and many specific service charters have been created.[3] Planning is well represented, with the *Development Control Charter* and the *Planning Charter* for the entire service. There are also *Local Environment Charters* that address the right of access to environmental information held by public authorities; the right to participate in decision-making on environmental issues; and the right to seek remedies in the event of shortcomings in environmental services.

It is not clear whether these charters are making any difference to the relationship between the public and the planning and environmental protection systems. The British Standards Institution emphasises that quality of service depends on management, personnel, training, organisational objectives, and resources. But the thrust of the charters is to focus on easily measurable facets of service delivery. In some cases it is reduced to rather shallow measurements and setting of targets for responding to inquiries, making decisions, and the like. The value of these is limited, and their impact will depend on how challenging they are.

Another question is the extent to which the charters extend the rights of the citizen in planning matters. The clear answer is 'very little'. The charters spell out existing rights and, in so far as this helps to increase understanding, they are to be welcomed. But the relationship between citizen and government is not fundamentally changed: indeed, in its original form it was defined in a one-dimensional way – 'the citizen as consumer' (LGMB 1992). The Labour administration consulted on the future of the Citizen's Charter in 1997 and subsequently has sought to establish a more bottom-up approach starting with the needs of users, though the basic idea remains much the same.

The new Charter programme is called *Service First* and adds three principles to the six noted above: constant search for innovation and improvement; working with other providers to increase effectiveness and coordination; and consulting and involving present and potential users. These achievements of the Service First principles will inform the Best Value initiative in local government. Anecdotal evidence suggests that there is considerable room for improvement in the planning service. The one-stop shop initiative (discussed in Chapter 5) is one way that the experience of the user of the service can be improved.

The most recent and important developments in public participation arise from the promotion of sustainable development through the Agenda 21 initiatives of local authorities (as described in Chapter 7). The local authority is recognised as the principal actor in translating and implementing global and national objectives for sustainable development at the local level, and participation is central to the formulation of strategies. LA21 embraces a far wider range of concerns than planning, including, for example, green accounting and purchasing, energy conservation, and recycling, but always at its heart is the objective of raising awareness and engaging citizens in promoting good environmental practice (Scott 1999). LA21 has become an important stimulus for promoting more creative ways of engaging local people in policy-making, some of which are finding their way into the separate arrangements for consultation on planning matters.

In local government generally there has been considerable innovation in participation methods since 1994, notably in the use of web sites and citizens' juries (De Montfort University and University of Strathclyde 1998). In the future, the government's aim to empower local community development through the *New Deal for Communities*, together with an emphasis on local action for sustainable development, will see many more experiments in community or neighbourhood action. They will seek to establish capacity for broader and long-term participation making use of visioning processes and more extensive use of ICT to engage a wider representation (Carley and Kirk 1998; Carley 1999). Nevertheless, there will remain a need for strong leadership from local authorities.

PUBLIC PARTICIPATION IN PLAN-MAKING

A major landmark in the growth of public participation was the 1968 Planning Act (and its Scottish equivalent of 1969) which made public participation a statutory requirement in the preparation of development plans.

The main stimulus for this came not from the grass-roots, but from central government. Under the old development plan system, the Department was becoming crippled by what a former permanent secretary called 'a crushing burden of casework'. The concept of ministerial responsibility was clearly shown to be inapplicable over the total field of development plan approval and planning appeals. A new system was therefore required which would remove much of the detailed work of planning, including approval of local plans, from central to local government. The decentralisation of powers was linked to the creation of procedures which ensured that other interests (predominantly development interests) had formal opportunities to participate and object during the plan-making and adoption process. Although ministers have consistently expressed their desire to see local communities making their own decisions on planning matters, the procedures also ensure that local choice, however democratically arrived at, does not transgress 'the general public interest' (or other private interests which central government deems to be important).

The procedural safeguards do this, however, at some cost of time and resources. The problem for government has been to balance the controls over local discretion with the need for a locally responsive and efficient development plan system. During the 1970s and 1980s many local authorities avoided preparing statutory development plans, in part because they believed the costs of taking a plan through the formal procedures of consultation and objection outweighed

any benefits (Bruton and Nicholson 1983). As a result, the legitimacy which plans provided to decision-making was limited. The failure of local authorities to keep plans up to date exacerbated this. When planning authorities did seek public 'involvement' they tended to adopt a 'prepare reveal and defend' strategy, or even 'attack and response' (Rydin 1999: 188 and 193).

The perverse outcome of all this was an over-reliance on informal policy (or in some cases no policy at all), *ad hoc* decision-making, and consequently considerably less accountability. Much of the debate during the 1980s and 1990s over 'the future of development plans' has turned on this issue. Planning practitioners have argued that in order to facilitate the production of plans, the procedures should be streamlined, for example by removing the requirement for a public inquiry. In fact, this was not the problem. Local authorities in general demonstrated little commitment to plan production until the adoption of statutory plans for the whole of their areas was made mandatory and their significance in development control increased.[4]

The formal opportunities for involvement in the plan-making process and the way they have changed over the past twenty years are set out in Chapter 4. It is useful to distinguish between opportunities for participation in the creation of planning policy and opportunities to formally object. When the public are given the opportunity to participate, the planning authority has not determined its final position and should be working cooperatively and creatively with other interests on shaping the plan. Its role is to mediate a wide range of interests, including those that are not directly represented in the process. When the authority offers the opportunity for the public to object, it has determined its position, and objections will seek to challenge that position. At this stage the authority is defending its preferred plan, and its role is as major interest.

Too often the formal objection and inquiry periods are confusingly described as opportunities to participate or be consulted. This is just not the case. During the objection and inquiry stages the planning authority is one side of an adversarial contest. This is not to suggest that the authority will not have 'an interest' or a view in the early stages. It is never a neutral observer or referee, as is sometimes thought. But it should be conducting an open debate on the merits or otherwise of particular policies and proposals.

The general effect of changes to the formal procedure since 1968 has been to reduce the emphasis on real public participation and consultation in the statutory procedure, while increasing opportunities for formal objection. In the early stages of the process, a statutory requirement for consultation before the planning authority has adopted its preferred view has been replaced with discretion to decide on the appropriate publicity for individual plans. Although the formal requirements have changed, the imperative for planning authorities to involve and consult local communities in plan preparation is in reality no less than before.

In contrast, the rights of formal objection after deposit of the plan have been extended. There is now a provision for objections where the local authority has not accepted the inspector's or panel's recommendations (introduced by the 1991 Act) and, more recently, a move to a two-stage deposit (introduced by the 1999 Regulations) and an end to the requirement for consultation with statutory consultees. The rationale is that two opportunities for formal objection will draw out significant problems earlier in the process and allow more time for their resolution. But it will also encourage local authorities to firm up their proposals very early in the process. Furthermore, central government is now playing a more active role in setting limits to local discretion, and in determining the content of plans. None of this can be conducive to effective real participation. PPG 12 (1999) talks of consultation 'based on a key issues approach which identifies the main matters and choices which need to be made by the plan, and which focuses on those local communities, businesses, organisations and individuals relevant to the proposals being put forward' (para. 2.8).

In Scotland, advice on consultation follows similar lines, but has for some time given greater emphasis to the benefits to be gained from early consultation with a wide range of interests. The position of central government could not be put more clearly. The emphasis is on sorting out problems that arise from conflicts between the proposals and interests who are directly affected. This has been reinforced by the increasing pressure to produce plans within a reasonable time scale, as well as the sheer size and complexity of plans. Inevitably, this means that the plan preparation process tends to focus on the concerns of those who have the inclination, skills, and resources to participate. Though the procedural safeguards are in principle open to all, it is only the better-organised and well-financed groups that are able to make the most use of them.

Nevertheless, consultation and participation in practice have been much more than adherence to formal procedures, and many planning authorities have been innovative in finding ways to involve a wider range of interests and concerns in plan making.

PUBLIC PARTICIPATION IN DEVELOPMENT CONTROL

Around a half a million applications for planning permission are made each year to local planning authorities in Britain, of which over four-fifths are granted. This enormous spate of applications involves great strains on the local planning machinery, which, generally speaking, is not adequately staffed to deal with them and at the same time undertake the necessary work involved in preparing and reviewing development plans. Yet full consideration by local planning staffs is needed if planning committees are to have the requisite information on which to base their decisions.

When a planning application is lodged, statutory consultees (listed in Chapter 5), other organisations, and neighbouring occupiers will be invited to comment. Others may also make representations if they are aware of the proposal. In England and Wales the system of notification in the case of 'bad neighbour' developments (such as sewage works, dance halls, and zoos) was replaced in 1992 with a requirement for notification of owners and other interests in land which is the subject of a planning application, but this does not necessarily extend to neighbours. Provision for notification of neighbours is dealt with under the provisions for publicity.[5]

Local authorities have the responsibility of deciding, on a case-by-case basis, what type of publicity to require. Major developments, as defined in the GDPO, require *either* site notices or neighbour notification, *and* a newspaper advertisement (see Chapter 5). A survey of London authorities found that direct contact by letter was most effective, although the letters used by about a third of authorities were written in a way which meant they were unlikely to be understood by two out of three people (Edmundson 1993: 13). The requirements of the GDPO have been described as 'overkill' and unnecessarily expensive, especially in the need for newspaper advertisements, which have questionable effect (Harrison 1994).

Moreover, in its Circular on the matter[6] the DoE stresses that obligations to publicise applications should not jeopardise the target of deciding 80 per cent of applications within a period of eight weeks. Speed is thus apparently to have higher priority than public participation.

Scotland and Northern Ireland have different arrangements for neighbour notification and publicity about planning applications. In the Scottish system, notification is the responsibility of the applicant, who certifies to the local authority that neighbours, as well as owners and lessees, have been notified. This can be problematic for the applicant, can lead to false certification (whether inadvertent or deliberate), and has been a constant source of complaint to the Ombudsman. A study of the Scottish system (Edinburgh College of Art and Allan 1995) found that policing the notification efforts of applicants would merely increase the workload for councils; the effectiveness could be improved only by transferring responsibility for notification to them. Overall, it advocated the Northern Ireland system of notification, where there is a two-tier approach. A non-statutory system requires the Planning Service to notify neighbours, but the identification of neighbours (through presentation of a list of 'notifiable interests') is undertaken by the applicant. The merits of the different approaches across the UK have been compared, with the Northern Ireland system coming out on top.

Once comments are gathered, it is up to the planning authority to consider them in making the decision. Should planning permission be granted and it later emerges that a notifiable neighbour has not been notified, the local authority cannot revoke the planning permission. The only recourse open to the third party is a private action against the applicant, but it has to be shown that failure to notify was carried out 'knowingly and with deceitful intention' (Berry *et al.* 1988: 806).

Over recent years there has been a considerable increase in opportunities for presentations to be made to planning committees, both by the applicant and by objectors, with evidence of very positive results in terms of the perception of the service by its 'customers' (Shaw 1998; Darke 1999). However, planning committees often have remarkably little time during a meeting in which to come to a decision. Agendas for meetings tend to be long: an average of five to six minutes for consideration of each application is nothing unusual, and in some cases the time spent on an application may be much less. It cannot, therefore, be surprising that in a large proportion of cases (in the bigger authorities at least) the recommendations of the planning officer are approved *pro forma*. Also, many applications are now dealt with through delegated powers and will not be discussed by committees (see Chapter 5). Limited open discussion of applications may be a reflection of harmonious relationships between councillors and officers, although there have been a number of well-publicised cases where elected members have consistently acted against the recommendations of officers.

Several important implications follow from this. First, and most obvious, is the danger that decisions will be given which are 'wrong' – that is, they do not accord with planning objectives. Second, good relationships with the public in general and unsuccessful applicants in particular are difficult to attain: there is simply not sufficient time. Third, this lack of time corroborates the view of many (unsuccessful) applicants that their case has never had adequate consideration – a view which is further supported by the manner in which refusals are commonly worded. Phrases such as 'detrimental to amenity' or 'not in accordance with the development plan', and so on, mean little or nothing to individual applicants. They suspect that their cases have been considered in general terms rather than in the particular detail which they naturally think is important. And they may be right: understaffed and overworked planning departments cannot give each case the individual attention which is desirable.

The value of public participation in development control needs to be considered in the light of the very different attitudes that people will take about a planning principle that they generally agree with (for example, preventing unnecessary development in the countryside) when it is applied to their own land. This natural human failing is encouraged by the curious compromise situation which currently exists in relation to the control of land. On the one hand, it is accepted in principle (and law) that there is no right to develop land, unless the development is publicly acceptable (as determined by a political instead of a financial decision). On the other hand, though the allocation of land to particular uses is determined by a public decision, the motives for private development are financial, and the financial profits which result from the development constitute private gain. This unhappy circumstance (which is discussed at length in Chapter 6) involves a clash of principles which the unsuccessful applicant for planning permission experiences in a particularly sharp manner. It follows that local planning officials may have a peculiarly difficult task in explaining to a landowner why, for example, a particular field needs to be 'protected from development'.

Nevertheless, the success which attends this unenviable task does differ markedly among different local authorities. The question is not simply one of the great variations in potential land values in different parts of the country or in the relative adequacy of planning staffs. Though these are important factors, there remains the less easily documented question of attitudes towards the public. Despite all the charters, there is considerable variation in the efforts that planning authorities make to assist and explain matters to the public.

RIGHTS OF APPEAL

An unsuccessful applicant for planning permission can, of course, appeal to the Secretary of State and, as already noted, a large number do so. Each case is considered by the Department on its merits. This allows a great deal of flexibility, and permits cases of individual hardship to be sympathetically treated. At the same time, however, it can make the planning system seem arbitrary, at least to the unsuccessful appellant. Although broad policies are set out in such publications as the *Planning Policy Guidance Notes*, the general view in the central departments is that a reliance on precedent could easily give rise to undesirable rigidities.

Other issues relevant to this view are the flexibility of the development plan, the wide area of discretion legally allowed to the planners in the operation of planning controls, and the very restricted jurisdiction of the courts. All these necessitate a judicial function for the Department. However, this function is only quasi-judicial: decisions are taken not on the basis of legal rules as in a court of law or in accordance with case-law, but on a judgement as to what course of action is, in the particular circumstances and in the context of ministerial policy, desirable, reasonable, and equitable. By its very nature this must be elusive, and the unsuccessful appellant may well feel justified in believing that the dice are loaded. The very fact that public inquiries on planning appeals are heard by ministerial 'inspectors' (and probably in the town hall of the authority whose decision is being appealed against) does not make for confidence in a fair and objective hearing.

Of course, part of the expressed dissatisfaction comes from those who are compelled to forgo private gain for the sake of communal benefit: the criticisms are not really of procedures, and they are not likely to be assuaged by administrative reforms or good public relations. Fundamentally, they are criticisms of the public control of land use – in particular, if not in principle.

THIRD PARTY INTERESTS

The rights of third parties (those affected by planning decisions, but having no legal interest, not being the applicant or the authority) are extremely limited. Over the years there have been calls to extend to them the right to appeal should planning permission be granted, as is already the case in Ireland. The rights of third parties were highlighted in the so-called Chalk Pit case (*Public Law*, Summer 1961: 121–8; Griffith and Street 1964). This, in brief, concerned an application to 'develop' certain land in Essex by digging chalk. On being refused planning permission, the applicants appealed to the minister, and a local inquiry was held. The inspector recommended dismissal partly because of the impact on the neighbouring property of a Major Buxton. The minister disagreed and allowed the appeal.

Major Buxton then appealed to the High Court, partly on the ground that in rejecting his inspector's findings of fact, the minister had relied on further information supplied by the Minister of Agriculture without giving the objectors any opportunity of correcting or commenting upon it. But Major Buxton now found that he had no legal right of appeal to the courts: indeed, he apparently had no legal right to appear at the inquiry. (He had only what the judge thought to be a 'very sensible' administrative privilege.) In short, Major Buxton was a 'third party': he was in no legal sense a 'person aggrieved'. Yet clearly in the wider sense of the phrase Major Buxton was very much aggrieved, and at first sight he had a moral right to object and to have his objection carefully weighed. But should the machinery of town and country planning be used for this purpose by an individual? Before the town and country planning legislation, landowners could develop their land as they liked, provided they did not infringe the common law, which was designed more to protect the right to develop than to restrain it. However, as the judge stressed, the planning legislation was designed 'to restrict development for the benefit of the public at large and not to confer new rights on any individual member of the public'.

This, of course, is the essential point. It is the job of the local planning authority to assess the public advantage or disadvantage of a proposed development, subject to a review by the Secretary of State if those having a legal interest in the land in question object. Third parties cannot usurp these government functions. Nevertheless, it is generally agreed that third parties should have a right to let their views be known to the planning committee.

THE USE OF PUBLIC INQUIRIES

Public inquiries into major planning appeals, and called-in planning applications, have had a stormy passage for many years, particularly those held in connection with highways and major developments such as Stansted, Windscale, and Sizewell. Similar difficulties are now being experienced with local plan inquiries, which, following the 1991 changes to the development plan regime, are bigger and more keenly contested affairs.

The planning inquiry is a microcosm of the land use planning system, and it reflects many of its competing positions and underlying conflicts of interest. It is perhaps in the inquiry where the clash of ideologies is most easily seen. McAuslan (1980: 72) used the example of road inquiries.

The disenchantment with public inquiries into road proposals is only the most public and publicised manifestation of a general disenchantment with the system of land use planning, to which the conflict of ideologies within and over the use of the law is an important contributor. What this use of the inquiry has shown is that the reforms introduced as a result of the Franks Report twenty years ago, based on the principle of openness, fairness, and impartiality, and concentrating on procedures, did not change (perhaps were not designed to change) the overriding purpose of the public local inquiry, which was and is to advance the administration's version of the public interest.

Using McAuslan's terminology, this is a triumph of 'the public interest ideology' over 'the ideology of public participation'. The important point here is that the inquiry is not an extension of public participation, but 'a limited and carefully controlled and confined discussion of specific proposals . . . inimical to the kind of wide ranging discussion that participators are demanding'. This applies equally to major planning inquiries and even, as is discussed later, examinations in public.

A difficulty with many inquiries is determining where the boundaries of discussion are to be drawn: there is always the danger that argument will spill over into a broader policy framework. It is common at inquiries into particular matters for the most general questions of policy to arise. This is hardly surprising, since typically the development being debated is, in fact, the application of one or more policies to a particular situation: this readily offers the opportunity for questioning whether the policy is intended to apply to the case at issue – or whether it should. Even wider issues arise, such as the desirability of supporting a particular way of generating nuclear power, or the need for more roads, or the role of the planning system in providing affordable housing. Pressure groups which, for example, may be opposed to the building of new roads or out-of-town shopping centres anywhere, irrespective of the merits (or otherwise) of particular projects, will want to use the inquiry as a platform on which to make their wider case. That they are able to do this is sometimes a reflection on the lack of national policy on certain issues. The Heathrow Terminal 5 inquiry is the latest in a long line of inquiries which have spent much time and public money debating what the national policy should be.

This raises the question as to whether the provisions for national policy debate are adequate. It makes sense, of course, to argue that Parliament should be the arena for the national policy debate, and the local authority for debate on local policies. It also seems reasonable to maintain that it is quite inappropriate for major issues of principle to be argued when they are simply being applied locally. But issues are not so easily packaged; in reality there is a sharing of competences among different jurisdictional levels. And in the UK, the lack of a regional tier has often meant that national and local levels have had to work cooperatively on strategic issues. Some site-specific proposals raise acute issues of national policy which have not been settled or adequately discussed; and sometimes government may

avert proper discussion at the national or regional levels because of the complexity and sensitivity of the issues involved. This may be the reason that one type of public inquiry, for which legislative provision was made in 1968, has never been used: the Planning Inquiry Commission (PIC).[7]

There have been several proposals for the funding of third parties at major public inquiries, though they differ on the form that this should take – and the difficulties to which it could give rise. The government has taken the narrow approach that 'most objectors participate in public inquiries to defend their own interests'. This is a perfectly proper activity, but there is no reason why it should be financed out of public funds. In Canada, however, the Berger Commission on the Mackenzie Valley Pipeline Inquiry arranged for a funding programme which cost nearly $2 million, for 'those groups that had an interest that ought to be represented, but whose means would not allow it'. The federal government has an 'intervener funding programme', which was used in the Beaufort Sea environmental assessment review (Cullingworth 1987).[8]

EXAMINATIONS IN PUBLIC

In the case of development plan inquiries, the separation of broad strategic policy and detailed site specific issues has been widely, if not completely, accepted. While delegating the adoption of local plans to local authorities, the 1968 Act confirmed that the public local inquiry would continue to be used for all development plans, and that objectors would maintain the statutory right to have their objections heard. The maintenance of the rights of those affected to obtain an independent hearing was thought to be particularly important given concerns raised in debate about the empowerment of local authorities to adopt their own plans (Bridges 1979). This argument quickly lost ground with the realisation of the practical consequences, evidenced in the inquiry into the Greater London Development Plan (GLDP). This inquiry considered 28,000 objections over twenty-two months in the years 1970 to 1972. The GLDP was not typical of the emerging notion of a structure plan and contained many detailed proposals, but the experience led to support for the introduction, in 1972, of the *examination in public* (EIP), a major departure from former practice. This involves a panel which considers only those matters that are selected for discussion. Objectors have no statutory right to be heard, 'effectively relieving the Secretary of State of any duty to inquire, in public, into objections made to a submitted development plan' (Dunlop 1976: 9). These changes produced a system which is the opposite of that intended by the Planning Advisory Group (PAG). The introduction of the EIP made the procedure for structure plans 'almost entirely administrative in character, being governed at almost every stage by discretion' exercised, until 1992, by central government (Bridges 1979: 246).

The rationale for EIPs, however, is far more than the negative one of avoiding lengthy, time-consuming, and quasi-judicial public inquiries: it is related essentially to the basic purpose and character of the plan. A structure plan does not set out detailed proposals and, therefore, does not show how individual properties will be affected. It deals with broad policy issues: the most common are green belts; the scale and general location of housing, industrial land, and commercial land; and the integration of land use and transport and sustainable development issues (Baker and Roberts 1999). Since the implications for particular sites should not be discussed, it is consistent that the procedure should not expend time on detailed objections.

An EIP is a required part of the adoption procedure for structure plans, unless the Secretary of State decides otherwise.[9] The issues selected are placed on deposit prior to the examination, together with a list of those selected to participate. There will usually be a preliminary meeting where the procedure and agenda can be discussed. This meeting, like the examination itself, is carried out by a panel with an independent chairman, supported by a panel secretary. The chairman has the discretion, both before and during an examination, to invite additional participants in addition to those selected by the authority, and to adjust the form of the proceedings if he or she considers this necessary. Though the Act made provision for regulations governing the conduct of the EIP, none has been made: instead, the DETR publishes a Code of Practice.[10] This stresses that the proceedings should be organised so as to promote 'intensive discussion without formality'. For example, 'objections to the general policies and proposals of the plan may be heard by means of a round-table session of the inquiry, chaired by the Inspector'.

The experience of EIPs has sometimes been quite different from this ideal, with excessively formal hearings involving senior counsel acting for local authorities and others. The proceedings certainly do not lend themselves to involvement by ordinary members of the public, and in fact this is not encouraged at any stage in the structure planning process. The main participants tend to be representative interest groups, notably the House Builders Federation and the Council for the Protection of Rural England, as well as local authorities.

This procedure clearly gives rise to difficulties, particularly since there is likely to be considerable criticism by any objectors who are excluded from participation in the examination. This is especially sensitive in England and Wales, where the planning authority itself is responsible for adopting the structure plan, holding and paying for the EIP, and selecting issues and participants. In this connection it is important to stress that the examination is envisaged as only one part of the process by which the plan is adopted (or, in Scotland, approved by the Secretary of State). Of crucial importance in this process is the extent to which effective citizen participation has taken place in the preparation of the plan. Baker and Roberts (1999) found some evidence of innovative practice on participation, for example in the use of video, the Internet, and targeted seminars, but this seems to be the exception rather than the rule, and they conclude by recommending that authorities should move to the formal deposit stage without delay.

LOCAL AND UNITARY DEVELOPMENT PLAN INQUIRIES

The precise role of the local plan inquiry has long been a subject of debate. The procedure is a long-standing feature of British government administration, with its origin in the parliamentary Private Bill procedure that provided an opportunity for objections to government proposals to be heard by a parliamentary committee (Wraith and Lamb 1971). The procedure has grown as much by accident as design since it has been successively amended to take into account changes elsewhere in the system. Its twofold purpose has continued: to gather information for government and to provide a route for individual redress. Like appeal inquiries, local plan inquiries involve the same balancing of private and public interests through a procedure which, although essentially administrative, has many of the hallmarks of judicial courtroom practice. However, the essential nature of the planning procedure is administrative. Final decisions are taken by government, at either the central or the local level. In making these decisions, account may be taken of matters not discussed at the inquiry. As the Franks Report (1957: para. 272) noted, the process 'must allow for the exercise of a wide discretion in the balancing of public and private interests'. The legitimacy of the decisions rests with the political accountability of the decision-maker (Parliament or the local council) rather than on the weighing and testing of evidence as in a court of law.

Though the Secretary of State has ultimate discretion, the procedure also attempts to safeguard the rights of the individual citizen. Some aspects of the inquiry procedure are much more akin to a judicial process. Objectors have a statutory right to appear, and the evidence is tested through a process of adversarial questioning before an independent party (a planning inspector). There is inherent ambiguity in a system which has as its main objective the gathering of evidence to assist in the making of a governmental decision, while at the same time operating in the manner of a judicial hearing (Wraith and Lamb 1971). The essential dilemma of this quasi-judicial process was described in the Franks Report (1957: paras. 273–5):

> If the administrative view is dominant the public enquiry cannot play its full part in the total process, and there is a danger that the rights and interests of the individual citizen affected will not be sufficiently protected. . . . If the judicial view is dominant there is a danger that people

will regard the person before whom they state their case as a kind of judge provisionally deciding the matter, subject to an appeal to the Minister.

The difficulties were increased by the 1991 reforms. In confirming the role of districts as the responsible authority for adopting local plans, they reinforced the administrative role of the inquiry. In introducing the provision which allows for objections where the local authority does not accept the recommendations of an inspector or panel, the changes lend weight to the judicial role. Thus, the central questions have changed little over the years following the Franks Committee, not least because while identifying the ambiguity in objectives, the Report found in favour of neither, and fell back instead on the need for balance, and consistent application in the inquiry procedure of the principles of 'openness, fairness and impartiality' (Bruton 1980: 377).

The three principles of 'openness, fairness and impartiality' have guided inspectors with some success, and the courts have played a relatively small part in the planning process (although this is growing). Nevertheless, each of the three principles requires some qualification. The Franks Report itself recognised that impartiality needed to be qualified since in some circumstances central government was both a party to the debate, perhaps putting forward a proposal, and at the same time the decision-maker. How, in this situation, can the procedure be impartial? This is the major complicating factor for local plan inquiries, EIPs, and major call-in inquiries. Here, one of the parties to the dispute will make the final decision, giving at least the appearance of being the judge and jury in its own court.[11]

The openness and fairness of the inquiry also need to be qualified. First, there is widespread misunderstanding of the procedure, especially the respective roles of inspector and local authority. The adversarial nature of the inquiry, with the inspector playing a passive role while objectors and the local authority exchange evidence and questions, has important implications for the way in which the agenda is structured; and it limits potential outcomes. In his case study of the Belfast Urban Areas Plan inquiry, Blackman (1991a) points out how an adversarial hearing focuses attention on the evidence brought forward to support the position of particular interests. The inquiry becomes moulded into a battle about which interest should prevail; and this precludes debate about alternative and potentially shared solutions, which may be in a 'common or generalisable social interest'.

All this has now to be considered in the context of the 'plan-led system'. More emphasis on statutory plans means that more development interests, neighbouring authorities, and service providers will be concerned to influence the content of plans and thus the outcome of inquiries. The number of objections to plans has increased dramatically over recent years.[12] Many have questioned the system's capacity to cope with this burden, particularly since the biggest test for contentious plans may be later in the modifications stages. One response to the increased scale of work at inquiries (and the burden on inspectors) has been to emphasise the potential that pre-deposit consultation and post-deposit negotiation might have in reducing the number of objections. After deposit of the plan, local authorities were encouraged to negotiate with objectors and publish agreed 'suggested changes' before the inquiry. The 1999 Regulations have formalised this approach by establishing a two-stage deposit. After the initial deposit the objections are discussed before a revised deposit stage (as explained in Chapter 4).

In response to widespread concerns about the mounting costs of inquiry work, the DoE commissioned research on the efficiency and effectiveness of local plan inquiries (Steel *et al*. 1994). The Inspectorate had already shown considerable flexibility in the way that it was prepared to innovate in the running of inquiries through round-table sessions and other less formal hearings for non-professionally represented objectors, but criticisms of the inquiry process grew.

The research concluded that the crux of the problem in the inquiry procedure was for local authorities to address. Many authorities were insufficiently prepared for the scale of the plan inquiry task and furthermore paid only passing attention to the advice that was then offered. Many were unable or unwilling to explore the possibility of compromise (or provide information and clarify uncertainties) with objectors.

Very few offered the support to objectors that would have reduced the effort needed from inspectors to clarify objections at the inquiry stage. The findings on the objectors' understanding of the procedure were very similar to those of Bruton *et al*. (1982), who had examined the issue ten years previously. In the main, local authorities took a very defensive attitude at the inquiry stage and were not well prepared. As a result, the Planning Inspectorate published *Development Plan Inquiries: Guidance for Local Planning Authorities* (1996), which brings the advice into line with the new demands of inquiries.

Research in Scotland has concluded similarly that management of the process has been poor in too many authorities. The *Review of Development Planning in Scotland* (Hillier Parker *et al*. 1998) makes similar points about a lack of focus to consultation, and the need for a greater sense of urgency and commitment to the process by local authorities. However, its recommendations concentrate on changing the procedures, in particular making the report of the inquiry binding on the authority.

The role of planning inspectors (and in Scotland reporters) is critical in the consideration of formal objections. Research on inquiries and the Inspectorate's own commissioned studies confirm that inspectors provide a high-quality service, although there are inconsistencies in practice arising from the wide discretion available to inspectors in managing proceedings. However, when problems arose in the procedure, inspectors had few powers and resources to put them right. The researchers' conclusion was that inquiry procedure rules were needed to provide inspectors with additional powers of sanction for those who did not follow the guidance, but this has not been acted upon. Instead, there have been revisions to the Code of Conduct.

One other issue that has been addressed is improving the business relationship between the Inspectorate and the local authorities who pay for their services on development plan inquiries. In 1994 Birmingham City Council successfully challenged the Inspectorate, saying that there was no legal basis for the Inspectorate to charge for inspectors' services (which had been the practice since 1968). This has been changed by the Town and Country Planning (Cost of Inquiries, Etc.) Act 1995, but not until millions of pounds had been returned (temporarily) to local authorities who had been wrongly charged. The Inspectorate has also now issued a model 'service agreement' which is used to outline the services that will be given and their costs.

Although this has not been fully examined in research, it is obvious that procedural problems are often compounded by the nature of the plans being examined, which are sometimes overly complex and detailed, and which are not conducive to effective communication in public consultation. An additional layer of complexity is added by the 'suggested changes' made prior to the inquiry, which in at least one case have been published in six separate documents. It is perhaps surprising therefore that attention is still primarily focused on the procedures. The form and content of plans have received much less attention, although their basic design is about as old as the Mini.

Those who want to contribute to shaping development plans will have to put considerable effort into the early stages of participation. This might now be more difficult in view of the increasing pressure on authorities to hasten plan preparation. This could be a recipe for conflict, frustration, and perhaps less credible plans.

THE HUMAN RIGHTS ACT

None of the discussion above takes account of the Human Rights Act 1998, which came into force on 2 October 2000. The Act enables citizens to take action under the European Convention of Human Rights (dating from 1950), through the domestic courts. The Convention was used in court cases involving planning well before the Act came into force, but the implications are still not clear. The Act does not confer new rights. However, its enactment raises awareness about the rights set out in the Convention and should ensure that they are taken into account by government and the courts. All public bodies are required to act in accordance with the Convention rights. Local planning authorities and other planning agencies, including the Planning Inspectorate, are thus bound by the Act.

The main issue for planning is the Convention right arising from Article 6 which guarantees that 'In the determination of his civil rights and obligations or of any criminal charge against him, everyone is entitled to a fair and public hearing within a reasonable time by an independent and impartial tribunal established by law.' This may be a problem for the UK since ministers make policy, apply it, and decide on appeals. While this duty is mostly undertaken through the Planning Inspectorate or Recorders' Office, their independence from ministers is in question (inspectors represent and 'stand in the shoes of the minister'). In any event, the minister has the right to 'recover' appeals for his or her decision.

The Scottish minister has already accepted that the neither ministers nor reporters are independent and impartial (in the meaning of the Convention), but has argued that the provision to further right to challenge decisions in the courts does meet the Convention. Although the right of challenge in the courts is severely restricted (being limited generally to procedural issues rather than the policy merits of the case), this argument has been accepted in at least one case so far. Nevertheless, the future of the Planning Inspectorate has been questioned.

A related issue is the idea of an environmental court. The merits or otherwise of an environmental court system, as currently practised, for example, in Australia and New Zealand, have been discussed in the DETR's *Environmental Court Project*, led by Professor Malcolm Grant. The report proposes a two-tier approach with a first-level court comprising a revised Planning Inspectorate as an independent tribunal and a second-level environmental court that would deal with the same sort of appeals now considered by ministers. At the time of writing, the government has confirmed its intention to continue with the current arrangements. The second stage of the review of the Inspectorate (discussed in Chapter 3) and more cases brought under the Human Rights Act will influence developments in the longer term.

INTERESTS IN PLANNING

'Interests' in planning are usually thought of in terms of the organisations, groups, and individuals who are actively engaged in the planning arena: they are identified by their participation in the land development and planning processes. These include land and property owners, developers, special interest groups, national government and its agencies, and local authorities themselves (in both their landowning and their regulatory capacities). Some organisations have become particularly skilled in presenting their views at both national and local levels. The House Builders Federation, for example, is an important national organisation that regularly presents evidence at public inquiries in support of the interests of the housebuilding industry.

Voluntary organisations operate in a variety of ways, from lobbying to education, and from active participation in planning processes to the ownership and management of protected land and property. Membership of such organisations has increased dramatically over the past twenty years.[13] National voluntary organisations obviously do not command the resources available to commercial interests, but they do employ experts and can be very effective in promoting their causes. Some of them have become increasingly sophisticated over the past decade or so, and can be appropriately described as 'major elites' (Goldsmith 1980). There are innumerable 'minor elites' of small groups who become involved in an *ad hoc* way with particular issues. The evolution of participation and consultation in planning has favoured these self-defined interest groups, and given them a relatively privileged position in the planning process.

It has long been appreciated that such groups are not necessarily representative of anything wider than the interests of their active supporters. Many people are not able or willing to take the time to engage in 'participation', and some groups who have a clear stake in planning outcomes (such as homebuyers or job-seekers) are too diffuse to have become effective participators, and 'rarely if ever emerge as definable actors in the development process' (Healey *et al*. 1988: chapter 7). Three 'interests' that permeate almost all

planning activity – race, gender, and disability – yet which are rarely represented in disputes over particular development projects are discussed briefly in the following sections. Other specific interests are beginning to receive special attention, notably those of older people (Gilroy 1999), while the social exclusion agenda of the Labour government will encourage local authorities to consider the needs of specific interests in the creation and implementation of planning policy. The revised PPG 12 (1999) states that when preparing plans, planning authorities

> should consider the relationship of planning policies and proposals to social needs and problems, including their likely impact on different groups in the population, such as ethnic minorities, religious groups, elderly and disabled people, women, single parent families, students, and disadvantaged people living in deprived areas. They should also consider the extent to which they can address issues of social exclusion though land use planning policies.
>
> (para. 4.13)

RACE AND PLANNING

Questions of equal opportunities for all racial groups have figured increasingly on the town planning agenda over the past thirty years, though with questionable impact on practice. The Race Relations Act 1976 places a duty on all local authorities to eliminate unlawful racial discrimination in their activities and to promote 'good relations between persons of different racial groups'. It was in 1978 that the RTPI established a joint working party with the Commission for Racial Equality (CRE) to investigate the multi-racial dimension of planning, and to make recommendations for any necessary changes in practice. Its report *Planning for a Multi-racial Britain* (RTPI and CRE 1983) was a frank assessment of the inadequacies of the then current thinking on race and planning. It is perhaps only a little less relevant today, and its recommendations apply to all planners, wherever they work.

The working party's deliberations were spurred by the deteriorating race relations in Britain's cities, and by numerous reports calling for more action from central government and other bodies (the Scarman Report 1981). These and other studies demonstrated that: 'Black people in Britain share all the problems of social malaise and multiple deprivation with their white neighbours. In addition, however, they experience the additional difficulties of racial discrimination and disadvantage, because of the colour of their skin' (CRE 1982: 9). The RTPI/CRE report argued that despite its record of innovation in participation, the profession had failed to address the issue of race. Indeed, the RTPI's own 1982 report on participation failed to consider the racial dimension explicitly. Ethnic minorities were mentioned only once, with reference to the need to provide interpretation at public meetings. *Planning for a Multi-racial Britain* sought to improve sensitivity to racial issues 'and show how planning practice can be modified to avoid racial discrimination, promote equality of opportunity, and improve race relations for the benefit of the whole community'. It identified three elements in the racial dimension of planning:

(1) reviewing the impact of current policies, practices and procedures upon different racial groups, with a view to ascertaining actual or potential racial discrimination;
(2) building racial distinctions into surveys, analyses and monitoring with a view to identifying the special needs of different racial groups; allowing the impact of policies to be assessed, and providing a basis for any appropriate positive action;
(3) positive action in planning policies, procedures, standards and decision making, partly by directing positive non-racial policies and actions towards groups containing high proportions of black people, and partly by taking special steps to ensure that black people have equal access to the benefits offered by town planning.

The report also considered how the profession could be made more representative in terms of its membership, and the implications of this for the education of planners. It became an important guide for committed practitioners and teachers, and formed the basis for other, more detailed guidance, foremost of which was the GLC's *Race and Planning Guidelines*. Publication of this was squeezed in just before abolition (when perhaps the most vigorous supporter of equal opportunities was lost).

The GLC report, however, has subsequently provided a rich source of material for guiding good practice in the London boroughs and elsewhere. Its recommendations illustrate the inadequacy of the 'colour-blind' approach identified as commonplace in the RTPI/CRE report. Such an approach, though holding that all people should be treated equally, is insensitive to the cultural traditions and needs of particular ethnic groups. Unintentional discrimination can therefore result. It is perhaps surprising, therefore, that Thomas and Krishnarayan (1993) are able to note the almost complete absence of reference to racial matters in major planning reports such as the Audit Commission's *Building in Quality* and the DoE's *Development Plans: A Good Practice Guide*. Despite evidence of innovative practice in a few places (Best and Bowser 1986), planning has made only a weak contribution to challenging racial discrimination and disadvantage. Thomas and Krishnarayan (1994a: 1891) explain this with reference to the 'socially conservative' nature of planning: the tendency to focus on technical problems of land use management rather than radical social goals, a tendency that has been strongly reinforced by professionalisation, the narrow interpretation of the objectives of planning by government and the courts, and the reproduction of existing social and spatial divisions in planning policy.

By the end of the 1990s little progress had been made. A survey for the Local Government Association revealed that, in general, equal opportunity issues now have a lower priority than in the 1980s; there is little policy guidance on the issue; and the colour-blind approach still dominates practice. The effect has been to institutionalise indirect discrimination within the planning system (Loftman and Beazley 1998a, b) and to stereotype different groups by oversimplifying their internal diversity (Ratcliffe 1998). However, the same survey also revealed examples of good practice, especially in the districts with a significant ethnic minority population. Many more authorities need to look towards the examples of good practice and past recommendations on how to address race and ethnic mix in planning.

WOMEN AND PLANNING

The absence of policy explicitly related to concerns of women has attracted little attention until recently. One of the many reasons for this is the inadequacy of the classic texts on social theory and urban studies. In reviewing these, Greed (1993: 233) writes:

> Women appear in studies of working class communities as a variety of oversimplified stereotypes, based on observing them as mono-dimensional residents tied to the area rather than as people with jobs, interests and aspirations beyond its boundaries. Young and Willmott give emphasis to women in their study, but their fondness for seeing them in the role of 'Mum', as virtually tea machines, and almost as wall paper to the main action of life, is open to question.

Greed has elsewhere (1991) described the male domination of the profession as 'only a temporary intermission'. Women were primary contributors to the social movement which promoted town planning early in the twentieth century. She argues that the professionalisation of planning, its institutionalisation within the government structure, and the limited access to qualifying courses have 'kept most women out'. The gender bias reflects, and in part perpetuates, the patriarchal structure of British society, and continues to influence the recruitment and education of planners. Women generally tend 'to under-achieve and under-aspire as regards a career', and they 'hesitate to embark upon a professional career', although once on a planning course, women tend to outperform men.

Awareness of women's issues in planning grew strongly during the 1980s and was reflected in important and influential reports from the GLC (1986) and the RTPI (1987, 1988). The negative impacts of increasing mobility and planning policies that encouraged this have come in for particular criticism (Hamilton 1999).[14] Though issues have been identified, their impact has been marginal. One difficulty (in addition to the power of traditional attitudes and ways of thinking and perceiving) is the general lack of explicit social policies in plans. Thus, the provision of sporting facilities and the open space standards applied to them are routinely regarded as legitimate land use matters. These predominantly male activities are contrasted by Greed (1993: 237) with crèches, which are commonly regarded as social issues, even though they 'may have major implications for central area office development' – a fact which is explicitly taken into account in some US cities (Cullingworth 1993: chapter 7). Another factor is the limited extent to which these issues have been addressed in the curricular of planning schools (Loevinger Rahder and O'Neill 1998).

Calder et al. (1993) have identified the scope of 'women's specific needs policies', and their research and that by L. Davies (1996) show that some specific policies are being incorporated in development plans, even weathering DETR scrutiny. However, this has not been taken up widely, and where relevant policies have been included, they have often been deleted following requests from the DETR after plan scrutiny, or in response to inspectors' reports after the inquiry. Davies (1996) notes the removal of a policy from the Hammersmith and Fulham UDP which required sheltered lockable spaces for buggies in new large-scale shopping developments.

Concerns for the needs of women are generally much less developed than those of race or disability, especially outside London, where some authorities simply 'didn't consider this an issue' (L. Davies 1996). Even where they are, 'child care' is the predominant 'women's issue'; reflecting the assumption of women's primary role as carers, and demonstrating 'the limited nature of the majority of planning initiatives which it may be argued are designed to ameliorate current constraints on women rather than to challenge the status quo in the drive for greater equality for women' (Little 1994b: 266).

PLANNING AND PEOPLE WITH DISABILITIES

One group that has received more attention from the planning system is people with disabilities. The 1990 Planning Act requires local authorities to draw to the attention of planning applicants the need to consider the requirements of the Chronically Sick and Disabled Persons Act 1970, and the Building Regulations require 'access provision' in new buildings, extended in 1992 to include access for people with sensory impairments. Disability in this sense has a broad definition including (as the RTPI's Practice Advice Note 3 points out) those who suffer breathlessness or pain, who need to walk with a stick, are partially sighted, have difficulty in gripping because of arthritis, or are pregnant. Access policies can also generally make life easier for parents with children and the elderly. As Gilroy and Marvin (1993: 24) point out, most people are or will be physically impaired at some time during their life. That impairment or disability can be turned into a handicap by the environment.

Despite this explicit concern, the needs of people with disabilities have been considered largely in terms of the design of the built environment. It is primarily the Building Regulations and not the planning system which are the means by which access requirements are enforced (L. Davies 1996). Important though this is, it leads to an overly simplistic stereotyping of the problems faced by individuals with disabilities. Difficulties in using public transport are also well known and now are covered by the Disability Discrimination Act 1995.

Thomas (1992: 25) has made a strong critique of typical attitudes:

The 'regs' can become a checklist which defines the needs of disabled people, ignoring, indeed disallowing, the possibility that individual professionals dealing with particular cases need to learn from the experience of disabled people themselves. The British legislation which relates specifically to planning with its references to practicality and reasonableness, reinforces a strand in planners' professional ideologies which emphasises the role of the planner in reaching optimum solutions in situations involving competing needs or interests. Thus might a fundamental right to an independent and dignified life be reduced to an 'interest' to be balanced against the 'requirements' of conservation or aesthetics.

L. Davies (1996) is critical of central government advice, which, although often very recently updated, fails to give the necessary impetus. The DoE *Good Practice Guide*, for example, is described as 'woefully lacking'. She also observes that whereas negotiations between interest groups and planners have often strengthened plans on these matters, further objection

by other interests and scrutiny by the DETR later in the process have resulted in a watering down of policies'. Above all, policies apply only to new building, whereas much of our environment, including the headquarters of the RTPI and universities (as Greed ruefully notes (1996c: 233)), presents major access problems for many people.

THE PLANNING PROFESSION AND PLANNING EDUCATION

Planning practice and education in the UK are strongly professionalised, more so than in any other European country. The Royal Town Planning Institute is incorporated by royal charter 'to advance the science and art of town planning for the benefit of the public'. Like other professional bodies, it upholds standards in practice by restricting membership to those who have approved educational qualifications and practical experience. Under the charter, the Institute has the power 'to devise standards of knowledge and skills for persons seeking corporate membership of the Chartered Institute.' It accredits planning courses in universities according to the specified standards. Working as a planner is not restricted to members of the RTPI, but many employers require membership.[15]

At the beginning of 2000 there were 17,500 members, of whom more than 1,000 were based overseas in more than ninety countries.[16] Previous surveys suggest that about half the professional staff working on planning in the public and private sectors are members of the RTPI (LGMB 1993c).[17] The overwhelming majority of the membership is young, white, and male. Women and ethnic minorities are under-represented. Even in the younger age groups, women account for only about 30 per cent of members, and at the higher levels women are very poorly represented.

The RTPI provides a range of services for its members and acts as the principal point of contact between the profession, the government, and other interests. The Institute also provides services for the consumer of the planning service, including information on where to get advice on planning, and the Planning Aid Service, which provides free independent advice on planning (through the voluntary effort of members) to people, communities and voluntary groups who cannot afford to pay for consultants.

The planning profession is thus well established in Britain, and the RTPI is easily the largest professional planning body in Europe. It has had a very strong influence on the development of planning practice in the UK and to some extent overseas. How this position has been reached in the years since the RTPI was established in 1913 is worthy of a note of explanation. The Institute held its own examinations for membership until 1992 but, from the earliest days, specific courses were set up to train planners, the first being at Liverpool University in 1909 followed by University College London in 1914 (Batey 1993; Collins 1989). By 1945 there were nine courses in town and country planning, all of which were postgraduate. Student numbers in town and country planning increased sharply to fifty-four separate courses in 1978 (Thomas 1990) and fifty-seven (twenty of them undergraduate) in 1981.

A government-inspired review, *Manpower Requirements for Physical Planning* (Amos *et al.* 1982), forced some courses to cease recruitment during the 1980s, although the impact on the number of planning students was reduced because of increased intakes in the remaining schools. In 1988 a total of 766 students were recruited into thirty-one courses. Student numbers have increased substantially since then, spurred on by the promotion of participation in higher education generally, although there have been fluctuations generally following the peaks and troughs of the economy. In the year 1998–99 there were over 3,000 students enrolled on planning courses in the twenty-four recognised schools and the Joint Distance Learning Diploma. This compares with 3,715 in 1991–92.

The Institute reviews the accreditation of planning schools at least once every five years and publishes guidelines setting out policy on the education of planners together with a core curriculum. Students completing accredited courses can (after two years of practical experience) apply for corporate membership of the Institute. Despite pessimistic forecasts in the 1980s, the demand for planning graduates has increased significantly over the past fifteen years. During the late 1980s and again in the late 1990s and into 2001, the supply of planning graduates is insufficient to meet demand. Current concerns in education centre on the increasing competition among planning schools for a smaller pool of applicants and the need to improve the quality of intakes.

The Town and Country Planning Association should also be mentioned here. It was established as

---

BOX 12.1 WORLD TOWN PLANNING DAY

World Town Planning Day was founded in 1949 by the late Professor Carlos della Paolera of the University of Buenos Aires. To promote the day, Paolera founded the International Organisation for the World Town Planning Day with the objective of advancing public and professional interest in town planning throughout the world. Arrangements should be made on or near 8 November each year for meetings, lectures, exhibitions, broadcasts, or celebrations relating to town planning.

Celebrations of the day were held in a different country every year (designated as president for that year) from 1950 until the late 1970s. Britain was the focus in 1965. Paolera designed a symbol for the day in the shape of a flag whose top half is blue and the bottom half green, respectively standing for air and land, and with a bright gold sun in the middle.

Source: *Town and Country Planning Association* (which occasionally flies the flag from its offices overlooking the Mall)

the Garden City Association by Ebenezer Howard in 1899 to realise the first garden city. It took on a wider remit to campaign for effective town planning in the early part of the twentieth century and took its present name in 1941. It is not a professional association and membership is open to anyone. The TCPA has been a consistent advocate for rational and humane forms of spatial development.[18] It is also a mine of critical yet constructive commentary on planning, notably through its journal *Town and Country Planning*.

ACCESS TO INFORMATION

Information is power; so it not surprising that un-democratic societies guard it jealously. It is surprising, however, that secrecy is so prevalent in Britain. The fiasco over the Crossman diaries revealed the absur-dities in striking detail. British secrecy is a legacy of old styles of government (Cullingworth 1993: 191). These persist in many ways, and government is still carried on in an elitist atmosphere of determining what actions are in the public interest, as viewed through the eyes of the government. Traditionally, partici-pation has been limited and information restricted, though there are clear signs of a significant change in attitudes.

Ironically, though traditional attitudes are more entrenched at the centre, it is central government that has forced local government to provide greater 'free-dom of information'. A major milestone in this is the Local Government (Access to Information) Act 1985, which imposes a duty on local authorities, the national parks, and other bodies such as PTAs 'to publish information . . . about the discharge of their func-tions and other matters'. The underlying principle of this Act is that all meetings of local authorities, their committees, and subcommittees should be open to the press and the public. There is a power to exclude only in narrowly defined circumstances. The Act also provides for public access to agendas, reports, and minutes of all meetings, and opens for public inspec-tion certain background papers which relate to the subject matter of reports to council, committee, or subcommittee meetings.

A review of the Act after ten years of operation (Steele 1995) was generally positive, finding that the Act had established 'minimum standards of openness and accountability'; that the provisions were exceeded by four out of five councils; and that this had been done at little extra cost. Planning figures prominently where rights are exercised, with 60 per cent of planning committees attended by more than ten people (against a 10 per cent average for council committees). Thus rights are being exercised, but take-up is limited to making information *available*, and most authorities do not actively promote information and thus make it truly *accessible*.

Similar conclusions were reached following a survey of public access to planning information in Scotland undertaken for the Scottish Office (University of Dundee 1997). The best-practice recommendations indicate the less than positive approach taken to access in some authorities – including having literature explaining how the system operates and ensuring that professional staff are available at suitable times. The Cabinet Office Service First initiative is going to require a much more thorough approach, including opening of offices beyond the usual working day.

A major influence on improving accessibility is the EU, particularly with its 1990 *Directive on Freedom of Access to Information on the Environment*. (Birtles 1991) This was a product of the commitment in the EC's Fourth Action Programme to enable groups and individuals to take a more effective part in protecting and promoting their interests. The Directive was brought into force in the UK by the Environmental Information Regulations 1992. The objective of the Directive is 'to ensure freedom of access to, and dissemination of, information on the environment held by public authorities and to set out the basic terms and conditions on which such information should be made available'.

The regulations define environmental information as 'the state of any water or air, the state of any flora or fauna, the state of any soil, or the state of any natural site or other land', together with activities or measures adversely affecting, or designed to protect, these states. Organisations affected are given two months to respond to any requests for information, and there are limited provisions for refusal to supply information where, for example, it is incomplete or subject to legal proceedings. An important provision in the regulations is the requirement for relevant organisations to produce a list of all information sources, and for this to be made publicly available. The Directive, however, does not address the two most important issues: 'knowing that there is someone to ask, and knowing that there is something to ask for' (Clabon and Chance 1992: 25). Once an issue is identified, the problem is mostly one of *navigation* around complex information systems (Moxen *et al*. 1995).

In order to encourage an active approach by governments, the Treaty on European Union also enables any citizen to make a complaint to the Commission that an EU provision is not being applied. Krämer (1991) argues that this casts the EC in the role of ombudsman for the environment. Additional rights to 'environmental information' arise from the Directives on environmental assessment, though the use of such measures to date is limited. Important factors are the limited knowledge of existence of information gaining access, the costs of retrieval, and understanding what is found.

In 1994 central government adopted a *Code of Practice on Access to Government Information* which, alongside the charter initiatives and requirements to justify expenditure more fully, has led to the publication of more detailed reports from departments and agencies. In 1999 the DETR also published a *Code of Practice on the Dissemination of Information during Major Infrastructure Developments*. The opening up of government has been increased following the Freedom of Information Act 2000.

The Act was preceded by the White Paper *Your Right to Know* (1997), explaining that the aim was to make all government information available 'unless it would clearly cause harm to national security, personal privacy, safety, business activities and law enforcement'. The Freedom of Information Act supersedes the Code and provides for a general right of access to information held by all public authorities except the secret services. As well as primary legislation requiring disclosure, it is also intended that much more government information will be routinely published, although the public authorities will be allowed to charge for it.

The cost issue is worth further note in respect of planning documents. Central government publications are now highly priced (presumably at market levels), as a result of which their cost is prohibitive to ordinary citizens. There is a striking contrast with many US government and European Union publications which the public can obtain free of charge. A good example has been set by the Scottish Office, which announced at the end of 1992 that it had decided to issue its planning guidance (NPPGs, Circulars, and PANs) generally free of charge, although retaining the discretion to make a charge for PANs where appropriate.

With local government, a bigger problem is the paucity of publications, but the cost of planning documents tends to be very high. No doubt these expensive reports are intended for a small, high-price market, but high quality (or quantity) of plan production is not important in relation to the need for public information and participation. A number of authorities have printed plans in newspaper or poster format. These are models well worth copying, as is the policy of North Kesteven District Council, whose plan is free, and those of the London Boroughs of Tower Hamlets and Barnet, which gave plans away free to residents.

Another means of providing information is via the Internet, and this is growing rapidly. Under the modernising government initiative the government aims to ensure that 25 per cent of all interactions with the citizen should be made available by electronic means by 2002, and 100 per cent by 2005.[19] Good examples of local authority web sites now contain details of planning applications and the local plan. Examples are listed with other web sites at the end of this book.

MALADMINISTRATION, THE OMBUDSMAN, AND PROBITY

Most legislation is based on the assumption that the organs of government will operate efficiently and fairly. This is not always the case, but even if it were, provision has to be made for investigating complaints by citizens who feel aggrieved by some action (or inaction). As modern post-industrial society becomes more complex, and as the rights of electors and consumers are viewed as important, pressures for additional means of protest, appeal, and restitution grow.

At the parliamentary level the case for an ombudsman was reluctantly conceded by the government, and a Parliamentary Commissioner for Administration was appointed in 1967. The Commissioner is an independent statutory official whose function is to investigate complaints of maladministration against central government departments and other government bodies acting on their behalf, and referred to him or her through Members of Parliament (Gregory and Pearson 1992). Powers of investigation extend over all central government departments, and there is an important right of access to all departmental papers. In 1994 the powers were extended to cover the *Code of Practice on Access to Government Information*.

Only a small fraction of the Parliamentary Commissioner's cases relate to planning matter, and of course the concern is with administrative procedures, not with the merits of planning decisions. The Commissioner's reports give full but anonymised texts of reports of selected cases which have been investigated. Illustrative cases include a complaint that the Secretary of State for the environment failed to understand the grounds on which a request had been made for intervention; a complaint that following a motorway inquiry, the inspector called for further evidence from the DoE on which objectors were not given the opportunity to cross-examine; and a complaint by a group of local residents that an appeal decision to allow a gypsy caravan site paid little heed to local residents' objections, ignored important relevant facts, and was taken on the basis of inconsistent attitudes. In all these cases the Parliamentary Commissioner concluded that the complaint could not be upheld. This is not always the case however; and the Commissioner's subsequent criticisms have led to changes in internal administrative procedures in the DETR.

The cases in which the Commissioner does find 'maladministration' are often of extraordinary

complexity, if not real confusion. Indeed, complexity and confusion can be major factors in the failures in communication and the misunderstandings which result in 'maladministration'.

The popularity of the Parliamentary Commissioner led to pressures for the establishment of a similar institution for local government. In the mid-1970s, *Commissioners for Local Administration* were set up for England, Scotland, and Wales. With good sense, they recently decided that they should be known as the *Local Government Ombudsmen* (using the Swedish word, where the office dates back to 1809 (Renton 1992). Not only is this their popular name, but it also makes explicit that their responsibilities are confined almost entirely to local government (though it is hardly gender sensitive!). The ombudsmen also deal with police authorities.

A high proportion of complaints concern planning matters: in 1998–99 about a fifth (3,290) of the record number of 15,869 complaints received in England. As with the Parliamentary Commissioner, complaints have to be referred via an elected member (although in exceptional circumstances the ombudsman can accept a complaint direct), a requirement concerning which there is considerable controversy. There is also concern about the situation which arises when a local authority refuses to 'remedy' a case in which maladministration or injustice is found by a Commissioner. To date, however, only limited legislative changes have been made. These include a power for local authorities to incur expenditure to remedy injustice without specific authorisation by the Secretary of State; a requirement that local authorities must notify the ombudsman of action taken in response to an adverse report; a power for the ombudsman to publish in a local newspaper a statement concerning cases in which a local authority has refused to comply with the ombudsman's recommendations; and a new responsibility for the ombudsmen to provide local authorities with advice on good practice, based on the experience of their investigations.

Not all have taken kindly to the 'interference' of the local ombudsmen, and their annual reports (while noting with satisfaction a general improvement in the handling of complaints by local authorities) often name authorities which have refused to remedy cases of maladministration and personal injustice. The irony is that the ombudsmen is wholly funded by contributions from local authorities.

The ombudsmen have constantly noted that aggrieved objectors to planning permissions (third parties, who account for about half of all cases) have little or no redress – unlike the aggrieved applicant, who can appeal to the Secretary of State. The ombudsmen have no power to deal with the merits of planning decisions, but they have difficulty in explaining the difference between a planning decision which constitutes maladministration and one which is simply disputed. Recent typical cases include complaints where the local authority approved applications under delegated powers even though the parish council had objected, and thus required discussion at committee; failure to make a condition on an approval to meet the concerns of objectors even though they were assured it would be made; failure to identify from site visits or to bring to the attention of the planning committee impacts on neighbours' amenity which may have led to refusal of applications; and failure to take enforcement action against uses not having planning permission. Where maladministration is found (about half the cases), the authority is usually asked to value the cost of the failure (for example, in loss of amenity) and to pay a sum for the trouble taken in bringing the case. The ombudsmen have certainly had an influence in encouraging local authorities to improve their planning procedures, and to go beyond minimum statutory requirements.

Despite improvements to procedure in many local authorities, there have been recent well-publicised cases of extreme maladministration leading to fraud and corruption. Although small in number such cases raise more general concerns about the integrity of officers and probity of councillors, especially in the national context, where the Scott Inquiry (and the dubious activities which gave rise to it) have brought these general questions to the attention of a very wide audience. The best-known case is that of North Cornwall District Council, though the London Borough of Brent, and Bassetlaw and Warwick District Councils, have also been subject to similar inquiries. Currently there

is considerable argument about the antics of some Doncaster councillors, seven of whom have been convicted of fraud. In North Cornwall, complaints were first taken up by the ombudsman, then by the district auditor, the police, and Channel 4 television. Finally, the DoE set up an official inquiry (Lees 1993), which unequivocally condemned the local councillors for granting permissions to local people for development in the open countryside.[20]

The 1992 Local Government Act and the Code of Conduct for Councillors establish the principle that councillors should not take part in proceedings if they have a direct or indirect pecuniary interest in the issue under discussion. But such a simple distinction does not cover the many ways in which councillors and officers can be influenced or themselves influence decisions. At the heart of the issue is lobbying. Applicants and objectors lobby councillors (and sometimes officers). Councillors may lobby colleagues (although not taking part in the decision themselves). Committee chairs can put pressure on officers, and so on. Whether or not such practices constitute improper activity is not always easy to discern.

Because of the number of cases of alleged impropriety or 'sleaze' in government generally during the Conservative administration, a committee was established to consider standards in public life (the Nolan Committee). Its 1997 report on local government included a chapter on planning. The report criticised the national code for being inadequate, complicated and, in parts, inconsistent and even impenetrable. Building on the report, the government proposed a 'new ethical framework' to govern the conduct of elected members and also local government employees (who were not covered by the code). Each local authority is required to produce its own code, and an independent Standards Board will have the responsibility of investigating alleged breaches of the local code.[21]

Planning was seen to require extra measures including the need for councillors to undertake training in planning because of the difficulties in dealing fairly with planning law and its implementation.[22] There should also be a greater degree of openness in the planning process; this would, among other things,

BOX 12.2 CODE OF BEST PRACTICE IN PLANNING PROCEDURES

Members and officers should avoid indicating the likely decision on an application or otherwise committing the authority during contact with applicants and objectors.

There should be opportunities for applicants and objectors, and other interested parties such as parish councils, to make presentations to planning committee.

All applications considered by planning committee should be subject to full, written reports from officers incorporating firm recommendations.

The reasons given by planning committee for refusing or granting and should be fully minuted, especially where these are contrary to officer advice or local plan.

Councillors and planning officers should make oral declarations at planning committee of significant contact with applicants and objectors, in addition to the usual disclosure of pecuniary and non-pecuniary interests.

No member should be appointed to planning committee without having agreed to undertake a period of training in planning procedures as specified by the authority.

Source: *Standards of Conduct in Local Government* (1997: 75)

---

assist in dealing with the problems facing local authorities in granting permission for their own proposed developments and 'the potential for planning permission being bought and sold'.

In coming to these conclusions, the Nolan Report noted that in 1947 'the need for postwar reconstruction was clear. Development enjoyed broad public support.' Things have now changed:

> Development is now a term which has a pejorative ring, and the planning system is seen by many people as a way of preventing major changes to cherished townscapes and landscapes. If the system does not achieve this (and it is a role which it was not originally designed to perform), then the result can be public disillusionment.
>
> (para. 277)

In Scotland a 1998 consultation paper on the Nolan Report, *A New Ethical Framework for Local Government in Scotland*, broadly accepted its recommendations, but took issue with a number of them. It proposed a single code for all local governments (instead of a model code) and it favoured a national Standards Commission instead of local authority standards committees. It also argued that reasons should not be required for the granting of planning permission since such decisions are not subject to any appeal process, and would not only add to the difficulties facing a planning com-

mittee but put permissions at increased risk of legal challenge on purely technical grounds.

IN CONCLUSION

The planning scene has been dominated for many years by a veritable orgy of institutional change. The pace of change has accelerated under the reforming Blair administration. Though all this was intended as a means of facilitating better planning, it is possible that it has had the opposite effect of restraining the development of policies appropriate to changing conditions and perceptions. If the filing cabinets are being constantly moved, it is difficult to bring their contents up to date. Furthermore, some of the institutional changes (even if promising in the longer run) may have added to the confusion over the role of 'town and country planning' in relation to regional and national economic planning, to the management of the economy, to the increasingly strident demands for environmental protection, to the place of public participation in the planning process, and to even more intractable issues such as 'the energy question', the distribution of incomes, and 'access to opportunity'. It is, however, a nice question as to whether a more stable institutional structure would have facilitated the formulation of more appropriate and effective policies in the context of the baffling economic and social problems of the time.

What does seem clear is that the faith in the efficacy of institutional change was misplaced. Consecutive attempts at the reorganisation of local government seem to have created as many problems as they solved. In the words of Matthew Arnold, 'faith in machinery is our besetting danger'.

The basic problems lie deeper: they relate to the functions, scope, and practicability of town and country planning'. The crucial issues with which 'planning' is concerned do not fall within the responsibility or competence of the planning authority, or even within that of local government – jobs and poverty being the two most obvious ones. Hence central government wrestles with the political pressures to which problems in such areas give rise, though typically with disappointing results.

From a cynical viewpoint, much effort is wasted at both local and central levels in attempting to control the uncontrollable. The proclamations of politicians are given a credibility which is unwarranted. It also has unfortunate consequences, since the illusion that problems can be 'solved' turns easily into a delusion, and constant failure debases the political process and breeds cynicism.

More positively, there has been a continued discussion of the limits, role, and purpose of planning. There has been a steady succession of reviews and studies: the 1976 RTPI discussion paper *Planning and the Future*; the Nuffield Foundation's 1986 Report *Town and Country Planning*; the TCPA's review of the planning system in the late 1990s, *Reinventing Planning* (2000); and most recently the review of environmental planning by the Royal Commission on Environmental Pollution which is due to report in 2001. It is interesting to speculate why this untypically deep questioning began when it did. Perhaps it was a sign of the coming of age of planning. Two factors were of particular importance: an awakening of concern for making government more responsive (what was inadequately termed 'public participation') and a sea change in the economy.

The first started with protests against unwanted developments, big and small, particular and generalised. Some of these protests led to gargantuan 'inquiries' – of which Roskill and Windscale were the epitome. Others were more modest and localised, but also much more numerous. With hindsight, the most important were those which in reality were protests not simply against a particular development (though that was the manifest objective), but against the policies which these represented.

Typically, these were the responsibility not of planners, but of other professions and, above all, of politicians who forged the policies. Politicians at both central and local levels perceived problems (understandably) in the terms in which they were presented. Problems labelled as housing shortages, road congestion, slum clearance, and redevelopment portrayed the obvious solutions: build houses quickly; build more roads; clear the slums; redevelop the worn-out parts of the inner city. The political responses were to 'solve' these clearly articulated problems. But policies involve choices and, again with hindsight, some of the choices had undesirable results: more houses involved high densities and few amenities; new roads increased the attraction of private transport and the decline of public alternatives; slum clearance destroyed communities; and so on.

The perceived 'failures' of planning – high-rise development, difficult-to-let council housing schemes, urban motorways, inner-city decline, and the like – added to the mounting concern about the role and character of planning. Whether, or to what extent, these were 'failures' and, if so, the degree to which 'planning' was to blame are questions which were seldom raised, let alone answered in their historical context. But they were seen to symbolise the inadequacy of planning.

An alternative interpretation would lay emphasis on the growth of real public participation. Public participation is not a subsidiary process which can be held in check: once it begins to work effectively, it transforms the nature of the planning process. On occasion it can get 'completely out of control', as it did in some well-publicised highway inquiries. Though disruptive, these led to a major reappraisal of both highway inquiry procedures and highway planning. Here the point is that the lesson was learned: it had become apparent that participation could work.

The professional acceptance of public participation (though by no means a unanimous) was a remarkable feature of the 1970s. That it came first in planning, rather than in other fields such as education or health, may be related to the transformed nature of planning education and the changed character of the 'intake' to the profession (Cherry 1974; Centre for Environmental Studies 1970). Indeed, it may be that it is this above all which explains the new humility, the introspective questioning, and the new intellectualism which were such marked features of the time. In this respect, planners departed from the norms of professionalism, though not without internal strife.

The profession's commitment to public participation continued into the 1980s, despite an increasingly hostile political framework. The growth of a participatory ethic, however, may have been of lesser importance than the impact of economic change. A new humility grew in response to a gradual realisation that changes in the economy were structural rather then cyclical. Policies based on the assumption that the task in hand was to channel the forces of economic growth were increasingly perceived to be misplaced. Planning was no longer to be preoccupied with controls over the location of growth: it was to be remoulded to assist in the actual promotion of growth.

Certainly the 1980s and 1990s saw a remarkable change in the political scene. A new and clear political philosophy emerged: a major objective of planning was now to facilitate enterprise with the minimum of constraints. Planning controls were reduced, most of the new town development corporations disbanded, the GLC and the metropolitan county councils abolished, and statutory requirements for public participation in the preparation of plans were reduced. Local government was increasingly bypassed in favour of *ad hoc* bodies designed to promote private-sector involvement.

Much of this dramatic change stemmed from the explicit political stance of the Thatcher government, and the belief that, somehow or other, planning itself was part of the problem – an attitude encapsulated by the remark that jobs were being locked up in the filing cabinets of planners. But there were also some deeper undercurrents. Above all, economic conditions highlighted the importance of the *promotion* of development in contrast to its *control*. Successive public expenditure crises also took their toll.

The advent of the Blair government in 1997 (following eighteen years of Conservative control) promised great changes, within a framework of what the election manifesto termed 'a new centre and centre-left politics'. Proposals for planning were largely subsumed under radical proposals for devolution, and 'good local government', but commitments were made in relation to a number of policy areas such as an integrated transport policy, 'life in our countryside', and a right to greater access. There followed an avalanche of publications, initiatives, and experiments, but implementation has been disappointing: there has been a great deal more talk than action.

More generally, there is increased confusion about the role of planning. The promises held out by the 'new' structure plan system failed to materialise. It appeared to be no more effective, speedy, flexible, or satisfying than the system it was designed to replace. Whether the latest changes will prove to be more effective remains to be seen. When the needs for economic growth and for planning clash, the former is likely to win.

A major unresolved issue is that of the allocation of land for housing. Public opinion is typically hostile to this, at least in the southern and more prosperous parts of the country, and neither statements of government commitment nor the role of the new regional development agencies has been positive. Indeed, in some areas, infrastructure inadequacies virtually preclude much-needed housing development. Unfortunately, the much-heralded integrated transport system is conspicuous by its absence.

It is always difficult to see current events in perspective, and there is abundant scope for debating whether the changes that have taken place are fundamental or not. More likely they will be overtaken by new problems, or by the redefinition of old problems which cannot readily be foreseen.

FURTHER READING

Planning and Politics

This is an enormous topic and many of the central texts on planning address it. Some recommendations for required reading on this topic are Ambrose (1986) *Whatever Happened to Planning?*; Reade (1987) *British Town and Country Planning*; Blowers (1980) *The Limits of Power*; Low (1990) *Planning, Politics and the State*; Thornley (1993) *Urban Planning under Thatcherism*; Healey (1997) *Collaborative Planning*; and Taylor (1998) *Urban Planning Theory since 1945* (chapter 5). See also the references for Chapter 1.

Participation in Planning

Overviews of the historical development of participation in planning are given by Rydin (1999) 'Public participation in planning' and Thomas (1996) 'Public participation in planning.' Early titles include Dennis (1970) *People and Planning*; Davies (1972) *The Evangelistic Bureaucrat: A Study of a Planning Exercise in Newcastle-upon-Tyne*; Broady (1968) *Planning for People*; Levin and Donnison (1969) 'People and planning'; and, from the USA, Gans (1968) *People and Plans* (and his more recent collection of essays (1991) *People, Plans and Policies*). Some titles give a flavour of their content: Wates (1977) *The Battle for Tolmers Square*; Forman (1989) *Spitalfields: A Battle for Land*; Anson (1981) *I'll Fight You for It: Behind the Struggle for Covent Garden*; and Christensen (1979) *Neighbourhood Survival*. More recent studies include Colenutt and Cutten (1994) 'Community empowerment', the DoE study *Community Involvement in Planning and Development Processes*; and the DETR study *Sustainable local Communities for the 21st Century: Why and How to Prepare an Effective LA21 Strategy* (1998).

Reviews of recent practice in participation are given by Reeves (1995) 'Developing effective public consultation: a review of Sheffield's UDP; D. Hall (1999) 'Town expansion: constructive participation' (on Stevenage); and Shaw (1998) 'Who's afraid of the double whammy'. For international comparisons, see Barlow (1995) and the report of the Council of

Europe's conference hosted by the Planning Inspectorate, *Public Participation in Regional/Spatial Planning in Different European Countries*. The DETR *Guidance on Enhancing Public Participation in Local Government* (1998) includes as an annexe a list of guides on public participation; and see also the Planning Officers' Society good practice guide, *Public Involvement in the Development Control Process* (1998).

Inquiries, Examinations in Public and Development Control

The DoE study *The Efficiency and Effectiveness of Local Plan Inquiries* (1997) contains an extensive bibliography. See also the Planning Inspectorate (1996) *Development Plan Inquiries: Guidance for Local Planning Authorities*. On EIPs, see Phelps (1995) 'Structure plans: the conduct and conventions of examinations in public' and Baker and Roberts (1999) *Examination of the Operation and Effectiveness of the Structure Planning Process*. On participation in development control, see Edinburgh College of Art and Allan (1995) *Review of Neighbour Notification* and Planning Aid for London (1995) *Publicity for Planning Applications*. The implications of the Human Rights Act is considered by Hart (2000).

The Profession

The intellectual development of the planning profession is explained by Healey (1985); and see also Taylor (1992) 'Professional ethics in town planning', Evans (1993) 'Why we no longer need a town planning profession'; and Grant (1999a) 'Planning as a learned profession'.

Race and Planning

Thomas (2000) provides a thorough examination of race and planning. In addition to RTPI (1993) *Ethnic Minorities and the Planning System*, see the DETR's *New Deal for Communities: Race Equality Guidance* (1999); and the earlier RTPI/CRE (1983) *Planning for a Multiracial Britain* is still relevant. See also Thomas and Krishnarayan (1994b) *Race, Equality and Planning:*

Policies and Procedures; but also see the reflections on this research by one of the researchers, H. Thomas (1997) 'Ethnic minorities and the planning system: a study revisited', and the LGA survey by Loftman and Beazley (1998a) *Race, Equality and Planning*.

Women and Planning

Two textbooks provide a comprehensive analysis of theory and practice. Greed (1994) *Women and Planning: Creating Gendered Realities* is a mine of interesting examples, and provides a guide to reading. Little (1994a) *Gender, Planning and the Policy Process* links the issue to wider debates. Other key sources are Gilroy and Marvin (1993) *Good Practices in Equal Opportunities* and Greed (1991) *Surveying Sisters: Women in a Traditional Male Profession*. See also London Women and Planning Group (1991) *Shaping our Borough: Women and Unitary Development Plans*; RTPI (1988a) *Planning for Choice and Opportunity* and Practice Advice Note no. 12 *Planning for Women* (1995); Little (1994b) 'Women initiatives in town planning in England'; and L. Davies (1996) 'Equality and planning: gender and disability'.

People with Disabilities

The British Standards Institution publishes the *Code of Practice for Access for the Disabled to Buildings* (BS 5810) and *Code of Practice for Design for the Convenience of Disabled People* (BS 5619). See also Development Control Policy Note 16 (1985) and RTPI (1988b, c) Planning Advice Notes: *Access for Disabled People* and *Access Policies in Local Plans*; and London Boroughs' Disability Resource Team (1991) *Towards Integration: The Participation of Disabled People in Planning*.

Access to Information

On the general topic, a starting point is the Policy Studies Institute report *Public Access to Information* (Steele 1995), updated with the White Paper *Your Right to Know: Explanatory Notes on the Freedom of Information Bill*; and University of Dundee (1997) *Public Access to Planning Information*. On access to

environmental information, see European Environmental Bureau (1994) *Your Rights under European Union Environment Legislation* and Moxen *et al.* (1995) *Accessing Environmental Information in Scotland*.

Maladministration, the Ombudsman and Probity

The annual reports of the separate Commissioners for England, Scotland, and Wales provide all the facts plus a flavour through thumbnail sketches of the cases being heard. See also CPRE (1999e) *The Local Government Ombudsman*.

NOTES

1 One successful example that is often quoted is Coin Street on the south bank of the Thames in central London, where the local community prepared their own plan for a highly valued commercial site including affordable housing and other amenities (Brindley *et al.* 1996). Other, less successful stories of community participation are *The Battle for Tolmers Square* (Wates 1976) and *Spitalfields: A Battle for Land* (Forman 1989).

2 Arnstein's (1969) ladder of participation has eight rungs: citizen control, delegated power, partnership, placation, consultation, informing, therapy, and manipulation.

3 Charters now include the *Travellers Charter*, the *Passenger's Charter*; the *Bus Passenger's Charter*: the *Rail User's Charter*; the *London Underground Customer's Charter*; and the *Victim's Charter*. There are many others. Details of current national government charters are available on the Modernising Public Services Group web site at <www.service first.gov.uk>. Note that the environmental charters, like many others, appear in different forms for the four countries of the UK.

4 The adoption of statutory local plans for the whole area of the authority has always been a requirement in Scotland, although this could be by a number of part area plans. This requirement explains in part the better performance in plan-making in Scotland.

5 Neighbours are defined as those having coterminous boundaries (either at the side or above and below) and within 4 m of the boundary. Most local authorities go beyond the statutory requirements and take a broader definition of neighbouring properties, and organisations that should be consulted (Spawforth 1995, cited in K. Thomas 1997).

6 Circular 15/92, *Publicity for Planning Applications*. See also the GDPO 1995: article 8.

7 This was heralded in the 1967 White Paper which preceded the 1968 reforms of the planning system. It was argued that for planning cases which raised wide or novel issues of more than local significance a Planning Inquiry Commission should be set up consisting of three to five members appointed by the Secretary of State to make recommendations.

8 On the basis of Canadian experience, Purdue and Kemp have advocated limited state funding on the basis that some objectors 'genuinely contribute to the wider understanding of the issues involved' (1985: 685).

9 In Baker and Roberts' (1999) survey, where very minor alterations were made only two of twenty-eight structure plan authorities had not undertaken an EIP as part of the process.

10 The code of practice which explains the procedures for examinations in public is given in the DETR booklet *Structure Plans: A Guide to Procedures* (2000), which is available free from the DETR publications centre and on the web site at {less}www.planning.detr.gov.uk/guides/structure {more}.

11 The findings of research (outlined in Chapter 3) show that local authorities make modest use of their power to reject inspectors' recommendations (only in one out of ten cases). However, it is the much smaller number of rejected recommendations that receive the most publicity.

12 The average number of objections in 1997–98 was 1,250, down from 1,400 in 1994–95. The Planning Inspectorate's work on local plan inquiries grew threefold between 1988 and 1991. This was largely due to an increase in the average

length of inquiries, from just over two weeks in 1988 to eight weeks in 1993 (sitting for three and half days per week). Since then the average length of inquiries has remained about the same, although there is great variation from one plan to another. The full duration from the opening of the inquiry to receipt of the inspector's report was forty-nine weeks in 1994–95. (This is still quite a bit less than the time the authority spends in preparing plans prior to deposit.)

13 Between 1971 and 1998, membership of the Royal Society for the Protection of Birds (established 1889) increased from 98,000 to over one million; the National Trust (established 1895) from 278,000 to 2.6 million; and the relative newcomer Friends of the Earth (established 1971) from a mere 1,000 to more than 114,000 in the UK and one million worldwide – though recent trends show a levelling off.

14 The impact of changes in transport and mobility has been identified as particularly important for women and a Women's transport network has been established within the DETR's Mobility Unit to secure wider understanding of women's specific transport needs.

15 'The [1998] Members Survey confirmed that 63.8 per cent of members said an RTPI qualification was unnecessary to gain employment in planning when they first entered the profession: and 54.4 per cent reported that it was not necessary in order for them to retain their present job' (Grant 1999a: 6).

16 Further information on the RTPI, its Charter, and its Code of Conduct is available from the Institute's headquarters at 26 Portland Place, London W1N 4BE, and on its web site: <www.rtpi.org.uk>.

17 Non-members working in planning include professionals who are not eligible for membership, such as architects who are working principally on planning matters, or students studying for qualifications that lead to membership, as well as those eligible who choose not to join. The total 'professional body' in its widest sense is thus more than 35,000 strong. In addition, there are support

staff, who in 1991 totalled over 8,000 in local government and an equal number elsewhere. In total, therefore, there is a planning workforce of some 50,000. Three-quarters work in the public sector (with about two-thirds in local government) and about a fifth were in private consultancies and the development industry. The dominant activities undertaken by planners are in development control and development planning, though they are engaged in a very wide range of other jobs.

18 For a review of the hundred year work of the TCPA, see the anniversary edition of its journal *Town and Country Planning* (68 (6)) and Hardy (1991a, b).

19 The ways in which the government intends to achieve these targets are set out in the Cabinet Office Channels Implementation Policy, available at <www.citu.gov.uk/moderngov/cppolicy.htm>.

20 In fact the rate of decisions going against officers'

recommendations was no higher and even less than that for many other authorities across the country. The publicity brought to North Cornwall has brought major changes and, for the new committee, extensive awareness-raising through training on both issues of good conduct in local government and the operation of the planning system.

21 *Modern Local Government: In Touch with the People* (Cm 4014, 1998), chapter 6; *Modernising Local Government: A New Ethical Framework* (1998); and *Local Leadership, Local Choice* (Cm 4298, 1999) chapter 4.

22 The DETR has subsequently published a suggested syllabus, *Training in Planning for Councillors* (1998), in cooperation with the RTPI, LGA, and IDEA. See also the LGA publication *Probity in Planning: The Role of Councillors and Officers* (1997) and Cowan (1999).

BIBLIOGRAPHY

Abel-Smith, B. and Townsend, P. (1965) *The Poor and the Poorest*, London: Bell

Abercrombie, P. (1945) *Greater London Plan*, London: HMSO

Adair, A. S., Berry, J. N., and McGreal, W. S. (1991) 'Land availability, housing demand and the property market', *Journal of Property Research* 8: 59–69

Adams, B., Okely, J., Mogan, D., and Smith, D. (1976) *Gypsies and Government Policy in Britain*, London: Heinemann

Adams, D. (1992) 'The role of landowners in the preparation of statutory local plans', *Town Planning Review* 63: 297–323

Adams, D. (1994) *Urban Planning and the Development Process*, London: UCL Press

Adams, D. (1996) 'The use of compulsory purchase under planning legislation', *Journal of Planning and Environment Law* 1996: 275–85

Adams, D. and Pawson, G. P. (1991) *Representation and Influence in Local Planning*, Manchester: Department of Planning and Landscape, University of Manchester

Adams, D., Russell, L., and Taylor-Russell, C. (1994) *Land for Industrial Development*, London: Spon

Adams, J. G. U. (1990) 'Car ownership forecasting: pull the ladder up, or climb back down?', *Traffic Engineering and Control* 31: 136–41

Adams, J. G. U. (1995) *Risk*, London: UCL Press

Adams, J. G. U. (1998) 'Carmageddon', in Barnett, A. and Scruton, R. (eds) *Town and Country*, London: Jonathan Cape

Adams, W. M. (1986) *Nature's Place: Conservation Sites and Countryside Change*, London: Allen & Unwin

Adams, W. M. (1996) *Future Nature: A Vision for Conservation*, London: Earthscan

Advisory Committee on Business and the Environment (1993) *Report of the Financial Sector Working Group*, London: Department of Trade and Industry

Advisory Committee on Business and the Environment (annual) *Progress Report*, London: Department of Trade and Industry

Albrechts, L. (1991) 'Changing roles and positions of planners', *Urban Studies* 28: 123–37

Albrechts, L., Moulaert, F., Roberts, P., and Swyngedouw, E. (eds) (1989) *Regional Policy at the Crossroads: European Perspectives*, London: Jessica Kingsley

Alden, J. (1992) 'Strategic planning guidance in Wales', in Minay, C. L. W. (ed.) 'Developing regional planning guidance in England and Wales: a review symposium', *Town Planning Review* 63: 415–34

Alden, J. and Boland, P. (eds) (1996) *Regional Development Strategies: A European Perspective*, London: Jessica Kingsley

Alden, J. and Offord, C. (1996) 'Regional planning guidance', in Tewdwr-Jones, M. (ed.) *British Planning Policy in Transition: Planning in the 1990s*, London: UCL Press

Alden, J. and Romaya, S. (1994) 'The challenge of urban regeneration in Wales: principles, policies and practice', *Town Planning Review* 65: 435–61

Aldous, T. (1989) *Inner City Urban Regeneration and Good Design*, London: HMSO

Aldridge, H. R. (1915) *The Case for Town Planning*, London: National Housing and Town Planning Council

Aldridge, M. (1979) *The British New Towns*, London: Routledge & Kegan Paul

Aldridge, M. (1996) 'Only demi-paradise? Women in garden cities and new towns', *Planning Perspectives* 11: 23-39

Aldridge, M. and Brotherton, C. J. (1988) 'Being a programme authority: is it worthwhile?', *Journal of Social Policy* 16: 349–69

Alexander, A. (1982) *Local Government in Britain since Reorganisation*, London: Allen & Unwin

Alexander, C. (1965) 'A city is not a tree', *Architectural Forum* 122: 1 (reprinted in Bell, G. and Thywitt, T. (eds) 1972) *Human Identity in the Urban Environment*, Harmondsworth: Penguin)

Alexander, E. R. (1981) 'If planning isn't everything, maybe it's something', *Town Planning Review* 52: 131–42

Alexander, E. R. (1992) *Approaches to Planning: Introducing Current Planning Theories, Concepts and Issues* (second edition), Philadelphia: Gordon & Breach

Alker, S., Joy, V., Roberts, P., and Smith, N. (2000) 'The definition of brownfield', *Journal of Environmental Planning and Management* 43: 49–69

Allaby, M. (1994) *Macmillan Dictionary of the Environment*, London: Macmillan

Allanson, P. and Whitby, M. (eds) (1997) *The Rural Economy and the British Countryside*, London: Earthscan

Allen, H. J. B. (1990) *Cultivating the Grass Roots: Why Local Government Matters*, The Hague: International Union of Local Authorities

Allen, J. and Hamnett, C. (eds) (1991) *Housing and Labour Markets: Building the Connections*, London: Unwin Hyman

Allen, K. (1986) *Regional Incentives and the Investment Decision of the Firm* (DTI), London: HMSO

Allen, R. (1995) 'Policy and grassroots action: a vital mix', in *First Steps: Local Agenda 21 in Practice*, London: HMSO

Allinson, J. (1999) 'The 4.4 million households: do we really need them anyway?', *Planning Practice and Research* 14: 107–13

Allison, L. (1986) 'What is urban planning for?', *Town Planning Review* 57: 5–16

Allmendinger, P. (1996a) 'Twilight zones', *Planning Week* 4 (29) (18 July 1996): 14–15

Allmendinger, P. (1996b) *Thatcherism and Simplified Planning Zones: An Implementation Perspective*, Oxford Planning Monographs vol 2, no. 1, Oxford Brookes University

Allmendinger, P. (1997) *Thatcherism and Planning: The Case of Simplified Planning Zones*, Aldershot: Avebury

Allmendinger, P. and Chapman, M. (eds) (1999) *Planning beyond 2000*, Chichester: Wiley

Allmendinger, P. and Tewdwr-Jones, M. (2000) 'New Labour, new planning? The trajectory of planning in Blair's Britain', *Urban Studies* 37: 1379 402

Allmendinger, P. and Thomas, H. (eds) (1998) *Urban Planning and the British New Right*, London: Routledge

Alonso, W. (1971) 'Beyond the inter-disciplinary approach to planning', *Journal of the American Institute of Planners* 37: 169–73 (reprinted in Cullingworth, J. B. (ed.) (1973) *Problems of an Urban Society*, vol. 3: Planning for Change. London: Allen & Unwin)

Alterman, R. and Cars, G. (eds) (1991) *Neighbourhood Regeneration: An International Evaluation*, London: Mansell

Alterman, R., Harris, D., and Hill, M. (1984) 'The impact of public participation on planning: the case of the Derbyshire structure plan', *Town Planning Review* 55: 177–96

Ambrose, P. (1974) *The Quiet Revolution: Social Change in a Sussex Village 1871–1971*, London: Chatto & Windus

Ambrose, P. (1986) *Whatever Happened to Planning?*, London: Methuen

Ambrose, P. (1994) *Urban Process and Power*, London: Routledge

Ambrose, P. and Colenutt, B. (1975) *The Property Machine*, Harmondsworth: Penguin

Amery, C. and Cruickshank, D. (1975) *The Rape of Britain*, London: Paul Elek

Amin, A. and Tomaney, J. (1991) 'Creating an enterprise culture in the North East? The impact of urban and regional policies of the 1980s', *Regional Studies* 25: 479–87

Amos, F. J. C., Davies, D., Groves, R., and Niner, P. (1982) *Manpower Requirements for Physical Planning*, Birmingham: Institute of Local Government Studies

Amundson, C. (1993) 'Sustainable aims and objectives: a planning framework', *Town and Country Planning* 62: 20–2

Anderson, M. A. (1981) 'Planning policies and development control in the Sussex Downs AONB', *Town Planning Review* 52: 5–25

Anderson, M. A. (1990) 'Areas of outstanding natural beauty and the 1949 National Parks Act', *Town Planning Review* 61: 311–39

Anderson, W. P., Kanaroglou, P. S., and Miller, E. J. (1996) 'Urban form, energy and the environment', *Urban Studies* 33: 7–35

Anon. (1956) 'Ye olde English green belt', *Journal of the Town Planning Institute* 42: 68–9

Anson, B. (1981) *I'll Fight for You: Behind the Struggle for Covent Garden*, London: Jonathan Cape

Archbishop of Canterbury's Commission on Urban Priority Areas (1985) *Faith in the City: A Call for Action by Church and Nation*, London: Church House Publishing

Archbishop of Canterbury's Commission on Urban Priority Areas (1990) *Living Faith in the City: A Progress Report*, London: General Synod of the Church of England

Archbishop's Commission on Rural Areas (1990) *Faith in the Countryside*, Stoneleigh Park, Warwickshire: Acora Publishing

Armitage Report (1980) *Lorries, People and the Environment* (Report of the Inquiry), London: HMSO

Armstrong, J. (1985) 'The Sizewell inquiry', *Journal of Planning and Environment Law* 1985: 686–9

Arnell, N. W., Jenkins, A. and George, D. G. (1994) *Implications of Climate Change for the NRA (National Rivers Authority)* London: HMSO

Arnold, C. (1997) 'Planning gain: how are off-site liabilities passed back to landowners', University of Cambridge: PhD dissertation, Department of Land Economy Library

Arnold, C. (1999) 'Planning gain: how are off-site liabilities passed back to land owners?', *Journal of Planning and Environment Law* 1999: 869–77

Arnstein, S. R. (1969) 'A ladder of citizen participation', *Journal of the American Institute of Planners* 35: 216–24

Arton Wilson Report (1959) *Caravans as Homes*, Cmnd 872, London: HMSO

Arup Economic Consultants (1990) *Mineral Policies in Development Plans* (DoE), London: HMSO

Arup Economic Consultants (1991) *Simplified Planning Zones: Progress and Procedures* (DoE), London: HMSO

Arup Economics and Planning (1994) *Assessment of the Effectiveness of Derelict Land Grant in Reclaiming Land for Development* (DoE), London: HMSO

Arup Economics and Planning (1995a) *Coastal Superquarries: Options for Wharf Facilities on the Lower Thames* (DoE), London: HMSO

Arup Economics and Planning (1995b) *Derelict Land Prevention and the Planning System* (DoE), London: HMSO

Arup Economics and Planning (1999) *Control of Outdoor Advertising: Fly-posting*, London: DETR

Ashby, E. (1978) *Reconciling Man with the Environment*, Stanford, CA: Stanford University Press

Ashby, E. and Anderson, M. (1981) *The Politics of Clean Air*, Oxford: Clarendon Press

Ashford, D. E. (ed) (1979) *The Politics of Urban Resources*, Chicago: Maaroufa

Ashworth, W. (1954) *The Genesis of Modern British Town Planning*, London: Routledge & Kegan Paul

Assembly of Welsh Counties (1992) *Strategic Planning Guidance in Wales*, Mold: Clwyd County Council

Association of Conservation Officers (1992) *Listed Buildings Repair Notices*, Brighton: ACO

Association of County Archaeological Officers (1993) *Archaeological Heritage*, London: Association of County Councils

Association of County Councils (1992) *National Parks: The New Partnership*, London: ACC

Association of County Councils (1993) *The Enabling Authority and County Government*, London: ACC

Association of London Government (1996a) *Paying the Wrong Fare: Affordable Public Transport*, London: ALG

Association of London Government (1996b) *Red Routes: Do They Work?*, London: ALG

Association of National Park Officers (1988) *National Parks: Environmentally Favoured Areas? A Proposal for a New Agricultural Policy to Achieve the Aims of National Park Designation*, Bovey Tracey: The Association

Association of Town Centre Management (1992) *Working for the Future of Our Towns and Cities*, London: ATCM

Association of Town Centre Management (1994) *The Effectiveness of Town Centre Management*, London: ATCM

Association of Town Centre Management (1997) *Managing Urban Spaces in Town Centres*, (DoE), London: The Stationery Office

Astrop, A. (1993) *The Trend in Rural Bus Services since Deregulation*, Crowthorne: Transport Research Laboratory

Atkinson, R. (1999a) 'Discourses of partnership and empowerment in contemporary British urban regeneration', *Urban Studies* 36: 59–72

Atkinson, R. (1999b) 'Urban crisis: new policies for the next century', in Allmendinger, P. and Chapman, M. (eds) *Planning beyond 2000*, Chichester: Wiley

Atkinson, R. and Moon, G. (1994) *Urban Policy in Britain: The City, The State and the Market*, London: Macmillan

Audit Commission: *see* section on Official Publications

Austin, J. (1995) 'Renewal areas: the research findings', London: Austin Mayhead

Austin Mayhead & Co. (1997) *Neighbourhood Renewal Assessment and Renewal Areas*, London: DETR

Automobile Association (1995a) *Transport and the Environment: The AA's Response to the Royal Commission on Environmental Pollution*, London: AA

Automobile Association (1995b) *Shopmobility: Good for People and Towns*, London: AA

Bailey, N. and Barker, A. (eds) (1992) *City Challenge and Local Regeneration Partnerships: Conference Proceedings*, London: Polytechnic of Central London

Bailey, N., Barker, A., and MacDonald, K. (1995) *Partnership Agencies in British Urban Policy*, London: UCL Press

Bailey, S. J. (1990) 'Charges for local infrastructure', *Town Planning Review* 61: 427–53

Bain, C., Dodd, A., and Pritchard, D. (1990) *RSPB Planscan: A Study of Development Plans in England and Wales*, Sandy, Beds: Royal Society for the Protection of Birds

Baker, M. (1998) 'Planning for the English regions: a review of the Secretary of State's regional planning guidance', *Planning Practice and Research* 13: 153–69

Baker, M. and Roberts, P. (1999) *Examination of the Operation and Effectiveness of the Structure Planning Process: Summary Report*, London: DETR

Baker, S., Kousis, M., Richardson, D., and Young, S. (1997) *The Politics of Sustainable Development*, London: Routledge

Baker Associates (1999) *Proposals for a Good Practice Guide on Sustainability Appraisal of Regional Planning Guidance*, London: Department of the Environment, Transport and the Regions

Balchin, P. (1995) *Housing Policy: An Introduction*, London: Routledge

Balchin, P. (1996) *Housing Policy in Europe*, London: Routledge

Balchin, P. (1999) 'Housing', in Cullingworth, J. B. (ed) *British Planning: 50 Years of Urban and Regional Policy*, London: Athlone

Balchin, P. and Rhoden, M. (eds) (1998) *Housing: The Essential Foundations*, London: Routledge

Balchin, P., Sýkora, L., with Bull, G. (1999) *Regional Policy and Planning in Europe*, London: Routledge

Baldock, D., Cox, G., Lowe, P., and Winter, M. (1990) 'Environmentally sensitive areas: incrementalism or reform?', *Journal of Rural Studies* 6: 143–62

Baldock, D., Bishop, K., Mitchell, K., and Phillips, A. (1996) *Growing Greener: Sustainable Agriculture in the UK*, London: Council for the Protection of Rural England

Balfour, J. (1983) *A New Look at the Northern Ireland Countryside*, Belfast: Department of the Environment, Northern Ireland

Ball, M. (1983) *Housing Policy and Economic Power*, London: Methuen

Ball, R. M. (1989) 'Vacant industrial premises and local development: a survey, analysis, and policy assessment of the problems in Stoke-on-Trent', *Land Development Studies* 6: 105–28

Ball, R. M. (1995) *Local Authorities and Regional Policy in the UK: Attitudes, Representations and the Local Economy*, London: Paul Chapman

Ball, S. and Bell, S. (1994) *Environmental Law* (second edition), London: Blackstone

Banham, R., Barker, P., Hall, P., and Price, C. (1969) 'Non-Plan: an experiment in freedom', *New Society* 26: 435–43

Banister, D. (1994a) *Transport Planning in the UK, USA and Europe*, London: Spon

Banister, D. (1994b) 'Reducing the need to travel through planning', *Town Planning Review* 65: 349–54

Banister, D. (ed.) (1995) *Transport and Urban Development*, London: Spon

Banister, D. (ed.) (1998) *Transport Policy and the Environment*, London: Spon

Banister, D. (1999) 'Review essay: the car is the solution, not the problem?', *Urban Studies* 36: 2415–19

Banister, D. and Button, K. (eds) (1993) *Transport, the Environment and Sustainable Development*, London: Spon

Banister, D. and Watson, S. (1994) *Energy Use in Transport and City Structure*, London: Planning and Development Research Centre, University College London

Banister, D., Capello, R., and Nijkamp, P. (eds) (1995) *European Transport and Communications Networks*, Chichester: Wiley

Bannon, M. J., Nowlan K. I., Hendry, J., and Mawhinney, K. (1989) *Planning: The Irish Experience 1920–1988*, Dublin: Wolfhound Press

Bardgett, L. (2000) *The Tourism Industry*, House of Commons Research Paper 00/66, London: House of Commons Library

Barker, G. M. A. (1998) 'Statutory local nature reserves in the United Kingdom', *Journal of Environmental Planning and Management* 41: 629–42

Barker, G. M. A. and Box, J. D. (1998) 'Statutory local nature reserves in the United Kingdom', *Journal of Environmental Planning and Management* 41: 629–42

Barker, T. C. and Savage, C. I. (1974) *An Economic History of Transport*, London: Hutchinson

Barkham, J. P., MacGuire, F. A. S., and Jones, S. J. (1992) *Sea-Level Rise and the UK*, London: Friends of the Earth

Barkham, R., Gudgin, G., Hart, M., and Harvey, E. (1996) *The Determinants of Small Firm Growth* (Regional Studies Association), London: Jessica Kingsley

Barlow, J. (1982) 'Planning practice, housing supply and migration', in Champion, A. G. and Fielding, A. J. (eds) *Migration Processes and Patterns*, vol 1: *Research Progress and Prospects*, London: Belhaven

Barlow, J. (1986) 'Landowners, property ownership, and the rural locality', *International Journal of Urban and Regional Research* 10: 309–29

Barlow, J. (1988) 'The politics of land into the 1990s: landowners, developers, and farmers in lowland Britain', *Policy and Politics* 16: 111–21

Barlow, J. (1995) *Public Participation in Urban Development: The European Experience*, London: Policy Studies Institute

Barlow, J. and Chambers, D. (1992) *Planning Agreements and Affordable Housing Provision*, Brighton: Centre for Urban and Regional Research, University of Sussex

Barlow, J. and Duncan, S. (1992) 'Markets, states and housing provision: four European growth regions compared', *Progress in Planning* 38: 93–177

Barlow, J. and Duncan, S. (1994) *Success and Failure in House Building: European Systems Compared*, Oxford: Elsevier Science

Barlow, J. and Gann, D. (1993) *Offices into Flats*, York: Joseph Rowntree Foundation

Barlow, J. and Gann, D. (1995) 'Flexible planning and flexible buildings: reusing redundant office space', *Journal of Urban Affairs* 17: 263–76

Barlow, J. and King, A. (1992) 'The state, the market, and competitive strategy: the housebuilding industry in the United Kingdom, France, and Sweden', *Environment and Planning A* 24: 381–400

Barlow, J., Cocks, R., and Parker, M. (1994a) *Planning for Affordable Housing* (DoE), London: HMSO

Barlow, J., Cocks, R., and Parker, M. (1994b) 'Delivering affordable housing: law, economics and planning policy', *Land Use Policy* 11: 181–94

Barlow Report (1940) *Report of the Royal Commission on the Distribution of the Industrial Population*, Cmd 6153, London: HMSO

Barnekov, T., Boyle, R., and Rich, D. (1989) *Privatism and Urban Policy in Britain and the United States*, Oxford: Oxford University Press

Barnekov, T., Hart, D., and Benfer, W. (1990) *US Experience in Evaluating Urban Regeneration* (DoE), London: HMSO

Barnett, A. and Scruton, R. (eds) (1998) *Town and Country*, London: Jonathan Cape

Barrett, S. and Fudge, C. (eds) (1981) *Policy and Action*, London: Methuen

Barrett, S. and Healey, P. (eds) (1985) *Land Policy: Problems and Alternatives*, Aldershot: Avebury

Barrett, S. and Whitting, G. (1983) *Local Authorities and Land Supply*, Bristol: School for Advanced Urban Studies, University of Bristol

Barrett, S., Stewart, M., and Underwood, J. (1978) *The Land Market and Development Process*, Bristol: School for Advanced Urban Studies, University of Bristol

Barton, H. (2000) *Sustainable Communities: The Potential for Eco-neighbourhoods*, London: Earthscan

Barton, H. and Bruder, N. (1995) *A Guide to Local Environmental Auditing*, London: Earthscan

Barton, H., Davis, G., and Guise, R. (1995) *Sustainable Settlements: A Guide for Planners, Designers and Developers*, Bristol: University of the West of England; and Luton: Local Government Management Board

Bassett, K. (1993) 'Urban cultural strategies and urban regeneration: a case study and critique', *Environment and Planning A* 25: 1773–88

Bassett, K. (1996) 'Partnerships, business elites and urban politics: new forms of governance in an English city?', *Urban Studies* 33: 539–55

Bastrup-Birk, H. and Doucet, P. (1997) 'European spatial planning from the heart', *Built Environment* 23: 307–14

Bateman, M. (1985) *Office Development: A Geographical Analysis*, New York: St Martin's Press

Bateman, M. and Riley, R. (eds) (1987) *The Geography of Defence*, London: Croom Helm

Batey, P. (1985) 'Postgraduate planning education in Britain', *Town Planning Review* 56: 407–20

Batey, P. (1993) 'Planning education as it was', *The Planner* 79 (4) (April 1993): 25–6

Batho Report (1990) *Report of the Noise Review Working Party*, London: HMSO

Batley, R. (1989) 'London docklands: an analysis of power relations between UDCs and local government', *Public Administration* 67: 167–87

Batley, R. and Stoker, G. (eds) (1991) *Local Government in Europe: Trends and Development*, London: Macmillan

Batty, M. (1984) 'Urban policies in the 1980s: a review of the OECD proposals for managing urban change', *Town Planning Review* 55: 489–98

Batty, M. (1990) 'How can we best respond to changing fashions in urban and regional planning?', *Environment and Planning B: Planning and Design* 17: 1–7

Baxter, J. D. (1990) *State Security, Privacy and Information*, Hemel Hempstead: Harvester Wheatsheaf

BDP Planning and Berwin Leighton Berwin (1998) *The Use of Permitted Development Rights by Statutory Undertakers*, London: The Stationery Office

BDP Planning and Oxford Institute of Retail Management (1994) *The Effects of Major Out-of-Town Retail Developments*, London: HMSO

Bean, D. (ed.) (1996) *Law Reform for All*, London: Blackstone

Beckerman, W. (1974) *In Defence of Economic Growth*, London: Jonathan Cape

Beckerman, W. (1990) *Pricing for Pollution: Market Pricing, Government Regulation, Environmental Policy* (second edition), London: Institute of Economic Affairs

Beckerman, W. (1995) *Small is Stupid: Blowing the Whistle on the Greens*, London: Duckworth

Bedfordshire County Council (1996) *A Step by Step Guide to Environmental Appraisal*, Sandy, Beds: Royal Society for the Protection of Birds

Beesley, M. E. and Kain, J. F. (1964) 'Urban form, car ownership and public policy: an appraisal of *Traffic in Towns*', *Urban Studies* 1: 174–203

Begg, H. M. and Pollock, S. H. A. (1991) 'Development plans in Scotland since 1975', *Scottish Geographical Magazine* 107: 4–11

Begg, I. (1991) 'High technology location and the urban areas of Great Britain: development in the 1980s', *Urban Studies* 28: 961–81

Begg, T. (1996) *Housing Policy in Scotland*, Edinburgh: John Donald

Bell, G. and Thywitt, T. (1972) *Human Identity in the Urban Environment*, Harmondsworth: Penguin

Bell, J. L. (1993) *Key Trends in Communities and Community Development*, London: Community Development Foundation

Bell, M. and Evans, D. (1998) 'The National Forest and Local Agenda 21: an experiment in integrated landscape planning', *Journal of Environmental Planning and Management* 41: 237–51

Bell, S. (1992) *Out of Order: The 1987 Use Classes Order: Problems and Proposals*, London: London Boroughs Association

Bell, S. (1997) *Design for Outdoor Recreation*, London: Spon

Bell, S. (1998) *Landscape: Patterns, Perception, Process*, London: Spon

Bendixson, T. (1989) *Transport in the Nineties: The Shaping of Europe*, London: Royal Institution of Chartered Surveyors

Bengs, C. and Böhme, K. (eds) (1998) *The Progress of European Spatial Planning*, Stockholm: Nordregio

Benington, J. (1994) *Local Democracy and the European Union: The Impact of Europeanisation on Local Governance*, London: Commission for Local Democracy

Bennett, R. and Errington, A. (1995) 'Training and the small rural business', *Planning Practice and Research* 10: 45–54

Bennett, R. J. (1990) 'Training and enterprise councils (TECs) and vocational education and training', *Regional Studies* 24: 65–82

Benson, J. F. and Willis, K. G. (1992) *Valuing Informal Recreation on the Forestry Commission Estate* (Forestry Commission Bulletin 104), London: HMSO

Bentham, C. G. (1985) 'Which areas have the worst urban problems?', *Urban Studies* 22: 119–31

Benyon, J. (ed.) (1984) *Scarman and After: Essays Reflecting on Lord Scarman's Report, the Riots, and Their Aftermath*, Oxford: Pergamon

Berry, J. and McGreal, S. (eds) (1995a) *European Cities, Planning Systems and Property Markets*, London: Spon

Berry, J. and McGreal, S. (1995b) 'Community and inter-agency structures in the regeneration of inner-city Belfast', *Town Planning Review* 66: 129–42

Berry, J., McGreal, W. S., and Deddis, W. G. (1993) *Urban Regeneration: Property Investment and Development*, London: Spon

Berry, J. N., Fitzsimmons, D. F., and McGreal, W. S. (1988) 'Neighbour notification: the Scottish and Northern Ireland models', *Journal of Planning and Environment Law* 1988: 804–8

Bertuglia, C. S., Clarke, G. P., and Wilson, A. G. (eds) (1994) *Modelling the City: Performance, Policy and Planning*, London: Routledge

Best, J. and Bowser, L. (1986) 'A people's plan for central Newham', *The Planner* 27 (11) (November 1986): 21–5

Best, R. (1981) *Land Use and Living Space*, London: Methuen

Beveridge Report (1942) *Social Insurance and Allied Services*, Cmd 6404, London: HMSO

Bibby, P. R. and Shepherd, J. W. (1990) *Rates of Urbanisation in England 1981–2001* (DoE), London: HMSO

Bibby, P. R. and Shepherd, J. W. (1993) *Housing Land Availability: The Analysis of PS3 Statistics on Land with Outstanding Planning Permission* (DoE), London: HMSO

Bibby, P. R. and Shepherd, J. W. (1997) 'Projecting rates of urbanization for England 1991–2016: method, policy, applications and results', *Town Planning Review* 68: 93–124

Bibby, P. R. and Shepherd, J. W. (1999) 'Refocusing national brownfield housing targets', *Town and Country Planning* 70: 302–5

Bingham, M. (1998) 'A plan-led system: the potential and actual role of development plans in development control', University of Cambridge: PhD dissertation (Department of Land Economy Library)

Biodiversity Group and Local Government Management Board (1997) *Guidance for Local Biodiversity Action Plans*, Luton: LGMB

Birch, A. H. (1998) *The British System of Government* (tenth edition), London: Routledge

Birkenshaw, P. (1990) *Government and Information*, London: Butterworths

Birtles, W. (1991) 'The European directive on freedom of access to information on the environment', *Journal of Planning and Environment Law* 1991: 607–10

Bishop, J. (1994) 'Planning for better rural design', *Planning Practice and Research* 9: 259–70

Bishop, K. (1992) 'Assessing the benefits of community forests: an

evaluation of the recreational use benefits of two urban fringe woodlands', *Journal of Environmental Planning and Management* 35: 63–76

Bishop, K. and Hooper, A. (1991) *Planning for Social Housing*, London: National Housing Forum (Association of District Councils)

Bishop, K., Tewdwr-Jones, M., and Wilkinson, D. (2000) 'From spatial to local: the impact of the European Union on local authority planning in the UK', *Journal of Planning and Environmental Management* 43: 309–34

Bishop, K. D. and Phillips, A. C. (1993) 'Seven steps to market: the development of the market-led approach to countryside conservation and recreation', *Journal of Rural Studies* 9: 315–38

Bishop, K. D., Phillips, A. C., and Warren, L. (1995a) 'Protected for ever? Factors shaping the future of protected areas policy', *Land Use Policy* 12: 291–305

Bishop, K. D., Phillips, A. C., and Warren, L. (1995b) 'Protected areas in the United Kingdom: time for new thinking', *Regional Studies* 29: 192–201

Blackaby, D. H. and Manning, D. N. (1990) 'The North–South divide . . .', *Papers of the Regional Science Association* 69: 43–65

Blackhall, J. C. (1993) *The Performance of Simplified Planning Zones*, Newcastle upon Tyne: Department of Town and Country Planning, University of Newcastle

Blackhall, J. C. (1994) 'Simplified planning zones (SPZs) or simply political zeal?', *Journal of Planning and Environment Law* 1994: 117–23

Blackman, T. (1985) 'Disasters that link Ulster and the North East', *Town and Country Planning* 54: 18–20

Blackman, T. (1991a) *Planning Belfast: A Case Study of Public Policy and Community Action*, Aldershot: Avebury

Blackman, T. (1991b) 'People-sensitive planning: communication, property and social action', *Planning Practice and Research* 6: 11–15

Blackman, T. (1995) *Urban Policy in Practice*, London: Routledge

Blackmore, R., Wood, C., and Jones, C. E. (1997) 'The effect of environmental assessment on UK infrastructure project planning decisions', *Planning Practice and Research* 12: 223–38

Blair, T. (1998) *Leading the Way: A New Vision for Local Government*, London: Institute for Public Policy Research

Blaker, G. (1995) 'Gypsy law', *Journal of Planning and Environment Law* 1995: 191–6

Blowers, A. (1980) *The Limits of Power: The Politics of Local Planning Policy*, Oxford: Pergamon

Blowers, A. (1983) 'Master of fate or victim of circumstance: the exercise of corporate power in environmental policy making', *Policy and Politics* 11: 375–91

Blowers, A. (1984) *Something in the Air: Corporate Power and the Environment*, London: Harper & Row

Blowers, A. (1986) 'Environmental politics and policy in the 1980s: a changing challenge', *Policy and Politics* 14: 11–18

Blowers, A. (1987) 'Transition or transformation? Environmental policy under Thatcher', *Public Administration* 65: 277–94

Blowers, A. (ed.) (1993) *Planning for a Sustainable Environment: A Report by the Town and Country Planning Association*, London: Earthscan

Blowers, A. and Evans, B. (eds) (1997) *Town Planning into the 21st Century*, London: Routledge

Blunden, J. and Curry, N. (1988) *A Future for Our Countryside?*, Oxford: Blackwell

Blunden, J. and Curry, N. (1989) *A People's Charter? Forty Years of the National Parks and Access to the Countryside Act 1949*, London: HMSO

Board on Sustainable Development, Policy Division (1999) *Our Common Journey: A Transition toward Sustainability*, Washington, DC: National Academy Press

Boddy, M., Lambert, C., and Shape, D. (1997) *City for the 21st Century: Globalisation, Planning and Urban Change in Contemporary Britain*, Bristol: Policy Press

Body, R. (1982) *Agriculture: The Triumph and the Shame*, London: Maurice Temple Smith

Bogdanor, V. (1993) *The Blackwell Encyclopaedia of Political Institutions*, Oxford: Blackwell

Bogdanor, V. (1999) *Devolution in the United Kingdom*, Oxford University Press (Opus Books)

Böhme, K. and Bengs, C. (eds) (1999) *From Trends to Visions: The European Spatial Development Perspective*, Stockholm: Nordregio

Bond, M. (1992) *Nuclear Juggernaut: The Transport of Radioactive Materials*, London: Earthscan

Bongers, P. (1990) *Local Government and 1992*, Harlow: Longman

Bonnel, P. (1995) 'Urban car policy in Europe', *Transport Policy* 2: 83–95

Bonyhandy, T. (1987) *Law and the Countryside: The Rights of the Public*, Oxford: Professional Books

Booker, C. and Green, C. L. (1973) *Goodbye London: An Illustrated Guide to Threatened Buildings*, London: Fontana

Booth, C., Darke, J., and Yeandle, S. (eds) (1996) *Changing Places: Women's Lives in the City*, London: Paul Chapman

Booth, P. (1996) *Controlling Development: Certainty and Discretion in Europe, the USA, and Hong Kong*, London: UCL Press

Booth, P. (1999a) 'From regulation to discretion: the evolution of development control in the British planning system', *Planning Perspectives* 14: 277–89

Booth, P. (1999b) 'Discretion in planning versus zoning', in Cullingworth, J. B. (ed.) *British Planning: 50 Years of Urban and Regional Policy*, London: Athlone

Booth, P. and Beer, A. R. (1983) 'Development control and design quality', *Town Planning Review* 54: 265–84 and 383–404

Borchardt, K. (1995) *European Integration: The Origins and Growth of the European Community* (fourth edition), Luxembourg: Office for the Official Publications of the European Communities

Borins, S. F. (1988) 'Electronic road pricing: an idea whose time may never come', *Transportation Research* A 22A: 37–44

Borraz, O. *et al.* (1994) *Local Leadership and Decision Study of France, Germany, the United States and Britain*, London: Local Government Chronicle/Joseph Rowntree Foundation

Bosworth, J. and Shellens, T. (1999) 'How the Welsh Assembly will affect planning', *Journal of Planning and Environment Law* 1999: 219–24

Boucher, S. and Whatmore, S. (1990) *Planning Gain and Conservation: A Literature Review*, Reading: Department of Geography, University of Reading

Boucher, S. and Whatmore, S. (1993) 'Green gains? Planning by agreement and nature conservation', *Journal of Environmental Planning and Management* 36: 33–50

Bourne, F. (1992) *Enforcement of Planning Control* (second edition), London: Sweet & Maxwell

Bovaird, T. (1992) 'Local economic development and the city', *Urban Studies* 29: 343–68

Bovaird, T., Gregory, D., Martin, S., Pearce, G., and Tricker, M.

(1990a) *Evaluation of the Rural Development Programme Process*, London: HMSO

Bovaird, T., Tricker, M., Martin, S. J., Gregory, D. G., and Pearce, G. R. (1990b) *An Evaluation of the Rural Development Programme Process*, London: HMSO

Bovaird, T., Tricker, M., Hems, L., and Martin, S. (1991a) *Constraints on the Growth of Small Firms: A Report on a Survey of Small Firms (Aston Business School)*, London: HMSO

Bovaird, T., Gregory, D., and Martin, S. (1991b) 'Improved performance in local economic development: a warm embrace or an artful sidestep?', *Public Administration* 69: 103–19

Bowers, J. (ed.) (1990a) *Agriculture and Rural Land Use*, Swindon: Economic and Social Research Council

Bowers, J. (1990b) *Economics of the Environment: The Conservationists' Response to the Pearce Report*, Newbury: British Association of Nature Conservationists

Bowers, J. (1992) 'The economics of planning gain: a reappraisal', *Urban Studies* 29: 1329–39

Bowers, J. (1995) 'Sustainability, agriculture, and agricultural policy', *Environment and Planning A* 27: 1231–43

Bowers, J. K. and Cheshire, P. C. (1983) *Agriculture, the Countryside and Land Use: An Economic Critique*, London: Methuen

Bowley, M. (1945) *Housing and the State 1919–1944*, London: Allen & Unwin

Bowman, J. C. (1992) 'Improving the quality of our water: the role of regulation by the National Rivers Authority', *Public Administration* 70: 565–75

Boyack, S. (1990) 'Speech of the Minister of Transport and the Environment to the Royal Town Planning Institute National Conference, 25 November 1999 (available on the Scottish Executive web site)

Boyack, S. (1999) Speech of the Minister for Transport and the Environment to the RTPI National Conference, 25 November 1999 (available on the Scottish Executive web site)

Boydell, P. and Lewis, M. (1989) 'Applications to the High Court for the review of planning decisions', *Journal of Planning and Environment Law* 1989: 146–56

Boyle, R. (ed.) (1985) 'Leveraging urban development: a comparison of urban policy directions and programme impact in the United States and Britain', *Policy and Politics* 13: 175–210

Boyle, R. (1993) 'Changing partners: the experience of urban economic policy in west central Scotland, 1980–90', *Urban Studies* 30: 309–24

Boyne, G., Jordan, G., and McVicar, M. (1995) *Local Government Reform: A Review of the Process in Scotland and Wales*, London: Local Government Chronicle Communications in association with the Joseph Rowntree Foundation

Boynton, J. (1986) 'Judicial review of administrative decisions: a background paper', *Public Administration* 64: 147–61

Bradbury, J. and Mawson, J. (eds) (1997) *British Regionalism and Devolution: The Challenges of State Reform and European Integration*, London: Jessica Kingsley

Bradford, M. and Robson, B. (1995) 'An evaluation of urban policy', in Hambleton, R. and Thomas, H. (eds) *Urban Policy Evaluation: Challenge and Change*, London: Paul Chapman

Bramley, G. and Watkins, C. (1995) *Circular Projections: Household Growth, Housing Development, and the Household Projections*, London: Council for the Protection of Rural England

Bramley, G. (1989) *Land Supply, Planning, and Private Housebuilding*, Bristol: School for Advanced Urban Studies, University of Bristol

Bramley, G. (1993a) 'The impact of land use planning and tax subsidies on the supply and price of housing in Britain', *Urban Studies* 30: 5–30

Bramley, G. (1993b) 'Land use planning and the housing market in Britain: the impact on housebuilding and house prices', *Environment and Planning A* 25: 1021–51

Bramley, G. (1996a) 'Impact of land use planning on the supply and price of housing in Britain: reply to comment by Alan W. Evans', *Urban Studies* 33: 1733–37

Bramley, G. (1996b) *Housing with Hindsight: Household Growth, Housing Need and Housing Development in the 1980s*, London: Council for the Protection of Rural England

Bramley, G. and Smart, G. (1995) *Rural Incomes and Housing Affordability*, Salisbury: Rural Development Commission

Bramley, G. and Watkins, C. (1996) *Steering the Housing Market: New Building and the Changing Planning System*, Bristol: Policy Press

Bramley, G., Bartlett, W., and Lambert, C. (1995) *Planning, the Market and Private House- Building*, London: UCL Press

Bramley, G., Munro, M., and Lancaster, S. (1997) *The Economic Determinants of Household Formation: A Literature Review*, London: Department for the Environment, Transport and the Regions

Bray, J. (1992) *The Rush for Roads: A Road Programme for Economic Recovery: A Report by Movement Transport Consultancy for ALARM UK and Transport 2000*, London: Transport 2000

Brearley, C. (1999) 'Integrated transport policy: the implications for planning', *Journal of Planning and Environment Law* 1999: 408–15

Breheny, M. J. (1983) 'A practical view of planning theory', *Environment and Planning B: Planning and Design* 10: 101–15

Breheny, M. J. (1989) 'Chalkface to coalface: a review of the academic–practice interface', *Environment and Planning B: Planning and Design* 16: 451–68

Breheny, M. J. (1991) 'The renaissance of strategic planning', *Environment and Planning B: Planning and Design* 18: 233–49

Breheny, M. J. (ed.) (1992) *Sustainable Development and Urban Form*, London: Pion

Breheny, M. J. (1995) 'The compact city and transport energy consumption', *Transactions of the Institute of British Geographers* NS 20: 81–101

Breheny, M. J. (1997) 'Urban compaction: feasible and acceptable?', *Cities* 14: 209–17

Breheny, M. J. (1999a) 'The people: where will they work?', *Town and Country Planning* 68: 174–5

Breheny, M. J. (ed.) (1999b) *The People: Where Will They Work? Report of the Town and Country Planning Association – Research into the Changing Geography of Employment*, London: Town and Country Planning Association

Breheny, M. J. and Congdon, P. (eds) (1989) *Growth and Change in a Core Region: The Case of South East England*, London: Pion

Breheny, M. J. and Hall, P. (eds) (1996) *The People: Where Will They Go? National Report of the Town and Country Planning Association: Regional Inquiry into Housing Need and Provision in England*, London: Town and Country Planning Association

Breheny, M. J. and Hooper, A. (1985) *Rationality in Planning: Critical Essays on the Role of Rationality in Urban and Regional Planning*, London: Pion

Breheny, M., Gent, T., and Lock, D. (1993) *Alternative Development Patterns: New Settlements* (DoE), London: HMSO

Brenan, J. (1994) 'PPG 16 and the restructuring of archaeological practice in Britain', *Planning Practice and Research* 9: 395–405

Brennan, A., Rhodes, J., and Tayler, P. (1998) *Evaluation of the Single Regeneration Challenge Fund Budget: A Partnership for Regeneration: An Interim Evaluation*, London: DETR

Bridge, G. (1993) *Gentrification, Class and Residence*, Bristol: School for Advanced Urban Studies

Bridges, L. (1979) 'The structure plan examination-in-public as an instrument of intergovernmental decision making', *Urban Law and Practice* 2: 241–64

Briggs, A. (1952) *History of Birmingham* (2 vols) Oxford: Oxford University Press

Brindley, T., Rydin, Y., and Stoker, G. (1996) *Remaking Planning* (second edition), London: Routledge

British Standards Institution *Code of Practice for Access for the Disabled to Buildings* (BS 5810), London: BSI

British Standards Institution *Code of Practice for Design for the Convenience of Disabled People* (BS 5619), London: BSI

British Waterways Board (1999) *Our Plan for the Future 2000–2004*, London: BWB

British Waterways Board (1993) *The Waterway Environment and Development Plans*, Watford: BWB

Britton, D. (ed.) (1990) *Agriculture in Britain: Changing Pressures and Policies*, Wallingford: CAB International

Broady, M. (1968) *Planning for People*, London: Bedford Square Press

Bromley, M. P. (1990) *Countryside Management*, London: Spon

Bromley, R. and Thomas, C. (eds) (1993) *Retail Change: Contemporary Issues*, London: UCL Press

Brooke, C. (1996) *Natural Conditions: A Review of Planning Conditions and Nature Conservation*, Sandy, Beds: Royal Society for the Protection of Birds

Brooke, R. (1989) *Managing the Enabling Authority*, Harlow: Longman

Brooks, J. (1999) '(Can) modern local government be in touch with the people?' *Public Policy and Administration* 14(1): 42–59

Broome, J. (1992) *Counting the Cost of Global Warming*, Cambridge: White Horse Press

Brotherton, D. I. (1989) 'The evolution and implications of mineral planning policy in the national parks of England and Wales', *Environment and Planning A* 21: 1229–40

Brotherton, D. I. (1992a) 'On the control of development by planning authorities', *Environment and Planning B: Planning and Design* 19: 465–78

Brotherton, D. I. (1992b) 'On the quantity and quality of planning applications', *Environment and Planning B: Planning and Design* 19: 337–57

Brotherton, I. (1993) 'The interpretation of planning appeals', *Journal of Environmental Planning and Management* 36: 179–86

Brown, C., Dühr, S., and Nadin, V. (forthcoming) *Sustainability, Development and Spatial Planning in Europe*, London: Routledge

Brownill, S. (1990) *Developing London's Docklands: Another Great Planning Disaster?*, London: Paul Chapman

Bruff, G. E. and Wood, A. P. (2000) 'Local sustainable development: land-use planning's contribution to modern local government', *Journal of Environmental Management and Planning* 43: 519–39

Brundtland Report (1987) *Our Common Future* (World Commission on Environment and Development), Oxford: Oxford University Press

Brunivells, P. and Rodrigues, D. (1989) *Investing in Enterprise: A Comprehensive Guide to Inner City Regeneration and Urban Renewal*, Oxford: Blackwell

Brunskill, I. (1989) *The Regeneration Game: A Regional Approach to Regional Policy*, London: Institute for Public Policy Research

Bruton, M. J. (ed.) (1974) *The Spirit and Purpose of Planning* (second edition), London: Hutchinson

Bruton, M. J. (1980) 'PAG revisited', *Town Planning Review* 48: 134–44

Bruton, M. J. (1985) *Introduction to Transport Planning*, London: UCL Press

Bruton, M. J. and Nicholson, D. J. (1983) 'Non-statutory plans and supplementary planning guidance', *Journal of Planning and Environmental Law* 432–43

Bruton, M. J. and Nicholson, D. J. (1985) 'Supplementary planning guidance and local plans', *Journal of Planning and Environment Law* 1985: 837–44

Bruton, M. J. and Nicholson, D. J. (1987) *Local Planning in Practice*, London: Hutchinson

Bryant, B. (1995) *Twyford Down: Roads, Campaigning and Environmental Law*, London: Spon

Bryant, S. (1999) '50 years of loss', *Town and Country Planning* 68: 16

Buchanan, C. D. (1958) *Mixed Blessing: The Motor Car in Britain*, London: Leonard Hill

Buchanan, C. D. (1963) *Traffic in Towns* (Buchanan Report), London: HMSO

Buckingham-Hatfield, S. and Evans, B. (eds) (1996) *Environmental Planning and Sustainability*, Chichester: Wiley

Budd, L. and Whimster, S. (eds) (1992) *Global Finance and Urban Living: A Study of Metropolitan Change*, London: Routledge

Building Design Partnership (BDP) (1996) *London's Urban Environment: Planning for Quality* (Government Office for London), London: HMSO

Bunce, M. (1994) *The Countryside Ideal: Anglo-American Images of Landscape*, London: Routledge

Bunnell, G. (1995) 'Planning gain in theory and practice: negotiation of agreements in Cambridgeshire', *Progress in Planning* 44: 1–113

Burbridge, V. (1990) *Review of Information on Rural Issues* (Central Research Unit Papers), Edinburgh: Scottish Office

Burchell, R. W. and Listokin, D. (eds) (1982) *Energy and Land Use*, New Brunswick, NJ: Center for Urban Policy Research, Rutgers University

Burchell, R. W. and Sternlieb, G. (eds) (1978) *Planning Theory in the 1980s*, New Brunswick, NJ: Center for Urban Policy Research, Rutgers University

Burgon, J. (2000) *An Assessment of Changes in the Developed and Undeveloped Coast of England and Wales Between 1965 and 1995*, Department of Geography, University of Reading

Burke, P. and Manta, M. (1999) *Planning for Sustainable Development: Measuring Progress in Plans*, Cambridge, MA: Lincoln Institute of Land Policy

Burke, T. and Shackleton, J. R. (1996) *Trouble in Store: UK Retailing in the 1990s*, London: Institute of Economic Affairs

Burns, D., Hambleton, R., and Hoggett, P. (1994) *The Politics of Decentralisation: Revitalising Local Democracy*, London: Macmillan

Burrows, R. (1997) *Contemporary Patterns of Residential Mobility in England*, York: Centre for Housing Policy, University of York

Burrows, R. and Rhodes, O. (1998) *Unpopular Places? Area Disadvantage and the Geography of Misery in England*, York: Joseph Rowntree Foundation

Burton, T. P. (1989) 'Access to environmental information: the UK experience of water registers', *Journal of Environmental Law* 1: 192–208

Burton, T. P. (1992) 'The protection of rural England', *Planning Practice and Research* 7 (1) (Spring): 37–40

Business in the Community (1990) *Leadership in the Community: A Blueprint for Business Involvement in the 1990s*, London: Business in the Community

Business in the Community (1992) *A Measure of Commitment: Guidelines for Measuring Environmental Performance*, London: Business in the Community

Butler, D. and Kavanagh, D. (1997) *The British General Election of 1997*, London: Macmillan

Butler, D., Adonis, A., and Travers, T. (1994) *Failure in British Government: The Politics of the Poll Tax*, Oxford: Oxford University Press

Butt, R. (1999) 'Beyond motherhood and apple pie', *Town and Country Planning* 68: 106

Byrne, D. (1989) *Beyond the Inner City*, Milton Keynes: Open University Press

Byrne, S. (1989) *Planning Gain: An Overview – A Discussion Paper*, London: Royal Town Planning Institute

Byrne, T. (1994) *Local Government in Britain* (sixth edition), London: Penguin

Cairncross, F. (1993) *Costing the Earth* (second edition), London: Business Books/Economist Books

Cairncross, F. (1995) *Green Inc: Guide to Business and the Environment*, London: Earthscan

Cairns, S. (1999) 'Redirecting the school run', *Town and Country Planning* 68: 300–1

Calder, N., Cavanagh, S., Eckstein, C., Palmer, J., and Stell, A. (1993) *Women and Development Plans*, Newcastle upon Tyne: Department of Town and Country Planning, University of Newcastle

Callander, R. F. (1987) *A Pattern of Landownership in Scotland*, Finzean: Haughend

Callander, R. F. (1998) *How Scotland is Owned*, Edinburgh: Canongate

Callies, D. (1999) 'An American perspective on UK planning', in Cullingworth, J. B. (ed.) *British Planning: 50 Years of Urban and Regional Policy*, London: Athlone

Callies, D. L. and Grant, M. (1991) 'Paying for growth and planning gain: an Anglo-American comparison of development conditions, impact fees and development agreements', *Urban Lawyer* 23: 221–48

Cambridge Policy Consultants (1998) *Regenerating London Docklands*, London: DETR

Cambridge University, Department of Land Economy (1999) *Environmental Court Project Final Report*, London: Department of the Environment, Transport and the Regions

Cameron, G. C. (1990) 'First steps in urban policy evaluation in the United Kingdom', *Urban Studies* 27: 475–95

Cameron, G. C., Monk, S., and Pearce, B. J. (1988) *Vacant Urban Land: A Literature Review*, London: Department of the Environment

Cameron, S., Baker, M., Bevan, M., Hull, A., and Williams, R. (1998) *Regionalisation, Devolution and Social Housing*, London: National Housing Forum

Campbell, M. (ed.) (1990) *Local Economic Policy*, London: Cassell

Campbell, M., Kearns, A., Wood, M. and Young, R. (2000) *Regeneration into the 21st Century: Policies into Practice – An Overview*, Bristol: Policy Press

Campbell, S. and Fainstein, S. S. (eds) (1996) *Readings in Planning Theory*, Oxford: Blackwell

Capita Management Consultancy (1996) *An Evaluation of Six Early Estate Action Schemes* (DoE), London: HMSO

Capita Management Consultancy (1997) *Housing Action Trusts: Evaluation of Baseline Conditions* (11 volumes), London: Department of the Environment, Transport and the Regions

Capita Management Consultancy (1998) *Housing Action Trusts: Evaluation of Baseline Conditions* (11 vols), London: Department of the Environment, Transport and the Regions

Capner, G. (1994) 'Green belt and rural economy', in Wood, M. (ed.) *Planning Icons: Myth and Practice* (Planning Law Conference, *Journal of Planning and Environment Law*), London: Sweet and Maxwell

Card, R. and Ward, R. (1996) 'Access to the countryside: the impact of the Criminal Justice and Public Order Act 1994', *Journal of Planning and Environment Law* 1996: 447–62

Cardiff University and Buchanan Partnership (1997) *Slimmer and Swifter: A Critical Examination of District Wide Local Plans and UDPs*, London: Royal Town Planning Institute

Carley, M., Chapman, M., Hastings, A., Kirk, K., and Young, R. (2000) *Urban Regeneration through Partnership: A Study of Nine Urban Regions in England, Scotland and Wales*, York: Policy Press

Carley, M. (1990a) 'Neighbourhood renewal in Glasgow: policy and practice', *Housing Review* 39: 49–51

Carley, M. (1990b) *Housing and Neighbourhood Renewal: Britain's New Urban Challenge*, London: Policy Studies Institute

Carley, M. (1991) 'Business in urban regeneration partnerships: a case study in Birmingham', *Local Economy* 6: 100–15

Carley, M. (1997) *Sustainable Transport and Retail Vitality: State of the Art for Towns and Cities*, Historic Boroughs Association of Scotland with Transport 2000

Carley, M. (1999) 'Neighbourhoods: building blocks of national sustainability', *Town and Country Planning* 69: 61–4

Carley, M. and Kirk, K. (1998a) *City-Wide Urban Regeneration* (Scottish Executive Central Research Unit), Edinburgh: The Stationery Office

Carley, M. and Kirk, K. (1998b) *Sustainable by 2020? A Strategic Approach to Urban Regeneration for Britain's Cities*, London: Policy Press

Carmichael, P. (1992) 'Is Scotland different? Local government policy under Mrs Thatcher', *Local Government Policy Making* 18 (5) (May): 25–32

Carmona, M. (1996) 'Controlling urban design: Part 1: A possible renaissance?', *Journal of Urban Design* 1: 47–73

Carmona, M. (1998) 'Residential design policy and guidance: prevalence, hierarchy and currency', *Planning Practice and Research* 13: 407–19

Carmona, M. (1999) 'Residential design policy and guidance: content, analytical basis, prescription and regional emphasis', *Planning Practice and Research* 14: 17–38

Carnie, J. K. (1996) *The Safer Cities Programme in Scotland* (Scottish Office, Central Research Unit), Edinburgh: HMSO

Carnwath, R. (1991) 'The planning lawyer and the environment', *Journal of Environmental Law* 3: 57–67

Carnwath, R. (1992) 'Environmental enforcement: the need for a specialist court', *Journal of Planning and Environment Law* 1992: 799–808

Carnwath Report (1989) *Enforcing Planning Control: Report by Robert Carnwath QC*, London: HMSO

Carole Millar Research (1999) *Perceptions of Local Government: A Report of Focus Group Research*, Edinburgh: Scottish Office Central Research Unit

Carson, R. (1951) *The Sea around Us*, New York: Oxford University Press

Carson, R. (1962) *Silent Spring*, Harmondsworth: Penguin edition (1965)

Carter, C. (1997) *Local–Central Government Relations: A Compendium of Joseph Rowntree Foundation Research Findings*, York: Joseph Rowntree Foundation

Carter, C. and John, P. (1992) *A New Accord: Promoting Constructive Relations between Central and Local Government*, York: Joseph Rowntree Foundation

Carter, N., Brown, T., and Abbott, T. (1991) *The Relationship between Expenditure-Based Plans and Development Plans*, Leicester: School of the Built Environment, Leicester Polytechnic

Carter, N., Oxley, M., and Golland, A. (1996) 'Towards the market: Dutch physical planning in a UK perspective', *Planning Practice and Research* 11: 49–60

Cartwright, L. (1997) 'The implementation of sustainable development by local authorities in the south east of England', *Planning Practice and Research* 12: 337–47

Cartwright, L. (2000) 'Selecting local sustainable development indicators: does consensus exist in their choice and purpose?', *Planning Practice and Research* 15: 65–78

Casellas, A. and Galley, C. (1999) 'Regional definitions in the European Union: a question of disparities', *Regional Studies* 33: 551–8

Castells, M. (1991) *The Informational City: Economic Restructuring and Urban Development*, Oxford: Blackwell

Catalano, A. (1983) *A Review of Enterprise Zones*, London: CES

CB Hillier Parker, Saxell Bird Axon (1998) *The Impact of Large Foodstores on Market Towns and District Centres* (DETR), London: The Stationery Office

Central Housing Advisory Committee (1967) *The Needs of New Communities*, London: HMSO

Central Office of Information: *see* section on Official Publications

Centre for Environmental Studies (1970) *Observations on the Greater London Development Plan*, London: CES (reprinted in Cullingworth, J. B. (ed.) (1973) *Problems of an Urban Society*, vol. 3: *Planning for Change*, London: Allen & Unwin)

Centre for Local Economic Strategies (1990) *Inner City Regeneration: A Local Authority Perspective*, Manchester: CLES

Centre for Local Economic Strategies (1991) *City Centres, City Cultures: The Role of the Arts in the Revitalisation of Towns and Cities*, Manchester: CLES

Centre for Local Economic Strategies (1992a) *Reforming the TECs: Towards a Strategy – Final Report of the CLES TEC/LEC Monitoring Project*, Manchester: CLES

Centre for Local Economic Strategies (1992b) *Social Regeneration: Directions for Urban Policy in the 1990s*, Manchester: CLES

Centre for Local Economic Strategies (1993) *The Green Local Economy: Integrating Economic and Environmental Development at the Local Level*, Manchester: CLES

Centre for Local Economic Strategies (1994) *Rethinking Urban Policy: City Strategies for the Global Economy*, Manchester: CLES

Centre for Local Economic Strategies (1996) *Regeneration through Work: Creating Jobs in the Social Economy*, Manchester: CLES

Centre for Local Economic Strategies (1998) *Joining Up? The New Deal, the Public Sector and the Employer Option*, Manchester: CLES

Centre for Local Economic Strategies (1999) *Strategies for Success: A First Assessment of Regional Development Agencies' Draft Regional Economic Strategies*, Manchester: CLES

Centre for Scottish Public Policy (1999) *Parliamentary Practices in Devolved Parliaments*, Edinburgh: The Stationery Office

Centre for Urban Policy Studies (1998) *The Impact of Urban Development Corporations in Leeds, Bristol and Central Manchester*, London: Department of the Environment, Transport and the Regions

Cervero, R. (1990) 'Transit pricing research: a review and synthesis', *Transportation* 17: 117–39

Cervero, R. (1995) 'Planned communities, self-containment and commuting: a cross-national perspective', *Urban Studies* 32: 1135–61

Chadwick, G. F. (1978) *A Systems View of Planning: Towards a Theory of the Urban and Regional Planning Process*, Oxford: Pergamon

Champion, A. G. (1989) *Counterurbanisation: The Changing Pace and Nature of Population Deconcentration*, London: Edward Arnold

Champion, A. G. (1996) *Urban Exodus: Migration between Metropolitan and Non-metropolitan Areas in Britain*, Swindon: Economic and Social Science Research Council

Champion, A. G. and Atkins, D. (1996) *The Counterurbanisation Cascade: An Analysis of the Census Special Migration Statistics for Great Britain*, Newcastle upon Tyne: University of Newcastle upon Tyne, Department of Geography

Champion, A. G. and Fielding, A. J. (eds) (1982) *Migration Processes and Patterns*, vol. 1: *Research Progress and Prospects*, London: Belhaven

Champion, A. G. and Green, A. E. (1992) 'Local economic performance in Britain during the late 1980s: the results of the third booming towns study', *Environment and Planning A* 24: 243–72

Champion, A. G., Atkins, A., Coombes, M., and Fotheringham, S. (1998) *Urban Exodus*, London: Council for the Protection of Rural England

Champion, T. and Watkins, C. (1991) *People in the Countryside: Studies of Social Change in Rural Britain*, London: Paul Chapman

Chandler, J. A. (1996) *Local Government Today* (second edition), Manchester: Manchester University Press

Chapman, D. and Larkham, P. (1992) *Discovering the Art of Relationship*, Birmingham: Faculty of the Built Environment, Birmingham Polytechnic

Chapman, D. and Larkham, P. (1999) 'Urban design, urban quality and the quality of life: reviewing the Department of the Environment's Urban Design Campaign', *Journal of Urban Design* 4: 211–32

Chapman, M. (1995) 'Urban policy and urban evaluation: the impact of the European Union', in Hambleton, R. and Thomas, H. (eds) *Urban Policy Evaluation: Challenge and Change*, London: Paul Chapman

Chapman, P. (1998) *Poverty and Exclusion in Rural Britain: The Dynamics of Low Income and Employment*, York: Joseph Rowntree Foundation

Charles, HRH The Prince of Wales (1989) *A Vision of Britain: A Personal View of Architecture*, London: Doubleday

Chartered Institute of Environmental Health (1995) *Travellers and Gypsies: An Alternative Strategy*, London: CIEH

Chartered Institute of Housing, National Housing Federation, Room, and Shelter (1999) *National Blueprint: The Delivery of Affordable Homes through the Planning System*, published jointly by the four organisations

Chartered Institute of Transport (1990) *Paying for Progress: A Report on Congestion and Road Use Charges*, London: CIT

Chartered Institute of Transport (1991) *London's Transport: The Way Ahead*, London: CIT

Chartered Institute of Transport (1992) *Paying for Progress: A Report on Congestion and Road Use Charges: Supplementary Report*, London: CIT

Checkland, S. G. (1981) *The Upas Tree: Glasgow 1875–1975 and after 1975–1980*, Glasgow: University of Glasgow Press

Cherry, G. E. (1974) *The Evolution of British Town Planning*, London: Leonard Hill

Cherry, G. E. (1975) *Environmental Planning 1939–1969*, vol. 2: *National Parks and Access to the Countryside*, London: HMSO

Cherry, G. E. (1981) *Pioneers in British Planning*, London: Architectural Press

Cherry, G. E. (1982) *The Politics of Town Planning*, London: Longman

Cherry, G. E. (1988) *People and Plans: The Shaping of Urban Britain in the Nineteenth and Twentieth Centuries*, London: Edward Arnold

Cherry, G. E. (1992) 'Green belt and the emergent city', *Property Review* (December): 97–101

Cherry, G. E. (1996) *Town Planning in Britain since 1900: The Rise and Fall of the Planning Ideal*, Oxford: Blackwell

Cherry, G. E. (1994a) *Rural Change and Planning: England and Wales in the Twentieth Century*, London: Spon

Cherry, G. E. (1994b) *Birmingham: A Study in Geography, History and Planning*, Chichester: Wiley

Cherry, G. E. and Penny, L. (1990) *Holford: A Study in Architecture, Planning and Civic Design*, London: Spon

Cherry, G. E. and Rogers, A. (1996) *Rural Change and Planning: England and Wales in the Twentieth Century*, London: Spon

Cheshire, P. and Sheppard, S. (1989) 'British planning policy and access to housing: some empirical estimates', *Urban Studies* 26: 469–85

Cheshire, P., D'Arcy, E., and Giussani, B. (1991) *Local, Regional and National Government in Britain: A Dreadful Warning*, Reading: Department of Economics, University of Reading

Cheshire County Council (1995) *Cheshire County Structure Plan: New Thoughts for the Next Century*, Chester: Cheshire County Environmental Planning Service

Chesman, G. R. (1991) 'Local authorities and the review of the definitive map under the Countryside and Wildlife Act 1981', *Journal of Planning and Environment Law* 1991: 611–14

Chesterton (1995) *Dwelling Provision through Planning Regeneration*, Hertford: Hertfordshire County Council

Chisholm, M. (1995) *Britain on the Edge of Europe*, London: Routledge

Christensen, T. (1979) *Neighbourhood Survival*, Dorchester: Prism Press

Christie, B. (2000) 'Finding a way forward for LA21 in Scotland', *EG* 6 (5): 5–8

Chubb, R. N. (1988) *Urban Land Markets in the United Kingdom* (DoE), London: HMSO

Church, A. (1988) 'Urban regeneration in London Docklands: a five-year review', *Environment and Planning C: Government and Policy* 6: 187–208

Church, C. and McHarry, C. (eds) (1999) *One Small Step: A Guide to Action on Sustainable Development in the UK*, London: Community Development Forum

Churchill, R., Warren, L. M., and Gibson, J. (eds) (1991) *Law, Policy and the Environment*, Oxford: Blackwell

Civic Trust (1991) *Audit of the Environment*, London: The Trust

Civic Trust (1999) *Brownfield Housing: 12 Years On*, London: The Trust

Clabon, S. and Chance, C. (1992) 'Legal profile: freedom of access to environmental information', *European Environment* 2 (3) (June): 24–5

Clapson, M. (1998) *Invincible Green Suburbs, Brave New Towns*, Manchester: Manchester University Press

Clark, A., Lee, M., and Moore, R. (1996) *Landslide Investigation and Management in Great Britain* (a support document for PPG 14) (DoE), London: HMSO

Clark, D. M. (1992) *Rural Social Housing – Supply and Trends: A 1992 Survey of Affordable New Homes*, Cirencester: Association of Community Councils in Rural England

Clark, G., Darrall, J., Grove-White, R., Macnaghten, P., and Urry, J. (1994) *Leisure Landscapes: Leisure, Culture and the English Countryside: Challenges and Conflicts*, London: Council for the Protection of Rural England

Clark, M., Smith, D., and Blowers, A. (1992) *Waste Location: Spatial Aspects of Waste Management, Hazards, and Disposal*, London: Routledge

Clawson, M. and Hall, P. (1973) *Land Planning and Urban Growth*, Baltimore: Johns Hopkins University Press

Claydon, J. (1998) 'Discretion in development control: a study of how discretion is exercised in the conduct of development control in England and Wales', *Planning Practice and Research* 13: 63–80

Claydon, J. and Smith, B. (1997) 'Negotiating planning gains through the British development plan system', *Urban Studies* 34: 2003–22

Clayton, A. M. H. and Radcliffe, N. J. (1996) *Sustainability: A Systems Approach*, London: Earthscan

Cleator, B. (1995) *Review of Legislation Relating to the Coasts and Marine Environment of Scotland*, Perth: Scottish Natural Heritage

Cloke, P. (ed.) (1987) *Rural Planning: Policy into Action?*, London: Paul Chapman

Cloke, P. (ed.) (1988) *Policies and Plans for Rural People*, London: Allen & Unwin

Cloke, P. (ed.) (1992) *Policy and Change in Thatcher's Britain*, Oxford: Pergamon

Cloke, P. (1993) 'On problems and solutions: the reproduction of problems for rural communities in Britain during the 1980s', *Journal of Rural Studies* 9: 113–21

Cloke, P. (1996) 'Housing in the open countryside: windows on "irresponsible" planning in rural Wales', *Town Planning Review* 67: 291–308

Cloke, P. and Little, J. (1990) *The Rural State: Limits to Planning in Rural Society*, Oxford: Clarendon Press

Cloke, P., Doel, M., Matless, D., Phillips, M., and Thrift, N. (1994) *Writing the Rural – Five Cultural Geographies*, London: Paul Chapman

Clotworthy, J. and Harris, N. (1996) 'Planning policy implications of local government reorganisation', in Tewdwr-Jones, M. (ed.) *British Planning Policy in Transition: Planning in the 1990s*, London: UCL Press

Coates, D. (1994) 'Wanted? More land, enquire within', *House Builder* (April): 29–30

Cobb, D., Dolman, P., and O'Riordan, T. (1999) 'Interpretations of sustainable agriculture in the UK', *Progress in Human Geography* 23: 209–35

Coccossis, H. and Nijkamp, P. (eds) (1995) *Sustainable Tourism Development*, Aldershot: Avebury

Cochrane, A. (1991) 'The changing state of local government: restructuring for the 1990s', *Public Administration* 69: 281–302

Cochrane, A. (1993) *Whatever Happened to Local Government?*, Buckingham: Open University Press

Cochrane, A. and Clarke, A. (1990) 'Local enterprise boards: the short history of a radical initiative', *Public Administration* 68: 315–36

Cocks, R. (1991) 'First responses to the new breach of condition notice', *Journal of Planning and Environment Law* 1991: 409–18

Cole, J. and Cole, F. (1993) *The Geography of the European Community*, London: Routledge

Cole, I. and Smith, Y. (1996) *From Estate Action to Estate Agreement: Regeneration and Change on the Bell Farm Estate, York*, Bristol: Policy Press

Coleman, A. (1990) *Utopia on Trial* (second edition), London: Hilary Shipman

Colenutt, B. and Cutten, A. (1994) 'Community empowerment in vogue or vain?', *Local Economy* 9: 236–50

Collar, N. A. (1999) *Green's Concise Scots Law: Planning* (second edition), Edinburgh: W. Green

Collins, L. (1996) 'Recycling and the environmental debate: a question of social conscience of scientific reason?', *Journal of Environmental Planning and Management* 39: 333–55

Collins, M. P. (1989) 'A review of 75 years of planning education at UCL', *Planner* 75: 18–22

Collins, M. P. and McConnell, S. (1988) *The Use of Local Plans for Effective Town and Country Planning: Report of a Research Project Funded by the Nuffield Foundation*, London: Bartlett School of Architecture and Planning, University College London

Colls, J. (1996) *Fundamentals of Air Pollution*, London: Spon

Comedia (1999) *Park Life: Urban Parks and Social Renewal*, Gloucestershire: Comedia

Commission on Social Justice (1994) *Social Justice: Strategies for National Renewal*, London: Vintage

Commission for Racial Equality (1982) *Local Government and Racial Equality*, London: CRE

Confederation of British Industry (1988) *Initiatives beyond Charity: Report of the CBI Task Force on Business and Urban Regeneration*, London: CBI

Confederation of British Industry (1995) *Missing Links: Setting National Transport Priorities*, London: CBI

Confederation of British Industry (1996) *Winning Ways: Developing the UK Transport Network*, London: CBI

Connal, R. C. and Scott, J. N. (1999) 'The new Scottish Parliament: what will its impact be?', *Journal of Planning and Environment Law* 1999: 491–7

Connolly, M. E. H. and Loughlin, S. (1990) *Public Policy in Northern Ireland: Adoption or Adaptation?*, Belfast: Policy Research Institute

Consortium Developments (1985) *New Country Towns*, London: Consortium Developments

Cook, A. J. and Davis, A. L. (1993) *Package Approach Funding: A Survey of English Highway Authorities*, London: Friends of the Earth

Cooke, P. and Morgan, K. (1993) 'The network paradigm: new departures in corporate and regional development', *Environment and Planning D* 11: 543–64

Coombes, M., Raybould, S., and Wong, C. (1992a) *Developing Indicators to Assess the Potential for Urban Regeneration* (DoE), London: HMSO

Coombes, T., Fidler, P., and Hathaway, A. (1992b) 'South West regional planning review: towards a regional strategy', in Minay, C. L. W. (ed.) 'Developing regional guidance in England and Wales: a review symposium', *Town Planning Review* 63: 415–34

Coon, A. (1988) 'Local plan provision: the record to date and prospects for the future', *The Planner* 74 (5) (May): 17–20

Coon, A. (1989) 'An assessment of Scottish development planning', *Planning Outlook* 32: 77–85

Cooper, A. J. (1998) 'The origins of the Cambridge green belt', *Planning History* 20: 6–19

Cooper, D. E. and Palmer, J. A. (eds) (1992) *The Environment in Question*, London: Routledge

Coopers & Lybrand (1985) *Land Use Planning and the Housing Market: An Assessment*, London: Coopers & Lybrand

Coopers & Lybrand (1987) *Land Use Planning and Indicators of Housing Demand*, London: Coopers & Lybrand

Coopers & Lybrand (1994) *Employment Department Baseline Follow-Up Studies: Final Overall Report to the Employment Department*, London: Coopers & Lybrand

Corner, T. (1999) 'Planning, environment, and the European Convention on Human Rights', *Journal of Planning and Environment Law* April: 301–4

Cornford, J. and Gillespie, A. (1992) 'The coming of the wired city? The recent development of cable in Britain', *Town Planning Review* 63: 243–64

Cornford, T. (1998) 'The control of planning gain', *Journal of Planning and Environment Law* 1998: 731–49

Corkindale, J. (1998) *Reforming Land Use Planning*, London: Institute of Economic Affairs

Costonis, J. J. (1989) *Icons and Aliens: Law, Aesthetics, and Environmental Change*, Urbana-Champaign: University of Illinois Press

Couch, C. (1990) *Urban Renewal: Theory and Practice*, London: Macmillan

Couch, C. and Gill, N. (1993) *Renewal Areas: A Review of Progress*, Working Paper 119, Bristol University School for Advanced Urban Studies

Couch, C., Eva, D., and Lipscombe, A. (2000) 'Renewal areas in north-west England', *Planning Practice and Research* 15: 257–68

Council for the Protection of Rural England (1988) *Welcome Homes: Housing Supply from Unallocated Land*, London: CPRE

Council for the Protection of Rural England (1990a) *From White Paper to Green Future*, London: CPRE

Council for the Protection of Rural England (1990b) *Our Finest Landscapes: Council for the Protection of Rural England Submission to the National Parks Review Panel*, London: CPRE

Council for the Protection of Rural England (1990c) *Future Harvest: The Economics of Farming and the Environment – Proposals for Action* (Jenkins, T. N.), London: CPRE and WWF

Council for the Protection of Rural England (1990d) *Planning Control over Farmland*, London: CPRE

Council for the Protection of Rural England (1991) *Energy Conscious Planning* (Owens, S.), London: CPRE

Council for the Protection of Rural England (1992a) *Campaigners Guide to Using EC Environmental Law* (Macrory, R.), London: CPRE

Council for the Protection of Rural England (1992b) *Campaigners Guide to Local Plans*, (Green Balance), London: CPRE

Council for the Protection of Rural England (1992c) *Our Common Home: Housing Development and the South East's Environment*, London: CPRE

Council for the Protection of Rural England (1992d) *Transport and the Environment: Council for the Protection of Rural England's Submission to the Royal Commission on Environmental Pollution*, London: CPRE

Council for the Protection of Rural England (1992e) *Where Motor Car is Master: How the Department of Transport Became Bewitched by Roads* (Kay, P. and Evans, P.), London: CPRE

Council for the Protection of Rural England (1992f) *The Lost Land: Land Use Change in England 1945–1990* (Sinclair, G.), London: CPRE

Council for the Protection of Rural England (1993a) *Index of National Planning Policies*, London: CPRE

Council for the Protection of Rural England (1993b) *Preparing for the Future: A Response by the Council for the Protection of Rural England to the Consultation Paper on the UK Strategy for Sustainable Development*, London: CPRE

Council for the Protection of Rural England (1993c) *Water for Life: Strategies for Sustainable Water Resource Management*, London: CPRE

Council for the Protection of Rural England (1994a) *Leisure Landscapes: Leisure, Culture and the English Countryside: Challenges and Conflicts* (Clark, G., Darrall, J., Grove-White, R., Macnaghten, P., and Urry, J.), London: CPRE

Council for the Protection of Rural England (1994b) *Environmental Policy Omissions in Development Plans*, London: CPRE

Council for the Protection of Rural England (1994c) *The Housing Numbers Game*, London: CPRE

Council for the Protection of Rural England (1994d) *Public Access to Planning Documents*, London: CPRE

Council for the Protection of Rural England (1995a) *Renewable Energy in the UK: Financing Options for the Future* (Mitchell, C.), London: CPRE

Council for the Protection of Rural England (1995b) *The End of Hierarchy: A New Perspective on Managing the Road Network* (Goodwin, P.), London: CPRE

Council for the Protection of Rural England (1995c) *Circular Projections: Household Growth, Housing Development, and the Household Projections* (Bramley, G. and Watkins, C.), London: CPRE

Council for the Protection of Rural England (1996a) *The Campaigners' Guide to Minerals*, London: CPRE

Council for the Protection of Rural England (1996b) *Growing Greener: Sustainable Agriculture in the UK* (Baldock, D., Bishop, K., Mitchell, K., and Phillips, A.), London: CPRE

Council for the Protection of Rural England (1997) *Planning More to Travel Less*, London: CPRE

Council for the Protection of Rural England (1998) *Rural Transport Policy and Equity*, London: CPRE

Council for the Protection of Rural England (1999a) *Fair Examination? Testing the Legitimacy of Strategic Planning for Housing*, London: CPRE

Council for the Protection of Rural England (1999b) *Hedging Your Bets: Is Hedgerow Legislation Gambling with Our Heritage*, London: CPRE

Council for the Protection of Rural England (1999c) *Transport: The New Regional Agenda*, London: CPRE

Council for the Protection of Rural England (1999d) *Plan, Monitor and Manage: Making it Work*, London: CPRE

Council for the Protection of Rural England (1999e) *The Local Government Ombudsman*, London: CPRE

Council for the Protection of Rural England (1999f) *Park and Ride: Its Role in Local Transport*, London: CPRE

Council of Europe (1989) *European Campaign for the Countryside: Conclusion and Declarations*, Strasbourg: The Council

Council of Europe (1991) *The Bern Convention of Nature Conservation*, Strasbourg: The Council

Council of Europe (1992) *The European Urban Charter*, Strasbourg: Council of Europe Standing Conference of Local and Regional Authorities of Europe

Counsell, D. (1998) 'Sustainable development and structure plans in England and Wales: a review of current practice', *Journal of Environmental Planning and Management* 41: 177–94

County Planning Officers Society (1991) *Regional Guidance and Regional Planning Conferences Progress Report*, CPOS

County Planning Officers Society (1992) Metropolitan Planning Officers Society, and District Planning Officers Society, *Planning in the Urban Fringe: Final Report of the Joint Special Advisory Group*, CPOS, Middlesbrough: Department of Environment, Development and Transportation, Cleveland County Council

County Planning Officers Society (1993) *Planning for Sustainability*, CPOS, Winchester: County Planning Department, Hampshire County Council

County Planning Officers Society (annual) *Opencast Coalmining Statistics*, CPOS (published annually by Durham County Council)

Cowan, R. (1995) *Planning Aid Handbook*, London: Royal Town Planning Institute

Cowan, R. (1999) *The Role of Planning in Local Government*, London: Royal Town Planning Institute

Cowell, R. and Owens, S. (1998) 'Suitable locations: equity and sustainability in the minerals planning process', *Regional Studies* 32: 797–811

Cox, A. (1984) *Adversary Politics and Land*, Cambridge: Cambridge University Press

Cox, G., Lowe, P., and Winter, M. (1986) *Agriculture, People and Policies*, London: Allen & Unwin

Cox, G., Lowe, P., and Winter, M. (1990) *The Voluntary Principle in Conservation*, Chichester: Packard

Crabtree, J. R. and Chalmers, N. A. (1994) 'Economic evaluation of policy instruments for conservation: standard payments and capital grants', *Land Use Policy* 11: 94–106

Craig, P. and de Búrca, G. (1999) *EU Law*, Oxford: Oxford University Press

Crawford, C. (1989) 'Profitability and its role in planning decisions', *Journal of Environmental Law* 1: 221–44

Cripps, J. (1977) *Gypsy Site Policy and Illegal Camping: A Report on the Working of the Caravan Sites Act 1968*, London: HMSO

Croall, J. (1995) *Preserve or Destroy: Tourism and the Environment*, London: Gulbenkian Foundation

Croall, J. (1997) *LETS Act Locally: The Growth of Local Exchange Trading Systems*, London: Gulbenkian Foundation

Cross, D. (1992) 'Regional planning guidance for East Anglia', in Minay, C. L. W. (ed.) 'Developing regional guidance in England and Wales: a review symposium', *Town Planning Review* 63: 415–34

Cross, D. and Bristow, R. (eds) (1983) *English Structure Planning*, London: Pion

Crouch, C. and Marquand, D. (eds) (1989) *The New Centrism: Britain out of Step with Europe?*, Oxford: Blackwell

Crow, S. (1996a) 'Lessons from *Bryan*', *Journal of Planning and Environment Law* 1996: 359–69

Crow, S. (1996b) 'Development control: the child that grew up in the cold', *Planning Perspectives* 11: 399–411

Crow, S. (1998a) 'Planning gain: there must be a better way', *Planning Perspectives* 13: 357–72

Crow, S. (1998b) 'Challenging appeal decisions, or the use and abuse of the "toothcomb"', *Journal of Planning and Environment Law* 1998: 419–31

Crow, S. (2000) 'The public examination of draft regional planning guidance: some reflections on the process', *Journal of Planning and Environment Law* (October): 990–1002

Cuddy, M. and Hollingsworth, M. (1985) 'The review process in land availability studies: bargaining positions for builders and planners', in Barrett, S. and Healey, P. (eds) *Land Policy: Problems and Alternatives*, Aldershot: Avebury

Cullinane, S. (1992) 'Attitudes towards the car in the UK: some implications for policies on congestion and the environment', *Transportation Research* 26A: 291–301

Cullingworth, J. B. (1954) *Town and Country Planning in England and Wales*, London: Allen & Unwin

Cullingworth, J. B. (1960a) 'Household formation in England and Wales', *Town Planning Review* 31: 5–26

Cullingworth, J. B. (1960b) *Housing Needs and Planning Policy: A Restatement of the Problems of Housing Need and 'Overspill' in England and Wales*, London: Routledge & Kegan Paul

Cullingworth, J. B. (ed.) (1973) *Problems of an Urban Society*, vol. 3: *Planning for Change*, London: Allen & Unwin

Cullingworth, J. B. (1975) *Environmental Planning 1939–1969*, vol. 1: *Reconstruction and Land Use Planning 1939–1947*, London: HMSO

Cullingworth, J. B. (1979) *Environmental Planning 1939–1969*, vol. 3: *New Towns Policy*, London: HMSO

Cullingworth, J. B. (1980) *Environmental Planning 1939–1969*, vol. 4: *Land Values, Compensation and Betterment*, London: HMSO

Cullingworth, J. B. (1987) *Urban and Regional Planning in Canada*, New Brunswick, NJ: Transaction

Cullingworth, J. B. (ed.) (1990) *Energy Policy Studies*, vol. 5: *Energy, Land, and Public Policy*, New Brunswick, NJ: Transaction

Cullingworth, J. B. (1993) *The Political Culture of Planning: American Land Use Planning in Comparative Perspective*, New York and London: Routledge

Cullingworth, J. B. (1997a) *Planning in the USA*, New York: Routledge

Cullingworth, J. B. (1997b) 'British land use planning: a failure to cope with change?', *Urban Studies* 34: 945–60

Cullingworth, J. B. (ed.) (1999) *British Planning: 50 Years of Urban and Regional Policy*, London: Athlone Press

Cullingworth, J. B. and Karn, V. A. (1968) *The Ownership and Management of Housing in the New Towns*, London: HMSO

Cullingworth Report (1967) *Scotland's Older Houses* (Report of the Scottish Housing Advisory Committee, Subcommittee on Unfit Housing in Scotland), Edinburgh: HMSO

Curry, N. (1992a) 'Nature conservation, countryside strategies and strategic planning', *Journal of Environmental Management* 35: 79–91

Curry, N. (1992b) 'Controlling development in the national parks of England and Wales', *Town Planning Review* 63: 107–21

Curry, N. (1993) 'Negotiating gains for nature conservation in planning practice', *Planning Practice and Research* 8 (2) (April): 10–15

Curry, N. (1994) *Countryside Recreation, Access and Land Use Planning*, London: Spon

Curry, N. (1997) 'Enhancing countryside recreation benefits through the rights of way system in England and Wales', *Town Planning Review* 68: 449–63

Curry, N. and Peck, C. (1993) 'Planning on presumption: strategic planning for countryside recreation in England and Wales', *Land Use Policy* 10: 140–50

Cyclists Public Affairs Group (1995) *Investing in the Cycling Revolution*, Godalming (Surrey): Cyclists Touring Club

Cyclists Touring Club (1993) *Cycle Policies in Britain: The 1993 CTC Survey*, Godalming: CTC

Dabinett, G. and Lawless, P. (1994) 'Urban transport investment and regeneration: researching the impact of South Yorkshire Supertram', *Planning Practice and Research* 9: 407–14

Dales, J. H. (1968) *Pollution, Property and Prices*, Toronto: University of Toronto Press

Dalziel, M. and Rowan-Robinson, J. (1986) 'Resurrecting the two price system for land', *Journal of Planning and Environment Law* 1986: 409–15

Damer, S. and Hague, C. (1971) 'Public participation in planning: evolution and problems', *Town Planning Review* 42: 217–24

Damesick, P. J. (1986) 'The M25: a new geography of development?', *Geographical Journal* 152: 155–60

Daniels, S. (1993) *Fields of Vision: Landscape Imagery and National Identity in England and the United States*, Oxford: Polity Press

Danson, M. (1999) 'Economic development: the Scottish Parliament and the development agencies', in McCarthy, J. and Newlands, D. (eds) *Governing Scotland: Problems and Prospects: The Economic Impact of the Scottish Parliament*, Aldershot: Ashgate

Danson, M. W., Lloyd, M. G., and Newlands, D. (1989) 'Rural Scotland and the rise of Scottish Enterprise', *Planning Practice and Research* 4: 13–17

Danson, M. W., Lloyd, M. G., and Newlands, D. (1992) *The Role of Regional Development Agencies in Economic Regeneration*, London: Jessica Kingsley

Darke, R. (1999) 'Public speaking rights in local authority planning committees', *Planning Practice and Research* 14: 171–83

Darley, G., Hall, P., and Lock, D. (1991) *Tomorrow's New Communities*, York: Joseph Rowntree Foundation

Dasgupta, M., Oldfield, R., Sharman, K., and Webster, V. (1994) *Impact of Transport Policies in Five Cities*, Crowthorne: Transport Research Laboratory

David, Tyldesely & Associates (1996) *Wildlife Impact: The Treatment of Nature Conservation in Environmental Assessment*, Sandy, Beds: Royal Society for the Protection of Birds

Davidoff, P. (1965) 'Advocacy and pluralism in planning', *Journal of the American Institute of Planners* 31 (reprinted in Faludi, A. (ed.) (1973) *A Reader in Planning Theory*, Oxford: Pergamon)

Davidoff, P. and Reiner, T. A. (1962) 'A choice theory of planning', *Journal of the American Institute of Planners* 28: 103–15 (reprinted in Faludi, A. (ed.) (1973) *A Reader in Planning Theory*, Oxford: Pergamon)

Davies, C., Pritchard, D., and Austin, L. (1992) *Planscan Scotland: A Study of Development Plans in Scotland*, Sandy, Beds: Royal Society for the Protection of Birds

Davies, H. W. E. (1992) 'Britain 2000: The impact of Europe for planning and practice', *The Planner, Town and Country Planning Summer School Proceedings* 78 (21) (November): 21–2

Davies, H. W. E. (1994) 'Towards a European planning system?', *Planning Practice and Research* 9: 63–9

Davies, H. W. E. (1996) 'Planning and the European question', in Tewdwr-Jones, M. (ed.) *British Planning Policy in Transition: Planning in the 1990s*, London: UCL Press

Davies, H. W. E. (1999) 'The planning system and the development plan', in Cullingworth, J. B. (ed.) *British Planning: 50 Years of Urban and Regional Policy*, London: Athlone

Davies, H. W. E. (1998) 'Continuity and change in the British planning system', *Town Planning Review* 69: 135–52

Davies, H. W. E. and Gosling, J. A. (1994) *The Impact of the European Community on Land Use Planning in the United Kingdom*, London: Royal Town Planning Institute

Davies, H. W. E., Edwards, D., and Rowley, A. R. (1984) 'The relevance of development control', *Town Planning Review* 51: 5–24

Davies, H. W. E., Edwards, D., and Rowley, A. R. (1986a) *The Relationship between Development Plans, Development Control, and Appeals*, Reading: Department of Land Management and Development, University of Reading

Davies, H. W. E., Rowley, A. R., Edwards, D., Blom-Cooper, A., Roberts, C., Rosborough, L., and Tilley, R. (1986b) *The Relationship between Development Plans and Development Control*, Reading: Department of Land Management and Development, University of Reading

Davies, H. W. E., Edwards, D., Roberts, C., Rosborough, L., and Sales, R. (1986c) *The Relationship between Development Plans and Appeals*, Reading: Department of Land Management and Development, University of Reading

Davies, H. W. E., Edwards, D., and Rowley, A. R. (1989a) *The Approval of Reserved Matters following Outline Planning Permission* (DoE), London: HMSO

Davies, H. W. E., Hooper, A. J., and Edwards, D. (1989b) *Planning Control in Western Europe*, London: HMSO

Davies, J. G. (1972) *The Evangelistic Bureaucrat: A Study of a Planning Exercise in Newcastle-upon-Tyne*, London: Tavistock

Davies, L. (1996) 'Equality and planning: race' and 'Equality and planning: gender and disability', in Greed, C. (ed.) *Implementing Town Planning: The Role of Town Planning in the Development Process*, Harlow: Longman

Davies, P. and Keate, D. (1995) *In the Public Interest: London's Civic Architecture at Risk*, London: English Heritage

Davies, R. (1995) *Retail Planning Policies in Western Europe*, London: Routledge

Davies, R. L. and Campion, A. G. (1983) *The Future of the City Centre*, Institute of British Geographers, London: Academic Press

Davis, K. C. (1969) *Discretionary Justice: A Preliminary Inquiry*, Baton Rouge: Louisiana State University Press

Davis, M. (1990) *City of Quartz: Excavating the Future in Los Angeles*, London: Verso

Davison, I. (1990) *Good Design in Housing: A Discussion Paper*, London: House Builders Federation/Royal Institute of British Architects

Davison, R. C. (1938) *British Unemployment Policy: The Modern Phase*, London: Longman

Davoudi, S. (1995a) 'City Challenge: a sustainable mechanism of temporary gesture' in Hambleton, R. and Thomas, H. (eds) *Urban Policy Evaluation: Challenge and Change*, London: Paul Chapman

Davoudi, S. (1995b) 'City Challenge: the three-way partnership', *Planning Practice and Research* 10: 333–44

Davoudi, S. (1999) 'A quantum leap for planners: the role of the planning system within changing approaches to waste management', *Town and Country Planning* 68: 20–3

Davoudi, S. and Atkinson, R. (1999) 'Social exclusion and the British planning system', *Planning Practice and Research* 14: 225–36

Davoudi, S. and Healey, P. (1994) *Perceptions of City Challenge Policy Processes: The Newcastle Case*, Newcastle upon Tyne: Department of Town and Country Planning, University of Newcastle upon Tyne

Davoudi, S. and Healey, P. (1995) 'City Challenge: sustainable process or temporary gesture?', *Environment and Planning C: Government and Policy* 13: 79–95

Davoudi, S., Healey, P., and Hull, A. (1997) 'Rhetoric and reality in British structure planning in Lancashire: 1993–95', in Healey, P., Khakee, A., Motte, A., and Needham, B. (1997) *Making Strategic Spatial Plans: Innovation in Europe*, London: UCL Press

Dawson, D. (1985) 'Economic change and the changing role of local government', in Lochlin, M., Gelfand, M. D., and Young, K. (eds) *Half a Century of Municipal Decline*, London: Allen & Unwin

Dawson, J. and Walker, C. (1990) 'Mitigating the social costs of private development: the experience of linkage programmes in the United States', *Town Planning Review* 61: 157–70

Dawson, J. A. (1994) *Review of Retailing Trends, with Particular Reference to Scotland*, Edinburgh: Scottish Office Central Research Unit Papers

de Groot, L. (1992) 'City Challenge: competing in the urban regeneration game', *Local Economy* 7: 196–209

De Montford University and University of Strathclyde (1998) *Enhancing Public Participation in Local Government*, London: Department of the Environment, Transport and the Regions

De Soissons, M. (1988) *Welwyn Garden City*, Cambridge: Publications for Companies

Deakin, N. and Edwards, J. (1993) *The Enterprise Culture and the Inner Cities*, London: Routledge

Dear, M. and Scott, A. J. (eds) (1981) *Urbanization and Urban Planning in a Capitalist Society*, London: Methuen

Deas, I. and Ward, K. (2000) 'The song has ended but the melody lingers: regional development agencies and the lessons of the urban development corporation experiment', *Local Economy* 14: 114–32

Debenham, Tewson & Chinnocks (1988) *Planning Gain: Community Benefit or Commercial Bribe?*, London: Debenham, Tewson & Chinnocks

Deimann, S. (1994) *Your Rights under European Union Environment Legislation*, Brussels: European Environmental Bureau

Delafons, J. (1990a) *Development Impact Fees and Other Devices*, Berkeley, CA: Institute of Urban and Regional Development, University of California at Berkeley

Delafons, J. (1990b) *Aesthetic Control: A Report on Methods Used in the USA to Control the Design of Buildings*, Berkeley, CA: Institute of Urban and Regional Development, University of California at Berkeley

Delafons, J. (1997) *Politics and Preservation: A Policy History of the Built Heritage 1882–1996*, London: Spon

Delafons, J. (1998) 'Reforming the British planning system 1964-5: the Planning Advisory Group and the genesis of the Planning Act of 1968', *Planning Perspectives* 13: 373–87

Denington Report (1966) *Our Older Homes: A Call for Action* (Central Housing Advisory Committee), London: HMSO

Dennington, V. N. and Chadwick, M. J. (1983) 'Derelict land and waste land: Britain's neglected land resource', *Journal of Environmental Management* 16: 229–39

Dennis, N. (1970) *People and Planning*, London: Faber

Dennis, N. (1972) *Public Participation and Planners' Blight*, London: Faber

Department of National Heritage (1996) *Protecting Our Heritage: A Consultation Paper on the Built Heritage of England and Wales*, London

Department of the Environment, Transport and the Regions and CABE (2000) *By Design: Urban Design in the Planning System – Towards Better Practice*, London: Thomas Telford

Derby City Council (1989) *Sir Francis Ley Industrial Park: Simplified Planning Zoning – The First Twelve Months*, Derby: The Council

Derby City Council (1991) *Sir Francis Ley Industrial Park: Simplified Planning Zoning – The First Three Years*, Derby: The Council

Derbyshire, A., McNaught, S., and Blatchford, C. (1999) 'Listing needs much bigger alterations', *Estates Gazette* (May): 145–7

Derthick, M. (1972) *New Towns In-Town: Why a Federal Program Failed*, Washington, DC: Urban Institute

Devon County Council (1991) *Traffic Calming Guidelines*, Exeter: The Council

Dewar, D. (1999) 'Doubts over designation plans' [for designation of the New Forest and the South Downs as national parks], *Planning* (8 October): 12

Diamond, D. (1979) 'The uses of strategic planning: the example of the National Planning Guidelines in Scotland', *Town Planning Review* 50: 18–25

Diamond, D. and Spence, N. (1989) *Infrastructure and Industrial Costs in British Industry*, London: HMSO

Dicken, P. (1992) *Global Shift: The Internationalization of Economic Activity*, London: Paul Chapman

Diefendorf, J. (1990) *Rebuilding Europe's Blitzed Cities*, London: Macmillan

Dinan, D. (1998) *Encyclopedia of the European Union*, London: Macmillan

Dinan, D. (1999) *Ever Closer Union: An Introduction to European Integration* (second edition), London: Macmillan

Distributive Trades Economic Development Committee (1988) *The Future of the High Street*, London: NEDO

District Planning Officers' Society (1992) *Affordable Housing*, DPOS

Doak, J. (1999) 'Planning for the reuse of redundant defence estate: disposal processes, policy frameworks and development impacts', *Planning Practice and Research* 14: 211–24

Dobry, G., Hart, G., Robinson, P., and Williams, A. (1996) *Blundell and Dobry: Planning Applications, Appeals and Proceedings*, London: Sweet and Maxwell

Dobry Report (1974a) *Control of Demolition*, London: HMSO

Dobry Report (1974b) *Review of the Development Control System: Interim Report*, London: HMSO

Dobry Report (1975) *Review of the Development Control System: Final Report*, London: HMSO

Dobson, A. (1990) *Green Political Thought: An Introduction*, London: HarperCollins

Dobson, A. (1991) *A Green Reader*, London: André Deutsch

Dobson, A. (1995) 'No environmentalism without democratisation', *Town and Country Planning* 64: 322–3

Docklands Consultative Committee (1991) *Ten Years of Docklands: How the Cake Was Cut*, London: DCC

Docklands Consultative Committee (1992) *All That Glitters is Not Gold: A Critical Assessment of Canary Wharf*, London: DCC

Dodd, A. M. and Pritchard, D. E. (1993) *RSPB Planscan Northern Ireland: A Study of Development Plans in Northern Ireland*, Sandy, Beds: Royal Society for the Protection of Birds

Doig, A. (1984) *Corruption and Misconduct in Contemporary British Politics*, Harmondsworth: Penguin

Doig, A. and Ridley, F. F. (1995) *Sleaze, Politicians, Private Interests and Public Reaction*, Oxford: Oxford University Press

Doling, J. (1997) *Comparative Housing Policy: Government and Housing in Advanced Industrialized Countries*, London: Macmillan

Donnison, D. and Middleton, A. (eds) (1987) *Regenerating the Inner City: Glasgow's Experience*, London: Routledge

Doogan, K. (1996) 'Labour mobility and the changing housing market' *Urban Studies* 33: 199–221

Dorling, D. and Atkins, D. (1995) *Population Density, Change and Concentration in Great Britain 1971, 1981, and 1991*, Studies on Medical and Population Subjects 58, Office of Population Censuses and Surveys, London: HMSO

Douglas, A. (1997) *Local Authority Sites for Travellers* (Scottish Office Central Research Unit), Edinburgh: The Stationery Office

Dower Report (1945) *National Parks in England and Wales*, Cmd 6628, London: HMSO

Dowling, J. A. (1995) *Northern Ireland Planning Law*, Dublin: Gill & Macmillan

Downey, J. and McGuigan, J. (1999) *Technocities*, London: Sage

Downs, A. (1992) *Stuck in Traffic: Coping with Peak-Hour Traffic Congestion*, Washington, DC: Brookings Institution/Lincoln Institute of Land Policy

Doxford, D. and Hill, T. (1998) 'Land use for military training in the UK: the current situation, likely developments and possible alternatives', *Journal of Environmental Planning and Management* 41: 279–97

Doyle, T. and McEachern, D. (1998) *Environment and Politics*, London: Routledge

Draper, P. (1977) *Creation of the DoE: A Study of the Merger of Three Departments to Form the DoE*, Civil Service Studies 4, London: HMSO

Drewry, G. and Butcher, T. (1991) *The Civil Service Today* (second edition), Oxford: Blackwell

Drivers Jonas (1992) *Retail Impact Assessment Methodologies*, Edinburgh: Scottish Office Central Research Unit

Drury, P. (1995) 'The value of conservation', *Conservation Bulletin* (English Heritage), July 1995: 20

Dubben, N. and Sayce, S. (1991) *Property Portfolio Management*, London: Routledge

Duffy, K. and Hutchinson, J. (1997) 'Urban policy and the turn to community', *Town Planning Review* 68: 347–62

Dungey, J. and Newman, I. (eds) (1999) *The New Regional Agenda*, London: Local Government Information Unit

Dunleavy, P. (1995) 'Policy disasters: explaining the UK's record', *Public Policy and Administration* 10: 52–70

Dunlop, J. (1976) 'The examination-in-public of structure plans: an emerging procedure', *Journal of Planning and Environmental Law* 8–17

Dunmore, K. (1992) *Planning for Affordable Housing*, London: Institute of Housing

Dunn, J. A. (1998) *Driving Forces: The Automobile, its Enemies and the Politics of Mobility*, Washington, DC: Brookings Institution Press

Durman, M. and Harrison, M. (1996) *Bournville 1895–1914*, Birmingham: Article Press, Department of Art, University of Central England

Durrant, K. (2000) 'Making design decisions', *The Planning Inspectorate Journal* 18: 7–10

Duxbury, R. (1999) *Telling and Duxbury's Planning Law and Procedure*, London: Butterworths

Dwyer, J. and Hodge, I. (1996) *Countryside in Trust: Land Management by Conservation, Recreation and Organisation*, Chichester: Wiley

Dynes, M. and Walker, D. (1995) *The Times Guide to the New British State: The Government Machine in the 1990s*, London: Times Books

Earp, J. H., Headicar, P., Banister, D., and Curtis, C. (1995) *Reducing the Need to Travel: Some Thoughts of PPG 13*, Oxford: School of Planning, Oxford Brookes University

Easteal, M. (1995) 'A thoroughly modern review', *Local Government Chronicle* (22 September): 14–15

Economist (1999) 'English devolution: regional awakening?', *The Economist* (30 January): 34

ECOTEC Research and Consulting (1990) *Dynamics of the Rural Economy*, London: DoE

ECOTEC Research and Consulting (1993a) *Review of UK Environmental Expenditure: A Final Report to the DoE*, London: HMSO

ECOTEC Research and Consulting (1993b) *Reducing Transport Emissions through Planning*, London: HMSO

Edinburgh College of Art (School of Planning and Housing) and Peter P. C. Allan (Chartered Town Planning Consultants) (1995) *Review of Neighbour Notification*, Edinburgh: Scottish Office Central Research Unit

Edinburgh College of Art, Brodies, W. S., and Halliday Fraser Munro Planning (1997) *Research on the General Permitted Development Order and Related Mechanisms*, Edinburgh: The Stationery Office

Edmundson, T. (1993) 'Public participation in development control', *Town and Country Planning Summer School Proceedings*, London: Royal Town Planning Institute

Edwards, G. and Spence, D. (1997) *The European Commission* (second edition), London: Cartermill International

Edwards, J. (1987) *Positive Discrimination, Social Justice and Social Policy: Moral Scrutiny of a Policy Practice*, London: Tavistock

Edwards, J. (1990) 'What is needed from public policy?', in Healey, P. and Nabarro, R. (eds) *Land and Property Development in a Changing Society*, Aldershot: Gower

Edwards, J. (1997) 'Urban policy: the victory of form over substance', *Urban Studies* 34: 825–43

Edwards, J. and Batley, R. (1978) *The Politics of Positive Discrimination*, London: Tavistock

Edwards, J. and Deakin, N. (1992) 'Privatism and partnership in urban regeneration', *Public Administration* 70: 359–68

Edwards Report (1991) *Fit for the Future: Report of the National Parks Review Panel*, Cheltenham: Countryside Commission

Ekins, P. (1986) *The Living Economy: A New Economics in the Making*, London: Routledge

Elcock, H. (1994) *Local Government*, London: Routledge

Elkin, T., McLaren, D., and Hillman, M. (1991) *Reviving the City: Towards Sustainable Development*, London: Policy Studies Institute

Elkin, S. H. (1974) *Politics and Land Use Planning: The London Experience*, Cambridge: Cambridge University Press

Elsom, D. (1996) *Smog Alert: Managing Urban Air Quality*, London: Earthscan

Elson, M. J. (1986) *Green Belts: Conflict Mediation in the Urban Fringe*, London: Heinemann

Elson, M. J. (1990) *Negotiating the Future: Planning Gain in the 1990s*, Gloucester: ARC

Elson, M. J. (1991) *Green Belts for Wales: A Positive Role for Sport and Recreation*, Cardiff: Sports Council for Wales

Elson, M. J. and Ford, A. (1994) 'Green belts and very special circumstances', *Journal of Planning and Environmental Law* 1994: 594–601

Elson, M. J., Walker, S., and Macdonald, R. (1993) *The Effectiveness of Green Belts* (DoE), London: HMSO

Elson, M. J., Macdonald, R., and Steenberg, C. and Broom G. (1995a) *Planning for Rural Diversification* (DoE), London: HMSO

Elson, M. J., Steenberg, C., and Wilkinson, J. (1995b) *Planning for Rural Diversification: A Good Practice Guide* (DoE), London: HMSO

Elson, M. J., Steenberg, C., and Mendham, N. (1996) *Green Belts and Affordable Housing: Can We Have Both?*, Bristol: Policy Press

Elvin, D. and Robinson, J. (2000) 'Environmental impact assessment', *Journal of Planning and Environment Law* 2000: 876–93

Emerson, M. (1998) *Redrawing the Map of Europe*, London: Macmillan

Emms, J. E. (1980) 'The Community Land Act: a requiem', *Journal of Planning and Environment Law* 1980: 78–86

Empty Houses Agency (1998) *Joined Up Thinking: A Directory of Good Practice for Local Authority Empty Property Strategies*, London: Empty Houses Agency

English Historic Towns Forum (1992) *Townscape in Trouble: Conservation Areas – The Case for Change*, Bath: The Forum

English Nature: *see* section on Official Publications

English Partnerships (2000) *Urban Design Compendium*, London: English Partnerships

English Tourist Board (1991) *Tourism and the Environment: Maintaining the Balance* (prepared in conjunction with the Employment Department Group), London: ETB

English Tourist Board (1995) *English Heritage Monitor* 1995 (published annually), London: ETB

English Regional Associations (1998) *Regional Governance: Statement of General Principles*, Taunton: ERA

English Tourist Board and the Civic Trust (1993) *Turning the Tide: A Heritage and Environment Strategy for a Seaside Resort*, London: ETB

Ennis, F. (1994) 'Planning obligations in development plans', *Land Use Policy* 11: 195–207

Ennis, F. (1997) 'Infrastructure provision, the negotiating process and the planner's role', *Urban Studies* 34: 1935–54

Environ (1994) *Parking Provision in City Centres: A Lot to be Desired?*, Leicester: Environ

Environ (1995) *Indicators of Sustainable Development in Leicester*, Leicester: Environ

Environment Agency: *see* section on Official Publications (under National Rivers Authority)

Environmental Resources Ltd (1992) *Economic Instruments and Recovery of Resources from Waste* (DoE), London: HMSO

Erikson, R. A. and Syms, P. M. (1986) 'The effects of enterprise zones on local property markets', *Regional Studies* 20: 1–4

Esher, L. (1981) *A Broken Wave: The Rebuilding of England 1940–1980*, London: Allen Lane (Penguin edition 1983)

Essex, S. (1996) 'Members and officers in the planning policy process', in Tewdwr-Jones, M. (ed.) *British Planning Policy in Transition: Planning in the 1990s*, London: UCL Press

Essex County Council (1973) *Design Guide for Residential Areas*, Chelmsford: The Council

Etzioni, A. (1967) 'Mixed-scanning: a third approach to decision-making', *Public Administration Review* (December) (reprinted in Faludi, A. (1973) *A Reader in Planning Theory*, Oxford: Pergamon)

European Conference of Ministers of Transport (1995) *Urban Travel and Sustainable Development*, Paris: Organisation for Economic Cooperation and Development

European Environmental Bureau (1994) *Your Rights under European Union Environment Legislation*, Brussels: EEB

European Ombudsman (1999) *The European Ombudsman Annual Report for 1998*, Luxembourg: OOPEC

European Spatial Planning and Information Network (ESPRIN) (2000) *Urban and Rural Relations: Final Report to the DETR*, School of Architecture and Planning, University of Newcastle upon Tyne

Evans, A. W. (1973) *The Economics of Residential Location*, London: Macmillan

Evans, A. W. (1983) 'The determination of the price of land', *Urban Studies* 20: 119–29

Evans, A. W. (1988) *No Room! No Room! The Costs of the British Town and Country Planning System*, London: Institute of Economic Affairs

Evans, A. W. (1991) 'Rabbit hutches on postage stamps: planning, development and political economy', *Urban Studies* 28: 853–70

Evans, A. W. (1995) 'The property market: ninety per cent efficient?', *Urban Studies* 32: 5–29

Evans, A. W. (1996) 'The impact of land use planning and tax subsidies on the supply and price of housing in Britain: a comment', *Urban Studies* 33: 581–5

Evans, A. W. and Eversley, D. (eds) (1980) *The Inner City: Employment and Industry*, London: Heinemann

Evans, B. (1993) 'Why we no longer need a town planning profession', *Planning Practice and Research* 8: 9–15

Evans, B. (1995) *Experts and Environmental Planning*, Aldershot: Avebury

Evans, D. (1992) *A History of Nature Conservation*, London: Routledge

Evans, R. (1997) *Regenerating Town Centres*, Manchester: Manchester University Press

Eve, G. and Department of Land Economy, University of Cambridge (1992) *The Relationship between House Prices and Land Supply* (DoE), London: HMSO

Eve, Grimley J. R. (1992) *Use of Planning Agreements*, London: HMSO

Everest, D. A. (1990) 'The provision of expert advice to government on environmental matters: the role of advisory committees', *Science and Public Affairs* 4: 17–40

Eversley, Lord (1910) *Commons, Forests and Footpaths: The Story of the Battle during the Past Forty-five Years for Public Rights over the Commons, Forests and Footpaths of England and Wales*, London: Cassell

Eversley, D. E. C. (1973) *The Planner in Society*, London: Faber

Fainstein, S. S. (1994) *The City Builders: Property, Politics, and Planning in London and New York*, Oxford: Blackwell

Fainstein, S. S. and Campbell, S. (1996) *Readings in Urban Theory*, Oxford: Blackwell

Fainstein, S. S., Gordon, I., and Harloe, M. (eds) (1992) *Divided Cities: New York and London in the Contemporary World*, Oxford: Blackwell

Faludi, A. (1973) *A Reader in Planning Theory*, Oxford: Pergamon

Faludi, A. (1987) *A Decision-Centred View of Environmental Planning*, Oxford: Pergamon

Faludi, A. (2000) 'The European Spatial Development Perspective: what next?', *European Planning Studies* 8: 237–50

Farnham, D. and Horton, S. (1992) 'Human resource management in the new public sector: leading or following private employer practice', *Public Policy and Administration* 7: 42–55

Farrington, J. (1995) 'Military land in Britain after the Cold War', *Geography* 80: 272–89

Farrington, J., Gray, D., Martin, S., and Roberts, D. (1998) *Car Dependence in Rural Scotland*, Scottish Office Central Research Unit, Edinburgh: The Stationery Office

Farthing, S. (1996) *Evaluating Local Environmental Policy*, London: Avebury

Feldman, P. (1999) *Closing Doors: Examining the Lack of Housing Choices in London*, London: National Housing Federation

Fergusson, A. (1973) *The Sack of Bath: A Record and an Indictment*, Salisbury: Compton Russell

Ferris, J. (1972) *Participation in Urban Planning: The Barnsbury Case – A Study of Environmental Improvement in London*, London: Bell

Field, B. and MacGregor, B. (1987) *Forecasting Techniques for Urban and Regional Planning*, London: Hutchinson

Fielden, G. B. R., Wickens, A. H., and Yates, I. R. (1994) *Passenger Transport after 2000 AD*, London: Spon (for the Royal Society)

Fielder, S. (1986) *Monitoring the Operation of the Planning System: The National Dimension*, Reading: Department of Land Management and Development, University of Reading

Fielding, T. and Halford, S. (1990) *Patterns and Processes of Urban Change in the United Kingdom* (DoE Reviews of Urban Research), London: HMSO

Fischer, M. and Black, M. (1995) *Greening Environmental Policy: The Politics of a Sustainable Future*, London: Paul Chapman

Fischer, F. and Forester, J. (eds) (1993) *The Argumentative Turn in Policy Analysis and Planning*, London: UCL Press

Fit, J. and Kragt, R. (1994) 'The long road to European spatial planning: a matter of patience and mission', *Tijdschrift voor Economische en Sociale Geografie* 85: 461–5

Fitzsimmons, D. S. M. (1995) 'Removal of agricultural occupancy conditions: the Northern Ireland controversy', *Journal of Planning and Environment Law* 1995: 670–8

Flowers Report (1981) *Coal and the Environment*, London: HMSO

Flynn, A. and Marsden, T. K. (1995) 'Rural change, regulation, and sustainability', *Environment and Planning A* 27: 1180–92

Flynn, A., Leach, S., and Vielba, C. (1985) *Abolition or Reform? The GLC and the Metropolitan County Councils*, London: Allen & Unwin

Foley, P. (1992) 'Local economic policy and job creation: a review of evaluation studies', *Urban Studies* 29: 557–98

Foley, P., Hutchinson, J., and Fordham, G. (1998) 'Managing the challenge: winning and implementing the Single Regeneration Budget Fund', *Planning Practice and Research* 13: 63–80

Fontaine, P. (1995) *Europe in Ten Lessons* (second edition), Luxembourg: Office for the Official Publications of the European Communities

Fordham, G. (1995) *Made to Last: Creating Sustainable Neighbourhood and Estate Regeneration*, York: Joseph Rowntree Foundation

Fordham, G., Hutchinson, J., and Foley, P. (1998) 'Strategic approaches to local regeneration: the Single Regeneration Budget Challenge Fund', *Regional Studies* 33: 131–41

Fordham, R. (1989) 'Planning gain: towards its codification', *Journal of Planning and Environment Law* 1989: 577–84

Fordham, R. (1990) 'Planning consultancy: can it serve the public interest?', *Public Administration* 68: 243–8

Forester, J. (1980) 'Critical theory and planning practice', *Journal of the American Planning Association* 46: 275–86

Forester, J. (1982) 'Planning in the face of power', *Journal of the American Planning Association* 48: 67–80

Forester, J. (1989) *Planning in the Face of Power*, Berkeley: University of California Press

Forman, C. (1989) *Spitalfields: A Battle for Land*, London: Hilary Shipman

Forsyth, J. (1992) 'Tower Hamlets: setting up a regeneration corporation in Bethnal Green', in Bailey, N. and Barker, A. (eds) *City Challenge and Local Regeneration Partnerships: Conference Proceedings*, London: Polytechnic of Central London

Fothergill, S. and Gudgin, G. (1982) *Unequal Growth. Urban and Regional Change in the UK*, London: Heinemann

Fothergill, S. and Guy, N. (1990) *Retreat from the Regions: Corporate Change and the Closure of Factories*, London: Jessica Kingsley/ Regional Studies Association

Fothergill, S., Kitson, M., and Monk, S. (1985) *Urban Industrial Change: The Causes of the Urban-Rural Contrast in Manufacturing Employment Change* (DoE), London: HMSO

Fothergill, S., Kitson, M., and Perry, M. (1987) *Property and Industrial Development*, London: Hutchinson

Fox, M. and Turner, S. (1994) 'Northern Ireland', in Freshfields Environment Group (ed.) *Tolley's Environmental Handbook*, Croydon: Tolley

Franks Report (1957) *Report of the Committee on Administrative Tribunals and Enquiries*, Cmnd 218, London: HMSO

Freestone, D. (1991) 'European Community environmental policy and law', in Churchill, R., Warren, L. M., and Gibson, J. (eds) *Law, Policy and the Environment*, Oxford: Blackwell

Freilich, R. H. (1999) 'Final report for compulsory purchase: an appropriate power for the 21st century', *Journal of Planning and Environmental Law* 1999: 1076–86

Freilich, R. H. and Bushek, D. W. (eds) (1995) *Exactions, Impact Fees and Dedications*, Chicago: American Bar Association

Frey, H. (1999) *Designing the City: Towards a More Sustainable Urban Form*, London: Routledge

Fried, M. (1969) 'Social differences in mental health', in Kosa, J., Antonovsky, A., and Zola, I. K. (eds) *Poverty and Health: A Sociological Analysis*, Cambridge, MA: Harvard University Press

Friends of the Earth (1989) *Action for People: A Critical Appraisal of Government Inner City Policy*, London: FoE

Friends of the Earth (1991) *Local Responses to 1989 Traffic Forecasts*, London: FoE

Friends of the Earth (1992) *Less Traffic, Better Towns*, London: FoE

Friends of the Earth (1994a) *Planning for the Planet: Sustainable Development Strategies for Local and Strategic Plans*, London: FoE

Friends of the Earth (1994b) *Working Future: Jobs and the Environment*, London: FoE

Friends of the Earth (1997) *Stopping the Sprawl: The Housing Campaign Guide*, London: FoE

Frost, M. and Spence, N. (1993) 'Global city characteristics and central London's employment', *Urban Studies* 30: 547–58

Froud, J. (1994) 'The impact of ESAs on lowland farming', *Land Use Policy* 11: 107–18

Fry, G. K. (1984) 'The attack on the civil service and the response of the insiders', *Parliamentary Affairs* 37: 353–63

Fudge, C., Lambert, C., and Underwood, J. (1982) 'Local plans: approaches, preparation and adoption', *The Planner* 68 (March/April): 52–3

Fudge, C., Lambert, C., and Underwood, J. (1983) *Speed, Economy and Effectiveness in Local Plan Preparation and Adoption*, Bristol: School of Advanced Urban Studies, University of Bristol

Fulford, C. (1998) *The Costs of Reclaiming Derelict Sites*, London: Town and Country Planning Association

Fuller Peiser and Reading University (1999) *Development of the Redundant Defence Estate*, London: Thomas Telford

Fulton Report (1968) *Report of the Committee on the Civil Service*, Cmnd 3638, London: HMSO

Fyson, A. (1999a) 'Iron out defence land policy to get the full benefits', *Planning* 30 July: 13

Fyson, A. (1999b) 'Inner cities will not hold growing housing numbers', *Planning* 28 May: 15

Gahagan, M. (1992) 'City challenge: a solution to regeneration through partnership?' in Bailey, N. and Barker, A. (eds) *City Challenge and Local Regeneration Partnerships: Conference Proceedings*, London: Polytechnic of Central London

Gallent, N. (2000) Planning and affordable housing: from old values to New Labour', *Town Planning Review* 71: 123–47

Gallent, N. and Bell, P. (2000) 'Planning exceptions in rural England: past, present and future', *Planning Practice and Research* 15: 375–84

Gans, H. J. (1968) *People and Plans: Essays on Urban Problems and Solutions*, New York: Basic Books

Gans, H. J. (1991) *People, Plans and Policies: Essays on Poverty, Racism and Other National Urban Problems*, New York: Columbia University Press

Gardner, B. (1996) *Farming for the Future: Policies, Production and Trade*, London: Routledge

Garmise, S. (1997) 'The impact of European regional policy on the development of the regional tier in the UK', *Regional and Federal Studies* 7:1–24

Garside, P. L. and Hebbert, M. (eds) (1989) *British Regionalism 1900–2000*, London: Mansell

Gatenby, I. and Williams, C. (1992) 'Section 54A: the legal and practical implications', *Journal of Planning and Environment Law* 1992: 110–20

Gatenby, I. and Williams, C. (1996) 'Interpreting planning law' in Tewdwr-Jones, M. (ed.) (1996) *British Planning Policy in Transition*, London: UCL Press

Geddes, M. (1995) *Poverty, Excluded Communities and Local Democracy*, London: Commission for Local Democracy

Geddes, P. (1915) *Cities in Evolution*, London: Benn

Gentleman, H. (1993) *Counting Travellers in Scotland*, Edinburgh: Scottish Office

Gentleman, H. and Swift, S. (1971) *Scotland's Travelling People*, Edinburgh: HMSO

Geraghty, P. J. (1992) 'Environmental assessment and the application of an expert systems approach', *Town Planning Review* 63: 123–42

Gerald Eve Research (1995) *Whither the High Street?*, London: Gerald Eve Research

Gibbs, D. (1994) 'Towards the sustainable city', *Town Planning Review* 65: 99–109

Gibbs, D., Longhurst, J., and Braithwaite, C. (1996) 'Moving towards sustainable development? Integrating economic development and the environment in local authorities', *Journal of Environmental Planning and Management* 39: 317–32

Gibson, J. (1991) 'The integration of pollution control', in Churchill, R., Warren, L. M., and Gibson, J. (eds) *Law, Policy and the Environment*, Oxford: Blackwell

Gibson, M. S. and Langstaff, M. J. (1982) *An Introduction to Urban Renewal*, London: Hutchinson

Giddens, A. (1998) *The Third Way: The Renewal of Social Democracy*, Cambridge: Polity Press

Gilbert, A. and Healey, P. (1985) *The Political Economy of Land: Urban Development in an Oil Economy*, Aldershot: Gower

Gilg, A. W. (ed.) (1983) *Countryside Planning Yearbook 1983*, Norwich: Geo Books

Gilg, A. W. (ed.) (1988) *The International Year Book of Rural Planning*, London: Elsevier

Gilg, A. W. (ed.) (1991–95) *Progress in Rural Policy and Planning* (5 vols), London: Belhaven

Gilg, A. W. (ed.) (1992) *Restructuring the Countryside: Environmental Policy in Practice*, Aldershot: Avebury

Gilg, A. W. (1996) *Countryside Planning: The First Half Century* (second edition), London: Routledge

Gilg, A. W. (1999) *Perspectives on British Rural Planning Policy, 1994–97*, Aldershot: Ashgate

Gillett, E. (1983) *Investment in the Environment: Planning and Transport Policies in Scotland*, Aberdeen: Aberdeen University Press

Gillingwater, D. (1992) 'Regional strategy for the East Midlands', in Minay, C. L. W. (ed.) 'Developing regional guidance in England and Wales: a review symposium', *Town Planning Review* 63: 415–34

Gilpin, A. (1995) *Environmental Impact Assessment: Cutting Edge for the Twenty-first Century*, Cambridge: Cambridge University Press

Gilpin, A. (1996) *Dictionary of Environmental and Sustainable Development*, Chichester: Wiley

Gilroy, R. (1999) 'Towards a gentle city?', *Town and Country Planning* 68: 334–5

Gilroy, R. and Marvin, S. (1993) *Good Practices in Equal Opportunities*, Aldershot: Avebury

Ginsburg, L. (1956) 'Green belts in the Bible', *Journal of the Town Planning Institute* 42: 129–30

Glaister, S. and Mulley, C. (1983) *Public Control of the British Bus Industry*, Aldershot: Gower

Glaister, S., Burham, J., Stevens, H., and Travis, T. (1998) *Transport Policy in Britain*, London: Macmillan

Glass, R. (1959) 'The evaluation of planning: some sociological considerations', in Faludi, A. (1973) *A Reader in Planning Theory*, Oxford: Pergamon

Glass, W. D. (1994) 'Regional rural planning policies', *Town and Country Planning Summer School 1994 Proceedings*: 21–4

Glasson, J. (1992a) 'The fall and rise of regional planning in the economically advanced nations', *Urban Studies* 29: 505–31

Glasson, J. (1992b) *An Introduction to Regional Planning*, London: UCL Press

Glasson, J. (1999) 'The first 10 years of the UK EIA system: strengths, weaknesses, opportunities and threats', *Planning Practice and Research* 14: 363–75

Glasson, B. and Booth, P. (1992) 'Negotiation and delay in the development control process', *Town Planning Review* 63: 63–78

Glasson, J., Therivel, T., and Chadwick, A. (1998) *Introduction to Environmental Impact Assessment*, (second edition) London: UCL Press

Glasson, J., Godfrey, K., and Goodey, B. with Absalom, H. and Borg, J (1995) *Toward Visitor Impact Management: Visitor Impacts, Carrying Capacity and Management Responses in Europe's Historic Towns and Cities*, Aldershot: Avebury

GMA Planning, P-E International, and Jacques & Lewis (1993) *Integrated Planning and Granting of Permits in the EC* (DoE Planning Research Programme), London: HMSO

Goldsmith, M. (1980) *Politics, Planning and the City*, London: Hutchinson

Goldsmith, F. B. and Warren, S. A. (eds) (1993) *Conservation in Progress*, Chichester: Wiley

Goldsmith, M. and Newton, K. (1986) 'Central–local government relations: a bibliographic summary of the ESRC research initiative', *Public Administration* 64: 102–8

Gomez-Ibanez, J. A. and Meyer, J. R. (1992) *Going Private: The International Experience with Transport Privatization*, Washington, DC: Brookings Institution

Goodchild, B. (1997) *Housing and the Urban Environment: A Guide to Housing Design, Renewal and Urban Planning*, Oxford: Blackwell

Goodchild, R. N. and Munton, R. J. C. (1985) *Development and the Landowner: An Analysis of the British Experience*, London: Allen & Unwin

Goodin, R. E. (1992) *Green Political Theory*, Oxford: Polity Press

Goodlad, R., Flint, J., Kearns, A., Keoghan, M., Paddison, R., and Raco, M. (1999) *The Role and Effectiveness of Community Councils with regard to Community Consultation*, Edinburgh: Scottish Office Central Research Unit

Goodman, R. (1972) *After the Planners*, Harmondsworth: Penguin

Goodwin, P., Hallett, S., Kenny, F., and Stokes, G. (1991) *Transport: The New Realism*, Oxford: Transport Studies Unit, University of Oxford

Goodwin, P. B. (1989) 'The *rule of three*: a possible solution to the political problem of competing objectives for road pricing', *Traffic Engineering and Control* 30: 495–7

Goodwin, P. B. (1992) 'A review of new demand elasticities with special reference to short and long run effects of price changes', *Journal of Transport Economics and Policy* 26: 155–69

Goodwin, P. B. (1995) *The End of Hierarchy? A New Perspective on Managing the Road Network*, London: Council for the Protection of Rural England

Gordon,, J. E. (ed.) (1994) *Scotland's Soils: Research Issues in Developing a Soil Sustainability Strategy*, Perth: Scottish Natural Heritage

Gore, T. and Nicholson, D. (1991) 'Models of the land development process: a critical review', *Environment and Planning A* 23: 705–30

Gosling Report (1968) *Report of the Footpaths Committee*, London: HMSO

Gowers Report (1950) *Houses of Outstanding Historic and Architectural Interest*, London: HMSO

Graham, S. and Marvin, S. (1996) *Telecommunications and the City: Electronic Spaces, Urban Places*, London: Routledge

Graham, S. and Marvin, S. (1999) 'Planning cybercities? Integrating telecommunications into urban planning', *Town Planning Review* 70: 89–114

Graham, T. (1991) 'The interpretation of planning permissions and a matter of principle', *Journal of Planning and Environment Law* 1991: 104–12

Graham, T. (1993) 'Presumptions', *Journal of Planning and Environment Law* 1993: 423–8

Graham, T. (1996) 'Contaminated land investigations: how will they work under PPG 23?' *Journal of Planning and Environment Law* 1996: 547–53

Grant, J. S. (1987) 'Government agencies and the Highlands since 1945', *Scottish Geographical Magazine* 103: 95–9

Grant, M. (1979) 'Britain's Community Land Act: a post mortem', *Urban Law and Policy* 2: 359–73

Grant, M. (1982) *Urban Planning Law* (supplement 1990), London: Sweet and Maxwell

Grant, M. (1986) 'Planning and land taxation', *Journal of Planning and Environment Law* 1986: 92–106

Grant, M. (1992) 'Planning law and the British planning system', *Town Planning Review* 63: 3–12

Grant, M. (1996) *Permitted Development* (second edition), London: Sweet and Maxwell

Grant, M. (ed.) (1997) *Encyclopedia of Planning Law and Practice* (6 vols), London: Sweet and Maxwell (looseleaf; updated regularly) [A supplementary updating *Monthly Bulletin* is issued to subscribers]

Grant, M. (1999a) 'Planning as a learned profession', Paper presented to the Royal Town Planning Institute Council, 20 January 1999, London: RTPI

Grant, M. (1999b) 'Compensation and betterment', in Cullingworth, J. B. (ed.) *British Planning: 50 Years of Urban and Regional Policy*, London: Athlone

Graves, G., Max, R. and Kitson, T. (1996) 'Inquiry procedure: another dose of reform?', *Journal of Planning and Environment Law* (February): 99–106

Gray, C. (1994) *Government beyond the Centre: Sub-national Politics in Britain*, London: Macmillan

Gray, T. S. (1995) *UK Environmental Policy in the 1990s*, London: Macmillan

Grayling, T. and Glaister, S. (2000) *A New Fares Contract for London*, London: Institute for Public Policy Research

Grayson, L. (1990) *Green Belt, Green Fields and the Urban Fringe: The Pressure on Land in the 1980s: A Guide to Sources*, London: London Research Centre

Greater London Council (1986) *Changing Places*, London: GLC

Greed, C. (1991) *Surveying Sisters: Women in a Traditional Male Profession*, London: Routledge

Greed, C. (1994) *Women and Planning: Creating Gendered Realities*, London: Routledge

Greed, C. (1993) *Introducing Town Planning* (first edition), London: Longman

Greed, C. (ed) (1996a) Implementing Town Planning: The Role of Town Planning in the Development Process, Harlow: Longman

Greed, C. (ed) (1996b) Investigating Town Planning: Changing Perspectives and Agendas, Harlow: Longman

Greed, C. (1996c) *Introducing Town Planning*, London: Addison Wesley Longman

Greed, C. (ed.) (1999) *Social Town Planning*, London: Routledge

Greed, C. and Roberts, M (eds) (1998) *Introducing Urban Design: Interventions and Responses*, Harlow: Addison Wesley Longman

Green, A., Hasluck, C., Owen, D., and Winnett, C. (1993) *Local Unemployment Change in Britain: Leaps and Lags in the Response to National Economic Cycles*, London: Jessica Kingsley

Green, C. (1999) 'Conflicting signals over cluster', *Planning* 27 August: 12.

Green, H., Thomas, M., Iles, N., and Down, D. (1996) *Housing in England 1994/95: A Report of the 1994/95 Survey of English Housing Carried Out by the Social Survey Division of ONS on Behalf of the DoE*, London: HMSO

Green, R. E. (ed.) (1991) *Enterprise Zones: New Directions in Economic Development*, Newbury Park, CA: Sage

Green Balance (1992) *Campaigners' Guide to Local Plans*, London: Council for the Protection of Rural England

Green Balance (1999) *Fair Examination? Testing the Legitimacy of Strategic Planning for Housing*, London: Council for the Protection of Rural England

Greenhalgh, L. and Worpole, K. (1995) *Park Life: Urban Parks and Social Renewal*, London: Comedia/Demos

Greenhalgh, L., Worpole, K., and Grove-White, R. (1996) *People, Parks and Cities: A Guide to Current Good Practice in Urban Parks* (DoE), London: HMSO

Greensmith, C. and Haywood, R. (1999) *Rail Freight Growth and the Land Use Planning System*, Sheffield: Centre for Regional, Economic and Social Research, Sheffield Hallam University

Greer, A. (1996) *Rural Politics in Northern Ireland: Policy Networks and Agricultural Development since Partition*, Aldershot: Avebury

Greer, P. (1992) 'The Next Steps initiative: an examination of the agency framework documents', *Public Administration* 70: 89–98

Gregory, R. and Pearson, J. (1992) 'The parliamentary ombudsman after twenty-five years', *Public Administration* 70: 469–98

Gregory, S. (1998) *Transforming Local Services: Partnership in Action*, York: Joseph Rowntree Foundation

Grieco, M. and Jones, P. M. (1994) 'A change in the policy climate? Current European perspectives on road pricing', *Urban Studies* 31: 1517–32

Griffith, J. A. G. and Street, H. (1964) *A Casebook on Administrative Law*, London: Pitman

Griffiths, D. (1996) *Thatcherism and Territorial Politics: A Welsh Case Study*, Aldershot: Avebury

Griffiths, R. (1993) 'The politics of cultural policy in urban regeneration strategies', *Policy and Politics* 21: 39–46

Griffiths, R. (1995) 'Eurocities', *Planning Practice and Research* 10: 215–21

Grove, G. A. (1963) 'Planning and the appellant', *Journal of the Town Planning Institute* 49: 128–33

Groves, P. (1992) 'A chairman's view of a TEC', *Local Government Policy Making* 19: 9–17

Guy, C. (1994) *The Retail Development Process*, London: Routledge

Guy, C. (1998) 'High street retailing in off-centre retail parks', *Town Planning Review* 69: 291–313

Guy, S. and Marvin, S. (1995) *Planning for Water: Space, Time and the Social Organisation of Natural Resources*, Newcastle upon Tyne: Department of Town and Country Planning, University of Newcastle upon Tyne

Gwilliam, M., Bourne, C., Swain, C., and Prat, A. (1998) *Sustainable Renewal of Suburban Areas*, York: York Publishing Services

Gyford, J. (1990) 'The enabling authority: a third model', *Local Government Studies* 17: 1–4

Gyford, J. (1991) *Citizens, Consumers and Councils: Local Government and the Public*, London: Macmillan

Gyford, J., Leach, S., and Game, C. (1989) *The Changing Politics of Local Government*, London: Unwin Hyman

Haar, C. M. (1951) *Land Planning in a Free Society*, Cambridge, MA: Harvard University Press

Hackett, P. (1995) *Conservation and the Consumer: Measuring Environmental Concern*, London: Routledge

Hagman, D. G. and Misczynski, D. J. (1978) *Windfalls for Wipeouts: Land Value Capture and Compensation*, St Paul, MN: West

Hague, C. (1984) *The Development of Planning Thought: A Critical Perspective*, London: Hutchinson

Hague, C. (ed.) (1996) *Planning and Markets*, Aldershot: Avebury

Haigh, N. (1990) *EEC Environmental Policy* (second edition), Harlow: Longman

Haigh, N. and Irwin, F. (eds) (1990) *Integrated Pollution Control in Europe and North America*, Bonn: Institute for European Environ-

mental Policy; and Washington, DC: Conservation Foundation

Hailsworth, A. G. (1988) *The Human Impact of Hypermarkets and Superstores*, Aldershot: Avebury.

Halcrow Fox & Associates and Birkbeck College, University of London (1986) *Investigating Population Change in Small to Medium-Sized Areas*, London: DoE

Hale, M. and Lachowicz, M. (1998) *The Environment, Employment and Sustainable Development*, London: Routledge

Hales, R. (2000) 'Land use development planning and the notion of sustainable development: exploring constraint and facilitation within the English planning system', *Journal of Environmental Planning and Management* 43: 99–121

Halkier, H., Danson, M., and Damburg, C. (eds) (1998) *Regional Development Agencies in Europe*, London: Jessica Kingsley

Hall, A. C. (1990) 'Generating urban design objectives for local areas: a methodology and case study application to Chelmsford, Essex', *Town Planning Review* 61: 287–309

Hall, A. C. (1996) *Design Control: Towards a New Approach*, Oxford: Heinemann

Hall, D. (1989) 'The case for new settlements', *Town and Country Planning* 58: 111–14

Hall, D. (1991) 'Regional strategies: groping for guidance', *Town and Country Planning* 80: 138–40

Hall, D. (1999) 'Town expansion: constructive participation', *Town and Country Planning* 68: 170–2

Hall, D., Hebbert, M., and Lusser, H. (1993) 'The planning background', in Blowers, A. (ed.) *Planning for a Sustainable Environment: A Report by the Town and Country Planning Association*, London: Earthscan

Hall, P. (1980) *Great Planning Disasters*, London: Weidenfeld & Nicolson

Hall, P. (1981a) *The Enterprise Zone Concept: British Origins, American Adaptations*, Berkeley, CA: Institute of Urban and Regional Development, University of California at Berkeley

Hall, P. (ed.) (1981b) *The Inner City in Context: The Final Report of the Social Science Research Council Inner Cites Working Party*, London: Heinemann (reprinted 1986 by Gower)

Hall, P. (1982) 'Enterprise zones: a justification', *International Journal of Urban and Regional Research* 6: 415–21

Hall, P. (1983) 'The Anglo-American connection: rival rationalities in planning theory and practice, 1955-1980', *Environment and Planning B: Planning and Design* 10: 41–6

Hall, P. (1988) *Cities of Tomorrow*, Oxford: Blackwell

Hall, P. (1989) *London 2001*, London: Unwin Hyman

Hall, P. (1991) 'The British enterprise zones', in Green, R. E. (ed.) *Enterprise Zones: New Directions in Economic Development*, Newbury Park, CA: Sage

Hall, P. (1992) *Urban and Regional Planning* (third edition), London: Routledge

Hall, P. (1993) 'Planning in the 1990s: an international agenda', *European Planning Studies* 1: 3–12

Hall, P. (1997a) 'The view from London Centre', in Blowers, A. and Evans, B. (eds) *Town Planning into the 21st Century*, London: Routledge

Hall, P. (1997b) 'Regeneration policies for peripheral housing estates: inward and outward looking approaches', *Urban Studies* 34: 873–90

Hall, P. (1999a) 'The regional dimension', in Cullingworth, J. B. (ed.) *British Planning: 50 Years of Urban and Regional Policy*, London: Athlone

Hall, P. (1999b) *Cities in Civilization: Culture, Innovation and Urban Order*, London: Weidenfeld & Nicolson

Hall, P. (1999c) *Sustainable Cities or Town Cramming*, London: Town and Country Planning Association

Hall, P. and Hass-Klau, C. (1985) *Can Rail Save the City?*, Aldershot: Gower

Hall, P. and Markussen, A. (eds) (1985) *Silicon Landscapes*, London: Allen & Unwin

Hall, P. and Ward, C. (1998) *Sociable Cities: The Legacy of Ebenezer Howard*, Chichester: Wiley

Hall, P., Gracey, H., Drewett, R., and Thomas, R. (1973) *The Containment of Urban England*, London: Allen & Unwin

Hall, P., Breheny, M., McQuaid, R., and Hart, D. (1987) *Western Sunrise: Genesis and Growth of Britain's Major High Tech Corridor*, London: Unwin Hyman

Hall, S. (1995) 'The SRB: taking stock', *Planning Week* 27 April: 16-17

Hall, S. and Nevin, B. (1999) 'Continuity and change: a review of English regeneration policy in the 1990s', *Regional Studies* 33: 447–91

Hall, T. (ed.) (1991) *Planning and Urban Growth in the Nordic Countries*, London: Spon

Hallett, G. (ed) (1977) *Housing and Land Policies in West Germany and Britain*, London: Macmillan

Hallsworth, A. G. (1992) *The New Geography of Consumer Spending*, London: Pinter

Hallsworth, A. G. (1994) 'Decentralization of retailing in Britain: the breaking of the third wave', *Professional Geographer* 46: 296–307

Halsey, A. J. (ed.) (1972) *Educational Priority*, vol. 1: *EPA Problems and Policies*, London: HMSO

Ham, C. and Hill, M. (1993) *The Policy Process in the Modern Capitalist State* (second edition), London: Harvester Wheatsheaf

Hambleton, R. (1986) *Rethinking Policy Making*, Bristol: School for Advanced Urban Studies, University of Bristol

Hambleton, R. (1990) *Urban Government in the 1990s: Lessons from the USA*, Bristol: School for Advanced Urban Studies, University of Bristol

Hambleton, R. (1991) 'The regeneration of US and British cities', *Local Government Studies* 17: 53–69

Hambleton, R. (1993) 'Issues for urban policy in the 1990s', *Town Planning Review* 64: 313–23

Hambleton, R. (1998) *Local Government Political Management Arrangements: An International Perspective*, Edinburgh: Scottish Office Central Research Unit

Hambleton, R. (2000) 'Modernising political management in local government', *Urban Studies* 37: 931–50

Hambleton, R. and Hoggett, P. (1990) *Beyond Excellence: Quality Government in the 1990s*, Bristol: School for Advanced Urban Studies, University of Bristol

Hambleton, R. and Sweeting, D. (1999a) 'Change for councils', *Planning* (22 October): 20

Hambleton, R. and Sweeting, D. (1999b) 'Restructuring our decision making', *Planning* (12 November): 16–17

Hambleton, R. and Taylor, R. (eds) (1993) *People in Cities: A Transatlantic Policy Exchange*, Bristol: School for Advanced Urban Studies, University of Bristol

Hambleton, R. and Thomas, H. (eds) (1995) *Urban Policy Evaluation: Challenge and Change*, London: Paul Chapman

Hambleton, R., Stewart, J., and Underwood, J. (1980) *Inner Cities: Management and Resources*, Bristol: School for Advanced Urban Studies, University of Bristol

Hamilton, K. (1999) 'Women and transport: disadvantage and the gender divide', *Town and Country Planning* 68: 318–19

Hamnett, C. and Randolph, B. (1991) *Cities, Housing and Profits: Flat Break-up and the Decline of Private Renting*, London: UCL Press

Hams, T. (2000) 'Local Agenda 21: integration, independence or extinction?', *EG* 6 (6): 6–8

Hanley, N., Whitby, M., and Simpson, I. (1999) 'Assessing the success of agri-environmental policy in the UK', *Land Use Policy* 16: 67–80

Hansen, A. (ed.) (1993) *The Mass Media and Environmental Issues*, Chichester: Belhaven Press

Hanson, J. (1999) *Decoding Homes and Houses*, Cambridge: Cambridge University Press

Harding, A. (1991) 'The rise of urban growth coalitions, UK-style?', *Environment and Planning C: Government and Policy* 9: 295–317

Harding, A., Evans, R., Parkinson, M., and Garside, P. (1996) *Regional Government in Britain: An Economic Solution?* (Joseph Rowntree Foundation), Bristol: Policy Press

Hardy, D. (1991a) *Campaigning for Town and Country Planning*, vol. 1: *From Garden Cities to New Towns 1899–1946*, London: Spon

Hardy, D. (1991b) *Campaigning for Town and Country Planning*, vol. 2: *1946–1990, From New Towns to Green Politics*, London: Spon

Hardy, S., Hart, M., Albrechts, L., and Katos, A. (1995) *An Enlarged Europe: Regions in Competition?*, London: Regional Studies Association and Jessica Kingsley

Harloe, M. (1995) *The People's Home? Social Rented Housing in Europe*, Oxford: Blackwell

Harman, R. (1995) *New Directions: A Manual of European Best Practice in Transport Planning*, London: Transport 2000

Harrap, P. (1993) *Charging for Road Use Worldwide: A Financial Times Management Report*, London: Financial Times

Harris, N. and Tewdwr-Jones, M. (1995) 'The implications for planning of local government reorganisation in Wales: purpose, process, and practice', *Environment and Planning C: Government and Policy* 13: 47–66

Harris, R. (1992a) 'The Environmental Protection Act 1990: penalising the polluter', *Journal of Planning and Environment Law* 1992: 515–24

Harris, R. (1992b) 'Integrated pollution control in practice', *Journal of Planning and Environment Law* 1992: 611–23

Harris, R. and Larkham, P. (1999) *Changing Suburbs: Foundation, Form and Function*, London: Spon

Harrison, A. (1992) 'What shall we do with the contaminated site?', *Journal of Planning and Environment Law* 1992: 809–16

Harrison, J. (1994) 'Who is my neighbour?', *Journal of Planning and Environment Law* 1994: 219–23

Harrison, M. (1992) 'A presumption in favour of planning permission?', *Journal of Planning and Environment Law* 1992: 121–9

Harrison, M. L. and Mordey, R. (eds) (1987) *Planning Control: Philosophies, Prospects and Practice*, London: Croom Helm

Harrison, P. (1983) *Inside the Inner City: Life under the Cutting Edge*, Harmondsworth: Penguin

Harrison, R. T. and Hart, M. (eds) (1992) *Spatial Policy in a Divided Nation*, London: Jessica Kingsley

Harrison, T. (1988) *Access to Information in Local Government*, London: Sweet and Maxwell

Harrop, D. O. and Nixon, J. A. (1999) *Environmental Assessment in Practice*, London: Routledge

Hart, D. (2000) 'The impact of the European Convention of Human Rights on planning and environmental law', *Journal of Planning and Environmental Law* 2000: 117–33

Hart, T. (1992) 'Transport, the urban pattern and regional change, 1960–2010', *Urban Studies* 29: 483–503

Hart, T. (1993) 'Transport investment and disadvantaged regions: UK and European policies since the 1950s', *Urban Studies* 30: 417–36

Hart, T., Haughton, G., and Peck, J. (1996) 'Accountability and the non-elected local state: calling training and enterprise councils to local account', *Regional Studies* 30: 429–41

Harte, J. D. C. (1990) 'The scheduling of ancient monuments and the role of interested members of the public in environmental law', *Journal of Environmental Law* 2: 224–49

Harte, J. D. C. (1989) 'The scope of protection resulting from the designation of Sites of Special Scientific Interest', *Journal of Environmental Law* 1: 245–54

Harvey, G. (1997) *The Killing of the Countryside*, London: Jonathan Cape

Hasegawa, J. (1992) *Replanning the Blitzed City Centre*, Buckingham: Open University Press

Hass-Klau, C. (1990) *The Pedestrian and City Traffic*, London: Belhaven

Hass-Klau, C., Nold, I., Bocker, G., and Crampton, G. (1992) *Civilised Streets: A Guide to Traffic Calming*, Brighton: Environmental and Transport Planning

Hassan, G. (ed.) (1999) *A Guide to the Scottish Parliament* (Centre for Scottish Public Policy), Edinburgh: The Stationery Office

Hastings, A. (1996) 'Unravelling the process of "partnership" in urban regeneration policy', *Urban Studies* 33: 253–68

Hastings, A. and McArthur, A. (1995) 'A comparative assessment of government approaches to partnership with the local community', in Hambleton, R. and Thomas, H. (eds) *Urban Policy Evaluation: Challenge and Change*, London: Paul Chapman

Hastings, A., McArthur, A., and McGregor, A. (1996) *Less than Equal? Community Organisations and Estate Regeneration Partnerships*, Bristol: Policy Press

Haughton, G. (1990) 'Targeting jobs to local people: the British urban policy experience', *Urban Studies* 27: 185–98

Haughton, G. and Hunter, C. (1994) *Sustainable Cities*, London: Jessica Kingsley

Haughton, G. and Lawless, P. (1992) *Policies and Potential: Recasting British Urban and Regional Policies*, London: Regional Studies Association

Haughton, G., Rowe, I., and Hunter, C. (1997) 'The Thames Gateway and the re-emergence of regional strategic planning', *Town Planning Review* 68: 407–22

Hausner, V. A. (1986) *Urban Economic Adjustment and the Future of British Cities: Directions for Urban Policy*, Oxford: Clarendon Press

Hausner, V. A. (ed.) (1987) *Critical Issues in Urban Economic Development* (2 vols), Oxford: Clarendon Press

Hausner, V. A. and Robson, B. (1986) *Changing Cities: An Introduction to the Economic and Social Research Council Inner Cities Research Programme*, Swindon: The Council

Hawes, D. and Perez, B. (eds) (1996) *The Gypsy and the State: The Ethnic Cleansing of British Society* (second edition), Bristol: Policy Press

Hawkins, K. (1984) *Environment and Enforcement Regulation and the Social Definition of Pollution*, Oxford: Oxford University Press

Hayton, K. (1990) *Getting People into Jobs* (DoE), London: HMSO

Hayton, K. (1992) 'Scottish Enterprise: a challenge to local land use planning?', *Town Planning Review* 63: 265–78

Hayton, K. (1996) 'Planning policy in Scotland', in Tewdwr-Jones, M. (ed.) *British Planning Policy in Transition: Planning in the 1990s*, London: UCL Press

Haywood, R. (1999) 'South Yorkshire Supertram: its property impacts and their implications for integrated land use–transportation planning', *Planning Practice and Research* 14: 277–99

Hazell, R. (ed.) (1999) *Constitutional Futures: A History of the Next Ten Years* (The Constitution Unit), Oxford: Oxford University Press

Healey, M. J. and Ibery, B. W. (1985) *The Industrialisation of the Countryside*, Norwich: Geo Books

Healey, P. (1979) *Statutory Local Plans: Their Evolution in Legislation and Administrative Interpretations*, Oxford: Department of Town Planning, Oxford Polytechnic

Healey, P. (1983) *Local Plans in British Land Use Planning*, Oxford: Pergamon

Healey, P. (1985) 'The professionalisation of planning in Britain', *Town Planning Review* 56: 492–507

Healey, P. (1986) 'The role of development plans in the British planning system: an empirical assessment', *Urban Law and Policy* 8: 1–32

Healey, P. (1988) 'The British planning system and managing the urban environment', *Town Planning Review* 59: 397–417

Healey, P. (1989) 'Directions for change in the British planning system', *Town Planning Review* 60: 125–49; 'Comments and response' (by Goodchild, R., Marwick, A., Grant, M., Jones, A., Lyddon, D., and Robinson, D.) *Town Planning Review* 60: 319–32

Healey, P. (1990) 'Places, people and politics: plan-making in the 1990s', *Local Government Policy Making* 17: 29–39

Healey, P. (1991a) 'Urban regeneration and the development industry', *Regional Studies* 25: 97–110

Healey, P. (1991b) 'The content of planning education programmes: some comments from recent British experience', *Environment and Planning B: Planning and Design* 18: 177–89

Healey, P. (1991c) 'Models of the development process', *Journal of Property Research* 8: 219–38

Healey, P. (1992a) 'Development plans and markets', *Planning Practice and Research* 7: 13–20

Healey, P. (1992b) 'The reorganisation of state and market in planning', *Urban Studies* 29: 411–34

Healey, P. (1992c) 'Planning through debate: the communicative turn in planning theory', *Town Planning Review* 63: 143–62

Healey, P. (1992d) 'An institutional model of the development process', *Journal of Property Research* 9: 33–44

Healey, P. (1993) 'The communicative work of development plans', *Environment and Planning B*, 20: 83–104

Healey, P. (1994) 'Development plans: new approaches to making frameworks for land use regulation', *European Planning Studies* 2: 39–57

Healey, P. (1997) *Collaborative Planning*, London: Macmillan

Healey, P. (1998) 'Collaborative planning in a stakeholder society', *Town Planning Review* 69: 1–21

Healey, P. and Barrett, S. M. (1990) 'Structure and agency in land and property development processes', *Urban Studies* 37: 89–104

Healey, P. and Nabarro, R. (1990) *Land and Property Development in a Changing Context*, Aldershot: Gower

Healey, P. and Shaw, T. (1993) 'Planners, plans and sustainable development', *Regional Studies* 27: 769–76

Healey, P., McDougall, G. and Thomas, M. (eds) (1982) *Planning Theory: Prospects for the 1980s*, Oxford: Pergamon

Healey, P., McNamara, P., Elson, M., and Doak, A. (1988) *Land Use Planning and the Mediation of Urban Change*, Cambridge: Cambridge University Press

Healey, P., Purdue, M., and Ennis, F. (1992a) 'Rationales for planning gain', *Policy Studies* 13: 18–30

Healey, P., Davoudi, S., O'Toole, M., Tavsanoglu, S., and Usher, D. (1992b) *Rebuilding the City: Property-Led Urban Regeneration*, London: Spon

Healey, P., Purdue, M., and Ennis, F. (1993) *Gains from Planning: Dealing with the Impacts of Development*, York: Joseph Rowntree Foundation

Healey, P. *et al.* (1994) *Trends in Development Plan Making in European Planning*, Newcastle upon Tyne: Department of Town and Country Planning, University of Newcastle upon Tyne

Healey, P., Purdue, M., and Ennis, F. (1995a) *Negotiating Development: Rationales and Practice for Development Obligations and Planning Gain*, London: Spon

Healey, P., Cameron, S., Davoudi, S., Graham, S., and Madani-Pour, A. (eds) (1995b) *Managing Cities: The New Urban Context*, Chichester: Wiley

Healey, P., Khakee, A., Motte, A., and Needham, B. (1997) *Making Strategic Spatial Plans*, London: UCL Press

Hebbert, M. (1992) 'The British garden city: metamorphosis', in Ward, S. (ed.) *The Garden City: Past, Present and Future*, London: Spon

Hebbert, M. (1998) *London: More by Fortune than Design*, Chichester: Wiley

Hedges, B. and Clemens, S. (1994) *Housing Attitudes Survey*, London: Department of the Environment

Heim, C. (1990) 'The Treasury as developer-capitalist? British new town building in the 1950s', *Journal of Economic History* 50: 903–24

Hendry, J. (1989) 'The control of development and the origins of planning in Northern Ireland', in Bannon, M. J., Nowlan, K.I., Hendry, J., and Mawhinney, K. (1989) *Planning: The Irish Experience 1920–1988*, Dublin: Wolfhound Press

Hendry, J. F. (1992) 'Plans and planning policy for Belfast: a review article', *Town Planning Review* 63: 79–85

Hendry, J. F. (1993) 'Conservation in Northern Ireland', in Royal Town Planning Institute, *The Character of Conservation Areas*, vol. 2: *Supporting Information*, London: Royal Town Planning Institute

Hennessy, P. (1989) *Whitehall*, London: Secker & Warburg [paperback edition (1990), Glasgow: Fontana]

Hennessy, P. (1992) *Never Again: Britain 1945–1951*, London: Jonathan Cape (Vintage edition 1993)

Herbert, D. T. (ed.) (1995) *Heritage, Tourism and Society*, London: Mansell

Herbert-Young, N. (1995) 'Reflections on section 54A and *plan-led* decision-making', *Journal of Planning and Environment Law* 1995: 292–305

Herington, J. (1990) *Beyond Green Belts: Managing Urban Growth in the 21st Century*, London: The Stationery Office (previously published by Jessica Kingsley)

Herington, J. (1984) *The Outer City*, London: Harper & Row

Heseltine, M. (1987) *Where There's a Will*, London: Hutchinson

Hewitt, P. (1989) *A Cleaner, Faster London: Road Pricing, Transport Policy and the Environment*, London: Institute for Public Policy Research

Heycock, M. (1991) 'Public policy, need and land use planning', in Nadin, V. and Doak, J. (eds) *Town planning Responses to City Change*, Aldershot: Gower

Heywood, F. (1990) *Clearance: The View from the Street: A Study of Politics, Land and Housing*, Birmingham: Community Forum

Hibbs, J. (1989) *The History of the British Bus Services*, Newton Abbot: David & Charles

Higgins, J., Deakin, N., Edwards, J., and Wicks, M. (1983) *Government and Urban Poverty: Inside the Policy-Making Process*, Oxford: Oxford University Press

Higgins, M. and Allmendinger, P. (1999) 'The changing nature of public planning practice under the New Right: the legacies and implications of privatisation', *Planning Practice and Research* 14: 39–67

Higman, R. (1991) *Local Responses to 1989 Traffic Forecasts*, London: Friends of the Earth

Hill, D. M. (1970) *Participating in Local Affairs*, Harmondsworth: Penguin

Hill, D. M. (1994) *Citizens and Cities: Urban Policy in the 1990s*, Hemel Hempstead: Harvester Wheatsheaf

Hill, D. M. (1999) *Urban Policy and Politics in Britain*, Basingstoke: Macmillan

Hill, L. (1991) 'Unitary development plans for the West Midlands: first stages in the statutory responses to a changing conurbation', in Nadin, V. and Doak, J. (eds) *Town Planning Responses to City Change*, Aldershot: Gower

Hill, M. (ed.) (1993) *The Policy Process*, London: Harvester Wheatsheaf

Hill, M. P. (1987) 'Housing land availability: some observations of the process of housing development in contrasting urban locations', *Land Development Studies* 4: 209–19

Hill, S. and Barlow, J. (1995) 'Single regeneration budget: hope for "those inner cities"?', *Housing Review* 44: 32–5

Hillier Parker, Dundas & Wilson, and Edinburgh School of Planning and Housing (1998) *Review of Development Planning in Scotland* (Scottish Office Central Research Unit), Edinburgh: The Stationery Office

Hillman, J. (1988) *A New Look for London* (Royal Fine Art Commission), London: HMSO

Hillman, J. (1990) *Planning for Beauty: The Case for Design Guidelines* (Royal Fine Art Commission) London: HMSO

Hillman, J. (1993) *Telelifestyles and the Flexicity: The Impact of the Electronic Home*, Dublin: European Foundation for the Improvement of Living and Working Conditions / Luxembourg: Office for Official Publications of the European Communities

Hillman, M. (1989) 'More daylight, less accidents', *Traffic Engineering and Control* 30: 191–3

Hillman, M. (1992) 'Reconciling transport and environmental policy objectives: the way ahead at the end of the road', *Public Administration* 70: 225–34

Hillman, M. (1993a) *Time for Change: Setting Clocks Forward by One Hour throughout the Year*, London: Policy Studies Institute

Hillman, M. (1993b) *Children's Helmets: The Case For and Against*, London: Policy Studies Institute

Hillman, M. (ed) (1993c) *Children, Transport and the Quality of Life*, London: Policy Studies Institute

Hillman, M. and Whalley, A. (1979) *Walking IS Transport*, London: Policy Studies Institute

Hillman, M., Adams, J., and Whitelegg, J. (1991) *One False Move: A Study of Children's Independent Mobility*, London: Policy Studies Institute

Hills, P. J. (1994) 'The car versus mixed use development', in Wood, M. (ed.) *Planning Icons: Myth and Practice* (Planning Law Conference, *Journal of Planning and Environment Law*), London: Sweet and Maxwell

Himsworth, C. M. G. (1994) 'Charging for inspecting the register', *Scottish Planning and Environment Law* 44: 59–60

Hirsch, D. (ed.) (1994) *A Positive Role for Local Government: Lessons for Britain from Other Countries*, London: Local Government Chronicle/Joseph Rowntree Foundation

Hobbs, P. (1992) 'The economic determinants of post-war British planning', *Progress in Planning* 38: 179–300

Hobhouse Report (1947) *Report of the National Parks Committee (England and Wales)*, Cmd 7121, London: HMSO

Hobley, B. (1987) 'Rescue archaeology and planning', *The Planner* 73 (4) (May): 25–7

Hobson, J., Hockman, S., and Stinchcombe, P. (1996) 'The future of the planning system', in Bean, D. (ed.) *Law Reform for All*, London: Blackstone

Hodge, I. (1989) 'Compensation for nature conservation', *Environment and Planning A* 21: 1027–36

Hodge, I. (1991) 'Incentive policies and the rural environment', *Journal of Rural Studies* 7: 373–84

Hodge, I. (1992) 'Supply control and the environment: the case for separate policies', *Farm Management* 8: 65–72

Hodge, I. (1995) *Environmental Economics: Individual Incentives and Public Choices*, London: Macmillan

Hodge, I. (1999) 'Countryside planning: from urban containment to sustainable development', in Cullingworth, J. B. (ed.) *British Planning: 50 Years of Urban and Regional Policy*, London: Athlone

Hodge, I. and Monk, S. (1991) *In Search of a Rural Economy: Patterns and Differentiation in Non-metropolitan England*, Cambridge: Department of Land Economy

Hogwood, B. W. (1995) *The Integrated Regional Offices and the Single Regeneration Budget*, London: Commission for Local Democracy

Hogwood, B. W. (1996) *Mapping the Regions: Boundaries, Coordination and Government* (Joseph Rowntree Foundation), Bristol: Policy Press

Hogwood, B. W. and Keating, M. (eds) (1982) *Regional Government in England*, Oxford: Clarendon Press

Holden, E. (1977) *The Country Diary of an Edwardian Lady*, London: Michael Joseph

Holford, W. G. (1953) 'Design in city centres', part 3 of MHLG, *Design in Town and Village*, London: HMSO

Hollis, G., Ham, G., and Ambler, M. (eds) (1992) *The Future Role and Structure of Local Government*, Harlow: Longman

Hollox, R. E. and Biart, S. W. (1982) 'Local plan inquiries: a case study', *Journal of Planning and Environment Law* 1982: 17–23

Holmans, A. (1995) *Housing Demand and Need in England 1991 to 2011*, York: Joseph Rowntree Foundation

Holmans, A. and Simpson, M. (1999) *Low Demand: Separating Fact from Fiction*, London: Chartered Institute of Housing

Holmans, A., Morrison, N., and Whitehead, C. (1998) *How Many Homes Will We Need?* London: Shelter

Home, R. (1989) *Planning Use Classes: A Guide to the 1987 Order* (second edition), Oxford: BSP Professional

Home, R. (1992) 'The evolution of the use classes order', *Town Planning Review* 63: 187–201

Home, R. H. (1982) *Inner City Regeneration*, London: Spon

Home, R. K. (1987) 'Planning decision statistics and the use classes debate', *Journal of Planning and Environment Law* 1987: 167–73

Home, R. K. (1993) 'Planning aspects of the government consultation paper on gypsies', *Journal of Planning and Environment Law* 1993: 13–18

Hood, C. (1995) 'Contemporary public management: a new global paradigm', *Public Policy and Administration* 10: 104–227

Hooper, A. (1979) 'Land availability', *Journal of Planning and Environment Law* 1979: 752–6

Hooper, A. (1980) 'Land for private housebuilding', *Journal of Planning and Environment Law* 1980: 795–806

Hooper, A. (1982) 'Land availability in South East England', *Journal of Planning and Environment Law* 1982: 555–60

Hooper, A. (1985) 'Land availability studies and private housebuilding', in Barrett, S. and Healey, P. (eds) *Land Policy: Problems and Alternatives*, Aldershot: Avebury

Hooper, A. (1999) *Design for Living: Constructing the Residential Built Environment in the 21st Century*, London: Town and Country Planning Association

Hooper, A., Pinch, P., and Rogers, S. (1988) 'Housing land availability: Circular advice, circular arguments and circular methods', *Journal of Planning and Environment Law* 1988: 225–39

Hooper, A., Pinch, P., and Rogers, S. (1989) 'Housing land availability in the South East', in Breheny, M. J. and Congdon, P. (eds) *Growth and Change in a Core Region: The Case of South East England*, London: Pion

Hooper, A., Dunmore, K., and Hughes, M. (1998) *Home Alone: The Housing Preferences of One-Person Households*, Amersham: Housing Research Foundation

Hoskyns, J. (1983) 'Whitehall and Westminster: an outsider's view', *Parliamentary Affairs* 36: 137–47

Hough, B. (1992) 'Standing in planning permission appeals', *Journal of Planning and Environment Law* 1992: 319–29

Hough, B. (1997) 'The erosion of the concept of development: problems in establishing a material change of use', *Journal of Planning and Environment Law* 1997: 895–903

Hough, B. (1998) 'Relevance and reasons in planning matters', *Journal of Planning and Environment Law* 1998: 625–34

House Builders' Federation (1987) *Private Housebuilding in the Inner Cities*, London: HBF

Housing Choice (1991) *Planning: A Citizen's Charter*, London: Housing Choice

Howard, E. B. and Davies, R. L. (1993) 'The impact of regional out-of-town retail centres: the case of the Metro Centre', *Progress in Planning* 40: 89–165

Howard, E. (1898) *Tomorrow: A Peaceful Path to Real Reform*, London: Swan Sonnenschein

Howard, E. (1902) *Garden Cities of Tomorrow*, London: Swan Sonnenschein (reprinted by Faber in 1946, with an introduction by F. J. Osborn, and an introductory essay by Lewis Mumford)

Howe, J. (1996) 'A case of inter-agency relations: regional development in rural Wales', *Planning Practice and Research* 11: 61–72

Hughes, D. (1996) *Environmental Law* (third edition), Oxford: Butterworths

Hughes, D. J. (1995) 'Planning and conservation areas: where do we stand following PPG 15, and whatever happened to Steinberg?', *Journal of Planning and Environment Law* 1995: 679–91

Hughes, J. T. (1991) 'Evaluation of local economic development: a challenge for policy research', *Urban Studies* 28: 909–18

Hughes, M., Clarke, M., Allen, H., and Hall, D. (1998) *The Constitutional Status of Local Government in Other Countries*, Edinburgh: Scottish Office Central Research Unit

Hull, A. and Vigar, G. (1998) The changing role of the development plan in managing spatial change, *Environment and Planning C* 16: 379–94

Hull, A. (1998) 'Spatial planning: the development plan as a vehicle to unlock development potential?', *Cities* 15: 327–35

Hull, A., Healey, P., and Davoudi, S. (1995) *Greening the Red Rose County: Working towards an Integrated Sub-regional Strategy*, Newcastle upon Tyne: Department of Town and Country Planning, University of Newcastle upon Tyne

Humber, J. R. (1980) 'Land availability: another view', *Journal of Planning and Environment Law* 1980: 19–23

Humber, R. (1990) 'Prospects and problems for private housebuilders', *The Planner/Town and Country Planning Summer School Proceedings* 76 (7) (23 February): 15–19

Hunt Report (1969) *The Intermediate Areas*, London: HMSO

Huppes, G. and Kagan, R. A. (1989) 'Market-oriented regulation of environmental problems in the Netherlands', *Law and Policy* 11: 215–39

Hurrell, A. and Kinsbury, B. (1992) *The International Politics of the Environment: Actors, Interests and Institutions*, Oxford: Clarendon Press

Hutter, B. M. (1989) 'Variations in regulatory enforcement styles', *Law and Policy* 11: 153–74

Hutton, N. (1986) *Lay Participation in a Public Inquiry: A Sociological Case Study*, Aldershot: Gower

Hutton, R. H. (1991) 'Local needs policy initiatives in rural areas: missing the target', *Journal of Planning and Environment Law* 1991: 303–11

Huxley Report (1947) *Conservation of Nature in England and Wales*, Cmd 7122, London: HMSO

IAURIF (Institut d'aménagement du territoire et d'urbanisme de la région d'Île de France) (1991) *La Charte de l'Île de France*, Paris: IAURIF

Ilbery, B. W. (1990) 'Adoption of the arable set-aside scheme in England', *Geography* 76: 69–73

Ilbery, B. W. (1998) *The Geography of Rural Change*, Harlow: Longman

Illich, I. (1971) *Deschooling Society*, New York: Harper & Row

Illich, I. (1977) *Disabling Professions*, London: Boyars

Improvement and Development Agency (1998) *Sustainability in Development Control*, Luton: IdeA

Imrie, R. (1996) *Disability and the City*, London: Paul Chapman

Imrie, R. and Thomas, H. (1993) *British Urban Policy and the Urban Development Corporations*, London: Paul Chapman

Imrie, R., Thomas, H., and Marshall, T. (1995) 'Business organisations, local dependence and the politics of urban renewal in Britain', *Urban Studies* 32: 31–47

Ince, M. (1984) *Sizewell Report: What Happened at the Inquiry*, London: Pluto

Inland Revenue (1992) *Company Cars: Reform of Income Tax Treatment – A Consultative Document*, London: Inland Revenue

Innes, J. E. (1995) 'Planning theory's emerging paradigm: communicative action and interactive practice', *Journal of Planning Education and Reseach* 14: 183–90

Innes, J. E. (1998a) 'Challenge and creativity in postmodern planning', *Town Planning Review* 69: v–ix

Innes, J. E. (1998b) 'Information in communicative planning', *Journal of the American Planning Association* 64: 52–63

Insight Social Research (1989) *Local Attitudes to Central Advice*, London: ISR

Institute of Logistics and Transport (1999) *Sustainable Distribution* (Discussion Paper), London: ILT

International Energy Agency (1989) *Energy and the Environment*, Paris: Organisation for Economic Cooperation and Development

Investment Property Databank (1993) *The Investment Performance of Listed Buildings*, London: RICS

Investment Property Databank (1999) *Investment Performance of Listed Buildings*, Swindon: English Heritage

Jackson, A., Morrison, N., and Royce, C. (1994) *The Supply of Land for Housing: Changing Local Authority Mechanisms*, Department of Land Economy, University of Cambridge

Jackson, A. R. W. and Jackson, J. M. (1996) *Environmental Science: The Natural Environment and Human Impact*, Harlow: Longman

Jackson, T. (1996) *Material Concerns: Pollution, Profit and Quality of Life*, London: Routledge

Jacobs, B. D. (1992) *Fractured Cities: Capitalism, Community and Empowerment in Britain and America*, London: Routledge

Jacobs, J. (1961) *The Death and Life of Great American Cities*, London: Cape/Penguin

Jacobs, J. (1985) 'Urban development grant', *Policy and Politics* 13: 191–9

Jacobs, M. (1990) *Sustainable Development: Greening the Economy*, London: Fabian Society

Jacobs, M. (1991) *The Green Economy: Environment, Sustainable Development, and the Politics of the Future*, London: Pluto Press

Jacobs, M. (1999) *Environmental Modernisation: The New Labour Agenda*, London: Fabian Society

James, S. (1990) 'A streamlined city: the broken pattern of London government', *Public Administration* 68: 493–504

James, S. (1997) *British Government: A Reader in Policy Making*, London: Routledge

James, S. (1999) *British Cabinet Government* (second edition), London: Routledge

Jansen, A. J. and Hetsen, H. (1991) 'Agricultural development and spatial organization in Europe', *Journal of Rural Studies* 7: 143–51

Jarvis, R. (1996) 'Structure planning policy and strategic planning guidance in Wales', in Tewdwr-Jones, M. (ed.) *British Planning Policy in Transition: Planning in the 1990s*, London: UCL Press

Jenkins, K., Oates, G., and Stott, A. (1985) *Making Things Happen: A Report on the Implementation of Government Efficiency Scrutinies: Report to the Prime Minister*, London: HMSO

Jenkins, M., Burton, E., and Williams, K. (eds) (1996) *Compact City: A Sustainable Urban Form?*, London: Spon

Jenkins, S. (1995, 1996) *Accountable to None: The Tory Nationalisation of Britain*, London: Hamish Hamilton (1995) (Penguin edition 1996)

Jenkins, T. N. (1990) *Future Harvest: The Economics of Farming and the Environment – Proposals for Action*, London: Council for the Protection of Rural England and World Wide Fund for Nature

Jenks, M., Burton, E., and Williams, K. (eds) (1996) *The Compact City: A Sustainable Urban Form?*, London: Spon

Jewell, T. (1995) 'Planning regulation and environmental consciousness: some lessons from minerals?', *Journal of Planning and Environment Law* 1995: 482–98

Jewell, D. and Raemaekers, J. (1999) 'Local authority archaeological services in Scotland two years after local government reorganisation', *Journal of Environmental Planning and Management* 42: 375–88

JMP Consultants (1995b) *PPG 13: A Guide to Better Practice: Reducing the Need to Travel through Land Use and Transport Planning*, London: HMSO

JMP Consultants (1995a) *Travel to Food Super-stores*, London: JMP Consultants

John, P. (1993) *Local Government in Northern Ireland*, York: Joseph Rowntree Foundation

Johnson, D., Martin, S., Pearce, G., and Simmons, S. (1992) *The Strategic Approach to Derelict Land Reclamation* (DoE), London: HMSO

Johnson, D. A. (1995) *Planning the Great Metropolis: The 1929 Regional Plan of New York and its Environs*, London: Spon

Johnson, P. (ed.) (1994) *20th Century Britain: Economic, Social and Cultural Change*, London: Methuen

Johnson, S. P. and Corcelle, G. (1989) *The Environmental Policy of the European Communities*, London: Graham & Trotman

Johnson, W. C. (1984) 'Citizen participation in local planning: a comparison of US and British experience', *Environment and Planning C: Government and Policy* 2: 1–14

Johnston, B. (1999a) 'New powers and policies for improved performance', *Planning* 24 September: 16–17

Johnston, B. (1999b) 'Know your rights: councils should not panic about the new human rights legislation', *Planning* 19 November: 17

Johnston, B. F. (1995) 'Commission makes case for coherence', *Planning* (20 October)

Johnston, R. J. and Gardiner, V. (eds) (1991) *The Changing Geography of the UK*, London: Routledge

Johnstone, R. J. (1991) *Environmental Problems: Nature, Economy and State*, Chichester: Belhaven Press

Jones, A. (1996) 'Local planning policy: the Newbury approach', in Tewdwr-Jones, M. (ed.) *British Planning Policy in Transition: Planning in the 1990s*, London: UCL Press

Jones, B. and Kavanagh, D. (1998) *British Politics Today* (sixth edition), Manchester: Manchester University Press

Jones, C. (1996) 'Property-led local economic development policies from advance factory to English Partnerships and strategic property investment?', *Regional Studies* 30: 200–6

Jones, C., Wood, C., and Dipper, B. (1998) 'Environmental assessment in the UK planning process: a review of practice', *Town Planning Review* 69: 315–39

Jones, C. E. and Wood, C. (1995) 'The impact of environmental assessment on public inquiry decisions', *Journal of Planning and Environment Law* 1995: 890–904

Jones, G. and Stewart, J. (1983) *The Case for Local Government*, London: Allen & Unwin

Jones, M. (1996) 'TEC policy failure: evidence from the baseline follow-up studies', *Regional Studies* 30: 509–32

Jones, M. R. (1995) 'Training and enterprise councils: a continued search for local flexibility', *Regional Studies* 29: 577–80

Jones, P. (1990) *Traffic Quotes: Public Perception of Traffic Regulation in Urban Areas: Report of a Research Study* (Department of Transport), London: HMSO

Jones, P. (1998) 'The Groundwork network', *Geography* 83: 189–93

Jones, P. (1999) 'Groundwork: changing places and agendas', *Town and Country Planning* 68: 315–17

Jones, P. M. (1991a) 'Gaining public support for road pricing through a package approach', *Traffic Engineering and Control* 32: 194–6

Jones, P. M. (1991b) 'UK public attitudes to urban traffic problems and possible countermeasures: a poll of polls', *Environment and Planning C: Government and Policy* 9: 245–56

Jones, R. (1982) *Town and Country Choice: A Critical Analysis of Britain's Planning System*, London: Adam Smith Institute

Joseph Rowntree Foundation (1994) *Inquiry into Planning for Housing*, York: JRF

Journal of Planning and Environment Law (1995) 'Are off-site effects material to material changes of use?' (and case of *Forest of Dean District Council v. Secretary of State for the Environment and Howells*), *Journal of Planning and Environment Law* 1995: 889 and 937–43

Jowell, J. (1975) 'Development control' [review article on the Dobry Report], *Political Quarterly* 46: 340–4

Jowell, J. (1977a) 'Bargaining in development control', *Journal of Planning and Environment Law* 1977: 414–33

Jowell, J. (1977b) 'The limits of law in urban planning', *Current Legal Problems* 1977 30: 63–83

Jowell, J. and Grant, M. (1983) 'Guidelines for planning gain?', *Journal of Planning and Environment Law* 1983: 427–31

Jowell, J. and Millichap, D. (1987) 'Enforcement: the weakest link in the planning chain', in Harrison, M. L. and Mordey, R. (eds) *Planning Control: Philosophies, Prospects and Practice*, London: Croom Helm

Jowell, J. and Oliver, D. (1994) *The Changing Constitution*, Oxford: Clarendon Press

JURUE: ECOTEC Research and Consultancy Ltd (1988) *Improving Urban Areas*, London: HMSO

JURUE: ECOTEC Research and Consultancy Ltd (1987) *Greening City Sites*, London: HMSO

Kain, J. F. and Beesley, M. E. (1965) 'Forecasting car ownership and use', *Urban Studies* 2: 163–85

Karn, V., Lucas, J. *et al.* (1996) *Home-Owners and Clearance: An Evaluation of Rebuilding Grants* (DoE), London: HMSO

Kay, G. (1998) 'The right to roam: a restless ghost', *Town and Country Planning* 67: 255–9

Kay, J. H. (1998) *Asphalt Nation: How the Automobile Took Over America and How We Can Take It Back*, Berkeley, CA: University of California Press

Kay, P. and Evans, P. (1992) *Where Motor Car is Master*, London: Council for the Protection of Rural England

Keating, M. and Jones, B. (1985) *Regions in the European Community*, Oxford: Oxford University Press

Keeble, D., Tyler, P., Broom, G., and Lewis, J. (1992) *Business Success in the Countryside: The Performance of Rural Enterprise* (DoE), London: HMSO

Keeble, L. (1969) *Principles and Practice of Town and Country Planning*, London: Estates Gazette

Keene, Hon. Justice (1999) 'Recent trends in judicial control', *Journal of Planning and Environment Law* (January): 30–37

Keith, M. and Rogers, A. (eds) (1991) *Hollow Promises: Rhetoric and Reality in the Inner City*, London: Mansell

Kellett, J. (1990) 'The environmental impact of wind energy developments', *Town Planning Review* 61: 139–55

Kelly, A. and Marvin, S. (1995) 'Demand-side management in the electricity sector: implications for town planning in the UK', *Land Use Policy* 12: 205–21

Kemeny, J. (1994) *From Public Housing to the Social Market: Rental Policy Strategies in Comparative Perspective*, London: Routledge

Kemp, R. (1980) 'Planning, legitimation and the development of nuclear energy: a critical theoretic analysis of the Windscale inquiry', *International Journal of Urban and Regional Research* 4: 350–71

Kenny, M. and Meadowcroft, K. (1999) *Planning Sustainability*, London: Routledge

Kent County Council (1995) *A Vision for EuroRegion: Towards a Policy Framework – The First Step*, Maidstone: Kent County Council

Kenyon, R. C. (1991) 'Environmental assessment: an overview on behalf of the RICS', *Journal of Planning and Environment Law* 1991: 419–22

Keogh, G. and Evans, A. W. (1992) 'The private and social costs of planning delay', *Urban Studies* 29: 687–99

Kerr, D. (1998) 'The Private Finance Initiative and the changing governance of the built environment', *Urban Studies* 35: 2277–01

Khan, M. A. (1995) 'Sustainable development: the key concepts, issues and implications', *Sustainable Development* 3: 63–9

Kidd, S. and Kumar, A. (1993) 'Development planning in the English metropolitan counties: a comparison of performance under two planning systems', *Regional Studies* 27: 65–73

King, A. D. (ed.) (1980) *Buildings and Society: Essays on the Social Development of the Built Environment*, London: Routledge

King, A. D. (1990a) *Global Cities: Post-imperialism and the Internationalisation of London*, London: Routledge

King, J. (1990b) *Regional Selective Assistance 1980-84: An Evaluation by DTI, IDS, and WOID*, London: HMSO

Kingdom, J. (1991) *Local Government and Politics in Britain*, London: Philip Allan

Kingdon, J. W. (1984) *Agendas, Alternatives, and Public Policies*, New York: HarperCollins

Kirby, A. (1985) 'Nine fallacies of local economic change', *Urban Affairs Quarterly* 21: 207–20

Kirk, G. (1980) *Urban Planning in a Capitalist Society*, London: Croom Helm

Kirkby, J., O'Keefe, P., and Timberlake, L. (1995) *The Earthscan Reader in Sustainable Development*, London: Earthscan

Kirkwood, G. and Edwards, M. (1993) 'Affordable housing policy: desirable but unlawful?', *Journal of Planning and Environment Law* 1993: 317–24

Kirwan, R. (1989) 'Finance for urban public infrastructure', *Urban Studies* 26: 285–300

Kitchen, T. (1996) 'A future for strategic planning policy: a Manchester perspective', in Tewdwr-Jones, M. (ed.) (1999) *British Planning Policy in Transition: Planning in the 1990s*, London: UCL Press

Kitchen, T. (1997) *People, Politics, Policies and Plans: The City Planning Process in Contemporary Britain*, London: Paul Chapman

Kitchen, T. (1999a) 'Consultation on government policy initiatives: the case of regional policy guidance', *Planning Practice and Research* 14: 5–15

Kitchen, T. (1999b) 'The structure and organisation of the planning service in English local government', *Planning Practice and Research* 14: 313–27

Kitson, T. (1999) 'The European Convention on Human Rights and local plans', *Journal of Planning and Environmental Law* April: 321–3

Kivell, P. (1987) 'Derelict land in England; policy responses to a continuing problem', *Regional Studies* 21: 265–9

Kivell, P. (1993) *Land and the City: Patterns and Processes of Urban Change*, London: Routledge

Klein, R. (1974) 'The case for elitism: public opinion and public policy', *Political Quarterly* 45: 406–17

Klosterman, R. E. (1985) 'Arguments for and against planning', *Town Planning Review* 56: 5–20

Knox, P. L. (1994) *Urbanization: An Introduction to Urban Geography*, Englewood Cliffs, NJ: Prentice-Hall

Knox, P. L. and Cullen, J. L. (1981) 'Planners as urban managers: an exploration of the attitudes and self-image of senior British planners', *Environment and Planning A* 13: 885–98

Knox, P. L. and Taylor, P. J. (eds) (1995) *World Cities in a World-System*, Cambridge: Cambridge University Press

Koslowski, J. and Hill, G. (1993) *Towards Planning for Sustainable Development*, Aldershot: Avebury

Kostof, S. (1991) *The City Shaped: Urban Patterns and Meaning through History*, London: Thames & Hudson

Kostof, S. (1992) *The City Assembled: The Elements of Urban Form through History*, London: Thames & Hudson

KPMG Consulting (1999a) *City Challenge: Final National Evaluation*, London: Department of the Environment, Transport and the Regions

KPMG Consulting (1999b) *Fiscal Incentives for Urban Housing: Exploring the Options*, London: Department of the Environment, Transport and the Regions

Kraan, D. J. and Veld, R. J. (eds) (1991) *Environmental Protection: Public or Private Choice?*, Dordrecht: Kluwer

Krämer, L. (1991) 'The implementation of Community environmental directives within member states: some implications of the direct effect doctrine', *Journal of Environmental Law* 3: 39–56

Krämer, L. (1992) *Focus on European Environmental Law*, London: Sweet and Maxwell

Krämer, L. (1999) *E C Environmental Law* (fourth edition), London: Sweet and Maxwell

Kromarek, P. (1986) 'The Single European Act and the environment', *European Environment Review* 1: 10–12

Kunzmann, K. R. and Wegener, M. (1991) 'The pattern of urbanisation in Western Europe', *Ekistics* 350 (September/October): 282–91

Labour Party (1995) *A Choice for England: A Consultation Paper on Labour's Plans for English Regional Government*, London: Labour Party

Lambert, A. J. and Wood, C. M. (1990) 'UK implementation of the European directive on EIA', *Town Planning Review* 61: 247–62

Lampe, D. and Kaplan, M. (1999) *Resolving Land-Use Conflicts through Mediation: Challenges and Opportunities*, Boston, MA: Lincoln Institute of Land Policy

Land Capability Consultants (1990) Cost Effective Management of Reclaimed Derelict Sites (DoE), London: HMSO

Land Use Consultants (1986) *Channel Fixed Link: Environmental Appraisal of Alternative Proposals* (Department of Transport), London: HMSO

Land Use Consultants (1991) *Permitted Development Rights for Agriculture and Forestry* (DoE), London: HMSO

Land Use Consultants (1993) *Local Moves: The Funding and Formulation of Local Transport Policy*, London: Council for the Protection of Rural England

Land Use Consultants (1995a) *The Effectiveness of Planning Policy Guidance Notes*, London: HMSO

Land Use Consultants (1995b) *Planning Controls over Agricultural and Forestry Development and Rural Building Conversions* (DoE), London: HMSO

Land Use Consultants (1996) *Reclamation of Damaged Land for Nature Conservation* (DoE), London: HMSO

Land Use Consultants (1999) *Review of National Planning Policy Guidelines*, Edinburgh: The Scottish Office

Landry, C., Montgomery, J., and Worpole, K. (1989) *The Last Resort: Tourism, Tourist Employment and 'PostTourism' in the South East*, Bournes Green, Gloucester: Comedia

Lane, J. and Vaughan, S. (1992) *An Evaluation of the Impact of PPG 16 on Archaeology and Planning*, London: Pagoda Associates

Lane, P. and Peto, M. (1995) *Blackstone's Guide to the Environment Act 1995*, London: Blackstone

Larkham, P. J. (1990) 'The use and measurement of development pressure', *Town Planning Review* 61: 171–83

Larkham, P. J. (1995) *Patterns in the Designation of Conservation Areas*, Birmingham: School of Planning, University of Central England in Birmingham

Larkham, P. J. (1996) *Conservation and the City*, London: Routledge

Larkham, P. J. (1999a) 'Preservation, conservation and heritage: developing concepts and applications', in Cullingworth, J. B. (ed.) *British Planning: 50 Years of Urban and Regional Policy*, London: Athlone

Larkham, P. J. (1999b) 'Tensions in managing the suburbs: conservation versus change', *Area* 31: 359–71

Larkham, P. J. and Chapman, D. W. (1996) 'Article 4 Directions and development control: planning myths, present uses, and future possibilities', *Journal of Environmental Planning and Management* 39: 5–19

Larkham, P. J. and Jones, A. (1993) 'Conservation and conservation areas in the UK: a growing problem', *Planning Practice and Research* 8 (2) (April): 19–29

Larkham, P. J. and Lodge, J. (1999) *Do Residents Understand What Conservation Means?*, Working Paper Series no. 71, Birmingham University of Central England

Larkham, P. J., Jones, A., and Daniels, R. (1992) *The Character of Conservation Areas*, London: Royal Town Planning Institute

Lash, S., Szerszynski, B., and Wynne, B. (1996) *Risk, Environment and Modernity: Towards a New Ecology*, London: Sage

Lasok, D. (1994) *Law and Institutions of the European Communities* (sixth edition), London: Butterworths

Last, K. V. (1999) 'Protection of archaeological sites: an outline of the law', *Scottish Planning and Environment Law* 73: 60–1

Lavers, A. and Webster, B. (1994) 'Participation in the plan-making process: financial interests and professional representation', *Journal of Property Research* 11: 131–44

Law, C. M. (1992) 'Urban tourism and its contribution to economic regeneration', *Urban Studies* 29: 599–619

Law Commission (1993) *Administrative Law: Judicial Review and Statutory Appeals*, Law Commission Consultation Paper 126, London: HMSO

Lawless, P. (1986) *The Evolution of Spatial Policy: A Case Study of Inner Urban Policy in the United Kingdom 1968–1981*, London: Pion

Lawless, P. (1989) *Britain's Inner Cities* (second edition), London: Paul Chapman

Lawless, P. (1991) 'Urban policy in the Thatcher decade: English inner-city policy, 1979–90', *Environment and Planning C: Government and Policy* 9: 15–30

Lawless, P. (1994) 'Partnership in urban regeneration in the UK: the Sheffield central area study', *Urban Studies* 31: 1301–24

Lawless, P. (1996) 'The inner cities: towards a new agenda', *Town Planning Review* 67: 21–43

Lawless, P. (1998) *Unemployment and Social Exclusion: Landscapes of Labour Inequality*, London: Jessica Kingsley

Lawless, P. and Dabinett, G. (1995) 'Urban regeneration and transport investment: a research agenda', *Environment and Planning A* 27: 1029–48

Lawless, P. and Robinson, D. (2000) 'Inclusive regeneration? Integrating social and economic regeneration in English local authorities', *Town Planning Review* 71: 289–310

Laws, F. G. (1991) *Guide to the Local Government Ombudsman Service*, Harlow: Longman

Lawton, R. and Pooley, C. G. (1992) *The New Geography of London*, London: Edward Arnold

Laxton (1996) *Laxton's Guide to Single Regeneration Budgets*, Oxford: Heinemann

Layfield, F. (1992) 'The Environmental Protection Act 1990: the system of integrated pollution control', *Journal of Planning and Environment Law* 1992: 3–13

Layfield, F. (1996) 'The planning inquiry: an inspector's perspective', *Journal of Planning and Environment Law* 1996: 370–6

Leach, N. (1997) *Rethinking Architecture: A Reader in Cultural Theory*, London: Routledge

Leach, R. (1994) 'The missing dimension to the local government review', *Regional Studies* 28: 797–802

Leach, S. and Game, C. (1991) 'English metropolitan government since abolition: an evaluation of the abolition of the English metropolitan councils', *Public Administration* 69: 141–70

Leach, S. and Stewart, M. (1992) *Local Government: Its Role and Function*, York: Joseph Rowntree Foundation

Leach, S., Davis, H., Game, C., and Skelcher, C. (1991) *After Abolition: The Operation of the Post-1986 Metropolitan Government System*, Birmingham: Institute of Local Government Studies, University of Birmingham

Leach, S., Stewart, J., Spencer, K., Walsh, K., and Gibson, J. (1992) *The Heseltine Review of Local Government: A New Vision or Opportunities Missed?*, Birmingham: Institute of Local Government Studies, University of Birmingham

Leach, S., Stewart, J., and Walsh, K. (1994) *The Changing Organisation and Management of Local Government*, London: Macmillan

Leather, P. (1997) *The State of UK Housing: A Factfile on Dwelling Conditions*, Bristol: University of Bristol

Leather, P. and Mackintosh, S. (1998) *Housing Renewal: Policy and Practice*, London: UCL Press

Leather, P. and Morrison, T. (1997) *The State of UK Housing*, York: Policy Press

Lederman, P. B. and Librizzi, W. (1995) 'Brownfields remediation: available technologies', *Journal of Urban Technology* 2: 21–29

Ledgerwood, G. (1985) *Urban Innovation: The Transformation of the London Docklands 1968–84*, Aldershot: Gower

Lee, P. and Murie, A. (1999) *Literature Review of Social Exclusion*, Scottish Office, Central Research Unit

Lees, A. (1993) *Enquiry into the Planning System in North Cornwall District*, London: HMSO

LeGates, R. (ed.) (1998) *Early Urban Planning 1870–1940* (9 vols), London: Routledge

LeGates, R. and Stout, F. (eds) (1996) *The City Reader*, London: Routledge

Leitch Report (1977) *Report of the Advisory Committee on Trunk Road Assessment*, London: HMSO

Leung, H. L. (1979) *Redistribution of Land Values: A Re-examination of the 1947 Scheme*, Cambridge: Department of Land Economy, University of Cambridge

Leung, H. L. (1987) 'Developer behaviour and development control', *Land Development Studies* 4: 17–34

Lever, W. F. (1991) 'Deindustrialisation and the reality of the post-industrial city', *Urban Studies* 28: 983–99

Lever, W. F. (1992) 'Local authority responses to economic change in west central Scotland', *Urban Studies* 29: 935–48

Levett, R. (1993) *Agenda 21: A Guide for Local Authorities in the UK*, Luton: Local Government Management Board

Levett, R. (2000) 'What counts for quality of life', *EG* 6(3): 6–8

Levett, R. and Christie, I. (1999) *The Richness of Cities: Urban Policies in a New Landscape*, Comedia in association with Demos (Leicester: Econ Distribution)

Levin, P. H. (1976) *Government and the Planning Process: An Analysis and Appraisal of Government Decision-Making Processes with Special Reference to the Launching of New Towns and Town Development Schemes*, London: Allen & Unwin

Levin, P. H. (1979) 'Highway inquiries: a study in government responsiveness', *Public Administration* 57: 21–49

Levin, P. H. and Donnison, D. V. (1969) 'People and planning', *Public Administration* 13: 473–9

Levy, F., Meltsner, A. J., and Wildavsky, A. B. (1974) *Urban Outcomes: Schools, Streets, and Libraries*, Berkeley: University of California Press

Lewis, J. and Townsend, A. (1989) *The North–South Divide: Regional Change in Britain in the 1980s*, London: Paul Chapman

Lewis, J. P. (1974) *A Study of the Cambridge Sub-region*, London: HMSO

Lewis, N. (1992) *Inner City Regeneration: The Demise of Regional and Local Government*, Buckingham: Open University Press

Lewis, N. C. (1994) *Road Pricing: Theory and Practice*, London: Thomas Telford

Lewis, R. (1995) 'Contaminated land: the new regime of the Environment Act 1995', *Journal of Planning and Environment Law* 1995: 1087–96

Lewis, R. P. (1992) 'The Environmental Protection Act 1990: waste management in the 1990s: waste regulation and disposal', *Journal of Planning and Environment Law* 1992: 303–12

Lichfield, N. (1992) 'From planning gain to community benefit', *Journal of Planning and Environment Law* 1992: 1103–18

Lichfield, N. (1995) *Community Impact Evaluation: Principles and Practice*, London: UCL Press

Lichfield, N. and Darin-Drabkin, H. (1980) *Land Policy in Planning*, London: Allen & Unwin

Lindblom, C. E. (1959) 'The science of muddling through', *Public Administration Review* (Spring) (reprinted in Faludi, A. (1973) *A Reader in Planning Theory*, Oxford: Pergamon)

Liniado, M. (1996) *Car Culture and Countryside Change*, Cirencester: National Trust

Lipman, C. (1999) 'Difficult decisions in a rural balancing act' [N. Ireland] (together with 'How the Province deals with countryside issues') *Planning* (29 October): 10–11

Little, J. (1994a) *Gender, Planning and the Policy Process*, Oxford: Pergamon

Little, J. (1994b) 'Women's initiatives in town planning in England', *Town Planning Review* 65: 261–76

Little, J., Peake, L. and Richardson, P. (eds) (1988) *Women in Cities*, New York: New York University Press

Littlewood, S. and Whitney, D. (1996) 'Re-forming urban regeneration? The practice and potential of English Partnerships', *Local Economy* 11: 39–49

Llewelyn-Davies (1994) *Providing More Homes in Urban Areas*, Bristol: School for Advanced Urban Studies, University of Bristol

Llewelyn-Davies (1996) *The Re-use of Brownfield Land for Housing*, York: Joseph Rowntree Foundation

Llewelyn-Davies Planning (1997) *Sustainable Residential Quality: New Approaches to Urban Living*, London: Llewelyn-Davies Planning

Llewelyn-Davies (1998a) *Urban Living: Perceptions, Realities and Opportunities*, Winchester: Hampshire County Council

Llewelyn-Davies (1998b) *Planning and Development Briefs: A Guide to Better Practice*, London: Department of the Environment, Transport and the Regions

Llewelyn-Davies (2000) *Millennium Villages and Sustainable Communities*, London: Department of the Environment, Transport and the Regions

Llewelyn-Davies and JMP Consultants (1998) *Parking Standards in the South East*, London: Department of the Environment, Transport and the Regions

Llewelyn-Davies *et al.* (1994) *The Quality of London's Residential Environment*, London: London Planning Advisory Committee

Lloyd, H. (1998) Are current visitor numbers at historic properties sustainable? A case study of Bateman's, the home of Rudyard Kipling, London, The National Trust (at <www.ntenvironment. com>)

Lloyd, H. (1999) 'Are current visitor numbers at historic properties sustainable? A case study from Batemans, the home of Rudyard Kipling', *Views* 30 (Spring) (Also at <www.ntenvironment. com/html/building/papers/visitor1.htm>)

Lloyd, M. G. (1990) 'Simplified planning zones in Scotland: government failure or the failure of government?', *Planning Outlook* 33: 128–32

Lloyd, M. G. (1992) 'Simplified planning zones, land development, and planning policy in Scotland', *Land Use Policy* 9: 249–58

Lloyd, M. G. and Livingstone, L. H. (1991) 'Marine fish farming in Scotland: proprietorial behaviour and the public interest', *Journal of Rural Studies* 7: 253–63

Lloyd, M. G. and Rowan-Robinson, J. (1992) 'Review of strategic planning guidance in Scotland', *Journal of Environmental Planning and Management* 35: 93–9

Local, D. (1999) 'The new Catch 22 on need', *Town and Country Planning* 68: 111

Local Government Association and the Improvement and Development Agency (1997) *Best Value and Sustainable Development Guidance*, Luton: IdeA

Local Government Management Board (1992) *Citizens and Local Democracy: Charting a New Relationship*, Luton: LGMB

Local Government Management Board (1993a) *Local Agenda 21 UK: A Framework for Local Sustainability*, Luton: LGMB

Local Government Management Board (1993b) *Local Agenda 21: Principles and Process – A Step by Step Guide*, Luton: LGMB

Local Government Management Board and Royal Town Planning Institute (1993c) *Planning Staffs Survey 1992*, Luton: LGMB

Local Government Management Board (1995a) *Sustainable Settlements: A Guide for Planners, Designers and Developers* (Barton, H., Davis, G., and Guise, R.), Luton: LGMB

Local Government Management Board (1995b) *Indicators for Local Agenda 21*, Luton: LGMB

Lock, D. (1989) *Riding the Tiger: Planning the South of England*, London: Town and Country Planning Association

Lock, D. (1994) 'Keynote address', in Wood, M. (ed.) *Planning Icons: Myth and Practice* (Planning Law Conference), London: Sweet and Maxwell

Lock (1999) 'The new catch-22 on need', *Town and Country Planning* 68(4): 111

Loevinger, Rahder, B., and O'Neill, K. (1998) 'Women and planning: education for social change', *Planning Practice and Research* 13: 247–65

Loftman, P. and Beazley, M. (1998a) *Race, Equality and Planning*, London: Local Government Association

Loftman, P. and Beazley, M. (1998b) 'Racial equality and the planning agenda', *Town and Country Planning* 67: 326–7

Loftman, P. and Nevin, B. (1992) *Urban Regeneration and Social Equity: A Case Study of Birmingham 1986–1992*, Research Paper, Birmingham: Faculty of the Built Environment, University of Central England

Loftman, P. and Nevin, B. (1995) 'Prestige projects and urban regeneration in the 1980s and 1990s: a review of the benefits and limitations', *Planning Practice and Research* 10: 299–315

Lomas, J. (1994) 'The role of management agreements in rural environmental conservation', *Land Use Policy* 11: 119–23

London and South-East Regional Planning Conference: *see* SERPLAN

London Boroughs Association (1990) *Road Pricing for London*, London: LBA

London Boroughs Association (1992) *Out of Order: The 1987 Use Classes Order: Problems and Proposals*, London: LBA

London Boroughs' Disability Resource Team (1991) *Towards Integration: The Participation of Disabled People in Planning*, London: Disability Resource Team

London Economics (1992) *The Potential Role of Market Mechanisms in the Control of Acid Rain* (DoE), London: HMSO

London Planning Advisory Committee (1998a) *Sustainable Residential Quality: New Approaches to Urban Living*, London: LPAC

London Planning Advisory Committee (1998b) *Dwellings over and in Shops in London*, London: LPAC

London Planning Advisory Committee (1991) *London: World City Moving into the 21st Century*, London: HMSO

London Research Centre (1991) *Much Ado about Nothing: An Examination of the Potential for the Planning System to Secure Affordable Housing with Special Reference to Planning Agreements*, London: LRC

London Research Centre (1992) *Paris, London: A Comparison of Transport Systems*, London: LRC

London Transport Planning (1997) *Review of Office and Retail Parking Standards in London Unitary Development Plans*, London: Transport

London Women and Planning Group (1991) *Shaping Our Borough: Women and Unitary Development Plans*, London: Royal Town Planning Institute

Longley, P., Batty, M., Shepherd, J., and Sadler, G. (1992) 'Do green belts change the shape of urban areas? A preliminary analysis of the settlement geography of south east England', *Regional Studies* 26: 437–52

Loughlin, M. (1994) *The Constitutional Status of Local Government*, Research Report no. 3, London: Commission for Local Democracy

Loughlin, M., Gelfand, M. D., and Young, K. (eds) (1985) *Half a Century of Municipal Decline*, London: Allen & Unwin

Low, N. (1991) *Planning, Politics and the State: Political Foundations of Planning Thought*, London: Unwin Hyman

Lowe, P. and Flynn, A. (1989) 'Environmental politics and policy in the 1980s', in Mohan, J. (ed.) *The Political Geography of Contemporary Britain*, London: Macmillan

Lowe, P. and Goyder, J. (1983) *Environmental Groups in Politics*, London: Allen & Unwin

Lowe, P., Murdoch, J., Marsden, T., Munton, R., and Flynn, A. (1993) 'Regulating the new rural spaces: the uneven development of land', *Journal of Rural Studies* 9: 205–22

Lowe, R. and Bell, M. (1998) *Towards Sustainable Housing: Building Regulation for the 21st Century*, York: Joseph Rowntree Foundation

Lowndes, V. (1998) *Guidance on Enhancing Public Participation in Local Government*, London: Department of the Environment, Transport and the Regions

Lutyens, E. and Abercrombie, P. (1945) *A Plan for the City and County of Kingston upon Hull*, Hull: Brown

Lyddon, D. and Dal Cin, A. (1996) *International Manual of Planning Practice* (third edition), The Hague: International Society of City and Regional Planners

Lynch, P. (1999) 'New Labour and the English regional development agencies: devolution as evolution', *Regional Studies* 33: 73–8

McAllister, D. (ed.) (1996) *Partnership in the Regeneration of Urban Scotland* (Scottish Office), Edinburgh: HMSO

McAllister, A. and McMaster, R. (1999) *Scottish Planning Law: An Introduction* (second edition), Edinburgh: Butterworths

McAuslan, J. P. W. B. (1975) *Land, Law and Planning*, London: Weidenfeld & Nicolson

McAuslan, J. P. W. B. (1980) *The Ideologies of Planning Law*, Oxford: Pergamon

McAuslan, J. P. W. B. (1991) 'The role of courts and other judicial type bodies in environmental management', *Journal of Environmental Law* 3: 195–208

McBride, J. (1973) 'The urban aid programme: is it running out of cash?', *Quest* 16 March

McCaig, E., Henderson, C., and MVA Consultancy (1995) *Sustainable Development: What it Means to the General Public*, Edinburgh: Scottish Office Central Research Unit

McCarthy, J. (1995) 'The Dundee waterfront: a missed opportunity for planned regeneration', *Land Use Policy* 12: 307–19

McCarthy, J. (1999) 'Urban regeneration in Scotland: an agenda for the Scottish Parliament', *Regional Studies* 33: 559–66

McCarthy, J. and Newlands, D. (eds) (1999) *Governing Scotland: Problems and Prospects – The Economic Impact of the Scottish Parliament*, Aldershot: Ashgate

McCarthy, J., Lloyd, G., and Fernie, K. (1999) 'Planning for social inclusion in Scotland', *Town and Country Planning* 68: 310–11

McCarthy, P. and Harrison, T. (1995) *Attitudes to Town and Country Planning* (DoE), London: HMSO

McClintock, H. (ed.) (1992) *The Bicycle and City Traffic*, London: Belhaven Press

McConville, J. and Sheldrake, J. (eds) (1995) *Transport in Transition: Aspects of British and European Experience*, Aldershot: Avebury

McCormick, J. (1991) *British Politics and the Environment*, London: Earthscan

McCormick, J. (1995) *The Global Environmental Movement*, Chichester: Wiley

McCrone, D. (1997) *Land, Democracy and Culture in Scotland*, Perth: Rural Forum

McCrone, G. (1991) 'Urban renewal: the Scottish experience', *Urban Studies* 28: 919–38

McCrone, G. and Stephens, M. (1995) *Housing Policy in Britain and Europe*, London: UCL Press

MacDonald, R. and Thomas, H. (1997) *Nationality and Planning in Scotland and Wales*, Cardiff: University of Wales Press

McDougall, G. (1993) *Planning Theory: Prospects for the 1990s*, Aldershot: Avebury

MacEwen, A. and MacEwen, M. (1982) *National Parks. Conservation or Cosmetics?*, London: Allen & Unwin

MacEwen, A. and MacEwen, M. (1987) *Greenprints for the Countryside? The Story of Britain's National Parks*, London: Allen & Unwin

MacEwen, M. (1973) *Crisis in Architecture*, London: Royal Institute of British Architects

MacEwen, M. and Sinclair, G. (1983) *New Life for the Hills*, London: Council for National Parks

MacFarlane, R. (1998) 'What – or who – is rural Britain?', *Town and Country Planning* 67: 184–8

McFarlane, R. (1993a) *Community Involvement in City Challenge: A Good Practice Guide*, London: National Council for Voluntary Organisations

McFarlane, R. (1993b) *Community Involvement in City Challenge: A Policy Report*, London: National Council for Voluntary Organisations

McGill, G. (1995) *Building on the Past: A Guide to the Archaeology and Development Process*, London: Spon

MacGregor, B. and Ross, A. (1995) 'Master or servant? The changing role of the development plan in the British planning system', *Town Planning Review* 66: 41–59

MacGregor, S. and Pimlott, B. (eds) (1990) *Tackling the Inner Cities: The 1980s Revisited, Prospects for the 1990s*, Oxford: Clarendon Press

McHarg, I. L. (1967/1992) *Design with Nature*, New York: Wiley

McIntosh Report (1999) *Moving Forward: Local Government and the Scottish Parliament* Edinburgh: Scottish Office

Mackay, D. (1995) *Scotland's Rural Land Use Agencies*, Aberdeen: Scottish Cultural Press

McKay, D. H. and Cox, A. W. (1979) *The Politics of Urban Change*, London: Croom Helm

Mackie, P. J. (1980) 'The new grant system for local transport: the first five years', *Public Administration* 59: 187–206

McKinnon, A. (1999) 'First steps: the DETR Sustainable Distribution document', *Town and Country Planning* 68: 257

McKinsey Global Institute (1996) *Driving Productivity and Growth in the UK Economy*, London: McKinsey

Mackintosh, M. (1992) 'Partnership: issues of policy and negotiation', *Local Economy* 3: 210–24

Mackintosh, S. and Leather, P. (1992) *Home Improvement under the New Regime*, Bristol: School for Advanced Urban Studies, University of Bristol

McLaren, D., Bullock, S., and Yousuf, N. (1998) *Tomorrow's World: Britain's Share in a Sustainable Future*, London: Earthscan

Maclennan, D. (1986) *The Demand for Housing: Economic Perspectives and Planning Practices*, Edinburgh: Scottish Development Department

Maclennan, D. (1989) 'Housing in Scotland 1977–1987', in Smith, M. E. H. (ed.) *Guide to Housing: Main Changes in Housing Law* (third edition), London: Housing Centre

McLoughlin, J. B. (1969) *Urban and Regional Planning: A Systems Analysis*, Oxford: Pergamon

McMaster, R. F. (1995) *Results of Royal Town Planning Institute Survey of Local Planning Authorities' Views on Local Government Re-organisation: A Mixed Picture*, Edinburgh: Royal Town Planning Institute in Scotland

Macnaghten, P., Grove-White, R., Jacobs, M., and Wynne, B. (1995) *Public Perceptions and Sustainability in Lancashire* (Report by the Centre for the Study of Environmental Change, Lancaster University), Preston: Lancashire County Council

McQuail, P. (1994) *Origins of the DoE*, London: Department of the Environment

McQuail, P. (1995) *A View from the Bridge*, London: Department of the Environment

Macrory, R. (1992) *Campaigners' Guide to Using EC Environmental Law*, London: Council for the Protection of Rural England

McSheffrey, G. (2000) *Planning Derry: Planning and Politics in Northern Ireland*, Liverpool University Press

Madanipour, A. (1996) *Design of Urban Space: An Inquiry into a Social-Spatial Process*, Chichester: Wiley

Madden, M. (1987) 'Planning and ethnic minorities: an elusive literature', *Planning Practice and Research* no. 2 (1987): 29–32

Maddison, D., Pearce, D., Johansson, O., Calthorpe, E., Litman, T., and Verhoef, E. (1996) *The True Costs of Road Transport*, London: Earthscan

Malbert, B. (1998) *Urban Planning Participation: Linking Practice and Theory*, Gothenburg: Department of Urban Design and Planning, University of Technology

Maloney, W. A. and Richardson, J. J. (1995) *Managing Policy Change in Britain: The Politics of Water*, Edinburgh: Edinburgh University Press

Malpass, P. (1994) 'Policy making and local governance: how Bristol failed to secure City Challenge funding (twice)', *Policy and Politics* 22: 301–12

Malpass, P. (1996) 'The unravelling of housing policy in Britain', *Housing Studies* 11: 459–70

Malpass, P. and Means, R. (eds) (1993) *Implementing Housing Policy*, Buckingham: Open University Press

Mandelker, D. R. (1962) *Green Belts and Urban Growth*, Madison, WI: University of Wisconsin Press

Mandelson, P. and Liddle, R. (1996) *The Blair Revolution*, London: Faber

Manley, J. (1987) 'Archaeology and planning: a Welsh perspective', *Journal of Planning and Environment Law* 1987: 466–84 and 552–63

Manley, S. and Guise, R. (1998) 'Conservation in the Built Environment' in Greed, C. and Roberts, M. (eds) *Introducing Urban Design: Interventions and Responses*, London: Addison Wesley Longman

Manning, P. K. (1989) 'Managing risk: managing uncertainty in the British nuclear installations inspectorate', *Law and Policy* 11: 350–69

Marquand, D. (1998) *Must Labour Win?* (9th ESRC Annual Lecture), Swindon: Economic and Social Research Council

Marriott, O. (1967) *The Property Boom*, London: Hamilton

Marris, P. (1982) *Community Planning and Conceptions of Change*, London: Routledge & Kegan Paul

Marsden, T. (1999) *Rural Sustainability*, London: Town and Country Planning Association

Marsden, T., Murdoch, J., Lowe, P., Munton, R. J. C., and

Flynn, A. (1993) *Constructing the Countryside: An Approach to Rural Development*, London: UCL Press

Marshall, J. N. and Wood, P. A. (1995) *Services and Space: Key Aspects of Urban and Regional Development*, Harlow: Longman

Marshall, R. (1988) 'Agricultural policy development in Britain', *Town Planning Review* 59: 419–35

Marshall, S. (1999) 'Restraining mobility while maintaining accessibility: an impression of the city of sustainable growth', *Built Environment* 25: 168–79

Marshall, T. (1991) *Regional Planning in England and Germany*, Oxford: School of Planning, Oxford Polytechnic

Marshall, T. (1995) *Clearing an Industrial Area: Collingdon Road and the Cardiff Bay Development Corporation 1988–1994*, Oxford: School of Planning, Oxford Polytechnic

Marte, L. (1994) *Ecology and Society: An Introduction*, London: Policy Press

Martin, L. R. G. (1989) 'The important published literature of British planners', *Town Planning Review* 60: 441–57

Martin, S. (1989) 'New jobs in the inner city: the employment impacts of projects assisted under the urban development grant programme', *Urban Studies* 26: 627–38

Martin, S. (1995) 'Partnerships for local environment action: observations of the first two years of Rural Action for the Environment', *Journal of Environmental Planning and Management* 38: 149–65

Martin, S. and Pearce, G. (1992) 'The internationalisation of local authority economic development strategies: Birmingham in the 1980s', *Regional Studies* 26: 499–509

Martin, S. J., Ticker, M. J., and Bovaird, A. G. (1990) 'Rural development programmes in theory and practice', *Regional Studies* 24: 268–76

Marvin, S. and Guy, S. (1997) 'Infrastructure provision, development processes and the co-production of environmental value', *Urban Studies* 34: 2023–36

Marvin, S. and Slater, S. (1996) *'Holes in the Road': Roads and Utilities in the 1990s*, Newcastle upon Tyne: Department of Town and Country Planning, University of Newcastle

Marwick, A. (1970) *Britain in the Century of Total War: War, Peace and Social Change 1900–1967*, Harmondsworth: Penguin

Masser, I. (1983) *Evaluating Urban Planning Efforts*, Aldershot: Gower

Matarosso, F. (1995) *Spirit of Place: Redundant Churches as Urban Resources*, Bournes Green, Gloucester: Comedia

Mather, A. S. (1991) 'The changing role of planning in rural land use: the example of afforestation in Scotland', *Journal of Rural Studies* 7: 299–309

Mather, A. S. (1999) 'Reducing the need to travel in the City of Bristol: promoting bus use through complementary measures', *Built Environment* 25: 94–105

Mather, G. (1988) *Paying for Planning*, London: Institute of Economic Affairs

Maunder, W. J. (1994) *Dictionary of Global Climate Change* (second edition), London: UCL Press

Mawson, J. (ed.) (1995) *The Single Regeneration Budget: The Stocktake*, Birmingham: Centre for Urban and Regional Studies, University of Birmingham

Mawson, J. (1996) 'The re-emergence of the regional agenda in the English regions: new patterns of urban and regional governance', *Local Economy* 10: 300–26

May, A. D. (1986) 'Traffic restraint: a review of the alternatives', *Transportation Research A* 20A: 109–21

Meadows, D. H., Meadows, D. L., and Randers, J. (1992) *Beyond the Limits: Global Collapse or a Sustainable Future*, London: Earthscan

Meadows, P. (ed) (1996) *Work Out – or Work In?: Contributions to the Debate on the Future of Work*, York: Joseph Rowntree Foundation

Meny, Y., Muller, P., and Quermonne, J.-L. (eds) (1996) *Adjusting to Europe: The Impact of the European Union on National Institutions and Policies*, London: Routledge

Meyerson, M. (1956) 'Building the middle-range bridge for comprehensive planning', *Journal of the American Institute of Planners* 22 (2) (reprinted in Faludi, A. (1973) *A Reader in Planning Theory*, Oxford: Pergamon)

Meynell, A. (1959) 'Location of industry', *Public Administration* 37: 9

Midwinter, A. (1995) *Local Government in Scotland: Reform or Decline?* London: Macmillan

Millar, S. (1999) 'Beauty spot bypass proving failure', *Guardian* (12 July): 3

Miller, C. (2000) *Planning and Pollution Revisited: A Background Paper for the Royal Commission on Environmental Pollution Review of Environmental Planning*, London: Royal Commission on Environmental Pollution

Millichap, D. (1995a) 'Law, myth and community: a reinterpretation of planning's justification and rationale', *Planning Perspectives* 10: 279–93

Millichap, D. (1995b) *The Effective Enforcement of Planning Controls* (second edition), London: Butterworths

Mills, E. S. and Hamilton, B. W. (1994) *Urban Economics* (fifth edition), New York: HarperCollins

Milner Holland Report (1965) *Report of the Committee on Housing in Greater London*, Cmnd 2605, London: HMSO

Milton, K. (1991) 'Interpreting environmental policy: a social scientific approach', in Churchill, R., Warren, L. M., and Gibson, J. (eds) *Law, Policy and the Environment*, Oxford: Blackwell

Milton Keynes Development Corporation (1992) *The Milton Keynes Planning Manual*, Chesterton Consulting: distributed by Keyne Creations, Milton Keynes

Minay, C. L. W. (ed.) (1992) 'Developing regional planning guidance in England and Wales: a review symposium', *Town Planning Review* 63: 415–34

Minhinnick, R. (1993) *A Postcard Home*, Llandysul, Dyfed: Gomer Press

Mishan, E. J. (1976) *Elements of Cost–Benefit Analysis*, London: Allen & Unwin

Mitchell, C. (1995) *Renewable Energy in the UK: Financing Options for the Future*, London: Council for the Protection of Rural England

Mitchell Report (1991) *Railway Noise and the Insulation of Dwellings: Report of the Committee to Recommend a National Noise Insulation Standard for New Railway Lines*, London: HMSO

Mogridge, M. J. H. (1990) *Travel in Towns*, London: Macmillan

Mohan, J. (ed.) (1989) *The Political Geography of Contemporary Britain*, London: Macmillan

Mole, D. (1996) 'Planning gain after the *Tesco* case', *Journal of Planning and Environment Law* 1996: 183–93

Möltke, K. von (1988) 'The *vorsorgerprinzip* in West German environmental policy' (printed as an appendix to the 12th Report of the Royal Commission on Environmental Pollution, 1991), *Best Practicable Environmental Option*, London: HMSO

Monk, S., Pearce, B. J., and Whitehead, C. M. E. (1996) 'Land-use planning, land supply, and house prices', *Environment and Planning A* 28: 495–511

Montanari, A. and Williams, A. M. (1995) *European Tourism: Regions, Spaces and Restructuring*, Chichester: Wiley

Montgomery, J. and Thornley, A. (eds) (1990) *Radical Planning Initiatives: New Directions for Urban Planning in the 1990s*, Aldershot: Gower

Moore, B. and Townroe, P. (1990) *Urban Labour Markets: Reviews of Urban Research* (DoE), London: HMSO

Moore, B., Rhodes, J., and Tyler, P. (1986) *The Effects of Government Regional Economic Policy* (DTI), London: HMSO

Moore, R. (ed.) (1995) *Soil Diversity: A Literature Review*, Perth: Scottish Natural Heritage

Moore, T. and Thorsnes, P. (1994) *The Transportation/Land Use Connection: A Framework for Practical Policy*, Chicago: American Planning Association

Moore, V. (2000) *A Practical Approach to Planning Law* (seventh edition), London: Blackstone Press

Morgan, K. (1994) *The Fallible Servant: Making Sense of the Welsh Development Agency*, Cardiff: Department of City and Regional Planning, University of Cardiff

Morgan, K. (1995) 'Reviving the valleys? Urban renewal and governance structures in south Wales', in Hambleton, R. and Thomas, H. (eds) *Urban Policy Evaluation: Challenge and Change*, London: Paul Chapman

Morgan, P. and Nott, S. (1995) *Development Control: Law, Policy and Practice* (second edition), London: Butterworths

Morphet, J. (1993a) *Greening Local Authorities: Sustainable Development at the Local Level*, Harlow: Longman

Morphet, J. (1993b) *Towards Sustainability: A Guide for Local Authorities*, Luton: Local Government Management Board

Morphet, J. and Hams, T. (1994) 'Responding to Rio: the local authority approach', *Journal of Environmental Planning and Management* 37: 479–86

Morris, A. E. J. (1994) *History of Urban Form: Before the Industrial Revolution* (third edition), Harlow: Longman

Morris, P. and Therivel, R. (eds) (1995) *Methods of Environmental Impact Assessment*, London: UCL Press

Morris, R. (1998) 'Gypsies and the planning system', *Journal of Planning and Environment Law* 1998: 635–43

Morrison, N. and Pearce, B. (2000) 'Developing indicators for evaluating the effectiveness of the UK land use planning system', *Town Planning Review* 71: 191–211

Morton, D. (1991) 'Conservation areas: has saturation point been reached?', *The Planner* 7 (17) (17 May): 5–8

Moseley, M. (1997) 'Parish appraisals as a tool of rural community development: an assessment of the British experience', *Planning Practice and Research* 12: 197–212

Moussis, N. (1999) *Access to European Union: Law, Economics, Policies*, Genval: Euroconfidential

Mowat, C. L. (1955) *Britain between the Wars 1918–1940*, London: Methuen

Moxen, J., McCulloch, A., Williams, D., and Baxter, S. (1995) *Accessing Environmental Information in Scotland* (Scottish Office Central Research Unit), Edinburgh: HMSO

Muchnick, D. M. (1970) *Urban Renewal in Liverpool*, Occasional Papers on Social Administration 33, London: Bell

Murdoch, J. and Marsden, T. (1994) *Reconstructing Rurality: The Changing Countryside in an Urban Context*, London: UCL Press

Murray, M. (1991) *The Politics and Pragmatism of Urban Containment: Belfast since 1940*, Aldershot: Avebury

Murray, M. R. and Greer, J. V. (1993) *Rural Planning and Development in Northern Ireland*, Aldershot: Avebury

MVA Consultancy (1995) *The London Congestion Charging Research Programme: Principal Findings* (Government Office for London), London: HMSO

Myerson, G. and Rydin, Y. (1996) 'Sustainable development: the implications of the global debate for land use planning', in Buckingham-Hatfield, S. and Evans, B. (eds) *Environmental Planning and Sustainability*, Chichester: Wiley

Mynors, C. (1992) *Planning Control and the Display of Advertisements*, London: Sweet and Maxwell

Mynors, C. (1993) 'The extent of listing', *Journal of Planning and Environment Law* 1993: 99–111

Mynors, C. (1998) *Listed Buildings, Conservation Areas and Monuments*, London: Sweet and Maxwell

Nadin, V. (1992a) 'Consultation by consultants', *Town and Country Planning* 61: 272–4

Nadin, V. (1992b) 'Local planning: progress and prospects', *Planning Practice and Research* 7: 27–32

Nadin, V. (1998a) 'Planning and the UK Presidency', *Town and Country Planning* 67: 60–3

Nadin, V. (1998b) UK Consults on the ESDP, *Town and Country Planning* 67: 67

Nadin, V. (1999) 'British planning in its European context', in Cullingworth, B. (ed.) *British Planning: 50 Years of Urban and Regional Policy*, London: Athlone

Nadin, V. and Daniels, R. (1992) 'Consultants and development plans', *The Planner* 78: 10–12

Nadin, V. and Doak, J. (eds) (1991) *Town Planning Responses to City Change*, Aldershot: Gower

Nadin, V. and Jones, S. (1990) ' A profile of the profession', *The Planner* 76: 13–24

Nadin, V. and Seaton, K. (eds) (2000) *Subsidiarity and Proportionality in Spatial Planning Activities in the European Union: Supplementary Volume*, Faculty of the Built Environment Working Paper 60, University of the West of England, Bristol

Nadin, V. and Shaw, D. (1998a) 'Promoting transnational planning', *Town and Country Planning* 67: 70–2

Nadin, V. and Shaw, D. (1998b) 'Transnational spatial planning in Europe: the role of Interreg IIc in the UK', *Regional Studies* 32: 281–99

Nadin, V. and Shaw, D. (1999) *Subsidiarity and Proportionality in Spatial Planning Activities in the European Union*, London: DETR

Nadin, V. and Brown, C. (1999) *The Management of the Interreg I and II Transnational Spatial Planning Initiatives*, Report for Kent County Council and DG Regio European Commission, Bristol: University of the West of England, Faculty of the Built Environment

Nadin, V., Cooper, S., Shaw, D., Westlake, T., and Hawkes, P. (1997) *The EU Compendium of Spatial Planning Systems and Policies*, Luxembourg: Office for the Official Publications of the European Communities

Nathan, M., Roberts, P., Ward, M. and Garside, R. (1999) *Strategies for Success?: A First Assessment of Regional Development Agencies' Draft Regional Economic Strategies*, Manchester: Centre for Local Government Economic Strategies

National Committee for Commonwealth Immigrants (1967) *Areas of Special Housing Need*, London: NCCI

National Farmers' Union (1994) *Real Choices: Report by the Long Term Strategy Group*, London: NFU

National Farmers' Union (1995a) *Taking Real Choices Forward*, London: NFU

National Farmers' Union (1995b) *The Rural White Paper: Submission of the National Farmers' Union of England and Wales*, London: NFU

National Housing Federation (1999) *Closing Doors: Examining the Lack of Housing Choices in London*, London: National Housing Federation

National Housing Forum (1996) *More than Somewhere to Live*, London: National Housing Forum

National Retail Planning Forum (1999) *A Bibliography of Retail Planning*, London: NRPF

National Society for Clean Air and Environmental Protection (1995) *1995 Pollution Handbook*, Brighton: NSCA

National Society of Allotment and Leisure Gardeners and Anglia Polytechnic University (1997) *English Allotments Survey*, NSALG

Neill, W. J. V., Fitzsimons, D.S., and Murtagh, B. (1995) *Reimaging the Pariah City: Urban Development in Belfast and Detroit*, Aldershot: Avebury

Nelson, B. F. and Stubb, C.-G. (1998) *The European Union: Readings on the Theory and Practice of European Integration* (second edition), London: Macmillan

Nettlefold, J. S. (1914) *Practical Town Planning*, London: St Catherine's Press

Nevin, B. and Murie, A. (1997) *Beyond a Halfway Housing Policy: Local Strategies for Regeneration*, London: Institute for Public Policy Research

Nevin, B. and Shiner, P. (1995) 'Community regeneration and empowerment: a new approach to partnership', *Local Economy* 9: 308–22

Nevin, M. and Abbie, L. (1993) 'What price roads? Practical issues in the introduction of road-user charges in historic cities in the UK', *Transport Policy* 1: 68–73

Newberry, D. (1990) 'Pricing and congestion: economic principles relevant to pricing roads' *Oxford Review of Economic Policy* 5: 22–38

Newberry, D. (1995) 'Royal Commission Report on Transport and the Environment: an economic critique', *Economic Journal* 105 (432): 1258–72

Newby, H. (1985) *Green and Pleasant Land: Social Change in Rural England* (second edition), London: Wildwood House

Newby, H. (1986) 'Locality and rurality: the restructuring of rural social relations', *Regional Studies* 20: 209–16

Newby, H. (1990) 'Ecology, amenity, and society', *Town Planning Review* 61: 3–20

Newby, H. (1991) *The Future of Rural Society*, Swindon: Economic and Social Research Council (unpublished mimeo)

Newman, P. (1995) 'The politics of urban redevelopment in London and Paris', *Planning Practice and Research* 10: 15–23

Newman, P. (2000) 'Changing patterns of regional government in the EU, *Urban Studies* 36: 895–908

Newman, P. and Thornley, A. (1996) *Urban Planning in Europe: International Competition, National Systems, and Planning Projects*, London: Routledge

Newman, P. W. G. and Kenworthy, J. R. (1989) *Cities and Automobile Dependence: A Sourcebook*, Aldershot: Gower

Newman, P. W. G. and Kenworthy, J. R. (1996) 'The land use–transport connection', *Land Use Policy* 13: 1–22

Newson, M. (ed.) (1992) *Managing the Human Impact on the Natural Environment: Patterns and Processes*, Chichester: Belhaven

Nicol, C. and Hooper, A. (1999) 'Contemporary change and the housebuilding industry', *Housing Studies* 14: 57–76

Nijkamp, P. (1993) 'Towards a network of regions: The United States of Europe' *European Planning Studies* 1: 149–68

Niner, P. (1999) *Insights into Low Demand for Housing*, Foundations Report 739, York: Joseph Rowntree Foundation

Noble, B. (1999) 'Walking and cycling in Great Britain', in *Transport Trends 1999* (DETR), London: The Stationery Office

Noise Review Working Party (1990) *Report of the Noise Review Working Party*, London: HMSO

Nolan Report (1995) *First Report of the Committee on Standards in Public Life*, Cm 2850, London: HMSO

Nolan Report (1997) *Third Report of the Committee on Standards in Public Life: Standards of Conduct in Local Government*, Cm 3702, London: The Stationery Office

Northcott, J. (1991) *Britain in Europe in 2010*, London: Policy Studies Institute

Norton, D. M. (1993) 'Conservation areas in an era of plan led planning', *Journal of Planning and Environment Law* 1993: 211–13

Nuffield Report (1986) *Town and Country Planning*, London: Nuffield Foundation

Nugent, N. (1999) *The Government and Politics of the European Community* (fourth edition), London: Macmillan

Oatley, N. (1995) 'Competitive urban policy and the regeneration game', *Town Planning Review* 66: 1–14

Oatley, N. (ed.) (1998) *Cities, Economic Competition and Urban Policy*, London: Paul Chapman

Oatley, N. (2000) 'New Labour's approach to age-old problems: renewing and revitalising poor neighbourhoods: the national strategy for neighbourhood renewal', *Local Economy* 15: 86–97

Oatley, N. and Lambert, C. (1995) 'Evaluating competitive urban policy: the City Challenge initiative', in Hambleton, R. and Thomas, H. (eds) *Urban Policy Evaluation: Challenge and Change*, London: Paul Chapman

Oc, T. and Tiesdell, S. (1991) 'The London Docklands Development Corporation 1981-1991: a perspective on the management of urban regeneration', *Town Planning Review* 62: 311–30

Oc, T. and Tiesdell S. (1998) 'City centre management and safer city centres: approaches in Coventry and Nottingham', *Cities* 15: 85–103

Oc, T., Tiesdell, S., and Moynihan, D. (1997) 'The death and life of City Challenge: the potential for lasting impacts in a limited life urban regeneration initiative', *Planning Practice and Research* 12: 367–81

Ofori, S. (1994) 'Urban policy and environmental regeneration in two Scottish peripheral estates', *Environmentalist* 14: 283–96

Ogden, P. (1992) *Update: London Docklands: The Challenge of Development*, Cambridge: Cambridge University Press

Ogilvie, W. (1997) *Birthright in Land*, London: Othila

Oliver, D. and Waite, A. (1989) 'Controlling neighbourhood noise: a new approach', *Journal of Environment Law* 1: 173–91

O'Neill, J. (1999) Moving and shaking: inquiries and hearings, *Planning Inspectorate Journal* 16 (Summer): 18–20

Organisation for Economic Co-operation and Development: *see* section on Official Publications

O'Riordan, T. (1985) 'What does sustainability really mean? Theory and development of concepts of sustainability', in *Sustainable Development in an Industrial Economy: Proceedings of a Conference*, Queen's College, Cambridge, UK Centre for Economic and Environmental Development

O'Riordan, T. (1992) 'The environment', in Cloke, P. (ed.) *Policy and Change in Thatcher's Britain*, Oxford: Pergamon

O'Riordan, T. (1997) *Ecotaxation*, London: Routledge

O'Riordan, T. and Cameron, J. (eds) (1994) *Interpreting the Precautionary Principle*, London: Earthscan

O'Riordan, T. and D'Arge, R. C. (1979) *Progress in Resource Management and Environmental Planning*, New York: Wiley

O'Riordan, T. and Voisey, H. (1997) 'The political economy of sustainable development', *Environmental Politics* 6: 1–23

O'Riordan, T. and Weale, A. (1989) 'Administrative reorganization and policy change: the case of Her Majesty's Inspectorate of Pollution', *Public Administration* 67: 277–94

Osborn, F. J. (1969) *Green Belt Cities*, London: Evelyn, Adams & Mackay

Osborne, D. and Gaebler, T. (1992) *Reinventing Government: How the Entrepreneurial Spirit is Transforming the Public Sector*, New York: Addison Wesley

Osborne, S. P. and Tricker, M. (2000) 'Village appraisals: a tool for sustainable community development in the UK', *Local Economy*, February: 347–58

Osmond, J. (1977) *Creative Conflict: The Politics of Welsh Devolution*, London: Routledge

Osmotherly, E. B. C. (1995) *Guide to the Local Ombudsman Service* (looseleaf; with updating service), London: Pitman

O'Toole, M. (1996) *Regulation Theory and the New British State: The Impact of Locality on the Urban Development Corporation 1987–1994*, Aldershot: Avebury

Ove Arup and University of Reading (1997) *Planning Policy Guidance on Transport (PPG 13): Implementation 1994–1996*, London: HMSO

Ove Arup Economics and Planning (1998) *Control of Outdoor Advertisements: Fly-Posting* (DETR), London: The Stationery Office

Ove Arup Economics and Planning (1999) *Implementation of PPG 13*, London: Ove Arup

Owen, S. (1991) *Planning Settlements Naturally*, Chichester: Packard

Owen, S. (1995) 'Local distinctiveness in villages', *Town Planning Review* 66: 143–61

Owen, S. (1996) 'Sustainability and rural settlement planning', *Planning Practice and Research* 11: 37–47

Owens, S. (1984) 'Energy and spatial structure: a rural example', *Environment and Planning A* 16: 1319–37

Owens, S. (1985) 'Potential energy planning conflicts in the UK', *Energy Policy* 13: 546–58

Owens, S. (1986a) 'Strategic planning and energy conservation', *Town Planning Review* 57: 69–86

Owens, S. (1986b) *Energy, Planning and Urban Form*, London: Pion

Owens, S. (1989) 'Integrated pollution control in the United Kingdom: prospects and problems', *Environment and Planning C: Government and Policy* 7: 81–91

Owens, S. (1990a) 'Land use for energy efficiency', in Cullingworth, J. B. (ed.) *Energy, Land, and Public Policy*, New Brunswick, NJ: Transaction Publishers

Owens, S. (1990b) 'The unified pollution inspectorate and best practicable environmental option in the United Kingdom', in Haigh, N. and Irwin, F. (eds) *Integrated Pollution Control in Europe and North America*, Bonn: Institute for European Environmental Policy, and Washington, DC: Conservation Foundation

Owens, S. (1991) *Energy Conscious Planning*, London: Council for the Protection of Rural England

Owens, S. (1994a) 'Land, limits and sustainability: a conceptual framework and some dilemmas for the planning system', *Transactions of the Institute of British Geographers* NS 19: 439–56

Owens, S. (1994b) 'Can land use planning produce the ecological city?', *Town and Country Planning* 63: 170–3

Owens, S. (1998) 'Better for everyone?', *Town and Country Planning* 67: 329–31

Owens, S. and Cope, D. (1992) *Land Use Planning Policy and Climate Change* (DoE), London: HMSO

Owens, S. and Cowell, R. (1996) *Rocks and Hard Places: Mineral Resource Planning and Sustainability*, London: Council for the Protection of Rural England

Oxford Retail Group (1995) *The Implementation of PPG 6*, Oxford: Oxford Institute of Retail Management, Templeton College

PA Cambridge Economic Consultants (1990b) *An Evaluation of Garden Festivals* (DoE), London: HMSO

PA Cambridge Economic Consultants (1990a) *Indicators of Comparative Regional–Local Economic Performance and Prospects* (DoE), London: HMSO

PA Cambridge Economic Consultants (1992) *An Evaluation of the Belfast Action Team Initiative*, Belfast: DoENI

PA Cambridge Economic Consultants (1987) *An Evaluation of the Enterprise Zone Experiment* (DoE), London: HMSO

PA Cambridge Economic Consultants (1995) *Final Evaluation of Enterprise Zones* (DoE), London: HMSO

PA Consulting Group (1999) *Interim Evaluation of English Partnerships: Final Report*, London: DETR

Pacione, M. (1990) 'Development pressure in the metropolitan fringe', *Land Development Studies* 7: 69–82

Pacione, M. (1991) 'Development pressure and the production of the built environment in the urban fringe', *Scottish Geographical Magazine* 107: 162–9

Pacione, M. (1995) *Glasgow: The Socio-spatial Development of the City*, Chichester: Wiley

Pacione, M. (1997) 'The urban challenge: how to bridge the great divide' in Pacione, M. (ed.) *Britain's Cities: Geographies of Division in Urban Britain*, London: Routledge

Pack, C. and Glyptis, S. (1989) *Developing Sport and Leisure* (DoE), London: HMSO

Paddison, R., Money, J., and Lever, B. (eds) (1993) *International Perspectives in Urban Studies*, London: Jessica Kingsley

Paddison, R., Money, J., and Lever, B. (eds) (1994) *International Perspectives in Urban Studies 2*, London: Jessica Kingsley

Paddison, R., Money, J., and Lever, B. (eds) (1995) *International Perspectives in Urban Studies 3*, London: Jessica Kingsley

Paddison, R., Money, J., and Lever, B. (eds) (1996) *International Perspectives in Urban Studies 4*, London: Jessica Kingsley

Page, S. (1995) *Urban Tourism*, London: Routledge

Pahl, R. E. (1970) *Whose City? and Other Essays in Sociology and Planning*, Harlow: Longmans

Painter, J. (1992) 'The culture of competition', *Public Policy and Administration* 7: 58–68

Pantazis, C. and Gordon, D. (eds) (2000) *Tackling Inequalities: Where are We Now and What Can be Done?*, Bristol: Policy Press

Parfect, M. and Power, G. (1997) *Planning for Urban Quality: Urban Design in Towns and Cities*, London: Routledge

Paris, C. (ed.) (1982) *Critical Readings in Planning* Theory, Oxford: Pergamon

Park, P. (2000) 'An evaluation of the landfill tax two years on', *Journal of Planning and Environment Law* (January): 3–13

Parker, H. R. (1954) 'The financial aspects of planning legislation', *Economic Journal* 64: 82–6

Parkhurst, G. (1995) 'Park and ride: could it lead to an increase in car traffic?', *Transport Policy* 2: 15–23

Parkinson, M. (1985) *Liverpool on the Brink*, Hermitage, Berkshire: Policy Journals Publishers

Parkinson, M. (1989) 'The Thatcher Government's urban policy 1979–1989', *Town Planning Review* 60: 421–40

Parkinson, M. (1998) *Combating Social Exclusion: Lessons from Area Based Programmes in Europe*, York: Policy Press

Parkinson, M. and Robson, B. (2000) *Urban Regeneration Companies: A Process Evaluation*, London: Department of the Environment, Transport and the Regions

Parry, M. and Duncan, R. (eds) (1995) *The Economic Implications of Climate Change in Britain*, London: Earthscan

Parsons, M. L. (1995) *Global Warming: The Truth behind the Myth*, New York: Insight/Plenum

Patten, C. (1989) 'Planning and local choice', *Municipal Journal* (13 October): 18–19

Paxton, A. (1994) *The Food Miles Report: The Dangers of Long Distance Food Transport*, London: SAFE Alliance

Peake, S. (1994) *Transport in Transition: Lessons from the History of Energy*, London: Royal Institute of International Affairs/ Earthscan

Peake, S. and Hope, C. (1994) 'Sustainable mobility in context: three transport scenarios for the UK', *Transport Policy* 1: 195–207

Pearce, B. J. (1992) 'The effectiveness of the British land use planning system', *Town Planning Review* 63: 13–28

Pearce, D. (ed.) (1991) *Blueprint 2: Greening the World Economy*, London: Earthscan (1994 edition)

Pearce, D. (ed.) (1993) *Blueprint 3: Measuring Sustainable Development*, London: Earthscan

Pearce, D. and Turner, R. K. (1992) 'Packaging waste and the polluter pays principle: a taxation solution', *Journal of Environmental Planning and Management* 35: 5–15

Pearce, D., Edwards, L., and Beuret, G. (1979) *Decision-Making for Energy Futures: A Case Study of the Windscale Inquiry*, London: Macmillan

Pearce, D., Markandya, A., and Barbier, E. B. (1989) *Blueprint for a Green Economy: Report for the UK DoE*, London: Earthscan

Pearce, G., Hems, L., and Hennessy, B. (1990) *The Conservation Areas of England*, London: Historic Buildings and Monuments Commission (English Heritage)

Pearlman, J. J. (1995) 'Modification orders made under Wildlife and Countryside Act 1981: an update', *Journal of Planning and Environment Law* 1995: 1106–13

Pendlebury, J. (1997) 'The statutory protection of historic parks and gardens: an exploration and analysis of "structure", "decoration" and "character"', *Journal of Urban Design* 2: 241–58

Pendlebury, J. (1999) 'The place of historic parks and gardens in the English planning system: towards statutory controls', *Town Planning Review* 70: 479–500

Pendlebury, J. and Townshend, T. (1999) 'The conservation of historic areas and participation', *Journal of Architectural Conservation* 2: 72–87

Penn, C. N. (1995) *Noise Control: The Law and Its Enforcement* (second edition), Crayford: Shaw

Pennington, M. (1996) *Conservation and the Countryside: By Quango or Market?*, London: Institute of Economic Affairs

Pepper, D. (1993) *Eco-Socialism: From Deep Ecology to Social Justice*, London: Routledge

Pepper, D. (1996) *Modern Environmentalism: An Introduction*, London: Routledge

Percy-Smith, J. (1994) *Submissions to the Commission on Aspects of Local Democracy*, London: Commission on Local Democracy

Perry, M. (1999) 'Clusters' last stand', *Planning Practice and Research* 14: 149–52

Perveen, F. (ed.) (1994) *Urban Environment: An Annotated Bibliography*, Manchester: British Council

Peter Scott Planning Services (1998) *Access to the Countryside in Selected European Countries: A Review of Access Rights, Legislation and Associated Arrangements in Denmark, Germany, Norway and Sweden*, Perth: Scottish Natural Heritage

Pezzey, J. (1989) *Economic Analysis of Sustainable Growth and Sustainable Development*, Washington, DC: World Bank

Pezzey, J. (1992) 'Sustainability', *Environmental Values* 4: 321–62

Pharoah, T. (1993) 'Traffic calming in west Europe', *Planning Practice and Research* 8: 20–8

Pharoah, T. (1996) 'Reducing the need to travel: a new planning objective in the UK?', *Land Use Policy* 13: 23–36

Pharoah, T. and Apel, D. (1995) *Transport Concepts in European Countries*, Aldershot: Avebury

Phelps, R. (1995) 'Structure plans: the conduct and conventions of examinations in public', *Journal of Planning and Environment Law* 1995: 95–101

Piatt, A. (1997) 'Public concern: a material consideration?', *Journal of Planning and Environment Law* 1997: 397–400

Pickard, R. D. (1996) *Conservation in the Built Environment*, Harlow: Addison Wesley Longman

Pickering, K. T. and Owen, L. A. (1997) *An Introduction to Global Environmental Issues*, London: Routledge

Pickvance, C. (1982) 'Physical planning and market forces in urban development', in Paris, C. (ed.) *Critical Readings in Planning Theory*, Oxford: Pergamon

PIEDA (1990) *Five Year Review of the Bolton, Middlesbrough and Nottingham Programme Authorities* (DoE), London: HMSO

PIEDA (1992) *Evaluating the Effectiveness of Land Use Planning* (DoE), London: HMSO

PIEDA (1995) *Involving Communities in Urban and Rural Regeneration* (DoE), London: Department of the Environment

Plan Local (1992) *The Character of Conservation Areas*, London: Royal Town Planning Institute

Plank, D. (ed.) (1998) *Joined Up Thinking: A Directory of Good Practice for Local Authority Empty Property Strategies*, London: Empty Homes Agency

Planning Aid for London (1986) *Planning for Women*, London: Planning Aid for London

Planning Aid for London (1995) *Publicity for Planning Applications*, London: Planning Aid for London

Planning and Environment Law Reform Working Group (1999) 'Planning obligations', *Journal of Planning and Environment Law* 1999: 113–33

Planning Exchange (1989) *Evaluation of the Use and Effectiveness of Planning Publications*, Edinburgh: Scottish Development Department

Planning Inspectorate (1996) *Development Plan Inquiries: Guidance for Local Authorities*, Bristol: PI

Planning Inspectorate (annual) *Annual Report* (previously published by the PI; now The Stationery Office)

Planning Officers' Society (1997) *Better Local Plans: A Guide to Writing Effective Policies*, Planning Officers' Society

Planning Officers' Society (1998) *Public Involvement in the Development Control Process: A Good Practice Guide*, London: Local Government Association

Planning Officers' Society Best Value Group (1999) 'Getting started on Best Value', Barnsley: Barnsley Metropolitan Borough Planning Department

Plowden, S. and Buchan, K. (1995) *A New Framework for Freight Transport*, London: Civic Trust

Plowden, S. and Hillman, M. (1984) *Danger on the Road: The Needless Scourge – A Study of Obstacles to Progress in Road Safety*, London: Policy Studies Institute

Plowden, S. and Hillman, M. (1996) *Speed Control and Transport Policy*, London: Policy Studies Institute

Plowden, W. (1971) *The Motor Car and Politics 1896–1970*, London: The Bodley Head

Plowden, W. (1994) *Ministers and Mandarins*, London: Institute for Public Policy Research

Plowden Report (1967) *Children and Their Primary Schools*, London: HMSO

Plummer, J. and Zipfel, T. (1998) *Regeneration and Employment: A New Agenda for TECs, Communities and Partnerships*, York: Policy Press

Pollock, A. (1999) 'New directions for strategic planning in Northern Ireland', *Town and Country Planning* 68: 337–9

Postle, M. (1993) *Development of Environmental Economics for the NRA* (National Rivers Authority), London: HMSO

Potter, S. (1992) 'New town legacies', *Town and Country Planning* 61: 298–302

Potter, S. (1999) 'Tax and green transport plans', *Town and Country Planning* 68: 48–9

Power, A. (1999) *Estates on the Edge*, London: Macmillan

Power, A, and Mumford, K. (1999) *The Slow Death of Great Cities? Urban Abandonment or Urban Renaissance?*, York: York Publishing Services

Power, A. and Tunstall, R. (1997) *Dangerous Disorder, Riots and Violent Disturbances in 13 Areas of Britain, 1991–92*, York: Joseph Rowntree Foundation

Poxon, J. (2000) 'Solving the development plan puzzle in Britain: learning lessons from history', *Planning Perspectives* 15: 73–89

Pressman, J. L. and Wildavsky, A. B. (1984) *Implementation: How Great Expectations in Washington are Dashed in Oakland; or Why It's Amazing that Federal Programs Work at All, This Being a Saga of the Economic Development Administration as Told by Two Sympathetic Observers who Seek to Build Morals on a Foundation of Ruined Hopes*, Berkeley: University of California Press

Pretty, J. N. (1995) *Regenerating Agriculture: Policies and Practice for Sustainability and Self-Reliance*, London: Earthscan

Pretty, J. N. (1999) *The Living Land: Agriculture, Food and Community Regeneration in the 21st Century*, London: Earthscan

Price Waterhouse (1993) *Evaluation of Urban Development Grant, Urban Regeneration Grant, and City Grant* (DoE), London: HMSO

Priemus, H. (1999) 'Coordination in planning public transport: evidence from the Netherlands', *Planning Practice and Research* 14: 97–106

Priemus, H. and Konings, R. (2000) 'Public transport in urbanised regions: the missing link in the pursuit of the economic vitality of cities', *Planning Practice and Research* 15: 233–45

Prior, A. and Allmendinger, P. (1999) 'Certainty of flexibility? The options for neighbour notification in development control in the UK', *Planning Practice and Research* 14: 185–98

Property Advisory Group (1980) *Structure and Activity of the Development Industry*, London: HMSO

Property Advisory Group (1981) *Planning Gain*, London: HMSO

Property Advisory Group (1983) *The Climate for Public and Private Partnerships in Property Development*, London: HMSO

Property Advisory Group (1985) *Report on Town and Country Planning (Use Classes) Order 1972*, London: DoE

Public Sector Management Research Centre (1992) *Parish Councils in England*, London: HMSO

Public Sector Management Research Unit (1988a) *An Evaluation of the Urban Development Grant Programme* (DoE), London: HMSO

Public Sector Management Research Unit (1988b) *Improving Inner City Shopping Centres: An Evaluation of Urban Programme Funded Schemes in the West Midlands* (DoE), London: HMSO

Pugh, C. (ed.) (1996) *Sustainability, the Environment and Urbanization*, London: Earthscan

Pugh-Smith, J. (1992) 'The local authority as a regulator of pollution in the 1990s', *Journal of Planning and Environment Law* 1992: 103–9

Pugh-Smith, J. and Samuels, J. (1993) 'PPG 16: two years on', *Journal of Planning and Environment Law* 1993: 203–10

Pugh-Smith, J. and Samuels, J. (1996a) *Archaeology in Law*, London: Sweet & Maxwell

Pugh-Smith, J. and Samuels, J. (1996b) 'Archaeology and planning: recent trends and potential conflicts', *Journal of Planning and Environment Law* 1996: 707–24

Punter, J. V. (1985) *Office Development in the Borough of Reading 1951–1984: A Case Study of the Role of Aesthetic Control*, Reading: Department of Land Management, University of Reading

Punter, J. V. (1986–7) 'A history of aesthetic control: the control of the external appearance of development in England and Wales' (parts 1 and 2), *Town Planning Review* 57: 351–81 and 58: 29–62

Punter, J. V. (1990) *Design Control in Bristol, 1940–1990: The Impact of Planning in the Design of Office Development in the City Centre*, Bristol: Redcliffe

Punter, J. V. (1992) 'Design control and the regeneration of docklands: the example of Bristol', *Journal of Property Research* 9: 49–78

Punter, J. V. (ed.) (1994) 'Design control in Europe', special issue of *Built Environment* 20 (2)

Punter, J. V. (1999) 'Design', in Cullingworth, B. (ed.) *British Planning: 50 Years of Urban and Regional Policy*, London: Athlone

Punter, J. V. and Carmona, M. (1997) *The Design Dimension of Planning: Theory, Content and Best Practice for Design Policies*, London: Spon

Punter, J. V., Carmona, M. C., and Platts, A. (1994) 'The design content of development plans', *Planning Practice and Research* 9: 199–220

Purdom, C. B. (1913) *The Garden City: A Study in the Development of a Modern New Town*, Letchworth: Temple Press

Purdom, C. B. (1925) *The Building of Satellite Towns*, London: Dent

Purdue, M. (1986) 'The flexibility of north American zoning as an instrument of land use planning', *Journal of Planning and Environment Law* 1986: 84–91

Purdue, M. (1989) 'Material considerations: an ever expanding concept?', *Journal of Planning and Environment Law* 1989: 156–61

Purdue, M. (1991) 'Green belts and the presumption in favour of development', *Journal of Environmental Law* 3: 93–121

Purdue, M. (1995) 'When a regulation becomes a taking of land: a look at two recent decisions of the United States Supreme Court', *Journal of Planning and Environment Law* 1995: 279–91

Purdue, M. (1999) 'The changing role of the courts in planning', in Cullingworth, J. B. (ed.) *British Planning: 50 Years of Urban and Regional Policy*, London: Athlone

Purdue, M. and Kemp, R. (1985) 'A case for funding objectors at public inquiries? A comparison of the position in Canada as opposed to the United Kingdom', *Journal of Planning and Environment Law* 1985: 675–85

Purdue, M., Healey, P., and Ennis, F. (1992) 'Planning gain and the grant of planning permission: is the United States' test of the *rational nexus* the appropriate solution?', *Journal of Planning and Environment Law* 1992: 1012–24

Quinn, M. J. (1996) 'Central government planning policy', in Tewdwr-Jones, M. (ed.) *British Planning Policy in Transition: Planning in the 1990s*, London: UCL Press

Rabe, B. G. (1994) *Beyond Nimby: Hazardous Waste Siting in Canada and the United States*, Washington, DC: Brookings Institution

Rackham, O. (1994) *The Illustrated History of the Countryside*, London: Weidenfeld & Nicolson (paperback edition published by Phoenix Illustrated 1997)

Radcliffe, J. (1991) *The Reorganisation of British Central Government*, Aldershot: Dartmouth

Raemaekers, J. (1995) 'Scots have way to go on strategic waste planning', *Planning* 1106 (17 February): 24–5

Raemaekers, J., Prior, A., and Boyack, S. (1994) *Planning Guidance for Scotland: A Review of the Emerging New Scottish National Planning Policy Guidelines*, Edinburgh: Royal Town Planning Institute in Scotland

Ramsay Report [1] (1945) *Report of the Scottish National Parks Survey Committee*, Cmd 6631, Edinburgh: HMSO

Ramsay Report [2] (1947) *National Parks and the Conservation of Nature in Scotland*, Cmd 7235, Edinburgh: HMSO

Ranson, S., Jones, G., and Walsh, K. (eds) (1985), *Between Centre and Locality: The Politics of Public Policy*, London: Allen & Unwin

Rao, N. (1990) *The Changing Face of Housing Authorities*, London: Policy Studies Institute

Ratcliffe, D. A. (1994) *Conservation in Europe: Will Britain Make the Grade? The Status of Nature Resources in Britain and the Implementation of the EC Habitats and Species Directive*, London: Friends of the Earth

Ratcliffe, P. (1998) 'Planning for diversity and change: implications of polyethnic society', *Planning Practice and Research* 13: 359–69

Ravaioli, C. (1995) *Economists and the Environment*, London: Zed Books

Ravenscroft, N. (2000) 'The vitality and viability of town centres', *Urban Studies* 37: 2533–49

Ravetz, A. (1980) *Remaking Cities*, London: Croom Helm

Ravetz, A. (1986) *The Government of Space: Town Planning in Modern Society*, London: Faber

Ravetz, A. (2000) *City Region 2000: Integrated Planning for a Sustainable Environment*, London: Earthscan

Ravetz, A. with Turkington, R. (1995) *The Place of Home: English Domestic Environments 1914–2000*, London: Spon

Rawcliffe, P. (1995) 'Making inroads: transport policy and the British environmental movement', *Environment* 37: 6–20 and 29–36

Read, L. and Wood, M. (1994) 'Policy, law and practice', in Wood, M. (ed.) *Planning Icons: Myth and Practice* (Planning Law Conference, *Journal of Planning and Environment Law*), London: Sweet and Maxwell

Reade, E. J. (1982) 'If planning isn't everything [.]', *Town Planning Review* 53: 65–78

Reade, E. J. (1983) 'If planning is anything, maybe it can be identified', *Urban Studies* 20: 159–71

Reade, E. J. (1985) 'Planning's usurpation of political choice', *Town and Country Planning* 54: 184–6

Reade, E. J. (1987) *British Town and Country Planning*, Milton Keynes: Open University Press

Reade, E. J. (1992) 'The little world of Upper Bangor', part 1: 'How many conservation areas are slums?'; part 2: 'Professionally prestigious projects or routine public administration?'; part 3: 'What is planning for anyway?', *Town and Country Planning* 61: 11–12, 25–7, and 44–7

Redclift, M. (1987) *Sustainable Development: Exploring the Contradictions*, London: Methuen

Redclift, M. and Sage, M. (1994) *Strategies for Sustainable Development*, Chichester: Wiley

Redman, M. (1990) 'Archaeology and development', *Journal of Planning and Environment Law* 1990: 87–98

Redman, M. (1991) 'Planning gain and obligations', *Journal of Planning and Environment Law* 1991: 203–18

Redman, M. (1999) 'Compulsory purchase, compensation and human rights', *Journal of Planning and Environment Law* 1999: 315–26

Redundant Churches Fund (1990) *Churches in Retirement: A Gazeteer*, London: HMSO

Reed, M. (1990) *The Landscape of Britain*, London: Routledge

Reed, R. (ed.) (1999) *Glasgow: The Forming of a City* (revised paperback edition), Edinburgh: Edinburgh University Press

Rees, W. E. (1990) 'Sustainable development as capitalism with a green face: a review article' [review of Pearce, D., Markandya, A., and Barbier, E. B. (1989) *Blueprint for a Green Economy: Report for the UK DoE*, London: Earthscan], *Town Planning Review* 61: 91–4

Reeve, A. (1998) 'Review article: recent literature in urban design', *Planning Perspectives* 13: 411–16

Reeves, D. (1995) 'Developing effective public consultation: a review of Sheffield's UDP process', *Planning Practice and Research* 10: 199–213

Regional Policy Commission (1996) *Renewing the Regions: Strategies for Regional Economic Development*, Sheffield: Sheffield Hallam University

Regional Studies Association (1990) *Beyond Green Belts*, London: Jessica Kingsley

Reid, A. and Murgatroyd, L. (1999) 'Variations in travelling time: the social context of travel', in DETR *Travel Trends 1999*: 3–13

Reid, C. (1994) *Nature Conservation Law*, Edinburgh: W. Green

Reid, D. (1995) *Sustainable Development: An Introductory Guide*, London: Earthscan

Reith Reports (1946) *Interim Report of the New Towns Committee*, Cmd 6759; *Second Interim Report*, Cmd 6794; and *Final Report*, Cmd 6876, London: HMSO

Rendel Geotechnics (1993) *Coastal Planning and Management: A Review* (DoE), London: HMSO

Rendel Geotechnics (1995) *Coastal Planning and Management: A Review of Earth Science Information Needs* (DoE), London: HMSO

Renton, J. (1992) 'The ombudsman and planning', *Scottish Planning Law and Practice Conference 1992*, Glasgow: Planning Exchange

Rice, D. (1998) 'National parks for Scotland: a major step forward?', *Town and Country Planning* 67: 159–61

Richardson, G., Ogus, A., and Burrows, A. (1982) *Policing Pollution: A Study of Regulation and Enforcement*, Oxford: Oxford University Press

Ridley, F. (1986) 'Liverpool is different: political struggles in context', *Political Quarterly* 57: 125–36

Ridley, F. F. and Doig, A. (eds) (1995) *Sleaze: Politicians, Private Interests and Public Reaction*, Oxford: Oxford University Press

Ridley, N. (1991) *The Local Right*, London: Centre for Policy Studies

Rietveld, P. and van Wissen, L. (1991) 'Transport policies and the environment: regulation and taxation', in Kraan, D. J. and Veld, R. J. (eds) *Environmental Protection: Public or Private Choice?*, Dordrecht: Kluwer

Rittel, H. W. J. and Webber, M. M. (1973) 'Dilemmas in a general theory of planning', *Policy Sciences* 4: 155–69

Rivlin, A. M. (1971) *Systematic Thinking for Social Action*, Washington, DC: Brookings Institution

Roberts, T. (1998) 'The statutory system of town planning in the UK: a call for detailed reform', *Town Planning Review* 69: iii–vii

Roberts, J., Cleary, J., Hamilton, K., and Hanna, J. (1992) *Travel Sickness: The Need for a Sustainable Transport Policy for Britain*, London: Lawrence & Wishart

Roberts, M. (1974) *An Introduction to Town Planning Techniques*, London: Hutchinson

Roberts, P. (1991) 'Environmental priorities and the challenges of environmental management', *Town Planning Review* 62: 447–69

Roberts, P. (1995) *Environmentally Sustainable Business: A Local and Regional Perspective*, London: Paul Chapman

Roberts, P. (1996) 'Regional planning guidance in England and Wales: back to the future?', *Town Planning Review* 67: 97–109

Roberts, P. (1999) *Urban Regeneration: A Handbook*, London: Sage

Roberts, P. and Lloyd, G. (1998) *Developing Regional Potential: Monitoring the Performance of the Regional Development Agencies*, Dundee: Centre for Planning Research, University of Dundee

Roberts, P. and Lloyd, G. (1999) 'Institutional aspects of regional planning, management, and development: models and lessons from the English experience', *Environment and Planning B: Planning and Design* 26: 517–31

Roberts, P., Thomas, K., and Williams, G. (eds) (1999) *Metropolitan Planning in Britain*, London: Jessica Kingsley

Robertson, D. J. (ed.) (1966) *The Lothians Regional Survey and Plan*, vol. 1: *Economic and Social Aspects*, Edinburgh: HMSO

Robertson, G. (1989) *Freedom, the Individual and the Law* (sixth edition), Harmondsworth: Penguin

Robins, N. (1991) *A European Environment Charter*, London: Fabian Society

Robinson, M. (1992) *The Greening of British Party Politics*, Manchester: Manchester University Press

Robinson, G. M. (1994) 'The greening of agricultural policy: Scotland's Environmentally Sensitive Areas', *Journal of Environmental Planning and Management* 37: 215–25

Robinson, P. C. (1990) 'Tree preservation orders: felling a dangerous tree', *Journal of Planning and Environment Law* 1990: 720–3

Robson, B. (1988) *Those Inner Cities: Reconciling the Social and Economic Aims of Urban Policy*, Oxford: Clarendon Press

Robson, B. (1999) 'Vision and reality: urban social policy', in Cullingworth, J. B. (ed.) *British Planning: 50 Years of Urban and Regional Policy*, London: Athlone

Robson, B. and Tomlinson, R. (1998) *Updating and Revising the Index of Local Deprivation*, London: Department of the Environment, Transport and the Regions

Robson, B. T., Bradford, M. G., Deas, I., Hall, E., Harrison, E., Parkinson, M., Evans, R., Garside, P., and Harding, A. (1994) *Assessing the Impact of Urban Policy* (DoE), London: HMSO

Robson, B., Parkinson, M., Boddy, M., and Maclennan, D. (2000) *The State of English Cities*, London: DETR

Roche, F. L. (1986) 'New communities for a new generation', *Town and Country Planning* 55: 312–13

Rodriguez-Bachiller, A., Thomas, M., and Walker, S. (1992) 'The English planning system: some insights from a more regulated system', *Town Planning Review* 63: 387–402

Roger, Tym & Partners (1982–84) *Monitoring Enterprise Zones: Year One Report* (1982); *Year Two Report* (1983); and *Year Three Report* (1984), London: Roger, Tym & Partners

Roger, Tym & Partners (1984) *Land Supply for Housing in Urban Areas*, London: Housing Research Foundation

Roger, Tym & Partners (in association with Land Use Consultants) (1987) *Evaluation of Derelict Land Grant Schemes* (DoE), London HMSO

Roger, Tym & Partners (1988) *An Evaluation of the Stockbridge Village Trust Initiative* (DoE), London: HMSO

Roger, Tym & Partners (1989a) *The Effect on Small Firms of Refusal of Planning Permission* (DoE), London: HMSO

Roger, Tym & Partners (1989b) *The Incidence and Effects of Planning Conditions* (DoE), London: HMSO

Roger, Tym & Partners (1990) *Development Control Performance*, London: National Planning Forum

Roger, Tym & Partners (1991) *Housing Land Availability* (DoE), London: HMSO

Roger, Tym & Partners (1995a) *The Use of Article 4 Directions* (DoE), London: HMSO

Roger, Tym & Partners (1995b) *Review of the Implementation of PPG 16, Archaeology and Planning*, London: English Heritage

Roger, Tym & Partners (1996a) *Enterprise Zones Monitoring 1994–95*, London: HMSO

Roger, Tym & Partners (in association with Oscar Faber TPA) (1996b) *The Gyle Impact Study* [retail impact], Edinburgh: HMSO

Roger, Tym & Partners (1998) *Urban Development Corporations: Performance and Good Practice*, London: DETR

Rogers, Lord (1999) *Towards an Urban Renaissance: Final Report of the Urban Task Force Chaired by Lord Rogers of Riverside*, London: Spon (for other publications, see Urban Task Force)

Rogers, A. (1985) 'Local claims on rural housing', *Town Planning Review* 56: 367–80

Rogers, A. (1999) *The Most Revolutionary Measure: A History of the Rural Development Commission*, Cheltenham: Countryside Agency

Rogers, C. and Stewart, M. (1998) *Sustainable Regeneration: Good Practice Guide*, London: Department of the Environment, Transport and the Regions

Roome, N. J. (1986) 'New directions in rural policy: recent administrative changes in response to conflict between rural policies', *Town Planning Review* 57: 253–63

Rose, C. (1990) *The Dirty Man of Europe: The Great British Pollution Scandal*, London: Simon & Schuster

Rosenbloom, S. (1992) 'Why working families need a car', in Wachs, M. and Crawford, M. (eds) *The Car and the City*, Ann Arbor: University of Michigan Press

Ross, A., Rowan-Robinson, J., and Walton, W. (1995) 'Sustainable development in Scotland: the role of Scottish Natural Heritage', *Land Use Policy* 12: 237–52

Ross, M. (1991) *Planning and the Heritage* (first edition), London: Spon

Ross, M. (1996) *Planning and the Heritage* (second edition), London: Spon

Ross Silcock Ltd and Social Research Associates (1999) *Community Impact of Traffic Calming Schemes*, Edinburgh: Scottish Executive Central Research Unit

Roth, G. (1996) *Roads in a Market Economy*, Aldershot: Avebury

Rowan-Robinson, J. (1998) *Quality Assessment in Development Control* (Scottish Office Central Research Unit), Edinburgh: The Stationery Office

Rowan-Robinson, J. and Durman, R. (1992a) *Section 50 Agreements*, Central Research Unit Papers, Edinburgh: Scottish Office

Rowan-Robinson, J. and Durman, R. (1992b) 'Conditions or agreements', *Journal of Planning and Environment Law* 1992: 1003–11

Rowan-Robinson, J. and Durman, R. (1992c) 'Planning agreements and the spirit of enterprise', *Scottish Geographical Magazine* 108: 157–63

Rowan-Robinson, J. and Durman, R. (1993) 'Planning policy and planning agreements', *Land Use Policy* 10: 197–204

Rowan-Robinson, J. and Lloyd, M. G. (1988) *Land Development and the Infrastructure Lottery*, Edinburgh: T. & T. Clark

Rowan-Robinson, J. and Lloyd, M. G. (1991) 'National planning guidelines: a strategic opportunity wasting', *Planning Practice and Research* 6: 16–19

Rowan-Robinson, J. and Ross, A. (1994) 'Enforcement of environmental regulation in Britain: strengthening the link', *Journal of Planning and Environment Law* 1994: 200–18

Rowan-Robinson, J. and Young, E. (1987) 'Enforcement: the weakest link in the Scottish planning control system', *Urban Law and Policy* 8: 255–88

Rowan-Robinson, J. and Young, E. (1989) *Planning by Agreement in Scotland*, Glasgow: The Planning Exchange/Edinburgh: Green

Rowan-Robinson, J., Ross, A., and Walton, W. (1995) 'Sustainable development and the development control process', *Town Planning Review* 66: 269–86

Rowan-Robinson, J., Ross, A., Walton, W., and Rothnie, J. (1996) 'Public access to environmental information: a means to what end?', *Journal of Environmental Law* 8: 19–42

Rowell, T. A. (1991) *SSSIs: A Health Check*, London: Wildlife Link

Rowthorn, R. E. (1999) *The Political Economy of Full Employment in Modern Britain* (Kalecki Memorial Lecture 1999), Cambridge: Faculty of Economics and Politics (mimeo)

Royal Commission on Environmental Pollution: *see* section on Official Publications

Royal Fine Art Commission (1997) *Design Quality and the PFI*, London: Thomas Telford

Royal Institute of British Architects (1995) *Quality in Town and Country: A Response to the Secretary of State by a Special Working Party*, London: RIBA

Royal Institution of Chartered Surveyors (1991) *Britain's Environmental Strategy: A Response by the RICS to the White Paper 'This Common Inheritance'*, London: RICS

Royal Institution of Chartered Surveyors (1994) *The Effects of Lasting Peace on Property and Construction in Northern Ireland*, Belfast: RICS Northern Ireland Branch

Royal Institution of Chartered Surveyors (1995a) *The Private Finance Initiative: The Essential Guide*, London: RICS

Royal Institution of Chartered Surveyors (1995b) *Considering Private Finance for Public Sector Works: How They Do It Over There*, London: RICS

Royal Town Planning Institute (1988a) *Planning for Choice and Opportunity*, London: RTPI

Royal Town Planning Institute (1988b) *Access for Disabled People* (Planning Advice Note), London: RTPI

Royal Town Planning Institute (1988c) *Access Policies in Local Plans* (Planning Advice Note), London: RTPI

Royal Town Planning Institute (1988d) *Managing Equality: The Role of Senior Planners*, London: RTPI

Royal Town Planning Institute (1992) *Planning Policy and Social Housing*, London: RTPI

Royal Town Planning Institute (1993) *Ethnic Minorities and the Planning System*, London: RTPI

Royal Town Planning Institute (1995) *Planning for Women* (Practice Advice Note 12), London: RTPI

Royal Town Planning Institute (1997a) *The Role of Elected Members in Plan Making and Development Control* (Darke, R. and Manson, R.), London: RTPI

Royal Town Planning Institute (1997b) *Slimmer and Swifter: A Critical Examination of District Wide Local Plans and UDPs* (Tewdwr-Jones, M. and Crow, S.), London Royal Town Planning Institute

Royal Town Planning Institute (1997c) *Regional Planning in England*, London: RTPI

Royal Town Planning Institute (1999a) *Radical Review of the Development Plan System in England: Process and Procedures*, London: RTPI

Royal Town Planning Institute (1999b) *Planning for Biodiversity*, London: RTPI

Royal Town Planning Institute (2001a) *Fitness for Purpose: Quality in Development Plans: A Guide to Good Practice*, London: RTPI

Royal Town Planning Institute (2001b) *Fitness for Purpose: Report of Research on the Quality of Development Plans*, London: RTPI

Royal Town Planning Institute and Commission for Racial Equality (1983) *Planning for a Multi-racial Britain*, London: RTPI

Rudlin, D. (1998) *Tomorrow: A Peaceful Path to Urban Reform: The Feasibility of Accommodating 75 per cent of New Homes in Urban Areas*, London: Friends of the Earth

Rural Voice (1990) *Employment of the Land*, Cirencester: Rural Voice

Rush, M. (1990) *Parliament and Pressure Politics*, Oxford: Clarendon Press

Russell, H., Dawson, J., Garside, P., and Parkinson, M. (1996) *City Challenge: Interim National Evaluation*, London: HMSO

Russell, S. (2000) 'Environmental appraisal of development plans', *Town Planning Review* 71: 529–46

Russell Barber, W. (2000) *Regional Government in England: A Preliminary Review of Literature and Research Findings*, London: DETR

Rutherford, L. A. and Peart, J. D. (1989) 'Opencast guidance: opportunities for green policies', *Journal of Planning and Environment Law* 1989: 402–10

Ryan, J. C. (1991) 'Impact fees: a new funding source for local growth', *Journal of Planning Literature* 5: 401–7

Ryder, A. A. (1987) 'The Dounreay inquiry: public participation in practice', *Scottish Geographical Magazine* 103: 54–7

Rydin, Y. (1984) 'The struggle for housing land: a case of confused interests', *Policy and Politics* 12: 431–46

Rydin, Y. (1986) *Housing Land Policy*, Aldershot: Gower

Rydin, Y. (1992) 'Environmental dimensions of residential development and the implications for local planning practice', *Journal of Environmental Planning and Management* 35: 43–61

Rydin, Y. (1993) *The British Planning System: An Introduction*, London: Macmillan

Rydin, Y. (1998) *Urban and Environmental Planning*, London: Macmillan

Rydin, Y. (1998a) 'The enabling local state and urban development resources, rhetoric and planning in east London', *Urban Studies* 35: 175–91

Rydin, Y. (1998b) 'Land use planning and environmental capacity: reassessing the use of regulatory policy tools to achieve sustainable development', *Journal of Environmental Planning and Management* 41: 749–65

Rydin, Y. (1999) 'Public participation in planning', in Cullingworth, J. B. (ed.) *British Planning: 50 Years of Urban and Regional Policy*, London: Athlone

Rydin, Y., Home, R., and Taylor, K. (1990) *Making the Most of the Planning Appeals System: Report to the Association of District Councils*, London: Association of District Councils

Safdie, M. and Kohn, W. (1998) *The City after the Automobile: An Architect's Vision*, Boulder, CO: Westview Press

Salter, J. R. (1992a) 'Environmental assessment: the challenge from Brussels', *Journal of Planning and Environment Law* 1992: 14–20

Salter, J. R. (1992b) 'Environmental assessment: the need for transparency', *Journal of Planning and Environment Law* 1992: 214–21

Salter, J. R. (1992c) 'Environmental assessment: the question of implementation', *Journal of Planning and Environment Law* 1992: 313–18

Salter, M. and Newman, P. (1992) 'Minding their own business in the planning department', *Municipal Journal* 50 (11–17 December): 28–9

Sandbach, F. (1980) *Environment, Ideology and Public Policy*, Oxford: Blackwell

Sanders, A.-M. and Rothnie, J. (1996) 'Planning registers: their role in promoting public participation', *Journal of Planning and Environment Law* 1996: 539–46

Sandford Report (1974) *Report of the National Park Policies Review Committee*, London: HMSO

Sassen, S. (1995) *The Global City: New York, London, Tokyo*, Princeton, NJ: Princeton University Press

Saunders, P. (1986) *Social Theory and the Urban Question* (second edition), London: Routledge

SAVE Britain's Heritage (1998) *Catalytic Conversion: Revive Historic Buildings to Regenerate Communities*, London: SAVE

Scanlon, K., Edge, A., and Willmott, T. (1994) *The Economics of Listed Buildings*, Cambridge: Department of Land Economy, University of Cambridge

Scargill, D. I. and Scargill, K. E. (1994) *Containing the City: The Role of Oxford's Green Belt*, Oxford: School of Geography, University of Oxford

Scarman Report (1981) *The Brixton Disorders 10–12 April 1981*, Cmnd 8427, London: HMSO

Schackleton, J. R. (1992) *Training Too Much? A Sceptical Look at the Economics of Skill Provision in the UK*, London: Institute of Economic Affairs

Schaffer, F. (1970) *The New Town Story*, London: MacGibbon & Kee

Schlesinger, A. (1999) 'New deal for communities – one year on', *Town and Country Planning* 68: 345–7

Schmidt-Eichstaedt, G. (1996) *Land Use Planning and Building Permission in the European Union*, Cologne: Deutscher Gemeindeverlag and Verlag W. Kölhammer

Schmitter, P. (1996) 'Some alternative futures for the Europen polity and their implications for European public policy', in Meny, Y., Muller, P., and Quermonne, J.-L. (eds) *Adjusting to Europe: The Impact of the European Union on National Institutions and Policies*, London: Routledge

Schofield, J. (1987) *Cost–Benefit Analysis in Urban and Regional Planning*, London: Allen & Unwin

Scholefield, G. P. (1990) 'Transport and society: the Rees Jeffreys discussion papers', *Town Planning Review* 61: 487–93

Schon, D. A. (1971) *Beyond the Stable State*, London: Temple Smith

Schubert, D. and Sutcliffe, A. (1996) 'The "Haussmannization" of London? The planning and construction of Kingsway–Aldwych, 1989–1935', *Planning Perspectives* 11: 115–44

Schuster Report (1950) *Report of the Committee on the Qualifications of Planners*, Cmd 8059, London: HMSO

Scott, A. (1999) 'Whose futures? A comparative study of Local Agenda 21 in mid Wales', *Planning Practice and Research* 14: 401–21

Scott, A. J. and Roweis, S. T. (1977) 'Urban planning in theory and practice: a reappraisal', *Environment and Planning A* 9: 1097–119

Scott, P. (1996) *The Property Masters: A History of the British Commercial Property Sector*, London: Spon

Scott Report (1942) *Report of the Committee on Land Utilisation in Rural Areas*, Cmd 6378, London: HMSO

Scottish Homes (1991) *Planning Agreements and Low Cost Housing in Scotland's Rural Areas*, Edinburgh: Scottish Homes

Scrase, T. (1991) 'Archaeology and planning: a case for full integration', *Journal of Planning and Environment Law* 1991: 1103–12

Scrase, T. (1999) 'The judicial review of local planning authority decisions: taking stock', *Journal of Planning and Environment Law* (August): 679–90

Scrase, T. (2000) 'Listed building controls: shifting ground on fixtures and fittings?', *Journal of Planning and Environment Law* 2000: 235–45

Seaton, K. and Nadin, V. (2000) *A Comparison of Environmental Planning Systems Legislation in Selected Countries*, Background Paper for the Royal Commission on Environmental Pollution, Occasional Paper 8: Faculty of the Built Environment, University of the West of England, Bristol

Seebohm Report (1968) *Report of the Committee on Local Authority and Allied Personal Social Services*, Cmnd 3703, London: HMSO

Segal Quince Wicksteed (1992) *Evaluation of Regional Enterprise Grants: Third Stage*, London: Department of Trade and Industry

Segal Quince Wicksteed (1996) *The Impact of Tourism on Rural Settlements*, Salisbury: Rural Development Commission

Self, P. (1982) *Planning the Urban Region: A Comparative Study of Policies and Organisations*, London: Allen & Unwin

Self, P. (1993) *Government by the Market? The Politics of Public Choice*, London: Macmillan

Sellgren, J. (1990) 'Development control data for planning research: the use of aggregated development control records', *Environment and Planning B: Planning and Design* 17: 23–7

Selman, P. (1988) *Countryside Planning in Practice: The Scottish Experience*, Stirling: Stirling University Press

Selman, P. (1999) *Environmental Planning: The Conservation and Development of Biophysical Resources* (second edition), London: Paul Chapman

Selman, P. (1995a) 'Theories for rural-environmental planning', *Planning Practice and Research* 10: 5–13

Selman, P. (1995b) 'Local sustainability: can the planning system help get us from here to there?', *Town Planning Review* 66: 287–302

Selman, P. (1996) *Local Sustainability: Managing and Planning Ecologically Sound Places*, London: Paul Chapman

Selman, P. (1998) 'Local Agenda 21: substance or spin?', *Journal of Environmental Planning and Management* 41: 533–53

Selman, P. and Wragg, A. (1999) 'Local sustainability planning: from interest-driven networks to vision-driven super-networks?', *Planning Practice and Research* 14: 329–40

SERPLAN (1992) *SERPLAN: Thirty Years of Regional Planning 1962–1992*, London: London and South East Regional Planning Conference

Sharland, J. (2000) 'Listed buildings: the need for a new approach', *Journal of Planning and Environmental Law* (November): 1093–110

Sharman, F. A. (1985) 'Public attendance at planning inquiries', *Journal of Planning and Environment Law* 1985: 152–8

Sharp, E. (1969) *The Ministry of Housing and Local Government*, London: Allen & Unwin

Sharp, T. (1940) *Town Planning*, Harmondsworth: Penguin

Sharp, T. (1947) *Exeter Phoenix*, London: Architectural Press

Shaw, D., Nadin, V., and Westlake, T. (1995) 'The compendium of spatial planning systems and policies', *European Planning Studies* 3: 390–5

Shaw, D., Nadin, V., and Westlake, T. (1996) 'Towards a supranational spatial development perspective: experience in Europe', *Journal of Planning Education and Research* 15: 135–42

Shaw, D., Roberts, P., and Walsh, J. (eds) (2000) *Regional Planning and Development in Europe*, Aldershot: Ashgate

Shaw, J. (1998) 'Who's afraid of the double whammy?', *Town and Country Planning* 67: 306–7

Shaw, K. (1995) 'Assessing the performance of urban development corporations: how reliable are the official government output measures?', *Planning Practice and Research* 10: 287–97

Shaw, K. and Robinson, F. (1999) 'Learning from experience? Reflections on two decades of British urban policy', *Town Planning Review* 70: 49–63

Shaw, T. (1992) 'Regional planning guidance for the North East: advice to the secretary of state for the environment', in Minay, C. L. W. (ed.) 'Developing regional guidance in England and Wales: a review symposium', *Town Planning Review* 63: 415–34

Sheaf Report (1972) *Report of the Working Party on Local Authority/Private Enterprise Partnership Schemes*, London: HMSO

Sheail, J. (1976) *Nature in Trust: The History of Nature Conservation in Britain*, Glasgow: Blackie

Sheail, J. (1983) 'Deserts of the moon: the Mineral Workings Act and the restoration of ironstone workings in Northamptonshire', *Town Planning Review* 54: 405–24

Sheail, J. (1992) 'The *amenity* clause: an insight into half a century of environmental protection in the United Kingdom', *Transactions of the Institute of British Geographers* NS 17: 152–65

Sheail, J. (1995) 'John Dower, national parks, and town and country planning in Britain', *Planning Perspectives* 10: 1–16

Shelbourn, C. (1996) 'Protecting the familiar and cherished scene', *Journal of Planning and Environmental Law* 1996: 463–9

Shelton, A. (1991) 'The well informed optimist's view', in Nadin, V. and Doak, J. (eds) *Town Planning Responses to City Change*, Aldershot: Gower

Shepherd, J. and Abakuks, A. (1992) *The National Survey of Vacant Land in Urban Areas of England 1990* (DoE), London: HMSO

Shepley, C. (1999) 'Decision-making and the role of the Inspectorate', *Journal of Planning and Environment Law* (May): 403–7

Shere, M. E. (1995) 'The myth of meaningful environmental risk assessment', *Harvard Environmental Law Review* 19: 409–92

Sherlock, H. (1991) *Cities are Good for Us*, London: Paladin

Shiva, V. (1992) 'Recovering the real meaning of sustainability', in Cooper, D. E. and Palmer, J. A. (eds) *The Environment in Question*, London: Routledge

Shoard, M. (1980) *The Theft of the Countryside*, London: Maurice Temple Smith

Shoard, M. (1987) *This Land is Our Land*, London: Paladin (updated edition, Gaia Books, 1997)

Shoard, M. (1999) *A Right to Roam*, Oxford: Oxford University Press

Short, J. R., Fleming, S., and Witt, S. (1986) *Housebuilding, Planning and Community Action*, London: Routledge & Kegan Paul

Shorten, J. and Daniels, I. (2001) *Farm Diversification and the Planning System, Final Report for the Planning Officers' Society*, Faculty of the Built Environment Occasional Paper, University of the West of England, Bristol

Shucksmith, M. (1983) 'Second homes: a framework for policy', *Town Planning Review* 54: 174–93

Shucksmith, M. (1988) 'Policy aspects of housebuilding on farmland in Britain', *Land Development Studies* 5: 129–38

Shucksmith, M. and Watkins, L. (1991) 'Housebuilding on farmland: the distributional effects in rural areas', *Journal of Rural Studies* 7: 153–68

Shucksmith, M., Henderson, M., Raybould, S., Coombes, M., and Wong, C. (1995) *A Classification of Rural Housing Markets in England* (DoE), London: HMSO

Sidaway, R. (1994) *Recreation and the Natural Heritage: A Research Review*, Perth: Scottish Natural Heritage

Sillince, J. (1986) *A Theory of Planning*, Aldershot: Gower

Sim, P. A. (1994) 'Mixed use development', in Wood, M. (ed.) *Planning Icons: Myth and Practice* (Planning Law Conference, *Journal of Planning and Environment Law*), London: Sweet and Maxwell

Simmie, J. (1981) *Power, Property and Corporatism*, London: Macmillan

Simmie, J. (1993) *Planning at the Crossroads*, London: UCL Press

Simmie, J. (ed.) (1994) *Planning London*, London: UCL Press

Simmie, J. and French, S. (1989) 'Corporatism, participation and planning: the case of London', *Progress in Planning* 31

Simmie, J. and King, R. (eds) (1990) The State in Action: Public Policy and Politics, London: Pinter

Simmonds, D. (1990) *The Impact of the Channel Tunnel on the Regions*, London: Royal Town Planning Institute

Simmons, I. G. (1993) *Interpreting Nature: Cultural Constructions of the Environment*, London: Routledge

Simpson, I. (1987) 'Planning gain: an aid to positive planning', in Harrison, M. L. and Mordey, R. (eds) *Planning Control: Philosophies, Prospects and Practice*, London: Croom Helm

Sinclair, G. (1992) *The Lost Land: Land Use Change in England 1945–1990*, London: Council for the Protection of Rural England

Sinfield, A. (1973) 'Poverty rediscovered', in Cullingworth, J. B. (ed.) *Problems of an Urban Society*, vol. 3: *Planning for Change*, London: Allen & Unwin

Skeffington Report (1969) *Report of the Committee in Public Participation in Planning*, London: HMSO

Skelcher, C. (1985) 'Transportation', in Ranson, S., Jones, G., and Walsh, K. (eds) *Between Centre and Locality: The Politics of Public Policy*, London: Allen & Unwin

Skelcher, C., McCabe, A., Lowndes, V., and Nanton, P. (1996) *Community Networks in Urban Regeneration: It All Depends Who You Know*, Bristol: Policy Press

Slater, S., Marvin, S., and Newson, M. (1994) 'Land use planning and the water sector', *Town Planning Review* 65: 375–97

Smallbone, D. (1991) 'Partnership in economic development: the case of UK local enterprise agencies', *Policy Studies Review* 10: 87–98

Smart, G. and Anderson, M. (1990) *Planning and Management of Areas of Outstanding Natural Beauty*, Cheltenham: Countryside Commission

Smeed Report (1964) *Road Pricing: The Economic and Technical Possibilities*, London: HMSO

Smith, A. G., Williams, G., and Houlder, M. (1986) 'Community influence on local planning policy', *Progress in Planning* 25: 1–82

Smith, D. (ed.) (1993) *Business and the Environment: Implications of the New Environmentalism*, London: Paul Chapman

Smith, D. L. (1974) *Amenity and Urban Planning*, London: Crosby Lockwood Staples

Smith, G. (1994) 'Vitality and viability of town centres', in Wood, M. (ed.) *Planning Icons: Myth and Practice* (Planning Law Conference, *Journal of Planning and Environment Law*), London: Sweet and Maxwell

Smith, N. (1996) *The New Urban Frontier: Gentrification and the Revanchist City*, London: Routledge

Smith, R. and Wannop, U. (1985) *Strategic Planning in Action: The Impact of the Clyde Regional Plan 1946–82*, Aldershot: Gower

Smith Morris, E. (1997) *British Town Planning and Urban Design: Principles and Policies*, Edinburgh: Addison Wesley Longman

Solesbury, W. (1976) 'The environmental agenda: an illustration of how situations may become political issues and issues may demand responses from government: or how they may not', *Public Administration* 54: 379–97

Solesbury, W. (1981) 'Strategic planning: metaphor or method?', *Policy and Politics* 9: 419–37

Solesbury, W. (1986) 'The dilemmas of inner city policy', *Public Administration* 64: 389–400

Solesbury, W. (1993) 'Reframing urban policy', *Policy and Politics* 21: 31–8

Solesbury, W. (1998) *Good Connections: Helping People to Communicate in Cities* [working paper of the Comedia/Demos project The Richness of Cities – see Worpole and Greenhalgh 1999]

Solesbury, W. (1999) 'Get connected', *Town and Country Planning* 68: 52–3

Sorafu, F. J. (1957) 'The public interest reconsidered', *Journal of Politics* 19: 616–39

Sorensen, A. D. and Day, R. A. (1981) 'Libertarian planning', *Town Planning Review* 52: 390–402

Southgate, M. (1994) 'The added value test of merger' [of English Nature and the Countryside Commission], *Town and Country Planning* 63: 134–5

Southgate, M. (1995) 'Nature conservation and planning', *Report for the Natural and Built Environment Professions*, issue 7 (August): 34–7

Sparks, L. (1987) 'Retailing in enterprise zones: the example of Swansea', *Regional Studies* 21: 37–42

Sparks, L. (1998) *Town Centre Uses in Scotland*, Edinburgh: The Stationery Office

Spawforth, P. (1995) 'Vox pop', *Planning Week* 3 (27): 16–17

Spencer, S. and Bynoe, I. (1998) *A Human Rights Commission: The Options for Britain and Northern Ireland*, London: Institute for Public Policy Research

Standing Advisory Committee on Trunk Road Assessment (Department of Transport): *see* section on Official Publications (Transport)

Starkie, D. N. M. (1982) *The Motorway Age: Road and Traffic Policies in Postwar Britain*, Oxford: Pergamon

Stead, D. and Nadin, V. (1999) 'Environmental resources and energy in the UK: the potential role of a national spatial planning framework', *Town Planning Review* 70: 339–62

Stead, D. and Nadin, V. (2000) *Urban and Rural Interdependencies in the West of England*, Faculty of the Built Environment Working Paper, University of the West of England, Bristol

Steel, J., Nadin, V., Daniels, R., and Westlake, T. (1995) *The Efficiency and Effectiveness of Local Plan Inquiries*, London: HMSO

Steele, J. (1995) *Public Access to Information: An Evaluation of the Local Government (Access to Information) Act 1985*, London: Policy Studies Institute

Steer Davies Gleave (1992) *Financing Public Transport: How Does Britain Compare?*, London: Transport 2000

Steer Davies Gleave (1994) *Promoting Rail Investment*, London: Transport 2000

Steer Davies Gleave (1995) *Alternatives to Traffic Growth: The Role of Public Transport and the Future of Freight*, London: Transport 2000

Stein, J. M. (ed.) (1995) *Classic Readings in Urban Planning*, New York: McGraw-Hill

Stevens Report (1976) *Report of the Committee on Planning Control over Mineral Working*, London: HMSO

Stewart, J. D. and Stoker, G. (eds) (1991) *The Future of Local Government*, London: Macmillan

Stewart, J. D. and Stoker, G. (eds) (1995) *Local Government in the 1990s*, London: Macmillan

Stewart, J. D., Walsh, K., and Prior, D. (1995) *Citizenship: Rights, Community, and Participation*, London: Pitman

Stewart, M. (1994) 'Value for money in urban public expenditure', *Public Money and Management* (October–December): 55–61

Stewart, M. and Taylor, M. (1995) *Empowerment and Estate Regeneration*, Bristol: Policy Press.

Stoker, G. (1991) *The Politics of Local Government*, London: Macmillan

Stoker, G. (ed.) (1999) *The New Management of British Local Governance*, London: Macmillan

Stokes, G., Goodwin, P., and Kenny, F. (1992) *Trends in Transport and the Countryside*, Cheltenham: Countryside Commission

Stone, P. A. (1973) *The Structure, Size and Costs of Urban Settlements* (National Institute of Economic and Social Research), Cambridge: Cambridge University Press

Storey, D. J. (1990) 'Evaluation of policies and measures to create local employment', *Urban Studies* 27: 669–84

Strange, I. (1999) 'Urban sustainability, globalisation and the pursuit of the heritage aesthetic', *Planning Practice and Research* 14: 301–11

Strathclyde Regional Council (1995) *Sustainability Indicators*, Glasgow: The Council

Stubbs, M. (1994) 'Planning appeals by informal hearing: an appraisal of the views of consultants', *Journal of Planning and Environment Law* 1994: 710–14

Stubbs, M. (1999) 'Informality and fairness: unlikely partners in the planning appeal?', *Journal of Planning and Environmental Law* 1999: 106–12

Stubbs, M. (2000) 'Informality and the planning appeal by hearing method: an appraisal of user satisfaction', *Town Planning Review* 71: 245–67

Suddards, R. W. and Hargreaves, J. M. (1996) *Listed Buildings* (third edition), London: Sweet and Maxwell

Suddards, R. W. and Morton, D. M. (1991) 'The character of conservation areas', *Journal of Planning and Environment Law* 1991: 1011–13

Susskind, L. and Consensus Building Institute (1999) *Using Assisted Negotiation to Settle Land Use Disputes*, Boston, MA: Lincoln Institute of Land Policy

Sustrans (1996) *The National Cycle Network: Guidelines and Practical Details*, Bristol: Sustrans

Sutcliffe, A. (ed.) (1981a) *British Town Planning: The Formative Years*, Leicester: Leicester University Press

Sutcliffe, A. (1981b) *Towards the Planned City: Germany, Britain, the United States and France 1780–1914*, Oxford: Blackwell

Sutcliffe, A. (ed.) (1984) *Metropolis 1890–1940*, London: Mansell

System Three (1999) *Research on Walking*, Edinburgh: Scottish Office Central Research Unit

Tait, M. and Campbell, H. (2000) 'The politics of communication between planning officers and politicians: the exercise of power through discourse', *Environment and Planning* A: 32: 489–06

Tant, A. P. (1990) 'The campaign for freedom of information: a participatory challenge to elitist British government', *Public Administration* 68: 477–91

Tarling, R., Hirst, A., Rowland, B., Rhodes, J., and Tyler, P. (1999) *An Evaluation of the New Life for Urban Scotland Initiative*, Edinburgh: Scottish Executive Development Department

Tate, J. (1994) 'Sustainability: a case of back to basics?', *Planning Practice and Research* 9: 367–79

Taussik, J. (1992) 'Pre-application enquiries', *Journal of Planning and Environment Law* 1992: 414–19

Taussik, J. and Smalley, J. (1998) 'Partnership in the 1990s: Derby's successful City Challenge bid', *Planning Practice and Research* 13: 283–97

Taylor, A. (1984) 'The planning implications of new technology in retailing and distribution', *Town Planning Review* 55: 161–76

Taylor, A. (1992a) *Choosing Our Future: A Practical Politics of the Environment*, London: Routledge

Taylor, M. (1995) *Unleashing the Potential: Bringing Residents to the Centre of Regeneration*, York: Joseph Rowntree Foundation

Taylor, N. (1992b) 'Professional ethics in town planning', *Town Planning Review* 63: 227–41

Taylor, N. (1998) *Urban Planning Theory since 1945*, London: Sage

Taylor, N. (1999) 'The elements of townscape and the art of urban design', *Journal of Urban Design* 4: 195–210

Teague, P. (1994) 'Governance structures and economic perfor-mance: the case of Northern Ireland', *International Journal of Urban and Regional Research* 18: 275–92

TEST (1984) *The Company Car Factor*, London: TEST

TEST (1991) *Wrong Side of the Tracks? Impacts of Road and Rail Transport on the Environment – A Basis for Discussion*, London: TEST

TEST (1992) *An Environmental Approach to Transport and Planning in Cardiff*, London: TEST

Tewdwr-Jones, M. (1994a) 'Policy implications of the plan-led system', *Journal of Planning and Environment Law* 1994: 584–93

Tewdwr-Jones, M. (1994b) 'The development plan in policy implementation', *Environment and Planning C Government and Policy* 12: 145–63

Tewdwr-Jones, M. (ed.) (1996) *British Planning Policy in Transition: Planning in the 1990s*, London: UCL Press

Tewdwr-Jones, M. (1997) 'Green belt or green wedges for Wales? A flexible approach to planning in the urban periphery', *Regional Studies* 31: 73–7

Tewdwr-Jones, M. (1998) 'Planning modernised?', *Journal of Planning and Environment Law* 1998: 519–28

Tewdwr-Jones, M. (1999) *British Planning Policy in Transition: Planning in the 1990s*, London: UCL Press

Tewdwr-Jones, M. and Allmendinger, P. (1998) 'Deconstructing communicative rationality: a critique of Habermasian collab-orative planning', *Environment and Planning A* 30: 1975–89

Tewdwr-Jones, M. and Crow, S. (1997) *Slimmer and Swifter: A Critical Examination of District Wide Local Plans and UDPs*, London: Royal Town Planning Institute

Thake, S. and Stauerbach, R. (1993) *Investing in People: Rescuing Communities from the Margin*, York: Joseph Rowntree Foundation

Thatcher, M. (1993) *The Downing Street Years*, London: HarperCollins

Therivel, R. (1995) 'Environmental appraisal of development plans: current status', *Planning Practice and Research* 10: 223–34

Therivel, R. and Partidario, M. R. (1996) *The Practice of Strategic Environmental Assessment*, London: Earthscan

Thirsk, J. (1997) *Alternative Agriculture: A History from the Black Death to the Present Day*, Oxford: Oxford University Press

Thomas, D. (1995) *Community Development at Work: A Case of Obscurity in Accomplishment*, London: Community Development Foundation

Thomas, H. (1992) 'Disability, politics and the built environment', *Planning Practice and Research* 7: 22–6

Thomas, H. (1996) 'Public participation in planning', in Tewdwr-Jones, M. (ed.) *British Planning Policy in Transition: Planning in the 1990s*, London: UCL Press

Thomas, H. (1997) 'Ethnic minorities and the planning system: a study revisited', *Town Planning Review* 68: 195–211

Thomas, H. (1999) 'Set the standard: how local authorities are handling the introduction of Best Value', *Planning* 17 September: 17

Thomas, H. (2000) *Race and Planning: The UK Experience*, London: UCL Press

Thomas, H. and Healey, P. (1991) *Dilemmas of Planning Practice: Ethics, Legitimacy, and the Validation of Knowledge*, Aldershot: Avebury

Thomas, H. and Krishnarayan, V. (1993) 'Race, equality and planning', *The Planner* 79: 17–20

Thomas, H. and Krishnarayan, V. (1994a) 'Race, disadvantage and policy processes in British planning', *Environment and Planning A* 26: 1891–1910

Thomas, H. and Krishnarayan, V. (eds) (1994b) *Race, Equality and Planning: Policies and Procedures*, Aldershot: Avebury

Thomas, H. and Lo Piccolo, F. (2000) 'Best value, planning and race equality', *Planning Practice and Research* 15: 79–94

Thomas, K. (1990) *Planning for Shops*, London: Estates Gazette

Thomas, K. (1997) *Development Control: Principles and Practice*, London: UCL Press

Thomas, R. (1996) 'The economics of the new towns revisited', *Town and Country Planning* 65: 305–8

Thomas, S. and Watkin, T. G. (1995) '"Oh, noisy bells, be dumb": church bells, statutory nuisance and ecclesiastical duties', *Journal of Planning and Environment Law* 1995: 1097–105

Thompson, H. (1992) 'Contaminated land: the implications for property transactions and the property market', *National Westminster Bank Quarterly Review* (November 1991): 20–33

Thompson, P. B. (1995) *The Spirit of the Soil: Agriculture and Environmental Ethics*, London: Routledge

Thomson, J. M. (1969) *Motorways in London*, London: Duckworth

Thornley, A. (1991) *Urban Planning under Thatcherism: The Challenge of the Market*, London: Routledge

Thornley, A. (1993) *Urban Planning under Thatcherism: The Challenge of the Market* (second edition), London: Routledge

Thornley, A. and Newman, P. (1996) *Replanning European Cities*, London: Routledge

Tiesdell, S. A., Oc, T., and Heath, T. (1996) *Revitalising Historic Urban Quarters*, Oxford: Butterworth-Heinemann

Tietenberg, T. (1990) 'Economic instruments for environmental regulation', *Oxford Review of Economic Policy* 6: 17–33

Till, J. E. (1995) 'Building credibility in public studies', *American Scientist* 83: 468–73

Tillotson, J. (1996) *European Community Law: Text, Cases and Materials* (second edition), London: Cavendish

Tillotson, J. (2000) *European Union Law*, London: Cavendish

Tinch, R. (1996) *The Valuation of Environmental Externalities* (DoE), London: HMSO

Tindall, F. (1998) *Memoirs and Confession of a County Planning Officer*, Ford, Midlothian: Pantile Press

Titmuss, R. M. (1958) 'War and social policy', in his *Essays on 'The Welfare State'*, London: Allen & Unwin

Todorovic, J. and Wellington, S. (2000) *Living in Urban England: Attitudes and Aspirations*, London: Department of the Environment, Transport and the Regions

Tolba, M. (1992) *Saving Our Planet: Challenges and Hopes*, London: Chapman & Hall

Tolley, R. (1990a) *Calming Traffic in Residential Areas*, Wales: Brefi Press

Tolley, R. (ed.) (1990b) *The Greening of Urban Transport: Planning for Walking and Cycling in Western Cities*, London: Belhaven

Tomaney, J. and Mitchell, M. (1999) *Empowering the English Regions*, London: Charter 88

Towers, G. (1995) *Building Democracy: Community Architecture in the Inner Cities*, London: UCL Press

Town and Country Planning Association (1989) *Bridging the North–South Divide*, London: Town and Country Planning Association

Town and Country Planning Association (1992) *New Settlements: Planning Policy Guidance* (spoof PPG), London: Town and Country Planning Association

Town and Country Planning Association (1993) *Strategic Planning for Regional Development*, London: Town and Country Planning Association

Town and Country Planning Association (1996) *The People: Where Will They Go?* (edited by Breheny, M. and Hall, P.), London: Town and Country Planning Association

Town and Country Planning Association (1999a) *The People: Where Will They Work? Report of Town and Country Planning Association Research into the Changing Geography of Employment* (edited by M. J. Breheny), London: Town and Country Planning Association

Town and Country Planning Association (1999b) 'Draft PPG3: Housing' (summary of Town and Country Planning Association response), *Town and Country Planning* 68: 244–5

Town and Country Planning Association (1999c) *Your Place and Mine: Reinventing Planning*, London: Town and Country Planning Association

Town and Country Planning Association (2000) *Reinventing Planning: The Report of the TCPA Inquiry into the Future of Planning*, London: TCPA

Townroe, P. and Martin, R. (1992) *Regional Development in the 1990s: The British Isles in Transition*, London: Jessica Kingsley

Townsend, A. R. (1993) 'The urban–rural cycle in the Thatcher growth years', *Transactions of the Institute of British Geographers* NS 18: 207–21

Townsend, P. (1976) 'Area deprivation policies', *New Statesman* 6 August

Townshend, T. and Pendlebury, J. (1999) 'Public participation in the conservation of historic areas: case studies from north-east England', *Journal of Urban Design* 4: 313–31

Transport 2000 (1995) *Moving Together: Policies to Cut Car Commuting*, London: Transport 2000 Trust

Transport 2000 (1997) *Changing Journeys to Work: An Employer's Guide to Green Commuter Plans* (supported by London First), London: Transport 2000

Travers, T., Jones, G., Hebbert, M., and Burnham, J. (1991) *The Government of London*, York: Joseph Rowntree Foundation

Travers, T., Biggs, S., and Jones, G. (1995) *Joint Working between Local Authorities: Experience for the Metropolitan Areas*, London: Local Government Chronicle and Joseph Rowntree Foundation

Treisman, M. (1995) *Traffic and Planning in Oxford: Commercial Development and the Environment*, Oxford: School of Planning, Oxford Brookes University

Trench, S. and Oc, T. (1990) *Current Issues in Planning*, Aldershot: Avebury

Trench, S. and Oc, T. (1995) *Current Issues in Planning*, vol. 2, Aldershot: Avebury

Trimbos, J. (1997) 'Planning in Northern Ireland', *Journal of Planning and Environment Law* (October): 904–7

Tromans, S. (1991) 'Roads to prosperity or roads to ruin? Transport and the environment in England and Wales', *Journal of Environmental Law* 3: 1–37

Tromans, S. and Clarkson, M. (1991) 'The Environmental Protection Act 1990: its relevance to planning controls', *Journal of Planning and Environment Law* 1991: 507–15

Tromans, S. and Grant, M. (1997) *Encyclopedia of Environmental Law*, London: Sweet and Maxwell

Tromans, S. and Turrall-Clarke, R. (1994 with 1996 supplement) *Contaminated Land*, London: Sweet and Maxwell

Tromans, S. and Turrall-Clarke (1999) *Contaminated Land: The New Regime*, London: Sweet and Maxwell

Tromans, S., Grant, M. and Nash, M. (eds) (1997) *Encyclopaedia of Environmental Law*, London: Sweet and Maxwell (looseleaf; updated regularly)

Truelove, P. (1992) *Decision Making in Transport Planning*, Harlow: Longman

Truelove, P. (1999) 'Transport planning', in Cullingworth, J. B. (ed.) *British Planning: 50 Years of Urban and Regional Policy*, London: Athlone

Tubbs, C. (1994) 'The New Forest: one step backwards', *Ecos* 15: 37–42

Tugnutt, A. (1991) 'Design: the wider aspects of townscapes', *Town and Country Planning Summer School 1991: Report of Proceedings*: 19–22

Turner, T. (1992) 'Open space planning in London: from standards per 1000 to green strategy', *Town Planning Review* 63: 365–86

Turner, T. (1996) *City as Landscape: A Post-postmodern View of Design and Planning*, London: Spon

Turok, I. (1988) 'The limits of financial assistance: an evaluation of local authority aid to industry', *Local Economy* 2: 286–97

Turok, I. (1989) 'Evaluation and understanding in local economic policy', *Urban Studies* 26: 587–606

Turok, I. (1990a) 'Public investment and privatisation in the new towns: a financial assessment of Bracknell', *Environment and Planning A* 22: 1323–36

Turok, I. (1990b) *Targeting Urban Employment Initiatives* (DoE), London: HMSO

Turok, I. (1992) 'Property-led urban regeneration: panacea or placebo?', *Environment and Planning A* 24: 361–79

Turok, I. and Edge, N. (1999) *The Jobs Gap in Britain's Cities: Employment Loss and Labour Market Consequences* (Joseph Rowntree Foundation), Bristol: Policy Press

Turok, I. and Shutt, J. (eds) (1994) *Urban Policy into the 21st Century*: special issue of *Local Economy* 9 (3): 211–304

Turok, I. and Wannop, U. (1990) *Targeting Urban Employment Initiatives* (DoE), London: HMSO

Turok, I. and Webster, D. (1998) 'The New Deal: jeopardised by the geography of unemployment?', *Local Economy* 12: 309–28

Tyldesley, D. (1999) *Planning for Biodiversity*, London: Royal Town Planning Institute

Tym, R. H. (1994) 'Planning in a rapidly changing economy', in Wood, M. (ed.) *Planning Icons: Myth and Practice* (Planning Law Conference, *Journal of Planning and Environment Law*), London: Sweet and Maxwell

Tyme, J. (1978) *Motorways versus Democracy*, London: Macmillan

UK Round Table on Sustainable Development (1997) *Housing and Urban Capacity*, UKRTSD, London: Department of the Environment, Transport and the Regions

UK Round Table on Sustainable Development (1998) *Aspects of Sustainable Agriculture and Rural Policy*, UKRTSD, London: Department of the Environment, Transport and the Regions

UK Round Table on Sustainable Development (2000) *Fifth Annual Report*, London: HMSO

Uman, M. F. (ed.) (1993) *Keeping Pace with Science and Engineering: Case Studies in Environmental Regulation*, Washington, DC: National Academy Press

Underwood, J. (1981a) 'Development control: a review of research and current issues', *Progress in Planning* 16: 175–242

Underwood, J. (1981b) 'Development control: a case study of discretion in action', in Barrett, S. and Fudge, C. (eds) *Policy and Action: Essays on the Implementation of Public Policy*, London: Methuen

United Nations (1992) *Environmental Accounting: Current Issues, Abstracts and Bibliography*, New York: Department of Economic and Social Development, United Nations

University of Edinburgh (1999) *Summary of Devolved Parliaments in the European Union*, Edinburgh: The Stationery Office

University of Liverpool, Environmental Advisory Unit (1986) *Transforming Our Waste Land: The Way Forward* (DoE), London: HMSO

Unwin, R. (1909) *Town Planning in Practice: An Introduction to the Art of Designing Cities and Suburbs*, London: T. Fisher Unwin

Upton, W. (1999) 'The European Convention on Human Rights and environmental law', *Journal of Planning and Environment Law* April: 315–20

Upton, W. and Harwood, R. (1996) 'The stunning powers of environmental inspectors', *Journal of Planning and Environment Law* 1996: 623–32

Urban Initiatives (1995) *Hertfordshire Dwelling Provision through Planning Regeneration*, Hertford: Hertfordshire County Environment Department

Urban Task Force (1999a) *Towards an Urban Renaissance: Final Report of the Urban Task Force*, London: Spon

Urban Task Force (1999b) *But Would You Live There? Shaping Attitudes to Urban Living*, London: DETR

Urban Task Force (1999c) *Fiscal Incentives for Urban Housing: Exploring the Options* (KPMG), London: DETR

Urban Task Force (1999d) *The Future Role of Planning Agreements in Facilitating Urban Regeneration* (Leslie Punter), London: Department of the Environment, Transport and the Regions

Urban Task Force (1999e) *Mock Planning Policy Guidance Note on Urban Development* (Tony Burton), London: Department of the Environment, Transport and the Regions

Urban Task Force (1999f) *Regional Centres for Urban Development: A Feasibility Study* (Price Waterhouse), London: Department of the Environment, Transport and the Regions

Urban Villages Group (1992) *Urban Villages: A Concept for Creating Mixed-Use Urban Developments on a Sustainable Scale*, London: Urban Villages Group

Urban Villages Group (1993) *Economics of Urban Villages*, London: Urban Villages Forum

URBED (1994) *Vital and Viable Town Centres: Meeting the Challenge* (DoE), London: The Stationery Office

Uthwatt Report (1942) *Final Report of the Expert Committee on Compensation and Betterment*, Cmd 6386, London: HMSO

Vale, B. (1995) *Prefabs: A History of the UK Temporary Housing Programme*, London: Spon

Varner, G. E. (1987) 'Do species have standing?', *Environmental Ethics* 9: 57–72

Veld, R. J. (1991) 'Road pricing: a logical failure', in Kraan, D. J. and Veld, R. J. (eds) *Environmental Protection: Public or Private Choice?*, Dordrecht: Kluwer

Vergara, C. J. (1995) *The New American Ghetto*, New Brunswick, NJ: Rutgers University Press

Verhoff, E. (1996) *The Economics of Regulating Road Transport*, Cheltenham: Edward Elgar

Verney Report (1976) *The Way Ahead: Report of the Advisory Committee on Aggregates*, London: HMSO

Vickerman, R. W. (1991) *Infrastructure and Regional Development*, London: Pion

Vigar, G. and Healey, P. (1999) 'Territorial integration and "plan-led" planning', *Planning Practice and Research* 14: 153–69

Vigar, G., Healey, P., Hull, A., and Davoudi, S. (2000) *Planning Governance and Spatial Strategy in Britain: An Institutionalist Analysis*, London: Macmillan

Vilagrasa, J. and Larkham, P. J. (1995) 'Post-war redevelopment and conservation in Britain: ideal and reality in the historic core of Worcester', *Planning Perspectives* 10: 149–72

Viner, D. and Hulme, M. (1994) *The Climate Impacts LINK Project*, Norwich: Climatic Research Unit, University of East Anglia

Vogel, D. (1986) *National Styles of Regulation: Environmental Policy in Great Britain and the United States*, Ithaca, NY: Cornell University Press

Vogel, D. (1995) 'The making of EC environmental policy', in Ugur, M. (ed.) *Policy Issues in the European Union*, London: Greenwich University Press

Wachs, M. (ed.) (1985) *Ethics in Planning*, New Brunswick, NJ: Center for Urban Policy Research

Wachs, M. and Crawford, M. (1992) *The Car and the City*, Ann Arbor: University of Michigan Press

Wade, E. C. S. and Bradley, A. W. (1993) *Constitutional and Administrative Law* (eleventh edition by Bradley, A. W. and Ewing, K. D.), London: Longman

Wägenbaur, R. (1991) 'The European Community's policy on implementation of environmental directives', *Fordham International Law Journal* 14: 455–77

Waite, R. (1995) *Household Waste Recycling*, London: Earthscan

Wakeford, R. (1990) *American Development Control: Parallels and Paradoxes from an English Perspective*, London: HMSO

Wakeford, R. (1993) 'Planning policy guidance: what's the use?', *Housing and Planning Review* (April/May): 14–18

Waldegrave, W., Byng, J., Paterson, T., and Pye, G. (1986) *Distant Views of William Waldegrave's Speech*, London: Centre for Policy Studies

Walker, M. and Reynard, M. V. (1990) *Costs in Planning Proceedings*, London: Longman

Walmsley, D. A. and Perrett, K. E. (1992) *The Effects of Rapid Transport on Public Transport and Urban Development*, Transport Research Laboratory, State of the Art Review 6, London: HMSO

Wannop, U. (1994) *Regional Planning and Governance in Britain in the 1990s*, Strathclyde Papers on Planning no. 27, Glasgow: University of Strathclyde, Centre for Planning

Wannop, U. (1995) *The Regional Imperative: Regional Planning and Governance in Britain, Europe and the United States*, London: Regional Studies Association and Jessica Kingsley

Wannop, U. (1990) 'The Glasgow Eastern Area Renewal (GEAR) project: a perspective on the management of urban regeneration', *Town Planning Review* 61: 455–74

Wannop, U. (1999) 'New towns', in Cullingworth, J. B. (ed.) *British Planning: 50 Years of Urban and Regional Policy*, London: Athlone

Warburton, D. (2000) 'Comment [on Blair and the environment]', *EG* 6 (10): 1–2

Ward, C. and Hardy, D. (1990) *Goodnight Campers! The History of the British Holiday Camp*, London: Spon

Ward, S. V. (1988) *The Geography of Interwar Britain: The State and Uneven Development*, London: Routledge

Ward, S. V. (ed.) (1992) *The Garden City: Past, Present and Future*, London: Spon

Ward, S. V. (1994) *Planning and Urban Change*, London: Spon

Ward, S. V. (1998) *Selling Places: The Marketing and Promotion of Towns and Cities*, London: Spon

Ward, S. V. (1999) 'Public–private partnerships', in Cullingworth, J. B. (ed) *British Planning: 50 Years of Urban and Regional Policy*, London: Athlone

Wardroper, J. (1981) *Juggernaut*, London: Temple Smith

Warren, H. and Davidge, W. R. (eds) (1930) *Decentralisation of Population and Industry: A New Principle in Town Planning*, London: King

Warwick Business School (1999) *Improving Local Public Services: Interim Evaluation of the Best Value Pilot Programme*, London: Department of the Environment, Transport and the Regions

Waste, R. J. (1998) *Independent Cities: Rethinking Urban Policy*, Oxford: Oxford University Press

Waters, G. R. (1994) 'Government policies for the countryside', *Land Use Policy* 11: 88–93

Wates, N. (1977) *The Battle for Tolmers Square*, London: Routledge & Kegan Paul

Watt, P. (1992) 'Publicity for planning applications', *Scottish Planning Law and Practice* 38: 15–16

Weale, A. (1992) *The New Politics of Pollution*, Manchester: Manchester University Press

Weale, A., O'Riordan, T., and Kramme, L. (1991) *Controlling Pollution in the Round: Change and Choice in Environmental Regulation in Britain and West Germany*, London: Anglo-German Foundation for the Study of Industrial Society

Weber, M. M. (1968–69) 'Planning in an environment of change', *Town Planning Review* 39: 179–95 and 277–95 (reprinted in Cullingworth, J. B. (ed.) (1973) *Problems of an Urban Society*, vol. 3: *Planning for Change*, London: Allen & Unwin)

Webster, B. and Lavers, A. (1991) 'The effectiveness of public local inquiries as a vehicle for public participation in the plan making process: a case study of the Barnet unitary development plan inquiry', *Journal of Planning and Environment Law* 1991: 803–13

Weir, S. (1995) 'Quangos: questions of democratic accountability', *Parliamentary Affairs* 48: 306–22

Welbank, M., Davies, N., and Haywood, I. (2000) *Mediation in the Planning System*, London: Department of the Environment, Transport and the Regions

Welford, R. (1995) *Environmental Strategy and Sustainable Development: The Corporate Challenge for the 21st Century*, London: Routledge

Wells, H. G. (1905) *A Modern Utopia*, London: Collins

Wenban-Smith, A. (1999) *Plan, Monitor and Manage: Making It Work*, London: Council for the Protection of Rural England

Wenban-Smith, A. and Pearce, B. (1998) *Planning Gains: Negotiating with Planning Authorities* (second edition), London: Estates Gazette

Westmacott, R. and Worthington, T. (1997) *Agricultural Landscapes: A Third Look*, Cheltenham: Countryside Commission

Weston, J. (2000) 'Reviewing environmental statements: new demands for the UK's EIA procedures', *Planning Practice and Research* 15: 135–42

Westwood, S. and Williams, J. (1996) *Imagining Cities: Scripts, Signs and Memories*, London: Routledge

Whatmore, K. and Boucher, S. (1993) 'Bargaining with nature: the discourse and practice of environmental planning gains', *Transactions of the Institute of British Geographers* NS 18: 166–78

Whatmore, S., Munton, R., and Marsden, T. (1990) 'The rural restructuring process: emerging divisions of agricultural property rights', *Regional Studies* 24: 235–45

Whitbread, M. and Marsay, A. (1992) *Coastal Superquarries to Supply South-East England Aggregate Requirements* (DoE), London: HMSO

Whitbread, M., Mayne, D., and Wickens, D. (1991) *Tackling Vacant Land: An Evaluation of Policy Instruments for Tackling Land Vacancy* (DoE), London: HMSO

White, P. (1986) 'Land availability, land banking and the price of land for housing: a review of recent debates', *Land Development Studies* 3: 101–11

White, P. (1995) *Public Transport: Its Planning, Management and Operation*, London: UCL Press

Whitehand, J. W. R. (1989) 'Development pressure, development control, and suburban townscape change', *Town Planning Review* 60: 403–20

Whitehand, J. W. R. and Larkham, P. J. (1991) 'Suburban cramming and development control', *Journal of Property Research* 8: 147–59

Whitehand, J. W. R. and Larkham, P. J. (1992) *Urban Landscapes: International Perspectives*, London: Routledge

Whitehead, A. (1999) 'From regional development to regional devolution', in Dungey, J. and Newman, I. (eds) *The New Regional Agenda*, London: Local Government Information Unit

Whitehead, C. (1999) *Housing Need in the South East*, Cambridge: University of Cambridge, Department of Land Economy

Whitehead, C., Holmans, A., Marshall, D., Royce, R., and Gordon, I. (1999) *Housing Needs in the South East*, Cambridge: Property Research Unit, Department of Land Economy, University of Cambridge

Whitelegg, J. (1993) *Transport for a Sustainable Future: The Case for Europe*, London: Belhaven

Whitelegg, J. (1994) *Roads, Jobs and the Economy*, London: Greenpeace

Whitelegg, J. (1997) *Critical Mass: Transport, Environment and Society in the 21st Century*, London: Pluto Press

Whitney, D. and Haughton, G. (1990) 'Structures for development partnerships in the 1990s: practice in West Yorkshire', *The Planner* 76 (21): 15–19

Widdicombe Report (1986) *The Conduct of Local Authority Business: Report*, Cmnd 9797; *Research Volumes*, Cmnd 9798, 9799, 9800, and 9801, London: HMSO. *Government Response to the Report*, Cm 433, London: HMSO, 1988

Wightman, A. (1996) *Who Owns Scotland?*, Edinburgh: Canongate

Wightman, A. (1999) *Scotland: Land and Power*, Luath Press in association with Democratic Left Scotland

Wilcox, S. (1995) *Housing Finance Review 1995/96*, York: Joseph Rowntree Foundation

Wildavsky, A. B. (1973) 'If planning is everything, maybe it's nothing', *Policy Sciences* 4: 127–53

Wildavsky, A. B. (1987) *Speaking Truth to Power: The Art and Craft of Policy Analysis* (second edition), London: Transaction Publishers

Wilding, R. (1990) *The Care of Redundant Churches: A Review of the Operation and Financing of the Redundant Churches Fund*, London: HMSO

Wilding, S. and Raemaekers, J. (2000) 'Environmental compensation for greenfield development: is the devil in the detail?', *Planning Practice and Research* 15: 211–32

Wilhelm, J. (1996) *Fax Messages from the Future*, London: Earthscan

Wilkes, S. and Peter, N. (1995) 'Think globally, act locally: implementing Agenda 21 in Britain', *Policy Studies* 16: 37–44

Wilkinson, D., Bishop, K., and Tewdwr-Jones, M. (1998) *The Impact of the EU on UK Planning Practice*, London: Department of the Environment, Transport and the Regions

Wilkinson, D. (1992) 'Maastricht and the environment: the implications for the EC's environmental policy of the Treaty on European Union', *Journal of Environmental Law* 4: 221–39

Wilkinson, D. and Appelbee, E. (1999) *Implementing Holistic Government: Joined-Up Action on the Ground*, Bristol: Policy Press

Wilkinson, D. and Waterton, J. (1991) *Public Attitudes to the Environment in Scotland*, Edinburgh: Scottish Office

Williams, G., Bell, P., and Russell, L. (1991) *Evaluating the Low Cost Rural Housing Initiative* (DoE), London: HMSO

Williams, G., Strange, I., Bintley, M., and Bristow, R. (1992) *Metropolitan Planning in the 1990s: The Role of Unitary Development Plans*, Manchester: Department of Planning and Landscape, University of Manchester

Williams, H. (1999) 'A need to clarify need', *Town and Country Planning* 68: 112

Williams, K. (1999) 'Urban intensification policies in England: problems and contradictions', *Land Use Policy* 16: 167–78

Williams, K., Burton, E., and Jenks, M. (2000) *Achieving Sustainable Urban Form*, London: Spon

Williams, N., Shucksmith, M., Edmond, H., and Gemmell, A. (1998) *Scottish Rural Life Update: A Revised Socio-economic Profile of Rural Scotland* (Scottish Office Central Research Unit), Edinburgh: The Stationery Office

Williams, P. (ed.) (1997) *Directions in Housing Policy: Towards Sustainable Housing Policies for the UK*, London: Paul Chapman

Williams, R. and Birch, N. (1994) 'The longer-term implications of national traffic forecasts and international network plans for local roads policy: the case of Oxfordshire', *Transport Policy* 1: 95–9

Williams, R. and Wood, B. (1994) *Urban Land and Property Markets in the UK*, London: UCL Press

Williams, R. H. (1996) *European Union Spatial Policy and Planning*, London: Paul Chapman

Williams, R. H. (2000) 'Constructing the European Spatial Development Perspective: for whom?', *European Planning Studies* 8: 357–65

Willis, K. and Garrod, G. (1993) *The Value of Waterside Properties*, Newcastle upon Tyne: Countryside Change Unit, University of Newcastle upon Tyne

Willis, K. G. (1995) 'Judging development control decisions', *Urban Studies* 32: 1065–79

Willmott, P. (1994) *Urban Trends 2: A Decade in Britain's Deprived Urban Areas*, London: Policy Studies Institute.

Wilson, A. and Charlton, K. (1997) *Making Partnerships Work: A Practical Guide for the Public, Private, Voluntary and Community Sectors*, York: Joseph Rowntree Foundation (York Publishing Services)

Wilson, D. and Game, C. (1998) *Local Government in the United Kingdom* (second edition), London: Macmillan

Wilson, E. (1993) *Strategic Environmental Assessment*, London: Earthscan

Wilson, G. K. (1983) 'Planning lessons from the ports', *Public Administration* 61: 265–81

Wilson Report (1963) *Committee on the Problems of Noise: Final Report*, Cmnd 2056, London: HMSO

Winpenny, J. T. (1991) *Values for the Environment*, London: HMSO

Winter, M. (1996) *Rural Politics: Policies for Agriculture, Forestry and the Environment*, London: Routledge

Winter, P. (1994) 'Planning and sustainability: an examination of the role of the planning system as an instrument for the delivery of sustainable development', *Journal of Planning and Environment Law* 1994: 883–900

Winter, P. (1998) *Legal and Liability Issues Preventing Urban Housing Development*, London: Town and Country Planning Association

Wolf, W. (1996) *Car Mania: A Critical History of Transport*, London: Pluto Press

Wolman, H. and Goldsmith, M. (1992) *Urban Politics and Policy: A Comparative Approach*, Oxford: Blackwell

Wolman, H. L., Cookford, C., and Hill, E. (1994) 'Evaluating the success of urban success stories', *Urban Studies* 31: 835–50

Womersley, J. R. (1994) 'The real EV [electric vehicle] challenge: reinventing an industry', *Transport Policy* 1: 266–70

Wong, C., Ravetz, J., and Turner, J. (2000) *The UK Spatial Planning Framework: A Discussion*, London: Royal Town Planning Institute

Wood, C. (1989) *Planning Pollution Prevention*, London: Heinemann

Wood, C. (1991) 'Urban renewal: the British experience', in R. Alterman and G. Cars (eds) *Neighbourhood Regeneration: An International Evaluation*, London: Mansell

Wood, C. (1994) 'Local urban regeneration initiatives: Birmingham Heartlands', *Cities* 11: 48–58

Wood, C. (1995) *Environmental Impact Assessment: A Comparative Review*, Harlow: Longman

Wood, C. and Bellanger, C. (1999) Directory of Environmental Impact Statements, Working Paper 179, Oxford Brookes School of Planning

Wood, C. (1996) 'Private sector housing renewal in the UK: progress under the "new regime" in Birmingham', Paper presented to the ACSP–AESOP Joint International Congress, Toronto

Wood, C. (1999) 'Environmental planning', in Cullingworth, J. B. (ed.) *British Planning: 50 Years of Urban and Regional Policy*, London: Athlone

Wood, C. (2000) 'Ten years on: an empirical analysis of UK environmental statement submissions since the implementation of the Directive 85/337/EEU *Journal of Environmental Planning and Management* 43: 721–47

Wood, C. and Jones, C. (1991) *Monitoring Environmental Assessment and Planning* (DoE), London: HMSO

Wood, C. and Jones, C. (1992) 'The impact of environmental assessment on local planning authorities', *Journal of Environmental Planning and Management* 35: 115–28

Wood, M. (ed.) (1994) *Planning Icons: Myth and Practice* (Planning Law Conference, *Journal of Planning and Environment Law*), London: Sweet and Maxwell

Wood, M. (1996) 'Local plans and UDPs: is there a better way?', *Journal of Planning and Environment Law* 1996: 807–15

Wood, R., Handley, J., and Bell, P. (1998) 'The character of countryside recreation and leisure appeals: an analysis of planning inspectorate records', *Journal of Planning and Environment Law* (November): 1007–27

Wood, W. (1949) *Planning and the Law*, London: Marshall

Woodward, G. D. and Larkham, P. J. (1994) *Ideal and Reality in Planning: Post-war Development in Worcester*, Birmingham: School of Planning, University of Central England

Woolmar, C. (1997) *Unlocking the Gridlock: The Key to a New Transport Policy*, London: Friends of the Earth

World Commission on Environment and Development: Brundtland Report (1987) *Our Common Future*, Oxford: Oxford University Press

World Health Organisation (1997) *City Planning for Health and Sustainable Development*, Copenhagen: WHO

Worpole, K. (ed.) (1999) *Richer Futures: Fashioning a New Politics*, London: Earthscan

Worpole, K. and Greenhalgh, L (1999) *The Richness of Cities: Urban Policy in a New Landscape*, Comedia in association with Demos (Leicester: Econ Distribution)

Worskett, R. (1969) *The Character of Towns*, London: Architectural Press

Wraith, R. E. and Hutchesson, P. G. (1973) *Administrative Tribunals*, London: Allen & Unwin

Wraith, R. E. and Lamb, G. B. (1971) *Public Inquiries as an Instrument of Government*, London: Allen & Unwin

Wrigley, M. and Seaton, K. (2001) 'Evaluating the National Land Use Database', *Journal of Property Research (forthcoming)*

Wrigley, N. (1998) 'Understanding store development programmes in post-property-crisis UK food retailing', *Environment and Planning A* 30: 15–35

Wrigley, N. and Lowe, M. (eds) (1996) *Retailing, Consumption and Capital*, Harlow: Longman

Wye College (1991) *Rural Society: Issues for the Nineties*, Wye, Kent: Wye College

Yardley, D. (1995) *Introduction to Constitutional and Administrative Law*, London: Butterworths

Yelling, J. (1999) 'The development of residential urban renewal policies in England: planning for modernization in the 1960s', *Planning Perspectives* 14: 1–18

Yelling, J. A. (1992) *Slums and Redevelopment: Policy and Practice in England 1918–1945*, London: UCL Press

Yeomans, D. (1994) 'Rehabilitation and historic preservation: a comparison of American and British approaches', *Town Planning Review* 65: 159–78

Yiftachel, O. (1989) 'Towards a new typology of urban planning theories', *Environment and Planning B: Planning and Design* 16: 23–39

Young, E. (1991) *Scottish Planning Appeals*, Edinburgh: Green/Sweet and Maxwell

Young, E. and Rowan-Robinson, J. (1985) *Scottish Planning Law and Practice*, Glasgow: Hodge

Young, K. (1984) 'Metropolitan government: the development of the concept of reality', in Leach, S. (ed.) *The Future of Metropolitan Government*, Birmingham: Institute of Local Government Studies, University of Birmingham

Young, K. (1986) 'Metropolis, R.I.P.', *Political Quarterly* 57: 36–46

Young, K. (1990) *The Politics of Local Government since Widdicombe*, York: Joseph Rowntree Foundation

Young, K. and Rao, N. (1997) *Local Government since 1945*, Oxford: Blackwell

Young, K., Gosschalk, B., and Hatter, W. (1996) *In Search of Community Identity* (Joseph Rowntree Foundation), York: York Publishing Services

Zonneveld, W. and Faludi, A. (1996) 'Cohesion versus competitive position: a new direction for European spatial planning', in Buunk, W. *et al.* (eds) *International Planning: A Dutch Perspective*, Amsterdam Study Centre for the Metropolitan Environment, University of Amsterdam

Zucker, P. C. (1999) *What Your Planning Professors Forgot to Tell You*, Chicago: Planners Press

OFFICIAL PUBLICATIONS

A NOTE ON OFFICIAL PUBLICATIONS

The student of town and country planning now has a very rich library of official publications to consult. This continues to grow at a rapid rate, and any list quickly becomes out of date. This appendix is a selective one. The first part lists the main sources of planning policy guidance. Other official documents are then presented by topic area.

Many official publications are published by The Stationery Office (TSO), formerly Her Majesty's Stationery Office (HMSO), but an increased number are now published by government departments and agencies. Some of these have the advantage of being free of charge, but there can be difficulties in determining both the publisher and the cost. Here the Internet has become invaluable, once one has identified the appropriate sites.

A few words of explanation about the mysteries of government publications may be helpful. *White Papers* are technically *Command Papers* – that is, they are 'presented to Parliament by Command of Her Majesty'. The reference to colour was at one time meaningful: a White Paper was a slim paper (often a policy statement or report) which had a white cover. It was therefore easily distinguishable from the more substantial Blue Books (which had blue covers). The Victorian Royal Commission reports were perhaps the best known of these.

Command papers are numbered sequentially and have a short prefix which varies, but which (so far) is always an abbreviation of the word 'command'. The earliest papers were prefixed C, but this was changed to Cd in 1900. This prefix lasted until 1918–19 when, on approaching the number 10,000, it was changed to Cmd. This served its purpose until 1956, in which year the prefix was changed again – and for the same reason – to Cmnd. The last (but presumably not the final) change was made in 1986, when the prefix became Cm. Why those responsible for numbering these papers are unable to go beyond 9,999 is not clear. What is clear is that authors and publishers alike have difficulty in getting the prefix correct every time.

Though still used to denote a statement of government policy, the term 'White Paper' now has no precise meaning. Colourful and graphic presentation has become common, and some White Papers bear a strong resemblance to company reports. Until relatively recently, it was reasonable to assume that if a White Paper was a policy document, it represented the government's view, or its fairly firm proposals. This distinguished it from a 'Green Paper', which was of the nature of a preliminary draft White Paper. Such papers began to appear in the late 1960s, and were popular for a time.

The distinction between white and green is now blurred and it is not uncommon to hear that a White Paper has a 'green tinge' (i.e. some of the proposals are still open for discussion). White Papers also review existing policy and present examples of good practice – it can be difficult to sort out what is new and what already is in place (which may be no accident). Similarly, Green Papers may have a 'white tinge' (i.e. certain issues have been firmly decided and are not open for further debate). Of course, circumstances change, and so do the minds of governments. (Thus the 1989 White Paper *The Future of Development Plans* announced the abolition of structure plans, but it was later decided to retain them.)

Green Papers have now largely been superseded by consultation papers, which, ironically, are often not published by TSO, but are available from the department concerned. In recent years there have been large numbers of these. Those concerned with planning are frequently printed in the monthly update to Grant's *Encyclopedia* and are listed on the relevant government web site.

Another recent innovation is the publication of annual reports of departments. Again these are Command Papers (with a range of colours). But they are both more and less than annual reports. They are more in the sense that they present details of recent and planned expenditure on the services which are covered by the budget of the Department(s) concerned. In this, they are parts of *The Government's Expenditure Plans* (which is their subtitle). But they are less than a full annual report in that they do not enter into discussion of the policies to which the figures relate. They merely state what these policies are (which is useful, but limited). As increasing sections of the executive parts of departments are hived off to agencies, the reports become of an even more summary nature (though the agencies themselves produce annual reports – of varied value).

Most government publications are published by TSO, but until 1996 the market for Northern Ireland, Scottish, and Welsh publications apparently did not justify the same treatment. As a consequence, many of their planning documents have had to be obtained direct (as is the case with most Department of the Environment Consultation Papers). Where publication is by the TSO, copies have to be obtained from the Belfast, Edinburgh, or Cardiff branches. Some publications of a number of English departments are also obtainable only from the departments concerned. To complicate matters further, recently some Scottish publications (both priced and free) have become obtainable from the Edinburgh TSO Bookshop. In short, a considerable amount of searching may be required before the source for a publication can be identified. In following lists, the publisher is TSO (for the respective countries) unless otherwise indicated. Names in parentheses relate to the author or agency concerned; names without parentheses relate to the publisher. Though we have attempted to be consistent on these matters, it is feared that we have not always succeeded.

PLANNING POLICY FOR ENGLAND

Planning Policy Guidance Notes

PPG 1   *General Policy and Principles*, 1997
PPG 2   *Green Belts*, 1995
PPG 3   *Housing*, 2000
PPG 4   *Industrial and Commercial Development and Small Firms*, 1992
PPG 5   *Simplified Planning Zones*, 1992
PPG 6   *Town Centres and Retail Developments*, 1996, revised consultation draft 1999
PPG 7   *The Countryside: Environmental Quality and Economic and Social Development*, 1997
PPG 8   *Telecommunications*, 1992, revised consultation draft 2000
PPG 9   *Nature Conservation*, 1994
PPG 10   *Planning and Waste Management*, 1999
PPG 11   *Regional Planning*, 1999
PPG 12   *Development Plans and Regional Planning Guidance*, 1999
PPG 13   *Transport* 1994, revised consultation draft 1999
PPG 14   *Development on Unstable Land*, 1990

PPG 15   *Planning and the Historic Environment*, 1994
PPG 16   *Archaeology and Planning*, 1990
PPG 17   *Sport and Recreation*, 1991; revised version in preparation
PPG 18   *Enforcing Planning Control*, 1991
PPG 19   *Outdoor Advertising Control*, 1992
PPG 20   *Coastal Planning*, 1992
PPG 21   *Tourism*, 1992
PPG 22   *Renewable Energy*, 1993
PPG 23   *Planning and Pollution Control*, 1994; revised version in preparation
PPG 24   *Planning and Noise*, 1994; revised version in preparation
PPG 25   *Development and Flood Risk*, 2001

Regional Planning Guidance Notes

RPG 1   *Strategic Guidance for Tyne & Wear*, 1989
RPG 2   *Strategic Guidance for West Yorkshire*, 1989
RPG 3   *Strategic Guidance for London Planning Authorities*, 1996
RPG 3   (Annex) *Supplementary Guidance for London on the Protection of Strategic Views*, 1991
RPG 3b/9b *Strategic Planning for the River Thames*, 1997
RPG 4   *Strategic Guidance for Greater Manchester*, 1989
RPG 5   *Strategic Guidance for South Yorkshire*, 1989
RPG 6   *Regional Planning Guidance for East Anglia*, 1991
RPG 7   *Regional Planning Guidance for the Northern Region*, 1993
RPG 8   *Regional Planning Guidance for the East Midlands Region*, 1994
RPG 9   *Regional Guidance for the South East*, 1994
RPG 9a   *The Thames Gateway Planning Framework*, 1995
RPG 10   *Strategic Guidance for the South West*, 1994
RPG 11   *Regional Planning Guidance for the West Midlands Region*, 1998
RPG 12   *Regional Planning Guidance for Yorkshire and Humberside*, 1996
RPG 13   *Regional Planning Guidance for the North West*, 1996

Minerals Policy Guidance Notes

MPG 1   *General Considerations and the Development Plan System*, 1996
MPG 2   *Applications, Permissions and Conditions*, 1998
MPG 3   *Coal Mining and Colliery Spoil Disposal*, 1999
MPG 4   *Revocation, Modification, Discontinuance, Prohibition and Suspension Orders*, 1997
MPG 5   *Stability in Surface Mineral Workings and Tips*, 2000
MPG 6   *Guidelines for Aggregates Provision in England*, 1994
MPG 7   *The Reclamation of Mineral Workings*, 1996
MPG 8   *Planning and Compensation Act 1991: Interim Development Order Permissions (IDOS) – Statutory Provisions and Proceedings*, 1991
MPG 9   *Planning and Compensation Act 1991: Interim Development Orders*, 1992
MPG 10   *Provision of Raw Material for the Cement Industry*, 1991
MPG 11   *Environmental Impacts and Mineral Working*, 1993

MPG 12   *Treatment of Disused Mine Openings and Availability of Information on Mined Ground*, 1994
MPG 13   *Guidelines for Peat Provision in England (including the Place of Alternative Materials)*, 1995
MPG 14   *Environment Act 1995: Review of Mineral Planning Permissions*, 1995
MPG 15   *Silica Sand*, 1996
MPG 17   *Oil, Gas and Coalbed Methane* (revised version in preparation)

Good Practice Guides

GPG 01A   *Design Bulletin 32: Residential Roads and Footpaths*
GPG 01B   *Places, Streets and Movement: A Companion Guide to Design Bulletin 32. Residential Roads and Footpaths*
GPG 05   *Telecommunications Prior Approval Procedures* (as applied to mast/tower development)
GPG 10   *Lighting in the Countryside: Towards Good Practice*
GPG 11   *European Spatial Development Perspective (ESDP)*
GPG 12   *Planning for Rural Diversification*
GPG 13   *Development Plans: A Good Practice Guide*
GPG 14   *Enforcing Planning Control: Good Practice Guide for Local Planning Authorities*
GPG 15   *Environmental Appraisal of Development Plans: A Good Practice Guide*
GPG 16   *Evaluation of Environmental Information for Planning Projects: A Good Practice Guide*
GPG 18   *Local Development Plans and Unitary Development Plans: Best Practice*
GPG 20   *Planning and Development Briefs: A Guide to Better Practice*
GPG 22   *Planning for Sustainable Development: Towards Better Practice*
GPG 23   *PPG 13: Guide to Better Practice*
GPG 24   *Preparation of Environmental Statements for Planning Projects That Require Environmental Assessment: A Good Practice Guide*
GPG 25   *Proposals for a Good Practice Guide on Sustainability Appraisal of Regional Planning Guidance*
GPG 26   *The One Stop Shop Approach to Development Consents*
GPG 27   *Vital and Viable Town Centres*

Research Reports

RES 01   *Structure Plans*
RES 02   *The Effectiveness of Planning Guidance on Sport and Recreation*
RES 09   *Impact of Large Foodstores on Market Towns and District Centres*
RES 14   *Environmental Geology in Land Use Planning: Advice for Planners and Developers*
RES 15   *Environmental Effects of Traffic Associated with Mineral Working*
RES 17   *Implementation of PPG 13 1994–96*
RES 18   *Operation of Compulsory Purchase Orders*
RES 22   *Application of Environmental Capacity to Land Use Planning*
RES 23   *Environmental Effects of Dust from Surface Mineral Workings*
RES 24   *Reclamation of Mineral Workings to Agriculture*
RES 25   *Reclamation of Damaged Land for Nature Conservation*
RES 29   *London in the UK Economy*
RES 33   *The Effectiveness of Planning Policy Guidance Notes*
RES 34   *Scoping Study: RPG Targets 1999 and Instructions*
RES 38   *Low Level Restoration of Sand and Gravel Workings*

RES 39   *Amenity Reclamation of Mineral Workings*

RES 40   *Landform Replication as a Technique for the Reclamation of Limestone Quarries: A Progress Report*

RES 41   *The Reclamation of Limestone Quarries Using Landform Replication*

RES 42   *Review of the Effectiveness of Restoration Conditions for Mineral Workings and the Need for Bonds*

RES 43   *The Potential for Woodland Establishment on Landfill Sites*

RES 44   *Tree Establishment on Landfill Sites: Research and Updated Guidance*

RES 45   *Landscaping and Revegetation of China Clay Wastes: Main Report*

RES 46   *The Reclamation and Management of Metalliferous Mining Sites*

RES 47   *Restoration of Damaged Peatlands*

RES 48   *Slate Waste Tips and Workings in Britain*

RES 49   *The Reclamation of Mineral Workings to Agriculture*

RES 50   *Guidance on Good Practice for the Reclamation of Mineral Workings to Agriculture*

RES 51   *Reclamation of Damaged Land for Nature Conservation*

RES 52   *Restoration and Revegetation of Colliery Spoil Tips and Lagoon*

RES 53   *Agricultural Quality of Restored Land at Bush Farm*

RES 54   *The Use of Soil-Forming Materials in the Reclamation of Older Mineral Workings and Other Reclamation Schemes*

RES 55   *Derelict Land Prevention and the Planning System*

RES 56   *Permitted Development Rights for Agriculture and Forestry*

RES 57   *Planning for Rural Diversification*

RES 60   *Subsidiarity and Proportionality in Spatial Planning Activities in the European Union: Final Report* 1999

RES 62   *Housing in the South East: The Inter-relationship between Supply, Demand and Land Use Policy*

RES 63   *Areas of Economic Pressure in the South East*

RES 64   *Examination of the Operation and Effectiveness of the Structure Planning Process: Summary Report*

RES 65   *The Effectiveness of PPG 13: A Pilot Study*

RES 66   *The Economic Consequences of Planning to the Business Sector*

RES 67   *Control of Outdoor Advertisements: Fly-Posting*

RES 68   *Parking Standards in the South East*

RES 69   *The Effectiveness of Planning Policy Guidance on Sport and Recreation*

RES 70   *The Impact of Large Foodstores on Market Towns and District Centres*

RES 71   *The Use of Density in Urban Planning*

RES 72   *Dwellings over and in Shops in London*

RES 73   *Town Centres: Defining Boundaries for Statistical Monitoring – Feasibility Study*

RES 74   *The Use of Permitted Development Rights by Statutory Undertakers*

RES 75   *Sustainable Residential Quality: New Approaches to Urban Living* (DETR, Government Office for London, London Planning Advisory Committee)

RES 76   *Implementation of PPG 13: Final Report*

RES 77   *The Application of Environmental Capacity to Land Use Planning*

RES 78   *The Operation of Compulsory Purchase Orders*

RES 79   *Mitigation Measures in Environmental Statements*

RES 80   *Departure Applications: The Effectiveness of the Town and Country Planning Act (Development Plans and Consultation Direction) 1992* (Joint DoE and DoT)

RES 81   *Analysis of Responses to the Discussion Document 'Quality in Town and Country'*

RES 82   *Sustainable Settlements and Shelter: The UK National Report on Habitat II*

RES 83   *Changes in the Quality of Environmental Statements for Planning Projects*

RES 84   *Planning Controls over Agricultural and Forestry Developments*
RES 85   *London's Urban Environment: Planning for Quality*
RES 86   *London in the UK Economy*
RES 87   *Attitudes to Town and Country Planning*
RES 88   *Monitoring Housing Land Supply: Calibrating Indicators of Constraint*
RES 89   *Review of the Land Use Planning Research Programme 1989/90–1993/94*
RES 90   *Use of Article 4 Directions*
RES 91   *Implementation of PPG 13: Interim Report* (with DoT)
RES 92   *Effectiveness of Planning Policy Guidance Notes*
RES 93   *Urbanization in England: Projections 1991–2016*
RES 94   *Hazardous Substances Consents*
RES 95   *Feasibility Study for Deriving Information about Land Use Stock*
RES 96   *Impact Fees for Planning*
RES 97   *Community Involvement in Planning and Development Processes*
RES 98   *Efficiency and Effectiveness of Local Plan Inquiries*
RES 99   *The Costs of Determining Planning Applications and the Development Control Service*
RES 100   *Analyses of Land Use Change Statistics*
RES 101   *Planning for Affordable Housing*
RES 102   *Integrated Planning and Granting of Permits in the EC*
RES 103   *Alternative Development Patterns: New Settlements*
RES 104   *Merry Hill Impact Study*
RES 105   *The Effectiveness of Green Belts*
RES 106   *Migration and Business Relocation: The Case of the South East*
RES 107   *Business Relocation: The Case of London and the South East Region*
RES 108   *Migration and the Metropolis: Patterns and Processes of Inter-regional Migration to and from South East England*
RES 109   *Housing Land Availability: Analysis of PS 3 Statistics on Land with Outstanding Planning Permission*
RES 110   *Reducing Transport Emissions through Planning*
RES 111   *East Thames Corridor – a Study of Development Capacity and Potential*
RES 112   *The National Survey of Vacant Land in Urban Areas of England 1990*
RES 113   *Evaluating the Effectiveness of Land Use Planning*
RES 114   *Use of Planning Agreements*
RES 115   *Planning, Pollution and Waste Management*
RES 116   *Residential Roads and Footpaths, Layout Considerations. Design Bulletin 32, (2nd edition)*
RES 117   *Effects of Major Out of Town Retail Development*
RES 118   *Relationship between House Prices and Land Supply*
RES 119   *Time for Design: Urban Design Practices in Six Urban Authorities*
RES 120   *Evaluating the Low Cost Rural Housing Initiative*
RES 121   *An Examination of the Effects of the Use Classes Order 1987 and the General Development Order 1988*
RES 122   *Monitoring Environmental Assessment and Planning*
RES 123   *Housing Land Availability*
RES 124   *Land Use Planning Policy and Climate Change*
RES 126   *Simplified Planning Zones: Progress and Procedures*
RES 127   *Rate of Urbanization in England 1981–2000*
RES 128   *Area Economic Studies: An Evaluation*

RES 129   *Approval of Reserved Matters following Outline Planning Permission*
RES 130   *Effect on Small Firms of Refusal of Planning Permission*
RES 131   *Incidence and Effects of Planning Conditions*
RES 132   *Study of Special Industrial Use Classes*
RES 133   *Kirklees, Wakefield and Doncaster Co-ordination Report*
RES 144   *The Process of Local Plan Adoption (vols 1 and 2)*

PLANNING POLICY FOR NORTHERN IRELAND

(Published by the Northern Ireland Planning Service)

Planning Policy Statements

PPS 1   *General Planning Principles*
PPS 2   *Planning and Nature Conservation*
PPS 3   *Development Control: Roads Considerations*
PPS 4   *Industrial Development*
PPS 5   *Retailing and Town Centres*
PPS 6   *Planning, Archaeology and Built Heritage*
PPS 7   *Quality Residential Environments*, issued for consultation
PPS 8   *Open Space, Sport, Recreation, Leisure and Commuity Facilities*, issued for consultation
PPS 9   *The Enforcement of Planning Control*
PPS 10   *Telecommunications*, issued for consultation

Development Control Advice Notes

DCAN 1   *Amusement Centres*
DCAN 2   *Multiple Occupancy*
DCAN 3   *Bookmaking Offices*
DCAN 4   *Hot Food Bars*
DCAN 5   *Taxi Offices*
DCAN 6   *Restaurants and Cafes*
DCAN 7   *Public Houses*
DCAN 8   *Small Unit Housing in Residential Areas*
DCAN 9   *Residential and Nursing Homes*
DCAN 10   *Environmental Impact Assessment*
DCAN 11   *Access for People with Disabilities*
DCAN 12   *Hazardous Substances*
DCAN 13   *Crèches, Day Nurseries and Pre-school Playgroups*
DCAN 14   *Telecommunications Prior Approval Procedures*
DCAN 15   *Vehicular Access Standards*

Other Recent Reports on Planning in Northern Ireland

1993   *A Planning Strategy for Northern Ireland* (HMSO, Belfast)

1996   *The Planning System in Northern Ireland*, HC Select Committee on Northern Ireland Affairs (HMSO, London)

1997   *Shaping Our Future: Towards a Strategy for the Development of the Region – Draft Regional Strategic Framework for Northern Ireland*, DoENI, Belfast

1998   *Shaping Our Future: Draft Regional Strategy for Northern Ireland* DoENI, Belfast

1999   *Proposals for Amendments for Planning Legislation in Northern Ireland: A Consultation Paper*

Annual   *The Planning Service: Annual Report and Accounts*

PLANNING POLICY FOR SCOTLAND

(Now published by the Scottish Executive)

Scottish National Planning Guidelines

NPG 1   *North Sea Oil and Gas: Coastal Planning Guidelines*, 1974

NPG 2   *Aggregate Working*, 1977

NPG 3   *Priorities for Development Planning; Land for Housing; Land for Large Industry; Land for Petrochemical Development; Rural Planning Priorities; National Scenic Areas; Nature Conservation; Forestry*, 1981

NPG 4   *Skiing Developments*, 1984

NPG 5   *High Technology; Individual High Amenity Sites*, 1985

NPG 6   *Location of Major Retail Development*, 1985

NPG 7   *Agricultural Land*, 1987

Scottish National Planning Policy Guidelines

NPPG 1   *The Planning System*, 2000

NPPG 2   *Business and Industry*, 1993

NPPG 3   *Land for Housing*, 1993; revised 1996

NPPG 4   *Land for Mineral Working*, 1994

NPPG 5   *Archaeology and Planning*, 1994

NPPG 6   *Renewable Energy*, 1994

NPPG 7   *Planning and Flooding*, 1995

NPPG 8   *Town Centres and Retailing*, 1996; revised 1998

NPPG 9   *The Provision of Roadside Facilities on Motorways and Other Trunk Roads in Scotland*, 1996

NPPG 10   *Planning and Waste Management*, 1996

NPPG 11   *Sport, Physical Recreation and Open Space*, 1996

NPPG 12   *Skiing Developments*, 1997

NPPG 13   *Coastal Planning*, 1997

NPPG 14   *Natural Heritage*, 1999

NPPG 15   *Rural Development*, 1999

NPPG 16   *Opencast Coal and Related Matters*, 1999

NPPG 17   *Transport and Planning*, 1999
NPPG 18   *Planning and the Historic Environment*, 1999

Scottish Planning Advice Notes

PAN 33   *Development of Contaminated Land*, 1988
PAN 36   *Siting and Design of New Housing in the Countryside*, 1991
PAN 37   *Structure Planning*, 1992; revised 1996
PAN 38   *Structure Plans: Housing Land Requirements*, 1993; revised 1996
PAN 39   *Farm and Forestry Buildings*, 1993
PAN 40   *Development Control*, 1993; Addendum 1997
PAN 41   *Development Plan Departures*, 1994; revised 1997
PAN 42   *Archaeology: The Planning Process and Scheduled Monuments Procedures*, 1994
PAN 43   *Golf Courses and Associated Developments*, 1994
PAN 44   *Fitting New Housing Development into the Landscape*, 1994
PAN 45   *Renewable Energy Technologies*, 1994
PAN 46   *Planning for Crime Prevention*, 1994
PAN 47   *Community Councils and Planning*, 1996
PAN 48   *Planning Application Forms*, 1996
PAN 49   *Local Planning*, 1996
PAN 50   *Controlling the Environmental Effects of Surface Mineral Workings*, 1996
PAN 50A   *The Control of Noise at Surface Mineral Workings*, 1996
PAN 50B   *The Control of Dust at Surface Mineral Workings*, 1998
PAN 50C   *The Control of Traffic at Surface Mineral Workings*, 1998
PAN 50D   *Control of Blasting at Surface Mineral Workings*, 1999
PAN 51   *Planning and Environmental Protection*, 1997
PAN 52   *Planning and Small Towns*, 1997
PAN 53   *Classifying the Coast for Planning*, 1998
PAN 54   *Planning Enforcement*, 1999
PAN 55   *The Private Finance Initiative and Planning*, 1999
PAN 56   *Planning and Noise*, 1999
PAN 57   *Transport and Planning*, 1999
PAN 58   *Environmental Impact Assessment*, 1999
PAN 59   *Improving Town Centres*, 1999

Other Recent Reports on Planning in Scotland

1997   *Review of Development Planning in Scotland* (SDD Research Findings no. 50)
1999   *Land Use Planning under a Scottish Parliament* and *Overview of Responses to Consultation*
The Scottish Executive Development Department also publishes a regular *Planning Bulletin*

PLANNING POLICY FOR WALES

(Now published by the Welsh Assembly)

Planning Policy Guidance – Wales

Planning Guidance Wales: Unitary Development Plans, 1996
Planning Guidance Wales: Planning Policy: First Revision, 1999

Technical Advice Notes (Wales)

TAN (W) 1	*Joint Housing Land Availability Studies*, 1997
TAN (W) 2	*Planning and Affordable Housing*, 1996
TAN (W) 3	*Simplified Planning Zones*, 1996
TAN (W) 4	*Retailing and Town Centres*, 1996
TAN (W) 5	*Nature Conservation and Planning*, 1996
TAN (W) 6	*Agriculture and Rural Development*, 2000
TAN (W) 7	*Outdoor Advertisement Control*, 1996
TAN (W) 8	*Renewable Energy*, 1996
TAN (W) 9	*Enforcement of Planning Control*, 1997
TAN (W) 10	*Tree Preservation Orders*, 1997
TAN (W) 11	*Noise*, 1997
TAN (W) 12	*Design*, 1997
TAN (W) 13	*Tourism*, 1997
TAN (W) 14	*Coastal Planning*, 1998
TAN (W) 15	*Development and Flood Risk*, 1998
TAN (W) 16	*Sport and Recreation*, 1998
TAN (W) 17	*Environmental Assessment*, 1998
TAN (W) 18	*Transport*, 1998
TAN (W) 19	*Telecommunications*, 1998
TAN (W) 20	*The Welsh Language: Unitary Development Plans and Development Control*, 2000

In preparation:
Development on Unstable Land
Development of Contaminated Land
Planning, Pollution Control; and Waste Management

Other Recent Documents on Planning in Wales

2000	*Wales Spatial Planning Framework Key Challenges for Wales*
2000	*Development of Planning Policy in Wales: Report of the Land Use Planning Forum*

EUROPEAN UNION PLANNING POLICY AND STUDIES

Reports and Papers

1990	*Green Paper for the Urban Environment*
1991	*Europe 2000: Outlook for the Development of the Community's Territory*
1992	*The Impact of Transport on the Environment: A Community Strategy for Mobility*
1992	*The Future Development of the Common Transport Policy*
1992	*Fifth Action Programme on the Environment, 1992–2000: Towards Sustainability*
1993	*Growth, Competitiveness and Employment*
1994	*Europe 2000+: Cooperation for European Territorial Development*
1995	*The Citizen's Network: Fulfilling the Potential of Public Passenger Transport in Europe*
1995	*Towards Fair and Efficient Policy in Transport*
1995	*The Trans-European Transport Network*
1996	*European Sustainable Cities Report*, EU Expert Group on the Urban Environment
1997	*Agenda 2000*
1997	*Towards an Urban Agenda in the European Union*
1998	*Community Policies and Spatial Planning*
1998	*Evaluation of the Performance of the EIA Process Final Report*
1999	*Sustainable Urban Development in the EU: A Framework for Action*
1999	*The European Spatial Development Perspective*, Committee on Spatial Development
2000	*TERRA: An Experimental Laboratory in Spatial Planning*
2000	*Sixth Periodic Report on the Social and Economic Situation of the Regions in the European Union*
2000	*Sixth Action Programme on the Environment*
2001	*Second Report on Social and Economic Cohesion*

Studies

4	*Urbanisation and the Function of Cities in the European Community*, 1992
8	*Study of the Prospects in the Atlantic Regions*, 1993
18	*The Prospective Development of the Northern Seaboard*, 1994
21	*The Regional Impact of the Channel Tunnel throughout the Community*, 1994
22	*The Prospective Development of the Central and Capital Cities and Regions*, 1994
28	*The EU Compendium of Spatial Planning Systems and Policies*, 1997
29	*Economic and Social Cohesion in the European Union: The Impact of Member States'own policies*, 1998
32	*Inclusive Cities: Building Local Capacity for Development*, 1999
36	*Spatial Perspectives for the Enlargement of the European Union*, 1999

AGENCIES OF PLANNING

Central Government

1970	*The Reorganisation of Central Government*, Cmnd 4506
1979	*Central Government Controls over Local Authorities*, Cmnd 7634
1984	*Progress in Financial Management in Government Departments*, Cmnd 9297

1988	*Civil Service Management Reform: The Next Steps*, Cm 524
1990	*Report of the Noise Review Working Party* (Batho Report)
1991	*Improving Management in Government: The Next Steps Agencies – Review 1991*, Cm 1730
1994	*Origins of the Department of the Environment* (P. McQuail), Department of the Environment
1995	*Review of the Department of the Environment*, Department of the Environment
1995	*View from the Bridge* (P. McQuail), Department of the Environment
1996	*Next Steps Agencies in Government, Review 1995*, Cm 3164
1999	*Modernising Government*, Cm 4310
2000	*Reinforcing Standards: Review of the First Report of the Committee on Standards in Public Life*, Cm 4557–1

Devolved and Regional Government

[*see also* list of Regional Planning Guidance Notes]

1997	*Scotland's Parliament*, Cm 3658
1997	*A Voice for Wales: The Government's Proposals for a Welsh Assembly*, Cm 3718
1998	*Building Partnerships for Prosperity*, DETR
1998	*The Future of Regional Guidance: Consultation Paper*, DETR
1998	*Guidance to the Regional Development Authorities on Rural Policy*, DETR
1999	*Guidance to the Regional Development Authorities on Sustainable Development*, DETR
1999	*Local Plans and Unitary Development Plans: A Guide to Procedures*, DETR
1999	*Quality of Life Counts: DETR Indicators for the UK Sustainable Development Strategy*, DETR
1999	*Regional Planning Consultation Draft*, PPG 11
1999	*Rural Economies*, Cabinet Office Performance and Innovation Unit
1999	*Structure Plans: A Guide to Procedures*, DETR
1999	*Supplementary Guidance to Regional Development Authorities*, DETR
1999	*Regional Strategies*, DETR
1999	*Regional Chambers*, DETR
1999	*Regional Development Agencies*, HC Environment, Transport and Regional Affairs Committee, 10th Report, Session 1998–99, HC 232
2000	*Regional Government in England: A Preliminary Review of the Literature and Research Findings*, DETR

Local Government

1979	*Central Government Controls over Local Authorities*, Cmnd 7634
1979	*Organic Change in Local Government*, Cmnd 7457
1979	*Central Government Controls over Local Authorities*, Cmnd 7634
1981	*Committee of Inquiry into Local Government in Scotland* (Stodart Report), Cmnd 8115
1983	*Rates: Proposals for Rate Limitation and Reform of the Rating System*, Cmnd 9008
1983	*Streamlining the Cities: Government Proposals for Reorganising Local Government in Greater London and the Metropolitan Counties*, Cmnd 9063
1984	*Report of the Committee on Inquiry into the Functions and Powers of the Islands Councils of Scotland* (Montgomery Report), Cmnd 9216
1986	*Paying for Local Government*, Cmnd 9714
1986	*The Conduct of Local Authority Business: Report of the Committee of Inquiry* (Widdicombe Report), Cmnd 9797

1992	*Functions of Local Authorities in England* (Local Government Review), TSO
1992	*The Role of Community and Town Councils in Wales* (Consultation Paper), Welsh Office
1993	*Local Government in Wales: A Charter for the Future*, Cm 2155
1997	*New Leadership for London: The Government's Proposals for a Greater London Authority*, Consultation Paper, Cm 3724
1997	*Standards of Conduct in Local Government*, 3rd Report of the Committee on Standards in Public Life (Nolan Reports), Cm 3702
1998	*Guidance on Enhancing Public Participation in Local Government*, DETR
1998	*Modern Local Government: In Touch with the People*, Cm 4014
1998	*A Mayor and Assembly for London*, Cm 3897
1998	*Modernising Local Government: Local Democracy and Community Leadership*, DETR
1998	*New Ethical Framework for Local Government in Scotland: A Consultation Paper*, Scottish Office
1999	*Local Government Political Management Arrangements: An International Perspective* (R. Hambleton), Scottish Office
1999	*Local Leadership, Local Choice*, Cm 4298
1999	*Report of the Community Planning Working Group*, Scottish Office
1999	*Report of the Commission on Local Government and the Scottish Parliament* (McIntosh Report), Scottish Office
2000	*Preparing Community Strategies: Draft Guidance to Local Authorities*, Scottish Executive

THE PLANNING POLICY FRAMEWORK AND DEVELOPMENT CONTROL

{*see also* Northern Ireland, Scottish, and Welsh publications}

1942	*Report of the Expert Committee on Compensation and Betterment* (Uthwatt Report), Cmd 6386
1942	*Report of the Committee on Land Utilisation in Rural Areas* (Scott Report), Cmd 6378
1944	*The Control of Land Use*, Cmd 6537
1947	*Town and Country Planning Bill 1947: Explanatory Memorandum*, Cmd 7006
1950	*Report of the Committee on the Qualifications of Planners* (Schuster Report), Cmd 8059
1951	*Town and Country Planning 1943–1951: Progress Report*, Cmd 8204
1965	*The Future of Development Plans: Report of the Planning Advisory Group*
1967	*Town and Country Planning*, Cmnd 3333
1977	*Memorandum on Structure and Local Plans*, DoE Circular 55/77
1978	*Planning Procedures: The Government's Response to the Eighth Report from the Expenditure Committee, Session 1976–77*, Cmnd 7056
1984	*Memorandum on Structure Plans*, DoE Circular 22/84
1984	*Enforcement of Planning Control in Scotland* (J. Rowan-Robinson, E. Young, and I. McLarty), Scottish Office
1985	*Lifting the Burden*, Cmnd 9571
1986	*Planning Appeals, Call-in and Major Public Inquiries: The Government's Response to the Fifth Report from the Environment Committee*, Cm 43
1989	*The Future of Development Plans*, Cm 569
1990	*Mineral Policies in Development Plans* (Arup Economic Consultants)
1991	*Enforcing Planning Control*, PPG 18
1991	*Examination of the Effects of the Use Classes Order 1987 and the General Development Order 1988* (Wootton Jeffreys Consultants and Bernard Thorpe)

1991   *Permitted Development Rights for Agriculture and Forestry* (Land Use Consultants)

1992   *Building in Quality: A Study of Development Control* (Audit Commission)

1992   *Development Plans: A Good Practice Guide*

1992   *Evaluating the Effectiveness of Land Use Planning* (Roger Tym & Partners *et al.*)

1992   *Industrial and Commercial Development and Small Firms*, PPG 4

1992   *Outdoor Advertising Control*, PPG 19

1992   *Publicity for Planning Applications*, DoE Circular 15/92

1992   *Simplified Planning Zones*, PPG 5

1992   *Use of Planning Agreements* (Grimley J. R. Eve *et al.*)

1993   *Integrated Planning and Granting of Permits in the EC* (GMA Planning, P-E International, and Jacques & Lewis)

1994   *Review of the Use Classes Order*, Scottish Office

1994   *Planning Out Crime*, DoE Circular 5/94

1994   *Costs of Determining Planning Applications and the Development Control Service* (Price Waterhouse)

1994   *Guidance Notes for Local Planning Authorities on the Methods of Protecting the Water Environment through Development Plans*, Environment Agency

1994   *Improving the Local Plan Process*, Consultation Paper, DoE

1994   *Mediation: Benefits and Practice*, Department of the Environment

1995   *Review of the Town and Country Planning System in Scotland: The Way Ahead*, Scottish Office

1995   *Planning Controls over Demolition*, DoE Circular 10/95

1995   *The Use of Conditions in Planning Permissions*, DoE Circular 11/95

1995   *Planning Controls over Agricultural Land and Rural Building Conversions* (Land Use Consultants)

1995   *Review of Neighbour Notification*, Scottish Office

1995   *Efficiency and Effectiveness of Local Plan Inquiries*

1995   *Effectiveness of Planning Policy Guidance Notes* (Land Use Consultants), Department of the Environment

1995   *Quality in Town and Country: Urban Design Guidelines*

1995   *Evaluation of Planning Enforcement Provisions* (Arup Economic Planning and Linklaters & Paines)

1996   *Planning System in Northern Ireland*, HC Northern Ireland Affairs Committee, 1st Report, Session 1995–96, HC 53

1996   *Speeding Up the Delivery of Local Plans and UDPs: Report of the Review*, DoE

1997   *Efficiency and effectiveness of Local Plan Inquiries*, DoE

1997   *Enforcing Planning Control: A Good Practice Guide for Local Planning Authorities*

1997   *General Policy and Principles*, PPG1

1997   *Speeding Up the Delivery of Local Plans and UDPs* Consultation Paper, DOE

1998   *Future of Regional Planning Guidance*, DETR

1998   *Impact of the EU on the UK Planning System* (D. Wilkinson, K. Bishop, and M. Tewdwr-Jones)

1998   *Modernising Planning: A Policy Statement*, DETR

1998   *Our Competitive Future: Building the Knowledge Driven Economy*, Cm 4176

1998   *Prevention of Dereliction through the Planning System*, DETR Circular 2/98

1999   *Approach to Future Land Use Planning Policy*, Welsh Assembly

1999   *Biotechnology Clusters: Report of a Team Led by Lord Sainsbury, Minister for Science*, DTE

1999   *Costs in the Planning Service*, Scottish Executive Central Research Unit

1999   *Development Control and Development Plan Preparation: Local Authority Concerns and Current Government Action*, DETR

1999   *Development Plans*, PPG 12
1999   *Examination of the Operation and Effectiveness of the Structure Planning Process* (M. Baker and P. Robert), DETR
1999   *Modernising Planning: A Progress Report*, DETR
1999   *Modernising Planning: Streamlining the Processing of Major Projects through the Planning System: Consultation Paper*, DETR
1999   *Planning Concordat between the Local Government Association and the DETR*, DETR
1999   *Planning for Telecommunications*, DETR Circular 1999/4
1999   *Planning Policy Guidance Note 11: Regional Planning – Public Consultation Draft*, DETR
1999   *Review of National Planning Policy Guidelines* (Land Use Consultants), Edinburgh: Scottish Office
1999   *Town Centres and Retail Developments: Consultation Draft*, PPG 6
1999   *Development Control and Development Plan Preparation: Local Authority Concerns and Current Government Action*
2000   *Planning Inspectorate Executive Agency Annual Report 1999–2000*
2000   *Training for Urban Design*
2000   *Survey of Urban Design Skills in Local Government*
2000   *By Design: Urban Design in the Planning System: Towards Better Practice*, DETR and CABE
2000   *The Control of Fly-Posting: A Good Practice Guide*, DETR
2000   *European Spatial Planning and Urban–Rural Relationships: The UK Dimension*, DETR
2000   *Quinquennial Review of the Planning Inspectorate (PINS)*

LAND POLICY

1940   *Report of the Royal Commission on the Distribution of the Industrial Population* (Barlow Report), Cmd 6153
1942   *Report of the Committee on Land Utilisation in Rural Areas* (Scott Report), Cmd 6378
1942   *Report of the Expert Committee on Compensation and Betterment* (Uthwatt Report), Cmd 6386
1944   *The Control of Land Use*, Cmd 6537
1946   *Reports of the New Towns Committee: Interim Report* Cmd 6759; *Second Interim Report* Cmd 6794; *Final Report* Cmd 6876 (Reith Report)
1947   *Town and Country Planning Bill 1947: Explanatory Memorandum*, Cmd 7006
1951   *Town and Country Planning 1943–1951: Progress Report*, Cmd 8204
1952   *Town and Country Planning Act 1947: Amendment of Financial Provisions*, Cmd 8699
1955   *Green Belts*, MHLG Circular 42/55
1957   *Green Belts*, MHLG Circular 50/57
1958   *Town and Country Planning Bill: Explanatory Memorandum*, Cmnd 562
1963   *London – Employment: Housing: Land*, Cmnd 1952
1965   *The Land Commission*, Cmnd 2771
1969   *Modifications in Betterment Levy*, Cmnd 4001
1972   *Development and Compensation: Putting People First*, Cmnd 5124
1972   *Local Authority/Private Enterprise Partnership Schemes*
1974   *Land*, Cmnd 5730
1975   *Development Land Tax*, Cmnd 6195
1977   *Statement on the Non-statutory Inquiry by the Baroness Sharp into the Continued Use of Dartmoor for Military Training*, Cmnd 6837

1981   *Planning Gain* (Property Advisory Group)

1984   *Green Belts*, DoE Circular 14/84

1984   *Land for Housing*, DoE Circular 15/84

1985   *Land Use Planning and the Housing Market* (Coopers & Lybrand), London: Coopers & Lybrand

1987   *Land Supply and House Prices in Scotland* (PIEDA), Scottish Office

1987   *Land Use Planning and Indicators of Housing Demand* (Coopers & Lybrand), London: Coopers & Lybrand

1987   *Evaluation of Derelict Land Schemes* (Roger Tym & Partners and Land Use Consultants)

1988   *Urban Land Markets in the UK* (R. N. Chubb)

1988   *Vacant Urban Land: A Literature Review* (G. E. Cameron, S. Monk, and B. Pearce) Department of the Environment

1990   *Development on Unstable Land*, PPG14

1990   *Contaminated Land*, HC Environment Committee, 1st Report, Session 1989–90, HC 170 [*Government Response*, Cm 1161]

1990   *Mineral Policies in Development Plans* (Arup Economic Consultants)

1990   *Rates of Urbanization in England 1981–2001* (P. R. Bibby and J. W. Shepherd)

1991   *Housing Land Availability* (Roger Tym & Partners)

1991   *Housing Land Availability: The Analysis of PS3 Statistics on Land with Outstanding Planning Permission* (P. Bibby and J. Shepherd)

1991   *Planning and Affordable Housing*, DoE Circular 7/91

1991   *Tackling Vacant Land: An Evaluation of Policy Instruments for Tackling Urban Land Vacancy* (M. Whitbread, D. Mayne, and D. Wickens, Arup Economic Consultants)

1992   *National Survey of Vacant Land in Urban Areas of England 1990* (J. Shepherd and A. Abakuks)

1992   *Relationship between House Prices and Land Supply* (Gerald Eve and Department of Land Economy, University of Cambridge)

1992   *Strategic Approach to Derelict Land Reclamation* (Public Sector Management Research Centre, Aston University)

1992   *Use of Planning Agreements* (Grimley J. R. Eve *et al.*)

1993   *Alternative Development Patterns: New Settlements*

1993   *Effectiveness of Green Belts* (M. Elson, S. Walker, and R. Macdonald)

1994   *Assessment of the Effectiveness of Derelict Land Grant in Reclaiming Land for Development*

1994   *Feasibility Study for Deriving Information about Land Use Stock* (A. R. Harrison), Bristol: Department of Geography, University of Bristol, for DoE)

1994   *Shopping Centres and their Future*, HC Environment Committee, Session 1993–94, HC 359 [*Government Response*, Cm 2767, 1995]

1995   *Derelict Land Prevention and the Planning System*

1995   *Derelict Land Survey* (Scotland), Scottish Office

1995   *Green Belts*, PPG 2

1995   *Reducing the Need to Travel though Land Use and Transport Planning*, DoE

1995   *Survey of Derelict Land in England 1993*

1996   *Household Growth: Where Shall We Live?*, Cm 3471

1996   *Minerals Planning Policy and Supply Practices in Europe*

1996   *Survey of Mineral Workings in England*

1997   *Enforcing Planning Control: Legislative Provisions and Procedural Requirements*, DoE Circular 1997/10

1997   *Operation of Compulsory Purchase Orders* (City University Business School)

1997   *Planning Obligations*, DoE Circular 1991/1

1997   *Shopping Centres* HC Environment Committee 4th Report, Session 1996–97, HC 210 [*Government Response*, Cm 3729]

1998   *Impact of Large Foodstores on Market Towns and District Centres* (CB Hillier Parker and Saxell Bird Avon)

1998   *Our Competitive Future: Building the Knowledge Driven Economy*, Cm 4176

1998   *Planning for the Communities of the Future*, Cm 3885

1998   *Planning for Sustainable Development*

1998   *Prevention of Dereliction through the Planning System*, DETR Circular 2/98

1999   *Biotechnology Clusters: Report of a Team Led by Lord Sainsbury, Minister for Science*, DTI

1999   *Contaminated Land: Draft Circular*, DETR

1999   *Green Belt Statistics: England 1997*, DETR Information Bulletin 1183

1999   *Housing: Consultation Draft PPG 3*

1999   *Land Reform: Proposals for Legislation* (Scottish Executive), SE 1999/1

1999   *Toward an Urban Renaissance* (Report of the Urban Task Force), London: Spon

2000   *Derelict Land and Section 215 Powers*, DETR

2000   *Scottish Vacant and Derelict Land Survey*, Scottish Executive

2000   *Town and Country Planning (Residential Development on Greenfield Land, England), Direction 2000*, DETR

HOUSING

1965   *Report of the Committee on Housing in Greater London* (Milner Holland Report), Cmnd 2605

1967   *Areas of Special Housing Need*, National Committee for Commonwealth Immigrants

1968   *The Older Houses in Scotland: A Plan for Action*, Cmnd 3598

1968   *Older Houses into New Homes*, Cmnd 3602

1971   *Fair Deal for Housing*, Cmnd 4728

1973   *Homes for People: Scottish Housing Policy in the 1970s*, Cmnd 5272

1973   *Widening the Choice: The Next Steps in Housing*, Cmnd 5280

1973   *Towards Better Homes: Proposals for Dealing with Scotland's Older Housing*, Cmnd 5338

1973   *Better Homes: The Next Priorities*, Cmnd 5339

1981   *Priority Estates Project 1981: Improving Problem Council Estates*

1985   *Development in the Countryside and Green Belts*, SDD Circular 24/85, Scottish Office

1985   *Home Improvement: A New Approach*, Cmnd 9513

1985   *Home Improvement in Scotland: A New Approach*, Cmnd 9677

1985   *Land Use Planning and the Housing Market* (Coopers & Lybrand), London: Coopers & Lybrand

1986   *Demand for Housing: Economic Perspectives and Planning Practices* (D. MacLennan), Scottish Office

1987   *Land Use Planning and Indicators of Housing Demand* (Coopers & Lybrand), London: Coopers & Lybrand

1990   *Area Renewal, Unfitness, Slum Clearance and Enforcement Action*, DoE Circular 6/90

1991   *Planning and Affordable Housing*, DoE Circular 7/91

1992   *Relationship between House Prices and Land Supply* (Gerald Eve and Department of Land Economy, University of Cambridge)

1992   *Scottish House Condition Survey 1991*

1993   *Rural Housing*, HC Welsh Affairs Committee, 3rd Report, Session 1992–93, HC 621 [*Government Response*, Cmnd 2375]

1996   *Housing in England 1994/95: A Report of the 1994/95 Survey of English Housing* (H. Green *et al.*)

1994   *Planning for Affordable Housing* (J. Barlow, R. Cocks, and M. Parker)

1995   *Projections of Households in England to 2016*

1996   *Home-Owners and Clearance: An Evaluation of Rebuilding Grants* (V. Karn, J. Lucas *et al.*)

1996   *Housing Need*, HC Environment Committee, 2nd Report, Session 1995–96, HC 22

1996   *Household Growth: Where Shall We Live?*, Cm 3471

1996   *Planning and Affordable Housing*, DoE Circular 13/96

1996   *Private Sector Renewal: A Strategic Approach*, DoE Circular 17/96

1997   *Economic Determinants of Household Formation: Where Will They Go?* (G. Bramley, M. Munro, and S. Lancaster), DETR

1997   *Planning Obligations*, DoE Circular 1/97

1998   *English House Condition Survey 1996*

1998   *Housing*, HC Select Committee on the Environment, Transport and Regional Affairs, 10th Report, Session 1997–98, HC [*Government Response*, Cm 4080]

1998   *Household Growth in Rural Areas: The Household Projections and Policy Implications*, RDC

1998   *Planning and Affordable Housing*, DETR Circular 6/98

1998   *Planning for the Communities of the Future*, Cm 3885

1999   *Housing: PPG 3*, HC Environment, Transport and Regional Affairs Committee, 17th Report, Session 1998–99, HC 490

1999   *Projections of Households in England to 2021*, DETR

1999   *Toward an Urban Renaissance* (Report of the Urban Task Force), London: Spon

2000   *Good Practice Guide on Monitoring the Provision of Housing through the Planning System*, DETR

2000   *Town and Country Planning (Residential Development on Greenfield Land, England), Direction 2000*, DETR

2000   *Quality and Choice: A Decent Home for All* (Green Paper) DETR

2000   *Conversion and Redevelopment: Process and Potential*, DETR

2000   *Monitoring Provision of Housing through the Planning System: Towards Better Practice*, DETR

2000   *The Way Forward for Housing*, DETR

THE ENVIRONMENT

Annual   *Digest of Environmental Protection and Water Statistics*, DETR

1970   *The Protection of the Environment: The Fight against Pollution*, Cmnd 4373

1977   *Nuclear Power and the Environment*, Cmnd 6820

1986   *Transforming Our Waste Land* (University of Liverpool, Environmental Advisory Unit)

1990   *This Common Inheritance: Britain's Environmental Strategy*, Cm 1200

1991   *Policy Appraisal and the Environment: A Guide for Government Departments*

1991   *Public Attitudes to the Environment in Scotland* (D. Wilkinson and J. Waterton), Scottish Office

1992   *Coastal Planning*, PPG 20

1992   *Economic Instruments and Recovery of Resources from Waste* (Environmental Resources Ltd)

1992   *Land Use Planning Policy and Climate Change*

1992   *Planning, Pollution and Waste Management* (Environmental Resources Ltd and Oxford Polytechnic School of Planning)

1992 *Potential Role of Market Mechanisms in the Control of Acid Rain* (London Economics)

1992 *Land Use Planning Policy and Climate Change* (S. Owens and D. Cope)

1993 *Eco-management and Audit Scheme for UK Local Government*, (Department of the Environment)

1993 *Environmental Appraisal of Development Plans: A Good Practice Guide*

1993 *Review of Environmental Expenditure* (ECOTEC)

1994 *Biodiversity: The UK Action Plan*, Cm 2428

1994 *Contaminated Land and the Water Environment*, Environment Agency

1994 *Climate Change: The UK Programme*, Cm 2427

1994 *Guidance Notes for Local Planning Authorities on the Methods of Protecting the Water Environment through Development Plans*, Environment Agency

1994 *Managing Demolition and Construction Waste* (Howard Humphreys & Partners)

1994 *Ozone: Report of the Expert Panel on Air Quality Standards*

1994 *Planning and Noise*, PPG 24 (revision in preparation)

1994 *Planning and Pollution Control*, PPG 23 (revision in preparation)

1994 *Protecting and Managing Sites of Special Scientific Interest in England*, National Audit Office

1994 *Sustainable Development: The UK Strategy*, Cm 2426

1995 *Air Quality: Meeting the Challenge: The Government's Strategic Policies for Air Quality Management*

1995 *Contaminants Entering the Sea*, Environment Agency

1995 *Environmental Agenda for Wales*, Welsh Office

1995 *Environmental Facts: A Guide to Using Public Registers of Environmental Information*, Department of the Environment

1995 *Guide to Risk Assessment and Risk Management for Environmental Protection*

1995 *Making Waste Work: A Strategy for Sustainable Waste Management in England and Wales*, Cm 3040

1995 *Ozone Layer*, Department of the Environment

1995 *Preparation of Environmental Statements for Planning Projects That Require Environmental Assessment: A Good Practice Guide*

1995 *Risk Assessment and Risk Management for Environmental Protection*

1995 *Sustainable Development*, HL Select Committee on Sustainable Development, Report, Session 1994–95, HL 72 [*Government Response*, Cm 3018]

1995 *Sustainable Development: What It Means to the General Public* (E. McCaig, C. Henderson, and MVA Consultancy), Scottish Office

1996 *Indicators of Sustainable Development for the UK*

1996 *Integrated Pollution Control: A Practical Guide*

1996 *Scottish Agriculture*, HC Scottish Affairs Committee Inquiry into the Future for Scottish Agriculture, Session 1995–96, HC 629 [*Reply*, Cm 3548, 1997]

1996 *This Common Inheritance: 1996 UK Annual Report*, Cm 3188

1996 *Waste Management: The Duty of Care – A Code of Practice*

1997 *Air Quality and Land Use Planning*

1997 *Environmentally Sensitive Areas and Other Schemes under the Agri-environmental Regulation*, HC Agriculture Committee, 2nd Report, Session 1996-97, HC 45 [*Government Response*, Cm 3707]

1997 *Mitigating Measures in Environmental Statements*, DETR

1997 *UK National Air Quality Strategy*, Cm 3587

1998 *Biodiversity in Scotland: The Way Forward*, Scottish Biodiversity Group

1998 *Local Biodiversity Action Plans: A Manual*, Scottish Biodiversity Group

1998 *Sustainable Development: Opportunities for Change – Consultation Paper on a Revised UK Strategy*, DETR

1998	*Sustainable Local Communities for the 21st Century: Why and How to Prepare an Effective LA21 Strategy*
1999	*A Better Quality of Life: A Strategy for Sustainable Development in the UK*, Cm 4345
1999	*Air Quality Strategy for England, Scotland, Wales and Northern Ireland*, DETR, SE, NAW, and DoENI, Cm 4548 (SE 2000/3; NIA 7)
1999	National Waste Strategy, Scottish Executive
1999	*Operation of the Landfill Tax*, HC Environment, Transport and Regional Affairs Committee, 13th Report, Session 1989–99, HC 150
1999	*Planning and Waste Management*, PPG 10
1999	*Reducing the Environmental Impact of Consumer Products*, HC Environment, Transport and Regional Affairs Committee, 11th Report, Session 1889–99, HC 149
1999	*State of the Environment of England and Wales: Coasts*, (Environment Agency)
1999	*Lessons from the European Commission's Demonstration Programme on Integrated Coastal Zone Management*, Luxembourg: OOPEC
1999	*Towards a European Integrated Coastal Zone Management (ICZM) Strategy: General Principles and Policy Options*, Luxembourg: OOPEC (1999)
2000	*Action for Scotland's Biodiversity* (Scottish Biodiversity Group, Scottish Executive)
2000	*Northern Ireland Waste Management Strategy*, DoE (NI)
2000	*Strategic Environmental Assessment of Plans and Programmes: Proceedings of the Intergovernmental Policy Forum Glasgow*
2000	*The Greening Government Initiative: First Annual Report from the Green Ministers Committee*

Royal Commission on Environmental Pollution

1st Report: *Report of the Royal Commission on Environmental Pollution*, Cmnd 4585, 1972

2nd Report: *Three Issues in Industrial Pollution*, Cmnd 4894, 1972

3rd Report: *Pollution in Some British Estuaries and Coastal Waters*, Cmnd 5054, 1972

4th Report: *Pollution Control: Progress and Problems*, Cmnd 5780, 1974

5th Report: *Air Pollution Control: An Integrated Approach*, Cmnd 6371, 1976

6th Report: *Nuclear Power and the Environment*, Cmnd 6618, 1976

7th Report: *Agriculture and Pollution*, Cmnd 7644, 1979

8th Report: *Oil Pollution of the Sea*, Cmnd 8358, 1981

9th Report: *Lead in the Environment*, Cmnd 8852, 1983

10th Report: *Tackling Pollution: Experience and Prospects*, Cmnd 9149, 1984

11th Report: *Managing Waste: The Duty of Care*, Cmnd 9675, 1985

12th Report: *Best Practicable Environmental Option*, Cm 310, 1988

13th Report: *The Release of Genetically Engineered Organisms to the Environment*, Cm 720, 1989

14th Report: *GENHAZ: A System for the Critical Appraisal of Proposals to Release Genetically Modified Organisms into the Environment*, Cm 1557, 1991

15th Report: *Emissions from Heavy Duty Diesel Vehicles*, Cm 1631, 1991

16th Report: *Freshwater Quality*, Cm 1966, 1992

17th Report: *Incineration of Waste*, Cm 2181, 1993

18th Report: *Transport and the Environment*, Cm 2674, 1994

19th Report: *Sustainable Use of Soil*, Cm 3165, 1996

20th Report: *Transport and the Environment: Developments since 1994*, Cm 3752, 1997

21st Report: *Setting Environmental Standards*, Cm 4053, 1998

22nd Report: *Energy: The Changing Climate*, HOC Papers 1999–2000

HERITAGE PLANNING

1950	*Houses of Outstanding Historic and Architectural Interest* (Gowers Report)
1979	*A National Heritage Fund*, Cmnd 7428
1987	*Historic Buildings and Conservation Areas: Policy and Procedures*, DoE Circular 8/87
1990	*Archaeology and Planning*, PPG 16
1990	*Conservation Areas of England*, EH
1992	*Buildings at Risk: A Sample Survey*, EH
1992	*Development Plan Policies for Archaeology*, EH
1992	*Managing England's Heritage: Setting Our Priorities for the 1990s*, EH
1993	*Conservation Area Management*, EH
1993	*Conservation Issues in Strategic Plans*, EH
1993	*Finding and Minding: A Report on the Archaeological Work of the Department of the Environment Northern Ireland*, DoENI
1994	*Ecclesiastical Exemption: What It Is and How It Works*, EH
1994	*Our Heritage: Preserving It, Prospering from It*, HC National Heritage Committee, 3rd Report, Session 1993–94, HC 139
1994	*Planning and the Historic Environment*, PPG 15
1995	*Conservation in London: A Study of Strategic Planning Policy in London*, EH
1995	*In the Public Interest: London's Civic Architecture at Risk*, EH
1995	*Local Government Reorganisation: Guidance to Local Authorities on Conservation of the Historic Environment*, EH
1996	*Conservation Areas in Strategic Plans*, EH
1996	*Conservation Issues in Local Plans*, EH
1996	*Protecting Our Heritage: A Consultation Paper on the Built Environment of England and Wales*, DNH
1996	*Something Worth Keeping: Post-war Architecture in England*, EH
1997	*Planning and the Historic Environment: Notifications and Directions by the Secretary of State*, DETR Circular 14/97
1998	*Preservation Orders: Draft Regulations: A Consultation Paper*
1998	*Monuments on Record: Celebrating 90 Years of the Royal Commissions on Historical Monuments*, CD-ROM, Royal Commission on Historical Monuments in England
1998	*Conservation-Led Regeneration: The Work of English Heritage*, English Heritage
1998	*Enabling Development and the Conservation of Historic Assets*, English Heritage
1998	*Tourism: Towards Sustainability*
1998	*Planning and the Historic Environment*, Scottish Office
1999	*Enabling Development and the Conservation of Heritage Areas*, EH
1999	*Heritage Dividend: Measuring the Results of English Heritage Regeneration*, EH
1999	*Historic Core Zones Project*, English Historic Towns Forum
1999	*Historic Royal Palaces: Annual Report and Accounts 1998–99*, HC 598, 1998–99
1999	*Historic Scotland: Annual Report and Accounts 1998–99, and Corporate Plan 1999–2000*, SE 1999/7 (also HC 639, Session 1998–99)
1999	*Investment Performance of Listed Office Buildings* (Investment Property Databank), English Heritage and RICS
1999	*Protecting and Conserving the Built Heritage in Wales* (National Audit Office Wales), TSO

1999	*Royal Parks Agency: Annual Report and Accounts 1998–99*, HC 600 1998–99
1999	*Conservation Plans in Action*, English Heritage
1999	*Tomorrow's Tourism: A Growth Industry for the New Millennium*
1999	Heritage Dividend: Measuring the Results of English Heritage Regeneration
2000	*Contemporary Issues in Heritage and Environment Interpretation*, DETR
2000	Environment and Heritage Service Annual Report (NI) 1999–2000 (HCP 655, 1998–99)
2000	*Power of Place*
2000	*The Stonehenge World Heritage Site Management Plan*
2000	*New Strategy for Scottish Tourism*
2000	*Consultation Paper on the Impact of the Shimuzu Judgement*
2000	*Seaside 2000: Consultation,* DCMS
2000	*World Heritage Sites: The Tentative List of the United Kingdom of Great Britain and Northern Ireland,* DCMS
2000	*British Government Panel on Sustainable Development: Sixth Report*
2000	*Indicators of Sustainable Development:* UK Round Table on Sustainable Development
2000	*Not Too Difficult: Economic Instruments to Promote Sustainable Development within a Modernised Economy,* UK Round Table on Sustainable Development

THE COUNTRYSIDE

Publications of the Countryside Commission (CC), the Rural Development Commission (RDC), and their successor the Countryside Agency (CA) are obtainable from Countryside Agency Postal Sales, PO Box 124, Walgrave, Northampton NN6 9TL.

1942	*Report of the Committee on Land Utilisation in Rural Areas* (Scott Report), Cmd 6378
1947	*Report of the National Parks Committee (England and Wales)* (Hobhouse Report), Cmd 7121
1947	*Report of the Special Committee on Footpaths and Access to the Countryside* (Hobhouse Subcommittee Report), Cmd 7207
1966	*Leisure in the Countryside*, Cmnd 2928
1975	*Food from Our Own Resources*, Cmnd 6020
1975	*Sport and Recreation*, Cmnd 6200
1979	*Farming and the Nation*, Cmnd 7458
1983	*New Look at the Northern Ireland Countryside* (Jean Balfour)
1986	*Review of Forestry Commission Objectives and Achievements*, National Audit Office
1987	*Annual Review of Agriculture 1987*, Cm 67
1990	*Dynamics of the Rural Economy* (ECOTEC), Department of the Environment
1991	*Fit for the Future: Report of the National Parks Review Panel*, CC
1991	*Permitted Development Rights for Agriculture and Forestry* (Land Use Consultants)
1992	*An Agenda for Investment in Scotland's Natural Heritage*, SNH
1992	*Business Success in the Countryside: The Performance of Rural Enterprise* (D. Keeble *et al.*)
1992	*Coastal Zone Protection and Planning*, HC Environment Committee, 2nd Report, Session 1991–92, HC 17 [*Government Response*, Cm 2011
1992	*Environmentally Sensitive Areas Wales: Aspects of Designation* (G. O. Hughes and A. M. Sherwood), Welsh Office
1992	*Indicative Forestry Strategies*, DoE Circular 29/92
1992	*Protected Landscapes in the United Kingdom*, CC

1992	*Rural Framework*, Scottish Office
1992	*Scottish Rural Life: A Socio-economic Profile of Rural Scotland*, Scottish Office
1992	*Tir Cymen: A Farmland Stewardship Scheme*, CCW
1993	*Conserving England's Marine Heritage: A Strategy*, EN
1993	*Countryside Survey 1990: Main Report*, Department of the Environment
1993	*Coastal Planning and Management: A Review* (Rendel Geotechnics), TSO
1993	*Development below Low Water Mark* (Discussion Paper), Department of the Environment
1993	*The Economy and Rural England* (R. Tarling, J. Rhodes, J. North, and Broom, G.) RDC
1993	*English Rural Communities* (A. Rogers) RDC
1993	*Forestry and the Environment*, HC Environment Committee, 1st Report, Session 1992–93, HC 257
1993	*Managing the Coast* (Discussion Paper), Department of the Environment
1993	*Role of the Countryside Commission in the Town and Country Planning System*, CC
1993	*Rural Development and Statutory Planning* (ARUP Economics and Planning), RDC
1993	*Waterway Environment and Development Plans*, British Waterways
1993	*Welsh Estuaries Review*, CCW
1994	*Countryside Stewardship: An Outline*, CC
1994	*Forestry and Woodlands*, HC Select Committee on Welsh Affairs, 1st Report, Session 1993–94, HC 35 [*Government Response*, Cm 2645]
1994	*Lifestyles in Rural England* (P. Cloke, P. Milbourne, and C. Thomas), RDC
1994	*National Forest: The Strategy*, CC
1994	*Nature Conservation*, PPG 9
1994	*Nature Conservation in Environmental Assessment*, EN
1994	*Our Forests: The Way Ahead*, Cm 2644
1994	*Sustainable Forestry: The UK Programme*, Cm 2429
1995	*Countryside Planning File*, CC
1995	*Countryside Stewardship Handbook*, CC
1995	*Environmental Impact of Leisure Activities*, HC Environment Committee, 4th Report, Session 1994–95, HC 246 [*Government Response*, HC 761]
1995	*Guide to Measures Available to Control the Recreational Use of Water* (Cobham Resources Consultants), Scottish Office
1995	*Policy Guidelines for the Coast*, DoE
1995	*Planning for Rural Diversification* (M. Elson, R. Macdonald, R. Steenberg, and G. Brown)
1995	*Planning for Rural Diversification: A Good Practice Guide* (M. Elson, C. Steenberg, and J. Wilkinson)
1995	*Programme for Partnership: Announcement of the Outcome of the Scottish Office Review of Urban Regeneration Policy*, Scottish Office
1995	*Protecting and Managing Sites of Special Scientific Interest in England*, HC Committee of Public Accounts, 11th Report, Session 1994–95, HC 252
1995	*Rebuilding the English Countryside: Habitat Fragmentation and Wildlife Corridors in Practical Conservation*, EN
1995	*Rural England: A Nation Committed to a Living Countryside*, Cm 3016
1995	*Rural Scotland: People, Prosperity and Partnership*, Cm 3041
1995	*Sustainable Rural Tourism*, CC
1995	*Tourism in the Countryside*, RDC
1996	*Impact of Tourism on Rural Settlements*, RDC
1996	*Partnership in the Regeneration of Urban Scotland* (D. McAllister), Scottish Office

1996	*People, Parks and Cities: A Guide to Current Good Practice in Urban Parks*, Department of the Environment
1996	*The Countryside: Environmental Quality and Economic Development*, Department of the Environment
1996	*Rural England: The Rural White Paper*, HC Environment Committee, 3rd Report, Session 1995–96, HC 163
1996	*Working Countryside for Wales*, Cm 3180
1997	*Countryside, Environmental Quality and Economic and Social Development*, PPG 7
1997	*Disadvantage in Rural Areas*, RDC
1997	*Economic Impact of Recreation and Tourism in the English Countryside*, RDC
1997	*Hedgerows Regulations 1997: A Guide to the Law and Good Practice*, DoE
1997	*Protecting Environmentally Sensitive Areas*, National Audit Office
1997	*Rural Traffic: Getting It Right*, CC
1997	*Survey of Rural Services*, RDC
1997	*Towards a Development Strategy for Rural Scotland: A Discussion Paper*, Scottish Office
1998	*Access to the Open Countryside in England and Wales: A Consultation Paper*, DETR
1998	*Access to the Countryside for Open-Air Recreation*, SNH
1998	*A Home in the Country? Affordable Housing in Rural England*, RDC
1998	*Guidance to the Regional Development Agencies on Rural Policy*, DETR
1998	*Household Growth in Rural Areas*, RDC
1998	*Investing in Quality: Improving the Design of New Housing in the Scottish Countryside – A Consultation Paper*, Scottish Office
1998	*National Scenic Areas: A Consultation Paper*, SNH
1998	*Natural Heritage Designations in Scotland*, SNH
1998	*Protection of Field Boundaries* (HC Environment, Transport and Regional Affairs Committee, HC 969, Session 1997/98) [*Government Response, 1999*]
1998	*Review of the Hedgerows Regulations 1997* (DETR)
1998	*Rural Development and Land Use Policies*, RDC
1998	*Rural Disadvantage: Understanding the Processes*, RDC
1998	*Sites of Special Scientific Interest: Better Protection and Management – Consultation Paper*, DETR
1998	*Towards a Development Strategy for Rural Scotland*, Scottish Office
1999	*Countryside Recreation: Enjoying the Countryside*, CC
1999	*Farming and Rural Conservation Agency Annual Report 1998–99*, HC 765 (1998/99)
1999	*National Forest: Guide for Developers and Planners*, National Forest Company
1999	*Our Plan for the Future 2000–2004*, British Waterways
1999	*Planning for the Quality of Life in Rural England*, CC
1999	*Rural Economies* (Cabinet Office Performance and Innovation Unit)
1999	*Rural Wales: A Statement by the Rural Partnership*, Welsh Assembly
1999	*State of the Environment of England and Wales: Coasts*, (Environment Agency)
1999	*Town and Country Parks*, 20th Report of the Environment, Transport and Regional Affairs Committee, Session 1998–99, HC 477 [*Government Response, Cm 4550, 2000*]
1999	*Unlocking the Potential: A New Future for British Waterways*, DETR
2000	*National Parks (Scotland) Act: Explanatory Notes*, TSO
2000	*Sites of Special Scientific Interest: Encouraging Positive Partnerships – Public Consultation Paper on Code of Guidance*, DETR

URBAN POLICIES

1977   *Policy for the Inner Cities*, Cmnd 6845

1981   *Priority Estates Project 1981: Improving Problem Council Estates*

1985   *Urban Programme*, National Audit Office, HC 513 (1984–85)

1986   *Evaluation of Environmental Projects Funded under the Urban Programme* (JURUE)

1986   *Evaluation of Industrial and Commercial Improvement Areas* (JURUE)

1986   *Urban Programme and the Young Unemployed* (C. Whitting)

1987   *Review of Data Sources for Urban Policy* (ECOTEC)

1988   *Action for Cities*, Cabinet Office

1988   *Evaluation of the Urban Development Grant Programme* (Public Sector Management Research Unit, Aston University)

1988   *Improving Inner City Shopping Centres: An Evaluation of Urban Programme Funded Schemes in the West Midlands* (Public Sector Management Research Centre, Aston University)

1988   *New Life for Urban Scotland*, Scottish Office

1988   *Socio-demographic Change and the Inner City* (M. Boddy *et al.*)

1988   *Stockbridge Village Trust: Building a New Community* (Roger Tym & Partners)

1988   *Urban Development Corporations*, National Audit Office, HC 492 (1987–88)

1990   *Evaluation of Garden Festivals* (PA Cambridge Economic Consultants)

1990   *Employment in the 1990s*

1990   *Green Paper on the Urban Environment*, CEC

1990   *Patterns and Processes of Urban Change in the UK* (T. Fielding and S. Holford)

1990   *Regenerating the Inner Cities*, National Audit Office, HC 169 (1989–90)

1990   *Targeting Urban Employment Initiatives* (I. Turok and U. Wannop)

1990   *Tourism and the Inner City: An evaluation of the Impact of Grant Assisted Tourism Projects* (Polytechnic of Central London School of Planning *et al.*)

1990   *Urban Labour Markets: Reviews of Urban Research* (B. Moore and P. Townroe)

1990   *Urban Scotland into the 90s: New Life: Two Years On*, Scottish Office

1990   *US Experience in Evaluating Urban Regeneration* (T. Barnekov, D. Hart, and W. Benfer)

1992   *Developing Indicators to Assess the Potential for Urban Regeneration* (M. Coombes, S. Raybould, and C. Wong)

1992   *Estate Action: New Life for Local Authority Estates: Guidelines for Local Authorities*, DoE

1993   *Evaluation of Urban Grant, Urban Regeneration Grant, and City Grant* (Price Waterhouse)

1993   *Future of Private Housing Renewal Programmes*, Consultation Paper, DoE

1993   *Progress in Partnership*, Scottish Office

1994   *Assessing the Impact of Urban Policy* (B. Robson *et al.*), DETR

1994   *Employment Department Baseline Follow-Up Studies* (Coopers & Lybrand), published by Coopers & Lybrand

1995   *Final Evaluation of Enterprise Zones* (PA Cambridge Consultants)

1995   *Impact of Environmental Improvements on Urban Regeneration* (PIEDA)

1995   *Involving Communities in Urban and Regional Regeneration*, DoE

1995   *Programme for Partnership: Announcement of the Outcome of the Scottish Office Review of Urban Regeneration Policy*, Scottish Office

1995   *Single Regeneration Budget*, HC Environment Committee, 1st Report, Session 1995–96, HC 26

1996   *City Challenge: Interim National Evaluation*

1996	*Evaluation of Six Early Estate Action Schemes* (Capita Management Consultancy), DoE
1996	*Government Response to the Environment Committee Report into the Single Regeneration Budget*, Cm 3178
1996	*Partnership in the Regeneration of Urban Scotland* (D. McAllister), Scottish Office
1996	*People, Parks and Cities* (L. Greenhalgh, K. Worpole, and R. Grove-White)
1997	*Housing Action Trusts: Evaluation of Baseline Condition* (Capita Management Consultants), DETR
1997	*Neighbourhood Renewal Assessment and Renewal Areas* (Austin Mayhead & Co.), DETR
1997	*Towards an Urban Agenda in the European Union*, CEC
1998	*Bringing Britain Together: A National Strategy for Neighbourhood Renewal* (Social Exclusion Unit), Cm 4045
1998	*Community-Based Regeneration Initiatives: A Working Paper*, DETR
1998	*Evaluation of the Single Regeneration Challenge Fund Budget: A Partnership for Regeneration – An Interim Evaluation* (A. Brennan, J. Rhodes, and P. Tyler), DETR
1998	*Impact of Urban Development Corporations in Leeds, Bristol and Central Manchester* (Centre for Urban Policy Studies, University of Manchester), DETR
1998	*New Deal for Communities: Phase I Proposals*, DETR
1998	*Regenerating London Docklands* (Cambridge Policy Consultants *et al.*), DETR
1998	*Regeneration Programmes: The Way Forward*, Discussion Paper, DETR
1998	*Sustainable Urban Development on the European Union: A Framework for Action*, CEC
1998	*Urban Development Corporations: Performance and Good Practice* (Roger Tym & Partners), DETR
1999	*City Challenge: Final National Evaluation* (KPMG Consulting), DETR
1999	*City-wide Urban Regeneration* (M. Carley and K. Kirk) Scottish Executive Central Research Unit
1999	*English Partnerships: Assisting Local Regeneration* (National Audit Office)
1999	*Evaluation of the New Life for Urban Scotland Initiative* (R. Tarling *et al.*), Scottish Executive
1999	*Interim Evaluation of English Partnerships: Final Report* (PA Consulting Group), DETR
1999	*Literature Review of Social Exclusion* (P. Lee and A. Murie), Scottish Office
1999	*Local Evaluation for Regeneration Partnerships: Good Practice Guide*, DETR
1999	*New Deal for Communities: Learning Lessons: Pathfinders' Experiences on NDC Phase I, DETR*
1999	Opportunity for All: Tackling Poverty and Social Exclusion (Department of Social Security), Cm 4445
1999	*Running and Sustaining Renewal Areas: Good Practice Guide*, DETR
1999	*Toward an Urban Renaissance* (Report of the Urban Task Force), London: Spon
1999	*Sustainable Urban Development in the European Union: A Framework for Action*, Luxembourg: OOPEC
2000	*Urban Regeneration Companies: A Process Evaluation* (M. Parkinson and B. Robson) DETR
2000	*Co-ordination of Area-based Initiatives: Research Working Paper*, DETR
2000	*Local Strategic Partnerships: Consultation Document*, DETR
2000	*Millennium Villages and Sustainable Communities: Final Report*, DETR
2000	*New Deal for Communities: Delivering* (and other guidance), DETR
2000	*City Challenge: Final National Evaluation*, DETR

TRANSPORT

1963	*Traffic in Towns* (Buchanan Report)
1964	*Road Pricing: The Economic and Technical Possibilities* (Smeed Report)
1966	*Transport Policy*, Cmnd 3057

1967   *Public Transport and Traffic*, Cmnd 3481

1968   *Transport in London*, Cmnd 3686

1969   *Roads for the Future: A New Inter-urban Plan*

1972   *New Roads in Towns*

1973   *Report of the Urban Motorways Project Team to the Urban Motorways Committee*

1976   *Transport Policy: A Consultative Document*

1977   *Transport Policy*, Cmnd 6836

1978   *Policy for Roads, England 1978*, Cmnd 7132

1978   *Report on the Review of Highway Inquiry Procedures*, Cmnd 7133

1980   *Policy for Roads, England 1980*, Cmnd 7908

1981   *Lorries, People and the Environment*, Cmnd 8439

1982   *Policy for Roads, England 1981*, Cmnd 8496

1982   *Public Transport Subsidy in Cities*, Cmnd 8735

1983   *Roads in Scotland: Report for 1982*, Cmnd 9010

1983   *Public Transport in London*, Cmnd 9004

1983   *Policy for Roads in England*, Cmnd 9059

1984   *Buses*, Cmnd 9300

1985   *Airports Policy*, Cmnd 9542

1986   *The Channel Fixed Link*, Cmnd 9735

1987   *Policy for Roads in England 1987*, Cm 125

1989   *National Road Traffic Forecasts (Great Britain) 1989*

1989   *Roads for Prosperity*, Cm 693

1989   *New Roads by New Means: Bringing in Private Finance*, Cm 698

1990   *Road Pricing for London*, London Boroughs Association

1990   *Trunk Roads, England: into the 1990s*, Department of Transport

1990   *Traffic Quotes: Public Perception of Traffic Regulation in Urban Areas* (P. Jones)

1991   *Cycling*, HC Transport Committee, Minutes of Evidence, 8 May 1991, Session 1990–91, HC 423

1991   *Railway Noise and the Insulation of Dwellings: Report of the Committee to Recommend a National Noise Insulation Standard for New Railway Lines* (Mitchell Report)

1991   *Urban Public Transport: The Light Rail Option*, HC Transport Committee 4th Report, Session 1990–91, HC 14

1992   *Developers' Contributions to Highway Works: Efficiency Scrutiny Report and the Department's Response*, Department of Transport

1992   *Developers' Contributions to Highway Works: Consultation Document*, Department of Transport

1992   *Role of Investment Appraisal in Road and Rail Transport*, Department of Transport

1992   *Traffic in London: Traffic Management and Parking Guidance* (DoT Local Authority Circular 5/92)

1993   *Paying for Better Motorways*, Cm 2200

1994   *Charging for the Use of Motorways*, HC Transport Committee, 5th Report, Session 1993–94, HC 376

1994   *Impact of Transport Policies in Five Cities* (M. Dasgupta *et al.*), Transport Research Laboratory

1994   *Trunk Roads, England: 1994 Review*, Department of Transport

1995   *Better Places through Bypasses: Report of the Bypass Demonstration Project*

1995   *Guidance on Decriminalised Parking Enforcement Outside London* (DoT Local Authority Circular 1/95)

1995   *Guidance on Induced Traffic* (DoT Guidance Note 1/95), Department of Transport

1995   *London Congestion Charging Research Programme: Principal Findings* (MVA Consultancy)

1995	*Managing the Trunk Road Programme*, Department of Transport
1995	*PPG 13: A Guide to Better Practice: Reducing the Need to Travel through Land Use and Transport Planning* (JMP Consultants)
1995	*Urban Road Pricing*, HC Transport Committee, 3rd Report, Session 1994–95, HC 104 [*Government Response*, Cm 3019]
1996	*Cycling in Great Britain* (Transport Statistics Report), Department of Transport
1996	*National Cycling Strategy*, Department of Transport
1996	*Transport Strategy for London*
1996	*Transport: The Way Forward*, Cm 3234
1997	*Keeping Buses Moving: A Guide to Traffic Management to Assist Buses in Urban Areas*, DoT Local Transport Note 1997/1
1997	*Keeping Scotland Moving: A Scottish Transport Green Paper*, Cm 3565
1997	*National Road Traffic Forecasts (Great Britain)*, DETR
1997	*Regulation of Heavy Lorries*, National Audit Office
1997	*Developing a Strategy for Walking*, DETR
1998	*Breaking the Logjam: The Government's Consultation Paper on Fighting Traffic Congestion and Pollution through Road User and Workplace Parking Charges*, DETR
1998	*Parking Standards in the South East* (Llewelyn-Davies and JMP Consultants), DETR
1998	*Guidance on Local Transport Plans*, DETR
1998	*Moving Forward: Northern Ireland Transport Policy Statement*, Northern Ireland Department of the Environment
1998	A New Deal for Transport: Better for Everyone, Cm 3950
1998	*New Deal for Trunk Roads in England*, DETR
1998	*Role of Technology in Implementing an Integrated Transport Policy*, Office of Science and Technology, DTI
1998	*Transporting Wales into the Future*, Welsh Office
1998	*Travel Choices for Scotland: The Scottish Integrated Transport White Paper*, Cm 4010
1999	*Benefits of Green Transport Plans*, DETR
1999	*Community Impact of Traffic Calming Schemes in Scotland*, Scottish Executive Central Research Unit
1999	*From Workhorse to Thoroughbred: A Better Role for Bus Travel*, DETR
1999	*Integrated Transport White Paper: Report of the HC Environment, Transport and Regional Affairs Committee*, Session 1998–99, HC 32 [*Government Response*, 3rd Special Report, Session 1998–99, HC 708
1998	*A Transport Statement for Northern Ireland*
1999	*Local Transport Strategies: Preliminary Guidance*, SDD
1999	*Research on Walking* (System Three), Scottish Executive Central Research Unit
1999	*Revision of Planning Policy Guidance Note 13, Transport: Public Consultation Draft*, PPG 13
1999	*Sustainable Distribution: A Strategy*, DETR
1999	*School Travel: Strategies and Plans: A Best Practice Guide for Local Authorities*, DETR
1999	*Travel Choices for Scotland: Strategic Roads Review*
1999	*Tackling Congestion: The Scottish Executive's Consultation Paper on Fighting Traffic Congestion and Pollution through Road User and Workplace Parking Charges* (Scottish Executive)
1999	*School Travel: Strategies and Plans – A Best Practice Guide for Local Authorities*, DETR
1999	*Transport: The Way Forward*, Cm 3234
1999	*Transport and the Economy*, SACTRA

2000   *Commission for Integrated Transport First Annual Report*
2000   *Guidance on Full Local Transport Plans*, DETR
2000   *Good Practice Guide for the Development of Local Transport Plans*, DETR
2000   *Road Charging Options for London*, DETR
2000   *Encouraging Walking: Advice to Local Authorities* DETR
2000   *Guidance on the New Approach to Appraisal* and *Understanding the New Approach to Appraisal*, DETR
2000   *Waterways for Tomorrow*, DETR

Standing Advisory Committee on Trunk Road Assessment

1977   *Trunk Road Assessment Report*
1979   *Trunk Road Proposals: A Comprehensive Framework for Appraisal*
1986   *Urban Road Appraisal*
1992   *Assessing the Environmental Impact of Road Schemes*
1994   *Trunk Roads and the Generation of Traffic*
1999   *Transport and the Economy*

Statistical and Research Reports on Transport

NB: A comprehensive list of transport statistics is available from Transport Statistics Publications, DETR, Great Minster House, 76 Marsham Street, London SW1P 4DR

Annual   *Transport Statistics, Great Britain*
            Transport Trends
            Focus on Public Transport
            Focus on Safety
            Focus on Roads
            Focus on Freight

Periodic *Focus on Personal Travel* (including the report of the National Travel Survey (NTS) (published every three years, with bulletins giving updates of the NTS in the intervening years, published by the DETR)
See also UN Economic Commission for Europe *Annual Bulletin of Transport Statistics for Europe and North America*

PLANNING, THE PROFESSION AND THE PUBLIC

1950   *Report of the Committee on the Qualifications of Planners* (Schuster Report), Cmnd 8204
1957   *Report of the Committee on Administrative Tribunals and Enquiries* (Franks Report), Cm 218
1969   *People and Planning (Report of the Committee on Public Participation in Planning* (Skeffington Report))
1969   *Information and the Public Interest*, Cmnd 4089
1978   *Report on the Review of Highway Inquiry Procedures*, Cmnd 7133
1978   *Planning Procedures: The Government's Response to the Eighth Report from the Expenditure Committee, Session 1976–77*, Cmnd 7056
1986   *Planning Appeals, Call-In and Major Public Inquiries: The Government's Response to the Fifth Report from the Environment Committee*, Cm 43

1991   *The Citizen's Charter: Raising the Standard*, Cm 1599
1992   *The Citizen's Charter: First Report*, Cm 2101
1992   *Green Rights and Responsibilities: A Citizen's Guide to the Environment*, Department of the Environment
1994   *Mediation: Benefits and Practice*, Department of the Environment
1995   *Accessing Environmental Information in Scotland* (J. Moxen *et al.*), Scottish Office
1995   *Attitudes to Town and Country Planning* (R. McCarthy and T. Harrison)
1995   *Community Involvement in Planning and Development Processes*
1995   *Environmental Facts: A Guide to Using Public Registers of Environmental Information*, DoE
1995   *Environmental Education; Is the Message Getting Through?* (MORI), Scottish Office
1997   *Rights Brought Home: The Human Rights Bill*, Cm 3782
1997   *Your Right to Know: The Government's Proposals for a Freedom of Information Act*, Cm 3818
1998   *Guidance on Enhancing Public Participation in Local Government* (V. Lowndes), DETR
1999   *Freedom of Information: Consultation on Draft Legislation*, Cm 4355
1999   *Freedom of Information Draft Bill*, HC Select Committee on Public Administration, 3rd Report, Session 1998–99, HC 570
1999   *An Open Scotland: Freedom of Information: A Consultation*, SE/19999/51
1999   *Improving Enforcement Appeal Procedures*, Consultation Paper, DETR
1999   *Role and Effectiveness of Community Councils with regard to Community Consultation* (R. Goodlad), Scottish Executive Central Research Unit

INDEX OF STATUTES

GENERAL INDEX